* TM 1-1520-238-10

AH-64 APACHE
Attack Helicopter
Pilot's Flight Operating
Instructions

by HEADQUARTERS, DEPARTMENT OF THE ARMY

©2011 Periscope Film LLC
ISBN #978-1-935700-67-8
www.PeriscopeFilm.com

NOTICE:

This manual is sold for historic research purposes only, as an entertainment. It is not intended to be used as part of an actual flight training program. No book can substitute for flight training by an authorized instructor. The licensing of pilots is overseen by organizations and authorities such as the FAA and CAA. Operating an aircraft without the proper license is a federal crime.

* TM 1-1520-238-10

TECHNICAL MANUAL

OPERATOR's MANUAL

FOR

HELICOPTER, ATTACK, AH-64A APACHE

- WARNING DATA
- TABLE OF CONTENTS
- INTRODUCTION
- DESCRIPTION AND OPERATION
- AVIONICS
- MISSION EQUIPMENT
- OPERATING LIMITS AND RESTRICTIONS
- WEIGHT/BALANCE AND LOADING
- PERFORMANCE DATA
- NORMAL PROCEDURES
- EMERGENCY PROCEDURES
- REFERENCES
- ABBREVIATIONS AND TERMS
- ALPHABETICAL INDEX

DISTRIBUTION STATEMENT A: Approved for public release; distribution is unlimited.

HEADQUARTERS, DEPARTMENT OF THE ARMY
31 AUGUST 1994

* This manual supersedes TM 55-1520-238-10, dated 28 June 1984, including all changes.

URGENT

TM 1-1520-238-10

C 7

CHANGE } HEADQUARTERS
NO. 7 } DEPARTMENT OF THE ARMY
WASHINGTON, D.C., **15** December 1999

Operator's Manual
for
AH-64A HELICOPTER

DISTRIBUTION STATEMENT A: Approved for public release; distribution is unlimited.

TM 1-1520-238-10, 31 August 1994, is changed as follows:

1. Remove and insert pages as indicated below. New or changed text material is indicated by a vertical bar in the margin. An illustration change is indicated by a miniature pointing hand.

Remove pages	Insert pages
A and B	A and B
----	C/(D blank)
9-13 and 9-14	9-13 and 9-14

2. Retain this sheet in front of manual for reference purposes.

By Order of the Secretary of the Army:

ERIC K. SHINSEKI
General, United States Army
Chief of Staff

Official:

[signature]
JOEL B. HUDSON
Administrative Assistant to the
Secretary of the Amy
9933402

DISTRIBUTION:
To be distributed in accordance with initial distribution No. (IDN 310293) requirements for
TM **1-1520-238-10**.

TM 1-1520-238-10
C 6

CHANGE }
NO. 6

HEADQUARTERS
DEPARTMENT OF THE ARMY
WASHINGTON, D.C., 4 June 1999

Operator's Manual
for
AH-64A HELICOPTER

DISTRIBUTION STATEMENT A: Approved for public release; distribution is unlimited.

TM 1-1520-238-10, 31 August 1994, is changed as follows:

1. Remove and insert pages as indicated below. New or changed text material is indicated by a vertical bar in the margin. An illustration change is indicated by a miniature pointing hand.

Remove pages	Insert pages
A and B	A and B
- - - -	C/(D blank)
2-55 and 2-56	2-55 and 2-56
2-87 and 2-88	2-87 and 2-88
2-93 and 2-94	2-93 and 2-94
2-97 and 2-98	2-97 and 2-98
3-29 through 3-32	3-29 through 3-32
3-63 and 3-64	3-63 and 3-64
3-64.27/(3-64.28 blank)	3-64.27/(3-64.28 blank)
4-49 and 4-50	4-49 and 4-50
5-15 and 5-16	5-15 and 5-16
8 0 through 8 12	8-9 through 8-12
8-15 and 8-16	8-15 and 8-16
9-1 and 9-2	9-1 and 9-2
9-11 through 9-14	9-11 through 9-14

2. Retain this sheet in front of manual for reference purposes.

By Order of the Secretary of the Army:

DISTRIBUTION:
To be distributed in accordance with initial distribution No. (IDN 310293) requirements for TM 1-1520-238-10.

TM 1-1520-238-10
C 6

By Order of the Secretary of the Army:

ERIC K. SHINSEKI
General, United States Army
Chief of Staff

Official:

JOEL B. HUDSON
Acting Administrative Assistant to the
Secretary of the Army
9914404

DISTRIBUTION:
To be distributed in accordance with IDN 310293, requirements for TM 1-1520-238-10.

TM 1-1520-238-10
C5

CHANGE

NO. 5

HEADQUARTERS
DEPARTMENT OF THE ARMY
WASHINGTON, DC 27 February 1998

Operator's Manual
for
HELICOPTER, ATTACK,
AH-64A APACHE

DISTRIBUTATION STATEMENT A: Approved for public release; distribution is unlimited.

TM 1-1520-238-10, 31 August 1994, is changed as follows:

1. Remove old pages and insert new pages as indicated below. New or changed text material is indicated by a vertical bar in the margin. An illustration change is indicated by a miniature pointing hand.

Remove pages	Insert pages
A and B	A and B
i and ii	i and ii
2-69 and 2-70	2-69 and 2-70
- - - - - -	2-70.1/(2-70.2 blank)
4-75 and 4-76	4-75 and 4-76
4-77/(4-78 blank)	4-77 and 4-78
Index 2.1 and Index 2.2	Index 2.1 and Index 2.2
Index 7 and Index 8	Index 7 and Index 8
Cover 1/(Cover 2 blank)	Cover 1/(Cover 2 blank)

2. Retain this sheet, in front of manual for reference purposes.

By Order of the Secretary of the Army:

Official:

JOEL B. HUDSON
Administrative Assistant to the
Secretary of the Army
04914

DENNIS J. REIMER
General, United States Army
Chief of Staff

DISTRIBUTION:
To be distributed in accordance with Initial Distribution No. (IDN) 310293 requirements for TM 1-1520-23310.

TM 1-1520-238-10
C4

CHANGE }   HEADQUARTERS
NO.   4   DEPARTMENT OF THE ARMY
                        WASHINGTON, D.C., 30 July 1997

Operator's Manual
for
AH-64A HELICOPTER
TM 1-1520-238-10, 31 August 1994, is changed as follows:

1. Remove and insert pages as indicated below. New or changed text material is indicated by a vertical bar in the margin. An illustration change is indicated by a miniature pointing hand.

Remove pages	Insert pages
A and B	A and B
i and ii	i and ii
1-1 and 1-2	1-1 and 1-2
2-7 and 2-8	2-7 and 2-8
2-45 through 2-54	2-45 through 2-54
2-71 through 2-74	2-71 through 2-74
2-85 through 2-88	2-85 through 2-88
3-7 and 3-8	3-7 and 3-8
3-18.3 and 3.18.4	3-18.3 and 3.18.4
3-18.9 through 3-18.11/(3-18.12 blank)	3-18.9 through 3-18.12
3-19 and 3-20	3-19 and 3-20
3-33 and 3-34	3-33 and 3-34
. . . .	3-34.1/(3-34.2 blank)
3-55 and 3-56	3-55 and 3-56
3-61 through 3-64	3-61 through 3-64
3-64.1 through 3-64.26	3-64.1 through 3-64.26
. . . .	3-64.27/(3-64.28 blank)
3-65 and 3-66	3-65 and 3-66
4-1 and 4-2	4-1 and 4-2
4-5 through 4-14	4-5 through 4-14
4-14.1/(4-14.2 blank)	4-14.1/(4-14.2 blank)
4-15 and 4-16	4-15 and 4-16
. . . .	4-16.1/(4-16.2 blank)
4-17 through 4-20	4-17 through 4-20
4-33 and 4-34	4-33 and 4-34
4-49 through 4-56	4-49 through 4-56
4-59 and 4-60	4-59 and 4-60
4-63 through 4-66	4-63 through 4-66
4-66.1 through 4-66.3/(4-66.4 blank)	4-66.1 through 4-66.3/(4-66.4 blank)
4-67 through 4-70	4-67 through 4-70
. . . .	4-70.1/(4-70.2 blank)
5-9 through 5-14	5-9 through 5-14
6-1 and 6-2	6-1 and 6-2
6-5 and 6-6	6-5 and 6-6
8-9 through 8-16	8-9 through 8-16

TM 1-1520-238-10
C4

Remove pages	Insert pages
9-3 and 9-4	9-3 and 9-4
9-9 and 9-10	9-9 and 9-10
9-13 through 9-16	9-13 through 9-16
9-21 and 9-22	9-21 and 9-22
Index 1 and Index 2	Index 1 and Index 2
Index 3 through Index 6	Index 3 through Index 6
Index 9 and index 10	Index 9 and index 10
Index 13 through Index 20	Index 13 through Index 20

2. Retain this sheet in front of manual for reference purposes.

By Order of the Secretary of the Army:

DENNIS J. REIMER
General, United States Army
Chief of Staff

Official:

JOEL B. HUDSON
Administrative Assistant to the
Secretary of the Army
04011

DISTRIBUTION:
To be distributed in accordance with DA Form 12-31-E, block no. 0293, requirements for TM 1-1520-238-10.

TM 1-1520-238-10
C3

CHANGE

NO. 3

HEADQUARTERS DEPARTMENTS OF
THE ARMY AND THE AIR FORCE
WASHINGTON, D.C., 20 September 1996

Operator's Manual
for
HELICOPTER, ATTACK, AH-64A APACHE

DISTRIBUTION STATEMENT A: Approved for public release; distribution is unlimited

TM 1-1520-238-10 ,31 August 1994, is changed as follows:

1. Remove and insert pages as indicated below. New or changed text material is indicated by a vertical bar in the margin. An illustration change is indicated by a miniature pointing hand.

Remove pages	Insert pages
A and B	A and B
2-1 and 2-2	2-1 and 2-2
2-15 and 2-16	2-15 and 2-16
2-45 through 2-48	2-45 through 2-48
2-55 and 2-56	2-55 and 2-56
2-67 through 2-70	2-67 through 2-70
3-3 through 3-6	3-3 through 3-6
--------	3-18.1 through 3-18.11/(3-18.12 blank)
3-19 and 3-20	3-19 and 3-20
3-29 through 3-64	3-29 through 3-64
3-64.1 through 3-64.26	
3-65 and 3-66	3-65 and 3-66
4-1 through 4-6	4-1 through 4-6
4-9 and 4-10	4-9 and 4-10
4-19 and 4-20	4-19 and 4-20
4-25 and 4-26	4-25 and 4-26
4-33 and 4-34	4-33 and 4-34
4-39 through 4-42	4-39 through 4-42
4-47 through 4-50	4-47 through 4-50
4-61 and 4-62	4-61 and 4-62
4-69 and 4-70	4-69 and 4-70
5-1 through 5-4	5-1 through 5-4
5-9 through 5-14	5-9 through 5-14
6-5 and 6-6	6-5 and 6-6
6-17 and 6-18	6-17 and 6-18
7-1 through 7-4	7-1 through 7-4

TM 1-1520-238-10

C3

Remove pages	Insert pages
7-11 and 7-12	7-11 and 7-12
7A-1 through 7A-4	7A-1 through 7A-4
7A-65 and 7A-66	7A-65 and 7A-66
8-3 through 8-6	8-3 through 8-6
8-11 and 8-12	8-11 and 8-12
8-15 and 8-16	8-15 and 8-16
9-3 and 9-4	9-3 lnd 9-4
9-7 through 9-22	9-7 through 9-22
B-1 through B-8	B-1 through B-8
B-11 and B-12	B-11 and B-12
B-15 and B-16	B-15 and B-16
Index 1 and Index 2	Index 1 and Index 2
----------	Index 2.1 and Index 2.2
Index 3 through Index 14	Index 3 through Index 14

2. Retain this sheet in front of manual for reference purposes.

By Order of the Secretary of the Army:

Official:
Official:

JOEL B. HUDSON
Administrative Assistant to the
Secretary of the Army
02728

DENNIS J. REIMER
General, United States Army
Chief of Staff

DISTRIBUTION:
To be distributed in accordance with DA Form 12-31-E, block no. 0293, requirements for TM 1-1520-238-10.

TM 1-1520-238-10

CHANGE

NO. 2

HEADQUARTERS
DEPARTMENT OF THE ARMY
WASHINGTON, D. C., 5 February 1996

Operator's Manual
for

HELICOPTER, ATTACK, AH-64A APACHE

DISTRIBUTION STATEMENT A Approved for public release; distribution is unlimited.

TM 1-1520-238-10, 31 August 1994, is changed as follows:

1. Distribution Statement is changed on the cover as shown above.

2. Remove and insert pages as indicated below. New or changed text material is indicated by a vertical bar in the margin. An illustration change is indicated by a miniature pointing hand.

Remove pages	Insert pages
A/(B blank)	A and B
2-11 through 2-16	2-11 through 2-16
2-23 through 2-26	2-23 through 2-26
2-29 through 2-30	2-28 through 2-30
2-33 through 2-36	2-33 through 2-36
2-65 and 2-66	2-65 and 2-66
2-79 and 2-80	2-78 and 2-80
- - - - -	2-80.1/(2-80.2 blank)
2-83 through 2-88	2-83 through 2-88
3-7 and 3-8	3-7 and 3-8
3-17 and 3-18	3-17 and 3-18
3-69 and 3-70	3-68 and 3-70
4-1 and 4-2	4-1 and 4-2
4-48 and 4-50	4-49 and 4-50
4-63 and 4-64	4-63 and 4-64
4-68 and 4-70	4-68 and 4-70
5-1 and 5-2	5-1 and 5-2
5-9 and 5-10	5-9 and 5-10
6-7 and 6-8	6-7 and 6-8
6-11 and 6-12	6-11 and 6-12
6-15 and 6-16	6-15 and 6-16
7-3 and 7-4	7-3 and 7-4
7-68 and 7-70	7-69 and 7-70
7A-3 and 7A-4	7A-3 and 7A-4
7A-65 and 7A-66	7A-65 and 7A-66
8-7 through 8-16	8-7 through 8-16

TM 1-1520-238-10
C2

Remove pages	Insert pages
9-3 and 9-4	9-3 and 9-4
9-9 and 9-10	9-9 and 9-10
9-13 through 9-22	9-13 through 9-22
Index 1 through Index 4	Index 1 through Index 4
Index 11 through Index 14	Index 11 through Index 14
Index 17 through Index 20	Index 17 through Index 20

3. Retain this sheet in front of manual for reference purposes.

By Order of the Secretaries of the Army:

DENNIS J. REIMER
General, United States Army
Chief of Staff

JOEL B. HUDSON
Acting Administrative Assistant to the
Secretary of the Army

DISTRIBUTION:
To be distributed in accordance with DA Form 12-31-E, block no. 0293, requirements for TM 1-1520-238-10.

TM 1-1520-238-10
C1

CHANGE

NO. 1

HEADQUARTERS
DEPARTMENT OF THE ARMY
WASHINGTON, D. C., 15 May 1995

Operator's Manual
for
HELICOPTER, ATTACK AH-64A APACHE

<u>DISTRIBUTION STATEMENT C</u>: Distribution is authorized to U.S. Government agencies and their contractors only to protect technical or operational information from automatic dissemination under the International Exchange Program or by other means. This determination was made on 1 July 1994. Other requests for this document will be referred to U.S. Army Aviation and Troop Command, ATTN: AMSAT-I-MT, 4300 Goodfellow Blvd., St. Louis, MO 63120-1798.

TM 1-1520-238-10, 31 August 1994, is changed as follows:

1. Remove and insert pages as indicated below. New or changed text material is indicated by a vertical bar in the margin. An illustration change is indicated by a miniature pointing hand.

Remove pages	Insert pages
- - - - -	A/(B blank)
1-1 and 1-2	1-1 and 1-2
2-7 and 2-8	2-7 and 2-8
4-7 through 4-10	4-7 through 4-10
4-13 and 4-14	4-13 and 4-14
- - - -	4-1 4.1/(4-1 4.2 blank)
4-15 through 4-20	4-15 through 4-20
4-49 through 4-60	4-49 through 4-60
4-65 and 4-66	4-65 and 4-66
- - - -	4-66.1 through 4-66.3/(4-66.4 blank)
4-67 and 4-68	4-67 and 4-68
B-3 and B-4	B-3 and B-4
Index 1 through Index 4	Index 1 through Index 4
Index 13 and Index 14	Index 13 and Index 14
Index 17 and Index 18	Index 17 and Index 18

2. Retain this sheet in front of manual for reference purposes.

TM 1-1520-238-10

INSERT LATEST CHANGED PAGES: DESTROY SUPERSEDED PAGES.

LIST OF EFFECTIVE PAGES

NOTE: The portion of the text affected by the changes is indicated by a vertical line in the outer margins of the page. Changes to illustrations are indicated by miniature pointing hands. Changes to wiring diagrams are indicated by shaded areas.

Date of issue for original and change pages are:

Original . . . . 0 . . . . . . 31 August 1994
Change . . . . 1 . . . . 15 May 1995
Change . . . . 2 . . . . 5 February 1996
Change . . . . 3 . . . 20 September 1996
Change . . . . 4 . . . . . . 30 July 1997
Change . . . . 5 . . . . 27 February 1998
Change . . . . 6 . . . . 4 June 1999
Change . . . . 7 . . . 15 December 1999

TOTAL NUMBER OF PAGES IN THIS PUBLICATION IS 533, CONSISTING OF THE FOLLOWING:

Page No.	*Change No.	Page No.	*Change No.
Title	5	2-80	2
A - B	7	2-80.1/(2-80.2 blank)	2
C/(D blank)	7	2-81 – 2-82	0
a - c/(d blank)	0	2-83 – 2-85	2
i	5	2-86	4
ii-iii/(iv blank)	0	2-87	6
1-1	0	2-88 – 2-92	0
1-2	4	2-93 – 2-94	6
2-1	3	2-95 – 2-96	0
2-2 – 2-6	0	2-97 – 2-98	6
2-7	4	3-1 – 3-3	0
2-8 – 2-11	0	3-4 – 3-5	3
2-12 – 2-13	2	3-6	0
2-14	0	3-7	4
2-15	3	3-18	0
2-16 – 2-23	0	3-17	ii
2-24	2	3-18	0
2-25	0	3-18.1 – 3-18.2	3
2-26	2	3-18.3 – 3-18.4	4
2-27 – 2-28	0	3-18.5 – 3-18.8	3
2-29	2	3-18.9	4
2-30 – 2-33	0	3-18.10	3
2-34 – 2-35	2	3-18.11 – 3-18.12	4
2-36 – 2-45	0	3-19	4
2-46 – 2-48	4	3-20 – 3-29	0
2-49	0	3-30	6
2-50 – 2-51	4	3-31	3
2-52	0	3-32	6
2-53 – 2-54	4	3-33 – 3-34	4
2-55	0	3-34.1/(3-34.2 blank)	4
2-56	6	3-35 – 3-54	3
2-57 – 2-65	0	3-55 – 3-56	4
2-66	2	3-57 – 3-61	3
2-67	3	3-62	4
2-68 – 2-69	0	3-63	6
2-70	5	3-64	4
2-70.1/(2-70.2 blank)	5	3-64.1 – 3-64.26	4
2-71	0	3-64.27/(3-64.28 blank)	6
2-72	4	3-65	4
2-73	0	3-66 – 3-68	0
2-74	4	3-69 – 3-70	2
2-75 – 2-79	0	3-71 – 3-72	0

*Zero in this column indicates an original page.

Change 7   A

TM 1-1520-238-10

INSERT LATEST CHANGED PAGES: DESTROY SUPERSEDED PAGES.

LIST OF EFFECTIVE PAGES

Page No.	*Change No.
4-1 – 4-2	4
4-3 – 4-4	3
4-5	4
4-6	0
4-7 – 4-10	4
4-11	0
4-12 – 4-14	4
4-14.1/(4-14.2 blank)	4
4-15 – 4-16	4
4-16.1/4-16.2 blank)	4
4-17 – 4-19	4
4-20	3
4-21 – 4-24	0
4-25 – 4-26	3
4-27 – 4-32	0
4-33	4
4-34	3
4-35 – 4-39	0
4-40	3
4-41	0
4-42	3
4-43 – 4-47	0
4-48	3
4-49	4
4-50	6
4-51 – 4-55	4
4-56 – 4-57	1
4-58 – 4-59	0
4-60	4
4-61	3
4-62	0
4-63	2
4-64	4
4-65	0
4-66	4
4-66.1–4-66.3/(4-66.4 blank)	4
4-67 – 4-70	4
4-70.1(4-70.2 blank)	4
4-71 – 4-74	0
4-75 – 4-78	5
5-1	2
5-2 – 5-3	3
5-4 – 5-8	0
5-9	4
5-10	0
5-11	4
5-12 – 5-13	0
5-14	4
5-15	0
5-16	6
5-17/(5-18 blank)	0

Page No.	*Change No.
6-1-6-5	0
6-6	4
6-7 – 6-8	2
6-9 – 6-10	0
6-11 – 6-12	2
6-13 – 6-15	0
6-16	2
6-17	0
6-18	3
7-1	0
7-2 – 7-3	3
7-4	2
7-5 – 7-11	0
7-12	3
7-13 – 7-68	0
7-69	2
7-70 – 7-73/(7-74 blank)	0
7A-1	0
7A-2 – 7A-3	3
7A-4	2
7A-5 – 7A-65	0
7A-66	3
7A-67 – 7A-70	0
8-1 – 8-3	0
8-4 – 8-6	3
8-7	2
8-8	0
8-9	2
8-10 – 8-12	6
8-13 – 8-14	4
8-15 – 8-16	6
8-10 – 8-16	4
8-17 – 8-22	0
9-1	0
9-2	6
9-3	0
9-4	4
9-5 – 9-6	0
9-7	3
9-8	0
9-9	4
9-10	3
9-11	6
9-12	6
9-13 – 9-14	7
9-15	4
9-16 – 9-20	3
9-21 – 9-22	4

*Zero in this column indicates an original page.

B Change 7

TM 1-1520-238-10

INSERT LATEST CHANGED PAGES: DESTROY SUPERSEDED PAGES.

LIST OF EFFECTIVE PAGES

Page No.	*Change No.	Page No.	*Change No.
A-I – A-2	0	Index 3	1
B-I – B-5	3	Index 4	4
B-6	0	Index 5	0
B-7	3	Index 6	4
B-8 – B-10	0	Index 7	0
B-11 – B-12	3	Index 8	5
B-13 – B-14	0	Index 9	4
B-15 – B-16	3	Index 10	0
B17/(B-18 blank)	0	Index 11 – Index 13	3
Index 1	4	Index 14 – Index 15	4
Index 2	3	Index 16 – Index 17	0
Index 2.1	3	Index 18 – Index 19	4
Index 2.2	5	Index 20 – Index 23/(Index 24)	0

*Zero in this column indicates an original page.

Change 7   C/(D blank)

TM 1-1520-238-10

WARNING

Personnel performing operations, procedures, and practices which are included or implied in this technical manual shall observe the following warnings. Disregard of these warnings and precautionary information can cause serious injury or loss of life.

WARNING

AVIATION LIFE SUPPORT EQUIPMENT

Aviation life support equipment shall be utilized in accordance with AR 95-1 and FM 1-302. Failure to do so may result in personal injury or loss of life.

WARNING

BATTERY ELECTROLYTE

Battery electrolyte is harmful to the skin and clothing. Neutralize any spilled electrolyte by thoroughly flushing contacted area with water.

WARNING

CANOPY JETTISON

Canopy jettison safety pins shall be installed in pilot, copilot/gunner, and external firing mechanisms when the helicopter is on the ground. The canopy jettison system is manually operated. The canopy can be jettisoned when no electrical power is on the helicopter. Pilot and copilot/gunner safety pins shall be removed before starting engines. Safety pins shall be Installed during engine shutdown check. Debris may be expelled 50 feet outward when system is actuated. Pilot and copilot/gunner helmet visor should be down to prevent eye injury.

WARNING

CARBON MONOXIDE

When smoke, suspected carbon monoxide fumes, or symptoms of anoxia exist, the crew should immediately ventilate the cockpit.

WARNING

ELECTROMAGNETIC INTERFERENCE (EMI)

No electrical/electronic devices of any sort, other than those described in this manual or appropriate maintenance manuals, are to be operated by crew members during operation of this helicopter. Flights near high power radio transmitters' high intensity radio transmission areas (HIRTA) may cause degraded system operation.

WARNING

FIRE EXTINGUISHER

Exposure to high concentrations of extinguishing agent or decomposition products should be avoided. The liquid should not be allowed to contact the skin; it may cause frostbite or low-temperature burns.

WARNING

GROUND OPERATION

Engines will be started and operated only by authorized personnel. Reference AR 95-1 and AR 95-13.

WARNING

HANDLING FUEL, OIL, AND HYDRAULIC FLUIDS

Turbine and lubricating oils contain additives which are poisonous and readily absorbed through the skin. Do not allow them to remain on skin longer than necessary. Prolonged contact may cause skin rash. Prolonged contact with hydraulic fluid may cause burns. Refer to TM 10-1101 and FM 10-68 when handling fuel.

WARNING

HIGH VOLTAGE

All ground handling personnel must be informed of high voltage hazards when working near Target Acquisition Designator Sight (TADS) and Pilot Night Vision Sensor (PNVS) equipment.

WARNING

laser light hazard

LASER LIGHT

The laser light beam is dangerous and can cause blindness if it enters the eye either directly or reflected from a surface. Personnel should wear approved laser protection whenever in a controlled area when laser rangefinder or laser target designators are being used. Laser shall be used only in controlled areas by qualified personnel.

TM 1-1520-238-10

WARNING

NOISE

Sound pressure levels around helicopters during some operating conditions exceed the Surgeon General's hearing conservation criteria as defined In TB MED 251. Hearing protection devices, such as the aviator helmet or ear plugs, are required to be worn by all personnel in and around the helicopter during Its operation.

WARNING

STARTING ENGINES AND AUXILIARY POWER UNIT

Be sure that the rotor and blast area is clear, and a fire guard is posted if available.

WARNING

VERTIGO

The anti-collision strobe lights should be off during fright through clouds to prevent sensations of vertigo as a result of reflections of the light on the clouds.

WARNING

WEAPONS AND AMMUNITION

Observe all standard safety precautions governing the handling of weapons and live ammunition. When not in use, point all weapons in a direction away from personnel and property in case of accidental firing. Do not walk in front of weapons. SAFE all weapons before servicing. To avoid potentially dangerous situations, follow the procedural warnings in this text.

WARNING

WING STORES JETTISON

All jettison safety pins shall be installed when the helicopter is on the ground. Safety pins shall be removed prior to fright. Failure to do so will prevent jettison of wing stores.

c/(d blank)

Technical Manual

No. 1-1520-238-10

TM 1-1520-238-10

HEADQUARTERS
DEPARTMENT OF THE ARMY
WASHINGTON, D. C.   31 August 1994

OPERATOR'S MANUAL

FOR

HELICOPTER, ATTACK,
AH-64A APACHE

REPORTING ERRORS AND RECOMMENDING IMPROVEMENTS

You can help improve this publication. If you find any mistakes, or if you know of a way to improve these procedures, please let us know. Mail your letter or DA Form 2028 (Recommended Changes to Publications and Blank Forms) directly to: Commander, US Army Aviation and Missile Command, ATTN: AMSAM-MMC-LS-LP Redstone Arsenal, AL, 35898-5230. You may also submit your recommended changes by E-mail directly to ls-lp@redstone.army.mil. A reply will be furnished directly to you. Instructions for sending an electronic 2028 may be found at the end of this TM immediately preceding the hard copy 2028.

DISTRIBUTION STATEMENT A: Approved for public release; distribution is unlimited.

TABLE OF CONTENTS

		Page
CHAPTER 1	INTRODUCTION	1-1
CHAPTER 2	AIRCRAFT AND SYSTEMS DESCRIPTION AND OPERATION	2-1
Section I.	Aircraft	2-1
Section II.	Emergency Equipment	2-19
Section III.	Engines and Related Systems	2-22
Section IV.	FuelSystem	2-34
Section V.	Flight Control System	2-42
Section VI.	Hydraulic and Pressurized Air Systems	2-50
Section VII.	Power Train System	2-56
Section VIII.	Rotors	2-59
Section IX.	Utility Systems	2-60
Section X.	Heating, Ventilation, Cooling, and Environmental Control Systems	2-64
Section XI.	Electrical Power Supply and Distribution Systems	2-66
Section XII.	Auxiliary Power Unit	2-72
Section XIII.	Lighting	2-74
Section XIV.	Flight Instruments	2-76
Section XV.	Servicing, Parking, and Mooring	2-86

* This manual supersedes TM 55-1520-238-10, dated 28 June 1984, including all changes.

Change 5   i

TABLE OF CONTENTS - continued

Page

CHAPTER 3	AVIONICS		3-1
Section I.		General	3-1
Section II.		Communications	3-7
Section III.		Navigation	3-30
Section IV.		Transponder and Radar	3-66
CHAPTER 4	MISSION EQUIPMENT		4-1
Section I.		Mission Avionics	4-1
Section II.		Armament	4-10
Section III.		Active and Passive Defense Equipment	4-69
CHAPTER 5	OPERATING LIMITS AND RESTRICTIONS		5-1
Section I.		General	5-1
Section II.		System Limits	5-2
Section III.		Power Limit	5-9
Section IV.		Loading Limits	5-10
Section V.		Airspeed Limits Maximum and Minimum	5-11
Section VI.		Maneuvering Limits	5-14
Section VII.		Environmental Restrictions	5-16
Section VIII.		Other Limits	5-17
CHAPTER 6	WEIGHT/BALANCE AND LOADING		6-1
Section I.		General	6-1
Section II.		Weight and Balance	6-3
Section III.		Fuel and oil	6-6
Section IV.		Personnel	6-10
Section V.		Mission Equipment	6-11
Section VI.		Cargo Loading	6-16
Section VII.		Center of Gravity	6-17
CHAPTER 7	PERFORMANCE DATA FOR AH-64A HELICOPTERS EQUIPPED WITH T700-GE-701 ENGINES		7-1
Section I.		Introduction	7-1
Section II.		Maximum Torque Available	7-4
Section III.		Hover Ceiling	7-9
Section IV.		Hover Limits	7-11
Section V.		Cruise	7-13
Section VI.		Drag	7-69
Section VII.		Climb-Descent	7-72

TABLE OF CONTENTS - continued

Page

CHAPTER 7A	PERFORMANCE DATA FOR AH-64A HELICOPTERS EQUIPPED WITH T700-GE-701C ENGINES	7A-1
Section I.	Introduction	7A-1
Section II.	Maximum Torque Available	7A-4
Section III.	Hover Ceiling	7A-10
Section IV.	Hover Limits	7A-13
Section V.	Cruise	7A-15
Section VI.	Drag	7A-66
Section VII.	Climb-Descent	7A-69
CHAPTER 8	NORMAL PROCEDURES	8-1
Section I.	Crew Duties	8-1
Section II.	Operating Procedures and Maneuvers	8-2
Section III.	Instrument Flight	8-17
Section IV.	Flight Characteristics	8-18
Section V.	Adverse Environmental Conditions	8-19
CHAPTER 9	EMERGENCY PROCEDURES	9-1
Section I.	Aircraft Systems	9-1
Section II.	Mission Equipment	9-21
APPENDIX A	REFERENCES	A-1
APPENDIX B	ABBREVIATIONS AND TERMS	B-1
ALPHABETICAL INDEX		Index 1

CHAPTER 1
INTRODUCTION

1.1 GENERAL.

These instructions are for use of the operators. They apply to AH-64A helicopters.

1.2 WARNINGS, CAUTIONS, AND NOTES.

Warnings, Cautions, and Notes are used to emphasize important and critical instruction and are used for the following conditions:

WARNING

An operating procedure, practice, condition or statement, which if not correctly followed, could result in personal injury or loss of life.

CAUTION

An operating procedure, practice, condition or statement, which if not strictly observed, could result in damage to or destruction of equipment, loss of mission effectiveness or long term health hazards to personnel.

NOTE

An operating procedure, condition or statement, which is essential to highlight.

1.3 DESCRIPTION.

This manual contains the best operating instructions and procedures for the AH-64A under most circumstances. The observance of limitations, performance, and weight balance data provided is mandatory. The observance of procedure is mandatory, except when modification is required because of multiple emergencies, adverse weather, terrain, etc. Basic flight principles are not included. *THIS MANUAL SHALL BE CARRIED IN THE HELICOPTER AT ALL TIMES.*

The AH-64A helicopter is designed as a weapons-delivery platform and is equipped with point target (Hellfire missile), area weapon (30mm chain gun), and aerial rocket (2.75-inch folding-fin type) systems. The AH-64A carries two crewmembers: a pilot and a copilot/gunner (CPG).

1.4 APPENDIX A, REFERENCES.

Appendix A is a listing of official publications cited within the manual applicable to, and available for, flight crews.

NOTE

Appendix A shall contain only those publications referenced in the manual, and shall not contain Department of the Army blank forms.

1.5 APPENDIX B, ABBREVIATIONS AND TERMS.

Definitions of all abbreviations and terms used throughout the manual are included in Appendix B.

1.6 INDEX.

The index lists, in alphabetical order, paragraphs, figures, and tables contained in this manual by page number.

1.7 ARMY AVIATION SAFETY PROGRAM.

Reports necessary to comply with the safety program are prescribed in AR 385-40.

1.8 DESTRUCTION OF ARMY MATERIAL TO PREVENT ENEMY USE.

For information concerning destruction of Army materiel to prevent enemy use, refer to TM 750-244-1-5.

1.9 FORMS AND RECORDS.

Army aviator's flight record and aircraft maintenance records, which are to be used by crewmembers, are described in DA PAM 738-751 and TM 55-1500-342-23.

1.10 EXPLANATION OF CHANGE SYMBOLS.

Changes to the text and tables, including new material on added pages, shall be identified by a vertical bar in the outer margin of the column of text in which the change appears, extending close to the entire area of the material affected. Change symbols for single column text shall be placed in the margin opposite the binding. Change symbols for double column text shall be placed in the margin adjacent to the binding for the columns of text nearest the binding. The change symbols shall be placed in the outer margin opposite the binding for the column of text farthest from the binding. Pages with emergency markings, which consist of black diagonal lines around three edges, shall have the vertical bar or change symbol placed in the margin between the text and the diagonal lines. Change symbols shall indicate the current changes only. A miniature pointing hand symbol shall be used to denote a change to an illustration. However, a vertical line in the outer margin (opposite the binding) rather than miniature pointing hands, shall be utilized when there have been extensive changes made to an illustration. Change symbols shall not be used to indicate changes in the following:

    a. Introductory material.

    b. Indexes and tabular data where the change cannot be identified.

    c. Correction of minor inaccuracies, such as spelling, punctuation, relocation of material, etc., unless such correction changes the meaning of the instructive information and procedures.

1.11 SERIES AND EFFECTIVITY CODES

All AH-64A helicopters have BUCS equipment installed. In most helicopters, the system is deactivated; in some it is operable. The designator symbol **B** indicates text headings, text contents and illustrations pertaining to helicopters with an operable BUCS.

Some AH-64A helicopters have T700-GE-701C engines installed. Those helicopters will have components, instrumentation, performance parameters, and procedures different from helicopters with T700-GE-701 engines installed. The designator symbols **701** and **701C** indicate material pertaining to those specific engines.

Some AH-64A helicopters have the 7-319200005-11 Fire Control Computer (FCC) with -51 software installed (EGI Mod); others have the 7-319200005-9A Fire Control Computer (FCC) with -49A software installed; others yet have the 7-319200005-5 FCC with -45 software. Because of differences in operation, displays, etc. designator symbols, **-49A**, and **-51** will indicate material peculiar to that software installation.

1.12 USE OF SHALL, SHOULD, AND MAY.

Within this technical manual, the word *shall* is used to indicate a mandatory requirement. The word *should* is used to indicate non-mandatory but preferred method of accomplishment. The word *may* is used to indicate an acceptable method of accomplishment.

CHAPTER 2
AIRCRAFT AND SYSTEMS DESCRIPTION AND OPERATION

Section I. AIRCRAFT

2.1 GENERAL.

The AH-64A helicopter is a twin engine, tandem seat, aerial weapons platform.

2.2 AIRCRAFT GENERAL ARRANGEMENT.

Figure 2-2 illustrates the general arrangement including accessing and some major exterior components.

2.2.1 Fuselage. The fuselage includes a forward, center, and aft section that employ aluminum alloy semi-monocoque construction. All major weight items (crew, fuel, and ammunition) are supported by bulkheads, frames, and a longitudinal support structure. The forward fuselage contains the copilot/gunner (CPG) station. There are also provisions for mounting the target acquisition and designation sight (TADS), pilot night vision sensor (PNVS), and a 30mm area weapon. The center section contains the pilot crew station and provides support for the oleo-damped main landing gear, main transmission, wings, fuel cells, and ammunition bay. The aft section includes the vertical stabilizer and has provisions for mounting the tail landing gear. The avionics bay and stowage compartments are contained in the aft section. The tail rotor, drive shafts, gearboxes, and stabilator are attached to the aft section.

2.2.2 Wings. Left and right wings are attached to the center fuselage. They are of aluminum cantilever, spar, and rib construction. Each wing provides two hardpoints for external stores and hydraulic and electrical quick disconnects.

2.2.3 Rotors. The helicopter has a fully articulated four-blade main rotor system equipped with elastomeric lead-lag dampers. The tail rotor is a semi-rigid design and consists of four blades.

2.2.4 Engines. The helicopter is powered by two horizontally-mounted turbo-shaft engines. Power is supplied to the main transmission through engine- mounted nose gearboxes, shafts, and overrunning clutches. The main transmission drives the main and X tail rotors and accessory gearbox.

2.3 SPECIAL MISSION KITS.

The helicopter can be equipped with an IR jammer kit, radar jammer kit, radar warning kit, winterization kit, chaff kit, and extended range kit. Refer to the applicable system for descriptive information.

2.4 PRINCIPAL DIMENSIONS.

Figure 2-3 illustrates principal helicopter dimensions.

2.5 TURNING RADIUS AND GROUND CLEARANCE.

Figure 2-4 illustrates helicopter turning radius and ground clearance.

2.6 DANGER AREAS.

2.6.1 Shaded Areas Illustrated. The illustrated shaded areas (fig 2-5) can be hazardous. Personnel approaching an operating helicopter must do so at a 45-degree angle from the front. The approach must be made from well outside the rotor disc area until recognition is received from the pilot. The pilot will then signal when closer approach is safe.

2.6.2 Air Flow. Air flow from the tail rotor and downwash from the main rotor are dangerous, even outside the turning radius of the helicopter when it is in hover or operating at takeoff power.

2.6.3 Exhaust Gases. Exhaust gases from the helicopter engines and auxiliary power unit (APU) can cause burns. Personnel should remain clear of these areas.

2.6.4 Canopy Jettison. During canopy jettison, acrylic fragments will be propelled approximately 50 feet from the helicopter. Personnel approaching a crash-damaged helicopter shall look for a signal from the crew that closer approach is safe.

2.6.5 Laser. The laser shall be given special safety considerations because of the extreme danger involved during its operation. Relatively low laser light levels can cause permanent damage to eyes and skin burns. There is an additional danger of electrical shock horn laser components.

2.7 EQUIPMENT STOWAGE COMPARTMENTS.

The aft storage bay (fig 2-2) is for the stowage of tie down devices, protective covers, and other helicopter equipment. The loading conditions for this bay are covered in Chapter 6, Weight/Balance and Loading. The survival equipment storage bay (fig 2-2) is large enough to store a combat helmet, an environmental survival kit, a survival weapon, and a box of field-type rations for each crewmember. The loading limitations for this bay are covered in Chapter 6, Weight/Balance and Loading.

2.8 WINDSHIELD AND CANOPY PANELS.

2.8.1 Windshield. The windshield consists of two heated laminated glass windshields. One is directly forward of the CPG; the other is directly above his head. The canopy consists of five acrylic panels: Two on each side of the crew stations and one directly above the pilot.

2.8.2 Canopy Panels. The two canopy panels on the right are independently hinged. They latch and unlatch separately by interior or exterior handles. They swing upward to provide entrance to, and exit from, the crew station. Failure to properly close either canopy causes the CANOPY caution light on the pilot caution/warning panel (fig 2-7) to illuminate. The two canopy panels on the left side are fixed and do not open.

2.8.3 Canopy Jettison System. The canopy jettison system provides rapid egress paths when the helicopter access door(s) are jammed or blocked. It consists of three CANOPY JETTISON handles and detonation cords installed around the periphery of each of the four acrylic side panels on the sides of the pilot and CPG stations. The pilot handle (fig 2-1) is located at the upper left corner of the pilot instrument panel (fig 2-7). The CPG handle (fig 2-1) is located at the upper left corner of the CPG panel (fig 2-8). The external ground crew handle is located under a quick-release panel directly forward of the CPG windshield (fig 2-2). When operated, the system severs the four side panels. To arm the system, a CANOPY JETTISON handle is rotated 90° left or right, which uncovers the word ARMED on both sides of the handle. To activate the system, the rotated CANOPY JETTISON handle is pushed in, detonating a primer/initiator within the handle. The primer/initiator ignites the detonation cord which, in turn, ignites and burns around the periphery of the side panels. The burning action cuts a fine line around the side panels, severing them from the fuselage.

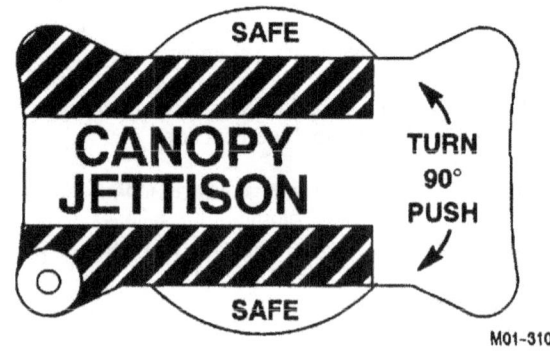

Figure 2-1. Canopy Jettison Handle

TM 1-1520-238-10

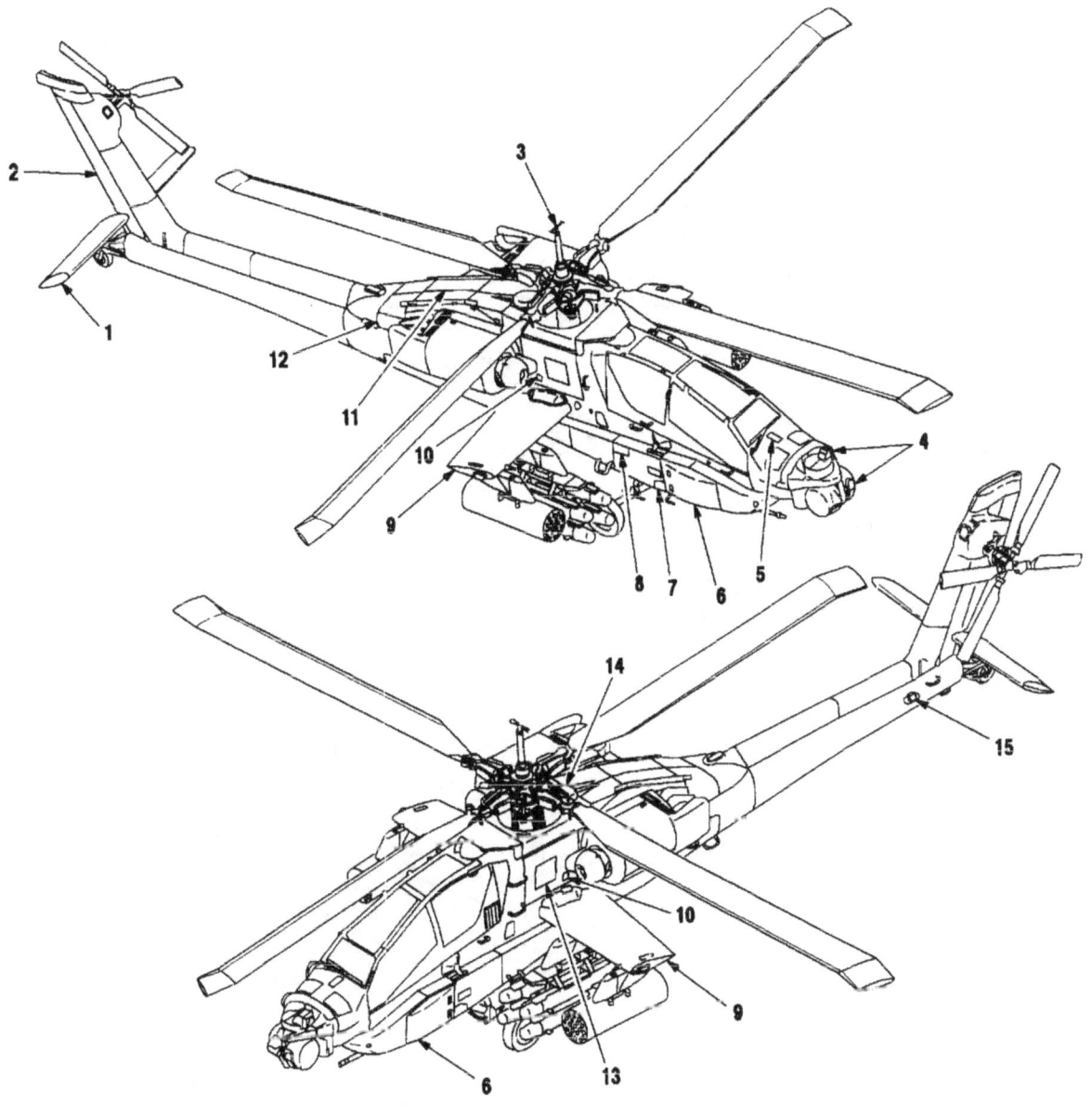

1. STABILATOR
2. VERTICAL STABILIZER
3. AIR DATA SENSOR
4. TADS AND PNVS TURRETS
5. CANOPY JETTISON HANDLE ACCESS DOOR
6. FORWARD AVIONICS BAY ACCESS DOOR
7. MOORING LUG ACCESS DOOR
8. FIRE EXTINGUISHER ACCESS DOOR
9. INTERCOMM ACCESS DOOR
10. MAIN TRANSMISSION OIL LEVEL SIGHT GAGE ACCESS DOOR
11. AFT EQUIPMENT BAY (CATWALK AREA) ACCESS DOORS
12. HYDRAULIC GROUND SERVICE PANEL ACCESS DOOR
13. HYDRAULIC OIL LEVEL SIGHT GAGE ACCESS DOOR
14. INFRARED COUNTERMEASURE DEVICE MOUNT
15. CHAFF PAYLOAD MODULE MOUNT

M01-094-1

Figure 2-2. General Arrangement (Sheet 1 of 2)

2-3

TM 1-1520-238-10

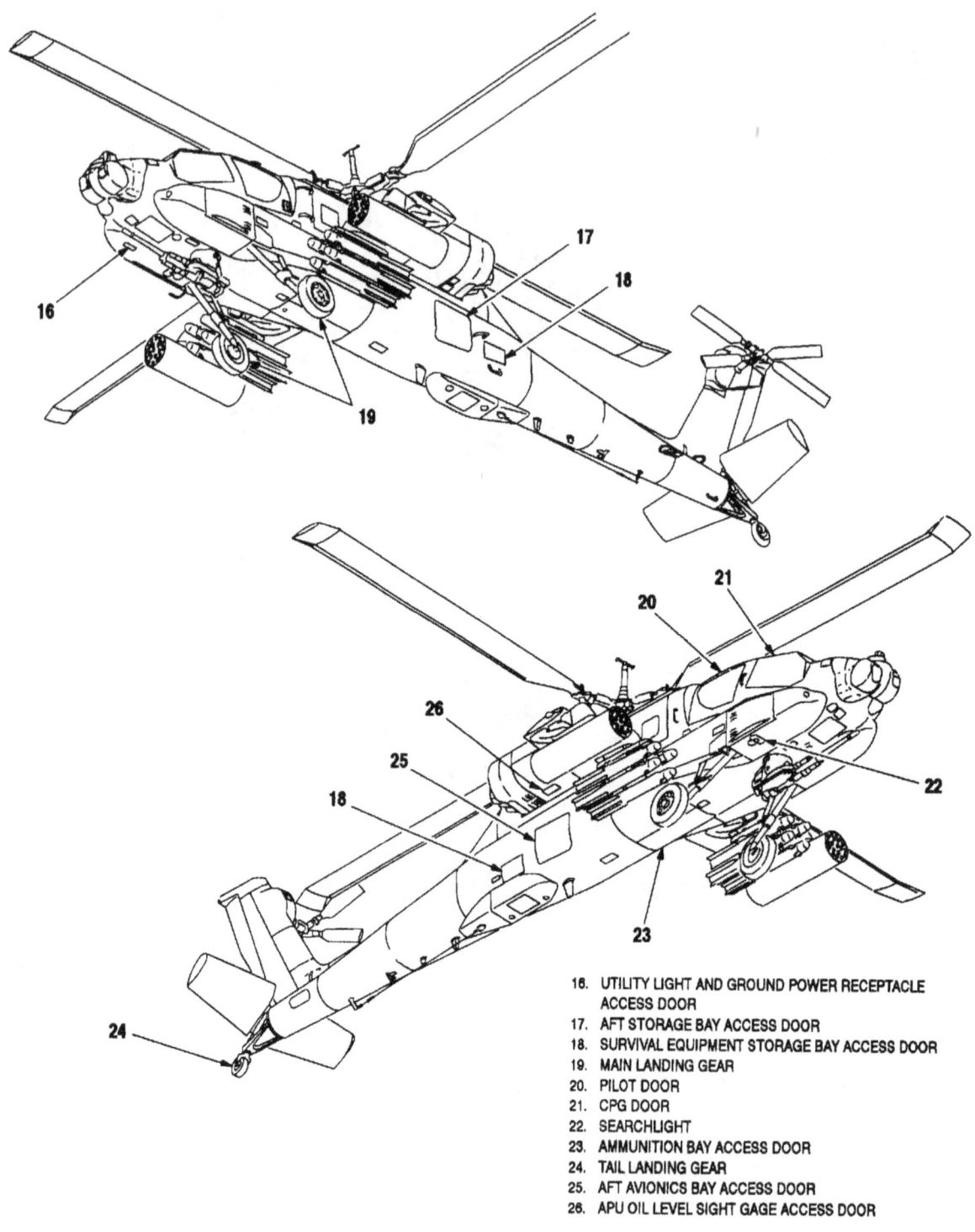

16. UTILITY LIGHT AND GROUND POWER RECEPTACLE ACCESS DOOR
17. AFT STORAGE BAY ACCESS DOOR
18. SURVIVAL EQUIPMENT STORAGE BAY ACCESS DOOR
19. MAIN LANDING GEAR
20. PILOT DOOR
21. CPG DOOR
22. SEARCHLIGHT
23. AMMUNITION BAY ACCESS DOOR
24. TAIL LANDING GEAR
25. AFT AVIONICS BAY ACCESS DOOR
26. APU OIL LEVEL SIGHT GAGE ACCESS DOOR

M01-094-2

Figure 2-2. General Arrangement (Sheet 2 of 2)

TM 1-1520-238-10

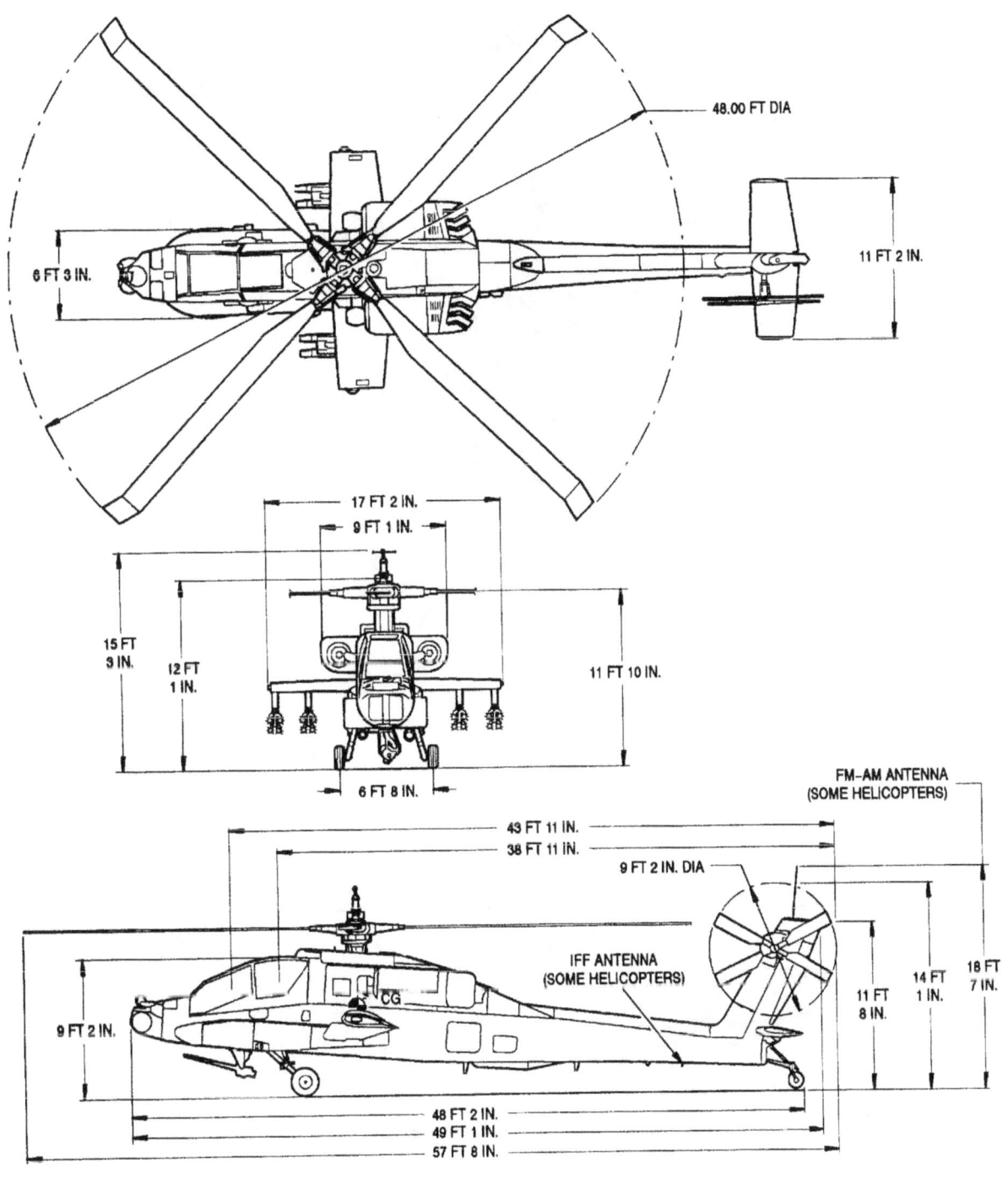

Figure 2-3. Principal Dimensions

TM 1-1520-236-10

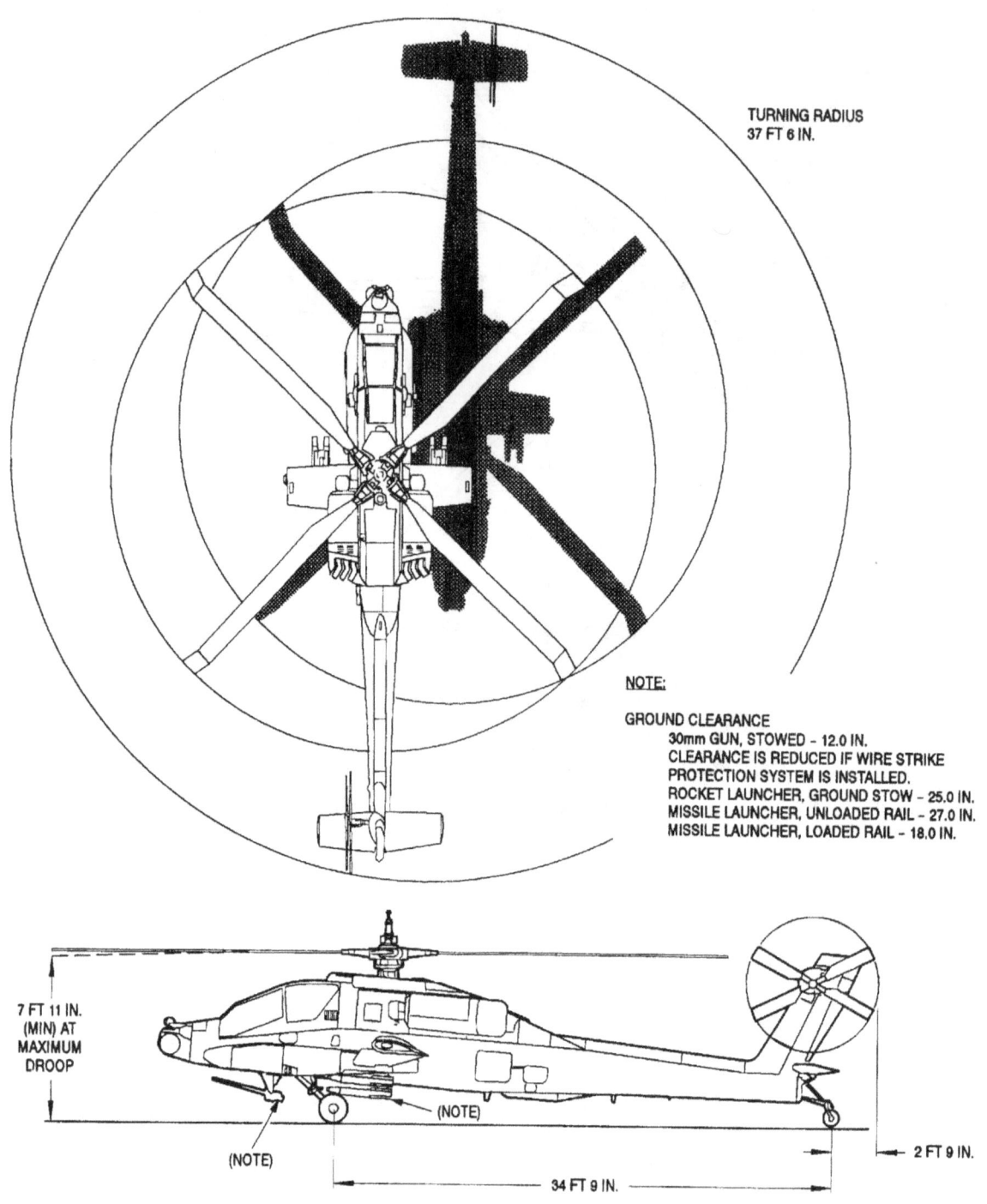

Figure 2-4. Turning Radius and Ground Clearance

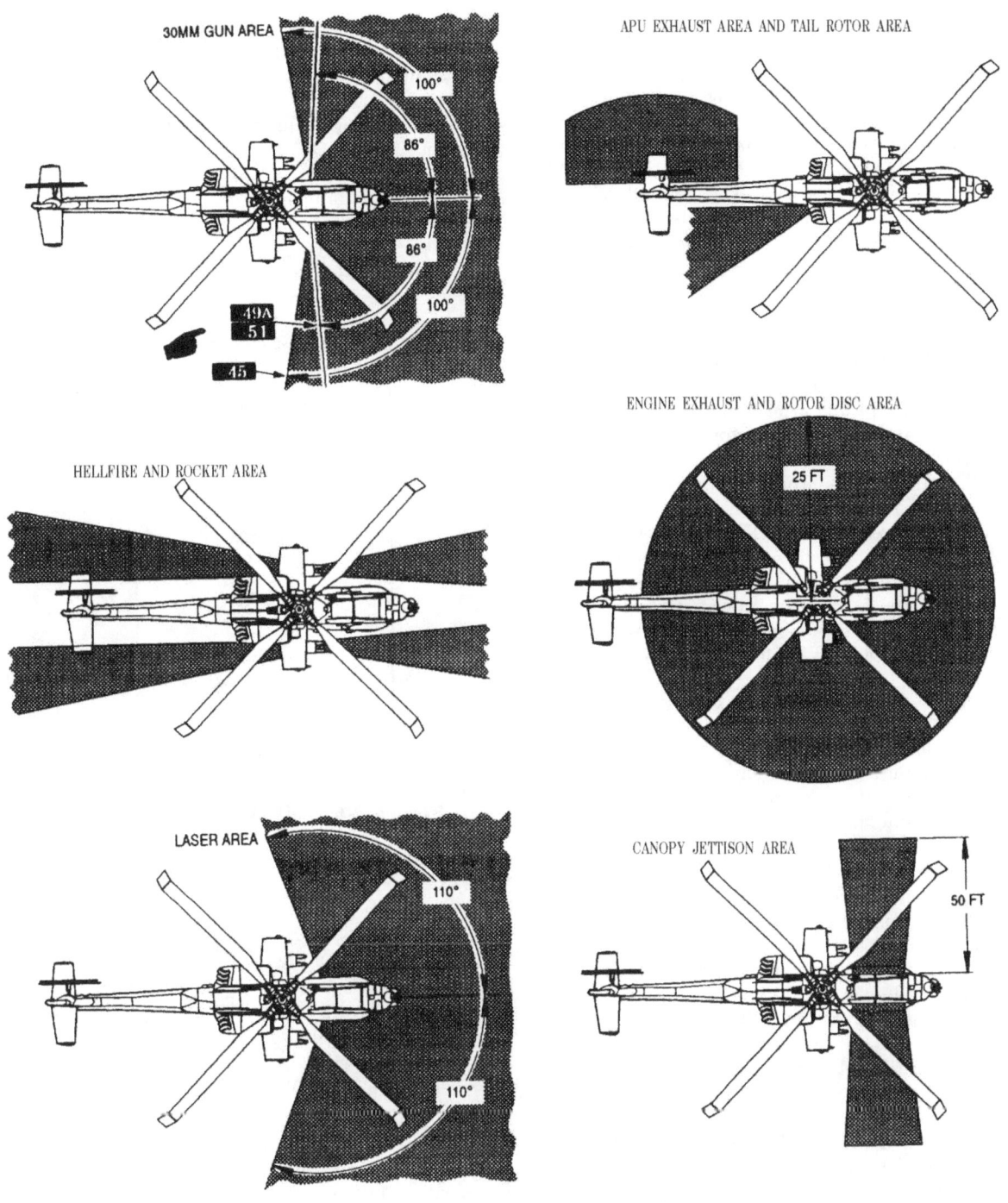

Figure 2-5. Danger Areas

2.9 LANDING GEAR.

The main landing gear (fig 2-2) supports the helicopter during ground operation (taxiing, take-off, and towing). The landing gear system is a three-point system consisting of the main landing gear, tail landing gear, and main landing gear brake system. The landing gear system provides for ease of maneuvering when taxiing and towing, has shock struts to absorb normal and high impact landings, and kneels to facilitate transport of the helicopter.

2.9.1 Main Landing Gear. Each main landing gear support consists of a trailing arm and a nitrogen/oil shock strut. The trailing arms transfer the helicopter landing and static loads to the airframe, and the shock struts absorb vertical loads. The upper ends of the left and right trailing arms attach to a cross tube which passes through the fuselage and is supported by fuselage-anchored pivot bearings. The upper ends of the shock struts are attached to mounts on the fuselage structure. In addition to its normal energy-absorbing function, each shock strut has a one-time high impact absorbing feature: shear rings are sheared and a rupture disk bursts causing a controlled collapse of the strut.

2.9.2 Tail Landing Gear. The tail landing gear consists of two trailing arms, nitrogen/oil shock strut, fork, axle, and wheel. The shock strut has an impact-absorbing capability similar to that of the main landing gear shock strut. The tail wheel is 360° free swiveling for taxiing and ground handling. The tail landing gear system incorporates a spring-loaded tail wheel lock. However, the tail landing gear is hydraulically unlocked from the pilot crew station or manually unlocked by a ground crewmember using a handle attached to the actuator. The tail wheel lock system is actuated by hydraulic pressure from the utility hydraulic system. Pressure is routed to the actuator through a control valve located in the tail boom. The valve is controlled by the tail wheel switch (fig 2-6) at the pilot station. When the tail wheel switch is placed in the **UNLOCK** position, pressure is applied to the actuator to retract the lock pin. A proximity switch will cause the advisory light above the switch to illuminate. When the tail wheel **LOCK/UNLOCK** switch is placed in the **LOCK** position, a valve shuts off hydraulic pressure and opens the line to the actuator. This relieves the pressure on the lock. Spring force will then move the lock pin to the lock position. If the tail wheel is unlocked manually, it can be locked from the pilot crew station by placing the

tail wheel switch in the **UNLOCK** position, then returning the switch to the **LOCK** position. The tail wheel shall be locked to:

    a. Absorb rotor torque reaction during rotor brake operation.

    b. Prevent shimmy during rolling takeoffs and landings.

    c. Prevent swivel during ground operation in high winds.

    d. Prevent swivel during operation on slopes.

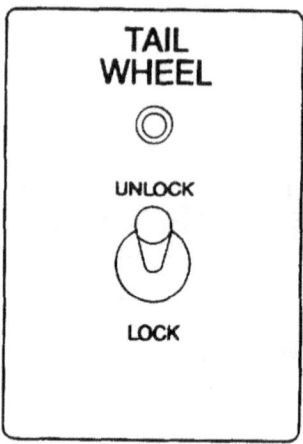

Figure 2-6. Tail Wheel Lock Panel

2.9.3 Landing Gear Brakes.

NOTE
It is necessary to maintain pressure on the brake until the **PARK BRAKE** handle is pulled out to lock the parking brakes. If the **PARK BRAKE** handle is pulled out without pressure applied to the brake pedals, the **PARK BRAKE** handle may remain out and the brakes will not be set.

The brake system affects only the main landing gear wheels. The main landing gear system consists of two independent hydromechanical systems, one left and one right. Braking action is initiated from either crew station by applying foot pressure at the top portion of the directional control pedals. This activates a master cylinder attached to each brake pedal (fig 2-7 and 2-8). The master cylinders pressurize hydraulic fluid in the master cylinder system components. This pressure is transmitted through tubing to the transfer valves, and the parking brake valve, to the wheel brake assemblies. It actuates pistons in each wheel brake assembly causing friction linings to move against a floating brake disk to stop wheel rotation. When the helicopter is parked, the pilot or CPG applies and maintains pressure on the brakes until the **PARK BRAKE** handle (fig 2-7) can be pulled out by the pilot to set the parking brakes. Hydraulic pressure is maintained in the system by the compensator valves mounted on the parking brake valve. Additional parking brake force may be achieved by holding the PARK BRAKE handle out and staging or pumping the brake pedals once or twice to maximize the holding force. Releasing the brake pedals before the PARK BRAKE handle will again lock the system and maintain the higher brake force. Either crewmember can release the parking brake by exerting pressure on the control pedal.

2.10 FLIGHT CONTROLS.

The flight control system consists of hydromechanical controls for the main and tail rotors and an electrical stabilator. A digital automatic stabilization system (DASE) is used to augment the controls, and provide a backup control system (BUCS). Refer to the appropriate system for complete descriptive information.

2.11 PILOT AND CPG INDICATORS, INSTRUMENT PANELS, CONSOLES, AND ANNUNCIATORS.

Figures 2-7 thru 2-12 provide an overview of instrumentation in both crew stations. Instruments will be discussed with their associated systems. Flight instruments are described in Section XIV of this chapter. Caution, warning, and advisory lights, as well as audio warning signals, are also discussed in Section XIV.

2.11.1 Indicators. Indicators for management of the helicopter systems are located on both pilot and CPG instrument and control panels. Refer to the applicable system for descriptive information.

2.11.2 Pilot Instrument Panel and Consoles. The pilot instrument panel is shown in figure 2-9 and the control consoles are shown in figure 2-11.

2.11.3 CPG Instrument Panel and Consoles. The CPG instrument panel is shown in figure 2-10 and the control consoles are shown in figure 2-12.

TM 1-1520-238-10

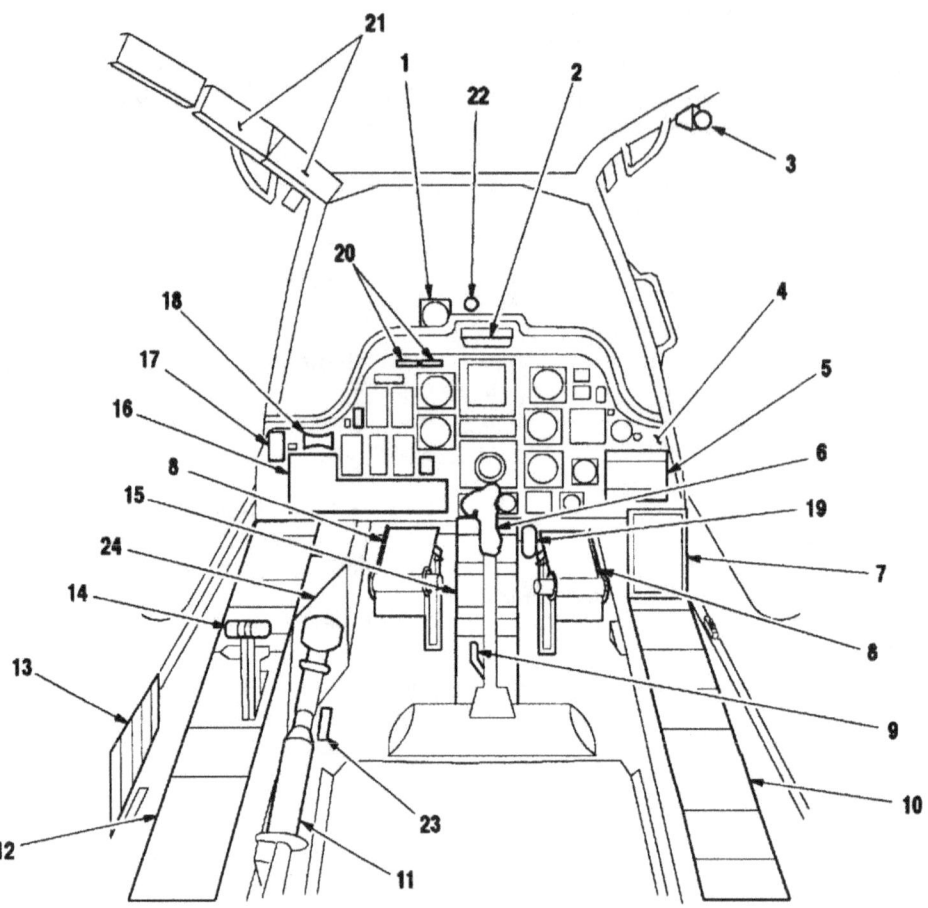

1. STANDBY COMPASS
2. MASTER CAUTION, WARNING PANEL
3. CANOPY DOOR RELEASE
4. INSTRUMENT PANEL
5. RDRCM, CHAFF, IRCM, AND AN/APR 39(VI) CONTROL PANELS
6. CYCLIC STICK
7. CAUTION, WARNING PANEL
8. DIRECTIONAL CONTROL AND BRAKE PEDALS
9. PEDAL ADJUST LEVER
10. RIGHT CONSOLE
11. COLLECTIVE STICK
12. LEFT CONSOLE
13. AUXILIARY AIR VENT
14. POWER LEVERS
15. CENTER CONSOLE
16. FIRE CONTROL PANEL
17. TAIL WHEEL LOCK PANEL
18. CANOPY JETTISON HANDLE
19. PARKING BRAKE HANDLE
20. ENGINE FIRE PULL HANDLES
21. CIRCUIT BREAKER PANELS
22. BORESIGHT RETICLE UNIT
23. STABILATOR MANUAL CONTROL PANEL
24. STOWAGE BOX

M01-006

Figure 2-7. Pilot Station Diagram

2-10

TM 1-1520-238-10

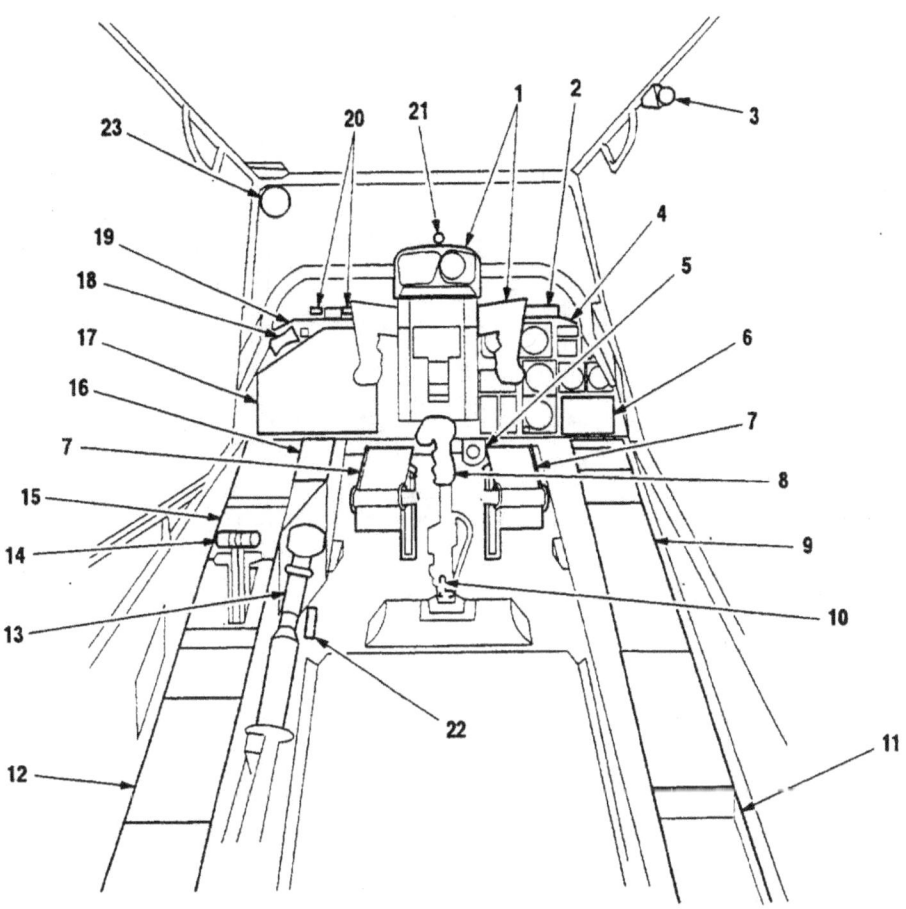

1. OPTICAL RELAY TUBE AND HANDGRIPS
2. MASTER CAUTION, WARNING PANEL
3. CANOPY DOOR RELEASE
4. RIGHT INSTRUMENT PANEL
5. CONDITIONED AIR OUTLET
6. CAUTION/ WARNING PANEL
7. DIRECTIONAL CONTROL AND BRAKE PEDALS
8. CYCLIC STICK
9. RIGHT CONSOLE
10. PEDAL ADJUST LEVER
11. MAP STORAGE COMPARTMENT
12. CIRCUIT BREAKER PANELS
13. COLLECTIVE STICK
14. POWER LEVERS
15. LEFT CONSOLE
16. DATA ENTRY KEYBOARD
17. FIRE CONTROL PANEL
18. CANOPY JETTISON HANDLE
19. LEFT INSTRUMENT PANEL
20. ENGINE FIRE PULL HANDLES
21. BORESIGHT RETICLE UNIT
22. STABILATOR MANUAL CONTROL PANEL
23. MIRROR

M01-007

Figure 2-8. CPG Station Diagram

TM 1-1520-238-10

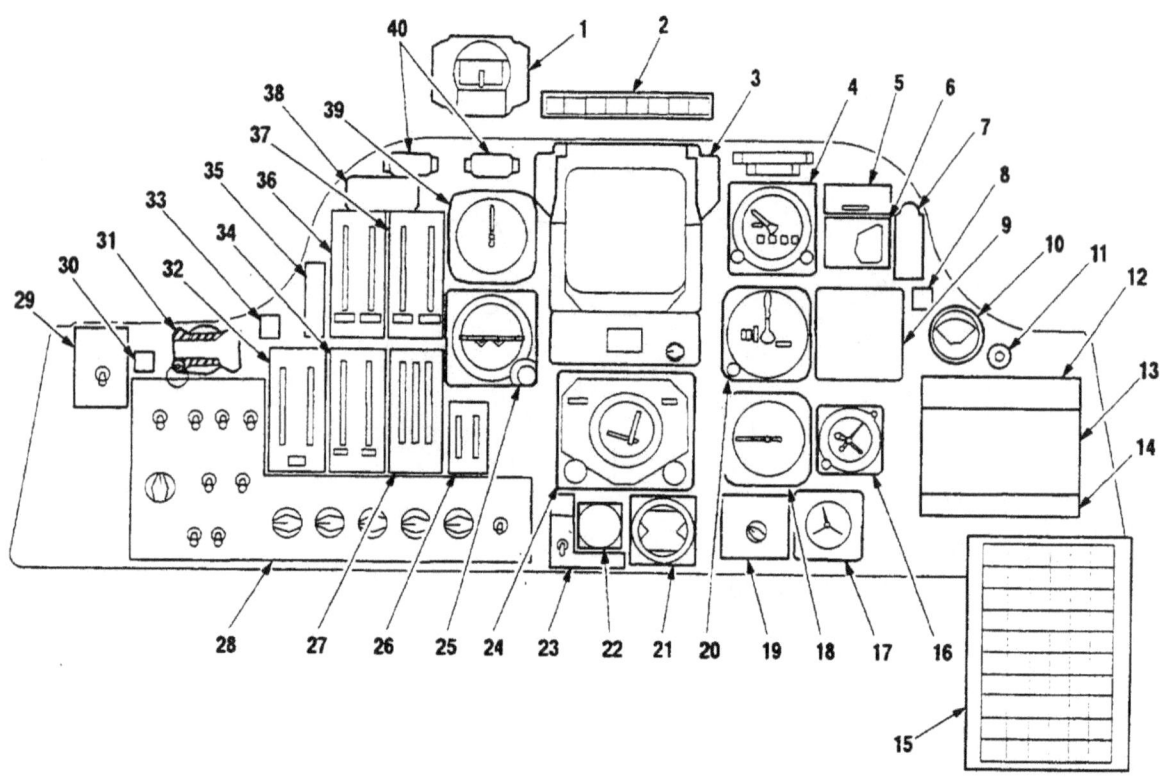

1. STANDBY MAGNETIC COMPASS
2. MASTER CAUTION, WARNING PANEL
3. VIDEO DISPLAY UNIT (VDU)
4. RADAR ALTIMETER
5. RADIO CALL PLACARD
6. STABILATOR POSITION INDICATOR
7. STABILATOR / AIRSPEED PLACARD
8. RADAR JAMMER INDICATOR
9. RADAR WARNING DISPLAY
10. ICING SEVERITY METER
11. PRESS-TO-TEST SWITCH
12. RADAR / INFRARED COUNTERMEASURES CONTROL PANEL
13. CHAFF DISPENSER CONTROL PANEL
14. RADAR WARNING CONTROL PANEL
15. CAUTION /WARNING PANEL
16. CLOCK
17. ACCELEROMETER
18. VERTICAL SPEED INDICATOR (VSI)
19. HARS CONTROL PANEL
20. ENCODING BAROMETRIC ALTIMETER
21. DUAL HYDRAULIC PRESSURE INDICATOR
22. UTILITY ACCUMULATOR PRESSURE GAGE
23. EMERGENCY HYDRAULIC PRESSURE SWITCH
24. HORIZONTAL SITUATION INDICATOR (HSI)
25. STANDBY ATTITUDE INDICATOR (SAI)
26. ENGINE OIL PRESSURE INDICATOR
27. ENGINE (Np) AND ROTOR (Nr) INDICATOR
28. FIRE CONTROL PANEL
29. TAIL WHEEL LOCK CONTROL PANEL
30. ARM SAFE INDICATOR
91. CANOPY JETTISON HANDLE
32. FUEL QUANTITY INDICATOR
33. FUEL TRANSFER INDICATOR (UNMODIFIED CAUTION/WARNING PANEL)
34. ENGINE GAS GENERATOR (NG) INDICATOR
35. INSTRUMENT DIM / TEST PANEL
36. ENGINE TURBINE GAS TEMPERATURE (TGT) INDICATOR
37. ENGINE TORQUE INDICATOR
38. FIRE EXTINGUISHER BOTTLE SELECT SWITCH
39. AIRSPEED INDICATOR
40. ENGINE FIRE PULL HANDLES

M01-008A

Figure 2-9. Pilot Instrument Panel

TM 1-1520-238-10

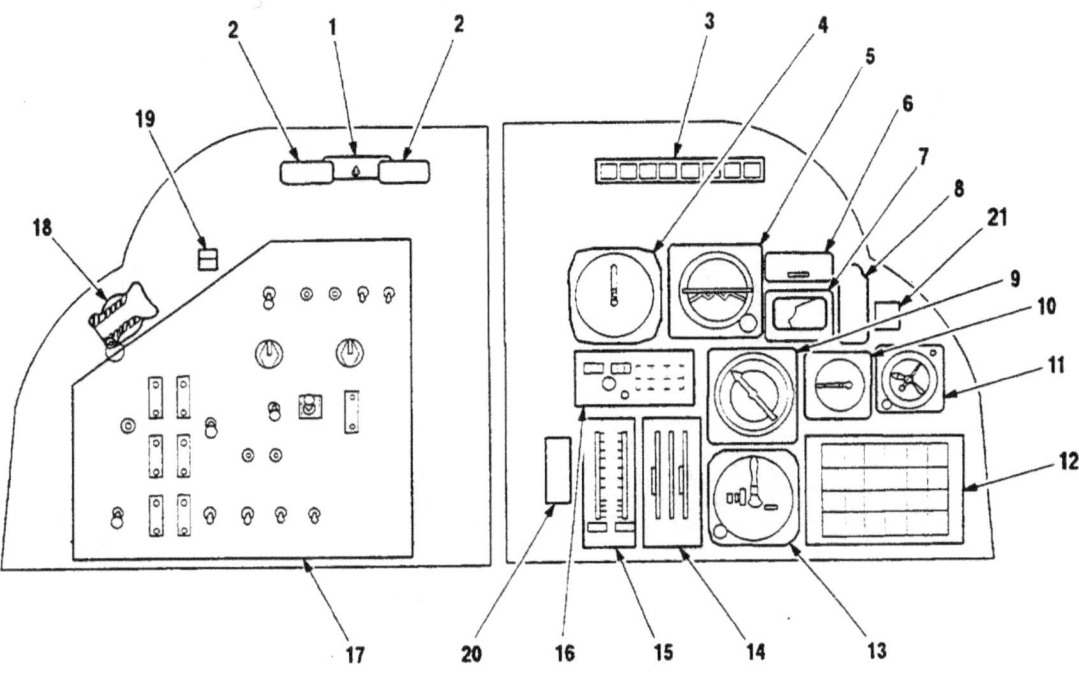

1. FIRE EXTINGUISHER BOTTLE SELECT SWITCH
2. ENGINE FIRE PULL HANDLES
3. MASTER CAUTION, WARNING PANEL
4. AIRSPEED INDICATOR
5. REMOTE ATTITUDE INDICATOR
6. RADIO CALL PLACARD
7. STABILATOR POSITION INDICATOR
8. STABILATOR/AIRSPEED PLACARD
9. RADIO MAGNETIC INDICATOR (RMI)
10. VERTICAL SPEED INDICATOR (VSI)
11. CLOCK
12. CAUTION/WARNING PANEL
13. BAROMETRIC ALTIMETER
14. ENGINE (N_p), ROTOR (N_r) INDICATOR
15. ENGINE TORQUE INDICATOR
16. SELECTABLE DIGITAL DISPLAY PANEL
17. FIRE CONTROL PANEL
18. CANOPY JETTISON HANDLE
19. ARM SAFE INDICATOR
20. ENGINE INSTRUMENT DIM/TEST PANEL
21. FUEL TRANSFER INDICATOR (UNMODIFIED CAUTION/WARNING PANEL)

M01-010A

Figure 2-10. CPG Instrument Panel

Change 2   2-13

TM 1-1520-238-10

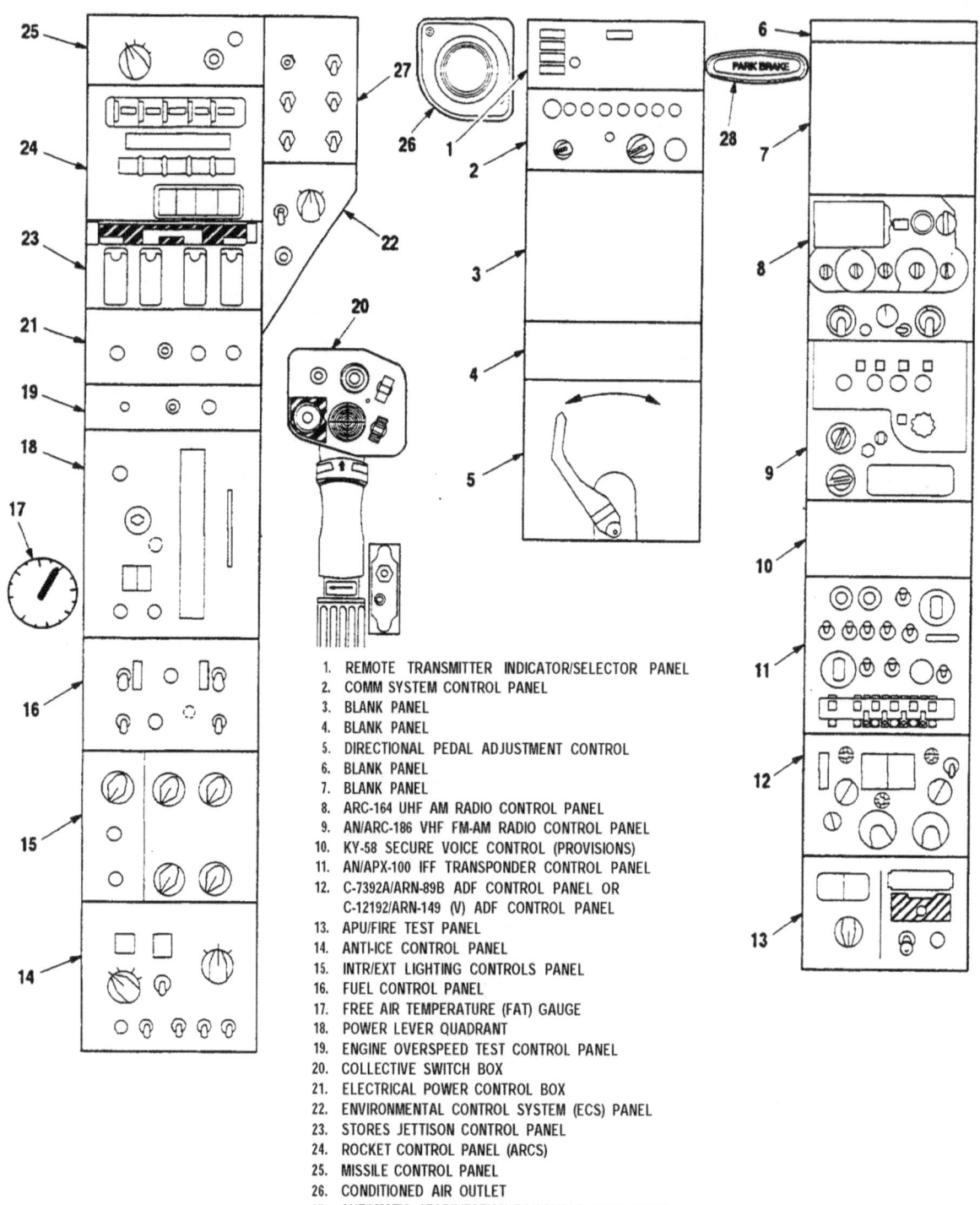

1. REMOTE TRANSMITTER INDICATOR/SELECTOR PANEL
2. COMM SYSTEM CONTROL PANEL
3. BLANK PANEL
4. BLANK PANEL
5. DIRECTIONAL PEDAL ADJUSTMENT CONTROL
6. BLANK PANEL
7. BLANK PANEL
8. ARC-164 UHF AM RADIO CONTROL PANEL
9. AN/ARC-186 VHF FM-AM RADIO CONTROL PANEL
10. KY-58 SECURE VOICE CONTROL (PROVISIONS)
11. AN/APX-100 IFF TRANSPONDER CONTROL PANEL
12. C-7392A/ARN-89B ADF CONTROL PANEL OR
    C-12192/ARN-149 (V) ADF CONTROL PANEL
13. APU/FIRE TEST PANEL
14. ANTI-ICE CONTROL PANEL
15. INTR/EXT LIGHTING CONTROLS PANEL
16. FUEL CONTROL PANEL
17. FREE AIR TEMPERATURE (FAT) GAUGE
18. POWER LEVER QUADRANT
19. ENGINE OVERSPEED TEST CONTROL PANEL
20. COLLECTIVE SWITCH BOX
21. ELECTRICAL POWER CONTROL BOX
22. ENVIRONMENTAL CONTROL SYSTEM (ECS) PANEL
23. STORES JETTISON CONTROL PANEL
24. ROCKET CONTROL PANEL (ARCS)
25. MISSILE CONTROL PANEL
26. CONDITIONED AIR OUTLET
27. AUTOMATIC STABILIZATION EQUIPMENT (ASE) PANEL
28. PARKING BRAKE

Figure 2-11. Pilot Control Consoles

TM 1-1520-238-10

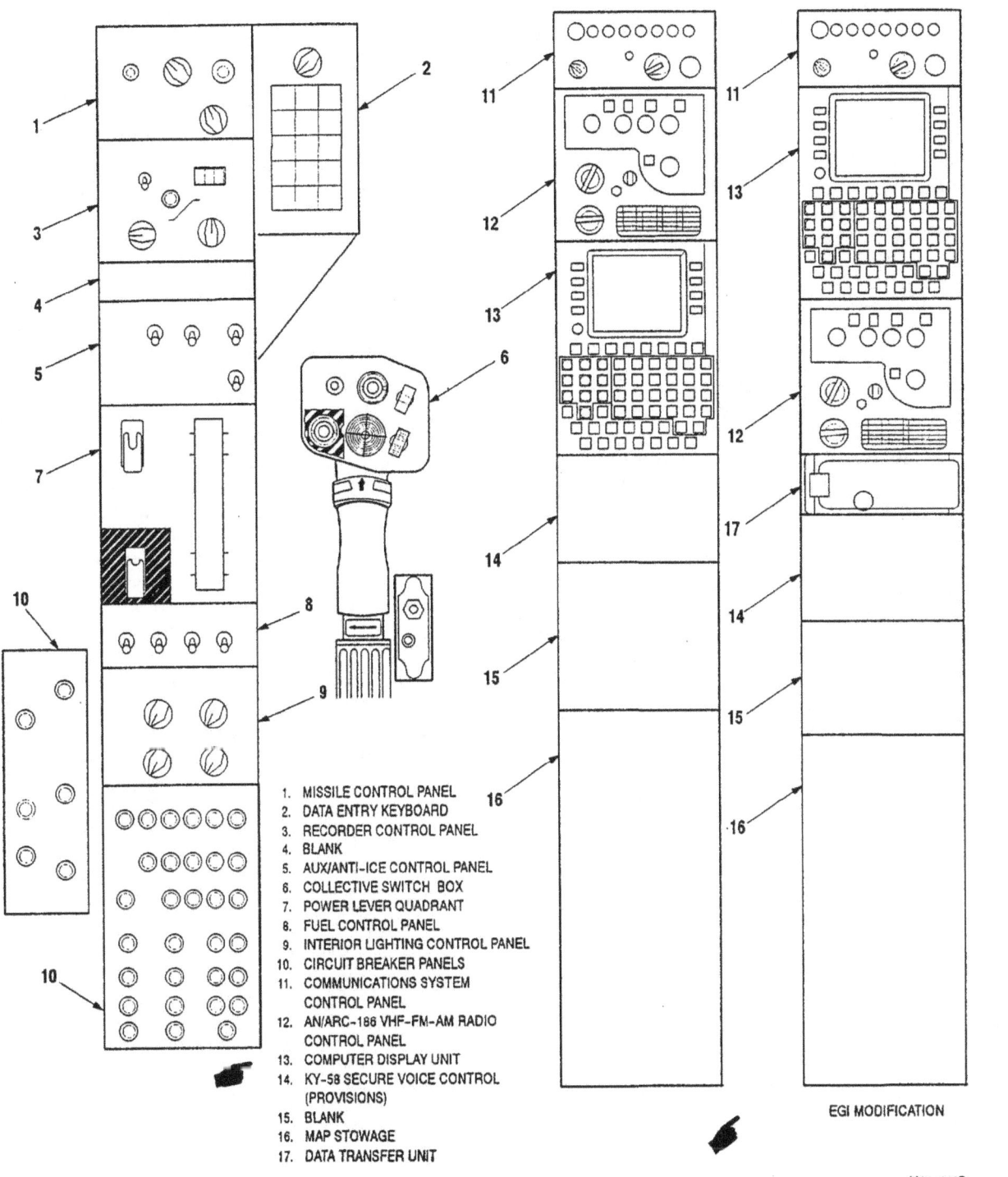

1. MISSILE CONTROL PANEL
2. DATA ENTRY KEYBOARD
3. RECORDER CONTROL PANEL
4. BLANK
5. AUX/ANTI-ICE CONTROL PANEL
6. COLLECTIVE SWITCH BOX
7. POWER LEVER QUADRANT
8. FUEL CONTROL PANEL
9. INTERIOR LIGHTING CONTROL PANEL
10. CIRCUIT BREAKER PANELS
11. COMMUNICATIONS SYSTEM CONTROL PANEL
12. AN/ARC-186 VHF-FM-AM RADIO CONTROL PANEL
13. COMPUTER DISPLAY UNIT
14. KY-58 SECURE VOICE CONTROL (PROVISIONS)
15. BLANK
16. MAP STOWAGE
17. DATA TRANSFER UNIT

EGI MODIFICATION

M01-011B

Figure 2-12. CPG Control Consoles

Change 3   2-15

2.12 CREW COMPARTMENTS.

The crew compartments are arranged in tandem and are separated by a ballistic shield (fig 2-13). The pilot sits aft of the CPG. Handholds and steps permit both crewmembers to enter and exit at the right side of the helicopter. A canopy covers both crew stations. The canopy frame and the transparent ballistic shield form a rollover structure. To provide for maximum survival and minimum vulnerability, armored seats are installed in both crew compartments.

2.13 CREWMEMBER SEATS.

WARNING

Seats stroke downward during a crash and any obstruction may increase the possibility of injury. Items should not be placed beneath seats.

The pilot and CPG seats (fig 2-14) provide ballistic protection and can be adjusted for height only. They are one-piece armored seats equipped with back, seat, and lumbar support cushions. Each seat is equipped with a shoulder harness lap belt, crotch belt, and inertial reel. The shoulder harness and belts have adjustment fittings and come together at a common attachment point. This provides a single release that can be rotated either clockwise or counterclockwise to simultaneously release the shoulder harness and all belts.

2.13.1 Seat Height Adjustment. Vertical seat adjustment is controlled by a lever on the right front of the seat bucket. When the lever is pulled out (sideways), the seat can be moved vertically approximately 4 inches and locked at any 3/4-inch interval. Springs counterbalance the weight of the seat. The lever returns to the locked position when released.

WARNING

With the collective in other than full-down position, the inertia reel control handle may be inaccessible.

2.13.2 Shoulder Harness Inertia Reel Lock Lever. A two-position shoulder harness inertia reel lock lever is installed on the lower left side of each seat (fig 2-14). When the lever is in the aft position, the shoulder harness lock will engage only when a forward force of 1-1/2 to 2Gs is exerted on the mechanism. In the forward position, the shoulder harness lock assembly is firmly locked. Whenever the inertia reels lock because of deceleration forces, they remain locked until the lock lever is placed in forward and then aft position.

2.13.3 Chemical, Biological, Radiological (CBR) Filter/Blower Mounting Bracket. The left side armored wing of each seat has a mounting bracket (fig 2-14) with an electrical interface connector for a CBR filter/blower.

TM 1-1520-238-10

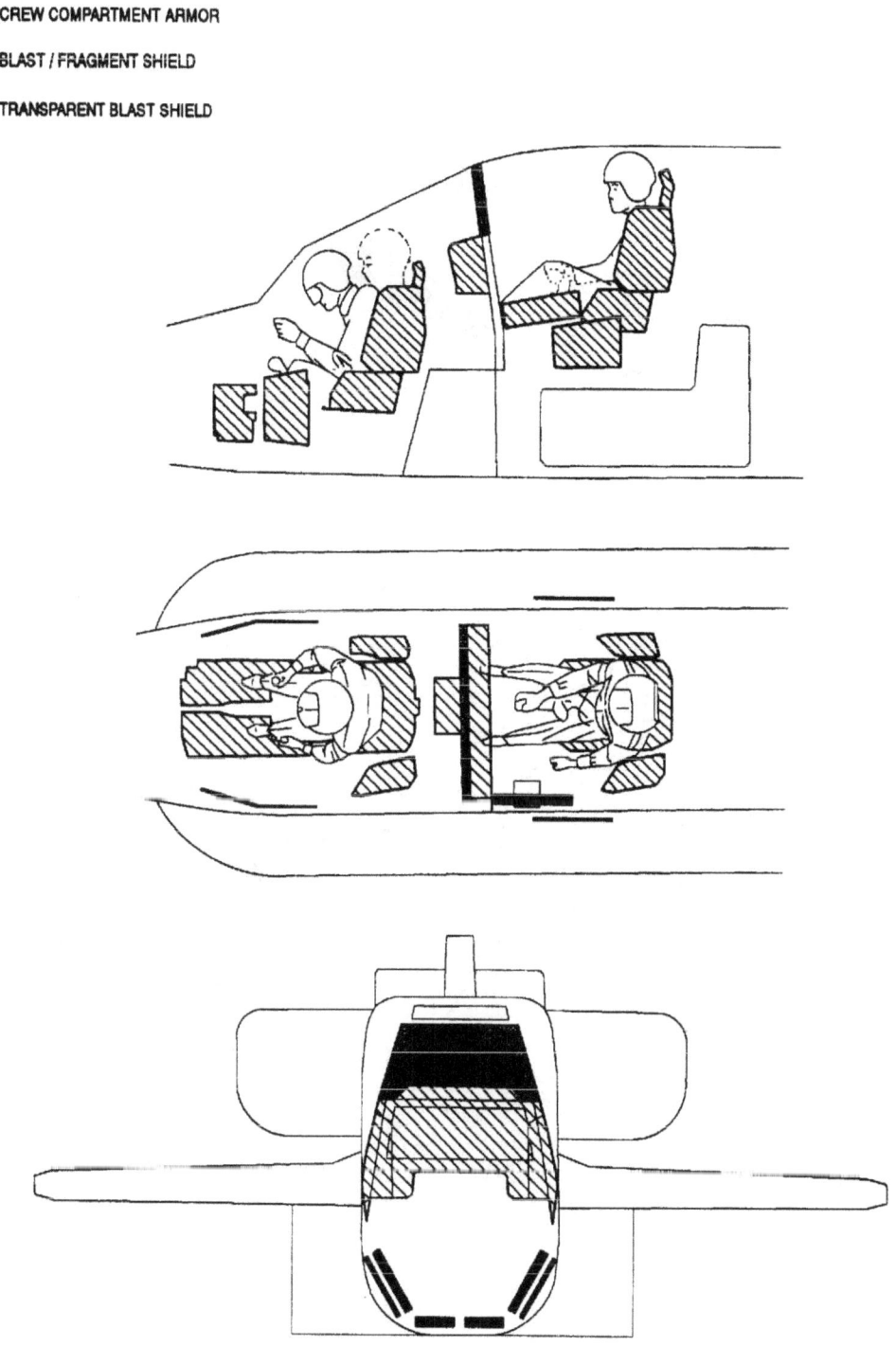

Figure 2-13. Armor Protection

2-17

TM 1-1520-238-10

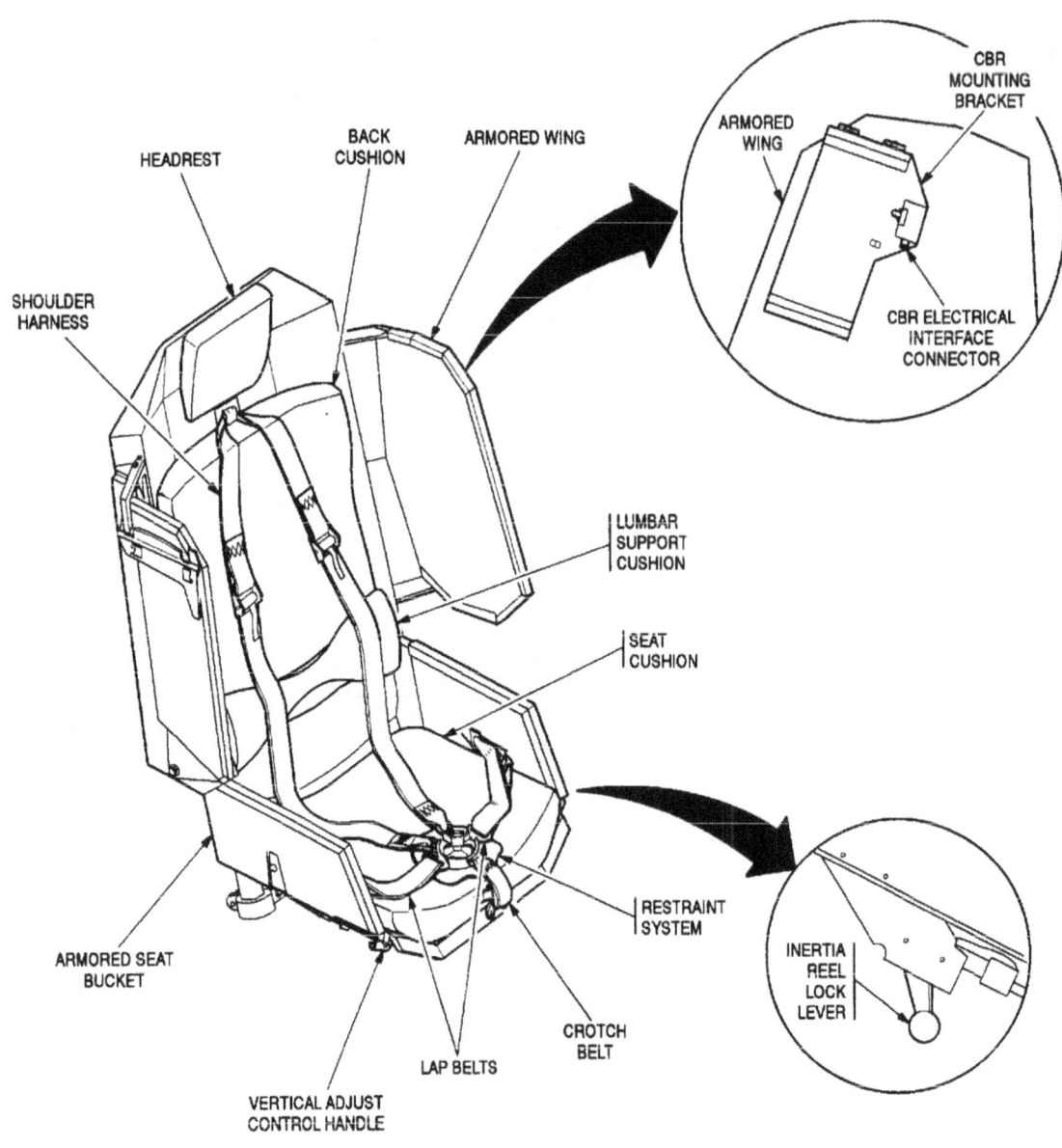

Figure 2-14. Crewmember Seat (Both Crew Stations)

Section II. EMERGENCY EQUIPMENT

2.14 GENERAL.

Emergency equipment on the helicopter consists of fire extinguishing equipment and two first aid kits (fig 9-1).

WARNING

Exposure to high concentration of fire extinguishing agent or decomposition products should be avoided. The gas should not be allowed to contact the skin; it could cause frostbite or low temperature burns. If agent comes in contact with skin, seek medical help immediately.

2.15 PORTABLE FIRE EXTINGUISHER.

A pressurized fire extinguisher is mounted on a quick release support located in the FAB fairing aft of the right FAB door and above the main landing gear wheel (fig 2-2). It is accessible through a hinged access panel and the panel is marked **FIRE EXTINGUISHER INSIDE**. The fire extinguisher compound is released by a hand-operated lever on top of the extinguisher. Inadvertent discharge of the bottle is prevented by a breakaway safety wire across the actuating lever. Operating instructions are printed on the extinguisher.

2.16 ENGINE AND AUXILIARY POWER UNIT (APU) FIRE DETECTION.

Two optical sensors, which react to visible flames, are located in each engine compartment and in the APU compartment. Three pneumatic fire/overheat detectors are located in the aft deck area. One detector is mounted on the main transmission support, and one on each of the two firewall louver doors. Amplified electrical signals from the sensors located in the engine compartment will illuminate the respective **ENG FIRE PULL** handle (fig 2-15) in both crew stations. The APU compartment sensors, or the aft deck fire/overheat detectors, will illuminate the **FIRE APU PULL** handle (fig 2-15) in the pilot station and the **FIRE APU** segment on the **MASTER CAUTION** panel in both crew stations. The engine fire pull handles (**ENG FIRE 1 PULL** and **ENG FIRE 2 PULL**) are located in similar positions in both crew stations; they are at the upper left corner of the instrument panel. The **FIRE APU PULL** handle is located on the APU panel on the pilot right console.

2.16.1 Engine and APU Fire Detector Testing.
FIRE PULL handle lamps are tested by pressing the **PRESS TO TEST** pushbutton on the **MASTER CAUTION** panel in either crew station. The fire detector circuits receive 28 vdc from the emergency dc bus through the **FIRE DETR** circuit breakers (**ENG 1, ENG 2** and **APU**) on the pilot overhead circuit breaker panel (fig 2-39). The fire detector circuits are tested by turning the **FIRE TEST DET** rotary switch located on the pilots aft right console (fig 2-15). The switch is spring-loaded to **OFF**. When set at **1**, the first sensor circuit for No. 1 engine, No. 2 engine, APU, and left and right firewall louver door fire detectors are tested, which causes all **FIRE PULL** handles to illuminate. The **FIRE APU** segments on both **MASTER CAUTION** panels also illuminate. When the **FIRE TEST DET** switch is set at 2, the second sensor circuit for No. 1 engine, No. 2 engine, APU, and main transmission support fire detector are tested; and all **FIRE PULL** handles and both **FIRE APU MASTER CAUTION** panel segments will illuminate. In either test, failure of a handle to light up indicates a fault in that particular sensor circuit.

2.17 ENGINE AND APU FIRE EXTINGUISHING SYSTEM

The fire extinguishing agent is stored in two spherical bottles, each containing a nitrogen precharge. The bottles, designated as primary (**PRI**) and reserve (**RES**) are mounted on the fuselage side of the No. 1 engine firewall. Tubing is installed to distribute the extinguishing agent to either of the engine nacelles or to the APU compartment. Bottle integrity maybe checked by inspecting the thermal relief discharge indicator which is viewed from below the left engine nacelle. A pressure gage on each bottle provides an indication of the nitrogen precharge pressure. Each bottle has three discharge valves that can be individually actuated by an electrically ignited pyrotechnic squib. There is one valve for each of the engine nacelles and one for the APU compartment. When a **FIRE PULL** handle is pulled, four events occur: fuel is shut off to the affected engine, engine cooling louvers to the affected engine close (not applicable to the APU), the appropriate squib firing circuit is armed, and the ENCU is shut off (not applicable to the APU). The extinguishing agent is not released to the fire area, however, until the **FIRE BTL** switch, located near each fire pull handle, is activated. The **FIRE BTL** switch is first set at PRI (primary bottle). When the fire is extinguished, the light in the

2-19

TM 1-1520-238-10

A PILOT

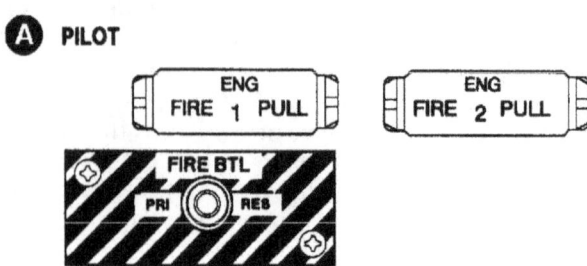

B APU/FIRE TEST

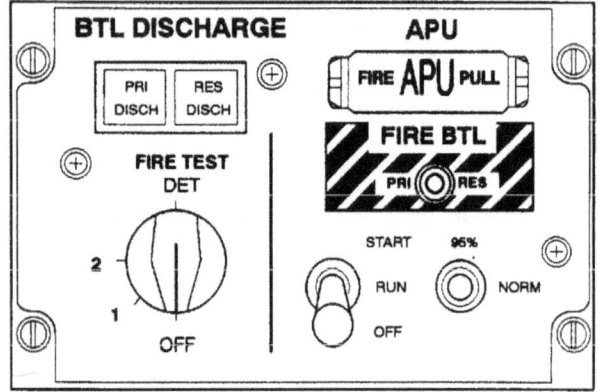

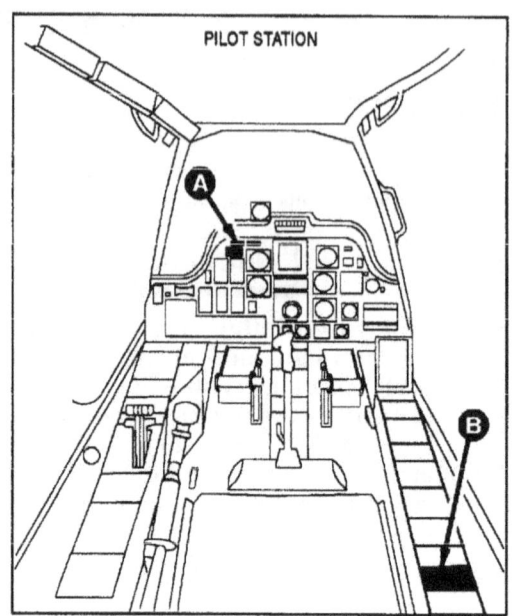

C CPG

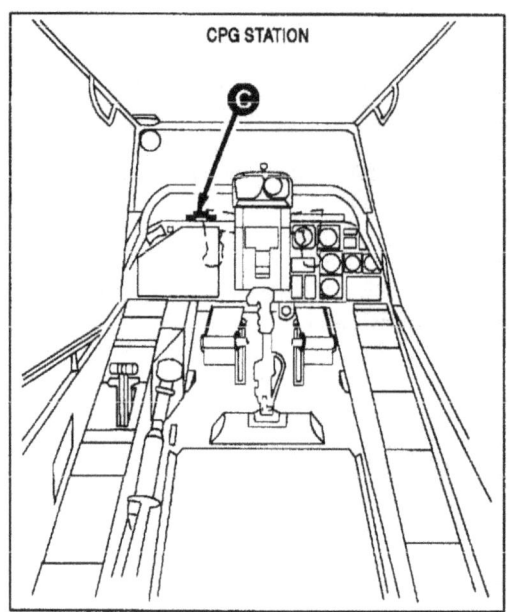

M01-124

Figure 2-15. Fire Detection and Extinguishing Controls

FIRE PULL handle will go out. If the fire is not extinguished, setting the switch at RES discharges the second bottle. Bottle discharge is indicated on the FIRE TEST panel by illumination of the PRI DISCH and RES DISCH displays. The fire extinguishing equipment receives 28 vdc from the emergency dc bus through the PILOT, CPG and APU FIRE EXTGH circuit breakers on the pilot overhead circuit breaker panel.

2.18 FIRST AID KITS.

Two first aid kits are provided: one on the inside, aft portion of the pilots right canopy panel and one on the lower side of the CPG left console.

2.19 CHEMICAL, BIOLOGICAL, AND RADIOLOGICAL (CBR) FILTER/BLOWER.

The CBR filter/blower provides filtered air to the flight crew when cockpit air is contaminated. Each crew member carries his own CBR blower/mask on board and connects it to the external power source, located on the left side armored wing of his seat (fig 2-14).

Refer to TM 3-4240-312-12&P for CBR filter blower operation, installation and maintenance instructions.

2.20 EMERGENCY PROCEDURES.

Chapter 9 describes emergency procedures.

Section III. ENGINES AND RELATED SYSTEMS

2.21 ENGINES.

The AH-64A helicopter can be equipped with either two T700-GE-701 **701** engines or two T700-GE-701C **701C** engines. The engines are mounted horizontally and housed in engine nacelles one on each side of the fuselage aft of the main transmission above the wing. The engines (fig 2-16) are a front-drive turboshaft engine of modular construction. The engines are divided into four modules: cold section, hot section, power turbine section, and accessory section.

2.21.1 Cold Section Module. The cold section module (fig 2-16) includes the main frame, diffuser and mid frame assembly, the inlet particle separator, the compressor, the output shaft assembly, and associated components. The compressor has five axial stages and one centrifugal stage. There are variable inlet guide vanes and variable stage-1 and stage-2 stator vanes. Components mounted on the cold section module are: the digital electronic control unit (DECU) **701C**, the electrical control unit (ECU) **701**, anti-icing and start bleed valve, history recorder/history counter, ignition system, and electrical cables as well as the accessory section module.

2.21.2 Hot Section Module. The hot section module (fig 2-16) consists of three subassemblies: the gas generator turbine, the stage one nozzle assembly, and the annular combustion liner.

2.21.3 Power Turbine Section Module. The power turbine module (fig 2-16) includes a two stage power turbine and exhaust frame. Mounted on the power turbine module is the thermocouple harness, the torque and overspeed sensor, and the Np sensor.

2.21.4 Accessory Section Module. The accessory section module (fig 2-16) includes the top mounted accessory gearbox and the following components: a hydromechanical unit (HMU), a fuel boost pump, oil filter, oil cooler, alternator, oil lube and scavenge pump, particle separator blower, fuel filter assembly, chip detector, oil/filter bypass sensors, oil/fuel pressure sensor, overspeed and drain valve (ODV), and an air turbine starter.

2.22 ENGINE COOLING.

Each engine is cooled by air routed through the engine nacelle. Airflow is provided by eductor pumping action of the infrared suppressor. Fixed louvers on the top and bottom of the aft portion of each nacelle and moveable doors in the bottom center forward portion of each nacelle accelerate convective engine cooling after shutdown. The moveable door is shut by engine bleed-air pressure during engine operation and is spring-loaded to open during engine shutdown.

2.23 ENGINE AIR INDUCTION.

The engines receive air through a bellmouth shaped nacelle inlet at the front of the engine. Air flows around the nose gearbox fairing before entering the engine nacelle inlet. From the inlet, air continues through canted vanes in the swirl frame where swirling action separates sand, dust, and other particles. Separated particles accumulate by centrifugal force in a scroll case. The particles are ejected overboard by a blower which forces them through a secondary nozzle of the infrared suppression device. Clean air, meanwhile, has passed through a deswirl vane which straightens the airflow and channels it into the compressor inlet.

2.24 INFRARED (IR) SUPPRESSION SYSTEM.

The IR suppression system consists of two assemblies: the primary nozzle and three secondary nozzles. The primary nozzle is mounted to the engine exhaust frame and directs exhaust gases into the secondary nozzle. The three secondary nozzles are attached and sealed to the engine nacelle with an adapter. During engine operation, exhaust gases are cooled by air drawn through the transmission area by a low pressure area created by the eduction action of the primary nozzle. The angles of the primary and three secondary nozzles prevents a direct view of the hot internal engine components. This also creates a low pressure area which causes an eduction action by drawing cooling air through the system. The cool air is mixed with the hot air in the three secondary nozzles which results in cooling the exhaust gases.

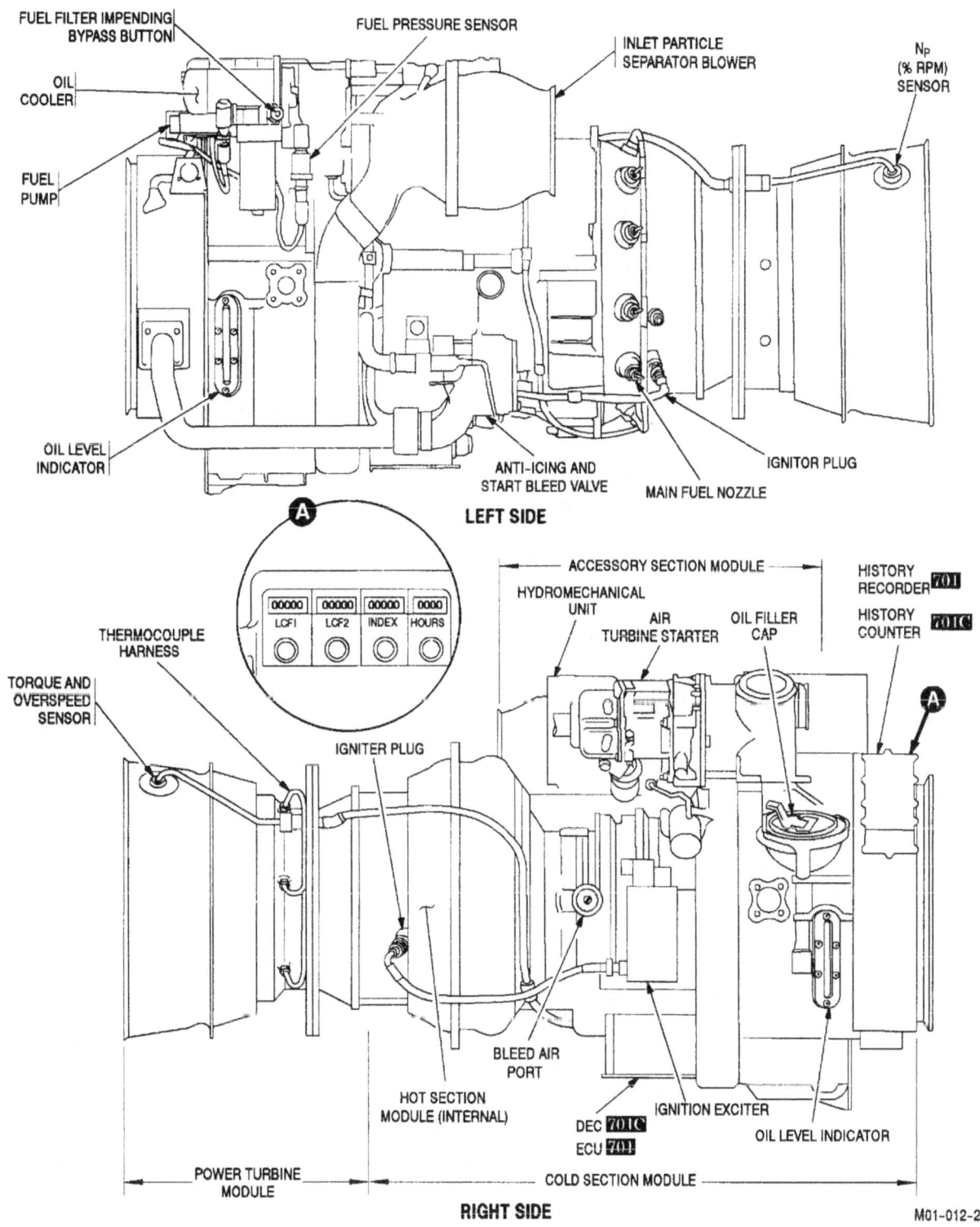

Figure 2-16. T700-GE-701/T700-GE-701C Engine (Sheet 1 of 2)

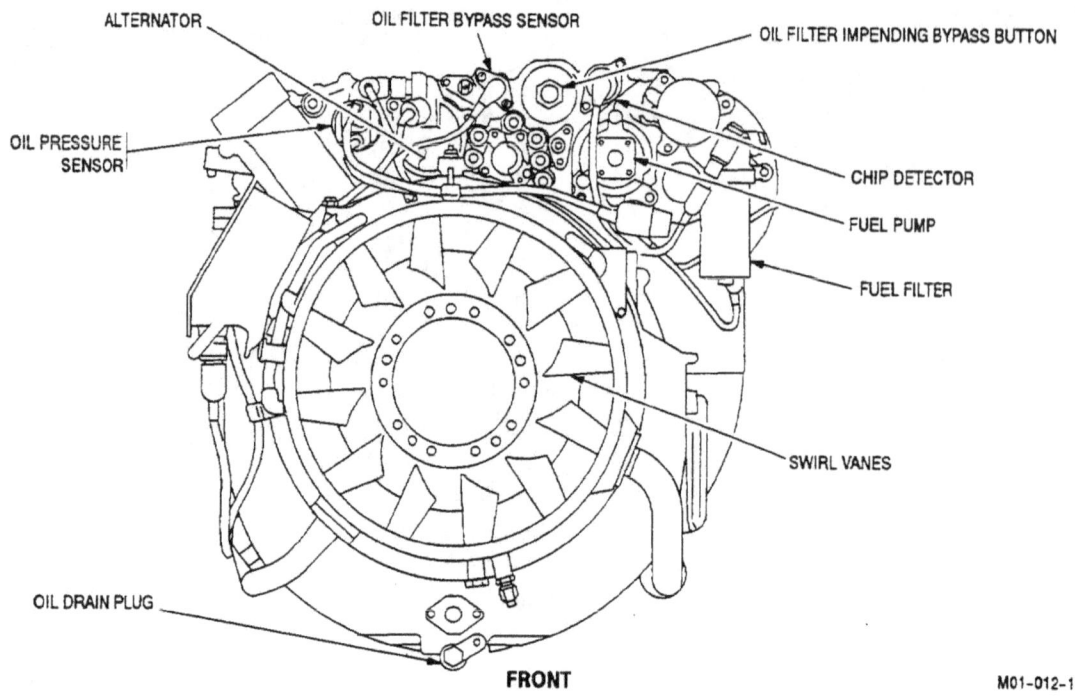

Figure 2-16 T700-GE-701/T700-GE-701C Engine (Sheet 2 of 2)

2.25 ENGINE AND ENGINE INLET ANTI-ICE SYSTEM.

To prevent damage to engines, the engine anti-ice system shall be activated when the aircraft is flown in visible moisture and the outside air temperature (OAT) is less than +5 °C (+41 °F).

The engine anti-ice system includes the engine inlet fairings, and the nose gearbox fairings. Engine fifth stage bleed air is used to heat the swirl vanes, nose splitter, and engine inlet guide vanes of each engine. The nose gearbox fairing is an electrically heated fairing to prevent the formation of ice on each engine nose gearbox fairing.

2.25.1 Engine and Engine Inlet Anti-Ice Operation. The engine anti-ice system is controlled by a two-position **ON/OFF** toggle switch located on the pilot **ANTI ICE** panel (fig 2-17). When the switch is placed to the ON position, the **ENG 1 ANTI ICE** and **ENG 2 ANTI ICE** advisory lights will illuminate and remain on during normal operation of the system. The **ENG 1 ANTI ICE** and **ENG 2 ANTI ICE** fail lights on the pilots caution/warning panel and the **ENG ANTI ICE** fail light on the CPG's caution/warning panel will illuminate until the nose gearbox fairing electrical heaters reach 205 °F and the temperature sensors within the engine inlets reach 150 °F. The time required to reach these temperatures will vary with OAT. When the nose gearbox fairings and the bleed air heated engine inlet reach operating temperatures, the caution/warning lights will extinguish. When the engine anti-ice system is turned off, the **ANTI ICE** panel advisory lights will extinguish and the caution/warning **ENG ANTI ICE** fail lights will illuminate and remain on until the engine inlet temperature sensors cool below 150 °F. The amount of time required for sensors to cool is dependent on OAT and the length of time the engine anti-ice system has been operating. The **ENG ANTI ICE** fail lights may remain illuminated for several minutes after the switch is placed in the **OFF** position. When the switch is **OFF**, the advisory lights on the pilot **ANTI ICE** panel indicate when the engine bleed-air valves are open. These lights will illuminate when engine speed is below 91% N_G (bleed valve open) and will extinguish when engine is above 91% N_G (bleed valve closed). When the switch is ON, the advisory lights will remain continuously illuminated unless the nose gearbox fairings overheat +121°C (250 °F) or under heat 96 °C (205 °F) or if the engine inlet section receives anti-ice air at less than 150 °F. If any of these three conditions occur, the advisory lights extinguish and the ENG **ANTI-ICE** fail lights illuminate.

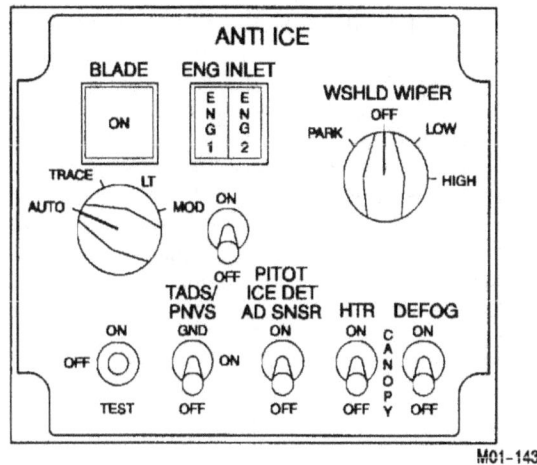

Figure 2-17. Pilot/Engine Anti-Ice Panel

2.26 ENGINE FUEL CONTROL SYSTEM.

The engine has a conventional fuel control system: **PWR** lever position and the degree of collective pitch basically establish the power output demands placed on the engines. Engine power is trimmed automatically through interaction of the engine HMU and the ECU **701** or DECU **701**. The ECU/DECU of each engine exchanges torque signals with the opposite engine to achieve automatic load-sharing between engines.

2.26.1 Fuel Boost Pump. A low-pressure suction fuel boost pump is installed on the front face of the engine accessory gearbox. It ensures that the airframe fuel supply system is under negative pressure, thus reducing the danger of fire in case of fuel system damage. If the **FUEL PSI ENG 1** or **FUEL PSI ENG 2** segment on the pilot caution/warning panel illuminates at idle speed and above, it could indicate a leak or restriction in the helicopter fuel system or a failed engine boost pump.

2.26.2 Fuel Filter. A fuel filter is located between the fuel boost pump and the high-pressure pump in the HMU. If this filter becomes clogged and impedes the passage of fuel, a bypass valve permits fuel to bypass the filter. The differential pressure initiating bypass actuates the fuel-pressure bypass sensor, thus causing the **FUEL BYP ENG 1** or **FUEL BYP ENG 2** segment on the pilot caution/warning panel to illuminate (fig 2-44). An impending filter bypass button on the filter housing pops out when filter element differential pressure indicates impending bypass.

2.26.3 Hydromechanical Unit (HMU). The HMU provides metered fuel to the combustor to control the gas generator (N_G) speed. The HMU contains a high pressure fuel pump to supply fuel to the metering section. The HMU responds to mechanical inputs from the crewmembers through the power available spindle (PAS) and the load demand spindle (LDS). The PAS is mechanically connected to the pilot **PWR** levers while the LDS is connected to a bellcrank attached to the collective servo. The HMU regulates fuel flow and controls positioning of the inlet guide vanes, variable compressor stage 1 and 2 vanes as well as the anti-ice and start bleed valve in response to engine inlet air temperature, compressor discharge air pressure, N_G speed, PAS and LDS positioning, and the ECU **701** or DECU **701C**. The torque motor feedback signals from the HMU to the ECU/DECU are provided by the linear variable displacement transducer (LVDT) to complete the control activated within the HMU at 100- 112% N_G speed. The HMU uses signals from the ECU/DECU to interpret fuel requirements and to vary fuel flow for automatic power control. The HMU will additionally provide N_G overspeed protection in the event the gas generator exceeds 108- 112% N_G. The reaction of the HMU to an N_G overspeed is the same as for an N_p overspeed. Overspeed protection protects the gas generator turbine from destructive overspeeds. When an N_G overspeed is sensed, fuel is directed to the MIN pressure valve of the HMU which causes it to close and shut off fuel to the engine.

2.26.4 Overspeed and Drain Valve (ODV). The ODV responds to a signal from the ECU/DECU. Under normal operation, fuel is routed from the HMU via the oil cooler and through the ODV to the combustor. When an overspeed condition is sensed, a signal from the ECU **701** or DECU **701** closes a solenoid in the ODV, thus routing fuel back into the HMU. All residual fuel is drained overboard. Fuel flow to the fuel manifold ceases, and the engine flames out.

NOTE

The ECU [701] and DECU [701C] are not interchangeable between -701 and -701C engines.

2.26.5 Electrical Control Unit (ECU) [701]. The ECU (fig 2-16) controls the engine and transmits operational information to the crew stations. It is a solid-state device mounted below the engine compressor casing. Powered by the engine alternator, the ECU receives inputs from the thermocouple harness, N_p sensor, torque and overspeed sensor, opposite engine torque for load sharing, N_G signal horn the alternator, N_p reference signal from the turbine speed control unit, and a feedback signal from the HMU for system stabilization. The torque-sharing system increases power on the lower-output engine to match it with the higher output engine. The ECU also receives opposite engine torque inputs to enable contingency power. When this input signal is 51% torque or below, contingency power is automatically enabled. However, contingency power is not applied until the flight crew pulls in collective above 867°C TGT. The ECU automatically allows the normally operating engine to increase its TGT limit, thereby increasing its torque output. The overspeed protection system senses a separate N_p signal independently of the governing channel. ECU also provides signals to the N_p indicator, TORQUE indicator, and history recorder. In case of the ECU malfunction, system operation maybe overridden by momentarily advancing the engine **PWR** lever to **LOCKOUT** and then retarding the lever past the **FLY** position to manually control engine power. This locks out the ECU from all control/limiting functions except N_p overspeed protection, which remains operational. To remove the ECU horn lockout operation, the engine **PWR** lever must be moved to **IDLE**, then back to **FLY**.

2.26.6 Digital Electronic Control Unit (DECU) [701C]. The DECU (fig 2-16) is mounted in the same location as the ECU. The DECU can be overridden like the ECU by momentarily advancing the engine **PWR** lever to **LOCKOUT**. The DECU, which incorporates improved technology, performs the same functions as the ECU except for the following functional and control improvements.

The DECU can be fully powered by either the engine alternator or by 400 Hz, 120 vac aircraft power. It incorporates logic which will eliminate torque spike signals during engine start-up and shutdown. The DECU contains an automatic hot start preventer (HSP). The DECU also provides signal validation for selected input signals within the electrical control system. Signals are continuously validated when the engine is operating at flight idle and above. If a failure has occurred on a selected input signal, the failed component or related circuit will be identified by a pre-selected fault code. Fault codes will be displayed on the engine torque meter (fig 2-18), which defines fault codes in terms of engine torque. Fault codes will be displayed starting with the lowest code for four seconds on/two seconds off, rotating through all codes and then repeating the cycle. The fault codes will be displayed on the engine torque meter only when all of the following conditions are met:

- N_c less than 20%
- N_P less than 35%
- Other engine shutdown
- Aircraft 400 Hz power available

The fault codes can be suppressed by pressing either **OVSP TEST** switch. The fault codes can be recalled by again pressing either **OVSP TEST** switch. Once a failure has been identified, the fault code will remain available for diagnostic indication until starter dropout on the next engine start.

a. Hot Start Prevention (HSP). The HSP system is a part of the DECU and prevents overtemperature during engine start, such as a compressor stall. The HSP system receives power turbine speed (N_P) signal, gas generator speed (N_G) signal, and TGT. When N_P and N_G are below their respective hot start reference, and TGT exceeds 900°C an output from the HSP system activates a solenoid in the ODV valve. This shuts off fuel flow and causes the engine to shut down. The HSP system will not operate if the aircraft 400 Hz power is not present at the DECU. The HSP system can be turned off by pressing and holding either **OVSP TEST** switch during the engine starting sequence.

TM 1-1520-238-10

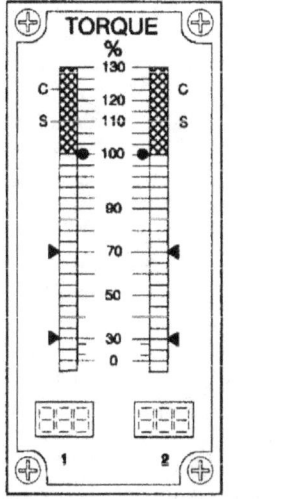

ENGINE TORQUE INDICATOR

SIGNAL FAILED	DIAGNOSTIC INDICATION ON TORQUE METER (+/-0% Tolerance)
DECU	15%
N_P DEMAND CHANNEL	25%
LOAD SHARE	35%
TGT CHANNEL	45%
ALTERNATOR POWER	55%
N_G CHANNEL	65%
N_P CHANNEL	75%
TORQUE AND OVERSPEED CHANNEL	85%
HOT START PREVENTION CHANNEL	95%
AIRCRAFT 400 HZ POWER	105%
COLLECTIVE CHANNEL	115%

M01-222

Figure 2-18. Signal Validation - Fault Codes

2.26.7 Fuel Pressure Warning System. The engine fuel pressure warning system for each engine consists of a pressure switch that illuminates the FUEL PSI segment on the pilot caution/warning panel. The **FUEL PSI ENG 1** and **FUEL PSI ENG 2** segments will illuminate when fuel pressure drops below 9 psi.

2.27 ENGINE ALTERNATOR.

2.27.1 Engine Alternator 701. The engine alternator (fig 2-16) supplies AC power to the ignition circuitry and the ECU. It also supplies a signal to the N_G speed indicator. All essential engine electrical functions are powered by the alternator. When the alternator power to the ECU is interrupted, a loss of Np and torque indications will occur on the affected engine(s), and the engine(s) will increase to the maximum power. Percent Np/Nr will increase above 100% and T4.5 limiting will be inoperative. When the alternator power providing the N_G signal is interrupted, a loss of N_G indications will occur with a corresponding engine out audio signal and warning light being activated. Actual engine operation is unaffected. A complete loss of engine alternator power results in affected engine(s) increasing to maximum power with a loss of of Np, N_G indication, torque and engine out audio signal and warning light being activated, and by an inability to start the engine.

2.27.2 Engine Alternator 701C. When the engine alternator (fig 2-16) power supply to the DECU is interrupted, 400 Hz, 115 vac aircraft power is used to power the DECU, therefore preventing an engine (high side) failure. Np/torque indications will not be affected. When alternator power supply for the N_G signal is interrupted, a loss of the associated engine N_G indication and an engine out audio signal and warning light will occur. Actual engine operation is unaffected. Complete failure of the alternator will cause loss of N_G indication, activation of an engine out audio and warning light, and inability to start the engine. Operation of the engine and all other indications will be normal.

2.28 ENGINE OIL SYSTEM.

Each engine is lubricated by a self-contained, pressurized, recirculating, dry sump system. Included are oil supply and scavenge pumps, an emergency oil system, an integral oil tank, a filter, an oil cooler, and seal pressurization and venting. A chip detector that illuminates the **CHIPS ENG 1** or **CHIPS ENG 2** segment on the pilot caution/warning panel is in line downstream of the scavenge pump.

2.28.1 Emergency Oil System. Small oil reservoirs, built into the engine oil sumps, are kept full during normal operation by the oil pump. If oil pressure is lost, oil will bleed slowly out of these reservoirs and be atomized by air jets thus providing an oil mist lubrication for the engine bearings for thirty seconds at 75% N_G. An **OIL PSI ENG 1** or **OIL PSI ENG 2** light on the pilot caution/warning panel will illuminate when oil pressure drops below 20-25 psi.

2.28.2 Oil Tank. Pertinent oil grades and specifications are in table 2-7. The filler port is on the right side of the engine (fig 2-16). The oil level is indicated by a sight gage on each side of the tank. Oil is supplied to the oil pump through a screen. The scavenge pump returns oil from the sumps to the oil tank through six scavenge screens.

2.28.3 Oil Cooler and Filter. Scavenge oil passes through an oil cooler (fig 2-16) before returning to the tank. It is cooled by transferring heat from the oil to fuel routed through the cooler. If the oil cooler pressure becomes too high, a relief valve will open to dump scavenge oil directly into the oil tank. Oil discharged from the oil pump is routed through a disposable element filter. As the pressure differential across the filter increases, the first indication will be a popped impending bypass button. As the pressure increases further, the **OIL BYP ENG 1** or **OIL BYP ENG 2** segment on the pilot caution/warning panel will illuminate. During engine starting, with oil temperature below the normal operating range, pressure may be high enough to close the oil filter bypass sensor switch. In this situation, the caution light or lights will remain on until the oil warms up and oil pressure decreases. The impending bypass indicator has a thermal lockout below +38 °C (100 °F) to prevent the button from popping.

2.28.4 Chip Detector. Each engine chip detector (fig 2-16) is mounted on the forward right side of the accessory gearbox. It consists of an integral magnet, electrical connector, and a housing. A removable screen surrounds the magnet. The detector attracts magnetic particles at a primary chip detecting gap. If chips are detected, a signal is sent to the pilot caution/warning panel to illuminate a **CHIPS ENG 1** or **CHIPS ENG 2** segment. These chip detectors are of the non-fizz burning type.

2.29 ENGINE IGNITION SYSTEM.

Each engine has an ignition exciter unit with two igniter plugs. The exciter unit receives power from its engine alternator. The MASTER IGN keylock switch on the pilot engine PWR lever quadrant (fig 2-19) is an enabling switch to the ENG START switches. When an

ENG START switch is placed to START, pneumatic motoring of the engine starter takes place and the ignition system is energized. Ignition cutout is automatic after the engine starts. Following aborted starts (Chapter 9, Emergency Procedures), the engine must be motored with the ignition system disabled. This is done by placing the ENG START switch to IGN OVRD. Chapter 5 contains the starting cycle limitations.

2.30 ENGINE STARTING SYSTEM.

The engine uses an air turbine starter (fig 2-16) for engine starting. System components for starting consist of the engine starter, a start control valve, an external start connector, check valves, controls, and ducting. Three sources may provide air for engine starts: The shaft driven compressor (SDC) (normally driven by the APU for engine starts), No. 1 *engine* compressor bleed air, or an externally connected ground source. In any case, the start sequence is the same. With the ENG FUEL switch ON and MASTER IGN switch ON, placing the ENG START switch momentarily to START will initiate an automatic start sequence. This will be evidenced by the illumination of the ENG 1 or ENG 2 advisory light above the ENG START switch on the pilot PWR lever quadrant. Compressed air is then directed through the start control valve to the air turbine starter. As the air turbine starter begins to turn, an overrun clutch engages which causes the engine to motor. The starter turbine wheel and gear train automatically disengage from the engine when engine speed exceeds starter input speed. At approximately 52% N_G, air to the starter shuts off, and ignition is terminated. If the engine does not start, the PWR lever must be returned to OFF and the ENG START switch must be placed at IGN OVRD (which aborts the automatic engine start sequence) momentarily before another start is attempted. Chapter 8 explains abort start procedures, and Chapter 5 contains the start cycle limits. If **the engine is equipped with a DECU 701C, fuel flow to** the engine will be automatically shut off if TGT exceeds 900 °C during the start sequence. If this occurs, the PWR lever must be returned to OFF and the ENG START switch must be placed in IGN OVRD (which aborts the automatic engine start sequence).

2.30.1 Engine Start Using APU. The APU provides on-board power for system check by ground personnel. The APU is capable of driving the main transmission accessories which include the ac generators, the hydraulic pumps, and the shaft driven compressor (SDC). The APU is normally left on during both engine starts, but it maybe shut down after one engine has reached 100% N_P. When N_P has reached 100% and the APU shut down, that engine may be used to drive the SDC through the transmission and accessory gear train for starting the second engine. A complete description of the auxiliary power unit appears in Section XII.

2.30.2 Engine Start Using External Source. An external air receptacle under the No. 1 engine nacelle provides an attachment point for an air line to start either engine from an external source. The assembly contains a check valve to prevent engine bleed air or SDC pressurized air from being vented overboard. An external air source of 40 psig and 30 pounds per minute (ppm) pressurizes the start system up to the engine start control valves which requires only that electrical power be applied for a normal start sequence.

2.30.3 Engine Start Using Engine Bleed Air Source. When the No. 1 engine is operating and it is necessary to start the No. 2 engine, bleed air may be used from the No. 1 engine compressor under certain circumstances. This technique is not normally used, however, it is fully automatic, and provides an alternative starting capability if the SDC shaft fails or if the SDC throttle valve or surge valve clogs (Section VI, Pressurized Air System). The starting sequence is the same as for APU starting, only the source of air to the start motor is different. When using this technique to ensure an adequate air pressure and flow to the No. 2 engine starter, collective pitch must be increased to a value that will increase the N_G of the No. 1 engine to a minimum of 95%.

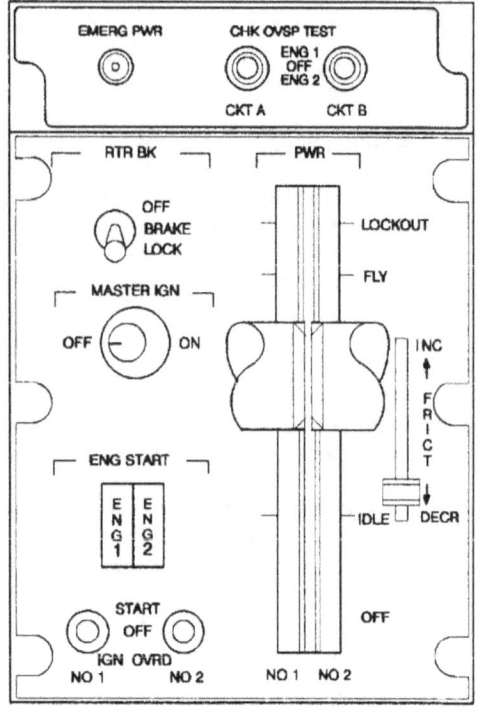
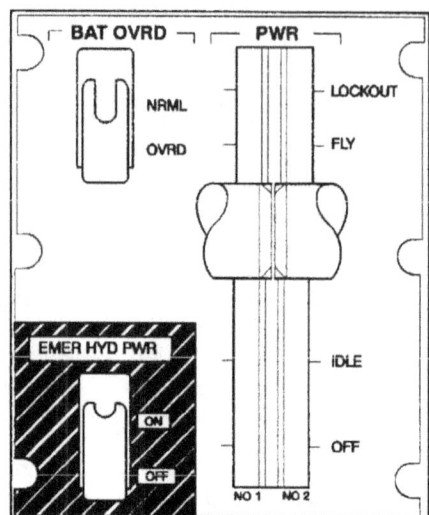

**PILOT PWR LEVER QUADRANT
AND
EMER PWR CHK OVSP TEST PANEL**

CPG PWR LEVER QUADRANT

M01-017

Figure 2-19. Pilot Emergency Check Overspeed Test Panel Power Lever Quadrant and CPG Power Lever Quadrant

2.31 ENGINE CONTROL SYSTEM.

The engine control system consists of the engine power lever quadrant, the engine chop controls, the load demand system, and the overspeed protection system.

2.31.1 Engine Power (PWR) Lever Quadrants. The **PWR** lever quadrants for the pilot and CPG (fig 2-19) allow either crewmember to manage engine power. The two quadrant control panels are different, although the **PWR** levers are identical. Friction however, can be set on only the pilot levers. The **PWR** levers have four detent positions: **OFF, IDLE, FLY** and **LOCKOUT**. The pilot detent override controls are mechanical while the CPG's are electrically operated. Movement of either **PWR** lever moves a cable to mechanically shut off fuel (stop cock) or to set NG speed. For flight, the lever is advanced to **FLY**. By moving the **PWR** lever momentarily to **LOCKOUT**, then retarding past **FLY**, NG speed may be manually controlled. When the **PWR** lever is in **LOCKOUT**, the TGT limiting system is deactivated,

and TGT must be closely monitored and controlled. The overspeed protection system is not disabled in the **LOCKOUT** position.

Physically confirm that engine chop collar is seated in its latched/centered position.

2.31.2 Pilot and CPG Engine Chop Control. The position of a knurled ring on each collective stick grip controls engine chop. The ring is placarded **UN-LATCH, CHOP** and **RESET**. When the ring is pushed forward **(UNLATCH)** and rotated 45° to the right to **CHOP**, the speed of both engines is reduced to idle. This is done by a switch in the grips that cuts out the power turbine speed reference signals for both engines. At this time, the **ENGINE CHOP** warning light will illuminate. The **ENG 1 OUT** and **ENG 2 OUT** lights and audio will not activate because engine chop activation disables this feature. The **LOW RPM ROTOR**

2-30

warning light and audio will activate. If the knurled ring is released after turning to **CHOP**, it will snap back to the center position, but the engines will remain at idle and the **ENGINE CHOP** warning light will remain illuminated. Retard both **PWR** levers to **IDLE**. If the knurled ring is then turned 45° to the left to **RESET** the engines will perform in accordance with **PWR** lever settings. When the knurled ring is released from RESET, it will snap back to the center position, but the circuit will remain in the **RESET** normal condition. Power for the engine-cut circuit is obtained by way of the **ENG CUT** circuit breaker on the pilot overhead circuit breaker panel.

2.31.3 Engine Load Demand System. When the engine **PWR** lever is in **FLY**, the ECU/DECU and HMU respond to collective pitch position to automatically control engine speed and provide required power. During emergency operations when the **PWR** lever is moved to **LOCKOUT** and then retarded to an intermediate position, between **IDLE** and **FLY**, the engine will respond to collective signals, but control of engine speed is no longer automatic and must be managed manually using the **PWR** lever.

> **WARNING**
>
> **The T700-GE-701 and T700-GE-701C engine is designed to shut down when an overspeed condition is sensed. The OVSP TEST circuit trips at 95 - 97% Np and should never be performed in flight. A power loss will result.**

a. Engine Overspeed Protection System. The engine overspeed protection system prevents destructive power turbine overspeeds. The system receives power turbine speed signals from the torque and overspeed sensor. When Np exceeds 119- 120% output, the ECU/DECU activates a solenoid in the ODV. This shuts off fuel flow, causing the engine to shut down. In order to test the system, two circuits are used: **CKT A** and **CKT B**. Both circuits must be closed before the overspeed protection test system is activated. When both **CKT A** and **CKT B** are closed, the system will trip. This will occur at 95- 97% Np. At this time a signal is sent to the ECU/DECU which closes a solenoid in the ODV and stops fuel flow to the combustor. During the test mode, automatic ignition is applied and remains on until 5 seconds after the test switches are released. This will ensure a positive relight when fuel flow resumes.

An advisory light on the **EMER PWR CHK OVSP TEST** panel indicates the availability of emergency electrical power. The system normally uses engine alternator power. The advisory light receives 28 vdc from the emergency dc bus through the **ENG START** circuit breaker. The No. 1 engine overspeed test receives 28 vdc from the No. 1 essential dc bus through the **ENG PWR 1** circuit breaker; for No. 2 engine, from the No. 2 essential dc. bus through the **ENG PWR 2** circuit breaker. The three circuit breakers are on the pilot overhead circuit breaker panel.

The overspeed protection system can be tested by the **EMER PWR CHK OVSP TEST** panel on the pilot left console as follows:

1. **CKT A** switch - **ENG 1**.

    a. N_G should remain stable. If N_G decreases, **CKT B** has malfunctioned.

2. **CKT A** switch - Release.

3. **CKT B** switch - **ENG 1**.

    a. N_G should remain stable. If N_G decreases, **CKT A** has malfunctioned.

4. **CKT A** switch - **ENG 1**.

    a. With both switches set to **ENG 1**, N_G should decrease immediately. Release both switches as soon as N_G begins to decrease.

5. Repeat steps 1 thru 4 for No. 2 engine.

2.32 ENGINE INSTRUMENTS.

The engine instruments are vertical scale type. A common feature of all vertical-scale instruments is the power-on indication: when electrical power is supplied to these instruments, the blue segment of each vertical-scale is illuminated. Another common feature is indicator-light dimming (not dimming of the edge-lighted panels but dimming of the vertical-scale segments and the digital displays). This dimming is accomplished for each crew station by the **DIM** control on the engine instrument test panels (fig 2-20) located in each crew station. An additional feature of the pilot engine instrument test panel is automatic dimming of the pilot engine instruments by a photoelectric cell located on the test panel. The cell dims or brightens in response to ambient crew station light but may be overridden by manual control. A third common feature of all engine instruments in each crew station is that they share power supplies. Each power supply energizes alternate lamp segments and one digital readout for all engine instruments in each respective crew station. If one power supply fails, every other lamp segment and all of ENG 1 or all of ENG 2 digits will extinguish. In addition, if one of the pilot power supplies fails, the **AUX PWR** light on the pilot engine instrument test panel illuminates. The CPG test panel does not have this feature. One of the pilot power supplies receives 28 vdc from the emergency dc bus through the ENG INST circuit breaker on the pilot circuit breaker panel. The other pilots power supply and both CPG power supplies receive 28 vdc from the emergency dc bus through the ENG INST circuit breaker on the CPG No. 1 circuit breaker panel. The illuminated segments at the vertical scale instruments are referenced to the adjacent instrument indices and utilize a technique called optimistic scaling. This means, for example, that for proper indication of 100% Np/Nr the segments immediately above the instrument index line for 100% should be at the threshold of illumination.

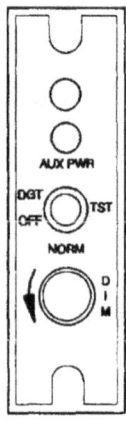

Figure 2-20. Engine Instrument Test Panel

2.32.1 Pilot Engine Instruments. All of the pilot engine instruments are on the left side of his instrument panel. These instruments are tested by the pilot engine instrument test panel (fig 2-20). When the switch on the test panel is set at NORM, all instrument lights are operated by their functional inputs. When the switch is set at DGT OFF, all digital displays turn off. When the switch is set at TST, all digital displays indicate 888, and all vertical-scale readings are illuminated sequentially from the bottom to full scale for 3 seconds, then extinguish. The following engine instruments are located on the pilot instrument panel:

a. Turbine Gas Temperature Indicator (TGT °C x 100). The TGT indicator is a dual, vertical-scale instrument with dual digital readouts beneath the scales. Temperature is sensed at the power turbine inlets by seven thermocouple probes. These signals are averaged and routed through the ECU/DECU, processed by the data signal converter, and transmitted to the crew station indicators. The DECU **701C** incorporates a 71 °C bias in its software which results in the indicated TGT reading cooler than the actual TGT. This permits use of the same TGT gauge for the -701 and -701C engines. When the helicopter is on battery power only and a given engine has not been started, the TGT signal is passed through the DECU to the TGT gauge without the 71 °C bias. When the helicopter has 115 vat, 400 Hz power applied, and a given engine has not been started, the DECU is powered and applies the 71 °C bias to the TGT signal for display. If when the helicopter has 115 vac, 400 Hz power applied prior to starting a given engine and the actual TGT minus the 71 °C bias results in a negative number, the indicated TGT maybe erroneous with a significant mismatch between pilot and CPG TGT gauges. During engine start, the TGT will increase until the sum of the actual TGT minus the 71 °C bias equals a positive number, then the indicated TGT values will be correct.

b. Torque Indicator (TORQUE %). The torque indicator is a dual vertical-scale instrument with dual digital readouts beneath the scales. The engine torque sensor sends pulsed signals to the engine ECU/DECU where torque signals are computed and sent to the crew station indicators.

c. Gas Generator Turbine Speed Indicator (NG RPM %). This is a dual, vertical-scale instrument with dual digital readouts beneath the scales. Speed signals are taken from the engine alternators and sent directly to the crew station indicators.

d. Power Turbine and Main Rotor Speed Indicator (ENG-RTR RPM %). This is a triple, vertical-scale instrument. The two outer scales (Np 1 and Np 2) indicate speeds of the power turbines. The engine power turbine driveshaft sensors transmit pulsed signals to the engine ECU/DECU where they are computed into speed signals and then sent to the crew station indicators. The middle scale registers main rotor speed (Np) and receives inputs directly from a magnetic pickup-type tachometer generator on the transmission.

e. Oil Pressure indicator (ENG OIL PSI x 10). This is a dual, vertical-scale instrument. Pressure signals are taken from the engine oil system transducers and are displayed on the crew station indicators.

2.32.2 CPG Engine Instruments. The CPG instrument panel has a power turbine and main rotor speed vertical-scale indicator and an engine torque vertical-scale indicator identical to those on the pilot instrument panel. To compensate for the absence of TGT, NG, and oil pressure indicators, a selectable digital display (SDD) panel (fig 2-21) is installed to the right of the CPG engine torque indicator. To obtain a readout, the **SELECT** knob is turned until an advisory light is displayed in the desired position on the right side of the panel. When the **SELECT** knob is rotated fully clockwise, no advisory light will be illuminated and the digital readout should be disregarded. Simultaneously, left engine and right engine digital readouts appear in the left upper corner of the panel for **FWD** and **AFT** cell fuel quantity. The CPG engine instruments are tested by pressing the **TEST** pushbutton on the SDD panel which causes all vertical-scale readings to be fully illuminated, and all digital displays to indicate **888**. The engine instrument test panel (fig 2-20) also has the capability of testing the CPG engine instruments by selecting the **TST** position of the **DGT/OFF/NORM/TST** switch. In this position all vertical-scale displays will be fully illuminated and the digital displays will indicate **888**. Selecting the **DGT OFF** position will blank all CPG digital displays.

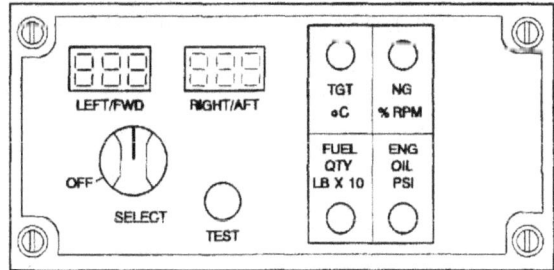

Figure 2-21. Selectable Digital Display Panel

2.33 ENGINE CAUTION/WARNING ANNUNCIATORS.

All cautions/warnings for the engines have been discussed in this section. These caution/warning segments, and their fault indicators, are summarized in Section XIV, tables 2-5 and 2-6. It should be noted that the CPG caution/warning panel has fewer segments than the pilot caution/warning panel and that specific faults are often not displayed. Illumination of any one of the pilot caution/warning segments that indicates an engine problem will simultaneously illuminate the CPG **ENG 1** or **ENG 2** segment.

2.34 ENGINE HISTORY RECORDER 701

An engine history recorder is mounted to the forward right side of the engine **701** (fig 2-16). Signals are sent to the history recorder by the ECU. The recorder displays two readouts of low cycle fatigue (LCF) events, a time temperature index, and engine operating hours. These readouts cannot be reset to zero. The engine history recorder is present only on the -701 engine.

2.35 ENGINE HISTORY COUNTER 701C

An engine history counter is mounted to the forward right side of the engine **701C** (fig 2-16). Signals are sent to the history counter by the DECU. The counter displays two readouts of LCF events, a time-temperature index and engine operating hours. These readouts cannot be reset to zero. The engine history counter is present only on the -701C engine.

Section IV. FUEL SYSTEM

2.35 FUEL SYSTEM.

The fuel system provides fuel and fuel management provisions (fig 2-22) to operate both engines and the APU. Fuel is stored in two crash-resistant, self-sealing fuel cells one forward and one aft of the ammunition bay (fig 2-23). Fuel may be transferred from either cell to the other. The system is also equipped to crossfeed (select which fuel cell supplies fuel to the engines). The helicopter has provisions for carrying either two or four external fuel tanks on the wing pylon attach points. Approved primary fuel grades and acceptable alternates are listed in Section XV, table 2-8.

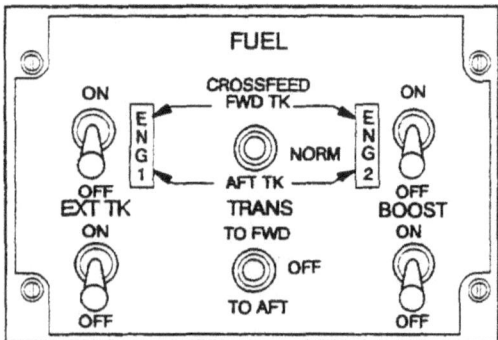

Figure 2-22 Pilot Fuel Control Panel

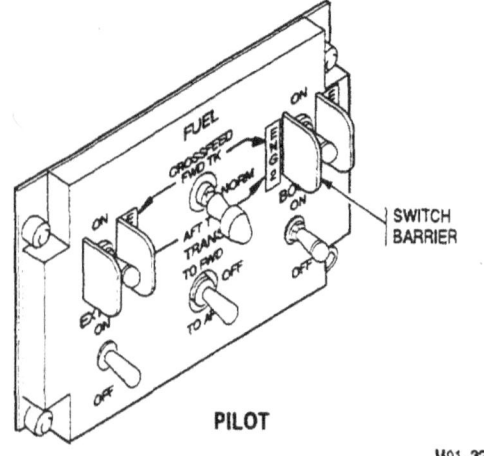

Figure 2-22.1 Pilot Fuel Control Panel (Modified)

2.36.1 Pilot Fuel Control Panel. Table 2-1 outlines the effects of switch settings on the pilot fuel control panel (fig 2-22). Some fuel control panels have been modified with switch barriers located on the ENG 1 and ENG 2 fuel switches (fig 2-22.1). The following paragraphs contain a description of fuel system provisions controlled by the panel.

a. Fuel Switch. Two switches, ENG 1 and ENG 2, control the activation of the crossfeed shutoff valves. In order for fuel to flow to the engines (or crossfeed to the opposite engine if selected), the respective fuel switch must be ON. When in the ON position, the crossfeed switch is enabled. This controls the positioning of the crossfeed valve.

WARNING

- The crossfeed switch shall be set to the NORM position at all times, in flight, unless executing emergency procedures for FUEL PSI ENG 1 and FUEL PSI ENG 2 warning advisory. A malfunctioning crossfeed valve could result in a single engine flameout.

- Do not switch directly from AFT TK to FWD TK crossfeed (or FWD TK to AFT TK) without pausing for at least 15 seconds in the NORM position to ensure both valves are sequencing to their proper positions. Failure to follow this procedure may result in a dual engine flameout if one of the crossfeed valves fails to properly position.

b. Crossfeed Switch. This switch simultaneously controls the position of both crossfeed/shutoff valves. Three discrete positions, FWD TK NORM and AFT TK allow the pilot to select the fuel cell(s) to feed both engines. The NORM setting feeds the No. 1 engine from the forward cell and the No. 2 engine from the aft cell. Both engines will feed from either the FWD TK or AFT TK when selected. This allows the pilot an emergency means to continue fight to a safe area after sustaining fuel system damage. The crossfeed switch can still be used on the ground to control fuel feed during hot refueling. CROSSFEED switch must be placed in the NORM position at least 30 seconds prior to takeoff. If the ENG 1 or ENG 2 fire pull handle is pulled, the selected engine crossfeed/shutoff valve automatically closes.

Some helicopters have been modified with green and amber crossfeed (**X FEED**) annunciator lights located on the pilot and CPG caution/warning panels (fig 2-44.1). The green **X FEED** annunciator light will illuminate on both caution/warning panels when crossfeed is selected by the pilot and both engine fuel valves are correctly positioned for the selection. The green **X FEED** annunciator light will also illuminate when override and crossfeed is selected by the CPG and both valves are correctly positioned. When the **CROSSFEED** switch on the pilot fuel control panel is placed in the **FWD TK** or **AFT TK** position the **X FEED** caution/warning lights will be inhibited for 3 ± 1 second while the fuel valves move to their selected position. The amber **X FEED** annunciator light and the **MASTER CAUTION** light (fig 2-43) will illuminate on both caution/warning panels when the pilot has selected crossfeed or the CPG has selected override and one or both fuel valves are incorrectly positioned. The amber **X FEED** caution/warning lights will also illuminate when a fuel valve is in the CLOSED position (in response to actuation of a fire handle or pilot fuel control panel **ENG 1** or **ENG 2** fuel switch in the **OFF** position.

NOTE

During engine start the amber **X FEED** caution/warning lights will illuminate if the **CROSSFEED** switch is not in the **AFT TK** position.

c. External Tank Switch (EXT TK). The pilot activates fuel transfer from as many as four external fuel tanks to the internal fuel cells by positioning the **EXT TK** switch to **ON**. The switch opens the auxiliary fuel tank shutoff valve and the two transfer shutoff valves. Opening the auxiliary fuel tank shutoff valve allows pressurized air to force fuel from the left auxiliary fuel tank into the forward Fuel cell and from the right auxiliary fuel tank into the aft fuel cell. When all auxiliary tanks are empty, the **EXT EMP** caution light on the pilot caution/warning panel illuminates.

NOTE

- The **REFUEL VALVE** switch (located on the external fuel servicing panel (fig 2-45) must be closed for operation of the **TRANS** switch to be effective. The transfer pump will not transfer fuel if the refuel valve is open.

- The fuel transfer system is the primary method of balancing loads.

d. Transfer Switch (TRANS). The **TRANS** switch controls the position of the fuel transfer air valve. The fuel transfer air valve directs pressurized air to the bi-directional transfer pump which allows fuel transfer between fuel cells. With the switch in the **TO AFT** or **TO FWD** position, pressurized air is directed to the air motor which turns the transfer pump and transfers fuel forward or aft. If the transfer pump is transferring fuel and the fuel cell to which it is transferring becomes full, a fuel level control valve will shut off the fuel flow through the pump. There is no need to stop the transferring operation until the fuel cells are at the desired level.

Some helicopters have been modified with green and amber transfer (**FUEL XFR**) annunciator lights located on the pilot and CPG caution/warning panels (fig 2-44.1). The green **FUEL XFR** annunciator light will illuminate on both caution/warning panels when the pilot or CPG fuel transfer switch is in the transfer position and transfer occurs. The amber **FUEL XFR** annunciator light and the **MASTER CAUTION** light (fig 2-43) will illuminate on both caution/warning panels when the pilot or CPG fuel transfer switch is in the transfer position and transfer does not occur.

WARNING

The CROSSFEED switch shall be set to the NORM position at all times unless executing emergency procedures for FUEL PSI ENG 1 and FUEL PSI ENG 2. Using the fuel system crossfeed may result in a dual engine flameout if one of the engine fuel valves fails to properly position.

e. BOOST Switch. The boost pump (located in the aft fuel cell) is used for starting the engines and when the **FUEL PSI ENG 1** and **FUEL PSI ENG 2** caution lights illuminate. The **BOOST** switch electrically opens the boost pump shutoff valve which directs pressurized air to the air driven boost pump. When the switch is in the **OFF** position, the engines receive fuel (through a suction feed system) via the engine-mounted fuel pumps. When the **BOOST** switch is in the **ON** position, and the **CROSSFEED** switch is in the **AFT TK** position, both engines feed from the aft fuel cell. The boost pump is automatically started and shut down during the engine start sequence. Shutdown occurs at approximately 52% N_G. It should be noted that the pilot **CROSSFEED** switch must be in the **AFT TK** position in order for the **BOOST** switch to latch on.

Table 2-1. Pilot Fuel Control Panel Switch Functions

Switch Position	Result	Indication
ENG 1 set OFF	Energizes fuel crossfeed/shutoff valve of the No. 1 engine to the closed position thus shutting off fuel flow to that engine.	None
ENG 1 set ON	Energizes fuel crossfeed/shutoff valve of the No. 1 engine to the position commanded by the CROSSFEED switch.	None
ENG 2 set OFF	Energizes fuel crossfeed/shutoff valve of the No. 2 engine to the closed position, thus shutting off fuel flow to that engine.	None
ENG 2 set ON	Energizes fuel crossfeed/shutoff valve of the No. 2 engine to the position commanded by the CROSSFEED switch.	None
CROSSFEED set FWD TK	Both engines feed from the forward fuel cell.	FWD cell fuel quantity indicators register fuel decrease while AFT cell indicators remain constant.
CROSSFEED set NORM	ENG No. 1 feeds from the forward fuel cell; ENG No. 2 feeds from the aft fuel cell.	Both fuel quantity indicators register fuel decrease from consumption.
CROSSFEED set AFT TK	Both engines feed from the aft fuel cell.	AFT cell fuel quantity indicators register fuel decrease while FWD cell indicators remain constant.
EXT TK switch set ON	Pressurized air is made available to all external tanks for transfer of fuel to the main fuel cells.	Both fuel quantity indicators register fuel increase during fuel transfer operation.
TRANS set TO FWD	Fuel is being pumped from AFT cell to FWD cell by the transfer pump.	AFT cell fuel quantity indicators register fuel decrease while FWD cell indicators register fuel increase.
TRANS set TO AFT	Fuel is being pumped from FWD cell to AFT cell by the transfer pump.	FWD cell fuel quantity indicators register fuel decrease while AFT cell indicators register fuel increase.
BOOST set ON	Pneumatically driven boost pump (in AFT cell) is delivering fuel to both engine CROSSFEED/ SHUTOFF VALVES.	BOOST PUMP ON segment of C/W panel illuminates. (Crossfeed must be set to AFT TK in order to latch BOOST switch.)

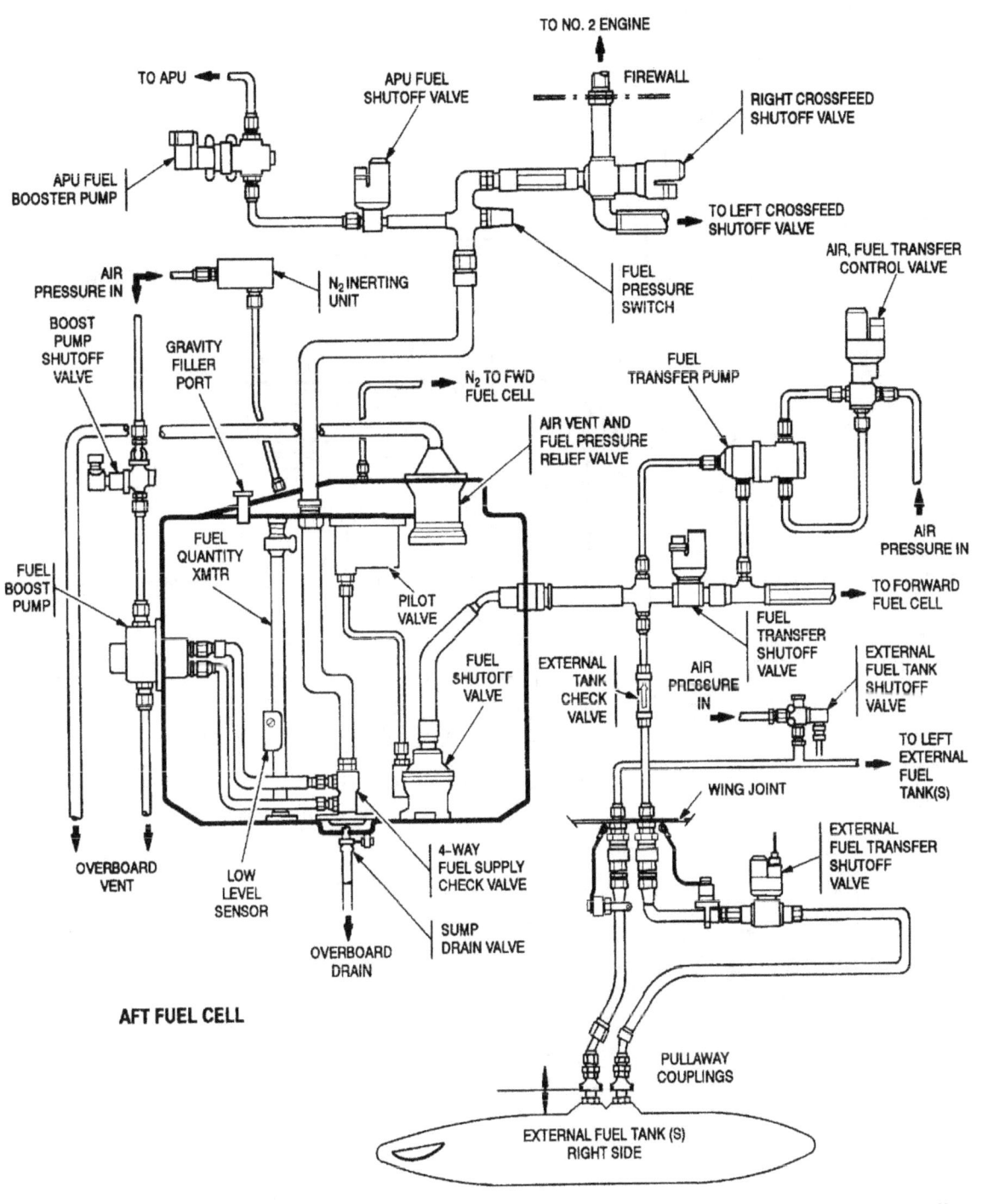

2-23. Fuel System (Sheet 1 of 2)

TM 1-1520-238-10

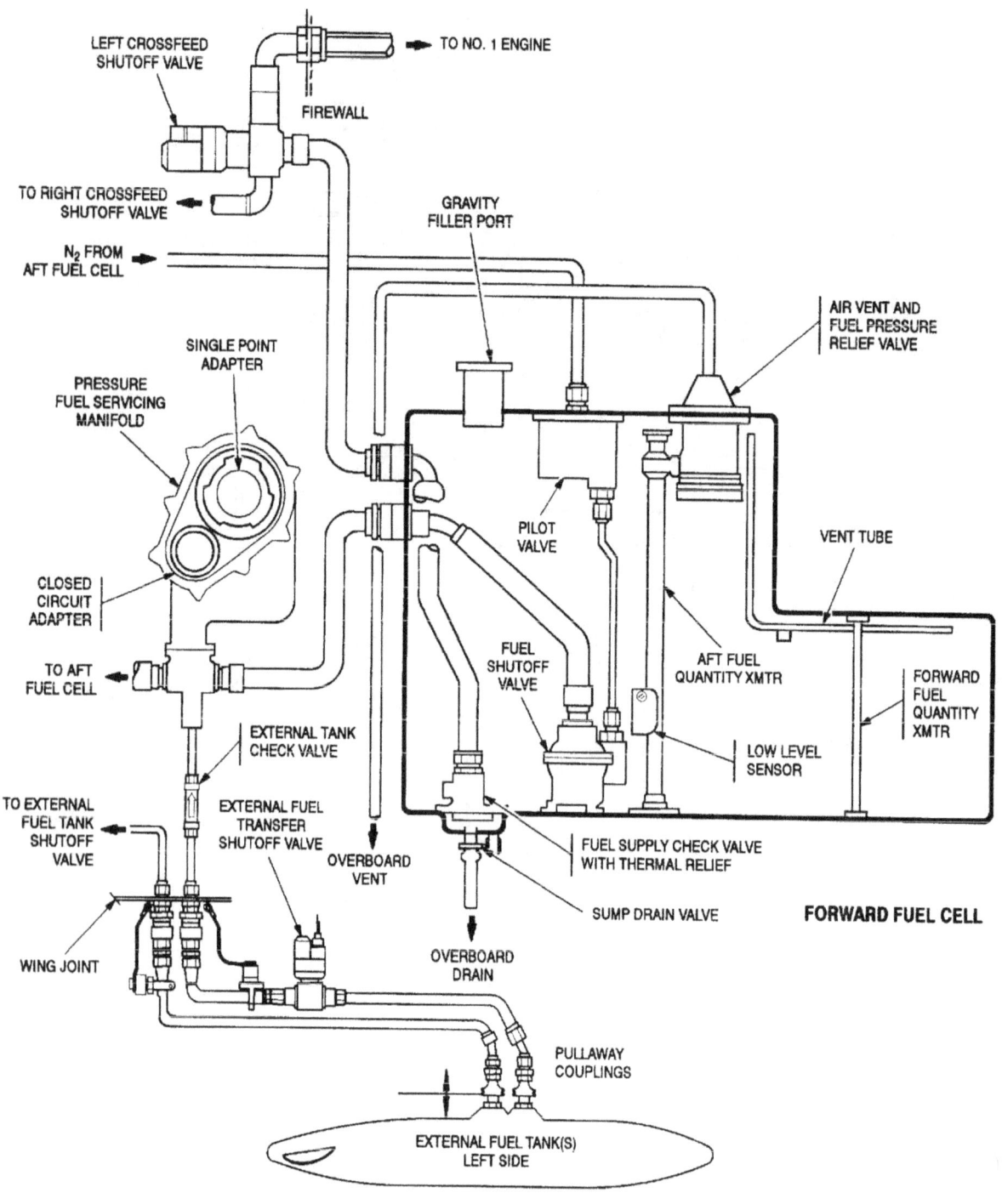

Figure 2-23. Fuel System (Sheet 2 of 2)

TM 1-1520-238-10

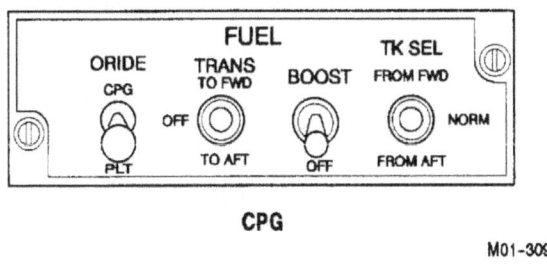

Figure 2-24. CPG Fuel Control Panel

2.36.2 CPG Fuel Control Panel. Table 2-2 outlines the effects of switch settings on the CPG fuel control panel (fig 2-24). The following paragraphs contain a description of fuel system provisions controlled by the panel:

a. Override Switch (ORIDE). With the **ORIDE** switch in the **PLT** position, only the pilot fuel control panel switches have control of fuel system operation. When the **ORIDE** switch is placed in the CPG position, The CPG fuel control panel switches are enabled. The switches that are not duplicated on the CPG **FUEL** panel are still active on the pilot **FUEL** panel, (i.e., **ENG 1**, **ENG 2** fuel switches, and **EXT TK** switch).

b. Transfer (TRANS) Switch. Operates the same as the pilot **TRANS** switch (paragraph 2.36.1 d).

c. BOOST Switch. Operates the same as the pilot **BOOST** switch (paragraph 2.36.1 e).

d. Tank Select (TK SEL) Switch. Operates the same as the pilot **CROSSFEED** switch (paragraph 2.36.1 b).

Table 2-2. CPG Fuel Control Panel Switch Functions

Switch Position	Result	Indication
ORIDE set PLT	Only the pilot fuel control panel is enabled.	None
ORIDE set CPG	CPG Fuel control panel is enabled.	None
TRANS switch	See Table 2-1.	
BOOST switch	See Table 2-1.	
TK SEL set FROM FWD	See Table 2-1, CROSSFEED set FWD TK.	
TK SEL set NORM	See Table 2-1, CROSSFEED set NORM.	
TK SEL set FROM AFT	See Table 2-1, CROSSFEED set AFT TK.	

2-39

2.36.3 Fuel Quantity Indicators. The pilot vertical-scale fuel quantity indicator (fig 5-1) indicates the quantity (in pounds X 10) of fuel remaining in the forward and aft fuel cells. A digital display at the bottom of the indicator shows the combined total of remaining fuel (in pounds X 10). The CPG selectable digital display (SDD) indicates the pounds of fuel remaining in the forward and aft fuel cells. The **SELECT** knob is rotated until the **FUEL QTY X 10** display is illuminated thus causing the pounds of fuel remaining in each cell to appear on the digital display. Combined total fuel is not indicated. The fuel quantity transmitters are capacitance-type transmitters. There are three fuel quantity transmitters: two are in the forward fuel cell because of the shape of the fuel cell, and one is in the aft fuel cell. The transmitters provide data to the fuel signal conditioner which, in turn, provides a readout to the fuel quantity indicators in the crew stations.

2.36.4 Fuel System Caution/Warning Lights. Fuel system caution/warning segments on the caution/warning panel are discussed in Section XIV, (tables 2-4 and 2-5). Their positions are shown on the caution/warning panel (fig 2-44).

2.36.5 Auxiliary Fuel Tank. For ferry missions, there are provisions for as many as four external fuel tanks to be carried on the wing pylon attachment points. The external tanks can be jettisoned in the same manner as any other externally mounted stores.

2.36.6 Refueling Provisions. All fuel service points are located on the right side of the helicopter. For all refueling operations, the **REFUEL valve** switch must be placed in the **OPEN** position and then **CLOSED** after completion of refueling. Section XV contains fuel specifications and weights. Chapter 6 describes the effects of fuel loading on weight and balance.

a. Gravity Filling. The forward and aft cells and the external auxiliary tanks are gravity-filled separately. Fuel cell filler ports are located forward and aft of the wing. Refueling time required for gravity refueling depends on the flow-rate capability of the servicing equipment. Section XV contains specific filling procedures.

NOTE

For gravity refueling, the closed vent system requires the refuel valve to be open to obtain maximum fill above the opening of the filler neck. When fuel has reached this level, the fuel flow rate will have to be decreased to accommodate the restricted vent flow rate.

b. Pressure Refueling. The pressure service manifold, located forward of the right wing, has two adapters for accommodating either a single point adapter (SPA) or a closed-circuit adapter (CGA). Using either nozzle, forward and aft cells may be filled separately or simultaneously dependent upon the refueling panel switch settings (fig 2-45). The shutoff valve at the bottom of each fuel cell responds to an adjoining float-type pilot valve at the top of the cell. This valve will automatically stop fuel flow when the cell is full. Shutoff is set to allow space for a 3% expansion in each cell. The **IND** switch to the left of the **FUEL QTY** indicator on the refueling panel energizes the quantity indicator and advisory lights. After refueling, the **REFUEL VALVE** switch must be returned to the **CLOSED** position, or fuel transfer from cell-to-cell cannot be accomplished by either crewmember. The inside panel of the refueling panel access door contains servicing instructions. Section XV contains complete pressure refueling procedures.

2.36.7 APU Fuel Supply. The APU receives fuel from the aft fuel cell. Fuel is routed from the aft fuel cell through the APU fuel shutoff valve to the APU boost pump and onto the APU.

a. APU Fuel Shutoff Valve. The APU fuel shutoff valve is controlled by the APU **OFF, -RUN,** and,- **START** switch When the switch is moved to the **RUN** position, the shutoff valve is electrically opened to allow fuel flow to the APU boost pump. If the **FIRE APU PULL** handle is pulled, the APU fuel shutoff valve will close.

b. APU Boost Pump. The electrically-driven APU boost pump is energized when the APU **START/RUN** switch is at the **START** or **RUN** position. The pump will continue to run until the APU is shut down. The pump delivers fuel to the APU integral fuel shutoff valve. Pulling the **FIRE APU PULL** handle will stop the pump.

2.38.8 Nitrogen Inerting Unit (NIU). The NIU reduces fire hazards associated with fuel cell ullages (airspace) by filling the ullage with oxygen-depleted air. The NIU is self-contained and automatically operated whenever pressurized air and 115 vac power is available. A go/no-go press-to-test monitor is located in the aft avionics bay. The NIU uses pressurized air from the pressurized air manifold and purges about 70% of the oxygen present. This air is then sent to the aft fuel cell and onward to the forward fuel cell.

2.38.9 **Fuel System Electrical Power Sources.** Electrical power for the fuel system is controlled by the following circuit breakers (fig 2-39):

a. FUEL FILL Circuit Breaker. The refuel panel receives 24 vdc power directly from the battery through the **FUEL FILL** circuit breaker on the pilot overhead circuit breaker panel.

b. FUEL BST, FUEL TRANS, and FUEL XFEED Circuit Breakers. Circuits controlled from the pilot and CPG **FUEL** control panels receive 28 vdc from the emergency dc bus through the **FUEL BST, FUEL TRANS,** and **FUEL XFEED** circuit breakers on the pilot overhead circuit breaker panel. The EXTEMP light on the pilot caution/warning panel receives 28 vdc from the No. 2 essential dc bus through the **FUEL TRANS** circuit breaker on the pilot overhead circuit breaker panel.

c. FUEL BST and APU HOLD Circuit Breakers. Fuel supply system components, automatically controlled by the APU control switch, receive 28 vdc from the emergency dc bus through the **FUEL BST** and **APU HOLD** circuit breakers on the pilot overhead circuit breaker panel and the **APU** circuit breaker in the aft avionics bay.

d. ENG INST Circuit Breakers. All fuel system gauges and caution/warning lights (except those on the refueling panel and the pilot **EXT EMP** caution light), receive 28 vdc from the emergency dc bus through the **ENG INST** circuit breakers on the CPG No. 1 circuit breaker panel and pilot overhead circuit breaker panel.

e. ENG START Circuit Breaker. Fuel Supply system components, automatically controlled by the automatic-engine start circuits, receive 28 vdc from the emergency dc bus through the **ENG START** circuit breaker on the pilot overhead circuit breaker panel.

f. JETT Circuit Breaker. External tank jettison circuits receive 28 vdc from the No. 2 essential dc bus through the **JETT** circuit breaker on the pilot overhead circuit breaker panel.

Section V. FLIGHT CONTROL SYSTEM

2.37 FLIGHT CONTROL SYSTEM.

The flight control system (fig 2-25) consists of hydromechanical flight controls, augmented by digital automatic stabilization equipment (DASE), and an automatically or manually controlled stabilator. The flight control system establishes vertical, longitudinal, lateral, and directional flight of the helicopter. The flight controls provide a cyclic stick, collective stick, and directional pedals in each crew station, connected in tandem, to provide control inputs to the main and tail rotor hydraulic servo actuators. A mixing unit combines inputs from the servoactuators, and transmits them to a non-rotating swashplate. The swashplate changes the linear motion from the mixer unit to rotating motion. The swashplate provides pitch changes for the four main rotor blades. Pedal inputs are transmitted in a similar manner to the tail rotor blades, except the mixer unit is not required. Description and operation of the main and tail rotor systems is in Section VIII.

NOTE

B Helicopters with operable BUCS have shear pins installed in the SPADS in place of steel pins. They also have servos equipped for BUCS operation.

B Each mechanical flight control linkage has a shear pin actuated decoupler (SPAD) installed. The SPADs allow backup control system (BUCS) engagement by means of a microswitch inside each SPAD if a jam occurs in the mechanical flight controls. The shear pins in the pilot SPADs shear at a force lower than those in the CPG SPADs. The SPADs are continuously monitored by the FD/LS. A single microswitch failure in one or more axes will cause a FD/LS message to appear.

CAUTION

- **B** Do not move flight controls without hydraulic power. You may damage or shear pins in the SPADs.
- Care shall be exercised in extending or folding down the CPG cyclic stick when the rotors are turning. Cyclic control system inputs may occur. It is recommended that the pilot hold his cyclic stick steady while the CPG is extending or folding his cyclic stick.

2.37.1 Cyclic Sticks. The cyclic sticks, one in each crew station, provide for helicopter movement about the pitch and roll axes. The CPG stick has a lockpin release mechanism at the base of the stick. This allows the CPG to fold the stick down while viewing the heads-down display and provides greater ease for ingress/egress. The cyclic stick remains functional in this position and is returned to the extended position by pulling aft on a lever in front of the stick grip. Both cyclic stick grips (fig 2-26) have switches for weapons firing, DASE disengagement, trim feel, radio and intercommunications, and flight modes symbology. The pilot grip also has a remote transmitter selector switch for radio selection. These switches will be described in more detail with their associated systems.

2.37.2 Collective Sticks. The collective sticks in both crew stations (fig 2-26) provide the crew with a means of adjusting pitch angle of the main rotor blades and fuel flow metering requirements of the gas generator turbine. Each collective stick has an engine chop collar just aft of the collective stick switch box (see Section III) to permit both engines to be reduced to idle without moving the **PWR** levers. A switch panel at the end of each collective stick contains a searchlight(**SRCH LT**) switch, an extend-retract (**EXT-RET**) momentary searchlight switch, a wing stores jettison guarded button (**ST JTSN**), a NVS switch, a **BRSIT HMD/PLRT** switch, and a radio frequency override (**RF OVRD**) switch. The **RF OVRD** switch is nonfunctional. The CPG collective stick has a BUCS select trigger switch. These switches will be discussed in more detail with their respective systems. A twist-type friction adjustment is installed on the collective assembly to prevent the collective stick from creeping during flight.

2.37.3 Directional Control Pedals. The directional control pedals, one set in each crew station, provide for helicopter movement about the yaw axis. Both sets of pedals are adjusted by applying foot pressure and moving a pedal adjust quick-release lever. Pressing the upper portion of either pedal actuates a master brake cylinder which delivers hydraulic power to a brake disc at the respective main landing gear wheel. Section I contains descriptions of the main landing gear and brake system.

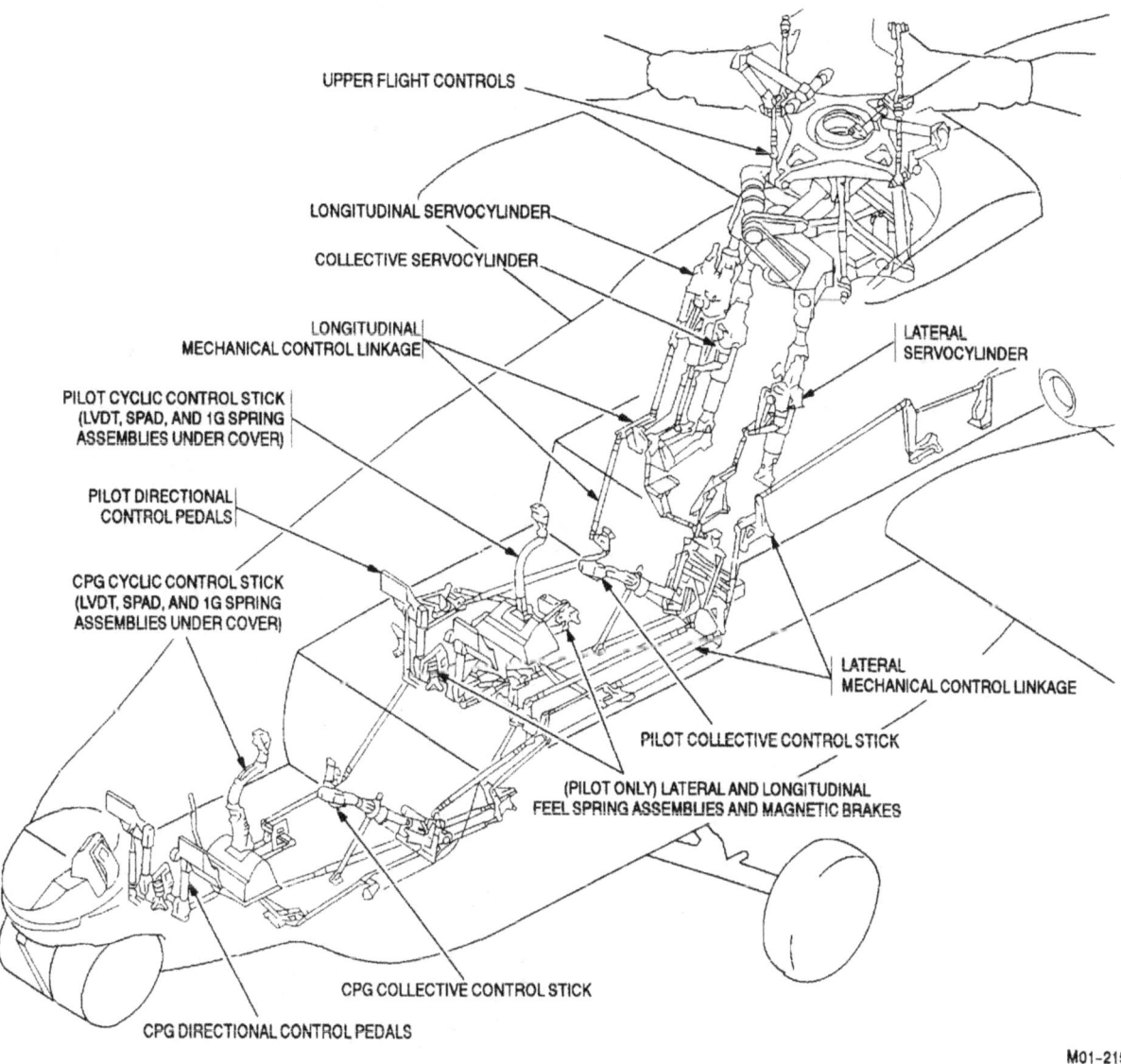

Figure 2-25. Primary Flight Control System

TM 1-1520-238-10

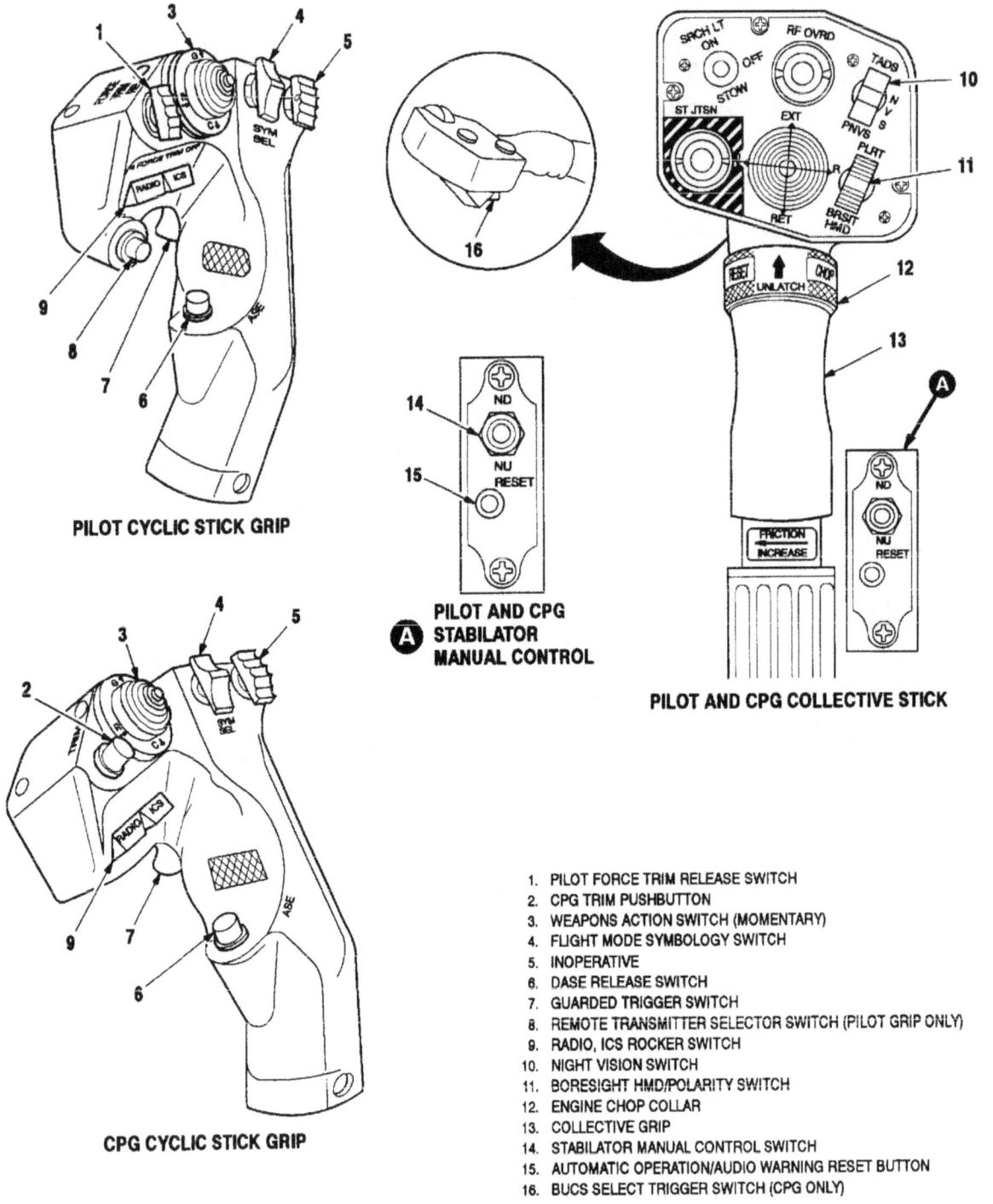

1. PILOT FORCE TRIM RELEASE SWITCH
2. CPG TRIM PUSHBUTTON
3. WEAPONS ACTION SWITCH (MOMENTARY)
4. FLIGHT MODE SYMBOLOGY SWITCH
5. INOPERATIVE
6. DASE RELEASE SWITCH
7. GUARDED TRIGGER SWITCH
8. REMOTE TRANSMITTER SELECTOR SWITCH (PILOT GRIP ONLY)
9. RADIO, ICS ROCKER SWITCH
10. NIGHT VISION SWITCH
11. BORESIGHT HMD/POLARITY SWITCH
12. ENGINE CHOP COLLAR
13. COLLECTIVE GRIP
14. STABILATOR MANUAL CONTROL SWITCH
15. AUTOMATIC OPERATION/AUDIO WARNING RESET BUTTON
16. BUCS SELECT TRIGGER SWITCH (CPG ONLY)

M01-014

Figure 2-26. Cyclic Stick Grip and Collective Stick Controls

2-44

2.37.4 Trim Feel.

> **CAUTION**
>
> **Use of the pilot FORCE TRIM REL switch in the force trim off position with the helicopter on the ground is not authorized. Inflight operation with the pilot FORCE TRIM REL switch in the force trim off position is authorized when briefed and the force trim off selection is acknowledged by both crewmembers.**

Either crewmember can trim the cyclic and pedal controls. A lateral, longitudinal, and directional trim feel magnetic brake and spring assembly is incorporated into each control system. Setting the pilot **FORCE TRIM REL** switch to the on position will engage the magnetic brakes in the longitudinal, lateral, and directional flight controls. The spring assemblies will hold the cyclic stick and directional pedals in trim. Movement of the cyclic or directional controls, by either the pilot or CPG, with **FORCE TRIM REL** switch on, will cause the spring assemblies to compress and provide feel to the controls. When control pressure is released, the controls will return to their trimmed position. Retrimming is accomplished by a **TRIM** pushbutton on the CPG cyclic stick **or** by a **FORCE TRIM REL** switch on the pilot cyclic stick grip. Pressing the button or pressing up on the switch releases the magnetic brake and allows the springs to travel to the new control position. Additionally, this action also allows the SAS actuators to recenter, if necessary. Releasing the button or switch will then allow the magnetic brake to engage and hold the springs at the new position. The pilot may press the **FORCE TRIM REL** switch to the full down position to disable trim feel entirely.

B For the CPG, trim feel is lost in the affected axis when the respective SPAD shear pin is broken or if mechanical linkage severance has occurred between the pilot and CPG stations.

With the pilot **FORCE TRIM REL** switch in the down position, the ATTD/HOVER HOLD and SAS actuator centering capabilities of the DASE are inoperative. The capabilities will return when the switch is returned to the center position. The full down position of the pilot **FORCE TRIM REL** switch may be used momentarily to unlatch the magnetic latch of the **ATTD/HOVER HOLD** switch on the **ASE** control panel (fig 2-27). The DASE capabilities of the pilot **FORCE TRIM REL** switch remain unaffected in the event of a trim system failure such as with the magnetic brake or spring capsule. Trim feel is operable throughout all ranges of cyclic stick or pedal travel. The trim system receives 28 vdc from the emergency dc bus through the **TRIM** circuit breaker on the pilot overhead circuit breaker panel.

2.37.5 Flight Control Servos. The four primary flight control servos are tandem units that use hydraulic pressure from both the primary and utility hydraulic systems. The primary and utility sides of the servos are independent of each other which provides redundancy in the servos. If one hydraulic system fails, the remaining system can drive the flight control servos. If both systems fail simultaneously during flight, either crewmember can use stored accumulator hydraulic pressure to provide limited hydraulics for safe landing. Section VI contains a detailed description of the emergency hydraulic system. The primary side of each flight control servo has two electrohydraulic solenoid valves. One responds to DASE computer signals for stability augmentation.

B The second electrohydraulic solenoid valve responds to DASE computer inputs for BUCS.

2.37.6 Digital Automatic Stabilization Equipment (DASE). The DASE augments stability and enhances maneuverability of the helicopter. DASE includes and/or controls the following: stability and command augmentation in pitch, roll, and yaw; attitude hold; heading hold; hover augmentation; turn coordination; and the BUCS. Major components include an **ASE** control panel (fig 2-27), a DASE computer, eight linear variable differential transducers (LVDT), two 26 vac transformers, and four hydraulic servo actuators.

    a. DASE Computer. The DASE computer receives 28 vdc from the No. 3 essential dc bus through the **ASE BUCS** circuit breaker and a stepped down 26 vac reference voltage from the No. 1 essential ac bus through the **ASE AC** circuit breaker. Both circuit breakers are on the pilot overhead circuit breaker panel. The DASE computer automatically disengages a mistrack of more than 35% and lights the **ASE FAIL** indicator on both caution/warning panels; concurrently, the **PITCH**, **ROLL**, and **YAW** switches on the **ASE** control panel will drop to **OFF** The squat switch on the left main landing gear disables the YAW CAS function during ground taxi to prevent overcontrol of the helicopter. The YAW channel circuitry in the DASE computer receives sideslip data from the air data sensor which provides turn coordination above 60 KTAS.

2-45

b. ASE Control Panel. The **ASE** control panel (fig 2-27) located in the pilot left console, has five single-throw, two-position magnetic switches labeled **PITCH, ROLL, YAW,** and **ATTD/HOVER HOLD** and a **BUCS TEST (BUCS TST)** switch.

B The **BUCS TST** switch on the **ASE** control panel permits the pilot to perform a go no-go check of the BUCS.

The **NOE/APRCH** switch is part of the stabilator system. The **ASE** release button at the base of each cyclic stick grip, when pressed, causes the three channel switches and the **ATTD/HOVER HOLD** switch to drop to **OFF**. The **ASE** control panel receives 28 vdc from the No. 3 essential dc bus through the **ASE DC** circuit breaker on the pilot overhead circuit breaker panel.

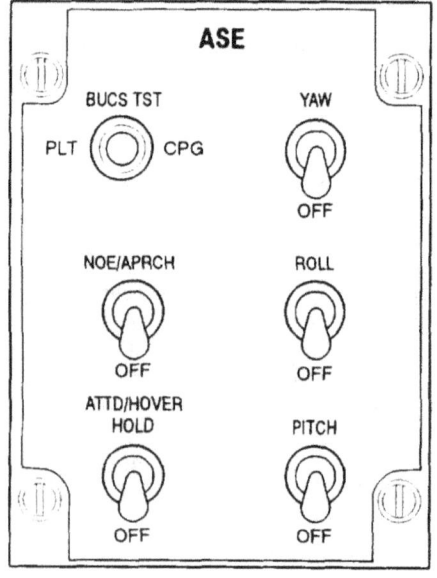

Figure 2-27. ASE Control Panel

c. Automatic Stabilization. The DASE has a stability augmentation system (SAS) and a command augmentation system (CAS). The SAS reduces pilot workload by dampening airframe movement caused by external forces such as in air turbulence and weapons recoil. The CAS augments helicopter response by mechanical control inputs and commands to the longitudinal (pitch), lateral (roll), and directional (yaw) flight control servoactuators. CAS signals are generated by movement of crew station flight controls which are sent to the DASE computer. The DASE computer sums the SAS/CAS information with inputs from the heading and attitude reference set (HARS) and the air data sensor system (ADSS). The DASE computer provides positioning commands to a two-stage electrohydraulic SAS servo valve on the primary side of the longitudinal, lateral, and directional flight control servoactuators. The position of the SAS servo valve determines the amount and direction of movement of the SAS actuators. The position of each of the SAS actuators is transmitted to the DASE computer by the LVDTs. The motion of each SAS actuator is summed with the mechanical input to each flight control servo, but the SAS actuator authority is limited to 10% bidirectional motion in all axes except the longitudinal where the authority is 10% aft and 20% forward. The DASE is engaged through the pilot ASE control panel.

d. Attitude/Hover Hold. The hover augmentation system (HAS) or HOVER HOLD mode of the DASE uses SAS actuators to maintain position and damp external disturbances to the helicopter. HAS is set by engaging the **ATTD/HOVER HOLD** switch or by using the momentary OFF (up) position of the pilot **FORCE TRIM REL** switch. HAS provides the pilot with limited station-keeping or velocity-hold during hover or low speed flight. Position-hold accuracy is a function of inertial velocity drift errors of the HARS which can vary with time. The SAS authority margin will be biased from its center position as these errors build with time. The attitude hold is a limited authority mode of the DASE in pitch and roll axes. This mode provides the pilot with limited hands-off flight capabilities in cruise flight. Attitude hold will only function if: (1) the **ATTD/HOVER HOLD, PITCH,** and **ROLL** switches on the **ASE** panel are engaged; (2) force trim is on, and (3) longitudinal airspeed of the helicopter is greater than 60 KTAS. CAS is removed when the attitude hold mode is engaged.

CAUTION

HAS should not be used as the sole method for station keeping. Cross-checking obstacle clearance using visual or NVS means shall be accomplished. Do not activate HAS on the ground or land with HAS on. Uncommanded aircraft attitude changes may result.

The initialization and stability of the HAS is affected by the HARS velocity drift. A HARS velocity error at HAS initialization will cause initial aircraft movement (at a rate proportional to the HARS velocity error) that may be trimmed using the cyclic force trim. Subsequent changes in the HARS velocity accuracy, during prolonged engagement of the HAS, will cause additional aircraft movement that may be re-trimmed.

NOTE

- **[51]** the GPS system is not keyed or is tracking less than 4 satellites, HAS performance may degrade over time. Hover Position Box and Velocity Vector accuracy will be degraded.

- Over time the SAS actuators may reach saturation limit if not centered by reinitializing. The ATTD/HOVER HOLD features in pitch, roll, or yaw will be lost if saturation occurs. The ATTD/HOVER HOLD feature is a limited hands off feature and the controls should always be monitored carefully.

Heading hold is a function of HAS. The HAS mode will function only if: (1) the ATTD/HOVER HOLD and all ASE channels are engaged; (2) force trim is on, and (3) airspeed is below 50 KTAS and ground speed is less than 15 knots. Degraded conditions will occur in attitude hold or HAS if individual ASE channel switches are not engaged. The **ATTD/HOVER HOLD** switch will disengage automatically when transitioning between modes; attitude hold will disengage at approximately 55 KTAS when decelerating. HAS will disengage when 15 knots ground speed or 50 KTAS is exceeded; or when ADSS fails or is turned off.

NOTE

Backup Control Systems (BUCS) are not incorporated on aircraft PV-529 (S/N 88-0199) and prior.

2.37.7 B Backup Control System (BUCS). The BUCS is a single-channel or non-redundant fly-by-wire flight control system which electronically operates all four flight controls. The BUCS provides full authority flight control with identical handling characteristics to DASE OFF flight in the affected axis. This is accomplished through the use of LVDTs attached to the flight controls in each crew station. The LVDT flight control inputs are made to the DASE, which electronically controls the BUCS servovalve on the primary side of each flight control actuator. For a given flight control, the BUCS will automatically engage when there is no mechanical connection between the flight control actuator and its flight control in **both** crew stations. The BUCS will not engage for any flight control which retains mechanical integrity between the flight control actuator and the associated flight control in **either** crewstation. The primary hydraulic system must be functional for the BUCS to operate. The **BUCS ON** caution light will illuminate when BUCS is engaged for any flight control.

WARNING

Illumination of the BUCS FAIL warning light while in flight shall be treated as a flight control system emergency

a. B Control Severance Engagement Logic.
The BUCS will automatically engage when a difference or mistrack exists between both LVDTs for a given flight control, one in each crewstation and both LVDTs for the corresponding flight control actuator. For example: if the lateral cyclic control was severed between the lateral main rotor actuator and both crewstations, the BUCS would engage when the crewmember flying made a lateral input greater than approximately 1.75 inches. For flight control severances, the required input to engage the BUCS for each flight control is approximately 1.75 inches, except longitudinal cyclic which is about 2.25 inches.

In the example, the logic the BUCS used to automatically engage was that the pilot and CPG lateral cyclic LVDTs were in agreement and the two LVDTs on the lateral actuator were in agreement, but that the difference or mistrack between the two sets was the equivalent of a lateral input greater than 1.75 inches. Therefore, the mechanical linkage must be severed between the actuator and the lateral cyclic controls in both crew stations.

CAUTION

The effect of the CPG assuming control with the BUCS select trigger switch is to transfer control of the helicopter from a flight control that still retains some integrity to a non-redundant electronic means of flight control. This shall only be done if the pilot is incapable of flying the helicopter.

If the lateral cyclic is severed between the pilot and CPG stations, the BUCS will *not* automatically engage because the pilot still has mechanical integrity between his lateral cyclic and the lateral actuator. The CPG is still capable of flying the helicopter with lateral cyclic by pressing the BUCS select trigger switch on his collective and making greater than a 1.75 inch lateral input. This will cause the BUCS to engage for the lateral flight control. The CPG shall assume control in this manner only in the most extreme of emergency conditions.

b. 🅑 Control Jam Engagement Logic. To engage the BUCS for a flight control jam, the crewmember flying the helicopter must exert enough force on the jammed flight control to break the pin in the SPAD. Each flight control in each crewstation has its own SPAD. The force required to break the pin in a SPAD is different for each of the flight controls; the required forces are less for the pilot than the CPG. Microswitches inside the SPAD cause the BUCS to engage when a pin is sheared. For example: if a lateral cyclic jam occurs, the pilot has to exert a force, left or right, to break the pin in the pilot lateral SPAD. The BUCS will then engage using the pilot lateral cyclic LVDT for control inputs. The CPG lateral cyclic may or may not move in response to the lateral main rotor actuator motion. If the jam is severe enough, the lateral cyclic controls may still be jammed from the CPG perspective. The mechanical flight control connections at the flight control actuators are designed to allow full actuator motion with the mechanical flight controls still jammed.

In the example, the pilot, now unjammed, makes lateral control inputs and the lateral main rotor actuator responds. If the act of breaking the pilot lateral SPAD allows freedom of motion for the lateral cyclic control linkage, the CPG lateral cyclic will follow the lateral actuator motion. If still jammed, the CPG lateral cyclic will not move. The CPG is still capable of flying the helicopter in either of these conditions. The CPG has to break his lateral SPAD to transfer control of the BUCS to his lateral cyclic LVDT. If the cyclic is following the motion of the actuator, the CPG only has to input a force opposite the motion until his lateral SPAD shears. If the control linkage is still jammed, the CPG exerts enough force in either direction until his lateral SPAD breaks. This shall be done only in the most extreme of emergency conditions and only if the pilot is incapable of flying the helicopter.

NOTE

To prevent an erroneous **BUCS FAIL** light during generator shutdown, turn **GEN 2** switch (fig 2-37) **OFF** prior to **GEN 1** switch. If **BUCS FAIL** light illuminates during generator shutdown, turn light off by toggling the **PITCH, ROLL,** or **YAW** switch on the **ASE** control panel.

C. 🅑 BUCS FD/LS. BUCS is monitored continuously. If a BUCS failure is detected, BUCS will not engage in the affected axis and the **BUCS FAIL** warning light in each crew station illuminates. To assure that BUCS is fully operational, a preflight self-test is provided in addition to the on-command DASE FD/LS test. The self-test verifies the integrity of BUCS before starting engines. The BUCS self-test requires primary hydraulic pressure within normal limits, **RTR BRAKE** switch set to **BRAKE,** flight controls centered, collective friction off, both **PWR** levers in the **IDLE** position or below, the helicopter on the ground and BUCS not engaged. All DASE channels must be off.

d. 🅑 Control Locks. The control locks protect the shear pins in the pilot and CPG cyclic longitudinal and lateral SPADS, and pedal directional SPADS (fig 2-25) from accidental breaking when the helicopter is on the ground without hydraulic power. They consist of two rig pins to prevent cyclic longitudinal and lateral movement and two pedal lock fixtures to prevent pedal movement, one set for the pilot station and one for the CPG station.

The two sets, comprising four rig pins and four pedal locks, are in a pouch located in the crew station. Each rig pin and pedal lock are attached together by a lanyard with a warning streamer.

CAUTION

When installing and removing the control locks, hydraulic power must be on the helicopter to prevent damage to the SPAD shear pins.

One rig pin is installed in the left side of the cyclic stick base to prevent longitudinal movement. The other rig pin, to prevent lateral movement, is installed in the lower right side of the cyclic stick shroud cover where it has been cut away to allow access for rig pin installation.

After each pedal is moved forward, the pedal lock fixture is installed through the brushes and cutout in the floor so it rests on the edge of a shelf. The pedal lock fixture is aligned so that when the pedal is moved aft, the pedal support fits into the fork of the pedal lock fixture. The pedal is then moved aft so the fork tightly engages the pedal support to prevent movement and the pedal adjust is tightened.

2.37.8 Stabilator System. A variable angle of incidence stabilator is installed to enhance helicopter handling characteristics. The stabilator is designed so that it will automatically be positioned by stabilator control units. These units determine stabilator position from airspeed, pitch rate, and collective stick position inputs. Automatic stabilator range of travel is from 5 ° trailing edge up through 25 ° trailing edge down. Manual control is from 10 ° trailing edge up to 35 ° trailing edge down. The stabilator is driven by two independent dc motors. Stabilator operation, in either manual or automatic mode, requires both ac and dc power supplies. Associated crew station controls and indicators are as follows:

a. Stabilator Circuit Breakers. The stabilator receives 115 vac from the No. 1 essential ac bus through the **STAB MAN AC** and **STAB AUTO AC** circuit breakers. It also receives 28 vdc from the No. 1 essential dc bus through the **STAB MAN DC** circuit breaker and from the No. 3 essential dc bus through the **STAB AUTO DC** circuit breaker. These circuit breakers are on the pilot overhead circuit breaker panel.

b. Stabilator Manual Control Switch. The stabilator manual control switch is located on the stabilator control panel (fig 2-26) installed inboard and forward of the friction grip on each collective stick. The switch, nose up (**NU**), and nose down (**ND**), permits crewmembers to control the stabilator angle of incidence. The manual mode will disengage both the normal auto and NOE/APRCH modes below 80 KTAS. Selection of the manual mode will cause the pilot and CPG **MAN STAB** caution lights to illuminate. Transitioning above 80 KTAS in the manual mode will result in automatic switchover to auto mode. Failure of the automatic switchover to auto mode will result in the following:

(1) Pilot and CPG **MASTER CAUTION** light starts flashing.

(2) Pilot and CPG caution/warning **MAN STAB** light starts flashing.

(3) Stabilator aural tone is heard.

Below 80 KTAS automatic mode is regained after selection of manual mode by momentarily pressing the automatic operation/audio tone **RESET** button on either collective stick.

C. NOE/APRCH Switch. The **NOE/APRCH** switch, located on the **ASE** control panel (fig 2-27), positions the stabilator at 25 degrees (trailing edge down) below 80 KTAS. If auto mode is on, this mode is selectable at any speed from a magnetically held switch. Transitioning past 80 knots will result in normal stabilator scheduling. Failure to revert to normal stabilator scheduling will result in the following:

(1) **NOE/APRCH** switch goes to **OFF**.

(2) Pilot and CPG **MASTER CAUTION** light starts flashing.

(3) Pilot and CPG caution/warning **MAN STAB** light starts flashing.

(4) Stabilator aural tone comes on.

Regaining auto mode after the NOE/APRCH mode is selected is accomplished by momentarily pressing the automatic operation/audio tone **RESET** button on either collective stick or setting the **NOE/APRCH** switch to **OFF** Manual mode may be regained by selecting manual mode with the **NU/ND** switch.

d. Stabilator Position Indicator. The **STAB POS DEG** indicator (fig 2-28), located in the upper right section of the pilot and CPG instrument panels, provides a visual indication of stabilator angle of incidence in either manual or automatic modes of operation. It is calibrated from 10 ° trailing edge up to 35 ° trailing edge down to reflect the position of the stabilator trailing edge. An **OFF** flag is displayed on the indicator face when the instrument is not operating and dc electrical power is applied to the indicator. With ac or dc power loss, both the **OFF** flag and pointer are not displayed.

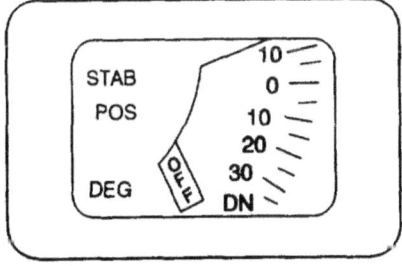

Figure 2-28. Stabilator Position Indicator

e. Stabilator Placard. A placard installed to the right of the position indicator in each crew station lists stabilator incidence angles for given airspeeds. Refer to Chapter 5 for limitations.

Section VI. HYDRAULIC AND PRESSURIZED AIR SYSTEMS

2.38 HYDRAULIC SYSTEMS.

Two independent hydraulic systems are installed so that failure of one system will not affect operation of the other. They are designated as the primary and utility hydraulic systems. They are similar but not identical; they have separate, as well as shared functions.

2.38.1 Primary Hydraulic System. The primary hydraulic system (fig 2-29) provides hydraulic power to the primary side of the lateral cyclic, longitudinal cyclic, collective, and directional servoactuators. Only the primary sides of these servoactuators (discussed in more detail in Section V) have electrohydraulic valves that allow the DASE and BUCS to affect the flight controls. Consequently, failure of the primary hydraulic system will result in the loss of DASE and BUCS. The primary hydraulic equipment includes the hydraulic pump, manifold, and servoactuators. The heat exchanger may still be installed in the primary system on some helicopters. The heat exchanger is obsolete and is being removed through attrition.

a. Servoactuators. The servoactuators (fig 2-29) can be commanded mechanically or electrically. Each actuator contains two hydraulic pistons on a common piston rod. One piston is driven by the primary hydraulic system, the other by the utility hydraulic system. The hydraulic provisions in the actuator are completely independent of each other; there is no exchange of fluid between systems.

B Each actuator is controlled by a common manual tandem servo valve which ports hydraulic pressure to each of the pistons. The servo valve spool is positioned by the associated mechanical control system to provide full authority control. Each actuator is also electrically controlled through the electrohydraulic valve sleeve to provide SAS or BUCS control. Each actuator is equipped with a hydraulically powered plunger which locks the manual servo valve spool at mid position when the DASEC powers the BUCS solenoid valve for BUCS engagement. Each actuator is also equipped with a DASEC controlled SAS solenoid valve that ports primary hydraulic pressure to the servo valve and the BUCS solenoid. Position transducer LVDT's measure the position of the servo valve sleeve and the actuator position. Each actuator incorporates a shear pin in the feedback linkage to decouple the actuator motion from a jammed mechanical control and prevent damage to the bellcrank attachments.

b. Primary Pump. The primary hydraulic pump is mounted on the accessory drive case of the main transmission (left side). The pump is of constant-pressure variable-displacement design driven by the transmission accessory gear train.

c. Primary Manifold. The primary manifold is installed on the left forward quadrant of the transmission deck. Its function is to store, filter, supply, and regulate the flow of hydraulic fluid. The manifold reservoir is pressurized on the return side by PAS air acting on the manifold reservoir piston. This prevents pump inlet cavitation. Servicing crews introduce fluid to the reservoir through ground support equipment (GSE) connections or the hand pump. Low pressure fluid entering the fill port is filtered by a 45 micron screen filter (before MWO 1-1520-238-50-52) or a 5 micron cartridge filter (after MWO 1-1520-238-50-52). The primary hydraulic system fluid capacity is six pints. The reservoir stores about one pint. Section XV contains specifications, capacities, and procedures for oil system servicing. Other provisions within the primary manifold are described below:

(1) **Air-Bleed Valve.** Used to deplete the pressurized air on the manifold for system repair or service.

(2) **High-Pressure and Low-pressure Relief Valves.** Regulates fluid pressure from the pump and the return to the manifold.

(3) **Reservoir Low-Level Indication Switch.** Switch is activated by the manifold reservoir piston. This illuminates an **OIL LOW PRI HYD** segment light on the pilot caution/warning panel to indicate minimum operating level.

(4) **Fluid Level indicator.** Located in the manifold reservoir housing allows visual inspection of the reservoir oil level.

(5) **Filters.** Filters on both manifold pressure and return sides have mechanical impending bypass indicators for visual inspection. These indicators operate on differential pressure. Both impending bypass indicators illuminate the **OIL BYP PRI HYD** segment on the pilot caution/warning panel. Only the return filter has bypass provisions.

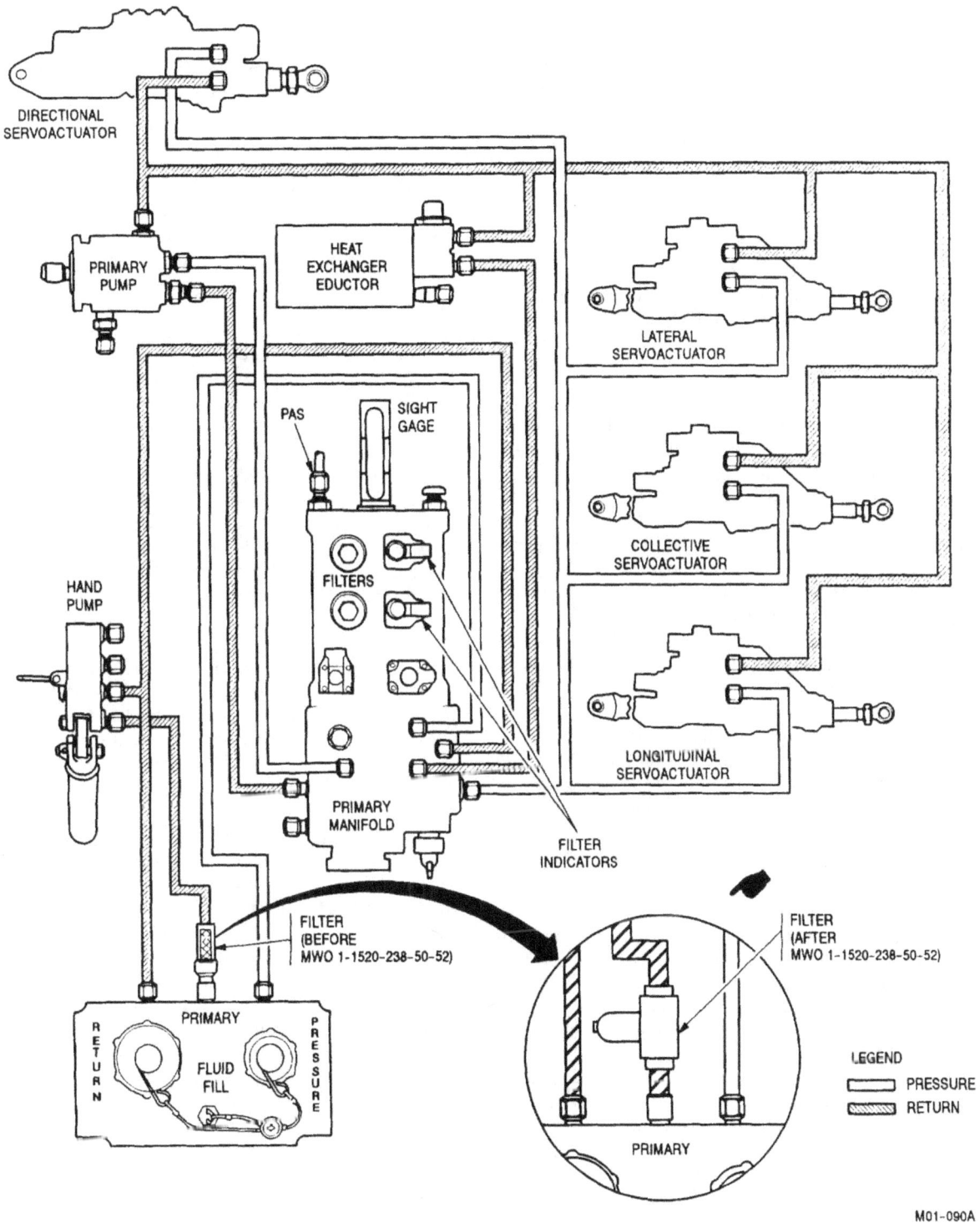

Figure 2-29. Primary Hydraulic System

(6) Pressure Switch. Senses primary system pressure and informs the pilot and CPG of a low oil pressure condition by illuminating the **PRI HYD PSI** segment on the pilot caution/warning panel and **PRI HYD** segment on the CPG panel. This switch illuminates the segments until the system pressure is above 2050 psi and will again illuminate the segments when pressure falls below 1250 psi.

(7) Pressure Transducer. Measures hydraulic pressure on the pressure side of the manifold and transmits this value to the left side of the pilot dual hydraulic gauge.

2.38.2 Utility Hydraulic System. The utility hydraulic system (fig 2-30) provides hydraulic power to the utility side of the lateral cyclic, longitudinal cyclic, collective, and directional servo actuators. This system also provides hydraulic power to the rotor brake, area weapon, external stores, tail-wheel lock, ammo carrier drive, and APU starter. The utility hydraulic pump (on the accessory drive case of the main transmission right side) is identical to that in the primary system. The hydraulic heat exchanger may still be installed on some helicopters. The heat exchanger is obsolete and is being removed through attrition. The only significant difference in the two systems is the manifold, with an associated accumulator and gas reservoir. Additional components in the system are the accumulator, rotor brake, and the utility hydraulic return accumulator that dampens hydraulic pressure surges caused by sudden actuation of the gun turret.

a. Utility Manifold. The utility manifold (fig 2-31) is installed on the aft main fuselage deck on the right side. It stores, filters, supplies, and regulates the flow of utility hydraulic fluid. Demands on the utility system are much greater than those on the primary system, and the utility manifold installation is therefore larger, with a reservoir capacity of 1.3 gallons. Total system capacity is about 2.6 gallons. An air-pressure reliefvalve, low-pressure reliefvalve, high-pressure relief valve, reservoir low-level indicating switch, fluid-level indicator, return filter, pressure filter, pressure switch, and pressure transducer, are identical in function to the primary system components. The pilot caution/warning panel provisions are also similar; but the lights are labeled **OIL LOW UTIL HYD, OIL BYP UTIL HYD,** and **UTIL HYD PSI.** Low utility hydraulic pressure also illuminates the CPG **UTIL HYD** light.

The utility manifold also incorporates several other features not duplicated on the primary manifold. These are the utility accumulator hydraulic pressure transducer and rotor brake solenoids.

(1) Low-Level and Auxiliary isolation Valves. Permit hydraulic fluid to flow to the external stores, ammo carrier drive, tail wheel lock actuator, and area weapon turret. If the reservoir fluid level decreases significantly, the reservoir piston, driven by PAS air, will contact and close the low-level valve. The auxiliary isolation valve, which normally requires two sources of pressure to permit fluid flow, will then close and deny hydraulic power to the area weapon turret, the external stores actuators, and the ammo carrier drive.

(2) Shutoff Valve. Located in the pressure line to the directional servo and TAIL WHL unlock actuator is actuated by the low-level switch in the utility system reservoir. The utility side of the directional servo actuator and the TAIL WHL unlock actuator will be inoperative if a low utility system fluid level is sensed.

(3) Accumulator isolation Valve. Normally isolates accumulator pressure from the rest of the utility system but allows system flow from the pump to pass through a portion of the valve and on to the utility side of the tandem servo actuators.

(4) Override Solenoid Valve. Normally de-energized closed, permits crew management of the accumulator reserve pressure. When the pilot or CPG places his **EMER HYD** switch ON, the override solenoid valve will energize open and accumulator fluid will pass to the accumulator isolation valve via emergency routing. In this case, another portion of the accumulator isolation valve will permit accumulator fluid to flow to the utility side of the servo actuators. The **EMER HYD** switch is powered through the **EMERG HYD** circuit breaker on the pilot overhead circuit breaker panel.

(5) Accumulator Hydraulic Pressure Transducer. Located in the manifold will provide the pilot with a continual indication of accumulator pressure on the accumulator hydraulic pressure indicator. During normal operation, the indicated pressure will be the same as that of the utility hydraulic system.

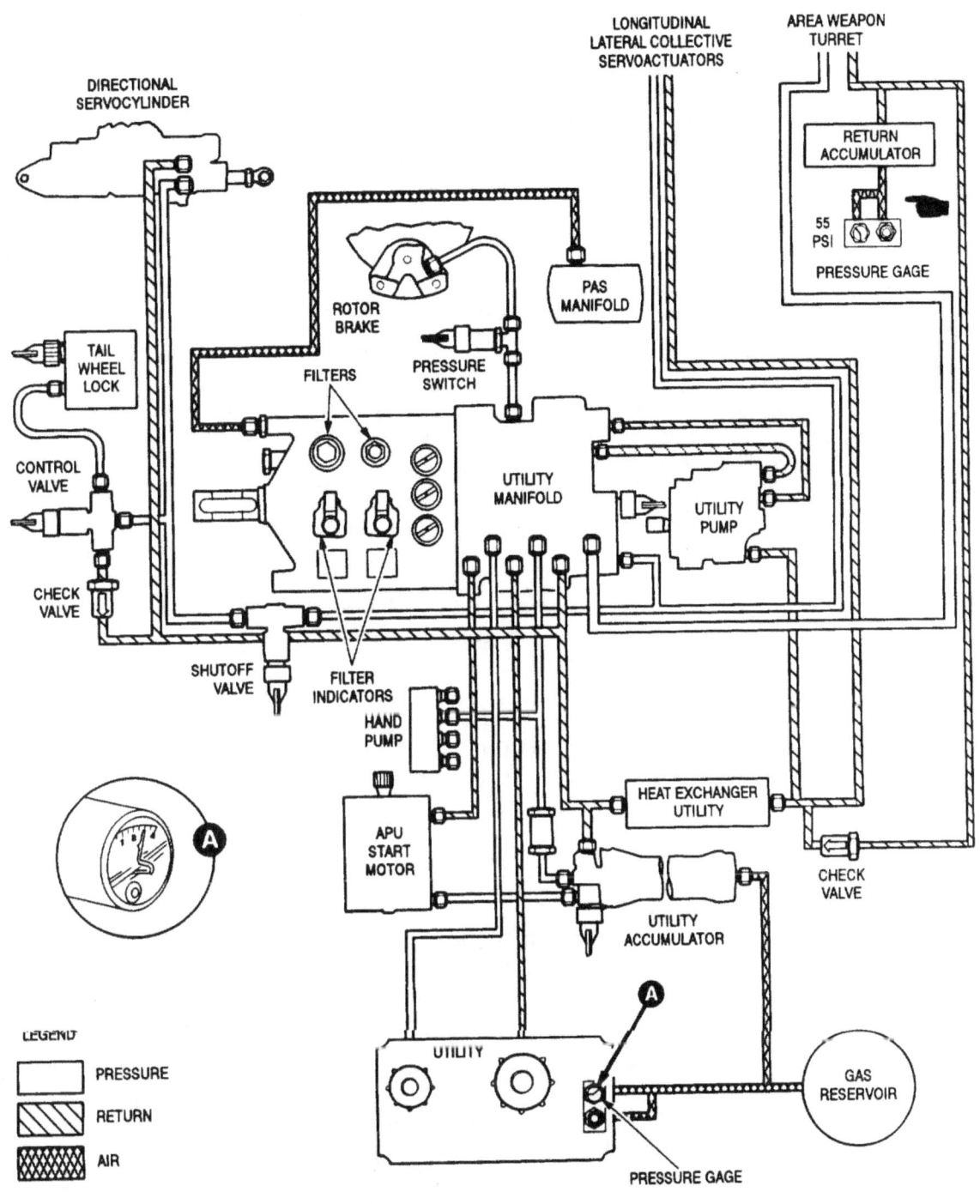

Figure 2-30. Utility Hydraulic System

(6) Rotor Brake Solenoid Valves. Valves are controlled by a three-position switch **OFF**, **BRAKE**, and **LOCK** on the pilot **PWR** lever quadrant adjacent to the **PWR** levers (fig 2-19). When the switch is positioned to **BRAKE**, utility system pressure regulated to 337 psi is applied to stop the rotor brake disc on the main transmission. When positioned to **LOCK**, the brake off solenoid valve traps 3000 psi pressure between the manifold and the rotor brake actuator.

b. Utility Accumulator. The accumulator is a multi-purpose installation unique to the utility hydraulic system. It is located on the right side of the helicopter directly beneath the APU. It stores hydraulic fluid at 3000 psi. The accumulator is charged by nitrogen gas. A gas storage tank supplies the charge and is serviced through a charging port (fig 2-31). The utility hydraulic manifold pressure is used for rotor brake application, APU starting, and emergency use of flight control operation.

2.38.3 Hand Pump. A hand pump is installed next to the primary system GSE panel on the right side of the helicopter. The pump provides one method of charging the fluid pressure in the utility accumulator as well as a method for the ground crew to fill both the primary and utility reservoirs. Low pressure fluid entering the fill port is filtered by a 45 micron screen filter (before MWO 1-1520-238-50-52) or a 5 micron cartridge filter (after MWO 1-1520-238-50-52). A lever may be moved to any of three positions. This, in turn, will open one of three mechanically operated check valves to the accumulator or to either reservoir. Section XV contains servicing instructions.

2.39 PRESSURIZED AIR SYSTEM (PAS).

The PAS cleans, pressurizes, regulates, and distributes air to the air turbine starters, fuel boost pump, fuel transfer pump, external fuel tanks, hydraulic reservoirs, heat exchangers, defog nozzles, engine cooling louver actuators, utility receptacle, ice detector sensor, and environmental control unit (ENCU). The pressurized air system has three sources of air: The primary source is the shaft driven compressor; the other sources are bleed air from the No. 1 engine and from an external air source.

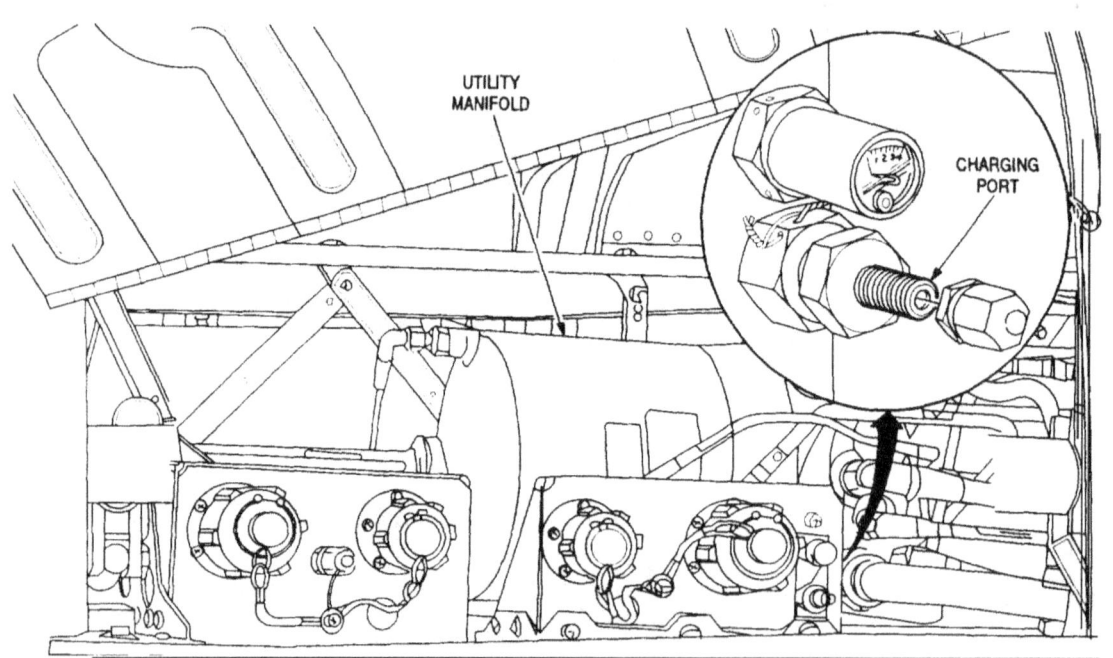

Figure 2-31. Utility Manifold

2.39.1 Shaft Driven Compressor (SDC). The SDC is mounted on the main transmission accessory case and uses outside air. The air passes through a particle separator and then through a butterfly-type throttle valve. This valve is closed by a time-delay solenoid when the APU START switch is engaged. The valve remains closed for 60 seconds after the APU **START** switch is released. This reduces APU loads during starting. The shaft driven compressor will compress air at 3:1 ratio and route it through a surge valve. This valve, mounted on top of the SDC, will open or close as necessary to keep SDC air pressure constant with variable system demand. Pressurized air then passes through a spring-operated check valve and into the PAS manifold. This check valve prevents pressurized air, from sources other than the SDC, from reverse routing back to the SDC. The SDC is lubricated by the main transmission accessory oil pump and has its own scavenge. If the SDC oil temperature exceeds 182 °C (360 °F), or if the SDC fails, the pilot **SHAFT DRIVEN COMP** caution light illuminates.

2.39.2 No. 1 Engine Bleed Air. The No. 1 engine provides an alternative source of pressurized air in case of SDC failure. The engine compressor bleed air selector valve opens automatically if SDC output falls below 10 psi. Simultaneously, a spring-operated check valve opens. When closed, this valve prevents pressurized air from going to the engine compressor when the SDC is operating. Each engine provides bleed air to its engine inlet to prevent ice formation and to keep the engine cooling louvers closed while the engine is operating. The bleed air used in these last two instances is independent of the pressurized air system.

2.39.3 External Air Source Receptacles. The receptacle for external air source connection (fig 2-46) is on the lower portion of No. 1 engine nacelle. When the air source is connected and operating, a normally closed check valve to the pressurized air system opens to allow air to pressurize the manifold.

Section VII. POWER TRAIN SYSTEM

2.40 POWER TRAIN.

The power train (fig 2-32) transmits engine power to the rotors and transmission-mounted accessories. The power train includes two engine nose gearboxes, two shafts to the main transmission, the main transmission, main rotor drive shaft, tail rotor drive shafts, intermediate gearbox, tail rotor gearbox, and APU drive shaft, and couplings.

2.40.1 Engine Nose Gearboxes. The engine nose gearboxes are mounted on the nose of each engine. They reduce nominal engine speed and change the angle of drive. Both nose gearboxes have self-contained pressurized oil systems with provisions to ensure limited operation if a total loss of pressurized lubrication occurs. These self-contained systems feature a pressure pump driven by the gearbox output shaft, filter with impending bypass filter indicator and filter bypass capability, high-pressure relief valve, and sump. Mounted on the nose gearbox output drive is an axial-type fan that draws air through the gearbox fairing and past cooling fins on the gearbox. The input drive shafts have flexible couplings that require no lubrication. Externally accessible accessories include a filler, breather, oil level sight gage, chip detector/temperature sensor, pressure transducer, and temperature probe.

a. Engine Nose Gearbox Caution Light Provisions. High oil temperature, low oil pressure, and the presence of chips are detected in each gearbox. The pilot is alerted to the condition by six caution segments: **OIL HOT NOSE GRBX 1**, **OIL HOT NOSE GRBX 2**, **OIL PSI NOSE GRBX 1**, **OIL PSI NOSE GRBX 2**, **CHIPS NOSE GRBX 1**, and **CHIPS NOSE GRBX 2**. Either the **ENG 1** or **ENG 2** segment will simultaneously illuminate on the CPG caution/warning panel (fig 2-44).

2.40.2 Main Transmission. The main transmission combines the two engine nose gearbox inputs and provides drive to the main rotor, tail rotor, accessories, and rotor brake disc. Two overrunning clutches at the main transmission permit either engine to be disengaged from the transmission during autorotation. The main transmission reduces the rpm input to the main rotor, tail rotor, and the rotor brake disc. The main transmission is mounted below the main rotor static mast base which allows its removal without removing the upper controls, mast, hub, or blades. The main rotor drive shaft is designed to carry torque loads only. The rotor hub is on a static mast which carries vertical and bending loads. The drive shaft rotates inside the static mast. The main transmission has primary and accessory drive trains. The primary drive train, through three stages, changes the angle and speed of the power drives to the main rotor, tail rotor, and rotor brake. Overrunning clutches allow the APU to drive the transmission accessory gearbox when the engines are not operating. The transmission accessories consist of two alternating-current generators, two hydraulic pumps, and a shaft driven compressor. A magnetic pickup measures main rotor rpm.

a. Main Transmission Lubrication. The main transmission has two independent oil systems. Each system has its own sump, pump, filter and heat exchanger. Oil level sight gauges are located in the transmission housing at each oil sump. These systems are not totally independent in the usual sense because during normal operation, the oils will mix. If oil loss occurs in either sump or in either heat exchanger, the diverter (float) valve will seal off that sump to prevent a total loss of oil. There are provisions throughout the transmission so that even with a total loss of oil there will be limited lubrication. Each oil filter has a bypass capability and an impending bypass capability and an impending bypass filter indicator. Each sump has a chip detector/temperature sensor and temperature transducer. Pressure is measured downstream of each heat exchanger. Oil passing through the heat exchanger. mounted inboard on each engine firewall, is air cooled. The heat exchangers have thermal bypass provisions for cold starts. A third oil pump, driven by the accessory drive, lubricates the accessory gears during APU operation. This pump draws oil from the right oil system sump.

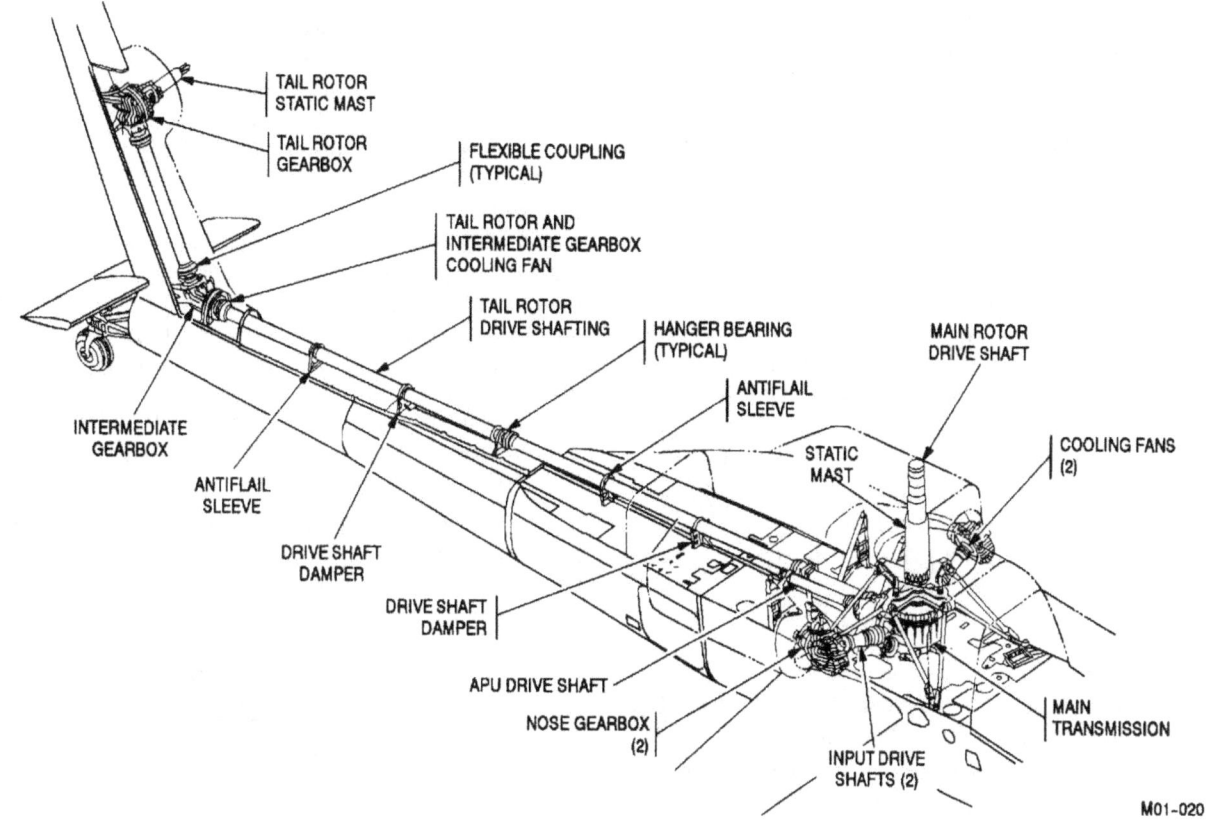

Figure 2-32. Power Train

b. Main Transmission Caution Lights. Sensors in the transmission sense adverse conditions which are displayed as caution lights (fig 2-44) in the crew stations The pilot station has six caution/warning lights: **OIL PSI MAIN XMSN 1, OIL PSI MAIN XMSN 2, OIL HOT MAIN XMSN 1, OIL HOT MAIN XMSN 2, CHIPS MAIN XMSN,** and **OIL PSI ACC PUMP.** Activation of any of these lights, except the **OIL PSI ACC PUMP,** simultaneously illuminates either **MAIN 1, MAIN XMSN 2,** or **CHIPS MAIN XMSN** on the CPG caution/warning panel.

2.40.3 Rotor Brake. The rotor brake reduces turn-around time for aircraft loading and servicing and prevents windmilling of the rotor system during gusty wind conditions. The rotor brake disc is visible at the aft end of the transmission.

CAUTION

With rotors turning, do not place the RTR BK switch in LOCK position.

NOTE

- When engaging rotor lock, pause in the **BRAKE** position until the **RTR BK** caution/warning light has illuminated prior to placing the switch in the lock position. The **PWR** levers will not advance past the ground idle detent with the rotor brake switch in the **LOCK** position.

- When operating engines with the rotor brake locked, monitor the main transmission temperature. If the transmission temperature reaches 130 °C (266 °F), secure operation, or release the rotor brake and turn the rotors until transmission temperature returns to normal.

a. Rotor Brake (RTR BK) Switch. The rotor brake is controlled by the **RTR BK** switch on the pilot **PWR** lever quadrant (fig 2-19). The switch has three positions: **OFF, BRAKE,** and **LOCK.** When set at **LOCK** with full utility hydraulic system pressure, the brake prevents the gas turbines from driving the power turbines when both engines are at idle. When the rotor is stopped, the switch may be set at **LOCK** which causes solenoid valves in the manifold to deenergize and all available utility hydraulic system or accumulator pressure to be applied to the brake. A system of interlocks prevents the rotor brake from being locked when the **PWR** levers are in any position except **IDLE** and **OFF.** When the switch is set at **BRAKE,** solenoid valves in the utility hydraulic manifold operate to actuate the brake. With the switch at **OFF,** the only hydraulic pressure to the brake is 30 psi from the pressurized air system which, when operating, pressurizes the return side of the utility hydraulic system.

b. Rotor Brake Solenoid and Pressure Switch. Rotor brake electrical solenoids and the pressure switch receive 28 vdc from the emergency dc bus through the **RTR BRK** circuit breaker on the pilot overhead circuit breaker panel. If this circuit breaker is open, or if helicopter emergency electrical power is lost for any reason, the rotor brake, if previously set at **LOCK,** will remain locked as long as accumulator pressure is available.

2.40.4 Tail Rotor Drive Shafts. There are four tail rotor drive shaft sections. Three tail rotor drive shafts lead from the transmission to the intermediate gearbox. Two are of equal length. The last shaft is installed on the vertical stabilizer between the intermediate and tail rotor gearboxes. Hanger bearings support the longer shafts. They are covered by aerodynamic fairings which may be opened for maintenance and inspection. The two equal-length shafts incorporate friction dampers and antiflails. Flexible couplings, attached to the shaft ends, are capable of accommodating shaft misalignments throughout the power range.

CAUTION

Prolonged out of ground effect hovering (20 - 30 min.) with outside air temperatures above 75 °F (24 °C) may cause the intermediate gearbox to overheat.

2.40.5 intermediate Gearbox. The intermediate gearbox, at the base of the vertical stabilizer, reduces the rpm and changes the angle of drive. A fan mounted on the gearbox input shaft draws air from an inlet on the vertical stabilizer. This air cools both the tail rotor gearbox and the intermediate gearbox. Four thermistors monitor temperature and an accelerometer measures vibration limits. The intermediate gearbox is a grease-lubricated sealed unit.

a. intermediate Gearbox Caution Light indicators. The four thermistors and the accelerometer provide crewmembers with temperature and vibration caution lights. Both crew stations have **TEMP INT** and **VIB GRBX** light segments.

CAUTION

Prolonged out of ground effect hovering (20 - 30 min.) with outside air temperatures above 75 °F (24 °C) may cause the tail rotor gearbox to overheat.

2.40.6 Tail Rotor Gearbox. The tail rotor gearbox, mounted on the vertical stabilizer, reduces the output rpm and changes the angle of drive. The tail rotor output shaft passes through the gearbox static mast. All tail rotor loads are transmitted to the static mast. The output shaft transmits only torque to the tail rotor. Lubrication of this gearbox is identical to that of the intermediate gearbox.

a. Tail Rotor Gearbox Caution Light indicators. The four thermistors and the accelerometer function in the same way as for the intermediate gearbox. The associated caution light on the pilot and CPG caution/warning panels are labeled **TEMP TR** and **VIB GRBX**. (Both the intermediate and tail rotor gearboxes activate the **VIB GRBX** light segment).

Section VIII. ROTORS

2.41 ROTOR SYSTEM.

The rotor system consists of a single, four-bladed, fully articulated main rotor and a four-bladed tail rotor assembly with two teetering rotor hubs.

2.41.1 Main Rotor. The main rotor has four blades. The head is a fully articulated system that allows the four blades to flap, feather, lead, or lag independent of one another. The head consists of a hub assembly, pitch housings, rotor dampers, and lead-lag links. Attached to the rotor head are four easily removable blades. The main rotor is controlled by the cyclic and collective sticks through a swashplate mounted about the static mast.

a. Hub Assembly. The main rotor hub is a steel and aluminum assembly that supports the main rotor blades; it is driven by the main rotor drive shaft. The hub rotates about a static mast, which supports it. This arrangement allows the static mast, rather than the main rotor drive shaft, to assume all flight loads. The hub is splined to the main rotor drive shaft by means of a drive plate adapter that is bolted to the hub. The hub is secured to the static mast by a large locknut secured by multiple bolts. The hub houses two sets of tapered roller bearings that are grease-lubricated and sealed. These bearings transfer hub loads to the static mast. Mechanical droop stops limit blade droop. When blade droop occurs, a striker plate on the pitch housing contacts a roller. The roller presses a plunger against a droop stop ring on the lower portion of the hub.

b. Pitch Housing. The pitch housing permits blade pitch changes in response to flight control movements transmitted through the swashplate. This is made possible within the four pitch housings by V-shaped stainless steel strap assemblies that are able to twist and flap to permit blade feathering and flapping. Cyclic and collective stick inputs are transmitted to the pitch housing horns by pitch links attached to the swashplate. Feather bearings are installed inboard on the pitch housing to allow vertical and horizontal loads to be transferred from the pitch housing to the hub. Centrifugal loads are transmitted by each strap assembly to the hub.

c. Lead-Lag Links. The lead-lag link for each blade is connected to the outboard end of each pitch housing and is secured in place by a pin and two bearings that allow the link to move horizontally. The pin goes through the V-portion of each strap within the pitch housing.

d. Damper Assemblies. Two damper assemblies control the lead-lag movement of each main rotor blade. Each damper attaches outboard to a link lug and inboard to a trunnion at the pitch housing. The damper contains elastomeric elements that distort to allow the blade to lead or lag.

e. Main Rotor Blades. Each main rotor blade is a constant-chord asymmetrical airfoil. The outboard tip is swept aft 20° and tapers to a thinner symmetrical section. The blade has a 21-inch chord. Tip weights are installed within the blade. Chord-wise, the leading-edge and forward half of the blade is a four-cell structural box of stainless steel and fiberglass with a stainless steel spar. The aft half of the blade has fiberglass skin with a nomex honeycomb core and a bendable trailing edge strip to aid in blade tracking. Each blade secured to its lead-lag link by two blade attachment pins. These pins can be removed without the use of tools and they pass vertically through the lead-lag link and blade root fittings which are both made of titanium. Five sets of stainless steel doublers are located on the upper and lower surfaces of the blade at the blade root. The blades may be folded by removing the appropriate blade attachment pin (one for each blade) and any two adjacent pitch link bolts and pivoting the blade to the rear position using a hand-held blade support device.

2.41.2 Tail Rotor. The tail rotor is of semirigid, teetering design. Two pairs of blades, each pair fastened to its own delta hinged hub, provide antitorque action and directional control. A titanium fork houses two elastomeric teetering bearings and drives the rotating swashplate through an attached scissors assembly. The tail rotor assembly is splined to, and driven by, the tail rotor gearbox drive shaft which passes through a static mast. Blade pitch changes when directional control inputs cause the non-rotating swashplate to act upon the rotating swashplate. One pitch link for each blade, attached to the rotating swashplate and pitch horn, causes blade movement about two pitch-change bearings in the blade root. Centrifugal forces are carried by stainless steel strap assemblies that attach outboard to the blade root and inboard at the hub center. An elastomeric bearing assembly positions the hub and strap pack in the tail rotor fork. Each blade has one stainless steel spar and two aluminum spars. Doublers and rivets attach the blade to the blade root. Brackets on the root fitting hold chord-wise balance weights. Spanwise balance weights are installed at the tip of each blade in an aluminum tip cap.

Section IX. UTILITY SYSTEMS

2.42 DEFOGGING SYSTEM.

The **CANOPY DEFOG** switch on the pilot **ANTI ICE** panel allows the pilot to direct pressurized air (with ECS on or off) to diffuser outlets at each canopy side panel. When the **CANOPY DEFOG** switch is set to ON, 28 vdc from the No. 1 essential dc bus through the **ECS CANOPY ANTI ICE** circuit breaker causes the defog valve to open. This allows hot air from the pressurized air system to flow through the shutoff valve to the air mixers. The hot pressurized air mixes with the crew station conditioned air in the air mixers. This partially cooled air is directed against the canopy side panels to defog them.

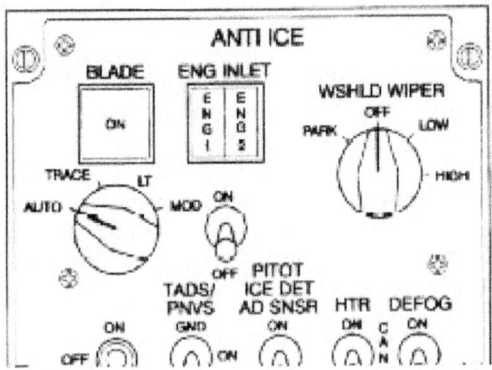

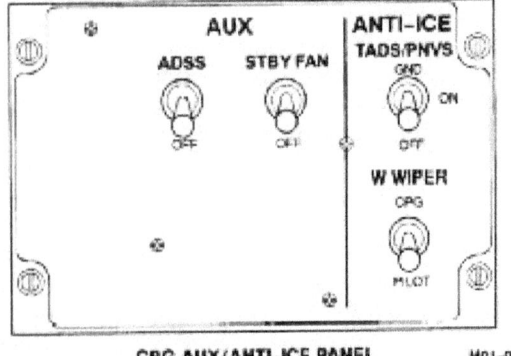

Figure 2-33. Pilot and CPG Anti-Ice Control Panels

2.43 ANTI-ICING AND DE-ICING.

Multiple anti-icing and de-icing protection is available for the windshields, pitot tubes, air data sensor, pilot night vision sensor (PNVS), target acquisition designation sight (TADS), engine inlets, nose gearboxes, and main and tail rotor blades. The 30mm weapon and the wing pylons are also susceptible to icing and maybe deiced by periodic flexing in azimuth or elevation. The pilot and CPG have **ANTI-ICE** panels on their left consoles (fig 2-33).

2.43.1 Windshield Anti-Ice/De-Ice. The two windshields, one in front, and the other directly above the CPG, have built-in transparent anti-ice elements that are heated by 115 vac to prevent ice accumulation. The pilot activates windshield heating by setting a two-position **CANOPY HTR** toggle switch on his **ANTI-ICE** panel to ON. In the ON position, the **CANOPY HTR** switch provides 28 vdc from the No. 1 essential dc bus through the **CANOPY ANTI-ICE CONTR** circuit breaker on the pilot overhead circuit breaker panel to turn on the windshield anti-ice control unit. When the control unit is operating, 115 vac from the No. 2 essential ac bus through the **CANOPY ANTI-ICE** circuit breaker on the pilot overhead circuit breaker panel is provided to the control unit. The 115 vac is distributed by the control unit to the window heater elements. The control unit is provided with temperature control and fail-safe features. When there is a failure in the windshield anti-ice system, the control unit will remove the 115 vac power from the unit. If the cause for the failure is an overheat or false temperature sensor signal, the control unit can be reset by turning the **CANOPY HTR** switch OFF and then back ON. Whenever the control unit, canopy heaters, canopy temperature sensors, or input power fails, or if the canopy overheats with the **CANOPY HTR** switch in the ON position, the **CANOPY ANTI-ICE FAIL** light on the pilot caution/warning panel will illuminate.

2.43.2 Pitot Tube and Air Data Sensor Anti-Ice/De-Ice. The pitot tubes, installed outboard on the leading edge of both wings, and the air data sensor, attached to the top of the main rotor de-ice distributor are protected from ice accumulation by internal heating elements. The **PITOT AD SNSR** switch, located on the pilot **ANTI-ICE** panel, is a dual fun4 Tc reswitch. In the ON position, the right hand pitot heater receives 28 vdc from the emergency dc bus through the **PITOT HTR** circuit breaker. The second fun4tion of the switch provides 28 vdc from the No. 3 essential dc bus through the

AIR DATA DC circuit breaker to the left wing pitot heater. Both circuit breakers are on the pilot overhead circuit breaker panel. The **AIR DATA DC** circuit breaker also provides 28 vdc to the ADS control relay. When energized, this relay sends 115 vac from the No. 1 essential ac bus through the **AIR DATA AC** circuit breaker to the air data sensor for anti-icing. In addition, the CPG **AUX** panel **ADSS** switch, when set to on, energizes the ADS control relay to provide the air data sensor with 115 vac from the No. 3 essential ac bus through the **AIR DATA AC** circuit breaker.

2.43.3 PNVS and TADS Anti-Ice/De-Ice. Three position TADS/PNVS toggle switches on the pilot **ANTI-ICE** panel and the CPG **AUX/ANTI-ICE** panel control the operation of the **TADS/PNVS** window heaters. These switches receive 28 vdc from the No. 1 essential dc bus through the **CANOPY ANTI-ICE CONTR** circuit breaker. Window heating is automatically controlled by a thermal control sensor in the respective heater power return circuits.

a. In Flight TADS/PNVS Anti-Ice/De-Ice. With the helicopter in flight and the pilot or CPG **TADS/PNVS** switch set to ON, the squat switch relay closes to provide 28 vdc from the pilot or CPG **TADS/PNVS** switch to the MRTU. This causes the MRTU to send a signal to energize anti-ice relays in the TADS power supply and PNVS electronic unit. The TADS and PNVS anti-ice relays then provide the TADS and PNVS window heaters with 115 vac from the No. 1 essential ac bus through the **TADS AC** on the CPG circuit breaker panel and the **PNVS AC** circuit breakers on the pilot overhead circuit breaker panel.

b. On Ground TADS/PNVS Anti-Ice/De-Ice. When the helicopter is on the ground, the squat switch opens removing power from the **TADS/PNVS** switches and the TADS and PNVS window heaters. When either **TADS/PNVS** switch is set to **GND**, a signal is sent to the MRTU which energizes the TADS/PNVS anti-ice relays. These relays then provide the TADS and PNVS window heaters with 115 vac from the No. 1 essential ac bus through the **TADS AC** and **PNVS AC** circuit breakers.

2.43.4 Engine Inlet and Nose Gearbox Anti-Ice. Engine bleed air is used to heat the engine air inlet, and heater blankets are used to anti-ice the nose gearbox. Engine anti-ice protection is controlled by a **ENG INLET ON/OFF** toggle switch on the pilot **ANTI-ICE** panel. In the **ON** position, **ENG 1** and **ENG 2** advisory lights located above the switch will illuminate when the system is functioning properly. The engine anti-ice valves and engine gearbox heater blanket controller receive 28 vdc from the No. 1 essential dc bus through the **ENG ANTI-ICE** circuit breaker on the pilot overhead circuit breaker panel. The **ENG INLET** switch in the OFF position energizes both engine anti-ice valves to the closed position. In the **ON** position, the **ENG INLET** switch deenergizes the anti-ice valves thus allowing them to open and furnish hot air to the engine inlet for anti-icing. It also energizes the engine gearbox heater blanket controller. The energized controller allows 115 vac to be supplied from the No. 1 essential ac bus through the **NOSE GRBX HTR** circuit breaker on the pilot overhead circuit breaker panel for the nose gearbox heater blankets.

2.43.5 Rotor Blade De-Icing System. The rotor blade de-ice three-position **ON, OFF,** and **TEST** toggle switch, located on the pilot **ANTI-ICE** panel receives 28 vdc from the No. 3 essential dc bus through the **BLADE DE-ICE CONTR** circuit breaker on the pilot overhead circuit breaker panel. In the **ON** position, the switch will provide 28 vdc power to the main rotor and tail rotor de-icing controller and turn on the controller. When the controller is turned on, it collects data from the signal processor unit, ice detector/rate sensor, and the outside air temperature sensor. If icing conditions are present, the controller will energize the blade de-ice control relay. The energized blade de-ice control relay provides a ground to the blade de-ice contactor control relay that receives 28 vdc from the No. 3 essential dc bus through the **BLADE DE-ICE** circuit breaker. With the blade de-ice contactor energized, 115 vat, 3 phase power is furnished from Gen 2 through the blade de-ice contactor to the de-ice controller. (Gen 1 will only furnish power when an overload is sensed on Gen 2, and the contactor is caused to trip). The de-ice controller rectifies the 115 vac to 268 vdc ($\pm$ 134 vdc) for the rotor blade heaters. The main rotor blades receive $\pm$ 134 vdc through slip rings at the main rotor distributor. The main rotor distributor provides a sequential delivery to the blade heater elements. In the **TEST** position, the rotor blade de-ice switch provides that the controller will complete one full cycle of the blade anti-ice circuits. The tail rotor blades receive $\pm$ 134 vdc from the de-icing controller through tail rotor slip rings. Power to the tail rotor blade elements and main rotor blade elements is controlled by the de-icing controller which times the amount of current allowed to the blades for heating. The de-icing controller will detect faults within the system. When a fault occurs, the controller shuts off power to the blades and illuminates the **BLADE ANTI ICE FAIL** segment on the pilot caution/warning panel.

a. **Rotor Blade De-Icing Rotary Switch Operation.** When the rotor blade de-ice system is operating, the **BLADE ON** advisory light on the pilot **ANTI-ICE** panel illuminates. The rotary switch above the **ON**, **OFF**, and **TEST** toggle switch on the pilot **ANTI-ICE** panel provides manual override. When the rotary switch is turned out of the **AUTO** position, it permits manual control to select conditions that provide manual heating to the rotor blades for **TRACE**, light **(LT)** and moderate **(MOD)** icing conditions.

2.43.6 Icing Severity Meter. An icing severity meter (fig 2-34) is provided in the right portion of the pilot instrument panel. An aspirated ice detector sensing head, located on the doghouse fairing assembly, provides input signals to a processor. These signals are proportional to liquid water content (LWC). The signals are sent to the icing severity meter to give the pilot a numerical indication of intensity. The meter is marked with both intensities and categories of ice accumulation. A **PRESS TO TEST** switch is provided adjacent to the meter. When this switch is pressed, the meter pointer will move to 1.5. Both the **BLADE** de-ice and the **PITOT AD SNSR** switches on the pilot **ANTI-ICE** panel (fig 2-33) must be on for the ice detector/rate sensor to operate properly.

2.44 RAIN REMOVAL.

CAUTION

Windshield wipers should not "be operated when canopies are dry. Scratches may result.

Two wipers are mounted on the canopy frame to wipe the two-windshields. Both wipers are electrically driven and are normally controlled by a four-position **WSHLD WIPER** rotary switch on the pilot **ANTI-ICE** panel. The pilot wiper moves horizontally and the CPG wiper moves vertically. They receive 28 vdc from the emergency dc bus through the **WSHLD WPR** circuit breaker on the pilot overhead circuit breaker panel. TWO speeds, **HIGH** and **LOW** may be selected by the pilot. To return the wipers to their static position adjacent to the canopy frame, the **WSHLD WIPER** knob is turned to **PARK** and held momentarily until the blades stop. The knob is spring-loaded from **PARK** to **OFF**. The CPG has limited control over the windshield wipers. Normally, the **W WIPER** switch on the CPG **ANTI-ICE** panel is left in the **PLT** position. When the switch is set at **CPG**, the CPG wiper blade moves at low speed.

2.45 WIRE STRIKE PROTECTION SYSTEM (WSPS).

The WSPS (fig 2-35) consists of six cutter assemblies and eleven deflectors. An upper cutter assembly is mounted on top of the pilot station canopy on the left side of the helicopter. A lower cutter assembly is mounted on the bottom of the helicopter, forward of the gun turret bay. A main landing gear cutter assembly is mounted on each main landing gear strut by the lower step on the forward side of the strut. A PNVS cutter assembly is mounted on the top of the TADS/PNVS. A gun turret deflector and cutter assembly is mounted on the forward side of the gun cradle. A forward and aft deflector are mounted on each of the pilot and CPG door hinges. An upper wiper deflector assembly is mounted by the upper windshield on the left forward side of the CPG station. A lower wiper deflector assembly is mounted by the lower windshield on the right forward side of the CPG station. A tailboom jack pad deflector assembly is mounted on the bottom of the tailboom just forward of the jack pad. A tail landing gear deflector assembly is mounted on the tail landing gear forward of the tail wheel. The wire strike protection system is designed to protect the helicopter from wire obstructions at low levels of flight.

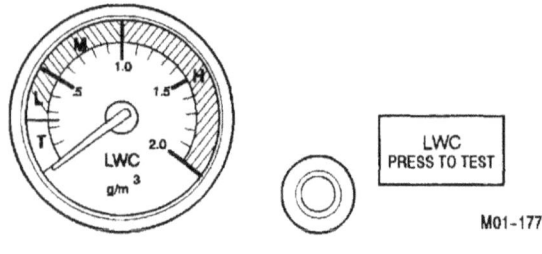

Figure 2-34. Icing Severity Meter and Press-to-Test Switch

TM 1-1520-238-10

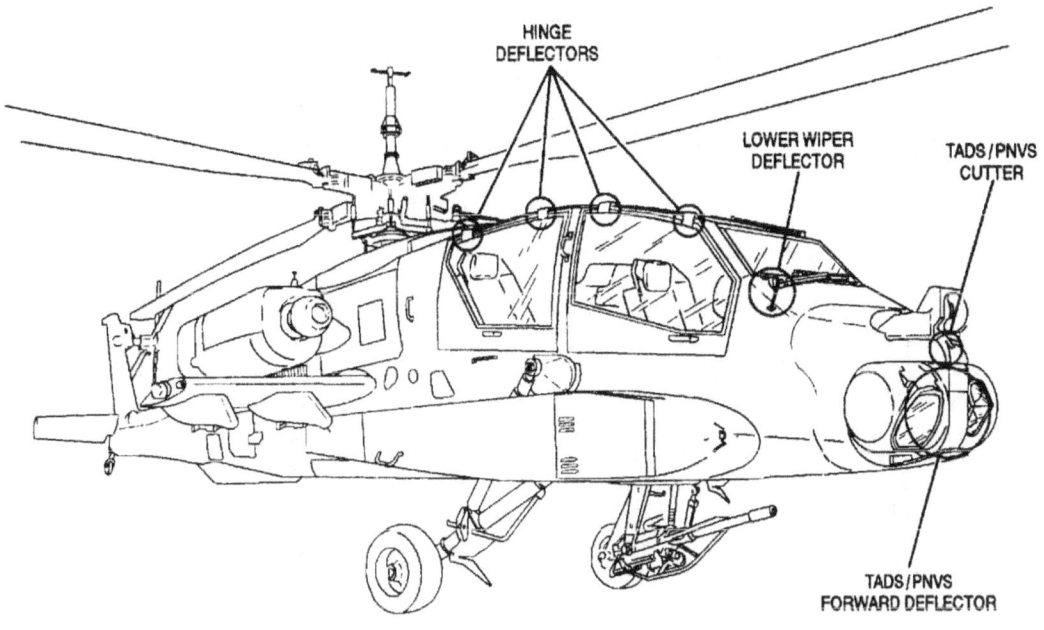

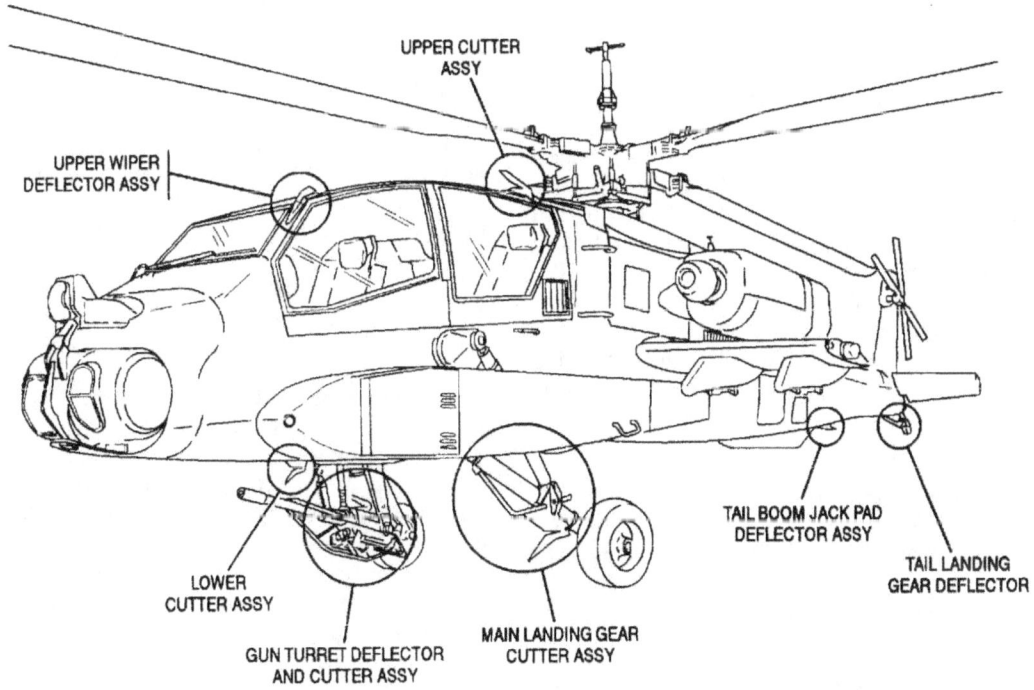

Figure 2-35. Wire Strike Protection System

2-63

Section X. HEATING, VENTILATION, COOLING, AND ENVIRONMENTAL CONTROL SYSTEMS

2.46 ENVIRONMENTAL CONTROL SYSTEM (ECS).

2.46.1 ECS Description. Crew compartment ventilating, heating, and air conditioning are provided by the ECS. In addition, the ECS is the primary source of conditioned air for the two forward avionics bays (FAB) and the TADS/PNVS turret. The ECS is comprised of the environmental control unit (ENCU), ECS shutoff valve, **ECS** control panel, and ducting to deliver pressurized air from the engines to the ENCU, and to route conditioned air from the ENCU to the crew stations, FABs, and TADS/PNVS turret. It includes an exhaust fan for the aft avionics bay, one fan in each FAB, and one fan in the electrical power center. The ENCU, located on the left side of the aft equipment bay, produces conditioned air with low moisture content. It takes hot pressurized air from the pressurized air system and cools it through a heat exchange and air expansion process. The ECS shutoff valve receives 28 vdc from the emergency dc bus through the **ECS CAB** circuit breaker. The FAB fans and aft avionics bay fan receive 28 vdc control voltage from the emergency dc bus through the **ECS CAB** circuit breaker and 115 vac from the No. 2 essential ac bus through the **ECS FAB FANS** circuit breaker. In addition, the electrical power center fan and aft avionics bay fan receive 115 vac from the No. 2 essential ac bus through the **ECS AFT FAN** circuit breaker. All circuit breakers are located on the pilot overhead circuit breaker panel. Thermal switches in the FABs will illuminate the **ECS** segment on the pilot caution/warning panel if the air temperature in either FAB is greater than 105 °F. The aft avionics bay fan is controlled by a separate automatic thermostat which turns the fan on and off depending on the temperature of the aft avionics bay.

2.46.2 ECS Normal Operation. Operation of the ECS is by the pilot **ECS** control panel (fig 2-36) located on the pilot left console (fig 2-11) and by the CPG AUX panel (fig 2-33) located on the CPG left console (fig 2-12). Normal operation is as follows:

   a. ECS Control Panel. The **NORM-STBY FAN** switch is normally set at **NORM.** When the switch is set to **STBY FAN,** the speed of the FAB fans is increased to draw additional air from the crew stations for avionics cooling. The CPG has a two-position **STBY FAN** switch, located on the **AUX** panel. This switch is normally in the OFF position to allow the pilot to control the FAB fan speed. When the CPG switch is set to **STBY FAN,** the speed of the FAB fans is increased.

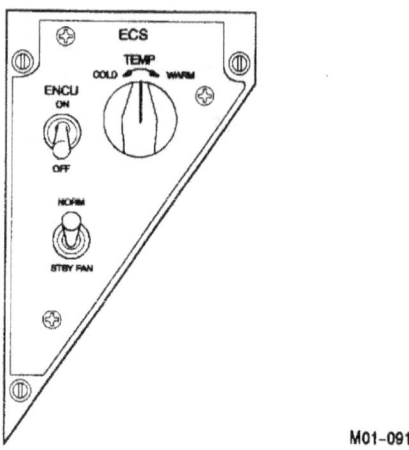

Figure 2-36. ECS Control Panel

The **TEMP** control on the pilot **ECS** panel controls the temperature of conditioned air to the crew stations, FABS, and TADS/PNVS turret. Moving the control between **COLD** and **WARM** operates the ENCU temperature control valve that determines the amount of hot pressurized air that will be allowed to mix with conditioned air. Setting the **ENCU** switch to **ON** opens the ECS shutoff valve and allows pressurized air to enter the ENCU. In addition, it provides 28 vdc to the temperature control sensor for operation of the pilot **TEMP** control and the ENCU temperature control valve.

   b. Temperature Control Sensor. The temperature sensor is located in the ENCU outlet duct. It positions the temperature control valve to maintain the selected temperature in the crew stations. It receives inputs from the **TEMP** control on the **ECS** panel, the FAB thermal switches and the outlet duct air temperature. An output voltage proportional to the difference between the selected temperature and the outlet duct air temperature is applied to the temperature control valve to position the valve to maintain the selected temperature. Thermal switches in the left FAB, in conjunction with the **ECS** panel circuitry, send signals to the temperature control sensor at 75 °F and 85 °F to reduce the temperature. When the left FAB temperature is below 75 °F, the **TEMP** control establishes the reference voltage sent to the temperature control sensor, thereby controlling the crew station temperature. If, during cold weather operations, the pilot adjusts the **TEMP** control to a position such that the temperature of the mixed air would be too hot for the FABs, the temperature sensor will automatically reduce the demanded air temperature. The amount that the temperature sensor can reduce the demanded air temperature is limited.

c. ECS Caution/Warning Light. If the air temperature in either FAB exceeds 105 °F, the **ECS** caution/warning light will illuminate. If this occurs, the pilot should reduce the demanded air temperature, but not to a level that adversely affects crew comfort, until the **ECS** caution/warning light extinguishes. During hot weather operations if the **ECS** caution/warning light illuminates, the pilot should adjust the **TEMP** control to full **COLD**. In all cases of **ECS** light illumination, the mission may be continued.

d. ECS Air Outlets. Conditioned air leaving the ENCU is routed to both crew stations. The crew stations each have four controllable crew air outlets. The CPG station has an additional four nonadjustable outlets to provide ECS air directly to each FAB, the TADS/PNVS turret, and the ORT. In the ECS cooling mode the two crew station floor outlets are closed and the waist and shoulder outlets are open. For heating mode the crew shoulder outlets are closed and the waist and floor outlets are open. The FAB and TADS/PNVS fans draw CPG cabin air forward and expel it overboard. Pilot crew station air is drawn through the electrical power center by a ventilation fan and then exhausted into the transmission bay.

2.46.3 ECS Emergency Operation. In the event of an ENCU failure, the actions of the pilot depend on the nature of the ambient temperature conditions. Crew station air flow with the ENCU inoperative is through the pilot's station into the CPG station, and out through the FABs. During cold weather operations, the pilot may, at his discretion, open the auxiliary ventilation door to periodically ventilate both crew stations. No other action is required. During hot or warm weather operations, the pilot should open the auxiliary ventilation door to provide crew station ventilation. If the ambient temperature is below 100 °F, no further action is recommended. If the ambient temperature is above 100 °F, the **NORM-STBY FAN** switch on the pilot ECS control panel should be set to the **STBY FAN** position. Depending on the ambient outside air temperature, the **ECS** caution/warning light may illuminate if either FAB temperature is greater than 105 °F.

TM 1-1520-238-10

Section XI. ELECTRICAL POWER SUPPLY AND DISTRIBUTION SYSTEMS

2.47 ELECTRICAL POWER SYSTEM.

All the helicopter electrical power requirements are supplied by two ac generators, two transformer/rectifiers (T/Rs) and in the case of complete failure, a 24-volt battery will supply flight critical systems. For ground operations, 115 vac external power can be supplied to the helicopter through the external power receptacle. The electrical power distribution system is shown in figure 2-38. The system caution lights are listed in tables 2-4 and 2-5.

CAUTION

In the event any circuit breaker opens for unknown reasons, do not attempt to reset the breaker more than one time. Repeated tripping of a circuit breaker is an indication of a possible problem with equipment or electrical wiring. Multiple attempts to reset the circuit breaker may result in equipment damage and/or an electrical fire.

2.47.1 DC Power Supply System. Two essential ac busses supply power to the two 350-ampere T/Rs through the **XFMR RECT 1** and **XFMR RECT 2** circuit breakers on the pilot overhead circuit breaker panel (fig 2-39). The No.1 and No. 2 T/Rs convert the ac input to 28 vdc. The 28 vdc power is applied to a dc contactor, which routes 28 vdc from T/R No. 1 to the No. 1 essential dc bus and from T/R No.2 to the No. 2 essential dc bus. The No. 1 and No. 2 essential dc buses each apply 28 vdc, in parallel, through isolation diodes to power the emergency dc bus during normal operation. For emergency operation, the battery powers the emergency dc bus, and the emergency dc bus diodes isolate the battery from the noncritical loads. The two essential dc buses also supply 28 vdc power to the No. 3 essential dc bus. The T/Rs are self-monitoring for overtemperature and will illuminate the **HOT RECT 1** and **HOT RECT 2** caution lights on the pilot caution/warning panel when an overheat condition exists. The dc contactor monitors the T/Rs for output loss or drop in voltage. If the dc contactor senses a fault in T/R No. 1, it disconnects T/R No. 1 from the No. 1 essential dc bus and illuminates the **RECT 1** caution light on the pilot caution/warning panel. At the same time, it connects No. 1 essential dc bus to No. 2 essential dc bus through

relay action inside the contactor. The same action takes place if T/R No. 2 fails. If both T/Rs fail, the **ELEC SYS FAIL** caution light on the CPG caution/warning panel will illuminate. The caution/warning lights receive 28 vdc from the emergency dc bus through the **CAUT** circuit breaker on the pilot overhead circuit breaker panel and the **EMERG BATT CAUT** circuit breaker on the CPG main circuit breaker panel (fig 2-40).

a. **Battery.** A 24-volt, 13 ampere-hour, 19 cell, nickel-cadmium battery provides emergency power. The battery is located in the aft avionics bay. The battery is charged by the battery charger which receives 28 vdc power from the No. 1 essential dc bus through the **BATT CHGR DC** circuit breaker and 115 vac from the No. 2 essential ac bus through the **BATT CHGR AC** circuit breaker. Both circuit breakers are on the pilot overhead circuit breaker panel. The battery charger will completely charge the battery and then maintain a trickle charge. The battery charger also contains fault sensing that monitors battery temperature, cell balance, and charger operation. If a fault occurs, either the **HOT BAT** or **CHARGER** caution light will illuminate on the pilot caution/warning panel. The helicopter is equipped with a battery heater which operates automatically when the battery is on.

b. **CPG Battery Override Switch.** A two-position guarded toggle **BAT OVRD** switch is located on the CPG **PWR** lever quadrant (fig 2-12). With the guard down, the switch is in the NRML position and enables the pilot EXT **PWR/BATT** switch on the pilot **ELEC PWR** control panel (fig 2-37) located on the left console (fig 2-11). **When the** guard is positioned up and the switch is set to **OVRD**, the pilot **EXT PWR/BATT** switch is inoperative, and the battery is disconnected from the emergency dc bus.

2.47.2 AC Power Supply System. The ac power supply system is the primary source of electrical power. It supplies 115 vac from two 35 kilovolt-ampere generators. Each generator and its associated components comprise an independent ac generating system that supplies about one-half of the total electrical requirements to the ac buses. The generators are mounted on, and driven by, the main transmission accessory gear box. The 115 vac power is monitored and regulated by the generator control unit (GCU). If the generator output is normal, the GCU applies voltage to the ac contactor which energizes and connects the generator output to the ac essential buses. The No. 1 ac contactor connects the No. 1 generator output to the No. 1 essential ac bus, and the No. 2 ac contactor connects the No. 2 generator to the No. 2 essential ac bus. If one generator fails, its load is automatically connected to the remaining generator.

a. Generators. Two generators, located on the main transmission accessory gearbox, supply 115 vac, three-phase power for operating the helicopter electrical equipment. Generator operation is controlled from the pilot **ELEC PWR** control panel (fig 2-37). When a generator control switch is in **GEN 1 or GEN 2** position, the selected generator(s) are brought on-line. When the switch is placed to **OFF/RESET**, the generator is taken off-line and fault sensing is reset. Two caution lights, **GEN 1 and GEN 2** on the pilot caution/ warning panel illuminate whenever the generator control unit senses a fault. These lights receive 28 vdc from the emergency dc bus through the **CAUT** circuit breaker on the pilot overhead circuit breaker panel (fig 2-39).

b. External Power Receptacle. The external power receptacle is located aft of the aft avionics bay. It provides a means of connecting external power to the helicopter. A micro switch, which is actuated by opening the

external power access door, informs the pilot when the door is open by illuminating the **EXT PWR** caution lamp on his caution/warning panel. The system is controlled from the **ELEC PWR** control panel. When the **BATT/EXIT PWR** switch is at **EXT PWR** position, the external power monitor checks the GPU for proper phase sequence, voltage, and frequency. The power monitor also inhibits connecting and charging of the battery. The **EXT PWR** caution light receives 28 vdc from the emergency dc bus through the **CAUT** circuit breaker on the pilot overhead circuit breaker panel (fig 2-39).

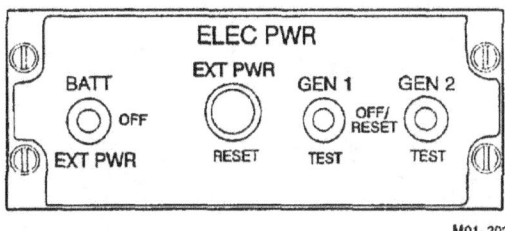

Figure 2-37. Pilot Electrical Power Control Panel

TM 1-1520-238-10

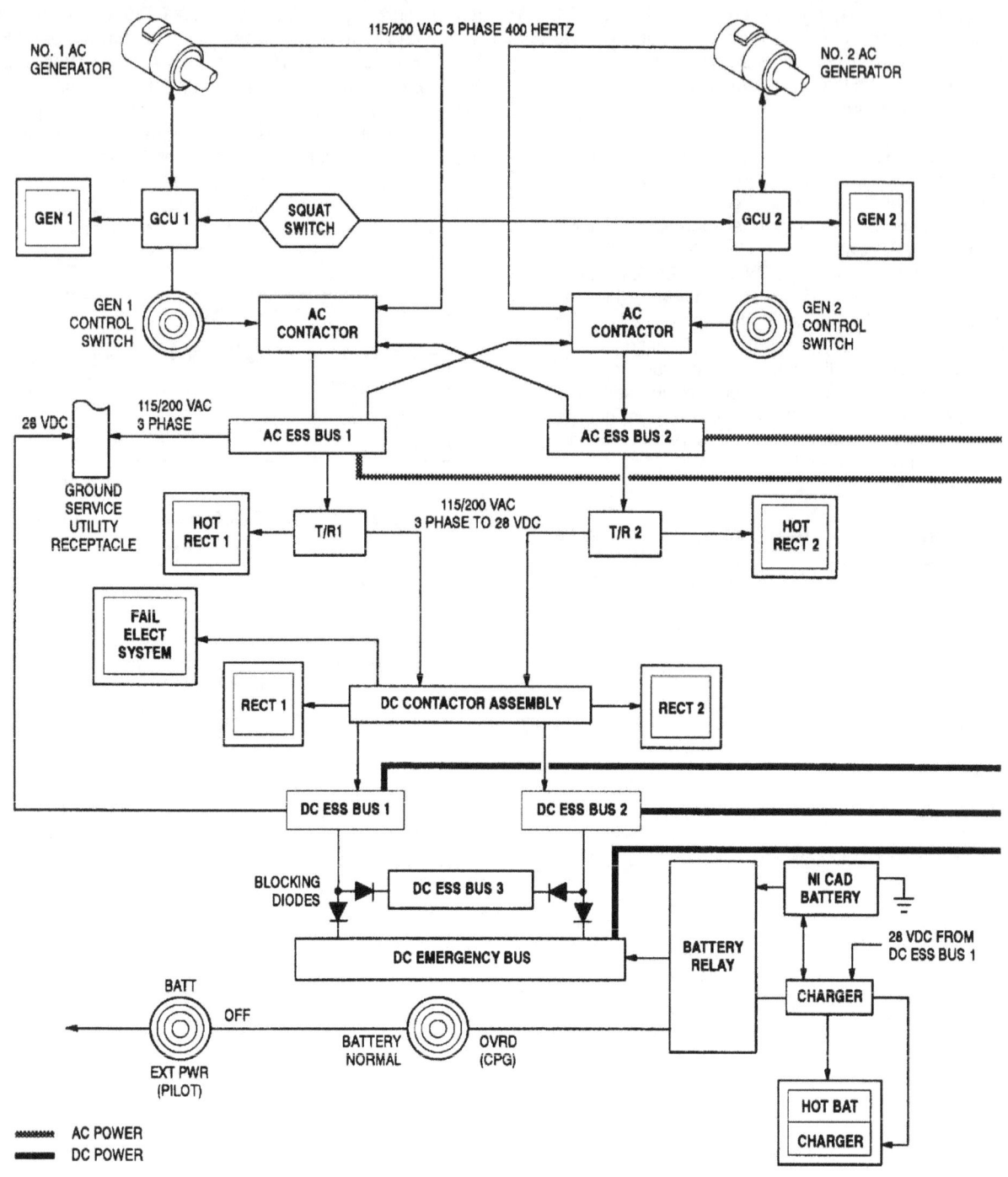

Figure 2-38. Electrical Power Distribution System (Sheet 1 of 2)

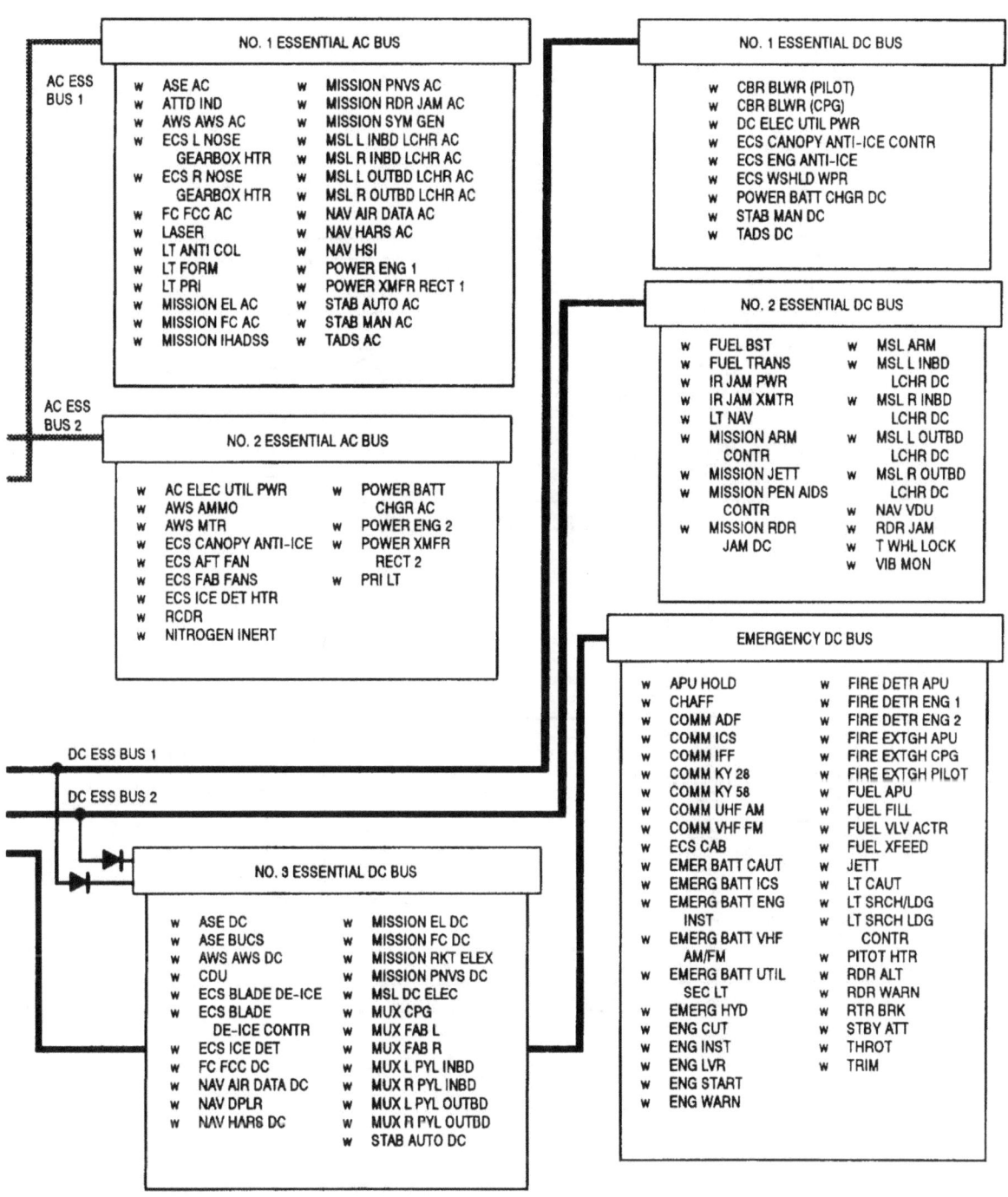

Figure 2-38. Electrical Power Distribution System (Sheet 2 of 2)

TM 1-1520-238-10

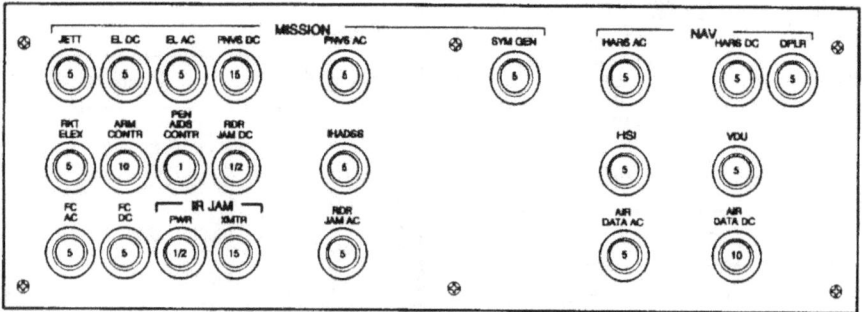

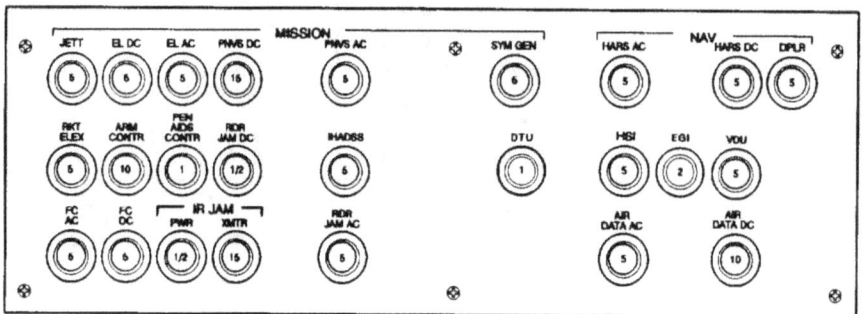

Figure 2-39. Pilot Overhead Circuit Breaker Panels (Sheet 1 of 2)

TM 1-1520-238-10

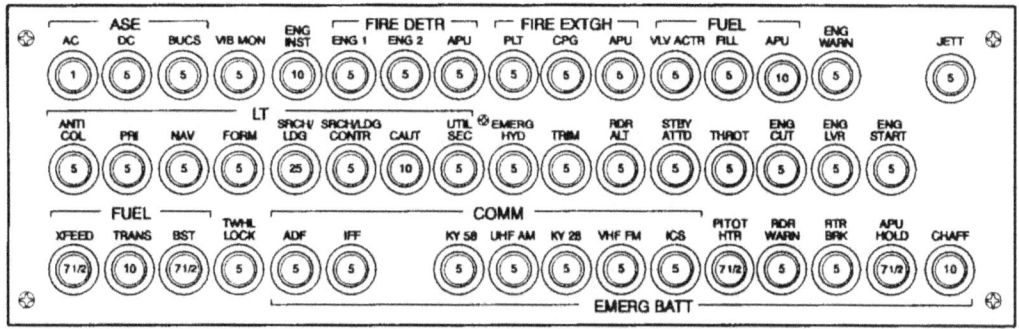
CENTER CIRCUIT BREAKER PANEL

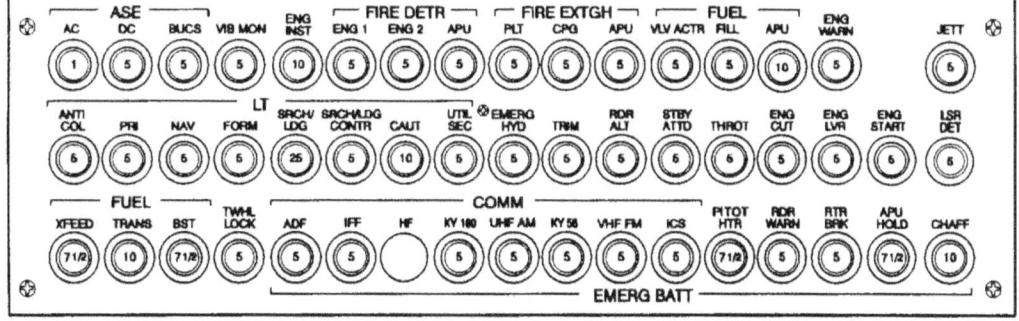
CENTER CIRCUIT BREAKER PANEL (LASER DETECTING SET MODIFICATION)

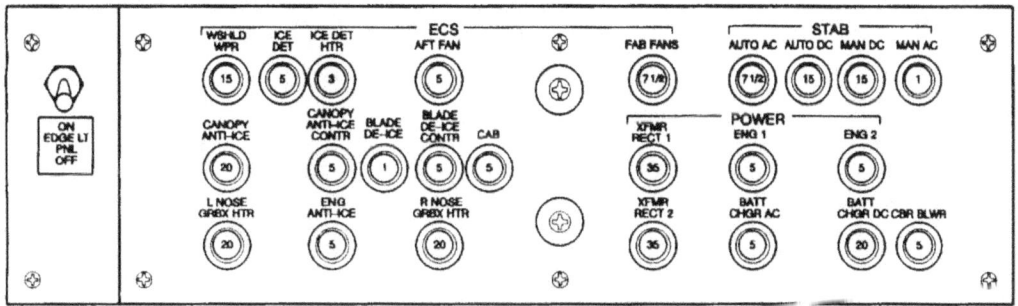
AFT CIRCUIT BREAKER PANEL

Figure 2-39 Pilot Overhead Circuit Breaker Panels (Sheet 2 of 2)

Change 5   2-70.1/(2-70.2 blank)

TM 1-1520-238-10

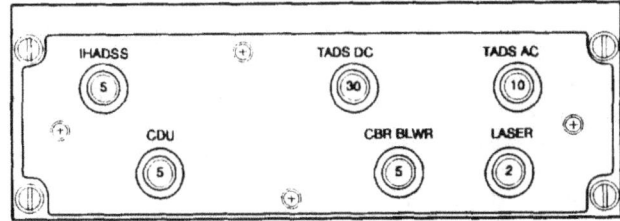

CPG NO. 2 CIRCUIT BREAKER PANEL

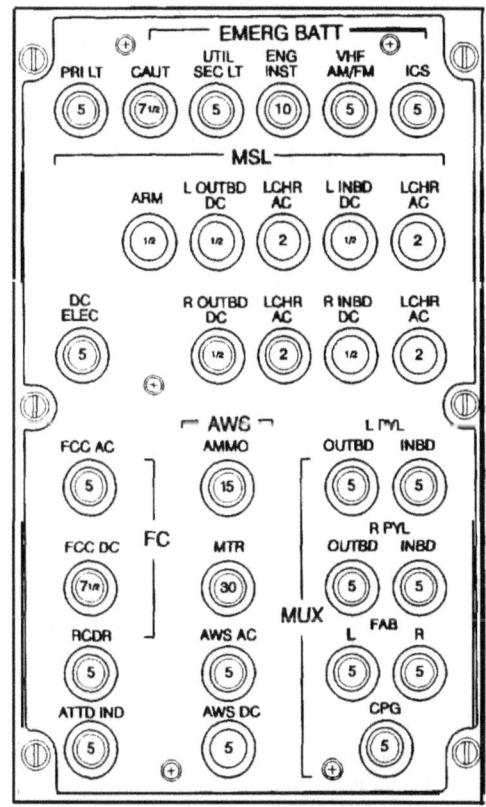

CPG NO. 1 CIRCUIT BREAKER PANEL

Figure 2-40. CPG Circuit Breaker Panels (Typical)

2-71

Section XII. AUXILIARY POWER UNIT

2.48 AUXILIARY POWER UNIT (APU).

WARNING

To prevent an accidental APU start, the APU circuit breaker in the aft avionics bay and the APU HOLD circuit breaker on the pilot overhead circuit breaker panel shall be out when battery or external electrical power is connected to the helicopter and unqualified personnel are in or around the pilot crew station.

The APU indirectly provides hydraulic, pneumatic, and electrical power for the operation of helicopter systems whenever the engines are not driving the main transmission accessory section. The APU provides the means of engine starting without the need for an aircraft ground power unit (AGPU). It is located just inboard of the right engine nacelle in the aft equipment bay. The APU consists of a gearbox, a compressor, and a turbine section, together with associated fuel, lubrication, and electrical systems. The APU controller and the APU control panel are installed separately from the APU. The utility hydraulic accumulator, located on the deck of the aft equipment bay below the APU, provides hydraulic pressure to the APU hydraulic starter through a solenoid-operated hydraulic start valve. The start valve opens when the APU **START** switch is positioned to **START** and closes automatically at 60% APU speed. The utility hydraulic accumulator is discussed in more detail in Section VI of this chapter.

2.48.1 Fuel System. The aft fuel cell provides fuel for APU operation and is discussed in more detail in Section IV of this chapter. The APU fuel control automatically regulates fuel flow. The APU burns approximately 135 pounds of fuel per hour.

2.48.2 Lubrication System. The APU has a self-contained oil system. An oil filler cap is located on the left side of the unit. Oil cooling is provided by airflow over cooling fins at the compressor inlet. The oil level sight gage (fig 2-45) is located on the right side and is an integral part of the oil sump. The sight gage can be inspected through an opening just below the right engine nacelle.

2.48.3 Electrical System. The APU electrical system requires dc power which is normally delivered by the helicopter battery when the APU control switch is set to **RUN**. This switch will apply dc power to energize the APU boost pump, open the APU fuel shutoff valve, and at the same time, send a signal to open the utility accumulator start solenoid valve to allow hydraulic pressure to turn the starter, and shut off the SDC inlet throttle valve to unload the SDC for an APU start. The **APU ON** caution light on the pilot caution/warning panel illuminates when the APU is operating.

2.48.4 APU Controller. The APU controller provides for automatic start and operation of the APU and power take-off clutch (PTO) engagement. The controller monitors the APU for loss of thermocouple, overtemperature, overspeed, overcurrent, low oil pressure, percent rpm, and exhaust gas temperature. The APU controller transmits a shutdown signal to the APU whenever the **APU** control panel switch is set to **OFF.** The controller also transmits the shutdown signal automatically for fault detection of any of the monitored functions. The **APU FAIL** caution light on the pilot caution/ warning panel is illuminated by the low oil pressure switch when automatic shutdown occurs. The PTO clutch actuation is also controlled by the APU controller. Normally, PTO clutch engagement occurs at 60% APU speed if the main rotor is below 90% speed; however, if the main rotor is above 90% speed, it will engage at 95% APU speed. For cold weather starts below 0° F (-18° C), PTO clutch engagement is at 95% APU speed using a cold start switch on the APU control panel.

2.48.5 APU Control Panel. The **APU** control panel (fig 2-41), on the pilot right console, has the following three distinct functions:

a. APU Control Switch. A three-position APU control switch allows the pilot to **START, RUN,** and shut down **(OFF)** the APU. This switch is spring-loaded from the **START** to **RUN** position. Pilot pressure at the **START** position is required for only one or two seconds to ensure electrical circuit latching.

CAUTION

Do not use APU 95% cold start switch when the ambient temperature is above 0° F C-18° C). Use of this switch will reduce the power takeoff (PTO) clutch life drastically, and could cause premature failure of the clutch duplex bearing/needle bearing as well as main transmission accessory gear case component failure.

b. 95% Cold Start Switch. A spring-loaded 95% switch allows the pilot to delay APU PTO clutch engagement during APU starts at temperatures below 0 °F (-18 °C). Delaying PTO clutch engagement until the APU is at 95% operating speed allows successful engagement in a cold weather environment. The switch is spring-loaded to the **NORM** position and must be held in the 95% position when used. PTO clutch engagement is inhibited until the switch is released and returned to the **NORM** position. The switch is used with the **APU ON** advisory light that indicates the APU is operating at or above 95% speed.

c. APU Fire Warning and Extinguishing Control. APU fire warning is provided by a warning light in the **FIRE APU PULL** handle and a **FIRE APU** warning light on the pilot and CPG **MASTER CAUTION** panels. Section II of this chapter describes emergency equipment operation. An APU fire can be extinguished by pulling the pilot **FIRE APU PULL** handle and placing the **FIRE BTL** selector switch either to **PRI** (primary) or **RES** (reserve). The fire test portion of the panel is discussed in detail in Section II of this chapter.

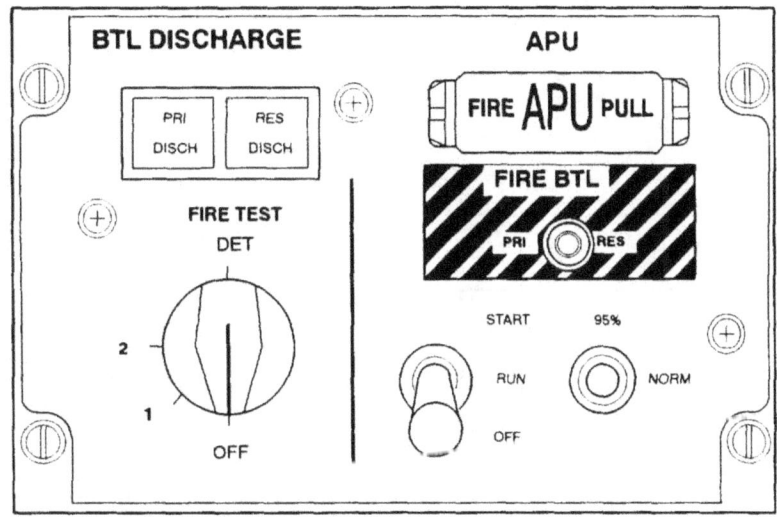

Figure 2-41. APU Control Panel

Section XIII. LIGHTING

2.49 LIGHTING EQUIPMENT.

The lighting equipment consists of exterior and interior lighting systems. The lighting equipment is controlled by the pilot and CPG lighting control panels (fig 2-42).

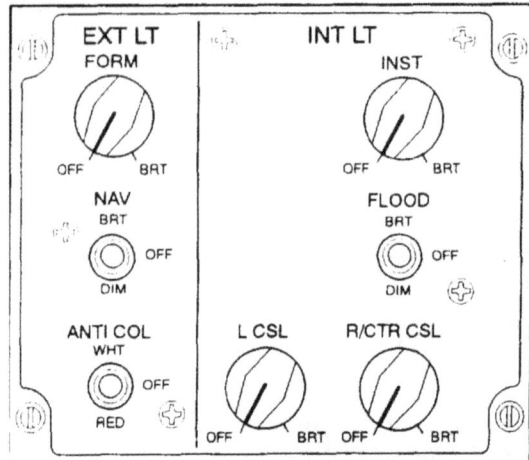

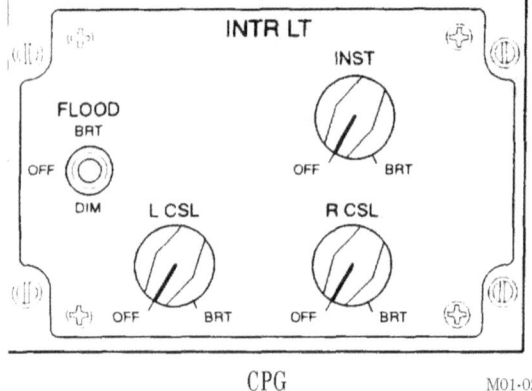

Figure 2-42. Pilot and CPG Lighting Control Panels

NOTE

The searchlight, crew station floodlights, and utility lights are the only lights available with battery power.

2.49.1 Exterior Lighting System.
The aircraft exterior lighting system consists off formation lights, navigation lights, anticollision lights, a searchlight, and an inspection maintenance light.

a. Formation Lights. The formation lights (green) are located on the upper surface of each wing, the upper centerline of the aft fuselage, and on the upper surface of the vertical stabilizer. Power for operation of the lights is 115 vac provided by the **FORM** circuit breaker on the pilot overhead circuit breaker panel (fig 2-39). Operation of the lights is controlled by the **FORM** rotary control on the pilot **EXT LT** panel.

b. Navigation Lights. The navigation lights are located on the wingtips (right side green, left side red, and the aft white) and on the top aft side of the vertical stabilizer. Power for operation of the lights is 28 vdc and is provided by the **NAV** circuit breaker on the pilot overhead circuit breaker panel. Operation of the lights is controlled by the **NAV** switch on the pilot **EXT LT** panel.

c. Anticollision Lights. High-intensity flashing red and white anticollision lights are located on each wingtip. Power for their operation is 115 vac through the **ANTI COL** circuit breaker and 28 vdc through the **LT NAV** circuit breaker on the pilot overhead circuit breaker panel. Operation of the lights is controlled by the **ANTI-COL** switch on the pilot **EXT LT** panel.

CAUTION

The searchlight can reach temperatures capable of igniting fires if contact is made with combustible/flammable materials. Do not land in areas such as high grassy meadows with the searchlight **ON**.

NOTE

Searchlight motion is inhibited for 60 seconds after the **SRCH LT** switch is placed in the **STOW** position.

d. Searchlight. The searchlight (fig 2-2) is located in a fairing under the forward end of the forward avionics bay just forward of the landing gear attachment and can be used as a landing light. Power for the searchlight is 28 vdc routed through a relay by the **SRCH LDG CONTR** and **SRCH LDG** circuit breakers on the pilot overhead circuit breaker panel. Operation of the light is controlled by the **SRCH LT** and **EXT-RET** switches on the pilot and CPG collective stick switchboxes (fig 2-26).

e. Inspection and Maintenance Light. The inspection and maintenance light is stored in the left equipment stowage compartment (fig 2-2). Two plug-in receptacle locations provide electrical power and will facilitate inspection and maintenance at all points on the helicopter. The receptacle is located adjacent to the CPG station on the underside of the right FAB, forward of the searchlight (fig 2-2). The second receptacle is located in the right aft avionics bay (fig 2-2). Power for operating the light is 24 vdc directly from the battery through the **MAINT LT** circuit breaker in the right aft avionics bay. Operation of the inspection and maintenance light is controlled by an **OFF**-**BRT** rheostat switch which is integral with the light.

2.49.2 Interior Lighting System. The helicopter interior lighting system consists of dimming lighting for engine instruments, flight instruments, avionics panels, console panels, and circuit breaker panels. All engine and flight instruments are equipped with red surface mounted edge lighting fixtures, and the avionics console and circuit breaker panel lights are integrally illuminated. Light is reflected to illuminate the panel markings and the clear edging around each switch or circuit breaker. In the event of total electrical power generation failure, the pilot and CPG flood lights, utility lights, and searchlight remain operable through the dc emergency bus.

a. Pilot Interior Lighting. Interior lighting for the pilot is controlled by three switch/rheostats and a three-position **FLOOD** toggle switch located on the pilot **INTR LT** control panel. The three switch/rheostats provide a detented **OFF** position and dim to **BRT** position. AC voltage is the primary power for the interior lights. The ac voltage is applied to a multichannel dimmer assembly through the **PRI** circuit breaker on the pilot overhead circuit breaker panel. The multichannel dimmer converts 115 vac to the proper dc levels for the three switch/rheostats. The **INST, L CSL,** and **R/CTR CSL** switch/rheostats vary the multichannel dimmer output. The **INST** switch/rheostat controls channels 1 and 2 of the multichannel dimmer. Channel 1 is for the right instrument panel, and channel 2 is for the left instrument panel. Both channels vary the multichannel output from **OFF** position to **BRT**. The **R/CTR CSL** control is split into two sections; A: **OFF** position, and **BRT** for APU **FIRE TEST** and APX100 control panel. B: **OFF** position, and **BRT** for the avionics control box lights in the center and right console. The **L CSL** controls the circuit breaker panels and the collective stick grip lights. The pilot also has an **EDGE LT PNL ON/OFF** switch located on the overhead circuit breaker panel that allows him to turn off circuit breaker edge lights independent of other console lights.

b. CPG Interior Lighting. The CPG interior lighting operation of the **INTR LT** switch/rheostats and the multichannel dimmer are the same as for the pilot operation. The **INST** switch/rheostat controls outputs to channel 1 (right side instruments), channel 2 (left side instruments), and the ORT. The **R CSL** controls outputs to channel 3 which is split into sections A and B. A: to the avionics lights; B: to the ASN 128 panel lights. The **L CSL** controls the circuit breaker and collective stick grip lights.

c. Emergency Floodlight System. The emergency floodlight system is installed under the glareshields of the pilot and CPG instrument panels. Power for the emergency floodlight system is emergency bus 28 vdc power through the **UTIL SEC** circuit breaker on the pilot overhead circuit breaker panel and the **UTIL SEC LT** circuit breaker on the CPG main circuit breaker panel (fig 2-40). Operation of the emergency floodlight system is controlled by the **FLOOD** three-position toggle switch on the pilot and CPG **INTR LT** control panel. When the switch is positioned to the **BRT** position, it turns on the blue green secondary lights to the brightest level. When in the dim position, the secondary lighting is dimmed.

d. Utility Light. A detachable utility light with a coiled extension cord is located to the left of the pilot and CPG seats. The utility light provides emergency red or white lighting in case instrument panel lighting fails. The light is operated by an **OFF BRT** rheostat switch integral with the light. Rotating the front section of the light selects white flood or red flood. The utility light receives 28 vdc from the emergency dc bus through the **UTIL SEC LT** circuit breaker on the pilot overhead circuit breaker panel and the **UTIL SEC LT** circuit breaker on the CPG No. 1 circuit breaker panel.

e. Dimming, MASTER CAUTION Panel, and Caution/Warning Panel Advisory Segment Lights. Dimming of all caution/warning lights is controlled by the **INST** control on both the pilot and CPG **INTR LT** control panels for the respective crew station. When the **INST** control is is in the **OFF** position, caution/warning lights are bright; any other position will cause the respective lights to go to a preset dim condition.

Section XIV. FLIGHT INSTRUMENTS

2.50 FLIGHT INSTRUMENTS.

The instruments discussed in this section are, for the most part, those that directly measure flight performance. Caution, warning, audio systems, and some flight instruments are common to both crew stations. The instruments are grouped as common, pilot, and CPG flight instruments.

2.50.1 Common Flight instruments. The flight instruments found in both the pilot instrument panel (fig 2-9) and the CPG instrument panel (fig 2-10) are the pressure (barometric) altimeter, instantaneous vertical speed indicator, airspeed indicator, attitude indicator, and the clock.

a. Pilot Barometric Altimeter. The pilot has an AAU-32/A encoding barometric altimeter. This altimeter is the same as the CPG's except the AAU-32/A interfaces with the IFF for Mode C operation.

b. CPG Barometric Altimeter. The CPG has an AAU-31/A barometric altimeter. The altimeter is graduated in 50-foot increments and marked at 100-foot intervals (0 - 9 x 100). Just left of center is a 100-foot drum and a 1000-foot drum to supplement the scale pointer. The scale window, at the lower right section of the instrument face, indicates barometric pressure setting in inches of mercury. It is adjustable by use of the barometric pressure set knob on the lower left corner of the indicator case. Maximum allowable altimeter error is 70 feet.

c. Vertical Speed indicator (VSI). The VSI measures the rates of change in static air pressure resulting from climbs and descents. An adjustment screw on the lower left corner is used to zero the pointer, if necessary, prior to flight.

d. Airspeed indicator. The airspeed indicator measures the difference between pitot pressure and static pressure. Instrument range markings and limitations are contained in Chapter 5, Section II, System Limits. At low airspeeds and high power settings, indicated airspeeds may be unreliable and fluctuate greater than 10 KIAS.

e. Pilot Standby Attitude indicator. The pilot standby attitude indicator provides an independent display of helicopter attitude. The indicator can display 360° of roll and ± 85° of pitch. A **PULL TO CAGE** knob at the lower right corner has two functions. Pulling it out with power applied to the instrument will cage the motor-driven internal gyroscope and level the background horizon line to 0° in pitch and roll. The knob may be turned to adjust the pitch of the artificial horizon relative to the fixed aircraft symbol. The indicator receives 28 vdc from the emergency dc bus through the **STBY ATTD** circuit breaker on the pilot overhead circuit breaker panel.

f. CPG Remote Attitude indicator. The CPG remote attitude indicator (RAI) displays helicopter attitude from information obtained from the heading and attitude reference system (HARS). The indicator can display 360° of roll and ± 90° of pitch. HARS input drives roll and/or pitch servos, which results in the appropriate roll or pitch of the artificial horizon. If HARS input ceases or becomes unreliable, 28-vdc power will cause an **OFF** flag to appear at the window on the left side of the instrument face. A pitch trim knob at the lower right corner may be turned to adjust the pitch of the artificial horizon relative to the fixed aircraft reference symbol. The RAI receives 115 vac from the No. 1 essential ac bus through the **ATTD IND** circuit breaker on the CPG No. 1 circuit breaker panel.

g. Clock. The clock combines the features of a standard clock and a stopwatch by displaying normal and elapsed time in hours, minutes, and seconds. Once wound, the clocks will run for eight days. The elapsed-time pushbutton control is on the upper right corner of the case. The clock is wound and set with a knob at the lower left corner of the case.

2.50.2 Pilot Flight instruments. The flight instruments in the pilot instrument panel (fig 2-9) are the video display unit, standby magnetic compass, free air temperature gage, accelerometer, and radar altimeter.

a. Video Display Unit (VDU). The VDU is a multipurpose instrument that provides the pilot with flight, navigation, and targeting information. A turn-and-slip indicator is located below the face of the cathode ray tube. The signals for the turn rate indicator are provided by the DASE.

b. Standby Magnetic Compass. The standby magnetic compass is attached to the pilot glareshield. Primary heading information is taken from the horizontal situation indicator, which is discussed in Chapter 3, Section III.

c. Free Air Temperature (FAT indicator. The FAT indicator (fig 2-11) is a self-contained unit mounted to the bulkhead adjacent to, and left of, the pilot **PWR** quadrant. A probe extends through the airframe to sense outside free air temperature. The dial, marked **FREE AIR,** indicates in degrees Celsius (°C).

d. Accelerometer. The accelerometer measures the g-forces during flight.

e. Radar Altimeter. For description refer to paragraph 3.18.

2.50.3 CPG Flight Instruments. The flight instruments unique to the CPG instrument panel (fig 2-10) are the radio magnetic indicator, airspeed indicator, and altimeter.

a. Radio Magnetic Indicator (RMI). The RMI, located at the bottom center of the CPG right instrument cluster, displays magnetic heading and ADF bearing to a selected station. The ADF will be discussed in Chapter 3 in conjunction with direction finder set AN/ARN-89 or the AN/ARN-149. Magnetic heading is provided through a synchro signal from the heading/attitude reference set (HARS) to a heading servo that positions the heading compass card within the instrument. When the HARS is aligning or failed, a heading warning signal will cause the **OFF** flag (right center on the instrument face) to appear. Internal power to the RMI is provided by 115 vac 400 Hz input power.

2.51 PITOT-STATIC SYSTEM.

Electrically-heated pitot tubes are installed at the outboard leading edge of both wings. Two associated static ports are installed flush with the fuselage: one on the right, and one on the left. The right wing pitot tube supplies ram air to the pilot airspeed indicator, and the left wing pitot tube supplies ram air to the CPG airspeed indicator. The static port on the right side of the fuselage provides static air pressure to the pilot barometric altimeter, rapid response vertical speed indicator, air speed indicator, No. 2 airspeed transducer, and the air data processor. The static port on the left side of the fuselage provides static air pressure to the CPG barometric altimeter, rapid response vertical speed indicator, airspeed indicator, and No. 1 airspeed transducer.

2.52 PILOT AND CPG MASTER CAUTION PANEL.

The master caution panels (fig 2-43), located at the center of the pilot and right center of the CPG glareshield (fig 2-9 and 2-10), are identical. When the **PRESS TO TEST** pushbutton is pressed, all caution/warning lights in that crew station illuminate. All segments have two lamps so that failure of a single lamp will produce only a dimming effect. If a segment fails to light, it indicates either the failure of both lamps or a defective circuit. The **MASTER CAUTION** display is also pushbutton operated; it is the only amber light on the master caution panel. Initially, when a caution/warning panel segment illuminates, it will flash. Simultaneously, the **MASTER CAUTION** will flash. Pressing the **MASTER CAUTION** light segment will extinguish the lamp and convert the illuminated fault segment to steady on. The **MASTER CAUTION** lights are individually reset in each crew station. and will remain on until the fault or condition is corrected. Warning lights are red, and when lighted, require immediate crewmember attention. The master caution panel segments are described in table 2-3.

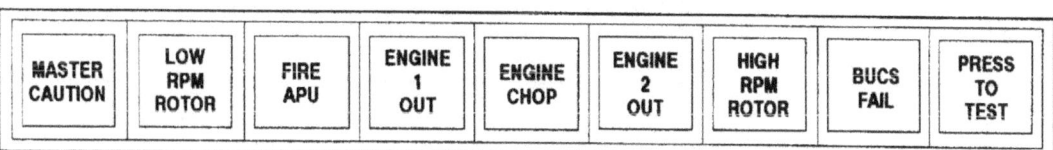

Figure 2-43. Pilot and CPG Master Caution Panel

Table 2-3. Master Caution Panel Indications

Word Segment	Color	Illumination Parameter or Fault
MASTER CAUTION	AMBER	Alerts both crewmembers to scan both their master caution/warning panels and caution/warning panels to identify fault condition.
LOW RPM ROTOR	RED	Nr is less than 94%.
FIRE APU	RED	Fire has been detected in the APU area or aft deck area.
ENGINE 1 OUT	RED	Np is below 94% or NG is below 63% on No. 1 engine. (Np disabled when the PWR lever is not in FLY).
ENGINE CHOP	RED	Both engines have been electrically retarded to idle power.
ENGINE 2 OUT	RED	Np is below 94% or NG is below 63% on No. 2 engine. (Np disabled when the PWR lever is not in FLY).
HIGH RPM ROTOR -701 ENGINE	RED	Nr is more than 104%.
HIGH RPM ROTOR -701C ENGINE	RED	Nr is more than 108%.
BUCS FAIL	RED	One of the following BUCS components has indicated a failure a. Control position transducer failure or misadjustment. b. Flight control actuator malfunction. c. BUCS tracer wire failure. d. DASEC NO-GO. e. BUCS select trigger pressed. f. BUCS circuit breaker out. g. No ac power to the DASEC. h. No primary hydraulic pressure.
PRESS TO TEST	WHITE	When depressed, all master caution panel and caution/warning panel segments, plus all other advisory lights, are illuminated within that crew station for test.

2.52.1 Pilot and CPG Caution/Warning Panels. The caution/warning panels, for the pilot and CPG (fig 2-44) are described in tables 2-4 and 2-5. Primarily cautionary in nature, the panels also include red warning segments. Often a specific fault indicated on the pilot panel will only light a general system segment on the CPG panel. Illumination of a segment on one crewmembers panel will not cause illumination of the **MASTER CAUTION** light in the other crew station.

a. **Warning and Caution Segment Dimming.** Dimming of all caution/warning light segments is controlled by the **INST** control on both the pilot and CPG **INTR LT** control panels (fig 2-42) for the respective crew station. When the **INST** control is in the **OFF** position, caution/warning lights are bright; any other position will cause the respective lights to go to a preset dim condition.

TM 1-1520-238-10

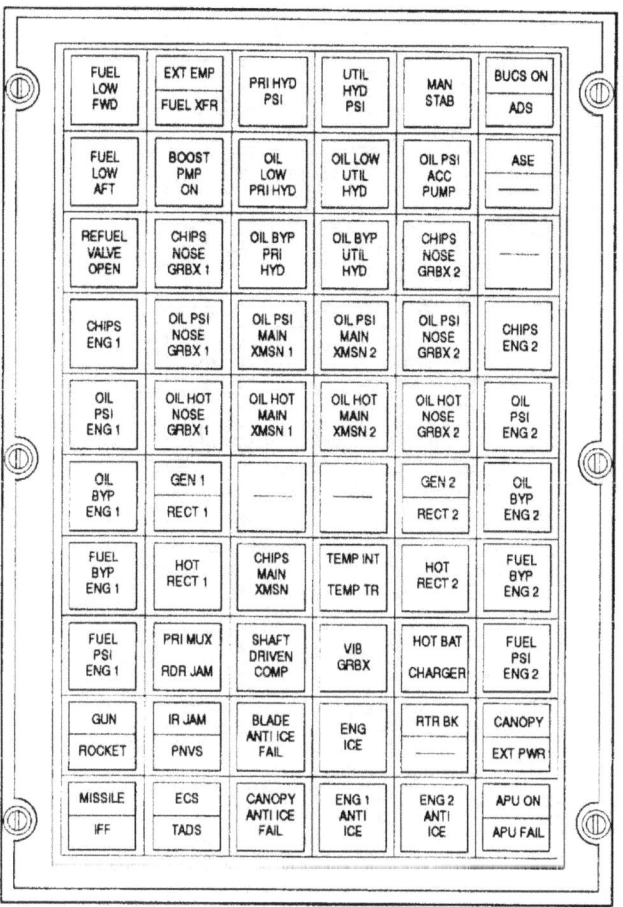

PILOT

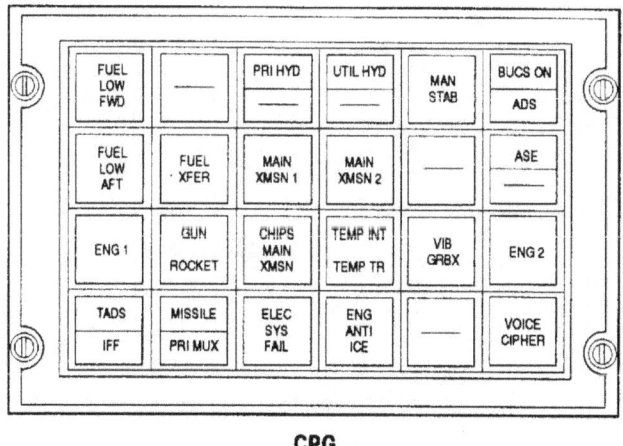

CPG

Figure 2-44. Pilot and CPG Caution/Warning Panels

2-79

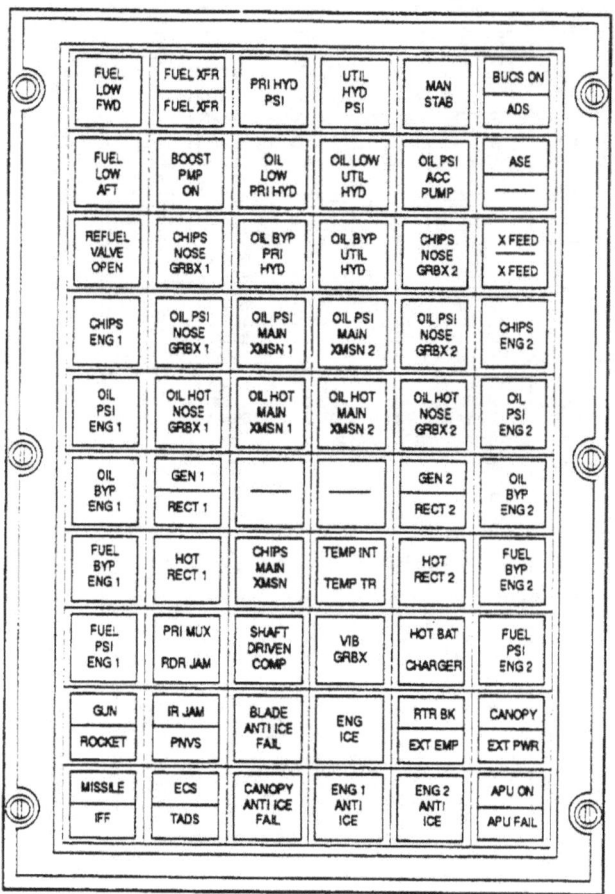

PILOT

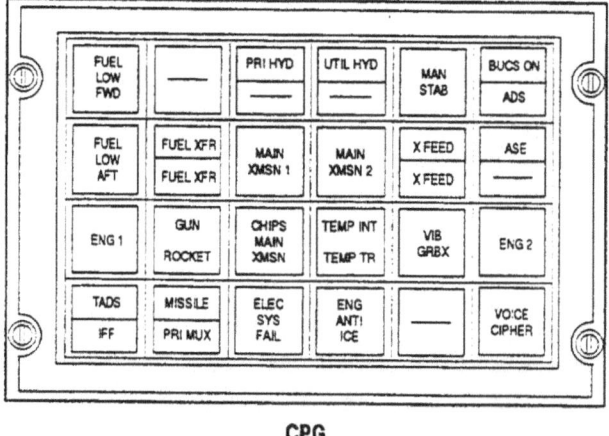

CPG

Figure 2-44.1 Pilot and CPG Caution/Warning Panels (Modified)

Table 2-4. Pilot Caution/Warning Light Segments

Word Segment	Color	Illumination Parameter or Fault
IFF	RED	Mode 4 is not able to respond to interrogation.
ENG ICE	AMBER	Ice detector probe indicates icing conditions are present.
BUCS ON	AMBER	Backup control system is activated in at least one control axis.
CHIPS ENG 1	AMBER	No. 1 engine scavenge oil contains metal fragments.
OIL PSI ENG 1	AMBER	No. 1 oil pressure is below 25 psi.
OIL BYP ENG 1	AMBER	No. 1 engine oil filter is clogged, and bypass has begun.
FUEL BYP ENG 1	AMBER	No. 1 engine fuel filter is clogged, and bypass has begun.
FUEL PSI ENG 1	RED	No. 1 engine fuel pressure is less than 9 psi.
	AMBER	Spare.
ROCKET	AMBER	Failed system.
MISSILE	AMBER	Failed system.
RTR BK	AMBER	Hydraulic pressure is being sensed at the rotor brake by the in-line pressure switch.
MAN STAB	AMBER	Automatic operation of the stabilator control unit has failed, or manual control has been selected.
CHIPS NOSE GRBX 1	AMBER	No. 1 engine nose gearbox oil contains metal fragments.
OIL PSI NOSE GRBX 1	AMBER	No. 1 engine nose gearbox oil pressure is below 26-30 psi.
OIL HOT NOSE GRBX 1	AMBER	No. 1 engine nose gearbox oil temperature is above 274 - 294°F (135 -145°C).
BLADE ANTI ICE FAIL	AMBER	Failed system.
CANOPY ANTI ICE FAIL	AMBER	Failed system or canopy temperature has exceeded limits.
ENG 1 ANTI ICE	AMBER	No. 1 engine nose gearbox heater and engine inlet temperature sensors are not at operating temperature. After reaching operating temperature, indicates No. 1 engine anti-ice subsystem has failed.
ENG 2 ANTI ICE	AMBER	No. 2 engine nose gearbox heater and engine inlet temperature sensors are not at operating temperature. After reaching operating temperature, indicates No. 2 engine anti-ice subsystem has failed.
GUN	AMBER	Failed system.
TADS	AMBER	Failed system.
PNVS	AMBER	Failed system.

Table 2-4. Pilot Caution/Warning Light Segments - continued

Word Segment	Color	Illumination Parameter or Fault
PRI HYD PSI	AMBER	Primary system hydraulic oil pressure is below 1250 psi.
OIL LOW PRI HYD	AMBER	Primary system hydraulic fluid is at minimum operating level.
OIL BYP PRI HYD	AMBER	A pressure differential of 60 to 80 psi has been detected in either the pressure or return filter of the primary hydraulic system. The return filter will bypass; the pressure filter will not.
OIL PSI MAIN XMSN 1	AMBER	Main transmission No. 1 system pressure is below 26-30 psi.
OIL HOT MAIN XMSN 1	AMBER	Main transmission No. 1 oil system temperature is above 274°- 294 °F (135° -145 °C).
GEN 1	AMBER	No. 1 generator is not on line.
RECT 1	AMBER	Transformer-rectifier has failed, or there is no ac input to the rectifier.
HOT RECT 1	AMBER	Temperature 190 °F (87 °C) at transformer-rectifier is excessive. Rectifier fan has probably failed.
SHAFT DRIVEN COMP	AMBER	Shaft driven compressor oil temperature is above 340°- 360 °F (171° -182 °C); or shaft driven compressor pressurized air output is less than 5-9 psi. No 1 engine bleed air will automatically supply the pressurized air system if No. 1 engine is operating.
PRI MUX	AMBER	The selected multiplex bus controller has malfunctioned and the other bus controller has assumed control of the multiplex bus. Light will extinguish when the operating bus controller is selected with the MUX switch on the CPG fire control panel.
UTIL HYD PSI	AMBER	Utility system hydraulic oil pressure is below 1250 psi.
OIL LOW UTIL HYD	AMBER	Utility hydraulic fluid is at minimum operating level.
OIL BYP UTIL HYD	AMBER	A pressure differential of 60 to 80 psid has been detected in either the pressure or return filter of the utility hydraulic system. The return filter will bypass; the pressure filter will not.
OIL PSI MAIN XMSN 2	AMBER	Main transmission No. 2 oil system pressure is below 26 - 30psi.
OIL HOT MAIN XMSN 2	AMBER	Main transmission No. 2 oil system temperature is above 274°- 294 °F (135° -145 °C).
GEN 2	AMBER	No. 2 generator is not on line.
RECT 2	AMBER	Transformer-rectifier has failed, or there is no ac input to the rectifier,
HOT RECT 2	AMBER	Temperature 190 °F (87 °C) at transformer-rectifier is excessive. Rectifier fan has probably failed.
CANOPY	AMBER	Canopy doors are not properly closed.

Table 2-4. Pilot Caution/Warning Light Segments - continued

Word Segment	Color	Illumination Parameter or Fault
EXT PWR	AMBER	Access door to external power receptacle is open.
HOT BAT	AMBER	Battery temperature is in excess of 134 °F (57 °C), or a defective cell has been detected. Battery charging is discontinued.
CHARGER	AMBER	Charger has failed to charge during a programmed charging cycle.
CHIPS MAIN XMSN	AMBER	Main transmission chip detector has detected metal fragments.
TEMP TR	AMBER	Tail rotor gearbox temperature is above 274°-294 °F (135° - 145 °C).
CHIPS NOSE GRBX 2	AMBER	No. 2 engine nose gearbox oil contains metal fragments.
OIL PSI NOSE GRBX 2	AMBER	No. 2 engine nose gearbox oil pressure is below 26 - 30 psi.
OIL HOT NOSE GRBX 2	AMBER	No. 2 engine nose gearbox oil temperature is above 274°- 294 °F (135° - 145 °C).
ASE	AMBER	One or more components of the automatic stabilization equipment is inoperative; one or more of the SAS channels is inoperative or has been selected OFF.

NOTE

Low fuel caution systems alert the crew that the fuel level in the tank has reached a specified level (capacity). Differences in fuel densities due to temperature and fuel type will vary the weight of the fuel remaining and the actual time the helicopter engine(s) may operate. Differences in fuel consumption rates, aircraft attitude, and operational conditions of the fuel subsystem will also affect the time the helicopter engine(s) may operate.

Word Segment	Color	Illumination Parameter or Fault
FUEL LOW FWD	AMBER	When light comes on, approximately 260 pounds of fuel remains in the forward fuel cell.
FUEL LOW AFT	AMBER	When light comes on, approximately 210 pounds of fuel remains in the aft fuel cell.
EXT EMP	AMBER	All external fuel tanks are empty.
REFUEL VALVE OPEN	AMBER	Refuel valve is open.
OIL PSI ACC PUMP	AMBER	Main transmission accessory gearbox oil pressure is below 26 - 30 psi.
TEMP INT	AMBER	Intermediate gearbox temperature is above 274°-294 °F (135° - 145 °C).
CHIPS ENG 2	AMBER	No. 2 engine scavenge oil contains metal fragments.
OIL PSI ENG 2	AMBER	No. 2 engine oil pressure is below 25 psi.
OIL BYP ENG 2	AMBER	No. 2 engine oil filter is clogged, and bypass has begun.
FUEL BYP ENGINE 2	AMBER	No. 2 engine fuel filter is clogged, and bypass has begun.

Table 2-4. Pilot Caution/Warning Light Segments - continued

Word Segment	Color	Illumination Parameter or Fault
FUEL PSI ENG 2	RED	No. 2 engine fuel pressure is less than 9 psi.
ECS	AMBER	The overheat sensor has sensed a mixed air FAB temperature greater than 105°F (41°C).
APU ON	AMBER	APU operation when APU NG speed is above 95%.
APU FAIL	AMBER	Indicates one of the following APU thermocouple malfunction, overtemperature, overspeed, low APU oil pressure, or overcurrent.
VIB GRBX	AMBER	Intermediate or tail rotor gearbox vibration level is excessive.
FUEL XFR	AMBER	TRANS switch is in the TO FWD or TO AFT position and fuel transfer is not occurring.
FUEL XFR Modified C/W panel	GREEN	TRANS selected, transfer is occurring.
FUEL XFR Modified C/W panel	AMBER	TRANS selected, transfer is not occurring.
X FEED Modified C/W panel	GREEN	CROSSFEED selected, fuel valves are correctly positioned.
X FEED Modified C/W panel	AMBER	CROSSFEED selected, fuel valves are incorrectly positioned.
ADS	AMBER	Failed system.
BOOST PUMP ON	AMBER	Boost pump operation is on and providing 8 to 10 psi in fuel line to fuel filter.
RDR JAM	AMBER	AN/ALQ-136 has failed, is off, or is in a warm-up cycle.
IR JAM	AMBER	AN/ALQ-144 has failed, or is in warm-up/cooldown cycle.

Table 2-5. CPG Caution/Warning Light Segments

Word Segment	Color	Illumination Parameter or Fault
IFF	RED	Mode 4 is not able to respond to interrogation.
ENG 1	AMBER	One or a combination of the following engine oil pressure low, engine oil filter in bypass, engine chips, engine fuel filter in bypass, engine fuel pressure low, nose gearbox chips, nose gearbox oil pressure low, nose gearbox oil temperature high.
MAIN XMSN 1	AMBER	One or a combination of the following: transmission oil temperature high, oil pressure low.
GUN	AMBER	Failed system.

Table 2-5. CPG Caution/Warning Light Segments - continued

Word Segment	Color	Illumination Parameter or Fault
ROCKET4	AMBER	Failed system.
ELEC SYS FAIL	RED	Complete dc electrical system failure. The battery is powering the dc emergency bus.
BUCS ON	AMBER	Backup control system is activated in at least one control axis.
MISSILE	AMBER	Failed system.
TADS	AMBER	Failed system.
PRI HYD	AMBER	Primary system hydraulic oil pressure is below 1250 psi.
MAN STAB	AMBER	Automatic operation of the stabilator control unit has failed, or manual control has been selected.
TEMP INT	AMBER	Intermediate gearbox temperature is above 274°- 294°F (135° - 145 °C).
UTIL HYD	AMBER	Utility system hydraulic oil pressure is below 1250 psi.
ENG ANTI ICE	AMBER	An engine anti-ice subsystem has failed.
VOICE CIPHER	AMBER	KY-58 is operating.
CHIPS MAIN XMSN	AMBER	Main transmission chip detector has detected metal fragments.
FUEL LOW FWD	AMBER	When light comes on, about 260 to 300 pounds of fuel remain in the forward fuel cell at cruise attitude.
FUEL LOW AFT	AMBER	When light comes on, about 210 to 270 pounds of fuel remain in the aft fuel cell at cruise attitude.
ASE	AMBER	One or more components of the automatic stabilization equipment is inoperative, or one or more of the SAS channels is inoperative or has been selected OFF.
ENG 2	AMBER	One or a combination of the following: engine oil pressure low, engine oil filter in bypass, engine chips, engine fuel filter bypass, engine fuel pressure low, nose gearbox chips, nose gearbox oil pressure low, nose gearbox oil temperature high.
MAIN XMSN 2	AMBER	One or a combination of the following: transmission oil temperature high, oil pressure low.
PRI MUX	AMBER	The selected bus controller has malfunctioned and the other bus controller has assumed control of the multiplex bus. Light will extinguish when the operating bus controller is selected with the MUX switch on the CPG fire control panel.
VIB GRBX	AMBER	Intermediate or tail rotor gearbox vibration level is excessive.
TEMP TR	AMBER	Tail rotor gearbox temperature is above 274°-294 °F (135° - 145°C).

Table 2-5. CPG Caution/Warning Light Segments - continued

Word Segment	Color	Illumination Parameter or Fault
FUEL XFER	AMBER	TRANS switch is in the TO FWD or TO AFT position and fuel transfer is not occurring.
FUEL XFR Modified C/W panel	GREEN	TRANS selected, transfer is occurring.
FUEL XFR Modified C/W panel	AMBER	TRANS selected, transfer is not occurring.
X FEED Modified C/W panel	GREEN	CROSSFEED selected, fuel valves are correctly positioned.
X FEED Modified C/W panel	AMBER	CROSSFEED selected, fuel valves are incorrectly positioned.
ADS	AMBER	Failed system.

2.53 HEADSET AUDIO WARNING SYSTEM.

In addition to the visual cues to help crewmembers identify faults, audio signals are provided as an aid in rapid recognition of critical conditions. These audio signals are described in table 2-6.

Table 2-6. Audio Warning Signals

Tone	Signal
Rising frequency	ENGINE OUT
Rising frequency	ROTOR RPM LOW
Steady tone	STAB FAIL
Steady, amplitude modulated	IFF
Falling frequency	MISSILE ALERT
Dependent on threat radar	RADAR WARNING ALERT
Intermittent	RADIO IN SECURE MODE

Section XV. SERVICING, PARKING, AND MOORING

2.54 SERVICING.

This section describes servicing information and procedures for various systems and components. Servicing points for fuel, engine oil, main transmission oil, nose gearbox, and APU oil are illustrated in figure 2-45. Fuel, lubricants, specifications, and fuel capacities are listed in table 2-7.

2.64.1 Fuel System Servicing. The helicopter has two crash-resistant self-sealing fuel cells located forward and aft of the ammunition bay in the center fuselage section. Each cell is serviced through gravity filler receptacles or pressure-filled through closed-circuit or single-point adapters (fig 2-45). Provisions are also made for as many as four external fuel tanks to be carried on the stores pylons. Table 2-7 lists individual tank capacities.

a. Fuel Types. Fuels are classified as primary, commercial equivalent, or emergency. Primary fuels are JP-4, JP-5, and JP-8. Commerical equivalent oils are listed in table 2-8. There are *no* emergency fuels authorized.

b. Use of Fuels. There is no special limitation on the use of primary fuel, but limitations in table 2-8 apply when commercial fuels are used. For the purpose of recording, fuel mixtures shall be identified as to the major component of the mixture.

c. Interchangeable Fuels. Fuels having the same NATO code number are interchangeable. Jet fuels (table 2-8) conforming to specification ASTM D-1655 may be used when MIL-T-5624 fuels are not available. This usually occurs during cross-country flights where aircraft using NATO F-40 (JP-4) are refueled with NATO F-44 (JP-5) or commercial ASTM type A fuels. Whenever this occurs, the engine operating characteristics may change because of lower operating temperatures. Slower acceleration, lower engine speed, harder starting, and greater range may be experienced. The reverse is true when changing from F-44 (JP-5) fuel to F-40 (JP-4) or commercial ASTM type B fuels.

d. Mixing of Fuels. When changing from one type of authorized fuel to another, (ie: JP-4 to JP-5), it is not necessary to drain the fuel system before adding new fuel.

e. Gravity Refueling. For gravity refueling, open fuel vent shutoff valve, remove filler cap, pull chain (opening anti-syphoning device), and service cells with fuel to the required level (table 2-7).

f. Closed-Circuit Pressure Refueling. Using service instructions printed on the inside panel of the refuel panel access door (fig 2-45), perform pressure refueling precheck. When closed-circuit pressure refueling do not exceed 15 psi fuel flow. Remove adapter cap, and using standard Army nozzle, service fuel cells with fuel to the required level (table 2-7). Using the standard Army nozzle, fuel flow at 15 psi is 56 gallons per minute.

g. Single-Point Pressure Refueling. Using service instructions printed on the inside panel of the refuel panel access door (fig 2-45), perform pressure refueling precheck. Remove adapter cap, and using an Army supplied SPA nozzle, fill fuel cells with correct fuel to the required level (table 2-7). When single-point pressure refueling do not exceed 50 psi fuel flow. With the SPA nozzle, fuel flow at 50 psi is at least 100 gallons per minute.

WARNING

- The pilot and CPG shall perform their armament safety check prior to entering the forward area refueling point (FARP).
- Radio transmissions shall be limited to EMERGENCIES ONLY until refueling has been completed.

h. Rapid (Hot) Refueling (Single Engine). For rapid turnaround the helicopter may be refueled with the rotors turning and the No. 2 engine shut down. The following procedures and steps shall be observed for rapid refueling:

1. **TAIL WHEEL** switch - **LOCK.**
2. **PARK BRAKE** - Set.
3. Weapons switches - Off.
4. **PLT/GND ARM/SAFE** switches - **OFF.**
5. **PLT/GND ORIDE** switch - **OFF.**
6. **HARS** switch - **NORM.**
7. **NO. 2 PWR** lever - **IDLE** for 2 minutes, then **OFF**
8. **ANTI-COL** switch - **OFF.**
9. Refueling - Monitor.
10. **REFUEL VALVE OPEN** caution light - Verify off when refueling is complete.
11. Fuel caps/grounding cables - Installed/removed.

2-86   Change 4

12. **ANTI-COL** switch – As Desired.

> **CAUTION**
>
> Do not hover with only one engine.

NOTE

- Upon completion of hot refueling ground taxi off refueling pad if terrain permits and accomplish steps 13 through 17.
- If terrain does not permit a safe ground taxi, perform the following on the refueling pad.

13. Engine 1 – Power lever to **FLY**.
14. Collective – Apply until 60% torque (#1 engine) is reached or aircraft is light on wheels.
15. **CROSSFEED** switch – **AFT TK**. Maintain power setting for 30 seconds.
16. **CROSSFEED** switch – **NORM**. Maintain power setting for 30 seconds.
17. Collective – Reduce to minimum torque.
18. **NO. 2** engine – Start.

   i. Rapid (Hot) Refueling (Dual Engine). For rapid turnaround the helicopter may be refueled, using the D-1 nozzle only, with the rotors turning and both engines at **IDLE**. The following procedures and steps shall be observed for rapid refueling:

   1. Repeat steps h.1 thru h.6 and steps h.8 thru h.17.

> **WARNING**
>
> When opening anti-syphoning device on external tank flapper valve, extreme care should be exercised when releasing air pressure to preclude venting of fuel overboard through filler neck.

NOTE

- Use of JP-4 is prohibited during auxiliary tank operation above 80 °F ambient.
- External fuel transfer is prohibited below minimum safe single engine airspeed
- Internal fuel transfer is prohibited during external fuel transfer.

   j. Refueling of External Auxiliary Tanks. The external auxiliary tanks are gravity-filled through filler receptacles. Remove the gravity filler caps and service the auxiliary tanks to the "AH-64 Fill line" marked on the flapper valve inlet to the tank.

   k. Fuel Sump Drains. Two fuel sump drains (detail K, fig 2-45), one for each tank, are located on the underside of the fuselage. They are used to drain fuel and check for fuel contamination. To actuate, press the plunger. Hold until sufficient fuel has been drained. The auxiliary fuel tanks also contain fuel drains. To actuate the auxiliary fuel tank drain valve, insert a phillips head screwdriver, push inward while twisting clockwise until sufficient fuel has been drained. Then reverse to stop fuel flow.

2.54.2 Oil System Servicing.

   a. Oil Types. Oils are classified as primary, commercial equivalent, or emergency. Commercial equivalent oils are listed in table 2-9. There are **no** emergency oils authorized.

   b. Nose Gearbox Servicing. The nose gearboxes are mounted on the front of each engine. When the oil falls below the proper level (fig 2-45), service with oil (table 2-7).

   c. Main Transmission Servicing. Access to the oil filler cap and the right sump oil level sight gage are through the transmission access panel on the right side of the fuselage (fig 2-2). The left sump oil level sight gage can be viewed through the transmission access door on the left side of the fuselage (fig 2-2). When the oil falls below the proper level, either side, (fig 2-45), service with oil (table 2-7).

> **CAUTION**
>
> Before beginning an extended flight with auxiliary fuel tanks installed, engine oil tanks will be filled to the "FULL" point on the sight glass.

   d. Engine Servicing. The engine oil tank is located in the engine frame. It is serviced through a gravity filler port. An oil level sight gage is located near the gravity filler port. When the oil falls below the proper level (fig 2-45), service with oil (table 2-7).

   e. APU Servicing. The APU oil filler cap and oil level sight gage are located on the oil reservoir on the APU gearbox. The sight gage can be viewed through an access panel under the No. 2 engine. When the oil falls below the proper level (fig 2-45), service with oil (table 2-7).

   f. Hydraulic System Servicing. The hydraulic system should only be serviced (fig 2-45) with approved fluids from table 2-7. Detailed hydraulic system servicing instructions are in TM 1-1520-238-23.

TM 9-1520-238-10

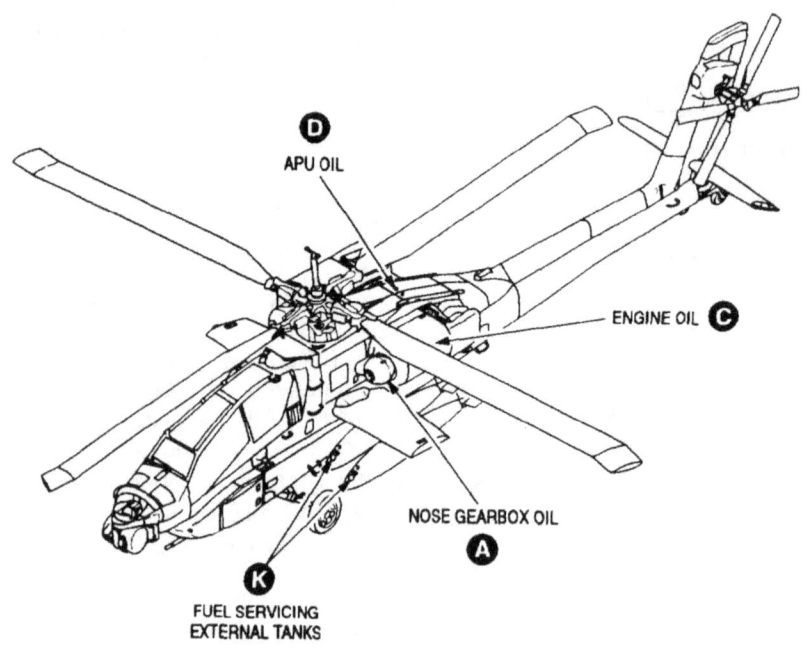

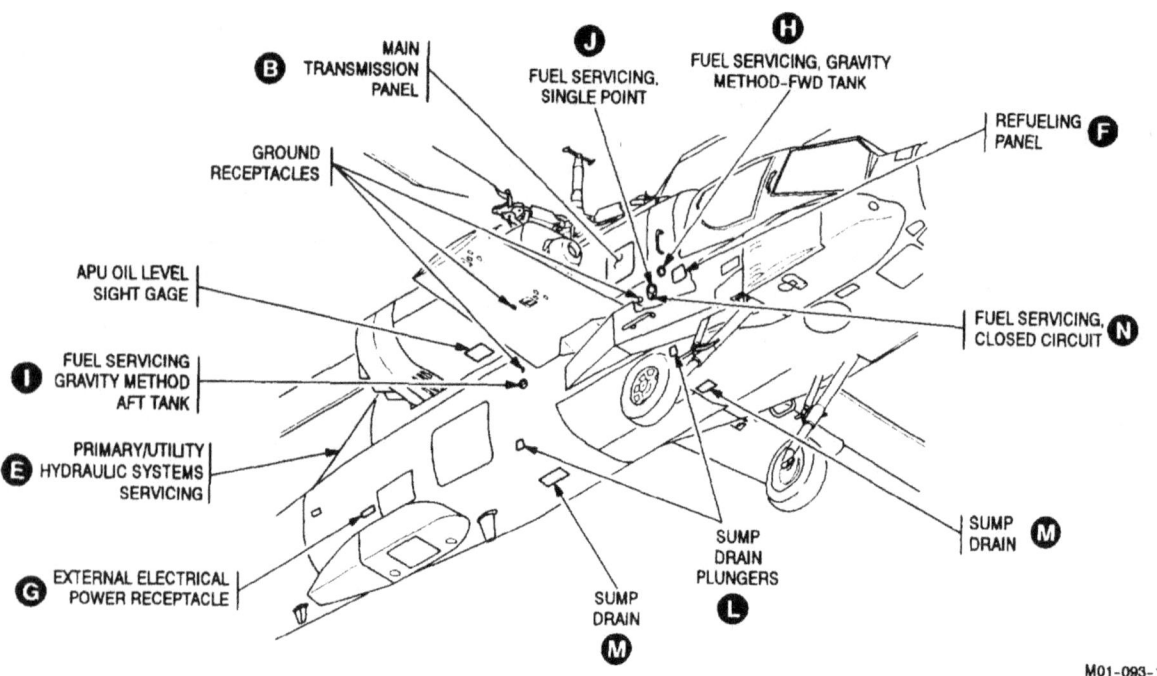

Figure 2-45. Servicing Diagram (Sheet 1 of 4)

TM 1-1520-238-10

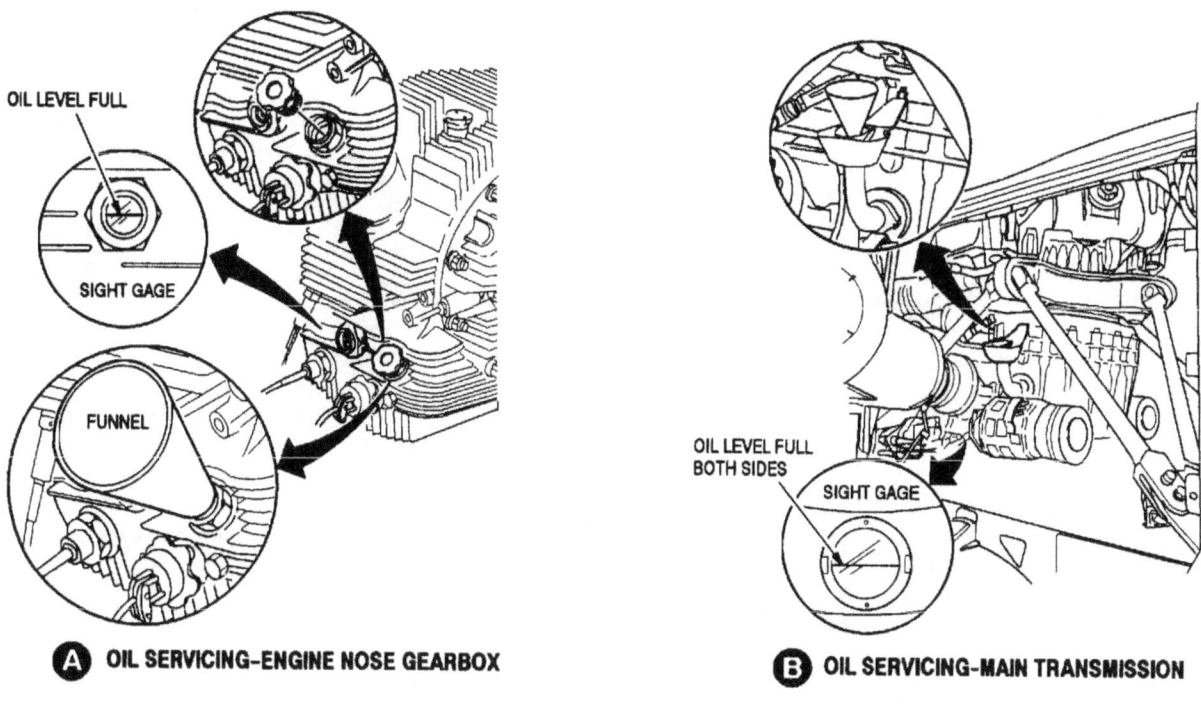

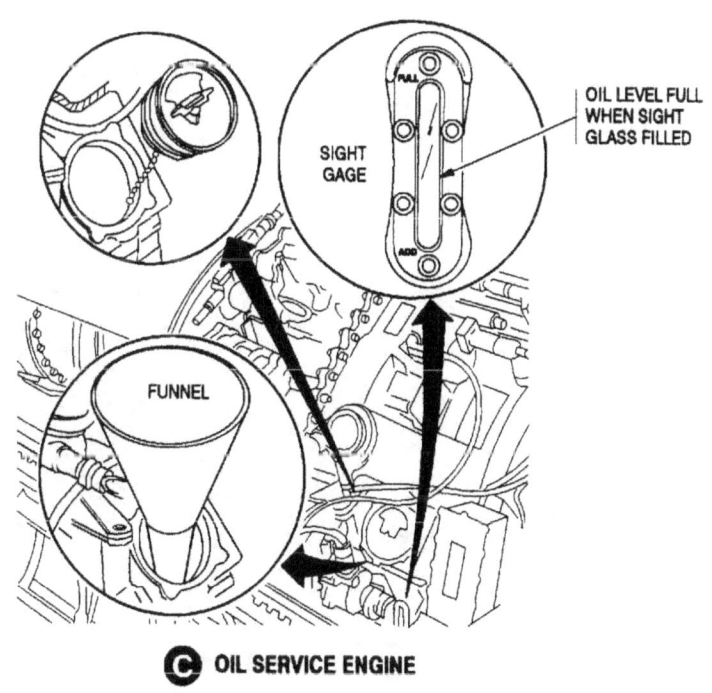

Figure 2-45 Servicing Diagram (Sheet 2 of 4)

2-89

TM 1-1520-238-10

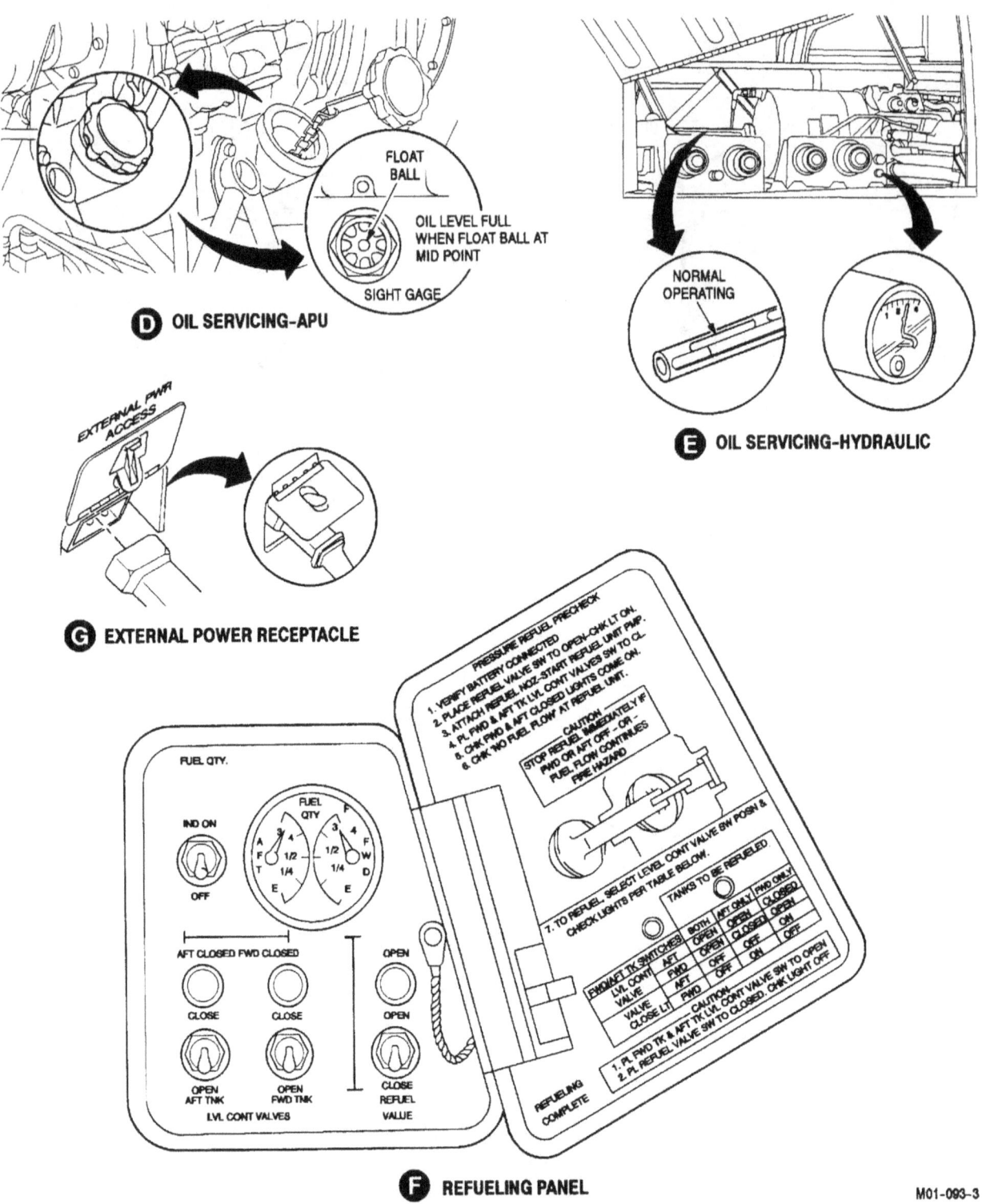

Figure 2-45 Servicing Diagram (Sheet 3 of 4)

2-90

TM 1-1520-238-10

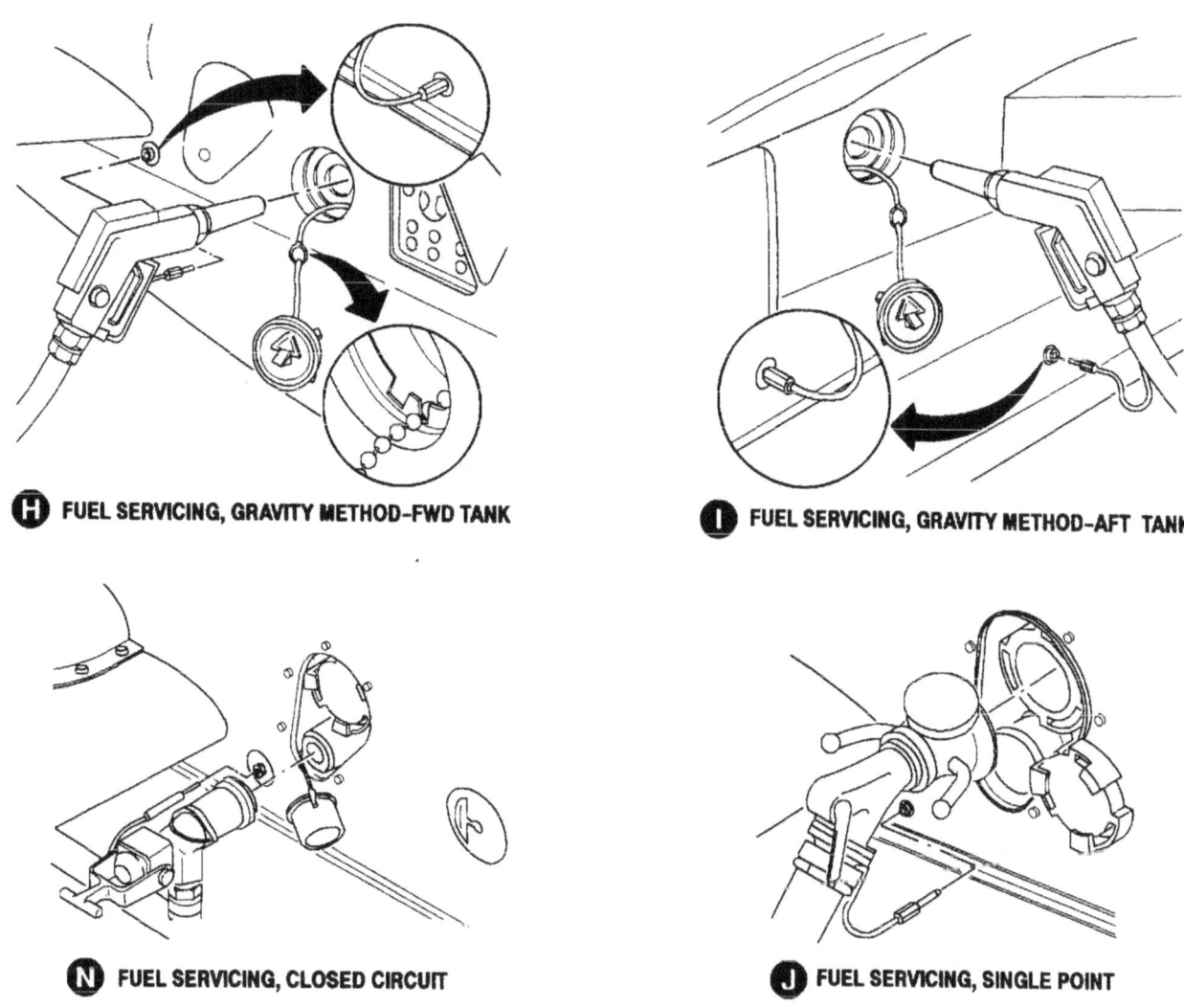

Figure 2-45 Servicing Diagram (Sheet 4 of 4)

2-91

Table 2-7. Fuel and Lubricant Specifications and Capacities

Tank or System	Capacity US	Name	Materials Spec	Grade
Forward Fuel Cell	156 gal usable 156 gal total	Turbine Fuel	MIL-T-5624 MIL-T-5624* MIL-T-83133	JP-4 JP-5 JP-8
Aft Fuel Cell	219 gal usable 220 gal total	Turbine Fuel	MIL-T-5624 MIL-T-5624* MIL-T-83133	JP-4 JP-5 JP-8
Auxiliary Fuel Tank	229 gal usable 230 gal total (each tank)	Turbine Fuel	MIL-T-5624 MIL-T-5624* MIL-T-83133	JP-4 JP-5 JP-8
Engine Oil		Lubricating Oil	MIL-L-23699* MIL-L-7808	
Main Transmission		Lubricating Oil	MIL-L-23699* MIL-L-7808	
Engine Nose Gearbox		Lubricating Oil	MIL-L-23699* MIL-L-7808	
Auxiliary Power Unit (APU)		Lubricating Oil	MIL-L-23699* MIL-L-7808	
Primary Hydraulic System		Hydraulic Fluid	MIL-H-83282* MIL-H-5606	
Utility Hydraulic System		Hydraulic Fluid	MIL-H-83282* MIL-H-5606	
Main Landing Gear Shock Strut		Hydraulic Fluid	MIL-H-5606	
Tail Landing Gear Shock Strut		Hydraulic Fluid	MIL-H-5606	
Brake System		Hydraulic Fluid	MIL-H-5606	

*Use in ambient temperatures of -25 °F (-32 °C) and above.
 Do not mix lubricating oils MIL-L-23699 and MIL-L-7808.
 Do not mix hydraulic fluids MIL-H-83282 and MIL-H-5606.

Table 2-8. Approved Fuels

US Military Fuel NATO Code No.	JP-4 (MIL-T-5624) F-40 (Wide Cut Type)	JP-5 (MIL-T-5624) or JP-8 (MIL-T-83133) F-44 or F-34 (High Flash Type)	
COMMERCIAL FUEL (ASTM-D-1655)	JET B	JET A	JET A-1 NATO F-34
American Oil Co.	American JP-4	American Type A	
Atlantic Richfield	Arcojet B	Arcojet A	Arcojet A-1
Richfield Div		Richfield A	Richfield A-1
B.P. Trading	B.P.A.T.G.		B.P.A.T.K.
Caltex Petroleum Corp.	Caltex Jet B		Caltex Jet A-1
Cities Service Co.		CITGO A	
Continental Oil Co.	Conoco JP-4	Conoco Jet-50	Conoco Jet-60
Gulf Oil	Gulf Jet B	Gulf Jet A	Gulf Jet A-1
EXXON Co. USA	EXXON Turbo Fuel B	EXXON A	EXXON A-1
Mobil Oil	Mobil Jet B	Mobil Jet A	Mobil Jet A-1
Phillips Petroleum	Philjet JP-4	Philjet A-50	
Shell Oil	Aeroshell JP-4	Aeroshell 640	Aeroshell 650
Sinclair		Superjet A	Superjet A-1
Standard Oil Co.		Jet A. Kerosene	Jet A-1 Kerosene
Chevron	Chevron B	Chevron A-50	Chevron A-1
Texaco	Texaco Avjet B	Avjet A	Avjet A-1
Union Oil	Union JP-4	76 Turbine Fuel	

NOTE: COMMERCIAL FUEL LIMITATIONS

Anti-icing and Biocidal Additive for Commercial Turbine Engine Fuel. The additive provides anti-icing protection and functions as a biocide to kill microbial growths in aircraft fuel systems. Icing inhibitor conforming to MIL-I-27686 shall be added to commercial fuel, not containing an icing inhibitor, during refueling operations, regardless of ambient temperatures. Refueling operations shall be accomplished in accordance with accepted commercial procedures. This additive (Prist or eq.) is not available through the Army Supply System, but is to be locally procured when needed.

Table 2-8. Approved Fuels – continued

US Military Fuel NATO Code No.	JP-4 (MIL-T-5624) F-40 (Wide Cut Type)	JP-5 (MIL-T-5624) or JP-8 (MIL-T-83133) F-44 or F-34 (High Flash Type)
FOREIGN FUEL	NATO F-40	NATO F-44
Belgium	BA-PF-2B	
Canada	3GP-22F	3-6P-24e
Denmark	JP-4 MIL-T-5624	
France	Air 3407A	
Germany (West)	VTL-9130-006	UTL-9130-0007/UTL 9130-010
Greece	JP-4 MIL-T-5624	
Italy	AA-M-C-1421	AMC-143
Netherlands	JP-4 MIL-T-5624	D ENG RD 2493
Norway	JP-4 MIL-T-5624	
Portugal	JP-4 MIL-T-5624	
Turkey	JP-4 MIL-T-5624	
United Kingdom (Britain)	D. Eng RD 2454	D.Eng RD 2498

NOTE: COMMERCIAL FUEL LIMITATIONS

Anti-icing and Biocidal Additive for Commercial Turbine Engine Fuel. The additive provides anti-icing protection and functions as a biocide to kill microbial growths in aircraft fuel systems. Icing inhibitor conforming to MIL-I-27686 shall be added to commercial fuel, not containing an icing inhibitor, during refueling operations, regardless of ambient temperatures. Refueling operations shall be accomplished in accordance with accepted commercial procedures. This additive (Prist or eq.) is not available through the Army Supply System, but is to be locally procured when needed.

Table 2-9. Approved Oils

Approved Domestic Commercial Oils for MIL-L-7808	
Manufacturer's Name	Manufacturer's Designation
American Oil and Supply Co.	PQ Turbine Oil 8365
Humble Oil and Refining Co.	ESSO/ENCO Turbo Oil 2389
Mobile Oil Corp.	RM-184A/RM-201A
Approved Domestic Commercial Oils for MIL-L-23699	
Manufacturer's Name	Manufacturer's Designation
Americal Oil and Supply Co.	PQ Turbine Lubricant 5247/6423/6700/7731/8878/9595
Bray Oil Co.	Brayco 899/899-G/899-S
Castrol Oil Co.	Castrol 205
Chevron International Oil Co., Inc.	Jet Engine Oil 5
Crew Chemical Corp.	STO-21919/STO-21919A/STD 6530
W.R. Grace and Co. (Hatco Chemical Div.)	HATCOL 3211/3611
Humble Oil and Refining Co.	Turbo Oil 2380 (WS-6000)/2395 (WS-6495)/2392/2393
Mobile Oil Corp.	RM-139A/RM-147A/Avrex S Turbo 260/Avrex S Turbo 265 Mobile 254
Royal Lubricants Co.	Royco 899 (C-915)/899SC/ Stauffer Jet II
Shell Oil Co., Inc.	Aeroshell Turbine Oil 500
Shell International Petroleum Co., Ltd.	Aeroshell Turbine Oil 550
Standard Oil Co., of California	Chevron Jet Engine Oil 5
Stauffer Chemical Co.	Stauffer 6924/Jet II
Texaco, Inc.	SATO 7377/7730 TL-8090
Approved Foreign Commercial Oils for MIL-L-7808	
Data not available at this time.	
Approved Foreign Commercial Oils for MIL-L-23699	
Data not available at this time.	

2.55 PARKING.

Helicopter parking shall be in accordance with local directives and the following minimum procedures: station helicopter on as level a surface as possible; set wheel brakes; lock tail wheel; turn all switches off; disconnect external power; chock wheels; secure rotor blades; attach static ground wire, and install engine inlet, exhaust, and pitot covers.

2.55.1 Protective Covers. Protective covers (fig 2-47) prevent damage from foreign objects and snow and water buildup to vital areas. All protective covers are part of the helicopter flyaway kit. The kit may be stored in the equipment stowage bay during flight. Covers are installed whenever the helicopter is on the ground for an extended period of time, or if severe environmental conditions such as ice or dust exist.

2.56 GROUND AIR SOURCE.

An external air receptacle (fig 2-46) under the No. 1 engine nacelle provides an attachment point for an external air line to start either engine or accomplish maintenance functions on the helicopter. An external air source that provides 40 psig and 30 pounds-per-minute air flow is required to pressurize the system for engine start. The maximum pressure from a ground source shall not exceed 50 psig.

2.57 TOWING.

The helicopter is towed by attaching a tow bar to the tail wheel fork. Towing the helicopter must be accomplished by trained personnel in accordance with instructions in TM 1-1520-238-23.

2.58 CANOPY AND WINDSHIELD CLEANING.

CAUTION

Do not attempt to clean either the PNVS or TADS turret windows or optics. These require special treatment by trained personnel.

The canopy and windshield shall be carefully cleaned, using aircraft cleaning practices, with clear water and a moist chamois or flannel cloth.

2.59 MOORING.

The helicopter is moored in accordance with TM 1-1500-250-23.

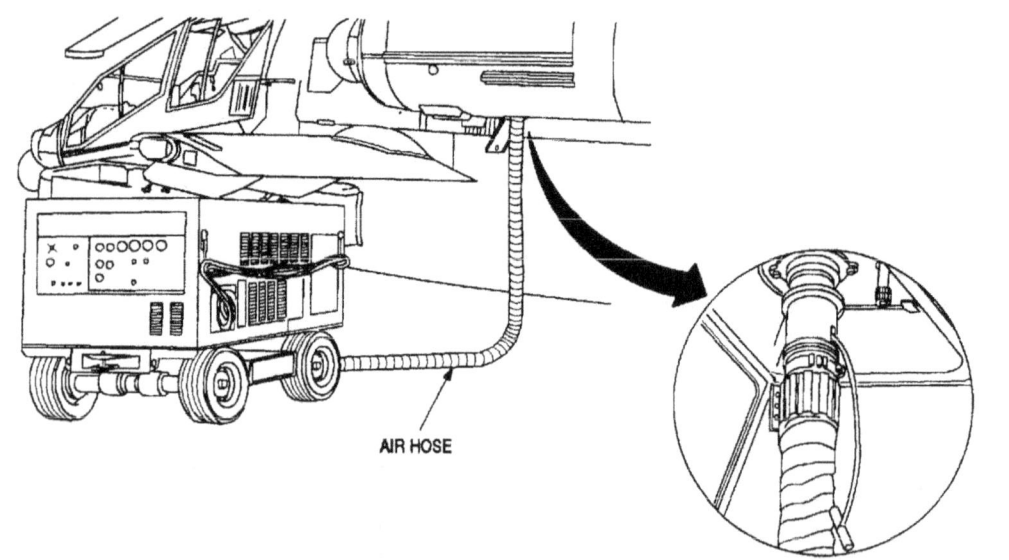

Figure 2-46. Ground Air Source

TM 1-1520-238-10

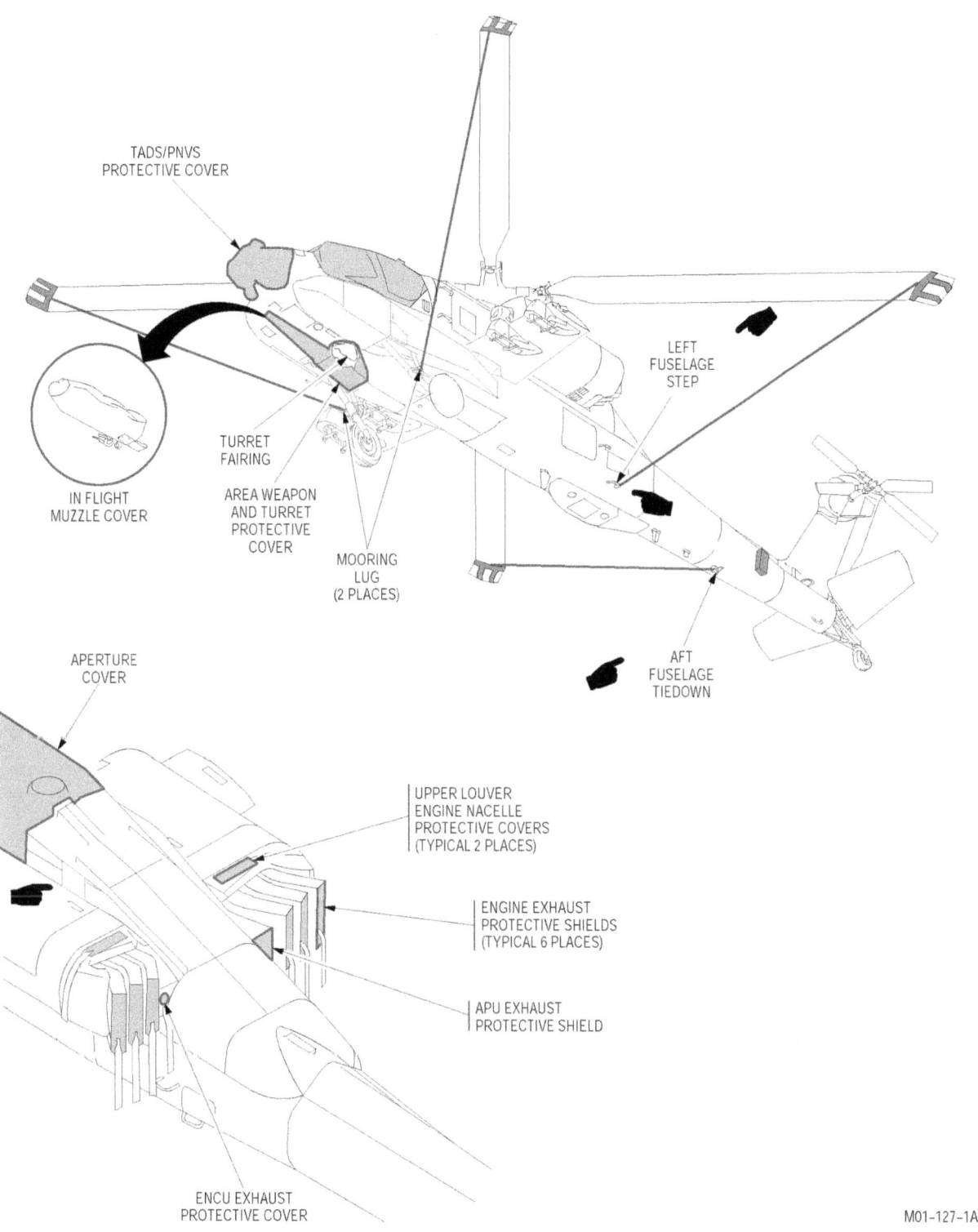

Figure 2-47. Protective Covers, Mooring, Towing (Sheet 1 of 2)

TM 1-1520-238-10

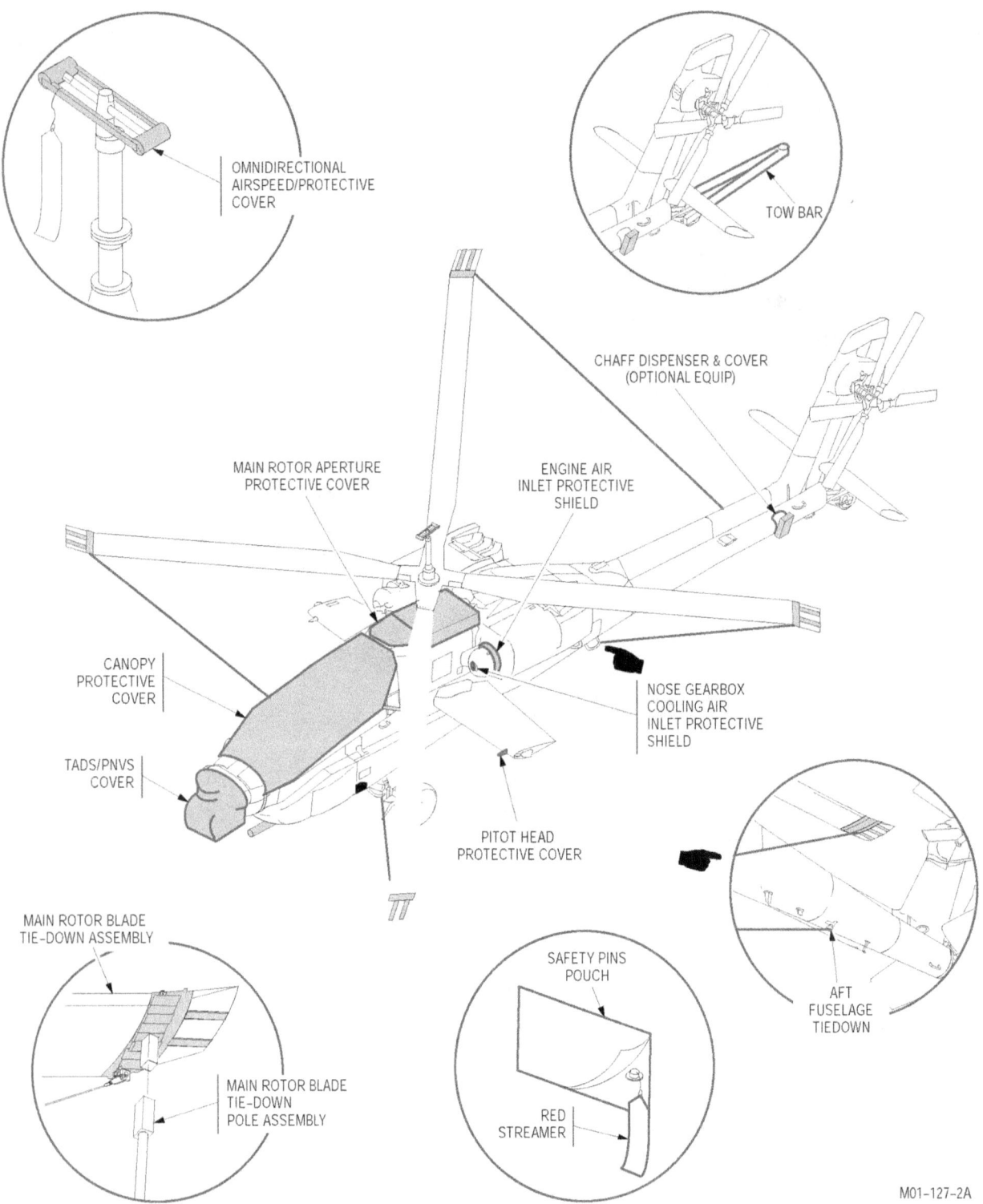

Figure 2-47. Protective Covers, Mooring, Towing (Sheet 2 of 2)

CHAPTER 3
AVIONICS

Section 1. GENERAL

3.1 DESCRIPTION.

This chapter covers the avionics equipment configuration installed in the AH-64A helicopter. It includes a brief description of the avionics equipment, its technical characteristics, capabilities, and locations. For mission avionics equipment, refer to Chapter 4, Mission Equipment.

The communications equipment provides intercommunication between crewmembers and VHF AM-FM and UHF AM radio communication. If installed, VHF FM Single Channel Ground and Airborne Radio Set (SINC-GARS) is also provided. The navigation equipment includes an Automatic Direction Finder (ADF), a Doppler Navigation System (DNS) and a Heading Attitude Reference Set (HARS). Transponder equipment consists of an Identification Friend or Foe (IFF) receiver transmitter. Height above ground level is provided by a radar altimeter. Each antenna is described with its major end item. Antenna locations on the airframe are shown (fig 3-1).

3.2 AVIONICS EQUIPMENT CONFIGURATIONS.

Avionics equipment configurations are shown in Table 3-1.

3.3 AVIONICS POWER SUPPLY.

Power to operate most of the avionics equipment is provided by the emergency dc bus. This allows for backup battery power that is used in the event of a complete electrical failure. External power may also be applied. Function selector should be OFF before applying helicopter power.

3.4 EMERGENCY OPERATION.

If both generators or both transformer-rectifiers (T/R's) fail, turn off all nonessential radio equipment to prevent excessive drain on the battery.

Table 3-1. Communication/Navigation Equipment

Equipment	Nomenclature	use	Range	Control Location	Remarks
Intercommunication	Intercommunication Control C-10414/ARC or C-11746/ARC	Intercommunication between crew members and control of navigation and communication radios.	Within crew stations and two external receptacles.	Pilot center console and CPG right console	
	Remote Transmitter Selector (used with C-10414) or Remote Transmitter Indicator ID-2403/ARC (used with C-11746)	Indicates control of communication and navigation equipment.	Not applicable	Pilot center console	
VHF FM-AM Communication	AN/ARC-186(V) VHF FM-AM No.1	Two-way FM voice communications. FM frequency range 30 to 87.975 MHz plus AM 116 to 151.975 MHz. "Receive Only" AM frequency range 108 to 115.975 MHz.	Line of sight	Pilot right console	
VHF FM-AM Communication	AN/ARC-186(V) VHF FM-AM No. 2	Same as VHF FM-AM No. 1 radio.	Line of sight	CPG right console	
VHF FM Communication (SINCGARS)	AN/ARC-201	Two-way voice communication in the frequency range of 30 to 87.975 MHz.	Line of sight	Pilot right console	
UHF AM Communication	AN/ARC-164	Two-way communication in the frequency range of 225 to 399.975 MHz.	Line of sight	Pilot right console	
Remote Control Unit (RCU) Voice Security System	Z-AHP KY-58	Secure communication for pilot AN/ARC-186 or AN/ARC-201 radio.	Not applicable	Pilot right console	Complete provisions

Table 3-1. Communication/Navigation Equipment - continued

Equipment	Nomenclature	Use	Range	Control Location	Remarks
(ADF)	**AN/ARN-89**	Radio range and broadcast reception. Automatic direction finding and homing in the frequency range of 100 to 3000 kHz.	50 NM	Pilot right console	
	C-7392/ARN-89	Provides controls for operation of AN/ARN-89 ADF.	Not applicable	Pilot right console	
(ADF)	AN/ARN-149 (V)	Provides relative bearing to the transmitting station being received. Includes standard commercial broadcast AM stations and nondirectional beacon (NDB) frequencies. Operates in the frequency range of 100 to 2199.500 kHz.	Line of sight for high-power NDB, 15 to 20 NM for low-power NDB	Pilot right console	
	C-12192/ARN-149	Provides controls for operation of AN/ARN-149 ADF.	Not applicable	Pilot right console	
(DNS)	AN/ASN-128	Provides present position or destination navigation information in latitude and longitude (degrees and minutes) or Universal Transverse Mercator (UTM) coordinates.	Not applicable	CPG right console	
Computer Display Unit (CDU)	CP-1252 CDU	Provides controls and indicators for operation of the AN/ASN-128 DNS.	Not applicable	CPG right console	

Table 3-1. Communication/Navigation Equipment - continued

Equipment	Nomenclature	Use	Range	Control Location	Remarks
(DNS)	AN/ASN-137	Provides present position or destination navigation information in latitude and longitude (degrees and minutes) or (UTM) coordinates.	Not applicable	CPG right console	
(CDU)	IP-1552/G CDU	Provides controls and displays for operation of the AN/ASN-137 DNS.	Not applicable	CPG right console	
(IFF)	AN/APX 100(V)	Transmits a specially coded reply to a ground based IFF radar interrogator system.	Line of sight	Pilot right console	
Absolute Altimeter	AN/APN-209	Measures absolute altitude.	0 to 1500 feet above ground level	Pilot right instrument panel	
(HARS)		Senses helicopter attitude and motion to define roll, pitch, heading, and flight path.	Not applicable	Pilot right instrument panel	
Data Transfer Receptacle (DTR)	DR-902B	Auto loading of mission data	Not applicable	CPG right console	
Embedded Global Positioning System Inertial Navigation Unit (EGI)	CN-1689(V)1/ASN	Provides present position or destination navigation information in latitude and longitude (degrees and minutes) or (UTM) coordinates.	Not applicable	CPG right console (CDU)	

3-4 Change 3

TM 1-1520-238-10

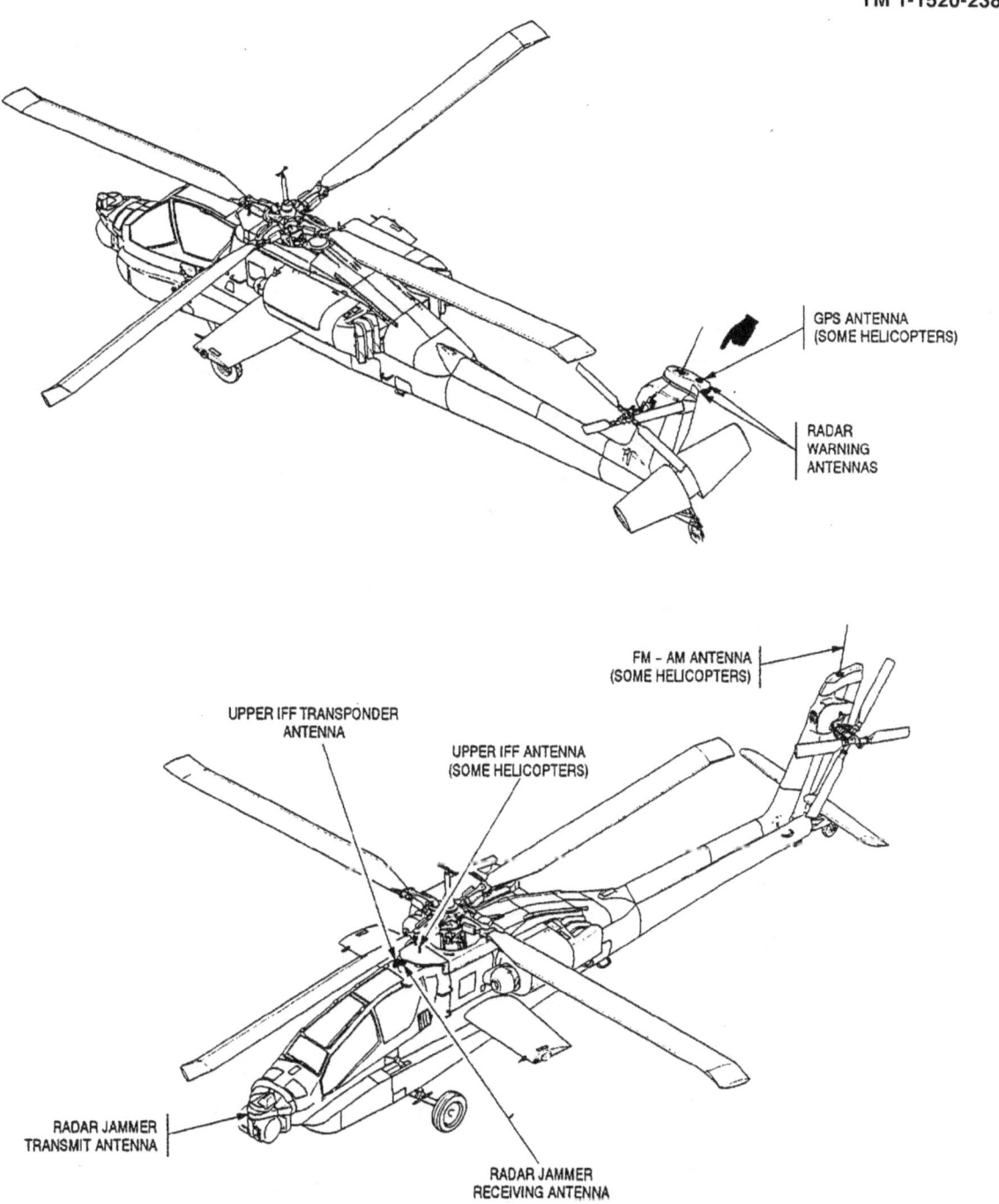

Figure 3-1. Antenna Arrangement (Sheet 1 of 2)

TM 1-1520-238-10

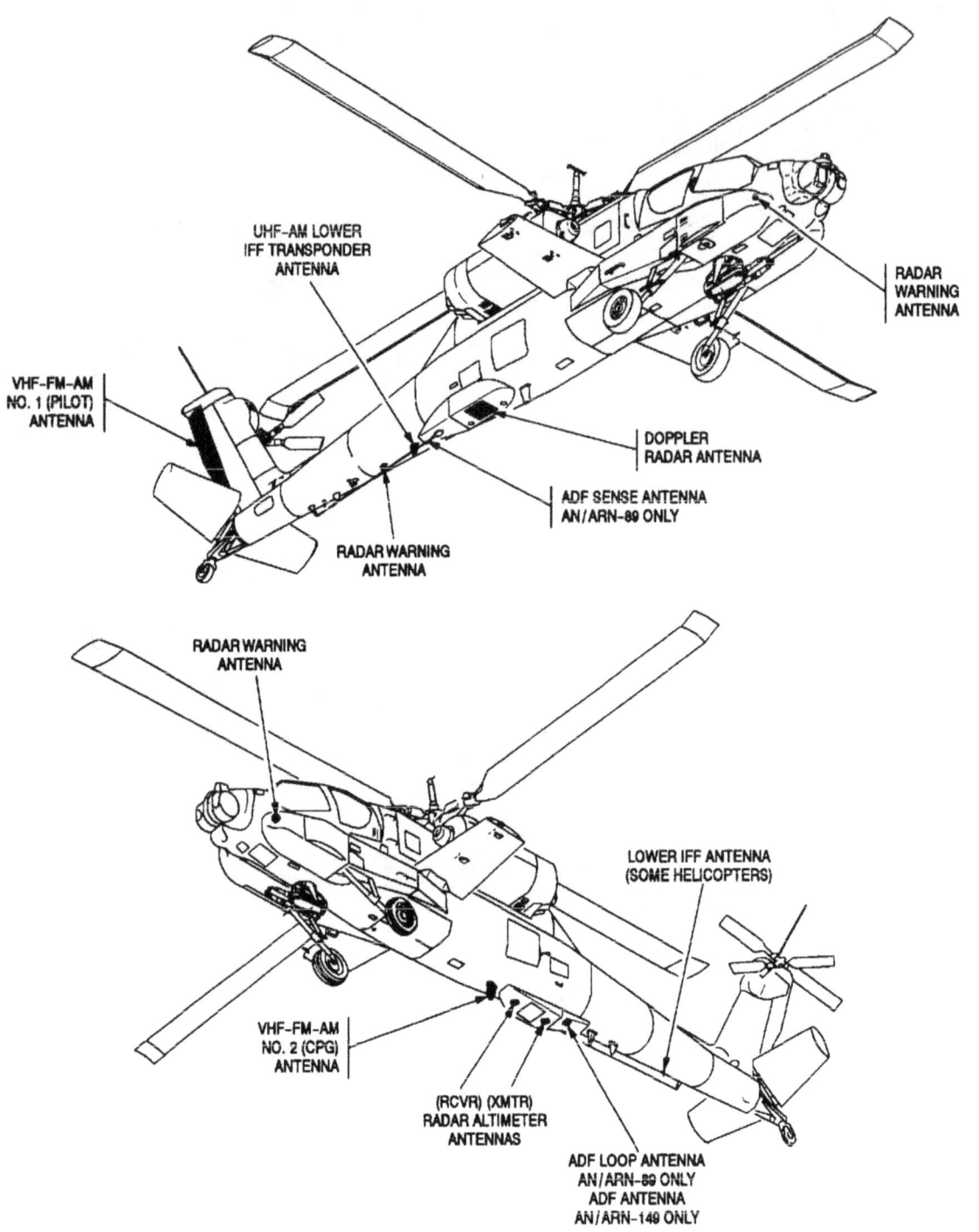

Figure 3-1. Antenna Arrangement (Sheet 2 of 2)

Section II. COMMUNICATIONS

3.5 INTERCOMMUNICATION SYSTEM (ICS) C-1041401)3/ARC OR C-11746(V)4/ARC.

TWO ICS control panels, placarded CSC (fig 3-2 and 3-3), one each for pilot and CPG, provide intercommunication capability between crewmembers. These also provide a means by which the pilot and CPG may select and control associated radio equipment for voice transmission and reception. Additionally, an external jack fitting on each service panel, located outboard on each wing, permits maintenance personnel to communicate with the crewmembers. A radio transmit/interphone rocker switch is installed on both cyclic stick grips (fig 2-26). A foot-operated radio transmit switch is installed on the right side of the aft cockpit deck. The forward crew station has two foot-operated switches: a radio transmit switch on the left side of the deck, and an interphone transmit switch on the right side. Hands free intercommunication is provided by a hot mike feature. The remote transmitter switch pushbutton on the pilot cyclic stick grip, directly below the **RADIO/ICS** switch, allows the pilot to remotely select the radio transmitter that he wishes to use. With the transmitter selector switch in position **5** (C-10414 configuration) or position **RMT** (C-11746 configuration), pressing the remote transmit select switch begins the transmitter select cycle. Identification of the selected transmitter is displayed by way of lamps above the pilot **CSC** panel. A placard is provided for each crewmember as ready reference to which receiver and transmitter selector switch position will enable a desired radio. Power for the (ICS) is provided from the emergency dc bus through the **ICS** circuit breaker in each crew station.

3.5.1 Controls and Functions. The function of the controls on the face of each **CSC** panel (fig 3-2 or 3-3) are as shown in table 3-2.

NOTE

- The **RADIO MON** switches are connected (ON) when pushed in fully and disconnected (OFF) when pulled out.

- The C-10414(V)3/ARC control panel is not illuminated.

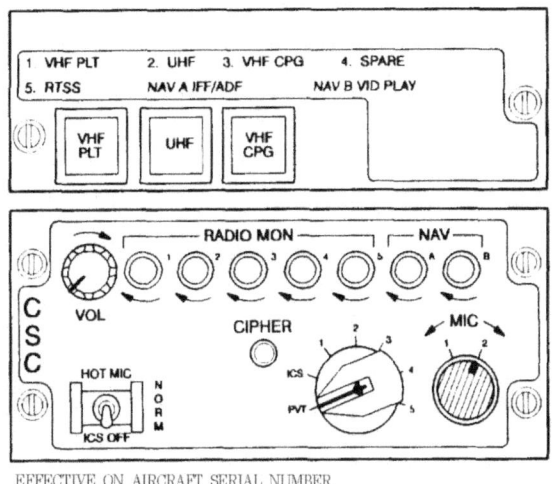

Figure 3-2. Control Panel C-10414(V)3/ARC and Remote Transmitter Selector

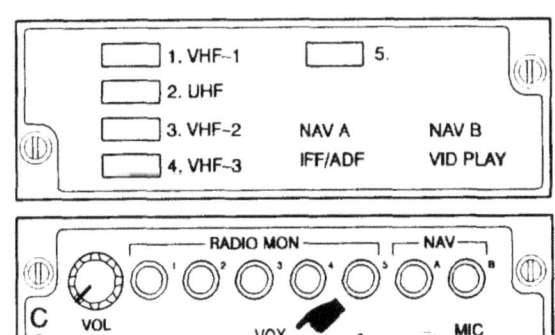

Figure 3-3. Control Panel C-11746(V)4/ARC and Remote Transmitter Indicator Panel ID-2403/ARC

Table 3-2. CSC Panel Control and Indicator Functions

Control/Indicator	Function
Receiver Selector Switches ON	
1	Connects pilot ARC-186 receiver to the headphone.
2	Connects pilot ARC-164 to the headphone.
3	Connects CPG ARC-186 receiver to the headphone.
4	Spare.
5	Spare.
NAV A	Connects the ARN-89 or the ARN-149 ADF receiver and the APX-100 IFF portion of the transponder to the headphone.
NAV B	Connects the audio output of the video recorder subsystem (VRS) to headphone (playback mode of the VRS ONLY).
Transmitter Selector Switch	
PVT	Not Used.
ICS	Enables ICS system when keyed.
1	Enables pilot ARC-186 transmission when keyed.
2	Enables pilot ARC-164 transmission when keyed.
3	Enables CPG ARC-186 transmission when keyed.
4	Spare.
5(C-11746)	Spare.
5(C-10414) RMT (C-11746)	Enables remote transmit select switch (pilot only).
ICS Switch HOT MIC	Interphone transmit on at all times.
VOX ON	Enables VOX.
NORM	Interphone transmit functions only when interphone transmit switch pressed.
ICS OFF	Disables interphone transmit.
VOL Control	Adjusts headset volume level.

TM 1-1520-238-10

Table 3-2. CSC Panel Control and Indicator Functions - continued

Control/Indicator	Function
MIC Switch	
1	Used when all maintenance headsets have dynamic microphones. (Pilot and CPG IHADSS helmets have linear microphones).
2	Used to communicate between pilot and CPG with IHADSS helmets; and between pilot/CPG with IHADSS helmets and maintenance personnel at wing stations. In the latter case, maintenance headsets must have a linear microphone or a compensating adapter.

NOTE

If the pilot communication control panel MIC switch is in the 2 position, headsets connected to the right wing station ICS connector must also have a linear microphone or compensating adapter. If the CPG communication control panel MIC switch in the 2 position, headsets connected to the left wing station ICS connector must also be configured correctly.

3.5.2 Modes of Operation. There are several methods of intercommunication operation. In all cases, no operator action is required to receive intercom signals other than adjusting the **VOL** control for a comfortable level at the headset.

3.5.3 Intercommunication for Crewmembers. Transmitting and receiving is accomplished by both crewmembers in the following manner.

1. Transmitter selector switch is set to **ICS** when the pilot is using the ICS floor switch. The transmitter selector switch may be set at any position when using either cyclic stick **RADIO/ICS** switch in the **ICS** position.

2. The **RADIO/ICS** switch on either cyclic stick grip should be pressed to ICS or the CPG foot-operated ICS switch should be pressed (ground crew must press a push-to-talk button on their ICS cord) to talk on the ICS. Speak into the microphone while pressing any of these switches. Release switch to listen.

3.5.4 External Radio Communication. Both crewmembers are able to communicate with external receiving station in the following manner:

1. Transmitter selector switch is set to the desired position, **1** thru 3, for either crew station. Pilot sets to **5** (C-10414) or **RMT** (C-11746) if he wishes to use the remote transmit select switch.

2. RADIO/ICS rocker switches- Press the cyclic stick grip RADIO rocker switch or the radio transmit floor switch. Speak into microphone while pressing switch. Release to listen.

3.5.5 Receiver Selection. Place RADIO MON switch(s) to ON as desired and adjust volume to a comfortable listening level.

3.6 RADIO SET AN/ARC-186(V).

Radio Set AN/ARC-186(V) is a VHF FM-AM transceiver that provides clear and secure voice communication capability at frequencies in the VHF AM and FM bands. The radio set has a guard frequency of 40.500 MHz for FM and a 121.500 MHz for AM. Over a frequency range of 108.00 MHz to 115.975 MHz, it functions as a receiver for the reception of (AM) transmissions. At frequencies in the range of 116.000 MHz to 151.975 MHz, the set operates both as an AM receiver and AM transmitter. From 108.000 MHz to 151.975 MHz, a total of 1760 AM voice communication channels spaced at 25 kHz is provided by the set. In a range of frequencies extended from 30.000 MHz to 87.975 MHz, it functions both as an FM receiver and FM transmitter. Operating in this frequency range, it provides 2320 FM voice communication channels with a spacing of 25 kHz. The radio set also provides 20 channel presets which can be any combination of AM or FM frequencies. Automatic tuning to both AM and FM emergency frequencies (121.5 MHz and 40.5 MHz, respectively) is provided by setting only one control. Power output of the transmitter is 10 watts.

3-9

3.6.1 Antennas. The FM-AM No. 1 (pilot) communication antenna (fig 3-1) is mounted on the vertical stabilizer as an integral part of the trailing edge assembly. Some helicopters have an FM-AM whip antenna mounted on top of the vertical stabilizer replacing the trailing edge antenna. The FM-AM No. 2 (CPG) communication blade antenna is mounted on the bottom center fuselage area directly forward of the doppler antenna fairing.

3.6.2 Controls and Functions. Controls for the AN/ARC-186(V) transceiver are on the front panel (fig 3-4) of the unit. The function of each control is as shown in table 3-3.

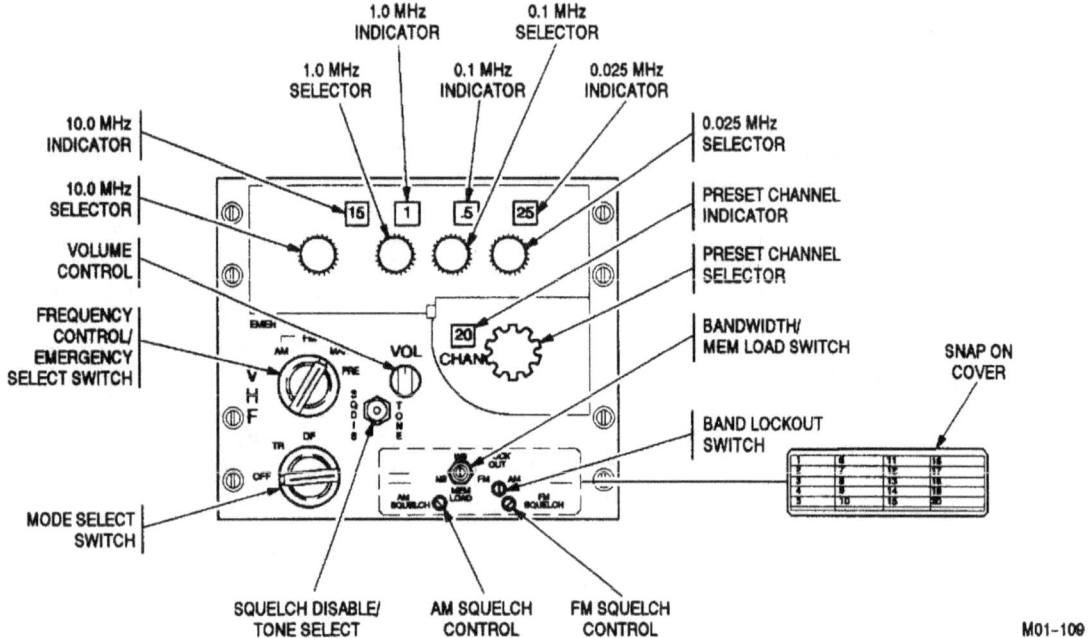

Figure 3-4. Control Panel AN/ARC-186

TM 1-1520-233-10

Table 3-3. AN/ARC-186(V) Controls and Functions

Control/Indicator	Function
0.025 MHz selector	Rotary switch. Selects r/t frequency in 0.025 MHz increments. Clockwise rotation increases frequency.
0.025 MHz Indicator	Indicates manually selected r/t frequency in 0.025 MHz increments.
0.1 MHz selector	Rotary switch. Selects r/t frequency in 0.1 MHz increments. Clockwise rotation increases frequency.
0.1 MHz indicator	Indicates manually selected r/t frequency in 0.1 MHz increments.
1.0 MHz selector	Rotary switch. Selects r/t frequency in 1.0 MHz increments. Clockwise rotation increases frequency.
1.0 MHz indicator	Indicates manually selected r/t frequency in 1.0 MHz increments.
10 MHz selector	Rotary switch. Selects r/t frequency in 10 MHz increments from 30 to 150 MHz. Clockwise rotation increases frequency.
10 MHz indicator	Indicates manually selected r/t frequency in 10 MHz increments from 30 to 150 MHz.
Present channel selector	Rotary switch. Selects preset channel from 1 to 20. Clockwise rotation increases channel number selected.
Preset channel indicator	Indicates selected preset channel.
Volume control	Potentiometer. Clockwise rotation increases volume.
Squelch disable/ tone select	Three-position switch. Center position enables squelch. SQ DIS position disables squelch. Momentary TONE position transmits tone of approximately 1000 Hz.
Frequency control/ emergency select switch	Four-position rotary switch. EMER AM/FM selects a prestored guard channel. MAN position enables manual frequency selection. PRE position enables preset channel selection.
Mode select switch	Three position rotary switch. OFF position turns transceiver off. TR position enables transmit receive modes. D/I? position is not used.
Bandwidth/memory load switch	Three-positions switch. NB position enables narrow band selectivity. WB enables wideband selectivity in the FM band. Momentary MEM LOAD allows manually selected frequency to go into selected preset channel memory.
AM squelch control	Screwdriver adjustable potentiometer. Squelch overridden at maximum counterclockwise position. Clockwise rotation increases input signal required to open the squelch.
FM squelch control.	Screw driver adjustable potentiometer. Squelch overridden at maximum counterclockwise position. Clockwise rotation increases input signal required to open the squelch.
Band lockout switch	Will lock out the AM or FM frequency of the band selected. Presently set to the center (LOCKOUT) position to receive both AM and FM bands.

3-11

3.6.3 Modes of Operation. Depending on the setting of the operating controls, the radio set can be used for the following controlled modes of operation:

a. Transmit/Receive (TR) Mode. Two-way in the clear and secure voice communication. Refer to paragraphs 3.8 through 3.8.2 for voice security system.

1. Set **OFF, TR, -D/F** mode select switch to **TR**.

2. Set **EMER AM/FM, -MAN, -PRE** frequency selector switch to **MAN** for manual frequency selection or to **PRE** for preset channel selection.

3. To manually select a frequency, rotate the four MHz selector switches until desired frequency is displayed at indicator windows.

NOTE

Rotating the MHz selector switches clockwise increases frequency. Frequencies can be manually selected in 0.025 MHz increments.

4. To select a preset channel, rotate preset channel selector switch until the number of the desired channel is displayed in the CHAN indicator window.

NOTE

Clockwise rotation of preset channel selector switch will increase the desired channel number **(1 to** 20). The radio set will automatically tune to the preset channel in TR mode.

b. DF Mode. Not Used.

c. AM Emergency (EMER AM) mode. Emergency two-way voice communication on selected guard channel.

1. Set mode select switch to **TR**.

2. Set frequency control/emergency select switch to **EMER AM**.

NOTE

Selecting the **EMER AM** mode will automatically disable the secure speech function and enable clear voice communication.

d. FM Emergency (EMER FM) mode. The FM emergency mode enables voice reception/transmission on a prestored guard frequency of 40.500 MHz.

1. Set mode select switch to **TR**.

2. Set **EMER AM/FM, -MAN, -PRE** frequency control/emergency select switch to **EMER FM**.

NOTE

Selecting the **EMER FM** mode will automatically disable the secure speech function and enable voice communication.

3.6.4 Operating Procedures, Radio Set AN/ARC-186.

a. Squelch Disable. To disable squelch, set squelch disable tone select **SQ DIS/TONE** switch to **SQ DIS**. Squelch will remain disabled (open) until switch is returned to center position.

b. Tone Transmission. To transmit (FM or AM) tone frequency of approximately 1000 Hz, set squelch disable tone select **SQ DIS/TONE** switch to the momentary **TONE** position. Releasing the switch removes the tone frequency.

c. Loading Preset Channels.

1. Set mode select switch to **TR**.

2. Set **EMER AM/FM-MAN-PRE** frequency control/emergency select switch to **MAN**.

3. Rotate the four MHz selector switches until desired frequency is displayed in indicator windows.

4. Rotate **CHAN** preset channel selector switch until the desired channel is displayed in the indicator window.

5. Remove snap-on cover.

6. Momentarily hold **WB-NB-MEM LOAD** switch to **MEM LOAD**. Preset frequency is now loaded into memory.

TM 1-1520-236-10

d. Wideband/Narrowband Selection

1. Remove snap-on cover.

2. For wideband operation, set **WB-NB-MEM LOAD** switch to **WB**.

3. For narrowband operation, set **WB-NB-MEM LOAD** switch to **NB**.

   NOTE

   The switch shall be placed in the **WB** position at anytime the **MEM LOAD** function is not being accomplished. The NB position is not used in this installation.

e. Band Lockout Selection.

1. Remove snap on cover.

   NOTE

   With the **LOCKOUT-AM-FM** switch set to **AM** or **FM**, the frequency of the band selected will be locked out. This will cause an audible warning to occur whenever a frequency in a locked out band is selected. For this installation, operational AM and FM bands are required and the **LOCKOUT-AM-FM** switch must be set to the **LOCKOUT** position.

2. Ensure **LOCKOUT-FM-AM** switch is in **LOCKOUT** position (indicated by a white dot on the switch).

3.7 RADIO SET RT-1167C/ARC-164(V) AND HAVE QUICK RADIOS

Receiver-Transmitter RT-1167/ARC-164 (fig 3-5) is an airborne, (UHF), (AM), radio transmitter-receiver (transceiver) set. It is a multi-channel, electronically-tunable transceiver with a fixed-tuned guard receiver. The transceiver operates on any one of 7,000 channels spaced in 0.025 MHz units in the 225.000 to 399.975 MHz frequency range. The guard receiver is on a permanent guard/emergency frequency of 243.000 MHz. The radio set primarily is used for voice communication. An additional radio set capability, although not fictional, is ADF. The radio set receives 28 vdc from the emergency dc bus through the UHF AM circuit breaker on the pilot overhead circuit breaker panel.

Receiver-Transmitter RT-1167C/ARC-164 has the same functions and capabilities as the RT-1167/ARC-164 plus a WWE-QUICK (HQ) mode of operation which provides an electronic counter counter measures (ECCM) frequency hopping capability. Radios with the HQ capability have frequency selector 1 on the extreme left of the front panel labeled A-3-2-T (fig 3-5). The single frequency mode of operation is referred to as the normal mode and the frequency hopping mode of operation is referred to as the anti-jam (AJ) or ECCM active mode.

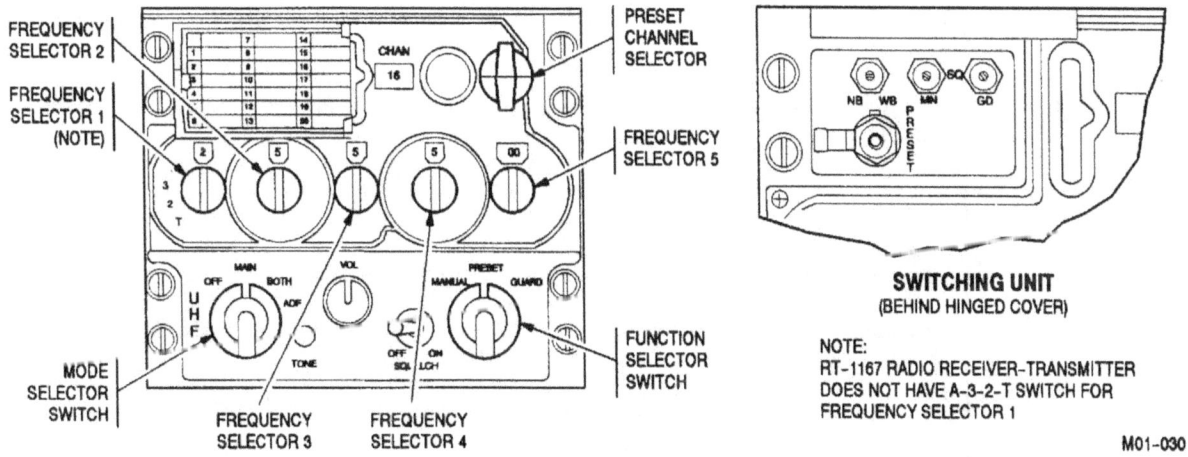

Figure 3-5. Receiver-Transmitter Radio, RT-1167/ARC-164(V)

3-13

3.7.1 Antenna. The UHF-AM antenna (fig 3-1) is located on the bottom center fuselage area and is mounted directly aft of the doppler fairing.

3.7.2 Controls and Functions. Controls for the ARC-164(V) are on the front panel of the unit (fig 3-5). The function of each control is described in table 3-4.

Table 3-4. RT-1167C/ARC-164(V) Controls and Functions

Control	Function
Frequency selector 1	For the RT-1167, manually selects 100's digit of frequency (either 2 or 3) in MHz.
A-3-2-T selector switch (RT-1167C radios only)	
A	Selects AJ mode.
3 and 2	Manually selects 100's digit of frequency (either 2 or 3) in MHz in normal mode or the first digit of the WOD in the AJ mode.
T	Momentary switch position allows the radio to receive a new TOD. Also used with TONE button for emergency clock start.
Frequency selector 2	Manually selects 10's digit of frequency (0 through 9) in MHz in normal mode, second digit of the WOD or first digit of net number in AJ mode.
Frequency selector 3	Manually selects unit digit of frequency (0 through 9) in MHz in normal mode, third digit of the WOD or second digit of net number in AJ mode.
Frequency selector 4	Manually selects tenths digit of frequency (0 through 9) in MHz in normal mode, fourth digit of the WOD or third digit of net number in AJ mode.
Frequency selector 5	Manually selects hundredths and thousandths digits of frequency (00, 25, 50 or 75) in MHz and the fifth and sixth digits of the WOD.
Preset channel selector	Selects and stores one of 20 preset channels in normal mode or the WOD in AJ mode.
CHAN indicator	Indicates selected preset channel. Indicator is blanked when radio is in MANUAL or GUARD mode.
Function selector switch	Selects method of frequency selection.
MANUAL	Any of 7,000 frequencies can be manually selected using the five frequency selectors.
PRESET	A frequency is selected using the preset channel selector switch for selecting any one of 20 preset channels as indicated on the CHAN indicator.
GUARD	The main receiver and transmitter are automatically tuned to the guard frequency and the guard receiver is turned off. Any manually set or preset frequency is blocked out.
SQUELCH ON-OFF switch	Turns squelch of main receiver on or off.

3.7.4 Warning Tones.

a. A steady warning tone will be heard in the headset when the AJ mode is selected and TOD, or a valid WOD, has not been entered.

b. A pulsating warning tone will be heard in the headset when an invalid operating net is selected.

3.7.5 Conference Capability.

NOTE

Conference communication is not possible when operating in the secure voice mode.

In the AJ mode, the radio can receive and process two simultaneous transmissions on the same net. This conference capability is available by selecting **00, 50,** or **75** with frequency selector **5** and then operating in the AJ mode. Conferencing is not possible when the net number ends in **25**.

In a conference net, the second transmitting radio will automatically shift its transmission frequency by 25 kHz when it monitors a transmission on the primary net frequency. The wide band receiver will read both transmissions without the interference normally associated with two radios transmitting on the same frequency simultaneously. Three simultaneous transmission will result in garbled reception.

3.7.6 Guard Operation.

NOTE

If the guard frequency is being jammed with the mode selector switch in the **BOTH** position and the radio in the AJ or normal mode, set the mode selector switch to **MAIN**.

The guard receiver is not affected by AJ mode operation. The guard frequency maybe monitored regardless of mode as long as the mode selector switch is set to **BOTH**. The **BOTH** position turns on the main transceiver and guard receiver. The guard receiver will remain tuned to 243.000 MHz regardless of manual or preset frequencies selected. When the function selector switch is set to **GUARD**, the AJ mode is removed, the guard receiver is turned off and the main transceiver is set at 243.00 MHz, making it possible to transmit and receive on the guard frequency.

3.7.7 Secure Communication. The HQ radio has a secure (cipher) communication capability. In either the normal or AJ mode, the radio can perform cipher mode operation. The effectiveness of communications in the AJ cipher mode depends on the frequency hop rate. It is recommended that the slowest hop rate (channel **20** WOD set at **00**) be used during AJ cipher mode operation.

3.7.8 Normal Mode Operating Procedures.

a. **Loading Preset Channels.**

1. Mode selector switch - **MAIN** or **BOTH**.

2. Frequency selectors - Select frequency to be set in memory.

3. Function selector switch - **PRESET**.

NOTE

For HQ radios, do not select channel 20.

4. Preset channel selector - Set to channel to be loaded.

5. Open the switching unit cover.

6. **PRESET** button - Press. The frequency is now loaded into memory.

7. Repeat steps 2, 3, 4 and 6 to complete presetting channels.

8. Close and latch switching unit cover.

b. **Transmit/Receive (MAIN) Operating Procedures.**

1. Mode selector switch - **MAIN**.

2. Function selector switch - **MANUAL**.

3. To manually select a frequency, set the five MHz selector switches until desired frequency is displayed in indicator windows.

4. To select a preset channel, rotate preset channel selector until desired channel is displayed in CHAN indicator window.

c. **Transmit/Receive/Guard Monitor Operating Procedures.**

1. Mode selector switch - **BOTH**.

NOTE

If reception on the selected frequency interferes with guard frequency reception, detune the radio set by selecting an open frequency or set function selector switch to **GUARD**.

2. Select the desired manual frequency or preset channel.

d. 1,020 Hz Tone Signal Transmission (MAIN) Procedures.

NOTE

The tone-modulated signal may be used to check out the radio receiver.

1. Mode selector switch - **MAIN**.

2. Select a desired frequency for tone signal transmission, either manually or with the preset channel selector.

3. **TONE** button - Press to transmit the 1,020 Hz signal.

3.7.9 HQ Mode (RT-1167C Only) Operating Procedures.

a. Activation of HQ Mode.

The HQ mode can be activated in one of two ways:

1. Function selector switch - **MANUAL**. A-3-2-T switch - A.

2. Function selector switch - **PRESET**. Preset channel selector - Channel **20**.

b. Entering the WOD.

NOTE

The WOD is normally entered before flight but can be entered in flight.

1. Mode selector switch - **MAIN**.

2. Function selector switch - **PRESET**.

NOTE

Channels containing WOD elements cannot be used in the normal preset function. Transmission can be made on WOD channels, but they will not be on the frequency stored in the channel memory.

3. Preset channel selector - Channel **20**.

4. Frequency selector 1- Set first WOD element.

5. Open switching unit cover.

6. PRESET button - Press.

NOTE

The WOD elements are entered into memory through channels 20 through 15. The WOD varies in length and will require from one to six of these channels.

7. Preset channel selector - Next lower channel.

8. Repeat steps 4, 6, and 7 until all elements of the WOD are loaded into preset memory.

9. Close and latch the switching unit cover.

NOTE

A single beep indicates that the WOD in channel 20 was entered into preset memory and that an additional WOD element is in the next lowest preset channel.

10. Preset channel selector - Set to channel 20. A single beep is heard.

NOTE

The double beep heard at the last channel signifies that the WOD elements have been transferred to volatile memory.

11. Preset channel selector - Set to progressively lower channels. A single beep is heard at each channel until the last channel; a double beep is heard at the last channel.

3.7.10 HQ Emergency Startup of TOD clock.

NOTE

This method of providing TOD is used when there is no TOD available from another HQ radio or external source. When using this method of synchronization, the flight commander or lead helicopter should emergency start his TOD clock. The lead helicopter then will transfer this TOD to other helicopters in flight. A valid TOD signal must be used by all helicopters before effective AJ communications can be achieved.

1. Function selector switch - **MANUAL** or **PRESET**, as desired.

NOTE

To preclude an inadvertent transmission of a new time, release the **TONE** button prior to releasing the **A-3-2-T** switch.

2. **A-3-2-T** switch - T and hold.

TM 1-1520-238-10

   3. **TONE** button - Press momentarily.
   4. A-3-2-T switch - Release.
   5. Return to designated frequency or channel.

a. **Transmitting TOD from Net Control Helicopter.**

   NOTE
   TOD transmission and reception is normally performed before flight, but it can be performed in flight.

   1. Mode selector switch - **MAIN** or **BOTH**, as desired.
   2. Function selector switch - **MANUAL** or **PRE- SET** as desired.
   3. Frequency selectors - Select predesignated frequency.

   or

   Preset channel selector - Set to predesignated channel.

   4. Call other helicopter(s) to send time.
   5. **TONE** button - Press momentarily to send TOD. Listen for 1,020 Hz tone.

b. **Receiving TOD.**

   NOTE
   *. After power-up and WOD transfer, the radio automatically accepts the first TOD message received.
   * To receive TOD in normal mode, after setting **A-3-2-T** switch to T, set the radio to the frequency being used.

   1. Mode selector switch - **MAIN.**
   2. Function selector switch - **MANUAL** or **PRE-SET**, as desired.
   3. Frequency selectors - Select predesignated frequency for reception of TOD.

   or

   Preset channel selector - Set to predesignated channel.

   4. Listen for net control helicopter state **STANDBY FOR TIME.**
   5. A-3-2-T switch Momentarily to T, then back to A. Listen for momentary two-tone signal followed by a single tone.

c. **Entering Net Number.**

   NOTE
   Setting the A-3-2-T switch to A overrides the hundredths digit in both manual and preset functions, puts the radio in AJ mode and programs the radio to use the net number in the three digits following the A.

   1. Function selector switch - **MANUAL** or **PRE-SET**, as desired.
   2. Frequency selectors - Set second, third and fourth switches to the desired net number.

   or

   Preset channel selector Set to a preset channel containing the desired net number in its second, third and fourth digits.

   3. A-3-2-T switch - A.

3.7A RADIO SET RT-1518/ARC-164 HAVEQUICK II (HQ II) RADIO

The HQ II system consists of a modification to AN/ARC-164 airborne, (UHF), (AM), radio transmitter-receiver (transceiver) set providing a frequency hopping or anti-jamming capability. Frequency hopping is a technique where the designated preset frequency being used for communication on a given link is automatically frequency-hopped many times per second. The purpose of this is to make it difficult for an adversary to jam the link since they cannot determine which channel is being used. By the time determination is made as to which frequency is being used, the communication link has changed to another frequency.

The frequency-hopping scheme is implemented in the equipment by storing or initializing every radio with a Word of Day (WOD), Time of Day (TOD), and a net number. The WOD programs the frequency-hopping rate and frequency-hopping pattern. The radio cannot function in the anti-jamming mode without a valid WOD. Up to six WODs may be entered at one time. The procedure for storing multiple WODs is called Multiple

Change 3   3-18.1

TM 1-1520-238-10

Word of Day (MWOD) loading. The TOD provides synchronization necessary for communicating in the anti-jamming mode by allowing frequency hopping at the same instant in time.

The RT-1518 HQ II radio (fig. 3-6) provides an analog display (Red Lt) for channel and frequency indicators, and functions/operates similar to the RT-1167C/ARC-164 radio described in paragraph 3.7 with seven additional features. Refer to table 3-4 for radio controls and functions. The additional features are:

    (1) Verify/Operate mode
    (2) Manual Single WOD Loading
    (3) Multiple WOD Manual Loading
    (4) Manual WOD Erase
    (5) Operational Date Loading
    (6) Training/Maintenance Nets
    (7) Loading/Changing FMT NETS

The HQ II Expanded Memory Board (EMB) capability is identified on the RT-1518 radio by an EMB sticker which appears next to the channel indicator display. If no sticker is displayed, the Multiple Word of Day (MWOD) must be entered to verify the HQ II radio capability.

3. 7A. 1 Antenna. The UHF-AM antenna (fig 3-1) is located on the bottom center fuselage area and is mounted directly aft of the Doppler fairing.

3. 7A. 2 Controls and Functions. Controls for the RT-1518/ARC-164 are on the front panel of the unit (fig 3-6). The function of each control is described in table 3-4.

CAUTION

The AN/ARC-164 radio must be placed in the VERIFY/OPERATE mode before operating in the Have Quick mode. This will prevent the radio from locking up and damaging the equipment.

3.7A.3 Modes of Operation.
    a. Verify/Operate.
      1. Mode Selector Switch - **MAIN** or **BOTH**
      2. CHAN switch - Set to channel 20.
      3. Frequency switches - Set to 220.000 on the frequency/status indicator.
      4. Function Selector switch - **PRESET**.
      5. Lift access cover and press **PRESET** button. The radio is now in the VERIFY/OPERATE mode.
      6. Function Selector switch - **MANUAL** or **PRESET**
      7. Select desired frequency or channel.
      8. Mode Selector Switch - OFF.

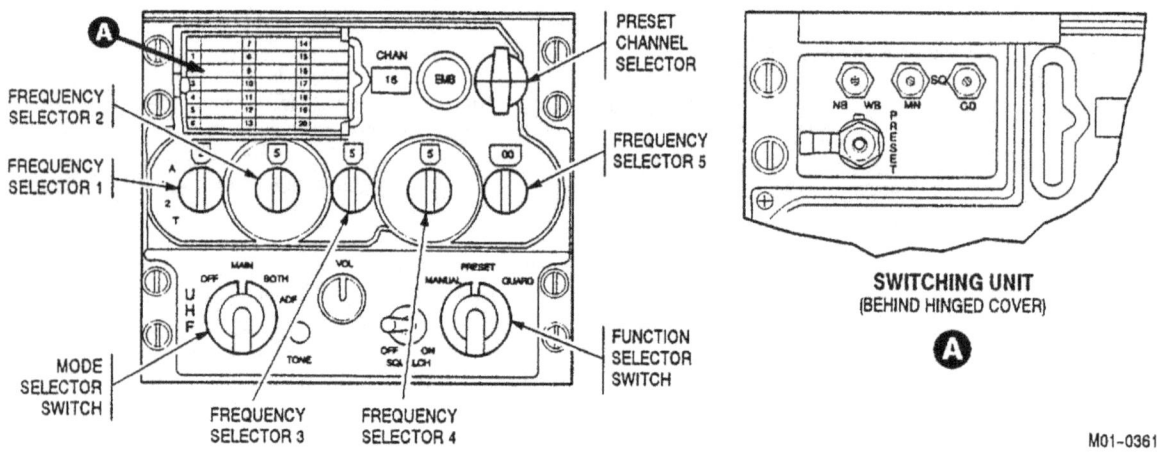

Figure 3-6. Receiver-Transmitter Radio, RT-1518/ARC-164

3.7A.4 Manual Single Word Of Day (SWOD) Loading.

1. Function Selector switch - **PRESET**.
2. **CHAN** switch - Set to Channel 20.
3. Frequency switches - Set to 220.000 MHz.
4. **PRESET** button - Lift preset frequency label cover and push **PRESET** button.
5. Frequency switches - Set to the first WOD segment frequency (300.050) of the Training/Maintenance WOD listed in table 3-5.

Table 3-5. Training/Maintenance WOD and Common Net

Preset Channel	WOD Segment	Common Net
20	300.500	300.000
19	376-000	300.000
18	359.100	300.000
17	314.300	300.000
16	297.600	300.000
15	287.400	300.000

6. Function Selector switch - **PRESET**.
7. CHAN switch - Set to Channel 20.
8. **PRESET** button - Lift preset frequency label cover and push **PRESET** button.

Repeat steps 5 through 8, setting channels 19 through 15 in order, substituting Training/Maintenance WOD for existing preset channels. After preset channel 15 frequency has been entered, the WOD must be initialized.

3.7A.5 Initializing the Word Of Day (WOD).
NOTE

If power is lost, or if preset channel 20 is selected when the **MANUAL-PRESET-GUARD** switch is in the **PRESET** position, the initialize procedure will be required again. To use this WOD, a TOD and NET must be entered.

1. **CHAN** switch - Set to Channel 20. A single beep should be heard.
2. **CHAN** switch - Set to Channel 19. A single beep should be heard.
3. **CHAN** switch - Set to Channel 18. A single beep should be heard.
4. **CHAN** switch - Set to Channel 17. A single beep should be heard.
5. **CHAN** switch - Set to Channel 16. A single beep should be heard.
6. **CHAN** switch - Set to Channel 15. A double beep should be heard.

3.7A.6A Multiple Word Of Day (MWOD) Loading.

1. CHAN switch - Set to Channel 20.
2. Function Selector switch - **PRESET**.
3. Frequency switches - Set to 220.025 (MWOD LOAD).
4. **PRESET** button - Lift preset frequency label cover and push **PRESET** button. Listen for single beep.

NOTE
- The radio is in the MWOD load mode.
- The radio will transmit and receive in the **VERIFY/OPERATE** mode only. The radio is disabled and will not transmit or receive in the **M-LOAD**, **ERASE**, or **FMT.CHG** modes.

5. Function Selector switch - **MANUAL**.
6. CHAN switch - Set to Channel 20.
7. Frequency switches - Set to the first WOD element.
8. **TONE** button - Press **TONE** and release. Listen for a wavering tone. The first WOD element is now entered.

Repeat steps 6 thru 8 for presets 19 thru 15, in that order, substituting WOD elements for existing preset channels.

9. CHAN switch - Set to Channel 14.

NOTE
- If two or more WODs have the same date code, the radio recognizes only the latest date entered.
- MWODs must be linked with an associated day-of-month code. This date code element has been added to every operational and training segment in HAVE QUICK and need only be loaded when MWOD is used.

10. Frequency switches - Set to the applicable day of the month code. Code format is XAB.XXX, where A and B are the day of the month the WOD is used. X can represent any value. For example, if today were 26 June, then select 226.025 or 326.475.

11. **TONE** button - Press **TONE** and release. Listen for a double beep. One complete WOD is now loaded. To load more WODs, repeat steps 5 thru 11.

3.7A.66 Multiple Word Of Day (MWOD) HaveQuick Operations.

NOTE

The operational date is the current Day of Month and must be entered into Channel 1 so the radio can select the proper WOD.

1. Mode Selector switch - **MAIN**.
2. **CHAN** switch - Set to Channel 20.
3. Function Selector switch - **PRESET**.
4. Frequency switches - Set to 220.025 MHz (MWOD LOAD).
5. **PRESET** button - Lift preset frequency label cover and push **PRESET** button. Listen for single beep.
6. Function Selector switch - **MANUAL**.
7. **CHAN** switch - Set to Channel 01.
8. Frequency switches - Set to the applicable Day of Month code. Code format is XAB.XXX, where A and B are the day of the month the WOD is used.
9. TONE button - Press TONE and release. Listen for a wavering tone.
10. Function Selector switch - **PRESET**.
11. **CHAN** switch - Set to Channel 20.
12. Frequency switches - Set to 220.000 MHz (Verify/Operate).
13. **PRESET** button - Lift preset frequency label cover and push **PRESET** button. Listen for single beep. Radio is now ready to receive TOD either via GPS or conventional means.
14. **HQ SINC** button - Press and hold for 3 seconds.
15. Function Selector switch - **MANUAL**.
16. Active Net - Select.

3.7A.7 Verifying MWOD is Loaded.

1. STATUS - Ensure radio is in the Verify/Operate mode.
2. Function Selector switch - **MANUAL**.
3. Frequency switches - Set to the day of the month to be verified. Code format is XAB.XXX, where A and B are the day of the month of the Date Tag associated with the WOD to check.

4. **CHAN** switch - Set to Channel 20. Switch momentarily to channel 19 and return to channel 20. A single beep verifies the **MWOD** with a matching day of the month code stored in memory. If a single beep is not heard, the code is not stored in memory. Repeat steps 3 and 4 to check other days of the month.

3.7A.8 MWOD Erase.

1. **CHAN** switch - Set to channel 20.
2. Function Selector switch - **PRESET**.
3. Frequency switches - Set to 220.050.
4. **PRESET** button - Lift preset frequency label cover and push **PRESET** button.
5. Function Selector switch - **MANUAL**.
6. **TONE** button - Press **TONE** and release. MWODs should now be erased.
7. Function Selector switch - **PRESET**.
8. Frequency switches - Set to 220.000.
9. **PRESET** button - Lift preset frequency label cover and push **PRESET** button. Close cover. The radio is now in the Verify/Operate model

3.7A.9 TOD Send.

1. Function Selector switch - **PRESET** or **MANUAL**. Rotate frequency switches or **CHAN** switch to a predesignated frequency for TOD transmission.
2. **TONE BUTTON** - Press **TONE** button for two seconds and release.
3. Loading Operational Date.
4. **CHAN** switch - Set to Channel 20.
5. Function Selector switch - **PRESET**.
6. Frequency switches - Set to 220.025.
7. **PRESET** button - Lift preset frequency label cover and push **PRESET** button.
8. **CHAN** switch - Set to Channel 01.
9. Frequency switches - Set to the operational date in the format of XAB.XXX, where A and B are the day of the month of the Date Tag. X can be any value.
10. **TONE** button - Press **TONE** and release. Listen for a wavering tone.
11. Function Selector switch - **PRESET**.

12. **CHAN** Switch - Set to Channel 20.
13. Frequency Switches - Set to 220.000.
14. **PRESET** Button - Lift preset frequency label cover and push **PRESET** button. Listen for a single beep.

3.7A.10 Net Numbers. The net number is used in the anti-jamming mode in the same fashion as a radio frequency in the normal mode of operation. The net number enables multiple users to operate simultaneously on a non-interfering basis with other users while sharing a common WOD and TOD. There are three nets available to the operator:

(1) Frequency Managed A-Nets (FMA-NET).
(2) Training Nets (T-NET).
(3) Frequency Managed Training Nets (FMT NETS).

a. Frequency Managed A-Nets (FMA-NET). The FMA NET provides four frequency tables or hop sets. There are 1000 possible nets for each hop set. The frequency table to be used is determined by the geographical area of operation. One large hop set has been coordinated for use in NATO-Europe and another large hop set for employment in non-NATO countries. The net number begins with an "A" and is followed by three digits between 000 to 999. The last two digits designate the frequency table to be used. Nets are selected IAW ABB.BCC where:

(1). A = A (Active)
(2). BB.B = Desired Net
(3). CC = 00 for basic HAVE QUICK A & B NETS.
= 25 for NATO-Europe
= 50 for non-NATO Europe
= 75 for future use.

b. **Training Nets (T-NETS).** Each Major Command is assigned a training WOD for daily training and radio maintenance. Training WODs may be loaded using a single WOD or MWOD methods. All training WODs are initialized with 300.0XX in channel 20. XX sets the frequency hop rate for the WOD in a SWOD only. In this training mode, the radio hops between the five frequencies loaded in with the WOD (locations 19 to 15) and five training nets are available. As shown below, a net number ending in 00 selects a training net.

(1). A00.000
(2). A00.100
(3). A00.200
(4). A00.300
(5). A00.400

c. **Frequency Managed Training Nets (FMT-NET).** To expand the number of training nets available to HAVE QUICK II users, 16 frequencies (Nets) have been loaded into the radio and are permanently stored in the radio memory. To use the FMT nets, a training WOD must be entered first. The FMT Nets are numbered A00.025 through A01.525. All six characters must be selected and the last two digits must be 25.

Selection of an FMT NET greater than A01.525 or ending in 50 or 75 will result in a pulsating tone.

d. Loading or Changing FMT NETS.

1. **CHAN** Switch - Set to Channel 20.
2. Function Selector switch - **PRESET**.
3. Frequency Switches - Set to 220.075.
4. **PRESET** Button - Lift preset frequency label cover and push **PRESET** button. Listen for a single beep.
5. Function Selector switch - **MANUAL.**
6. CHAN Switch - Set to appropriate memory location (Channels 20 to 5).
7. Frequency Switches - Set to select first frequency.
8. **TONE** Button - Press TONE and release.

Repeat steps 6 through 8 until all frequencies are loaded.

9. Frequency Switches - Set to 220.000 MHz when all frequencies have been loaded.
10. CHAN Switch - Set to Channel 20.
11. Function Selector switch - **PRESET.**
12. **PRESET** Button - Lift preset frequency label cover and push **PRESET** button.

TM 1-1520-238-10

3.7B RADIO SET RT-1518C/ARC-164 HAVEQUICK II (HQ II) RADIO

The RT-1518C/ARC-164 HQ II radio provides a Liquid Crystal Display (LCD) (Green Lt) for channel, frequency indicators, and operator prompts; a zeroize switch feature; and an electronic Fill Port data loading capability.

3.7B.1 Antenna. The UHF-AM antenna (fig 3-1) is located on the bottom center fuselage area and is mounted directly aft of the Doppler fairing.

3.7B.2 Controls and Functions. Controls for the RT-1518C/ARC-164 are on the front panel of the unit (fig 3-7). The function of each control is described in table 3-6.

CAUTION]

The AN/ARC-164 radio must be placed in the VERIFY/OPERATE mode before operating in the Have Quick mode. This will prevent the radio from locking up and damaging the equipment.
3.7B.3 Modes of Operation.

a. Power Up. All segments of both displays light up momentarily on power up and a series of beeps may be heard. If the frequency/status indicator displays a frequency after power up, the radio is in the **VERIFY/OPERATE** mode. Proceed to step 5. If **M-LOAD**, FMT.CHG or ERASE is displayed, proceed as follows:

1. **CHAN** switch - Set to channel 20.
2. Frequency switches - Set to 220.000 on the frequency/status indicator.
3. Function Selector switch - **PRESET.**
4. Lift access cover and press **LOAD** button. The radio is now in the **VERIFY/OPERATE** mode.
5. Function Selector switch - **MNL** or **PRESET**
6. Select desired frequency or channel.
7. Mode Selector Switch - **OFF**

b. Normal Operation. The radio set has the following modes of operation:
(1) MAIN mode: Two-way voice communication.
(2) BOTH mode: Allows using the main transceiver with constant monitoring of the guard receiver.
(3) GUARD (GRD) mode: Allows transmission and reception on guard frequency.
(4) Manual (MNL) Mode: Single frequency used.

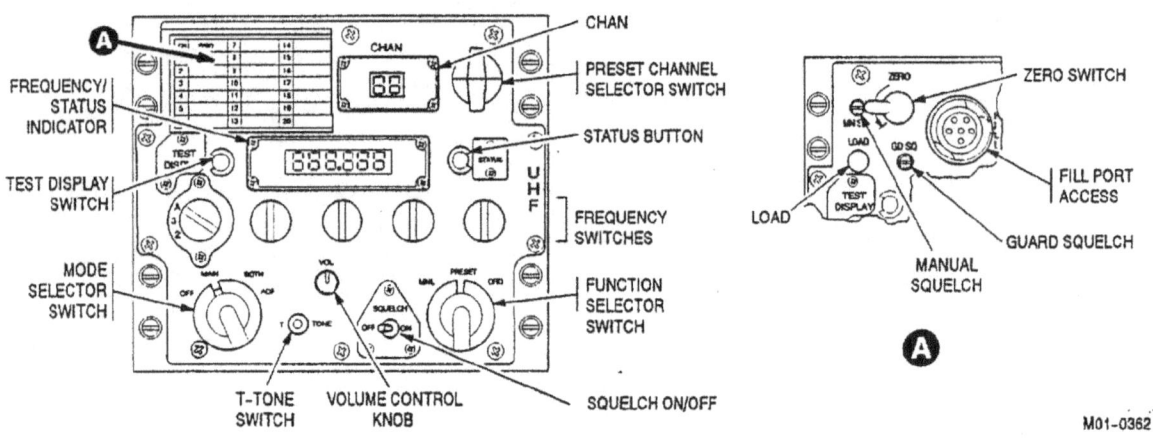

Figure 3-7. Receiver-Transmitter Radio, RT-1518C/ARC-164 3-18.6 Change 3

Table 3-6. RT-1518C/ARC-164 Controls and Functions

Control	Function
Frequency/Status Indicator	Displays individual frequency switch settings or any of the following operator prompts: **VER/OP**- indicates radio is in normal operating mode. **M-LOAD** -indicates radio is in MWOD load mode. **ERASE** - indicates radio is in MWOD erase mode. **FMT.CHG** - radio is in Frequency Management Training Change mode. **FILL** - indicates a keyfill device is connected to the **FILL** port. **WOD OK** - indicates valid WOD was received from keyfill device. **BAD** - indicates bad WOD or parity was received from keyfill device.
CHAN Indicator	Channel display indicator
Preset Channel Selector Switch	Selects one of 20 preset channels.
STATUS Switch	When depressed, an alternate display is shown on the frequency/status indicator for five seconds.
Frequency Switches	Rotary switches that select the corresponding hundreds, tens, units, tenths, and thousandths digits for the desired frequency in the normal mode, and the desired WOD elements or net number in the anti-jamming mode. The A position puts the radio in to the anti-jamming frequency-hopping mode of operation when selected.
Function Selector Switch	Three position switch **(MNL, PRESET, GRD)** which selects the main transmitter and receiver frequency. In Manual **(MNL)** frequency is manually selected using five rotary switches. The **PRESET** position allows selection of one of twenty preset frequencies using the rotary channel selector switch. The GRD position tunes the receiver and transmitter to emergency frequency 243.000MHz.
SQUELCH ON/OFF	Enables and disables squelch of the main receiver.
VOL Control Knob	Rotary knob used to set receiver volume. Does not control transmitter output.
T - TONE Switch	Three position toggle switch with two position being spring loaded; the middle position being off. When TONE is selected, a tone is transmitted on the selected frequency. When placed in the T position, reception of the TOD is enabled for one minute.
Mode Selector Switch	In **MAIN** the main receiver and transmitter are operational. In **BOTH**, the main receiver and transmitter and guard receiver are operational. **ADF** is not used. **OFF** turns the radio off.
TEST DISPLAY Switch	Lights all segments of the frequency status and channel indicators when depressed. Also used with the **T-TONE** switch for emergency clock start.
Manual Squelch **(MN SQ)**	Adjusts the threshold level of squelch for the main receiver.
ZERO Switch	Erases all MWOD elements when rotated to ZERO position.
FILL Port Access	**FILL** port access for loading MWOD segments.
Guard Squelch (GD SQ)	Adjusts the threshold level of squelch for the guard receiver.
LOAD Button	Stores selected frequency in preset channels 1 through 20 in normal mode. In ECCM mode, channel 20 is reserved for entry of WODs.

c. HaveQuick (HQ) Operation. The HQ mode of operation provides a jam resistant capability by means of a frequency hopping technique that changes frequency many times per second. Automatic frequency changing in an apparently random manner provides the jam resistance of the radio. The HQ radios permit communication in radio-jamming environments. Three elements are required for uninterrupted, successful communication. Radios must use the word-of-day (WOD), be time-synchronized with each other with a time-of-day (TOD), and share a common net.

(1) Word Of Day (WOD). The WOD programs the frequency hopping rate and frequency hopping pattern. The radio cannot function in the anti-jamming mode without a valid WOD The WOD does not take up preset memory, but the WOD memory is accessed through preset locations 20 through 15. The seventh location, accessed through channel 14, stores the Day of the Month information. This date code works in conjunction with the TOD and specifies the day a specific WOD is to be utilized when Multiple Word of Day (MWOD) are entered. At midnight (GMT) transitions, the radio automatically generates a new frequency hopping pattern based on the new days WOD. Up to six WODs may be entered at one time, allowing for multiple day use of the radio set without installing another WOD.

(2) Time Of Day (TOD). Time synchronization is necessary for communicating in the anti-jamming mode to allow frequency hopping at the same instant in time. TOD information is obtained from the UHF radios that have been modified to receive the Universal Time Coordinated (UTC) signal. UTC is a worldwide standard and is available from a variety of sources, such as the Command Post, Global Positioning System (GPS), and AWACS. A valid TOD signal will be heard as a two-beat frequency tone. Once all radios are operating on UTC, uninterrupted voice communications are insured. The radio automatically accepts the first TOD signal after power up. The first reception must occur in the normal mode. Updates of the TOD may be performed in the anti-jamming or normal mode. Subsequent TOD transmissions are ignored unless the operator enables the radio to receive a new TOD. If communications during anti-jamming operations become slightly garbled it is an indication of drift in TOD synchronization...

d. HaveQuick II Command Codes. There are four separate Command Code functions associated with loading the HaveQuick II radio. The operator enters a six digit command code into preset 20 to begin the unique initialization procedure. The Command Codes, their function and Frequency/Status indication are shown in table 3-7.

Table 3-7. HAVEQUICK II Command Codes

Command Code	Function	Frequency/Status Indication
220.000	Verify/Operate	VER/OP
220.025	MWOD Load	M-LOAD
220.050	MWOD Erase	ERASE
220.075	FMT.CHG Frequency	FMT.CHG

3.7B.4 Manual Single Word Of Day (SWOD) Loading.

NOTE

The below procedures step through entering the Training/Maintenance WOD for performing checks and maintenance. An actual WOD may be entered as a substitute in the below steps.

1. Function Selector switch - **PRESET**.
2. **CHAN** Switch - Set to Channel 20.
3. Frequency Switches - Set to 220.000 MHz.
4. **LOAD** Button - Lift preset frequency label cover and push **LOAD** button. To check that radio is in the **VER/OP** mode, press the **STATUS** button and the frequency/status LCD should display **VER/OP** for five seconds.
5. Frequency Switches - Set to the first WOD segment frequency (300.050) of the Training/ Maintenance WOD listed in table 3-8.

Table 3-8. Training/Maintenance WOD and Common Net

Preset Channel	WOD Segment	Common Net
20	300.050	300.000
19	376.000	300.000
18	359.100	300.000
17	314.300	300.000
16	297.600	300.000
15	287.400	300.000

6. Function Selector switch - **PRESET**.

7. **CHAN** switch - Set to Channel 20.

8. **LOAD** button - Lift preset frequency label cover and push **LOAD** button.

Repeat steps 5 through 8, setting channels 19 through 15 in order, substituting Training/Maintenance WOD for existing preset channels. After preset channel 15 frequency has been entered, the WOD must be initialized.

3.78.5 Initializing the Word Of Day (WOD).

NOTE

If power is lost, or if preset channel 20 is selected when the **MANUAL-PRESET-GUARD** switch is in the PRESET position, the initialize procedure will be required again. To use this WOD, a TOD and NET must be entered.

1. **CHAN** switch - Set to Channel 20. A single beep should be heard.

2. **CHAN** switch - Set to Channel 19. A single beep should be heard.

3. **CHAN** switch - Set to Channel 18. A single beep should be heard.

4. **CHAN** switch - Set to Channel 17. A single beep should be heard.

NOTE

If TOD is being automatically beaconed from another station, the first TOD message received within one minute of selected T position will be accepted.

5. **CHAN** switch - Set to Channel 16. A single beep should be heard.

6. **CHAN** switch - Set to Channel 15. A double beep should be heard.

7. Function Selector switch - **MNL**. This will prevent accidental erasure of the initialized WOD.

3.78.6 Multiple Word Of Day (MWOD) Loading.

1. **CHAN** switch - Set to Channel 20.

2. Function Selector switch - **PRESET**.

3. Frequency switches - Set to 220.025 (MWOD LOAD).

4. **LOAD** button - Lift preset frequency label cover and push **LOAD** button. Listen for single beep.

NOTE

- If radio is now in the MWOD load mode. **M-LOAD** will be displayed on the frequency/status indicator for five seconds by pressing the **STATUS** switch.

- The radio will transmit and receive in the **VERIFY/OPERATE** mode only. The radio is disabled and will not transmit or receive in the **M-LOAD**, **ERASE**, or **FMT.CHG** modes.

5. Function Selector switch - **MNL**.

6. **CHAN** switch - Set to Channel 20.

7. Frequency switches - Set to the first WOD element.

8. **T-TONE** switch - Toggle to **TONE** and release. Listen for a wavering tone. The first WOD element is now entered.

Repeat steps 5 thru 8 for presets 19 thru 15, in that order, substituting WOD elements for existing preset channels.

9. **CHAN** switch - Set to Channel 14.

NOTE

- If two or more WODs have the same date code, the radio recognizes only the latest date entered.

- MWODs must be linked with an associated day-of-month code. This date code element has been added to every operational and training segment in HAVE QUICK and need only be loaded when MWOD is used.

10. Frequency switches - Set to the applicable day of the month code. Code format is XAB.XXX, where A and B are the day of the month the WOD is used. X can represent any value. For example, if today were 26 June, then select 226.025 or 326.475.

11. **T-TONE** switch - Toggle to **TONE** and release. Listen for a double beep. One complete WOD is now loaded. To load more WODs, repeat steps 5 thru 11.

NOTE

The operational date is the current Day of Month and must be entered into Channel 1 so the radio can select the proper WOD.

Change 4   3-18.9

12. **CHAN** switch - Set to Channel 01.

13. Frequency switches - Set to the current Day of Month code. Code format is XAB.XXX, where A and B are the day of the month the WOD is used.

14. **T-TONE** switch - Toggle to **TONE** and release. Listen for a wavering tone.

15. Function Selector switch - **PRESET**.

16. **CHAN** switch - Set to Channel 20.

17. Frequency switches - Set to 220.000 MHz (Verify/Operate).

18. **LOAD** button - Lift preset frequency label cover and push **LOAD** button. Listen for single beep. Radio is now ready to receive TOD and then operate in the active mode.

3.7B.7 MWOD Loading using KYK-13 Keyfill Device

The KYK-13 has six channels and can hold up to six WODs. MWOD keying is supplied through cryptologic sources. Load the radio as follows:

1. **STATUS** button - Press; ensure the frequency/status LCD displays **VER/OP**.

2. Function Selector switch - **MNL**.

3. Preset Frequency Label Cover - Lift to reveal **FILL** Port Access connector.

NOTE

The fill cable for the KYK-13 may be used while loading MWOD information into the radio, but is not required.

4. KYK-13 - Set mode switch to **OFF/CHECK**.

5. KYK-13 - Install on **FILL** Port Access connector.

6. KYK-13 - Set mode switch to **ON**.

7. **STATUS** button - Press; ensure the frequency/status LCD displays **FILL**.

8. KYK-13 - Set address switch to applicable channel (1 through 6).

9. **LOAD** button - Press for two seconds. Listen for a series of beeps and confirm frequency/status indicator displays WOD OK. If frequency/status indicator displays **BAD**, the KYK-13 must be reloaded prior to repeating steps 4 through 9.

NOTE

The **CHAN** indicator steps down from memory location 20 to 14, then displays 01 while the KYK-13 is connected and turned on. This allows entry of the operational date information if required.

10. KYK-13 - Set to next applicable channel and repeat step 9. Observe that WOD OK is displayed on the frequency/status indicator after each WOD is loaded.

NOTE

If operational date is desired, proceed to step 11; if not desired, proceed to step 12.

11. **STATUS** button - Depress, if operational date entry is desired. Set date on frequency switches in XAB.XXX format. Toggle **T-TONE** switch to **TONE** position and release.

12. KYK-13 - Set mode switch to **OFF/CHECK** and remove KYK-13.

13. Preset Frequency Label Cover - Close.

The radio will return to its' previous mode prior to loading and both displays will return to the previous settings.

3.7B.8 Verifying MWOD is Loaded.

1. **STATUS** button - Press; ensure the frequency/status LCD displays **VER/OP**.

2. Function Selector switch - Set to **MNL**.

3. Frequency switches - Set to the day of the month to be verified. Code format is XAB.XXX, where A and B are the day of the month of the Date Tag associated with the WOD to check.

4. **CHAN** switch - Set to Channel 20. Switch momentarily to channel 19 and return to channel 20. A single beep verifies the MWOD with a matching day of the month code stored in memory. If a single beep is not heard, the code is not stored in memory. Repeat steps 3 and 4 to check other days of the month.

3.7B.9 MWOD Erase.

1. **CHAN** switch - Set to channel 20.

2. Function Selector switch - **PRESET**.

3. Frequency switches - Set to 220.050.

4. **LOAD** button - Lift preset frequency label cover and push **LOAD** button.

5. Function Selector switch - **MNL**.

6. **T-TONE** switch - Toggle to **TONE** and release. MWODs should now be erased.

7. Function Selector switch - **PRESET**.

8. Frequency switches - Set to 220.000.

9. **LOAD** button - Lift preset frequency label cover and push **LOAD** button. Close cover. The radio is now in the Verify/Operate mode/

3.7B.10 Alternate MWOD Erase.

1. Preset Frequency Label Cover - Open to reveal **ZERO** switch.

2. ZERO switch - Rotate counterclockwise, then return to normal position. **ERASE** should be displayed on the frequency/status indicator. All MWODs are now erased.

3.7B.11 Receiving Time of Day (TOD) and TOD Update 45 / 49A.

NOTE

A steady warning tone will be heard when the anti-jamming mode is selected and the TOD or a valid WOD has not been entered. A pulsating tone will be heard if an invalid net is selected.

1. Function Selector switch - **PRESET** or **MNL**.

2. **T-TONE** switch - Toggle to **T** and release.

3. TOD - Request from another station.

NOTE

If TOD is being automatically beaconed from another station, the first TOD message received within one minute of selected T position will be accepted.

a. TOD Send.

1. Function Selector switch - **PRESET** or **MNL**. Rotate frequency switches or **CHAN** switch to a predesignated frequency for TOD transmission.

2. **T-TONE** switch - Toggle to **TONE** position for two seconds and release.

3. Loading Operational Date.

4. **CHAN** switch - Set to Channel 20.

5. Function Selector switch - **PRESET**.

6. Frequency switches - Set to 220.025.

7. **LOAD** button - Lift preset frequency label cover and push **LOAD** button.

8. **CHAN** switch - Set to Channel 01.

9. Frequency switches - Set to the operational date in the format of XAB.XXX, where A and B are the day of the month of the Date Tag. X can be any value.

10. **T-TONE** switch - Toggle to **TONE** and release. Listen for a wavering tone.

11. Function Selector switch - **PRESET**.

12. **CHAN** switch - Set to Channel 20.

13. Frequency switches - Set to 220.000.

14. **LOAD** button - Lift preset frequency label cover and push **LOAD** button. Listen for a single beep.

b. Clock Start.

Press **TEST DISPLAY** switch while simultaneously pressing the **T-TONE** switch to the **T** position. To check if TOD is loaded, press the **T-TONE** switch to the **TONE** position for two seconds; two tones should be heard.

3.7B.12 Loading GPS Time with Single Word of Day (WOD) ▌51▐.

NOTE

The EGI must be operational to use GPS time via the EGI. The time on CDU line 1 of the ADMIN page cannot have an arrow by it. The absence of the arrow indicates that it is GPS time.

The following procedure is used for loading GPS time from the EGI to the Havequick radio with a single Word of Day (WOD).

1. Insert Word of Day.
2. Function Selector switch - **PRESET**.
3. Preset Channel Selector - 20. A single beep should be heard.
4. Preset Channel Selector - 19 - 15. A single beep should be heard 19 -16 and double beep at 15.
5. **HQ SYNC VAB** (on Data page) - Depress and release.
6. Function Selector switch - **Manual**.
7. Frequency Selector(s) - A000.0, A000.1, A000.2, A000.3, OR A000.4.

3.7B.13 Loading GPS Time with Multiple Word of Day (MWOD) ▌51▐.

The following procedure is used for loading GPS time from the EGI to the Havequick radio with a multiple word of Day (MWOD).

1. Insert Word of Day.
2. Mode Selector switch - **MAIN**.
3. **CHAN** switch - Set to 20.
4. Function Selector switch - **PRESET**.
5. Frequency switches - Set to 220.025 (MWOD LOAD).
6. **PRESET** button - lift preset frequency label cover and push **PRESET** button. Listen for single beep.
7. Function Selector switch - **Manual**.
8. **CHAN** switch -Set to 01.
9. Frequency switches - Set to applicable day of month code. Code format is XAB.XXX, where A and B are the day of the month the WOD is used. X can represent any value. For example, if today is 26 June, enter X26.XXX.
10. **TONE** button - Press **TONE** and release. Listen for a wavering tone.
11. Function Selector switch - **PRESET**.
12. **CHAN** switch - Set to 20.
13. Frequency switches - Set to 220.000 (Verify/ Operate).
14. **PRESET** button - lift preset frequency label cover and push **PRESET** button. Listen for single beep. Radio is now ready to receive TOD either via GPS or conventional means.
15. **HQ SYNC VAB** (on Data page) - Depress and release.
16. Function Selector switch - **Manual**.
17. Active Net - Select.

The **TONE** button can be pressed to verify radio has accepted GPS time, but time will be transmitted on tuned frequency. If the radio does not accept GPS time, verify the EGI is operational and the time on the **ADMIN** page is GPS time (i.e. no arrow by the time).

3.7B.14 Net Numbers. The net number is used in the anti-jamming mode in the same fashion as a radio frequency in the normal mode of operation. The net number enables multiple users to operate simultaneously on a non-interfering basis with other users while sharing a common WOD and TOD. There are three nets available to the operator:

(1). Frequency Managed A-Nets (FMA-NET).

(2). Training Nets (T-NET).

(3). Frequency Managed Training Nets (FMT NETS).

a. Frequency Managed A-Nets (FMA-NET). The FMA NET provides four frequency tables or hopsets. There are 1000 possible nets for each hopset. The frequency table to be used is determined by the geographical area of operation. One large hopset has been coordinated for use in NATO-Europe and another large hopset for employment in non-NATO countries. The net number begins with an "A" and is followed by three digits between 000 to 999. The last two digits designate the frequency table to be used. Nets are selected LAW ABB.BCC where:

(1). A= A (Active)

(2). BB.B= Desired Net

(3). CC = 00 basic HAVE QUICK A & B NETS.

= 25 for NATO-Europe

= 50 for non-NATO Europe

= 75 for future use.

b. Training Nets (T-NETS). Each Major Command is assigned a training WOD for daily training and radio maintenance. Training WODs may be loaded using a single WOD or MWOD methods. All training WODs are initialized with 300.0XX in channel 20. XX sets the frequency hop rate for the WOD in a SWOD only. In this training mode, the radio hops between the five frequencies loaded in with the WOD (locations 19 to 15) and five training nets are available. As shown below, a net number ending in 00 selects a training net.

(1). A00.000

(2). A00.100

(3). A00.200

(4). A00.300

(5). A00.400

c. Frequency Managed Training Nets (FMT-NET). To expand the number of training nets available to Have Quick users, 16 nets are available. To use the FMT nets, 16 frequencies have been loaded into the radio and are permanently stored in the radio memory. To use the FMT nets, a training WOD must be entered first. The FMT Nets are numbered A00.025 through A01.525. All six characters must be selected and the last two digits <u>must</u> be 25. Selection of an FMT NET greater than A01.525 or ending in 50 or 75 will result in a pulsating tone.

3.7B.15 Loading or Changing FMT NETS.

1. **CHAN** Switch - Set to Channel 20.

2. Function Selector Switch - **PRESET**.

3. Frequency Switches - Set to 220.075.

4. **LOAD** Button - Lift preset frequency label cover and push **LOAD** button. Listen for a single beep.

5. Function Selector Switch - **MNL**.

6. **CHAN** Switch - Set to appropriate memory location (Channels 20 to 5).

7. Frequency Switches - Set to select first frequency.

8. **T-TONE** Switch - Toggle to **TONE** and release.

Repeat steps 6 through 8 until all frequencies are loaded.

9. Frequency Switches - Set to 220.000 MHz when all frequencies have been loaded.

10. **CHAN** Switch - Set to Channel 20.

11. Function Selector Switch - **PRESET**.

12. **LOAD** Button - Lift preset frequency label cover and push **LOAD** button.

3.8 VOICE SECURITY SYSTEM TSEC/KY-58.

The TSEC/KY-58 (RCU), located in the right console, interfaces with the AN/ARC-186 and AN/ARC-201 radios to provide secure voice (ciphony) for these radios. The TSEC/KY-58 receives 28 vdc from the 28 vdc emergency bus through the **KY-28** circuit breaker on the pilot overhead circuit breaker panel.

Two operating modes are available: PLAIN mode for in-the-clear voice transmission or reception, and **C/RAD 1** (cipher) mode for secure radio transmission or reception.

NOTE

The RCU **POWER** switch must be set to **ON** and the **KY-28** circuit breaker must be IN before AN/ARC-136 or AN/ARC-201 radio communication (plain or ciphered) is possible.

3.8.1 Controls and Functions. Voice security system controls that require adjustment by the pilot include those on the Z-AHQ Power Interface Adapter, TSEC/KY-58 (located in the aft avionics bay), and the Z-AHP KY-58 RCU (located in the pilot right-hand console). Each of these devices are shown in figure 3-6. The function of each control and indicator is described in tables 3-5 through 3-7.

Table 3-5. Z-AHQ Power Interface Adapter Control and Indicator Functions

Control or Indicator	Function
BBV, DPV, BBN, DPN 4-position switch	Set according to type of radio being secured. Set to BBV for pilots VHF FM radio.
PTT button (push-to-talk)	Clears crypto alarm that occurs upon power up. Alarm can also be cleared by pressing any push-to-talk switch in the pilot compartment.
FILTER IN/OUT Selector	Prevents adjacent channel interference when using radios with channel spacing of 25 kHz. Must be set to IN for pilots AN/ARC-186(V) VHF-FM radio.
REM/LOC Switch	Sets the Z-AHQ to the local mode. Switch returns to REM (remote) position upon release, but equipment remains in local mode until any PTT is keyed.

TM 1-1520-238-10

Table 3-6. Z-AHP Remote Control Unit Control and Indicator Functions

Control or Indicator	Function
ZEROIZE switch (two-position toggle switch housed under a spring-loaded cover)	Use in an emergency to delete all crypto-net variables (CNVs) from KY-58 registers. Renders KY-58 unusable until new variables are loaded.
DELAY switch	Not used.
C/RAD 1/PLAIN Switch	Set switch to C/RAD 1 (cipher radio 1) to use secure voice. Set switch to PLAIN when operating radio in the clear.
Switch guard	Rotate to the left to prevent C/RAD1/PLAIN switch from accidentally being set to PLAIN.
MODE Switch	Set to OP (operate) to use pilots VHF radio in either the Ciphered or Plain mode. Set to LD (load) when installing (CNV) in the TSEC/KY-58 (TM 11-5810-262-12&P). Set to RV (receive variable) during manual remote keying (TM 11-5810-262-12&P).
POWER switch	Turns KY-58 on and off. Must be on (up) for operation in either plain or cipher mode.
FILL switch	Selects desired (CNV).

Table 3-7. TSEC/KY-58 Control and Indicator Functions

Control or Indicator	Function
VOLUME control	Sets audio level of pilots VHF-FM radio.
MODE switch	Set to P (Plain) to operate pilots VHF-FM radio in the clear. Set to C (Cipher) to operate pilot's VHF-FM radio in the ciphered (secure speech) mode. Set to LD (load) when installing (CNVs) in the TSEC/KY-58 (TM 11-5810-262-12&P). Set to RV (receive variable) during manual remote keying (TM 11-5810-262-12&P).
FILL connector	Used to load (CNVs) into the TSEC/KY-58 registers (TM 11-5810-262-12&P).
FILL switch	Pull knob and set to Z1-5 to zeroize (delete) (CNVs) in TSEC/KY-58 registers 1-5. Set to 1, 2, 3, 4 or 5 to select desired CNV. Pull knob and set to Z-ALL to zeroize (delete) (CNVs) in all TSEC/KY-58 registers. Zeroizing all registers renders TSEC/KY-58 unusable.
Power switch	Set to OFF to turn off both the TSEC/KY-58 and the pilots VHF-FM radio. Set to ON to operate the TSEC/KY-58 and pilots VHF-FM radio. The TD (time delay) position is not used.

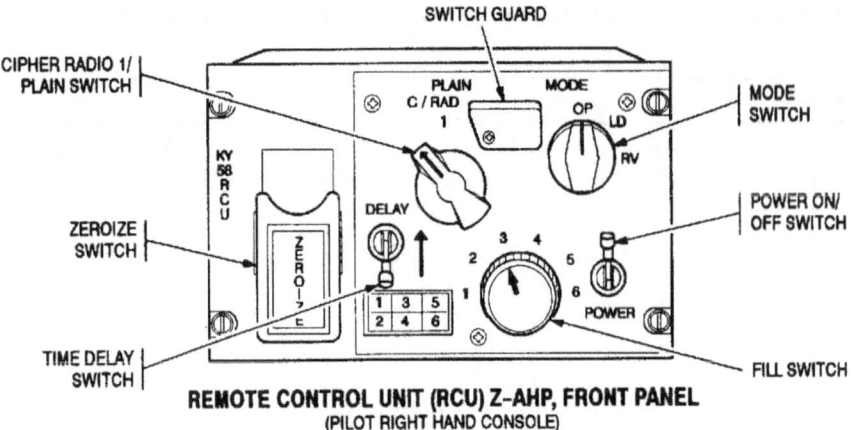
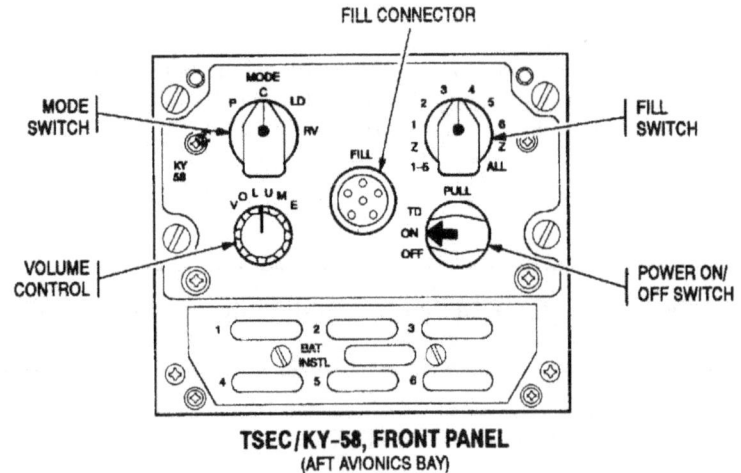
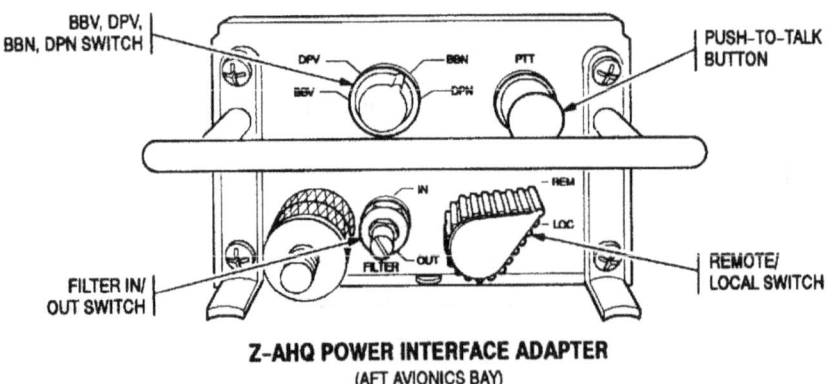

Figure 3-6. Voice Security System Equipment

TM 1-1520-238-10

3.8.2 Operating Procedures.

a. Preliminary Operation.

NOTE

Before the pilot VHF-FM radio may be operated in ciphered (secure voice) mode, it must be loaded with one or more (CNVS). Refer to TM 11-5810-262-12&P for complete details on loading these variables.

1. KY-58 **POWER** switch - **ON.**

2. KY-58 **MODE** switch - **C** (cipher).

3. KY-58 **VOLUME** control - As desired.

4. KY-58 Fill switch - Any numbered storage register position **(1-6).**

5. Power Interface Adapter 4- Position switch to **BBV**

6. Power Interface Adapter **FILTER** selector - **IN.**

7. RCU **DELAY** switch - Down (off) position.

8. RCU **C/RAD 1-PLAIN** switch - **C/RAD 1.**

9. RCU **MODE** switch - **OP.**

10. RCU Fill switch - Set to the proper (CNV).

11. RCU **POWER** switch - **On.**

NOTE

At this time you should hear an intermittent tone and background noise. The background noise is normal. The tone is a crypto alarm that must be cleared before the radio can be used. If step 12 does not clear the intermittent tone, double check steps 1 through 11. If necessary, refer to TM 11-5810-262-12&P.

12. **RADIO/ICS** switch - **RADIO** press and release (or the pilot push-to-talk floor switch). This should clear the crypto alarm.

b. Cipher Mode.

1. After steps 1 through 12 are complete, the radio is ready to transmit and receive secure speech in ciphered mode.

2. To transmit, press any push-to-talk switch. You may begin speaking following the beep.

c. Plain Mode.

1. **C/RAD 1/PLAIN - PLAIN.**

2. To transmit, press any push-to-talk switch. You may begin speaking immediately.

d. Automatic Remote Keying (AK).

NOTE

AK causes an old CNV in one of the registers to be replaced by a new one, or an empty register to be filled. Your net controller simply transmits the new CNV to your TSEC/KY-58.

1. Your net controller will contact you by using a secure voice channel, and tell you to wait for an AK transmission. You must not transmit during this period.

2. You will hear one or two beeps in your headset when the AK occurs.

e. Manual Remote Keying (MK).

NOTE

MK requires you to use the RCU to change CNVs.

1. The net controller will contact you by using a secure voice channel. He will tell you to stand by for a new or replacement CNV and that you will use an MK action.

2. Fill switch -6. Notify the net controller by radio when you have done this and stand by.

3. The net controller will tell you to set the **MODE** switch to **RV**. Notify the net controller when you have done this and stand by.

4. When notified by the net controller, set the fill switch to the storage position selected to receive the CNV. Notify the net controller when you have done this and stand by.

3-23

5. The net controller will ask you to listen for a beep. Wait two seconds.

6. **MODE** switch - **OP**.

If the MK operation was successful, the net controller will contact you via the new CNV. If the MK operation was not successful, the net controller will contact you by a clear voice (plain) transmission, tell you to set your fill switch to position 6, and stand by while the MK operation is repeated.

3.9 VOICE SECURITY SYSTEM TSEC/KY-28.

Not installed.

3.10 RADIO SET AN/ARC-201.

Radio set AN/ARC-201, consists of a panel mounted RT-1476/ARC-201 (transceiver) (fig 3-7). The receiver-transmitter is an airborne, (VHF) (FM) transceiver used as part of a series of (SINCGARS). The radio set does not have a guard channel. It has an (ECCM) frequency hopping mode (FH) of operation. The radio set provides communications of secure or plain voice over the frequency range of 30 to 87.975 Mz at 25 kHz intervals. A frequency offset tuning capability of -10 kHz, -5 kHz, +5 kHz and +10 kHz is provided for transmitting and receiving in the non-ECCM mode only. When used with the TSEC/KY-58 voice security system, the radio set is capable of transmitting and receiving clear voice or cipher mode communications. In the installed configuration there is no retransmission or homing capability. A memory holding battery is used to retain stored FH parameters, time and preset frequencies when primary power is removed from the transceiver. The radio set receives 28 vdc from the emergency dc bus through the **VHF FM** circuit breaker on the pilot overhead circuit breaker panel.

3.10.1 Antennas. The antennas used are the same as the AN/ARC-186 antennas.

3.10.2 Controls and Functions. Controls for the AN/ARC-201 are on the front panel of the unit (fig 3-7). A display provides operator interface, depending on switch positions and keyboard entries, for displaying manual and preset frequencies, offset frequencies, time, CUE, and transceiver status during self test. The function of each switch is described in table 3-8.

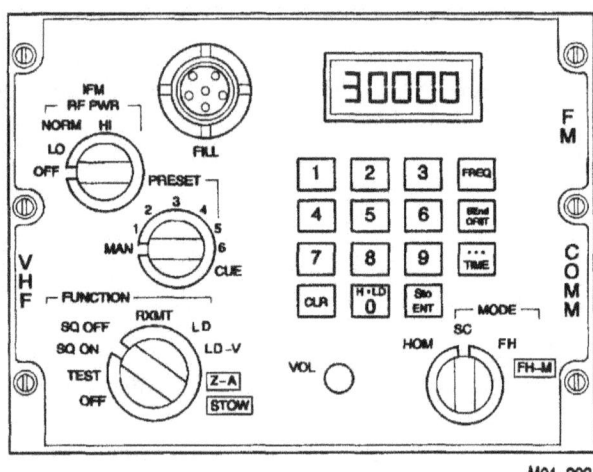

Figure 3-7. Control Panel AN/ARC-201

TM1-1520-238-10

Table 3-8. AN/ARC-201 Control Functions

Control	Function
FUNCTION switch	Selects basic operational condition of the transceiver.
OFF	Primary power is off. Memory holding battery power is on. Used during limited periods of inactivity when reloading FH parameters and preset frequencies is not desirable, e.g., between missions or overnight.
TEST	Self test of the transceiver and ECCM function.
SQ ON (Squelch On)	Power is applied to the transceiver and the squelch is enabled.
SQ OFF (Squelch Off)	Power is applied to the transceiver and the squelch is inoperative.
RXMT (Retransmit)	Not used.
LD (Load)	To load preset frequencies by normal keyboard entry, to fill ECCM net parameters and lockout channels, and to reset time.
LD-V (Load - variable)	To load the TRANSEC variable into the transceiver for use with ECCM.
Z-A (Zero All)	Pull and turn switch. Not an operational position. Zeros ECCM variables. Clears the TRANSEC variable to avoid a security compromise.
STOW	Pull and turn switch. Removes all power from the transceiver, including memory holding battery power. Used for extended storage.
MODE switch	Selects operational mode.
HOM (Homing)	Not used.
SC (Single Channel)	Selects single channel mode of operation. Operating frequency is selected by PRESET switch or keyboard entry.
FH	Selects (FH) mode of operation. PRESET switch positions 1 to 6 select frequency hopping net parameters.
FH M (Frequency Hopping - Master)	Pull and turn switch. Designates control station as the time standard for all communicating radios within a common net.

3-25

TM 1-1520-238-10

Table 3-8. AN/ARC-201 Control Functions - continued

Control	Function
PRESET switch	Selects specific predetermined operating conditions within the transceiver.
MAN (Manual)	In single channel mode, selects any operating frequency within the prescribed band, in 25 kHz increments, using the keyboard. Inoperative in FH modes.
Positions 1 through 6	In SC mode, preset frequencies are selected or loaded. In FH or FH-M mode, net parameters are selected.
CUE	Used by a non-ECCM radio to signal an ECCM radio. In SC mode, this is a seventh preset frequency.
FM RF POWER switch	Not used. Leave switch in OFF position.
Keyboard	To enter or display data, depending on the key switch actuated and the positions of the MODE and FUNCTION switches.
Numbers 1 through 9	To key in frequency, and load time information and frequency offsets.
FREQ	To display the current operating frequency during single channel manual or preset operation. To load manual and preset frequencies.
SEnd OFST (Offset)	To modify a single channel operating frequency which has been manually selected or preset for offsets of ± 5 kHz or ± 10 kHz or to initiate a transmission if a hopset or lockout set is in the holding memory and the MODE switch is set to FH-M.
TIME	To display or change the time setting maintained within each transceiver.
Sto ENT (Store Enter)	Initiates entry into transceiver of all valid complete entries by keyboard entry. Stores a received hopset or lockout set held in holding memory.
HoLD (Hopping Load) / O	Initiates transfer of ECCM parameters to transceiver. Enter O in the same manner as other keyboard numbers.
CLR (Clear)	Clears partial or erroneous entries.
VOL Control	Adjusts receiver volume.
FILL Connector	To enter ECCM variables from an external fill device.

3.10.3 Modes of Operation.

a. ECCM Mode. The ECCM mode provides a jam resistant capability by means of a (FH) technique that changes frequency many times a second. Automatic frequency changing in an apparently random manner provides the jam resistance of the radio. This capability permits communications in radio jamming environments. For uninterrupted successful communications, radios must be time synchronized with each other and share a common net.

(1) **Time Synchronization.** Time synchronization is provided by a clock inside the radio set. The operation of the (FH) net is time dependent and the accuracy of time, as it is registered in each radio set on a common net, is of significant importance. A relative standard or master must be established to prevent gradual time creepage during normal communications. One radio set will be designated as master. The time difference will be accommodated in the other radios, not the master radio. The net control station, or master, is designated as the time standard for all radio sets in the net.

(2) **Loading ECCM Parameters.** To use the ECCM function, TRANSEC variables and hopsets or lockouts are loaded into the transceiver by an external fill device connected to the **FILL** connector.

(3) **Signalling an ECCM Radio.** In a typical ECCM signaling operation, an operator with a non-ECCM radio set attempting a contact within an ECCM net places the **PRESET** switch to **CUE**. The ECCM radio display indicates **CUE** for 7 seconds and the ECCM radio operator hears a tone in his headset for 2 seconds each time the non-ECCM radio is keyed. The non-ECCM radio must be keyed for at least 4 seconds to insure reception by the ECCM radio. The ECCM radio is then switched to the single channel **CUE** frequency or some other predetermined frequency to establish contact.

b. Other Operational Modes. Depending on the settings of the operational controls, the radio set can be used for the following modes of operation:

(1) Two way clear voice (SC).

(2) Two way secure voice with TSEC/KY-58 installed (SC).

(3) Two way frequency hopping voice (FH or FH-M).

(4) Two way frequency hopping secure voice with TSEC/KY-58 installed (FH or FH-M).

3.10.4 Operating Procedures.

a. Manually Entering Frequency. When using the nine digits on the keyboard, entry of information is normally displayed digit-by-digit left to right on the display. Operation of the transceiver is altered only after complete, valid data is registered on the display and the **Sto ENT** button pressed within the 7 second time-out period. Illegal entries will not register on the display. Incomplete entries will not be accepted when the **Sto ENT** button is pressed.

Acceptance of valid data is signaled by a momentary blink of the display when the **Sto ENT** button is pressed. At this time, the transceiver acts upon the entry.

Partial or erroneous data can be erased at any time by pressing the **CLR** button, at which time the last readout will be cleared.

1. **FUNCTION** switch - **SQ ON** or **SQ OFF**, as desired.

2. **MODE** switch - **SC**.

3. **PRESET** switch - **MAN**.

4. **FREQ** button - Press. Frequency is displayed.

5. **CLR** button - Press. Display shows all bottom dashes.

6. Keyboard - Enter 5 digits for frequency.

7. **Sto ENT** button - Press. Display blinks momentarily.

b. Loading Preset Channels. Seven preset channels are available (**PRESET** positions **1** through **6** and **CUE**). These channels select discrete frequencies in non-ECCM modes and provide an analogous function in the (FH) ECCM mode. The function of the **PRESET** switch is two-fold. In a normal single-channel mode (MODE switch set to SC), either manual or preset frequencies are selected. In an ECCM frequency hopping mode (MODE switch set to **FH** or **FH-M**), nets are selected according to predetermined preloaded data. When a (FH) net is selected, the display will indicate **FH** followed by the corresponding valid (FH) data number. The CUE frequency can serve as a seventh preset channel in single-channel operation or as a special signaling frequency in the FH modes.

1. **FUNCTION** switch - **LD.**

2. **PRESET** switch - Number 1 through 6, as desired.

3. **MODE** switch - **SC.**

4. **FREQ** button - Press. Frequency is displayed.

5. **CLR** button - Press. Display shows all bottom dashes.

6. Keyboard - Enter 5 digits for frequency.

7. **Sto ENT** button - Press. Display blinks momentarily.

8. Repeat steps 1 through 7 for each desired preset channel.

c. **Setting a Frequency Offset (Non-ECCM Mode Only).**

NOTE

Frequency offset tuning is provided for transmitting and receiving in the non-ECCM mode only.

A single channel operating frequency, either manually selected or preset, can be offset by +10 kHz, +5 kHz, -5 kHz, or -10 kHz. The offsets are shown in the two right hand digits of the display.

1. **MODE** switch - **SC.**

NOTE

Negative offsets are indicated on the display by a negative sign (-) appearing in the center digit position. Positive offsets are indicated by no prefix. No offset is indicated by 00 in the two right side digits of the display.

2. **SEnd OFST** button - Press. Any current valid offset applied to the selected single channel frequency will be displayed.

3. **CLR** button - Press. Display shows all bottom dashes.

NOTE

Repeated pressing of **SEnd OFST** button will alternate between plus and minus offsets.

4. **SEnd OFST** button - Press if a negative offset is desired. A negative sign (-) will appear in the center digit position.

5. Keyboard - Enter **5** or **1** and **0**. The display indicates **05** or **10**.

6. When a valid offset is shown on the display, along with a minus or plus (no indication) sign, press **Sto ENT** button within 7 seconds of keyboard entry. Display momentarily blinks.

7. **FREQ** button - Press. The original operating frequency with the offset added or subtracted will be displayed.

8. To cancel an offset:

   a. **SEnd OFST** button - Press. Existing offset will be displayed.

   b. **CLR** button - Press. Display shows all bottom dashes.

   c. **HoLD** button - Press. 00 is displayed.

   d. **Sto ENT** button - Press. Operating frequency is returned to its non-offset condition.

d. **Setting Frequency Hopping.**

1. **MODE** switch - **FH** or **FH-M,** depending on radio set which is to be the master station.

2. **PRESET** switch - Select frequency hopping net parameters 1 through 6, as desired.

3. **FUNCTION** switch - **SQ ON** or **SQ OFF**, as desired.

e. **Setting Time for Frequency Hopping.**

1. **FUNCTION** switch - **LD.**

NOTE

- If an error occurs during any entry sequence, pressing the **CLR** button will erase the display and start the sequence over.

- Pressing the **TIME** button repeatedly will cause the display field to change as described.

2. **TIME** button - Press. Days are displayed.

3. **CLR** button - Press. Display shows all bottom dashes.

4. Keyboard - Enter new days digits.

5. **Sto ENT** button - Press. Display blinks momentarily.

6. **TIME** button - Press. Hours and minutes are displayed.

7. **CLR** button - Press. Display shows all bottom dashes.

NOTE

When hours and minutes are entered, and the **Sto ENT** button is pressed, minutes and seconds are displayed with seconds zeroed and new time started. This is to accommodate presetting and display of time prior to a time mark.

8. Keyboard - Enter new hours and minutes.

9. **Sto ENT** button - Press. Display blinks momentarily.

f. Loading ECCM Net Parameters and Lockout Channels.

1. **MODE** switch - **LD** or **LD-V**.

2. Connect fill device to the **FILL** connector.

3. **HoLD** button - Press.

g. Continous Self Test.

NOTE

The tests will be continually repeated until terminated by changing the **FUNCTION** switch position.

1. **FUNCTION** switch - **TEST**. Display shows E. After 3 seconds display changes to all **8s**.

2. After 3 seconds, display indicates GOOD or **FAIL** followed by a number to indicate the failed component as follows:

   a. **1** indicates failure of the transceiver.

   b. **3** indicates failure of the ECCM module.

   c. **7** indicates failure of interface to the transceiver.

   d. **8** indicates internal failure of control panel.

h. Other Continous Tests.

NOTE

The following tests are performed in any operational mode on a continuous basis with the results of the test either audible or visible.

(1) **Voltage Standing Wave Ratio (VSWR) Test.** This test is performed each time the transmitter is keyed and during the transmission. If VSWR exceeds 5 to 1, the audio sidetone will be inhibited.

(2) **Secure Mode Test.** This test is performed each time the transmitter is keyed if a TSEC/KY-58 is installed. A short beep indicates the TSEC/KY-58 is operating properly. A continuous tone indicates a faulty TSEC/KY-58.

(3) **ECCM Test.** When the **PRESET** switch is changed while operating in a (FH) mode, the fill data is examined and either FXXX, if a valid hopset has been loaded, or FILLn, if no hopset or an invalid hopset has been loaded, will be displayed. A built-in-test (BIT) is also run on the non-volatile random-access memory with the **FUNCTION** switch in the **Z-A** position.

Section III. NAVIGATION

3.11 INTRODUCTION.

The navigation systems of the AH-64 are divided into 3 major groupings: Stand Alone Radio Navigational Aids; a non-integrated navigation system; and a integrated navigation system. The stand alone radio navigation aids consist of the AN/ARN-89 or the AN/ARN 149(V)3 Automatic Direction Finder (ADF) Sets. The non integrated navigation system consists of the AN/ASN-128 or AN/ASN-137 Doppler Navigation Sets (DNS), the IP-1552G Computer Display Unit (CDU) and the Heading Attitude Reference System (HARS). In aircraft equipped with the non-integrated navigation system, the installed Doppler Navigation Set performs the navigation calculations. The integrated navigation system consists of the Embedded Global Positioning System (GPS) Inertial (EGI) unit, the Air Data Sensor Subsystem (ADSS), the HARS, the AN/ASN-137 DNS, the IP-1552G Computer Display Unit (CDU) and the navigation software module in the Fire Control Computer (FCC). In aircraft with the integrated navigation system, all navigation calculations are performed by the navigation software module in the FCC. A Data Transfer Unit (DTU) is used in the integrated navigation system to provide bulk loading of navigational data.

NOTE

During an electrical system malfunction and operating on **EMERG BATT** power, the HSI/RMI will not provide adequate indications to the station.

3.12 AUTOMATIC DIRECTION FINDER SET AN/ARN-89

Direction finder set AN/ARN-89 is an airborne, Low Frequency (LF), Automatic Direction Finder (ADF) radio that provides an automatic or manual compass bearing on any radio signal within the frequency range of 100 to 3,000 kHz. The ADF displays helicopter bearing relative to a selected radio transmission. On the pilot instrument panel, it is shown by the No. 2 bearing pointer of the Horizontal Situation Indicator (fig 2-9) On the CPG panel, it is shown on the bearing pointer of the Radio Magnetic Indicator (RMI) (fig 2-10). The ADF has three modes of operation that permit it to function as a Continuous Wave (CW) Automatic Direction Finder, a (CW) Manual Direction Finder, or as an Amplitude-Modulated (AM) broadcast receiver. Power to operate the set is provided through the **ADF** circuit breaker on the pilot overhead circuit breaker panel.

3.12.1 Antennas. The ADF antennas (fig 3-1) are located on the bottom center fuselage area, aft of the Doppler/Radar Altimeter antenna fairing. The ADF loop antenna is mounted under the aft fairing. The ADF sense wire antenna is supported between the aft fairing and a 7-inch standoff 12 feet aft of the fairing.

3.12.2 Controls and Function. Controls for the AN/ARN-89 are on the front panel of the C-7392 (fig 3-8) installed on the pilot right console. The function of each control is described in table 3-9.

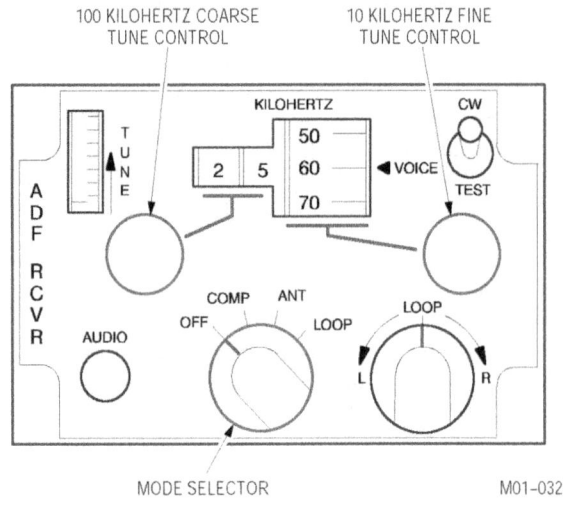

Figure 3-8. Control Panel AN/ARN-89

Table 3-9. AN/ARN-89 Control Functions

Control	Function
Mode selector switch	
OFF	Turns power to the set OFF.
COMP	Provides for operation as an ADF.

Table 3-9. AN/ARN-89 Control Functions - continued

Control	Function
ANT AM	Provides for operation as an receiver using the sense antenna.
LOOP	Provides for receiver operation as a manual direction finder using only the loop antenna.
-LOOP L-R control knob	Manually provides for left and right control of the loop when the mode selector switch is in the LOOP position. This control knob is spring-loaded to return to center.
AUDIO	Adjusts audio volume for station identification.
KILOHERTZ tune controls	
100 KHz coarse tune control knob	Tunes receiver in 100 KHz steps as indicated by first two digits of the KILOCYCLES indicator.
10 KHz fine tune control knob	Tunes receiver in 10 KHz steps as indicated by last two digits of KILOCYCLES indicator.
CW, VOICE, TEST switch	
CW (in the COMP mode)	Enables the tone oscillator to provide an audible tone for tuning to a (CW) station when the mode selector switch is at COMP.
CW (ANT or LOOP mode)	Enables the beat frequency oscillator to permit tuning to a CW station when the mode selector switch is at ANT or LOOP.
VOICE	Permits (LF) receiver to operate as a receiver when the mode selector switch is at any position.
TEST (in the COMP mode)	Provides for slewing of the loop through 180 degrees to check operation of the receiver in the COMP mode. This switch position is inoperative in LOOP and ANT modes.
TUNE meter	Indicates relative signal strength while tuning the set to a specific radio signal.
KILOCYCLES indicator	Indicates operating frequency to which receiver is tuned.

3.12.3 Operating Procedures.

a. Starting Procedure.
1. Mode selector switch - **COMP, ANT** or **LOOP.**
2. Frequency - Select.
3. CW-VOICE-TEST switch - CW or VOICE as desired.
4. CSC panel NAV A receiver select switch ON.
5. Fine tune control - Adjust for maximum upward indication on the TUNE meter.
6. AUDIO control - Adjust as desired.

b. **COMP Mode Operation.**
1. Mode selector switch - COMP. The horizontal situation indicator No. 2 bearing pointer and the radio magnetic indicator bearing pointer will display the magnetic bearing to the ground station from the helicopter as read against the integral compass card on each instrument.

TM 1-1520-238-10

2. Test the ADF as follows:

a. **CW-VOICE-TEST** switch – Set to **TEST** position. Check that both bearing pointers associated with the ADF rotate approximately 180°.

b. **CW-VOICE-TEST** switch – Release.

c. **ANT Mode Operation.**

 1. Mode selector switch – **ANT**

 2. Monitor receiver in headset.

d. **LOOP Mode Operation (manual direction finding).**

 1. Mode Selector switch – **LOOP**.

 2. Turn the **LOOP L-R** control to **L** (left) or **R** (right) to obtain an audio null and a **TUNE** indicator null. Check either ADF bearing pointer for a display of magnetic bearing to or from the station. The two null positions may be as much as 180° apart in this mode.

NOTE

During an electrical system malfunction and operating on **EMERG BATT** power, the HSI/RMI will not provide adequate indications to the station.

3.13 AUTOMATIC DIRECTION FINDER SET AN/ARN-149(V)3.

ADF set AN/ARN-149 is an airborne, LF, ADF radio that provides an automatic compass bearing on any radio signal within the frequency range of 100 to 2199.5 kHz. The ADF displays helicopter bearing relative to a selected radio transmission. On the pilot instrument panel, it is shown by the horizontal situation indicator (HSI) No. 2 bearing pointer (fig 2-9). On the CPG panel, it is shown on the RMI bearing pointer (fig 2-10). The ADF operates in the ANT (audio only) and ADF modes. It also has a self test mode. In addition, a submode provides the capability of identifying keyed CW signals. The ADF receives 28 VDC from the emergency DC bus through the ADF circuit breaker on the pilot overhead circuit breaker panel. The **HARS AC** circuit breaker on the pilot overhead circuit breaker panel provides power for the 26 VAC reference voltage used by the ADF receiver.

3.13.1 Antenna. The ADF antenna (fig 3-1) is located on the bottom center of the fuselage under the Doppler/Radar Altimeter (DRA) fairing.

3.13.2 Controls and Functions. Controls for the AN/ARN-149 are on the front panel of the ADF control panel (fig 3-9) located on the pilot right console. The function of each control is described in table 3-10.

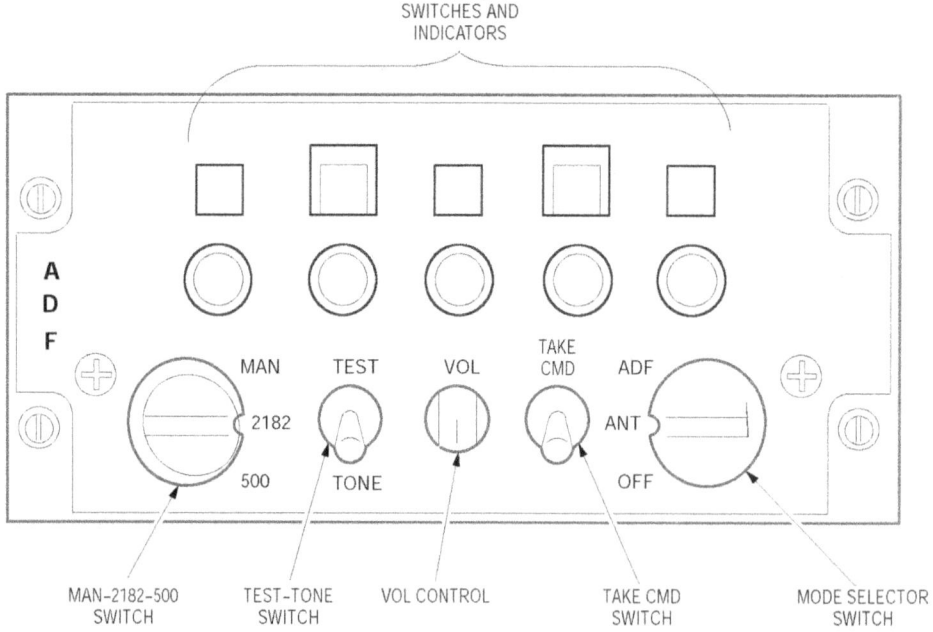

Figure 3-9. AN/ARN-149(V)3 ADF Control Panel

3-32 Change 6

Table 3-10. AN/ARN-149(V)3 ADF Control Panel Control Functions

Control	Function
Mode selector switch	
OFF	Removes power from the ADF.
ANT	Provides for operation as an AM receiver.
ADF	Provides both ADF and AM receiver operations.
TARE CMD switch	Not used.
VOL control	Controls the audio volume in 12 discrete steps.
TEST-TONE switch	Selects the submode of operation. If TEST is selected, the HSI No. 2 bearing pointer and RMI pointer momentarily move 90 degrees as a self-test. If TONE is selected, the normal audio is replaced by a 1 kHz tone for tuning to a (cw) station.
Frequency switches and indicators	Control and indicate the selected operating frequency when the MAN-2182-500 switch is in the MAN position. 2182 kHz and 500 kHz are international distress frequencies.

NOTE

2182 kHz and 500 kHz are international distress frequencies.

MAN-2182-500 switch	
MAN	Enables the frequency switches for manual frequency selection.
2182	2182 kHz is the operating frequency.
500	500 kHz is the operating frequency.

3.13.3 Modes of Operation. The AN/ARN-149 ADF system has two functional modes and a submode as follows:

  a. ADF Mode. In this mode, the system functions as an ADF that provides a relative bearing-to-station to the HSI No. 2 bearing pointer and the RMI pointer.

b. ANT Mode. In this mode, the system functions as an audio receiver, providing only an audio output of the received signal.

c. Tone Submode. This submode may be chosen in either ADF or ANT mode of operation. In the tone submode, the system provides a 1 kHz audio output tone when a signal is being received to identify keyed cw signals. This submode can be used when normal audio reception is insufficient to identify the station; the cw coded signal can be used to identify the station.

3.13.4 Operating Procedures.

a. ANT (audio only) Operation.

1. Mode selector switch - **ANT**.

2. **MAN-2182-500** switch - As desired. If **MAN** is selected, set the desired frequency with the five frequency switches.

3. **VOL** control - As desired.

4. If cw operation is desired, **TEST-TONE** switch - **TONE**.

b. ADF Operations.

1. Mode selector switch - **ADF**.

2. **MAN-2182-600** switch - As desired. If **MAN** is selected, set the desired frequency with the five frequency switches.

3. **VOL** control - As desired. The HSI No. 2 bearing pointer and the RMI pointer will indicate relative bearing to the selected ground station.

4. If cw operation is desired, **TEST-TONE** switch - **TONE**.

c. Self-Test Operation.

NOTE

The ADF system must be in ADF mode and receiving a valid ground station signal to perform the self-test.

1. Set to ADF mode (para. d.2). Note position of HSI No. 2 bearing pointer and RMI pointer.

2. **TEST-TONE** switch - Momentarily **TEST**.

3. HSI No. 2 bearing pointer and the RMI pointer rotate 90° from their original positions and then return to their original positions.

3.14 NON-INTEGRATED NAVIGATION SYSTEM

WARNING

The non integrated navigation system does not meet FM requirements for use as a primary navigation system for IFR operations in IMC.

The non-integrated navigation system consists of a Heading Attitude Reference System (HARS) and either the AN/ASN-128 or AN/ASN-137 Doppler Navigation Set (DNS). The non-integrated navigation system is not connected to, and operates independently of, the Mission System 1553 multiplex (MUX) bus. The MUX bus is responsible for passing the DNS velocity data from the ARINC bus to the HARS through special hardware interfaces in the CPG and DASE MRTUs. The MUX bus is also responsible for tracking/filtering MAGVAR changes that are calculated in the DNS and passing these changes to the HARS via the same special interfaces. In this configuration the HARS is dependent upon receiving DNS velocity for internal velocity damping and maintaining proper heading and/or attitude, and the DNS is dependent upon receiving HARS heading/attitude in order to perform the navigation calculations. Additionally, the HARS is dependent upon proper MAGVAR updates for driving the HSI, RMI, and DNS magnetic heading inputs. The MUX bus controller (FCC or BBC) uses the HARS inertial data outputs to drive all the inertial symbology presented on the video displays. This includes the heading tape, velocity vector, acceleration cue, horizon line, hover position box and trim ball. The Hover Position Box will drift during a stationary hover, and conversely, the aircraft will drift if the aircraft is flown to hold the Hover Position Box centered on the display. The amount of drift may reach as much as 21 feet per minute, and the drift may be random in nature. This drift is caused by HARS velocity errors, which are strongly influenced by the DNS velocity characteristics at these hover velocities. The following paragraphs provide a description of the HARS and DNS internal operations and the controls and displays used by the crew to operate the non-integrated navigation system.

3.14.1 Heading Attitude Reference System (HARS).

a. System Description

The HARS is a doppler aided strapdown inertial reference system. The HARS provides all attitude, velocity, and acceleration data for the helicopter. The HARS aligns itself by adjusting its own vertical axis to coincide with the earth's gravity vector and by measuring the rotational speed of the earth about its inertial axis.

By aligning the vertical axis with the earth's gravity, the HARS is able to determine the UP direction. In measuring the earth's rotational speed the HARS is able to determine which direction is East; North is then 90 degrees counter-clockwise from East. Any change of heading or movement of the helicopter during alignment while on the ground will disrupt accurate measurements and calculations resulting in an alignment error. The only corrective action is for the helicopter to be returned to a stationary condition and the HARS realigned. Rotor vibrations at 100% Nr have little effect on the accuracy of the HARS alignment. Inflight, the HARS maintains alignment by continuously gyrocompassing and estimating system errors. Inflight alignments or restarts require valid doppler data. The HARS may be aligned using one of four methods. The accuracy of the HARS alignment is dependent on the method chosen. The methods and accuracies are discussed in Table 3-11.

The HARS uses doppler velocities to damp the drift in the inertial data. Both the doppler and inertial velocities are combined by the Kalman filter program in the HARS to take advantage of the best of each. If doppler velocities are not valid (memory or malfunction), the HARS will reject the doppler data and function in a free inertial mode. In free inertial, the heading and velocities will drift in a sinusoidal oscillation called a Schuler period; approximately 84 minutes. The HARS free inertial condition is signaled to the flight crew by velocity vector flashing. The most likely cause for the HARS to reject doppler data is that the doppler is in memory. The doppler normally returns valid data over flat terrain. Over areas of tall grass or water, doppler data inconsistencies can develop resulting in a memory condition and invalid doppler data.

Over mountainous terrain, doppler data is better than over grass or water, but not as good as over flat terrain. The HARS will reject doppler velocities as long as the memory or malfunction condition exists. In addition, if the free inertial condition exists too long, the HARS inertial velocities will drift enough so that when the doppler data again becomes valid, the HARS will continue to reject the doppler velocities because they are no longer within the capture window of the Kalman filter. If this occurs, the available corrective actions are limited. The pilot can slow the helicopter to less than 40 KTAS in an attempt to let the HARS Kalman filter recapture the doppler velocities. If this fails, the only remaining corrective options are either to land the helicopter and when stationary place the HARS control switch in NORM to cage the HARS inertial velocities to zero, or to attempt an in-flight alignment (restart) of the HARS. There is always the option to continue flight in the free inertial condition, realizing that everything that uses the HARS data (HAS, flight symbology, navigation, and fire control) will be degraded to the extent that the HARS has drifted and will continue to drift.

The HARS computes error estimates and accelerometer biases during flight. This data is stored by the HARS as mission data memory on shutdown. The mission data memory allows the HARS to maintain a running calibration of its internal instruments. If, however, the HARS has experienced more than 12 minutes total of free inertial since it was turned on it will not update the mission data memory on shutdown. The mission data memory can be most easily corrupted by moving the helicopter during alignment and not realigning before flight. Extended free inertial and corrupt mission data memory are the two primary causes for inaccurate navigation in the non integrated system.

HARS accomplishes internal bit and temperature stabilization (for approximately 90 seconds) prior to initiating alignment. The status of the HARS is continuously monitored by the FD/LS; the on-command FD/LS test (test 05 HARS) will fault isolate.

The HARS receives 28 vdc from the No. 3 essential dc bus through the **HARS DC** circuit breaker and 115 vac from the No. 1 essential ac bus through the HARS AC circuit breaker; both circuit breakers are on the pilot overhead circuit breaker panel.

b. Controls and Functions. Control of the HARS is provided by the HARS mode selector switch. Control and indicator functions of the HARS mode selector switch are described in table 3-11.

The HARS control panel (fig 3-10) is located on the pilot lower right instrument panel. The control panel has a mode selector switch with four positions: OFF, OPR and two ALIGN positions: FAST and NORM. Signals are sent to the MUX and doppler for use by the fire control computer and other systems such as DASE, navigation, stabilator, and symbology. The VDU (fig 4-2) and the HDU (fig 4-9) display this information to the pilot.

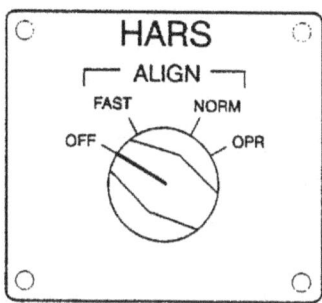

Figure 3-10. HARS Control Panel
NOTE
*. Loss of the heading tape during HARS alignment indicates a fault occurred with the HARS during alignment. Check the FD/LS.
*. If any helicopter movement occurs (heading or position) with the HARS switch in either of the ALIGN positions, the HARS shall be turned off and realigned without moving the helicopter.
*. Whenever the helicopter is on the ground and will not be moved for longer than one minute, set the HARS switch to NORM. This will prevent the HARS inertial velocities from drifting. The HARS switch shall be set to OPR prior to moving the helicopter.

TM 1-1520-238-10

Table 3-11. HARS Control Functions and Alignment Methods

Method/Control	How to Initiate	Requirements/Function	Effects On Accuracy
Off		Turns HARS off	
Normal	Switch to NORM	PPOS, MV and spheroid data entered through SP1 on data entry keyboard (DEK) prior to initiating alignment	Best accuracy attained by permitting completion of HARS alignment prior to starting the engines. Some degradation to alignment accuracy will occur if the engines are started prior to the HARS completing its alignment. This degradation will not occur if one engine is started and the rotor speed established at 100% within one minute and forty-five seconds after placing the HARS switch to NORM.
Stored Heading	Switch to FAST	Helicopter shall not have been moved since HARS was shut down	Uses data from previous HARS alignment and flight. Accuracy is dependent on that data. Alignment time is the shortest of all methods (approximately 90 to 120 seconds). Accuracy is not affected by engine starts.
FAST	Switch to FAST	Same as normal alignment	Least accurate method. Fastest method when stored heading alignment is not possible.
Inflight Restart	Switch to OFF, then to OPR Recycle doppler if necessary	AN/ASN-128 doppler MAL light must be OFF OFF, then to OPR.	If the HARS loses alignment in flight, the DASE will disengage. Switch the HARS After approximately 90 seconds, the HSI HDG flag will disappear and the DASE may be reengaged. The heading will appear to oscillate as the HARS attempts to align. The heading accuracy should be within 3 degrees after five minutes. Navigation performance will be degraded significantly.
OPR	Switch to OPR	Operating mode. Switch shall be in OPR prior to moving the helicopter.	

3.14.2 Required Navigation Data. The non integrated navigation system uses two components of data for proper operation: magnetic variation (MAGVAR) and spheroid (SPH). The HARS aligns with inertial North. MAGVAR is used to correct the inertial heading to one referenced to magnetic North. The symbolic heading tape, HSI, and RMI will all indicate magnetic heading. MAGVAR is entered in two places. Using the DEK, the present position MAGVAR on page 2 of Spare Position 1 (**SP1**) shall be checked and entered, if necessary, during the AFTER STARTING APU-CPG checklist. The second place that MAGVAR is input is in the DNS whenever coordinate data is entered. The spheroid is used to define the geographic reference frame that is used for the coordinate data. Spheroid is entered in 2 places. Using the DEK, the spheroid on page 2 of SP1 shall be checked and entered, if necessary during the AFTER STARTING APU-CPG checklist. The second

3-36 Change 3

place that spheroid is input is the DNS. Only 1 spheroid can be utilized by the DNS. The DNS PPOS and destination spheroids must be the same. The DEK SP1 and DNS PPOS spheroid must be the same.

3.14.3 Doppler Navigation Sets.

NOTE

The non integrated navigation system uses one of two Doppler Navigation Sets (DNS): AN/ASN-128 or AN/ASN-137 for navigation. The DNS which is installed may be determined by which Computer Display Unit (CDU) is installed in the CPG right-hand console. The CDU (fig 3-11) identifies those helicopters equipped with the AN/ASN-128 DNS. The CDU (fig 3-12) are those helicopters equipped with the AN/ASN-137 DNS. The CDU used with the AN/ASN-137 DNS is not part of the DNS, but is used to communicate with the DNS.

The DNS provides doppler velocity, and navigational position and steering information for the helicopter. The DNS is the navigator in the non integrated system.
The HARS provides attitude reference (heading, pitch, and roll) to the DNS. Navigational accuracy may degrade at altitudes above 10,000 feet AGL, and at pitch and roll angles greater than 30 degrees. The DNS is capable of providing position readouts in either the Military Grid Reference System (MGRS) form of Universal Transverse Mercator (UTM) or in latitude and longitude (LAT/LONG) coordinates. Navigation calculations are accomplished by the DNS in LAT/LONG and then converted to UTM (if necessary) for display purposes. Coordinate data may be entered in either coordinate format. The AN/ASN-128 DNS has 10 internal storage locations for coordinate data. The AN/ASN-137 DNS has 20 storage locations. Primary power to operate the DNS is provided through the **DPLR** circuit breaker on the pilot overhead circuit breaker panel. In addition, the AN/ASN-137 DNS also uses power provided through the **MUX FAB R** circuit breaker on the CPG No. 1 circuit breaker panel.

3.14.4 DNS Antenna. The AN/ASN-128 and AN/ASN-137 DNS use a common antenna (RT-1193A). The DNS antenna is located in a combined antenna/ra-dome and (RT) housing protected by the fairing on the bottom center fuselage area as shown in figure 3-1.

3.14.5 AN/ASN-128 DNS Controls and Displays. The controls and displays are on the front of the CDU CP1252 (fig 3-11) located on the CPG right console. The function of each control is described in table 3-12.

3.14.6 Modes of Operation. The three basic modes of operation are as follows:

a. Test Mode. The **TEST** mode contains two functions: **LAMP TEST** mode, in which all display segments are lit, and **TEST** mode, in which system operation is verified. In the **LAMP TEST** mode. system operation is identical to that of navigate mode except that all lamp segments and the **MEM** and **MAL** indicator lamps are lighted to verify their operation. In **TEST** mode, the system antenna no longer transmits or receives electromagnetic energy; instead, self-generated test signals are inserted into the electronics to verify operation. System operation automatically reverts into the backup mode during test mode. Self-test of the DNS is done using Built-In-Test Equipment (BITE) and all units connected and energized for normal operation. Self-test isolates failures to one of the three units of the DNS. The CDU (except for the keyboard and display) is on a continuous basis, and any failure is displayed by turn-on of the **MAL** indicator lamp on the CDU. The signal data converter and (RTA) are tested by turning the **MODE** selector switch to **TEST**. Failure of these components is displayed on the CDU by turn-on of the MAL indicator lamp. Identification of the failed unit is indicated by a code on the display panel of the CDU.
Continuous monitoring of the signal data converter and **RTA** is provided by the **MEM** indicator lamp. The **MEM** indicator lamp will light in normal operation when flying over smooth water. However, if the lamp remains lighted beyond 10 minutes over land or rough water, there is a malfunction in the doppler set. Then, to determine the nature of the malfunction, the operator should turn the **MODE** selector switch to **TEST**. As the keyboard is used, operation is verified by observing the alphanumeric readout.

TM 1-1520-238-10

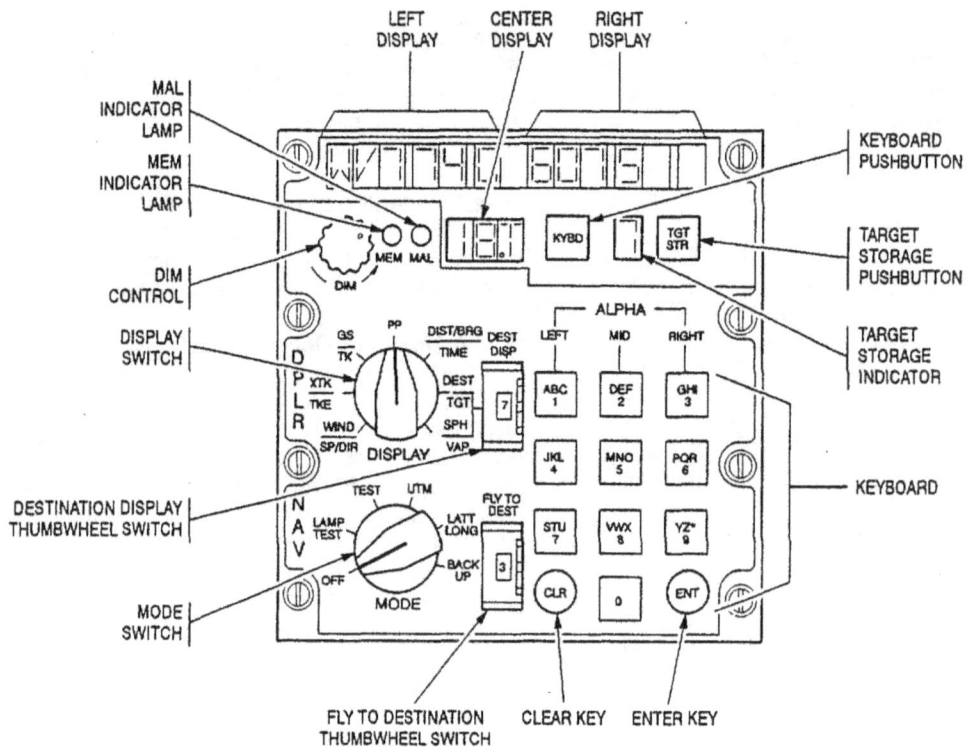

Figure 3-11. CDU CP1252, Used With AN/ASN-128 DNS

Table 3-12. AN/ASN-128 Control and Indicator Functions

Control/Indicator	Function
MODE Selector	
OFF	Turns off electrical power to the set.
LAMP TEST	Checks the operation of all lamps.
TEST	Initiates a (BIT) exercise for the DNS.
UTM	Selects (UTM) as the navigational mode of operation.
LATT/LONG	Selects LAT/LONG as the navigational mode of operation.
BACKUP	Places navigation set in true airspeed plus remembered wind mode of operation. If true airspeed is not available, places navigation set in remembered velocity mode of operation.
DISPLAY selector	Selects navigation data for display.
WIND SP/DIR (Left Display)	Windspeed in kilometers per hour (km/h).

3-38 Change 3

Table 3-12. AN/ASN-128 Control and Indicator Functions - continued

Control/Indicator	Function
(Right Display)	Wind direction relative to true north (degrees) All references to headings or track angle degrees are referred to magnetic north.
XTK-TKE (Left Display)	Distance crosstrack (XTK) of initial course to destination in km and tenths of a km.
(Right Display)	Track angle error (TKE) in degrees displayed as right or left of bearing to destination.
GS-TK (Left Display)	Ground speed (GS) in km/hr.
(Right Display)	Track angle (TK) in degrees.
PP (MODE switch set to UTM) (Center Display)	Present position UTM zone.
(Left Display)	Present position UTM area square designator and easting in km to nearest ten meters.
(Right Display)	Present position UTM area northing in km to nearest ten meters.
PP (MODE switch set to LATT/LONG) (Left Display)	Present position longitude in degrees, minutes, and tenths of minutes.
(Right Display)	Present position latitude in degrees, minutes, and tenths of minutes.
DIST/BRG-TIME (Center Display)	Time to destination selected by FLY TO DEST (in minutes and tenths of minutes).
(Left Display)	Distance to destination selected by FLY TO DEST (in km and tenths of a km).
(Right Display)	Bearing to a destination selected by FLY TO DEST (in degrees).
DEST-TGT (Mode switch set to UTM) (Center Display)	UTM zone of destination selected by DEST DISP thumbwheel.
(Left Display)	UTM area and easting of destination set on DEST DISP thumbwheel.
(Right Display)	Northing of destination set on DEST DISP thumbwheel.
DEST-TGT (Mode switch set to LATT/LONG) (Left display)	Latitude (N 84° or S 80° max.) of destination set on DEST DISP thumbwheel.

Change 3 3-39

Table 3-12. AN/ASN-128 Control and Indicator Functions - continued

Control/Indicator	Function
Right Display)	Longitude of destination set on DEST DISP thumbwheel.
SPH-VAR	
(left Display)	Spheroid code of destination set on DEST DISP thumbwheel.
(Right Display)	Magnetic variation (in degrees and tenths of degrees) of destination set on DEST DISP thumbwheel.
MEM Indicator Lamp	Lights when radar portion of navigation set is in nontrack condition.
MAL Indicator Lamp	Lights when navigation set malfunction is detected by built in self-test.
DIM Control	Control light intensity of display characters.
Left, Right, and Center Display Lamps	Lights to provide data in alphanumeric and numeric characters, as determined by setting of DISPLAY switch, MODE switch, and operation of keyboard.
Target Storage Indicator	Displays destination number (memory location) in which present position will be stored when TGT STR pushbutton is pressed.
TGT STR Pushbutton	Stores present position data when pressed.
KYBD Pushbutton	Used in conjunction with the keyboard to allow data to be entered into the computer. Also lights up keyboard when pushed the first time.
DEST DISP Thumbwheel switch DISPLAY	Destination display thumbwheel switch is used along with DEST-TGT and SPH-VAR position of switch to select destination whose coordinates or magnetic variation are to be displayed or entered Destinations are 0 though 9, P (Present Position) and H (Home).
Keyboard	Used to set up data for entry into memory. When the DISPLAY switch is tuned to the position in which new data is required and the KYBD pushbutton is pressed, data may be displayed on the appropriate left, right, and center display. To display a number, press the corresponding key or keys (1 through 0). To display a letter, first depress the key corresponding to the desired letter Then depress a key in the left, middle, or right column, corresponding to the position of the letter on the key. Example: to enter an L, first depress L, then 3, 6, or 9 in the right column.
FLY-TO DEST Thumbwheel switch information is desired	Selects the destination for which XTK/TKE and DIST/BRG/TIME are displayed when the DISPLAY switch is turned to either of these positions from which steering Destinations are 0 through 9, and H (Home).
ENT key	Enters data set up on keyboard into memory when pressed.
CLR key	Clears last entered character when pressed once. When pressed twice, clears entire display panel under keyboard control.

b. **Navigate Mode.** In the navigate mode **(UTM** or **LATT/LONG** position of the **MODE** selector switch), power is applied to all system components, and all required outputs and functions are provided. Changes in present position are computed and added to initial position to determine the instantaneous latitude/longitude of the helicopter. Destination and present position coordinates can be entered and displayed in **UTM** and latitude/longitude. At the same time, distance, bearing and time-to-go to any one of ten preset destinations are computed and displayed as selected by the **FLY TO DEST** thumbwheel.

c. **Backup Mode** In this mode, remembered velocity data are used for navigation. The operator can insert ground speed and track angle with the keyboard and the display in **GS-TK** position. This remembered velocity data can be manually updated through use of the keyboard and CDU **DISPLAY** selector switch in **GS-TK** position. When **GS-TK** values are inserted under these conditions, navigation continues using only these values.

3.14.7 Methods of Operation. Methods of operation are as follows

a **Window Display and Keyboard Operation.** In all data displays except UTM coordinates, the two fields are the left and right display windows. In UTM coordinates displays, the first field of control is the center window and the second field is the combination of the left and right displays. When pressing the **KYBD** pushbutton, one or other of the fields described above is under control. If it is not desired to change the display in the panel section under control, the pilot can advance to the next field of the display panel by pressing the **KYBD** pushbutton again. The last character entered may be cleared by pressing the **CLR** key. That character may be a symbol or an alphanumeric character. However, if the **CLR** key is pressed twice in succession, all characters in the field under control will be cleared, and that field will still remain under control.

b. **Data Entry**. To enter a number, press the corresponding key. To enter a letter, first press the key corresponding to the desired letter. Then press a key in the left, middle, or right column corresponding to the position of the letter on the pushbutton. For Example: To enter an **L**, first press **L**, then either **3**, **6**, or **9** in the right column. The computer program is designed to reject unacceptable data (for example, a UTM area of WI does not exist and will be rejected). If the operator attempts to insert unacceptable data, the display will be blank after **ENT** is pressed.

c. **Starting Procedure**.

(1) **Lamp Test**
1. **MODE** selector switch **LAMP TEST**. All display segments and indicator lamp should be on.

2 **DIM** control Turn fully clockwise, then fully counterclockwise, and return to full clockwise. All segments of the display should alternately glow bright, go off, and then glow bright.

(2). **Tes**t.

1. **MODE** selector switch **TEST**.

NOTE
• Ignore the random display of alpha and numeric characters which occurs during the first 15 seconds. Also ignore test velocity and angle data displayed after the display has frozen. A successful test cannot be accomplished until HARS has completed BIT.

• If the **MAL** lamp lights during any mode of operation except **LAMP TEST**, the CDU **MODE** switch should be turned first to **OFF** and then to **TEST** to verify the failure. If the **MAL** lamp remains on after recycling to **TEST**, enter the failure of the DNS on DA Form 2408-13-1 in the aircraft log book.

• After approximately 15 seconds, one of the following displays in table 3-13 will be observed in the left and right displays:

TM 1-1520-238-10

Table 3-13. AN/ASN-128 Window Displays

DISPLAY		REMARKS
LEFT	RIGHT	
GO	No display Display blank is (normal).	If right display is blank, system is operating satisfactorily.
GO	P	If right display is P, then pitch or roll data is missing, or pitch exceeds 90 deg. In this case, pitch and roll in the computer are both set to zero and navigation continues in a degraded operation. Problem may be in the HARS or helicopter cabling.
		NOTE If the TEST mode display is BU, MN or NG, the MODE switch should be recycled through OFF to verify that the failure is not a momentary one. If the TEST mode display is BU or MN, the data entry may be made in the UTM or LATT/LONG mode: but any navigation must be carried on with the system in the BACKUP mode
BU	C,R,S, or H followed by a numeric code	A failure has occurred and the system has automatically switched to a BACK-UP mode of operation as follows: 1. If no true airspeed is available, last remembered velocity is being used for navigation. 2. The operator has the option of turning the MODE switch to BACKUP and entering the best estimate of ground speed and track angle. 3. If true airspeed is available, true airspeed plus remembered wind is used for navigation. **NOTE** The operator has the option of turning the MODE switch to BACKUP and entering his best estimate of wind speed, direction, ground speed, and track angle The operator should update present position immediately because it is possible that significant navigation errors may have accumulated.
MN	C,R,S, or H followed by a numeric code	A failure has occurred and the BACKUP mode, used for manual navigation (MN), is the only means of valid navigation. The operator may use the computer as a dead reckoning device by entering ground speed and track data. The operator should update present position immediately because it is possible significant navigation errors may have accumulated.
NG	C,R,S, or H followed by a numeric code	A failure has occurred in the system and the operator should not use the system.
EN		The 9 V battery has failed All stored data must be reentered. This display may be cleared by pressing the KYBD pushbutton.

3-42 Change 3

d Entering UTM DATA. Enter the following initial data (para 3.14.3.b) before navigating with the doppler:

(1). Spheroid of operation, when using UTM coordinates.

(2). UTM coordinates of present position zone, area, easting (four significant digits), and northing (four significant digits). Latitude/ longitude coordinates may be used.

(3). Variation of present position to the nearest one-tenth degree.

(4). Coordinates of desired destination 0 through 5 and H (6 through 9 are normally used for target store locations but may also be used for destinations). It is not necessary to enter all destinations in the same coordinate system.

NOTE

It is not necessary to enter destinations unless steering information is required; unless, it is desired to update present position by overflying a destination; or, unless a present position variation computation is desired (paragraph 3.14.7.g.). If a present position variation running up-date is desired, destination variation must be entered. The operator may enter one or more destination variations to effect the variation update. It is not necessary for all destinations to have associated variations entered.

(5). Variations of destinations to be nearest one-tenth degree.

e. Entering Spheroid and/or Variation.

1. **MODE** selector switch - **UTM (LATT/ LONG**, or **BACKUP** may also be used).

2. **DISPLAY** selector switch - **SPH-VAR**.

3. **DEST DISP** thumbwheel - **P**, numeral, or **H**, as desired.

4. **RYBD** pushbutton Press. Observe display freezes and **TGT STR** indicator blanks. Press **KYBD** pushbutton again and observe left display blanks. If no spheroid data is to be entered, depress **KYBD** pushbutton again and proceed to step 6.

5. Enter Spheroid data. (Example: **INO**) Press keys 3 (left window blanks), **3, 5, 5,** and **0**. Left display should indicate **INO**.

6. **KYBD** pushbutton Press. Observe right display blanks. If no variation data is to be entered, press **ENT** key.

7. Enter Variation data. (Example: **E 01.2**.) Press keyboard **2** (right window blanks), **2, 0, 0, 1**, and 2 display indicates **NO E 01.2**. Press ENT key. The entire display will blank and **TGT STR** number will reappear. Display should indicate **INO E 01.2**.

f. Entering Present Position or Destination In UTM.

1. **MODE** selector switch - **UTM**.

2 **DISPLAY** selector switch - **DEST-TGT**

3. **DEST DISP** thumbwheel - **P**, numerical, or **H,** as desired.

4. Enter present position and destination. (Example: entry of zone 31T, area CF, easting 0958, and northing 3849.)

   a. **KYBD** pushbutton Press. Observe that display freezes and **TGT STR** indicator blanks

   b. To enter zone **31T KYBD** button Press. Observe that center display blanks.

   c. Keys **3, 1, 7**, and **8** - Press.

   d. To enter area CF, easting 0958, and northing 3849, KYBD button - Press. Observe left and right displays blank.

   e. Keys **1, 3, 2, 3, 0, 9, 5, 8, 3, 8, 4**, and **9** - Press.

   f. **ENT** pushbutton - Press. Left, right, and center displays will momentarily blank and **TGT STR** number will appear. Displays should indicate **31T CF 09583849**.

TM 1-1520-238-10

g. Entering Present Position or Destination Variation In LAT/LONG. The variation of a destination must be entered after the associated destination coordinates are entered (since each time a destination is entered, its associated variation is deleted). The order of entry for present position is irrelevant

NOTE
If operation is to occur in a region with relatively constant variation, the operator enters variation only for present position, and the computer will use this value throughout the flight.

1. **MODE selector** switch - **LATT/LON**

2. **DISPLAY** selector switch - **DEST-TGT**.

3. **DEST DISP** thumbwheel - **P**, numerical, or **H**, as desired.

4. Present position or destination - Enter. (Example: Entry of N41 degrees 10.1 minutes and EO 35 degrees 50.2 minutes.) Press **KYBD** pushbutton. Observe that display freezes and **TGT STR** indicator blanks. Press **KYBD** pushbutton again and observe left display blanks. To enter latitude press keys 5, 5, 4, 1, 1, 0 and 1. To enter longitude press **KYBD** pushbutton (right display should clear) and keys **2, 2, 0, 3, 5, 5, 0**, and **2**

5 **ENT** pushbutton - Press. Entire display will blank and **TGT STR** number will reappear. Display should indicate **N41** degrees **10.1 EO 35** degrees **50.2**.

h. Entering Ground Speed and Track.

1. **MODE** selector switch - **BACK UP**.

2. **DISPLAY** selector switch - **GS-TK**.

3. Ground speed and track - Enter. (Example: Enter 131 km/h and 024 degrees). Press **KYBD** pushbutton. Observe that left display freezes and **TGT STR** indicator blanks. To enter ground speed press keys **1, 3** and 1. Left display blanks: To enter track angle press keys **0, 2**, and **4**.

4. **ENT** pushbutton - Press. The entire display will blank, and **TGT STR** number will reappear. Display should indicate **131 024** degrees.

I. Initial Data Entry. Initial data entry variation coordinates are normally done prior to takeoff. To make the initial data entry, do the following:

1 Present position variation - Enter (para 3.14.7.f).

2. **DISPLAY** selector switch - **DEST-TGT**.

3. **DEST DISP** thumbwheel - **P**. Do not press **ENT** key now.

4. **ENT** pushbutton - Press as helicopter is sitting over or overflies initial fix position.

5. **FLY-TO DEST** thumbwheel - Desired destination location.

3.14.8 Operating Procedures.

a. Update of Present Position From Stored Destination. The helicopter is flying to a destination set by the **FLY TO DEST** thumbwheel. When the helicopter is over the destination, the computer updates the present position when the **KYBD** pushbutton is pressed. This is accomplished by using stored destination coordinates for the destination number shown in **FLY TO DEST** window and adding to them the distance traveled between the time the **KYBD** pushbutton was pressed and the **ENT** key was pressed.

1. **DISPLAY** selector switch - **DIST/BRG-TIME**.

2. **KYBD** pushbutton - Press when helicopter is over the destination. Display freezes.

NOTE
If a present position update is not desired as indicated by an appropriately small value of distance to go on overflying the destination, set the DISPLAY selector to some other position. This aborts the update mode.

3. **ENT** key - Press.

b. **Update of Present Position From Landmark.** There are two methods for updating present position from a landmark. Method 1 is useful if the landmark comes up unexpectedly and the operator needs time to determine the coordinates. Method 2 is used when a landmark update is anticipated.
1. Method 1.
   a. **DISPLAY** selector switch **PP**.
   b. **KYBD** pushbutton Press as landmark is overflown. Present position display will freeze. Compare landmark coordinates with those on display.

   (1) Landmark coordinates Enter. If difference warrant an update.
   (2) **ENT** key Press if update is required.
   3) **DISPLAY** selector switch Set to some other position to abort update.
2. Method 2
   a. **DISPLAY** selector switch **DEST/TGT**.
   b. **DEST DISP** thumbwheel **P** Present position coordinates should be displayed.
   c. **KYBD** pushbutton Press. Observe that display freezes.
   d. Landmark coordinates Manually enter via keyboard.
   e. **ENT** key Press when overflying landmark.
   f. **DISPLAY** selector switch Set to some other position to abort update.

c. **Left-Right Steering Signals.** Flying shortest distance to destination from present position:
1. **DISPLAY** selector switch **XTK-TKE**.
2. **MODE** selector switch UTM.

3. To center the pointer, fly the helicopter in the direction of the course deviation bar on the HSI.

d. Target Store (TGT STR) Operation. Two methods may be used for target store operation. Method 1 is normally used when time is not available for preplanning a target store operation. Method 2 is used when time is available and it is desired to store a target in a specific DEST DISP position.
1. Method 1.
   a. **TGT STR** pushbutton Press when flying over target. Present position is automatically stored and the destination location is that which was displayed in the target store indicator (position 6, 7, 8, or 9) immediately before pressing the TGT STR pushbutton.
2. Method 2.
   a. **MODE** selector switch **UTM** or **LATT/ LONG**, depending on coordinate format desired.
   b. **DISPLAY** selector switch DEST-TGT.
   c. **DEST DISP** thumbwheel **P**.
   d. **KYBD** pushbutton Press when overflying potential target. Display should freeze.

   NOTE

Do not press **ENT** key while **DEST DISP** thumbwheel is at P.

   e. If it is desired to store the target, turn **DEST DISP** thumbwheel to destination location desired and press **ENT** key.
   f. If it is not desired to store the target, momentarily place **DISPLAY** selector switch to another position.

o. **Transferring Stored Target Coordinates From One Location to Another.** The following procedure allows the operator to transfer stored target coordinates from one thumbwheel location to another. For example: it is assumed that the pilot wants to put the coordinates of stored target 7 into location of destination 2.

TM 1-1520-238-10

NOTE

Throughout this procedure, range, time-to-go, bearing, and left/right steering data are computed and displayed for the destination selected via the **FLY TO DEST** thumbwheel.

1. **DISPLAY** selector switch **DEST-TGT**.
2. **DEST DISP** thumbwheel 7.
3. **KYBD** pushbutton Press.
4. **DEST DISP** thumbwheel 2.
5. **ENT** key Press.

f. **Transferring Variation From One Location to Another.** The procedure to transfer variation data to the same location where the associated stored target coordinates have been transferred is the same as in para 3.14.8.e, except that the DISPLAY selector switch is placed at SPH-VAR.

g. **Dead Reckoning Navigation**. As an alternate BACKUP mode, dead reckoning navigation can be done using ground speed and track angle estimates provided by the operator.

1. **MODE** selector switch **BACKUP**.
2. **DISPLAY** selector switch **GS-TK**.
3. Best estimate of ground speed and track angle - enter via keyboard.
4. **MODE** selector switch Set to any other position to abort procedure.

h. **Operation During and After Power Interruption**.

During a dc power interruption, the random access memory (RAM) (stored destination and present position) data is retained by power from a 9-volt dc dry nicad battery. This makes it unnecessary to reenter any navigational data when power returns or before each flight. If the battery does not retain the stored destination data during power interruption, the display will indicate an EN when power returns. This indicates to the pilot that previously stored data has been lost and present position (spheroid/variation) and destinations must be entered. The computer, upon return of power, resets present position variation to EO 00.0-degree destination, and associated variations to a non-entered state, and remembers wind to zero and spheroid to CL6. The following data must be entered following battery failure:

1. Enter spheroid.
2. Enter present position variation.
3. Enter present position.
4. Enter each destination and its associated variation.

i. **Stopping Procedure.**

1. **MODE** selector switch **OFF**.

3.15 AN/ASN-137 DNS CONTROLS AND DISPLAYS.

The AN/ASN-137 DNS uses the IP-1552G CDU to communicate for data entry and display. The CDU (fig 3-12) has a full alphanumeric keyboard for data entry, 13 special purpose fixed action buttons (FAB), and 8 variable action buttons (VAB). In the non integrated navigation system only 6 of the FABs are active: **NAV, FDLS, STR, SPC, CLR and FLPN**. The CDU display architecture is organized in page formats. Each page can display information on 8 separate lines. The bottom line is always the data entry scratchpad, which is 22 characters wide. The 8 VABs are arranged 4 on each side of the CDU display. The function performed by the VAB depends on the particular page displayed. VABs may be used to transfer data to or from the scratchpad to a specific location on the display or select another page to perform other functions. The CDU IP-1552G is located in the CPG right-hand console. The function of each is described in table 3-14.

3-46 Change 3

TM 1-1520-238-10

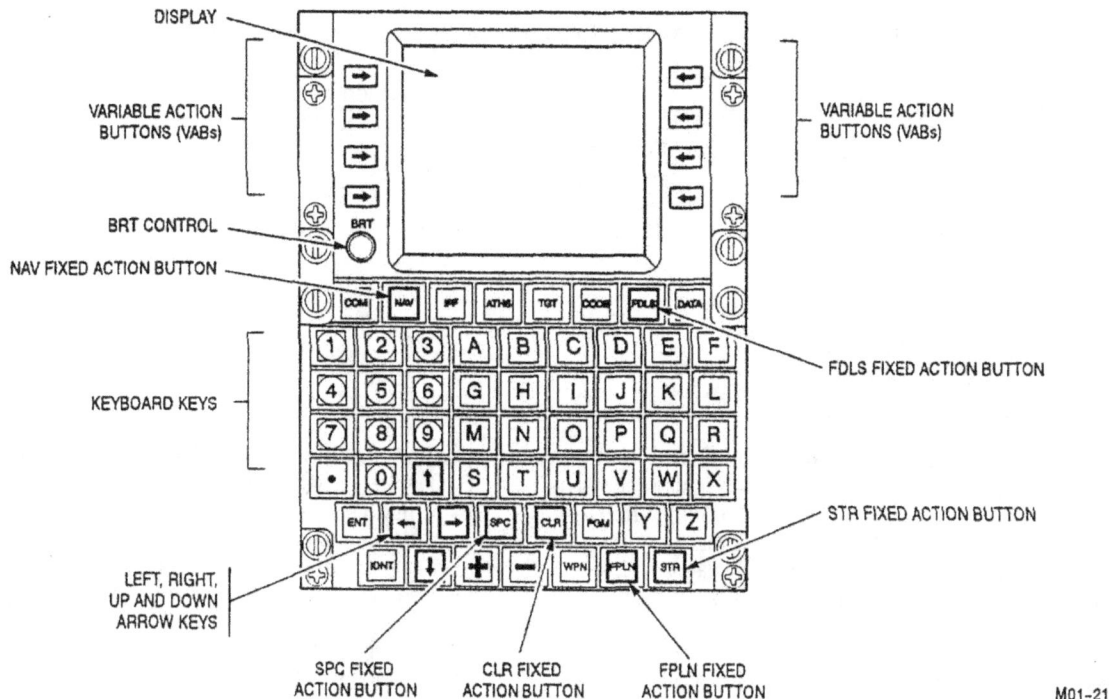

Figure 3-12. CDU IP-1552G, Used with AN/ASN-137 DNS

Table 3-14. AN/ASN-137 Control and Display Functions

Control/Display	Function
Display Screen	Display formats are organized as pages with 8 display lines; the bottom line is always used as the scratchpad for entering or editing data.
Keyboard	Alphanumeric keyboard is used for entering data.
Left/Right Arrows	Move the cursor one space left or right, as appropriate, per keystroke. When pressed constantly, will move the cursor at about 4 character positions per second.
Up/Down Arrows	Scroll through the waypoint dictionary pages, one page per keystroke. When constantly pressed, will scroll through all dictionary pages at about one page per second.
BRT Control	Adjust display brightness.
CLR	If pressed in response to an ERROR prompt in the scratchpad, will clear the error prompt and position the cursor at the first detected data entry error. If pressed a second time, will clear the entire scratchpad. If no ERROR prompt is present when CLR is pressed, it will clear the entire scratchpad.
SPC	Enters a blank space character.
STR	Stores DNS present position in storage locations 16 through 19 in circular rotation.

Change 3 3-47

Table 3-14. AN/ASN-137 Control and Display Functions - continued

Control/Display	Function
FLPPN	Displays the first (H-1-2) of 7 waypoint dictionary pages Dictionary pages may then be scrolled using the UP/DOWN ARROW keys.
NAV	Displays NAV top level page. Pressing the NAV FAB will override all other display pages and return the CDU display to the NAV top level page.
FDLS	Used to access the FDLS Test page. DNS and CDU FDLS tests are initiated from this page.
COM	Not active.
IFF	Not active.
ATHS	Not active.
TGT	Not active.
CODE	Not active.
DATA	Not active.

3.15.1 Modes of Operation. There are four modes c operation: Navigation, backup, hover bias calibration (HBCM), and test (On Command and Continuous).When electrical power is first applied to the helicopter, the CD1 is powered and the NAV top level page is displayed a shown in paragraph a. The DNS is not powered at this time.

NOTE

All data and messages shown in displays are typical representations.

a. Navigation Mode. This is the initial powerup mode of the DNS. In navigation mode, power is applied to all DNS components. Computed present position data is derived by the DNs computer from doppler radar data. This is the most accurate navigation mode.

b. Backup Mode. The backup mode is manually selected by the CPG. Backup mode is usually selected for only one of two reasons: DNS FDLS has detected a failure of the doppler radar ground speed for navigation calculations. Dynamic doppler radar velocities will not be available. The (HARS) will go into free inertial when the DNS is in backup mode. The CPG can change the ground speed used by entering the computed or best estimation of ground speed on the NAV top level page. Ground speed can only be entered by the CPG when the DNS is in backup mode.

c. HBCM. The HBCM calibrates the DSN system for small velocity errors that may be present in the doppler/transmitter subsystem. The velocity bias corrections are computed by DNS computer, and are applied to all subsequent doppler radar velocities. The velocity bias (error), when present, will be most noticeable when the helicopter is at a hover or slow speeds. The reason is that at a hover the velocity error, although small, is about the same amount as the actual helicopter velocity. When HCBM is selected, the CPG can manually start and stop the calibration. If a calibration NO-GO status is displayed at the conclusion, a recalibration may be restarted by the CPG. If a GO calibration is computed, the bias velocities will automatically be stored and applied continuously to all subsequent navigtion computations.

d. Test Mode. The DNS and CDU have both Continuous and On Command FDLS tests. The DNS continuous test checks the doppler antenna (RTA), and doppler signal data converter computer (SDCC). The CDU monitors its own internal functions and the dedicated 1553B mux bus between the CDU and DNS SDCC. If a failure is detected by the Continuous test, the CDU will display the prompt v FDLS on line 6 of the NAV top level page. The CPG can then display system status by pressing the **FDLS** fixed action button (FAB). The CPG can initiate the ON Command test from the FDLS page. The On Command test initiates a full DNS subsystem component test. If no failures are detected, the CDU will display the pitch, roll, and inertial heading, in degrees, currently being used by the DNS. The pitch and roll angles will reflect the actual orientation of the helicopter on level ground, pitch will be approximately 4.5 to 5.5°, and roll should be about 0.0. If the helicopter has a lean due to cyclic or pedal inputs, this will be reflected in the roll angle displayed. The heading value displayed is the inertial heading of the helicopter; HSI/RMI heading plus or minus the magnetic variation.

3.15.2 CDU Displays. The CDU display architecture is organized into page formats. Each page consists of 8 display lines. The first seven are 22 characters wide for displaying data. The bottom (eighth) line is 21 characters wide and is a scratchpad for entering or editing data. Individual pages may display navigation data, display failure messages, or indicate selectable functions. The CDU displays are categorized into 5 basic page formats: **NAV** top level, **ADMIN, FDLS, HBCM,** and **FPLN.** The CDU display will automatically revert back to the NAV top level page based on two basic rules:

Rule 1.-The CDU will revert to the **NAV** top level page automatically from the **ADMIN** or any of the **FPLN** (waypoint dictionary pages after 30 seconds if there are no characters in the scratchpad. The 30 second timer is reset each time a FPLN page is scrolled with the Up/Down arrow when the scratchpad contains no data. Remember: If the scratchpad contains any data the display will NOT automatically revert back to the **NAV** top level page.

Rule 2.-The CDU will revert to the **NAV** top level page automatically from the **ADMIN** page 3 seconds after pressing any of the following (VAB)s: **PWR ON** or **PWR OFF, BACKUP, MODE UTM** or **MODE L/L,** and **DISPL KPH** or **DISPLKTS.**

If the **HBCM** VAB on the **ADMIN** page or **FDLS** FAB is pressed the CDU will NOT revert automatically back to the **NAV** top level page. The **NAV** top level page can be manually selected at any time from any page by pressing the **NAV** FAB.

a. **NAV Top Level Page.** The **NAV** top level page (fig 3-13) will be displayed approximately 30 seconds after aircraft electrical power is applied. No data will be displayed until the DNS power is turned on. The DNS is powered by pressing the **ADMIN** VAB and then pressing the **PWR OFF** VAB on the **ADMIN** page. The DNS indicates a failure if it is powered and the HARS is OFF or in the first 90 seconds of alignment. When the DNS is powered, the NAV top level page (fig 3-14) will display the following:

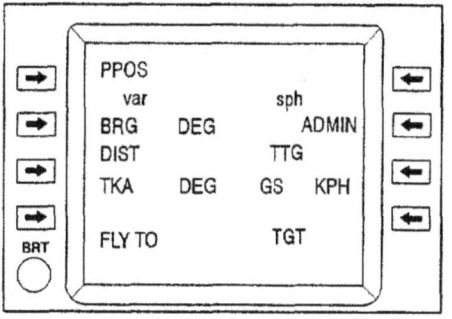

Figure 3-13. Power Up Display, NAV Top Level DNS PWR OFF

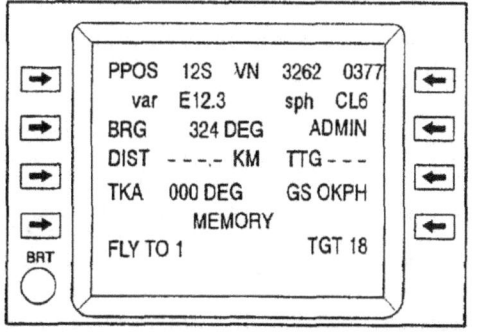

Figure 3-14. NAV Top Level Page PWR ON

TM 1-1520-238-10

NOTE
'Refer to Data Entry and Operating Procedures (para 3.14.7.f and 3.14.8) as necessary.

Line 1.-**PPOS** represents the current DNS computed present position in either UTM or LAT/LONG. The PPOS data may be changed or edited.

Line 2. **var** and **sph** are the magnetic variation and spheroid associated with the **PPOS**. **var** and **sph** may be changed or edited.

Line 3.-**BRG** is the direct bearing to the destination specified in the **FLY TO** position on line 7. **ADMIN** is a VAB label that selects the page for the various administrative functions associated with the DNS.

Line 4.-**DIST** and **TTG** are the distance and time-to go (at current ground speed) to the destination specified in the **FLY TO** position. **DIST** will display in either kilometers or nautical miles, independent of the PPOS data, as selected on the **ADMIN** page.

Line 5.-**TKA** and **GS** are the computed track angle and ground speed. Ground speed may be displayed in either kilometers per hour (**KPH**) or knots as selected on the **ADMIN** page. The units for both **DIST** and **GS** are same for a given selection: **KM** and **KPH**, or **NM** and **KTS**.

Line 6.-This is the CDU and DNS status line. The line may be blank or display **MEMORY, BACKUP,** or **vFDLS**. A display on this line indicates that the DNS has either lost radar lock: **MEMORY**; the CPG has manually selected **BACKUP;** or the Continuous FDLS test has detected a fault in the **CDU/DNS** system.

Line 7. **FLY TO** shows the selected destination to which the DNS is navigating. The displayed value can be any of the 20 locations in; the FPLN dictionary; **0** (or **H**) through 19. The **FLY TO** number can be changed.
TGT displays the storage location that will be used next When the **STR** FAB is pressed; locations 16 to 19 are used repetitively.
Line 8.-This is the scratchpad line for data entry.

    b. **ADMIN Page**. The **ADMIN** page (fig 3-15 and 3-16), is used to select other modes of DNS operation and administrative DNS management and display functions. The **ADMIN** page has the following:

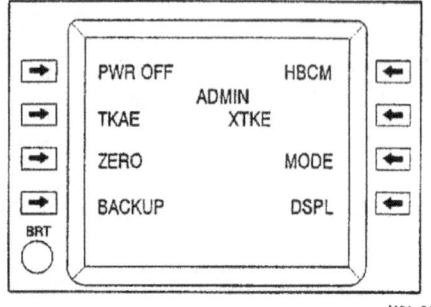

Figure 3-15. **ADMIN Page PWR OFF**

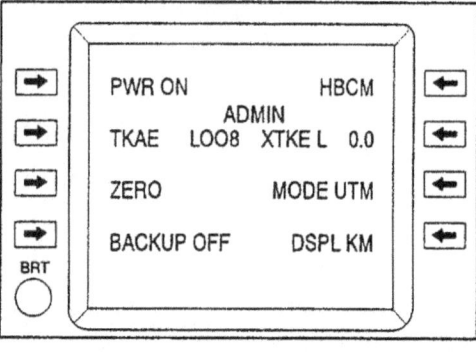

Figure 3-16. **ADMIN Page PWR ON**

Line 1. **PWR OFF** or **PWR ON** is an alternate action VAB powering the DNS. **HBCM** selects the Hover Bias Calibration Mode page.
Line 2.-Identifies the **ADMIN** page.
Line 3.-**TKAE** is a data display only showing the Track Angle Error. The **L** or **R** character specifies which direction to steer to reduce the **TKAE** to zero. **XTKE** is a data display only showing the Cross Track Error in kilometers or nautical miles.
Line 4.-Blank.
Line 5.-**ZERO** is a function which permits zeroizing the CDU or DNS memory. The specific system to zero is accomplished by entering either CDU or DNS in the scratchpad and pressing the **ZERO** VAB. Zeroizing the CDU will FPLN dictionary storage locations to default values. Zeroing the DNS will cause loss of HBCM data.
MODE L/L or **MODE UTM** is an alternate action VAB selecting the coordinate system for display and data entry of coordinate data.

Line 6 Blank.

Line 7. **BACKUP ON** or **BACKUP OFF** is an alternate action VAB selecting DNS **BACKUP** mode. **DSPL KM** or **DSPL NM** is an alternate action VAB that permits switch of **DIST** and GS only from **KM/KPH** to **NM/KTS** and back.
Line 8. Scratchpad.
c. **HBCM Page**. The **HBCM** page, (fig 3-17), is used to **START** and **STOP** the bias calibration of the DNS. The 30-second automatic reversion to **NAV** top level page is disabled when **HBCM** is selected. Refer to paragraph 3.15.c for HBCM operating procedures. The display has the following:

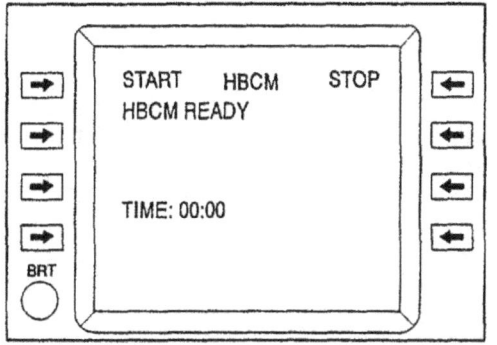

Figure 3-17. HBCM Page

Line 1.-**START** is used to activate the HBCM calibration. **HBCM** identifies the page. **STOP** is used to terminate the calibration.
Line 2.-**HBCM READY** indicates the DNS is ready to start the calibration sequence. **HBCM ACTIVE**, **HBCM GO**, and **HBCM NO-GO** will also be displayed on the line as appropriate.
Line 3.-Error messages pertinent loan HBCM condition are displayed on this line.
Line 4.-Blank.
Line 5.-**TIME XX:XX** indicates elapsed time in minutes/seconds for DNS computations of the bias calibration. The timer will stop while the DNS is in memory during the bias calibration.
Line 6.-Blank.
Line 7.-Blank.

Line 8. Scratchpad. Not used in HBCM.

d. **FDLS Page.** The **FDLS** page, (fig 3-18), is used to initiate the On Command FDLS test function. The **FDLS** page is selected by pressing the **FDLS** FAB. The **FDLS** page has the following display.

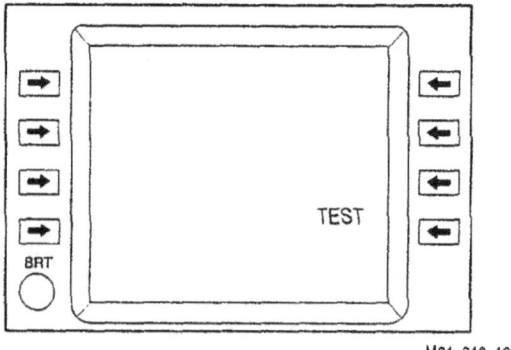

Figure 3-18. FDLS Page

Line 1 through Line 6. Blank.
Line 7.-**TEST** initiates the On Command FDLS test for the CDU and DNS.
Line 8. Scratchpad. Not used in FDLS.

e. **FPLN Page**. The **FPLN** FAB selects the waypoint coordinate dictionary pages. The first **FPLN** page (fig 3-19), displayed when the **FPLN** FAB is pressed is determined by the **FLY TO** selection on the **NAV** top level page.

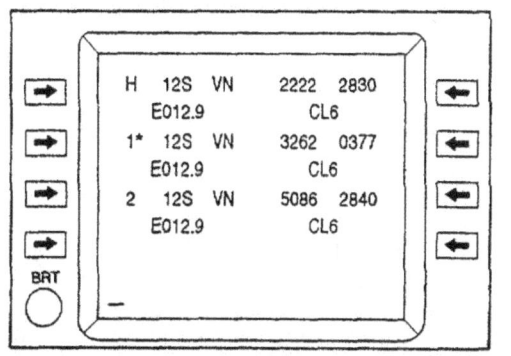

Figure 3-19. FPLN Dictionary Page

The waypoints are organized on pages as shown in table 3-15.

Table 3-15. Waypoints

Page	Waypoints
1	Home 1, 2
2	3,4,5
3	6,7,8
4	9,10, 11
5	12, 13, 14
6	15, 16, 17
7	18 and 19

The Up/Down arrow keys can be used to scroll through the FPLN dictionary pages. An * is shown immediately after the storage location currently being used for **FLY TO** computations and it cannot be changed until some other location is chosen to **FLY TO**. The page has the following data organization:

Line 1. - Displays the coordinate location data.

Line 2. - Displays the magnetic variation and spheroid associated with the coordinate data above it.

Line 3. - Displays the next coordinate location data.

Line 4. - Displays the magnetic variation and spheroid associated with the coordinate data above it.

Line 5. - Displays the next coordinate location data.

Line 6. - Displays the magnetic variation and spheroid associated with the coordinate data above it.

Line 7 - Blank.

Line 8. - Scratchpad for data input.

3.15.3 Data Entry Procedures. The CDU display present position can be manually updated, waypoint locations stored, **FLY TO** destination selected and other data entries made via the CDU keyboard. Keyboard entries appear on the scratchpad. The CDU checks the validity of all data entered before it transfers the data to the appropriate location on the display.

a. General Data Entry and Data Correction.
General data entry and data correction procedures follow the example of **PPOS**, **var**, **sp**h, and **MODE UTM** on the **NAV** top level page.

DATA ENTRY EXAMPLE

1. Use the keyboard keys, (fig 3-11), enter the following into the scratchpad:

12SVN32620377E12C5
NOTE
- The last character (5) is deliberately erroneous.
- Observe that each character appears on the scratchpad, left justified, as it is entered.

2. **PPOS** Press. Observe that the scratchpad display alternates between the data entered and the prompt **ERROR**.

3. **CLR FAB** - Press. Error prompt is cleared and the data in the scratchpad has the cursor positioned under the error as shown below:

12SVN32620377E12C5_

NOTE
The cursor is positioned under the first error detected. If more than one error is detected, the cursor moves left to right to the next error. Continue editing until all data is corrected.

4. Enter **6**. Observe that **5** changes to **6**.

5. **PPOS** - Press. Observe that the scratchpad is now blank and lines 1 and 2 of the CDU display appear as follows:

PPOS 12S VN 3262 0377
var E012.0   sph CL6

3.15.4 Power Up Procedures. The steps in this procedure list actions required (and expected CDU display results) to ready the DNS for flight. Refer to Data Entry (para 3.15.3) for detailed techniques and parameters of such actions as:

- UTM or L/L Coordinate Entries
- Hover Bias Calibration
- General Data Entry and Error Correction

a. Display Check.

   1. **BRT** - Rotate control fully clockwise. Observe brightness change.

   2. **BRT** - Adjust control for best visibility of display.

b. DNS Power On.

   1. **ADMIN** - Press. Observe that the ADMIN Page is displayed

   2. **PWR OFF** - Press. Observe **PWR OFF** changes to **PWR ON** for three seconds then CDU display changes to **NAV** top level present position page.

   NOTE
   - If PWR OFF changes to PWR --- or the NAV top level present position page is not displayed after approximately three seconds, the DNS has failed.
   - If the RAM battery has failed, all navigation data in the SDCC memory is lost and the following prompts are automatically displayed on the FDLS page (fig 3-22).

   DNS RAM ERASED
   DNS IN REMEMBERED VEL
   END OF LIST

c. FDLS Test.

   1. **FDLS** - Press. Observe that the Continuous Test Results - **FDLS** Page is displayed, as shown in figure 3-20.

   2. **TEST** - Press. Observe that NAV SYS TEST IN PROG is displayed until the test is complete and data is available for DNS test status.

   3. If a GO message is displayed (fig 3-21), press the NAV FAB.

   NOTE
   The heading, pitch and roll values are calculated and displayed approximately five seconds after the GO message.

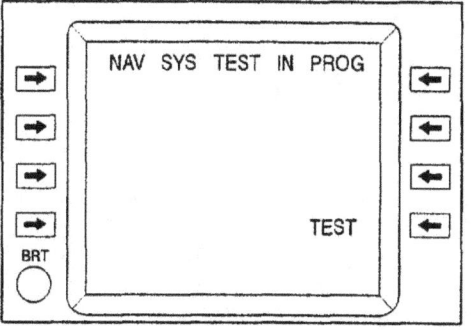

Figure 3-20. FDLS Test in Progress (Normal Mode)

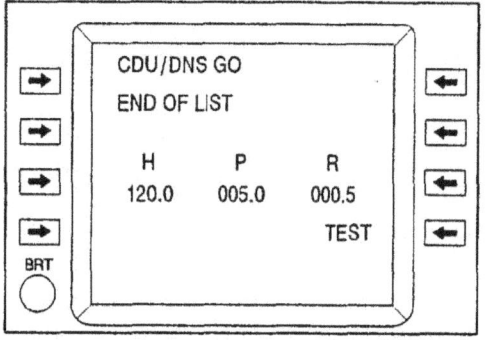

Figure 3-21. FDLS Status Page - On Command Test Results GO

d. **Selecting MODE UTM or L/L and DSPL KM or KTS.** If the operator desires to switch between **MODE UTM** and **MODE L/L** or **DSPL KM** and **DSPL ITS**, perform the following:

   1. **ADMIN** - Press. Observe that the ADMIN page is displayed.

   2. **MODE UTM** - Press to change to **MODE L/L** or vice versa.

   3. **DSPL KM** - Press to change to **DSPL KTS** or vice versa.

e. **Coordinate Entries for FPLN dictionary.**

   1. FPLN FAB - Press. Observe that **FPLN** dictionary page is displayed.

TM 1-1520-238-10

2. Use up and down arrow keys to scroll to **FPLN** dictionary page 1 (locations H, 1 and 2).
3. Use the keyboard to enter H coordinate data into scratchpad.

UTM EXAMPLE
11SGQ52184911E14.8C6
LAT/LONG EXAMPLE
N7425.9W12057.6E10.5

4. Press left hand VAB of desired storage location (H). Observe that the scratchpad is now blank and the entered data now appears on lines land 2 (H location) of CDU display.

5. Repeat steps 3 and 4, entering data and pressing appropriate VAB for each waypoint to be entered. Use Down and Up arrow keys to scroll through FPLN dictionary pages as desired.

6. V FAB - Press to display the NAV top level age or allow time out to automatically revert o NAV top level page.

f. Coordinate Entries for PPOS. Coordinate entries for PPOS data (including magnetic variation and spheroid) may be entered by two different methods.

(1). Method 1 - Manual Keyboard Entry.

1. Use the keyboard to enter coordinate data into the scratchpad.
2. VAB 1 - Press. Observe that the coordinate data now appears on lines land 2 of NAV top level page display.

(2). Method 2 - Updating to a Stored Waypoint.

1. Enter the number of the stored waypoint into the scratchpad.
2. **FLYTO** - Press. Observe that waypoint number now appears after **FLY TO**

b. Press top right VAB. Observe that **FLY TO** changes to **UPDATE TO**.

NOTE

The procedure can be aborted by pressing top right VAB.
d. **FLYTO** - Press. Observe that the present position coordinate data is now updated to that of the stores waypoint.

3.15.5 Operating Procedures. When first powered up the DNS is in navigation mode. Most functions and modes of operation are accessed via the NAV top level present position page. Exceptions are the On Command Test and FPLN dictionary Pages.

1. **ADMIN** - Press. Observe that the ADMIN page is displayed.

2. **PWR OFF** - Press. Observe **PWR OFF** changes to **PWR ON** and CDU display changes to **NAV** top level present position page after 3 seconds.

NOTE

If **PWR OFF** changes to **PWR ---** or the NAV top level present position page is not displayed after approximately three seconds, the DNS has failed.

If **FDLS** Status Page (fig 3-22) is displayed, the RAM battery has failed. The DNS RAM battery test is only performed at **PWR ON**. The On Command FDLS test can be initiated from this page by pressing **TEST**. Refer to On Command Test (FDLS) part of this paragraph for detailed procedures.

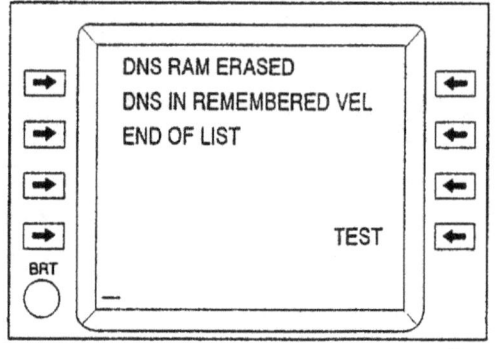

Figure 3-22. FDLS Page - Continuous Test Results

3-54 Change 3

a. CDU or DNS Memory Zeroize. The DNS memory zeroize function is used for maintenance purposes only after the CDU or signal data converter has been replaced. The CDU memory zeroize resets all flight plan dictionary page waypoints to bogus default values. This is done from the NAV top level present position page as follows:

1. **ADMIN** - Press. Observe that the **ADMIN** page is displayed.

2. Enter CDU on the scratchpad.

3. **ZERO** - Press.

4. FPLN - Press. Observe that all locations on the **FPLN** dictionary page displayed contain the similar default values:

   31T BV 6355 8711
   E 000.0   sph CL6

5. Using the up and down arrow keys, scroll through all **FPLN** dictionary pages, observing that all location contain the default value.

6. **NAV** - Press. Observe that the **NAV** top level present position page is displayed with default data next to the captions.

b. BACKUP Mode. In the **BACKUP** mode, the last remembered radar velocities become the source for navigating.

This mode provides the capability to the operator of manually entering **GS** and **TKA** via the **NAV** top level present position page

The **BACKUP** mode must be manually accessed. To operate the **BACKUP** mode do the following:

1. **ADMIN** - Press. Observe that the **ADMIN** page is displayed.

2. **BACKUP OFF** - Press. Observe that **BACKUP OFF** changes to **BACKUP ON**.

3. Enter GS data into the scratchpad and press VAB 7. Observe the data is now displayed on line 5 next to **GS**.

4. Enter TKA data into the scratchpad and press VAB 3. Observe that the data is now displayed on line 5 next to **TKA**.

C. HBCM. In **HBCM** the DNS determines the hover bias velocities in both the longitudinal and lateral axis of the DNS. The computed bias velocity corrections are applied to all future DNS hover velocity computations. **HBCM** is accessed via the **NAV** top level present position and **ADMIN** pages.

NOTE

Hover bias calibration only needs to be performed when the signal data converter or RTA has been replaced.

1. **ADMIN** - Press. Observe that the **ADMIN** page is displayed.

2. **HBCM** - Press. Observe that the **HBCM** page (fig 3-17) is displayed.

NOTE

- If line 2 of the **HBCM** page does not display **HBCM READY**, the DNS has failed.

- The time indicated on line 5 of the **HBCM** page increments only when DNS is performing hover bias computations. if the DNS goes into memory mode, the time display halts until the signal from the RTA is sufficient for DNS computations.

3. **START** - Press. Observe that on line 2 **HBCM READY** changes to **HBCM ACTIVE** and on line 5 the time displayed begins incrementing.

NOTE

The calibration continues until the CPG presses **STOP** to stop it. When the calibration is stopped, the mode status will be displayed. Allow the calibration to run for at least two minutes but not more than 8 minutes.

4. **STOP** - Press. Observe that the time display freezes and a mode status message is displayed on line 2. If applicable, DNS error messages will be displayed on lines 3 and 4. The DNS indicates the HBCM update was accepted when an **HBCM GO** is displayed or not accepted when an **HBCM NO-GO** message is displayed along with a description of the exceeded limit(s) on other lines. Possible DNS error messages are:

TIME BELOW 2 MIN
TIME EXCEEDED 8.2 MIN
BIAS EXCEEDS 0.3 KTS
EXCESSIVE MEM CONDITION

NOTE

If a NO-GO condition is detected, a recalibration can be initiated by pressing **START** again. This will zero the timer and allow for additional bias velocities calculations. If either TIME message is displayed, re-initiate the calibration. If the BIAS message is displayed, re-initiate calibration one time. If the BIAS message is displayed again, write up the system. If the excessive memory message is displayed, attempt to re-locate the aircraft over a surface area that is a non-reflective surface (short grass/coarse surface area) and re-initiate the calibration.

5. **NAV FAB** - Press to exit the **HBCM**. Observe the CDU displays the **NAV** top level page.

d. MODE UTM or L/L. This is an alternate action function which toggles **PPOS** and waypoint dictionary coordinate displays between UTM and LAT/LONG. This is done from the **NAV** top level present position page as follows:

1. **ADMIN** - Press. Observe that the **ADMIN** page is displayed.

2. **MODE UTM** - Press to change to **MODE L/L** or vice versa.

e. DISPLAY KM or KTS. This is an alternate action function which toggles the **DIST** and **GS** displays between **KM** and **KPH** or **NM** and **KTS** respectively. This is done from the **NAV** top level present position page as follows:

1. **ADMIN** - Press. Observe that the **ADMIN** page is displayed.

2. **DSPL KM** - Press to change to **DSPL KTS** or vice versa.

f. PPOS Updates to a Stored Waypoint Coordinate. **PPOS** updates to a stored waypoint coordinate are implemented from the **NAV** top level present position page. When overflying a waypoint, do the following:

NOTE

A flashing **UPDATE SPHEROID** message will appear on the scratchpad if the **FLY TO** waypoint selected has a different spheroid than the current present position data.

1. VAB 5 - Press. Observe **PPOS** data freezes and **FLY TO** changes to **UPDATE TO**. The DNS continues to compute aircraft present position.

2. **FLY TO** - Press to complete an update to the position of the waypoint plus any offset distance traveled following the initiation of a **PPOS** update.

3. To reject an update (abort), press either top right VAB again or the **NAV** FAB.

NOTE

Following either an acceptance or rejection of a **PPOS** update, the CDU displays **FLY TO** and resumes displaying DNS computed PPOS and status data dynamically.

g. PPOS Updates to an Unstored Waypoint. PPOS updates to an unstored waypoint are implemented from the **NAV** top level present position page as follows:

1. **PPOS** - Press. Observe PPOS changes to **UPDATE** and display freezes. This allows for comparison of DNS computed present position location with a known present position location.

NOTE

This type of PPOS update does not require that the aircraft remain in a hover to update the present position. This type of update also provides for position compensation for any distance traveled following a PPOS update.

2. To complete the PPOS update to a known value, the operator enters the desired present position into the scratchpad and presses **UPDT** or:

3-56   Change 4

3. Rejects (aborts) the PPOS update by pressing the **NAV** FAB.

NOTE

Following either an acceptance or rejection of a PPOS update, the CDU resumes displaying DNS computed PPOS and status data dynamically.

h. FLY TO Destination Selection. The number next to the **FLY TO** caption on line 7 specifies which prestored waypoint the DNS is to navigate to. Valid **FLY TO** entries include values **1-19** and 0 or **H**. **H** and 0 are synonymous. Enter a **FLY TO** destination as follows:

1. Enter desired prestored waypoint value on scratchpad.
2. **FLY TO** Press. Observe that the entered value now appears' next to **FLY TO** and the scratchpad is now blank.

i. UTM Coordinate Entries for Waypoint Dictionary. UTM coordinate entries, including magnetic variation and spheroid, may be entered into the waypoint coordinate. The following is a typical format to be used when entering UTM coordinates, magnetic variation and spheroid data in the CDU scratchpad:

11SGQ52184911E14.8C6

Where		
	11S	=Grid zone identifier
	GQ	=UTM 100 KM-square identifier
	5218	52.18 KM Easting within GQ
	4911	=49.11 KM Northing within GQ
	E14.8	=Magnetic variation
	C6	= Spheroid identifier of CL6

A complete entry of UTM coordinates, magnetic variation and spheroid data is not required under all circumstances. Refer to table 3-16 for partial data entry. When an incomplete set of data is entered into the CDU scratchpad:

Rule 1 Order of Entry. The order of entry for UTM coordinates, magnetic variation and spheroid remains the same regardless of which parameters are or are not entered. They are: grid zone identifier, 100 KM-square identifier, Easting/Northing; magnetic variation; and spheroid. Data entered into the scratchpad is always justified on the left side.

Rule 2-Spheroid Change. Table 3-17 shows character string entries that are required for sph data.

Table 3-16. Waypoint Dictionary - Partial Entry Rules

Partial Data Entries		Required Changes to Other Data				
		Easting/ Northing	Magnetic Variation	Spheroid		
Grid Zone	100KM ID					
Grid Zone	Not Affected	Change	Note 1	Note 1		
100KM ID	Note 1	Change	Note 1	Note 1		
Easting/	Note 1	Note 1	Change Northing	Note 1	Note 1	
Northing	Note 1	Note 1	Change Easting	Note 1	Note 1	
Magnetic Variation	Not Affected	Not Affected	Not Affected	Note 2	Not Affected	
Spheroid	Not Affected	Not Affected	Not Affected	Not Affected		

Note 1 If data for this parameter is not entered, its value defaults to the values associated with PPOS.
Note 2 The character E or W must precede numeric values for magnetic variation. Leading and trailing zeros are not required for magnetic variation. Entering a hyphen (-) sets the magnetic variation to no computed data.

Table 3-17. Spheroid String Entries

Spheroid	Character String Entry	Automatic Display
AUSTRALIAN	AU	AUO
BESSEL	BE	BEO
EVEREST	EV	EVO
INTERNATIONAL	IN	INO
CLARKE 1880	CO	CLO
CLARKE 1866	C6	CL6
MALAYAN*	MA	MAO

*Currently not utilized

J. UTM Coordinate Entries for Present Position. When an incomplete set of UTM coordinate, magnetic variation and spheroid are entered into PPOS, refer to table 3-18 for **exceptions** to waypoint dictionary entries.

Table 3-18. PPOS - Partial Entry Rules

Partial Data Entries	Required Changes to Other Data				
	Grid Zone	100KM ID	Easting/ Northing	Magnetic Variation	Spheroid
Grid Zone		Change	Change	Not Affected	Not Affected
100KM ID	Not Affected	Change	Change	Not Affected	Not Affected
Easting/ Northing	Not Affected	Not Affected	Change Northing Change Easting	Not Affected	Not Affected
Magnetic Variation	Not Affected	Not Affected	Not Affected	Note 1	Not Affected
Spheroid		Not Affected	Not Affected	Not Affected	Not Affected

Note 1: The character E or W must precede numeric values for magnetic variation. Leading and trailing zeros are not required for magnetic variation. Entering a hyphen (-) sets the magnetic variation to no computed data.

TM 1-1520-238-10

k. CDU Validity Checks For UTM Coordinate Data. The CDU checks the parameters of all data entered through the scratchpad. Any non-valid characters are flagged as erroneous. Valid parameters are as shown in table 3-19.

Table 3-19. UTM Coordinate Data - Valid Entries

Parameter	Valid Entries
Grid Zone	1 - 60 for the first two characters. C - H, J - N and P - X for the third character.
UTM Identifier	A - H, J - N and P - V for the most significant character.
Magnetic Variation	E or W and 0 - 180.0 degrees.
Spheroid	AU, BE, EV, IN, CO, C6 and MA. MA is not currently utilized.
Fly-To	1 - 19 and H and 0. H and O. are synonymous.

l. LATT/LONG Coordinate Entries for Waypoint Dictionary. LATT/LONG coordinates, including magnetic variation, may be entered into the waypoint coordinates as a single character string. The following is a typical format used when entering **LATT/LONG** coordinates and magnetic variation.

N7425.9W12057.6E10.5

Where N7425.9 = 74 degrees 25.9 minutes North Latitude
W12057.6 = 120 degrees 57.6 minutes West Longitude
E10.5 = 10.5 degrees East Magnetic Variation

When **LATT/LONG** and magnetic variation are all entered, the required sequence is latitude, longitude, then magnetic variation.

A complete entry of LATT/LONG coordinates and magnetic variation data is not required under all circumstances. Data entered in the scratchpad is always left justified. The following rules apply when an incomplete set of LATT/LONG coordinates and magnetic variation data is to be entered into the CDU scratch-pad.

NOTE
Leading zeros are required when entering degrees or minutes of latitude and longitude, i.e., N0905.5W01010.3 represents an entry of 9 degrees 5.5 minutes North latitude and 10 degrees 10.3 minutes West longitude.

Rule 1 - Latitude Range of Degrees. South latitude range for entering degrees is 00 - 80 North latitude range for entering degrees is 00 - 84.
Rule 2 - Longitude Range of Degrees. The range for entering degrees is 000 - 180.
Rule 3 - Latitude Change. A change in latitude must be accompanied by a change in longitude. Entry of magnetic variation is optional. If magnetic variation is not entered, its value defaults to the value associated with present position.
Rule 4 - Magnetic Variation Change. A change can be made to magnetic variation without changing LATT/LONG coordinate data or affecting their values.

m. LATT/LONG Coordinate Entries for PPOS. When an incomplete set of **LATT/LONG** coordinates and magnetic variation is entered into **PPOS**, the following exception applies:

NOTE
A change in latitude must be accompanied by a change in longitude. Magnetic variation remains unchanged.

Change 3 3-59

TM 1-1520-238-10

n. CDU Validity Checks for LATT/LONG Coordinate Data. The CDU checks the parameters of all data entered into the scratchpad. Any nonvalid characters are flagged as erroneous. Valid parameters are shown in table 3-20.

Table 3-20. LAT/LONG Coordinate Data

Parameter	Valid Entries
PPOS Latitude 1st Character	N or S only.
PPOS Latitude Degrees	00 - 84 for North latitude degrees. 00 - 80 for South latitude degrees.
PPOS Minutes	00.0 - 59.9.
PPOS Longitude 1st Character	E or W only.
PPOS Longitudinal Degrees	000 - 180.
PPOS Longitude Minutes	00.0 - 59.9.
Magnetic Variation	E or W and 0 - 180.0 degrees.
FLY TO	1 - 19 and H or 0 H and 0 are synonymous.

o. Waypoint Dictionary Pages and Transfer Functions. Up to 20 waypoints can be entered into the waypoint dictionary at one time. The waypoint dictionary pages are accessed via the **FPLN** FAB. If the CPG desires to change display modes (**UTM** or **LATT/LONG**) before making entries into the waypoint dictionary, perform steps 1 and 2. If the mode does not need to be changed, go to step 3.

1. **ADMIN** - Press. Observe that the **ADMIN** page is displayed.

2. **MODE UTM** - Press. Observe that **MODE UTM** changes to **MODE L/L** or vice versa and after three seconds the **NAV** top level present position page is displayed.

3. FPLN FAB - Press. Observe that FPLN waypoint dictionary page 1 is displayed (Waypoints home, 1 and 2).

NOTE

Refer to UTM or LATT/LONG coordinate entries for waypoint dictionary paragraphs for entry details. Refer to CDU validity checks for UTM or LATT/LONG coordinate data for data entry parametersm

4. Enter the desired coordinates into the scratchpad.

3-60 Change 3

5. **VAB** - Press adjacent to the desired location (waypoint number). Observe that the scratchpad is now blank and the data appears at the desired waypoint location.

The CDU allows for the transfer of coordinate data from one waypoint to another. To transfer waypoints, set the CDU to **MODE UTM** and follow the example.

EXAMPLE

6. **FPLN** FAB - Press. Observe that waypoint dictionary page 1 (home, 1 and 2) is displayed.

7. Enter **12SWN00000000E12C6** into the scratchpad.

8. VAB **1** - Press. Observe that the home waypoint location now displays:

```
12S WN        0000    0000
  E012.0        CL6
```

and the scratchpad is now blank.

9. VAB **5** - Press. Observe that the data in the home location is unchanged and the scratchpad now displays the data entered in step 2.

10. Press down arrow key. Observe that waypoint dictionary page 2 (locations 3, 4, and 5) are now displayed.

11. VAB **1** - Press. Observe that location 3 now contains the same data as the home location.

p. Target Storage. The DNS target store function operates similarly to the AN/ASN-128 target function. To perform this function, press the **STR FAB.** This will store DNS PPOS coordinate data, magnetic variation and spheroid data values into the waypoint number designated on line 7 of the **NAV** top level present position page. The **TGT STR** number automatically increments to the next higher number after each storage function until it reached 19, then the counter will reset to **16.**

For example: **TGT 17** specified that at the time the **STR FAR** was pressed, the instantaneous DNS PPOS data was stored into waypoint number **17.**

q. On Command Test (FDLS). The On Command Test is the manually initiated portion of the CDU and DNS FDLS (para 3.15.c). If the continuous test causes an FDLS message to be displayed on line 7 of the **NAV** top level present position page do the following:

1. **FDLS** FAB - Press. Observe that the CDU/DNS system status is displayed as shown in figure 3-23.

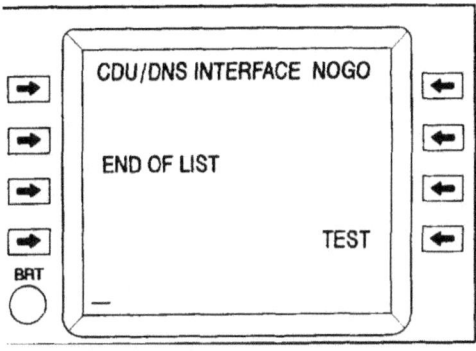

Figure 3-23. FDLS Status Page On-Command Test Results NO-GO

NOTE

An **END OF LIST** message is displayed following the last message (**GO** or **NO-GO**). If there are more than three malfunctions detected, use the Down and Up arrow keys to scroll through the display pages.

2. One or more of the following continuous test messages may be displayed:

- DNS RAM ERASED
- PROG PLUG NOT CONNECTED or PROPERLY PROGRAMMED
- RTA NOT CONNECTED
- DNS IN REMEMBERED VEL
- GS/TKA ENTRY ENABLED
- CDU/DNS INTERFACE NO-GO

3. The on command test GO/NO-GO self test can be initiated any time the FDLS page is displayed. It is not necessary to wait for an √**FDLS** error message to be displayed before pressing the **FDLS FAB** to display the **FDLS** page. The GO/NO-GO self test is comprised of the following internal tests:

   a. **PROM** Check sum verification.

   b. **RAM** read/write test.

   c. Limited non-volatile memory retention check.

   d. Display hardware BIT.

   e. 1553B terminal BIT.

4. Initiate the On Command test as follows:

   a. **TEST** - Press. Observe that DNS SYSTEM **TEST IN PROG** is displayed until the test is complete and data is available for DNS test status. The GO or NO-GO status is then displayed. The following is a list of possible NO-GO messages:

   (1) **PITCH/ROLL FAIL**

   (2) **HDG** or **A.C. REF FAIL**

   (3) **SDCC PWR SUPPLY FAIL**

   (4) **SDCC CPU/MEMORY FAIL**

   (5) **SDCC WRAPAROUND FAIL**

   (6) **SDCC A/D FAIL**

   (7) **SDCC 1553 I/O FAIL**

   (8) **SDC FAIL**

   (9) **RTA FAIL**

   (10) **SDCC CMD FAIL**

   If no failures are detected, the CDU displays the **FDLS** page as shown in figure 3-24.

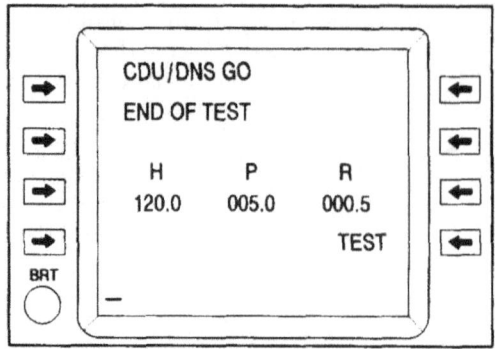

Figure 3-24. FDLS Status Page - On-Command Test Results

NOTE

The heading, pitch and roll values are calculated and displayed no more than five seconds after the GO message.

   b. **NAV** FAB - Press. Observe that the **NAV** top level present position page is displayed.

3.15.6 Stopping Procedures.

1. **ADMIN** - Press. Observe that the **ADMIN** page is displayed.

2. **PWR ON** - Press. Observe that **PWR ON** changes to **PWR OFF**.

3.16 INTEGRATED NAVIGATION SYSTEM [51]

NOTE

- The integrated navigation system (EGI) does not meet US ARMY requirements to fly FM approved GPS approaches.

- All data contained paragraph 3.16 through paragraph 3.16.16 is applicable to [51] FCC software only.

The Integrated Navigation System is a software module in the FCC that uses data from all of the navigation sensors. The data for the Navigation System comes from the Embedded GPS Inertial (EGI) sensor, a five-channel GPS receiver, the AN/ASN-137 Doppler Navigation System, the Heading Attitude Reference System and the Fire Control Computer. The Navigation System software resides in the FCC only. In the event of an FCC failure, the helicopter will automatically configure to the Non-Integrated Navigation System as described in paragraph 3.14, with the BBC as the bus controller.

NOTE

- Waypoints stored via the Data Transfer Unit (DTU) will not be accessible if the BBC is the bus controller. The CPG can store up to 20 waypoint coordinates in the CDU with the **MUX** switch in **SEC** (Emergency return routes to be used as a backup in the event of a FCC failure).

- If HARS is free inertial for 20 or more seconds, the **HARSVEL**? prompt is displayed on the CDU. If HARS is free inertial for 20 or more seconds and groundspeed is less than 10 knots, the electronic trim ball will flash. Care should be taken when engaging HAS as uncommanded aircraft drift may occur. Aircraft position should be carefully monitored using visual cues and symbology.

- If the GPS system is not keyed or is tracking less than 4 satellites, HAS performance may degrade over time. Hover Position Box and Velocity Vector accuracy will be degraded.

In normal operation, the Navigation System redundantly integrates data from the sensors. The EGI is the primary sensor for inertial data with the HARS as the backup. The GPS is the primary sensor for helicopter position. The Doppler provides helicopter velocity data. The GPS, if keyed and verified, will also provide velocity data. The fire control system and flight symbology use the redundant data. There are three instruments and two subsystems that receive dedicated inputs from the sensor array. The HSI, RMI, and RAI receive data directly from the HARS for heading and attitude. The DNS provides data directly to the distance-to-go and No.1 needle displays on the HSI. The DASE and DNS receive data directly from the HARS. For those instruments, including the DASE and DNS, that receive data directly from a sensor, there is no redundancy. If the sensor providing the data has failed, then the affected instrument or subsystem will not function. The distance-to-go and steering information is always displayed on the CDU and HMD symbology. The velocity vector, acceleration cue, vertical speed indicator, and heading tape are driven by the navigation system and not by any single sensor. This allows redundancy in the displays in the event of invalid data or failure of any sensor. No one velocity sensor can cause the error. However, because the Hover Augmentation System (HAS) of the DASE receives velocity data directly from the HARS, it is possible for a velocity error in the HARS to affect HAS and not affect the Navigation System. In the event that the HARS has been in a free inertial condition for longer than 20 seconds the **HARSVEL?** message is displayed in the CDU, NAV status page. If the HARS has been in a free inertial condition for longer than 20 seconds and the groundspeed of the helicopter is less than 10 knots, the pilot HMD trim ball will flash. The purpose of the flashing trim ball is to alert the pilot that if the HAS is engaged there may be an initial drift of the helicopter.

With the integrated navigation system, the Hover Position Box drift varies according to whether the EGI or HARS is in use as the inertial sensor, and if using the EGI, whether or not the GPS is operating in PPS (keyed) or SPS (not keyed) mode or operating at all. Using the EGI with the GPS keyed produces optimal HAS/Hover Position Box drift performance. The amount of drift may be up to 6 feet the first minute, and as much as 23 feet after 5 minutes. All other modes of operation (EGI or HARS) may produce Hover Position Box drift that is random and unpredictable. In these modes of operation, the HAS/Hover Position Box drift performance may be similar to that of the non-integrated navigation system (up to 21 feet per minute).

3.16.1 Embedded GPS Inertial (EGI).

a. System Description. The EGI is a velocity aided, strapdown, ring laser gyro based inertial unit. The EGI unit also houses the 5-channel GPS receiver. The ring laser inertial unit and GPS receiver are treated as separate sensors by the Navigation System and have separate FD/LS indications. The ring laser gyro operates on an optical principle called the Sagnac Effect which deals with the properties of light beams traveling in opposite directions around a closed loop. In operation, two laser beams are directed around the ring in opposite directions; clockwise and counter-clockwise. If the ring rotates in the clockwise direction while the light beams are in transit then the clockwise beam will seem to travel a shorter distance than the counter-clockwise beam. This is used as a measure of rotation by observing the interference pattern created at the end of the ring when the two laser beams mix together; the faster the rotation, the greater the interference pattern.

b. Operation. The EGI begins alignment whenever 28 vdc power is available and aligns in the same manner as the HARS. The vertical axis is found by aligning the vertical ring laser gyro with earth's gravity vector. Inertial north is found by measuring the eastward rotation of the earth about its axis. The time to align for the

EGI is approximately 4 minutes. The time is not significantly affected by temperature. The heading tape symbology will be displayed when the first platform, EGI or HARS aligns. There is no effect on Navigation System accuracy if the engines are started or the main rotor is turning during alignment. There is no reason, other than normal checklist items, to delay engine start.

3.16.2 Global Positioning System (GPS).

a. System Description. The GPS receiver installed in the EGI is a 5-channel receiver. The receiver is capable of operating in either C/A code or encrypted P/Y code. The Group User Variable (GUV) is the normal encryption key used. The GUV key is loaded into the EGI using a KYK-13 or equivalent device. When keyed the GPS receiver will automatically use anti-spoof/jam capabilities when they are in use. The EGI keying connector is located on the aft portion of the right-hand FAB. The EGI will retain the key through power ON/OFF/ON cycles. Because of safeguards built into the EGI, it is not considered classified when keyed. The antenna for the GPS receiver is located on the top of the vertical stabilizer.

b. Operation. The operation of the GPS receiver is entirely automatic. The GPS receiver is powered when the EGI is powered. If the GPS is keyed it may take as long as 12 minutes to verify the key with the satellites. The GPS receiver will provide position and time. The Navigation System automatically configures to handle the GPS and DNS velocities for velocity aiding of both the EGI and HARS. When the GPS has tracked the first satellite, it will provide the date and time to the Navigation System. The date and time (ZULU) is displayed at the top of the ADMIN page of the CDU. There are no specific operator actions that may be taken with the GPS.

3.16.3 Data Transfer Unit (DTU).

The Data Transfer Unit (DTU) is located in the aft portion of the CPG right-hand console. The cartridge is programmed using the Aviation Mission Planning System (AMPS). The AMPS writes the coordinate files for waypoints, targets, present position and laser codes to the cartridge. In the helicopter, the FCC handles all reading and writing to the cartridge. Whenever the CPG makes changes to any of the data, the FCC will automatically update the cartridge with the changes. This will preserve any changes made by the CPG in the event of a power transient. On power-up, either initial or following a transient, the FCC will check for copies of the files it has written out to the data cartridge. If it finds the saved files present, it will automatically load all of the saved files. It the saved files are not found, the FCC will load the AMPS files into the system. This has the effect of automatically loading the AMPS files when a "new" cartridge is used. The CPG has the capability of manually loading and saving files by using the pages on the CDU. FCC memory data may only be "dumped" by powering down the FCC.

3.16.4 Required Navigation Data.

The Integrated Navigation System uses two components of data for proper operation: map datum and altitude. Magnetic variation is automatically calculated by the Navigation System and requires no action by the flightcrew. Spheroid is not used in any portion of the Integrated Navigation System.

a. Altitude. Altitude is used with all coordinate data to permit the accurate prepointing or cueing to any of the coordinate data in the Navigation System. If altitude is not entered and the coordinate data is used to prepoint, the current helicopter altitude is used. The range of altitude entry is -900 to +20000 ft. MSL.

b. Map datums. Map datums are mathematical models of the Earth used to calculate the coordinates on maps and charts. Currently, many datums are used throughout the world to produce maps. The standard for US Forces is World Geodetic System 1984 (WGS 84). However, many US Military Grid Reference System (MGRS) and foreign maps are still based on other datums. Not correcting for different datums can cause significant errors. The coordinates for a point on the Earth's surface in one datum will not match the coordinates from another datum for the same point. The Integrated Navigation System requires that the datum be entered with the coordinate data. If it is not entered, it will default to the present position datum. The datum included with each set of coordinate data allows the Navigation System to calculate the compensations for different datums. Multiple datums can be used and the Navigation System will accurately generate the compensated navigation or prepointing data. The datum used is identified by a 2 digit number. Table 3-21 lists all of the datums that may be used with the Integrated Navigation System.

Table 3-21. Datum Names and Codes

Datum Name/Description	Datum Code Used By Navigation System
Adindan	01
ARC 1950	02
Australian 66	03
Bukit Rimpah: Indonesia	04
Camp Area Astro	05
Djarkarta: Indonesia	06
European 1950	07
Geodetic Datum 1949	08
Ghana	09
Guam 1963	10
Gunung Segara: Indonesia	11
Gunung Serindung: Indonesia	12
Herat North	13
Hjorsey	14
Hu-Tzu-Shan	15
Indian: Mean Value For Thailand And Viet-Nam	16
Ireland 1965	17
Kertau: West Malaysia	18
Liberia	19
Other International	20
Luzon	21
Merchich	22
Montjong Lowe: Indonesia	23
Minna	24
North American Datum of 1927	25
NAD 27 Alaska: Canada	26
Old Hawaii: Maui	27
Old Hawaii: Oahu	28
Old Hawaii: Kauai	29
Ordnance Survey of Great Britain 1936	30
Qornoq	31
Sierra Leone 1960	32
Campo Inchauspe: Argentina	33
Chua Astro: Paraguay	34

Table 3-21. Datum Names and Codes (cont)

Datum Name/Description	Datum Code Used By Navigation System
Corrego Alegre: Brazil	35
Provisional South American 1956	36
Yacare: Uruguay	37
Tananarive: South American 1925	38
Timbali: East Malaysia	39
Tokyo: Japan: Korea: Okinawa	40
Voirol	41
Special Datum: Indian Special	42
Special Datum: Luzon	43
Tokyo Datum	44
WGS 84 Special Datum	45
Sino-Soviet Bloc	46
WGS 84 Standard Datum*	47

* All FAA Sectionals and FLIP use this datum for coordinate data.

3.16.5 COMPUTER DISPLAY UNIT (CDU) IP-1552G
The CDU IP-1552G (fig 3-24.1) used in the integrated navigation system equipped aircraft has five additional FABs to assist the operator with aircraft systems functions. The additional FABs are: **TGT, CODE, DATA, PGM,** and **WPN**. Data upload can be accomplished manually through the CDU alphanumeric keyboard or electronically through the Data Transfer Unit (DTU). Functions of each FAB is described in table 3-22. VAB functions are described with each page/sub page display description.

3.16.6 Navigation System Initialization. When electrical power is applied to the helicopter, the CDU is powered and the **NAV** top level page is displayed as shown in figure 3-25.

NOTE
If DTC overwrites active FLY-TO or TGT, it is necessary to de-select and re-select active FLY-TO or TGT.

a. At initial power-up mode of the aircraft, power is automatically applied to the EGI. The HARS is powered when an alignment mode is selected, and the DNS is automatically powered during the HARS startup sequence. The CDU will default to the top level **NAV** page and will be displayed as shown in figure 3-25.

TM 1-1520-238-10

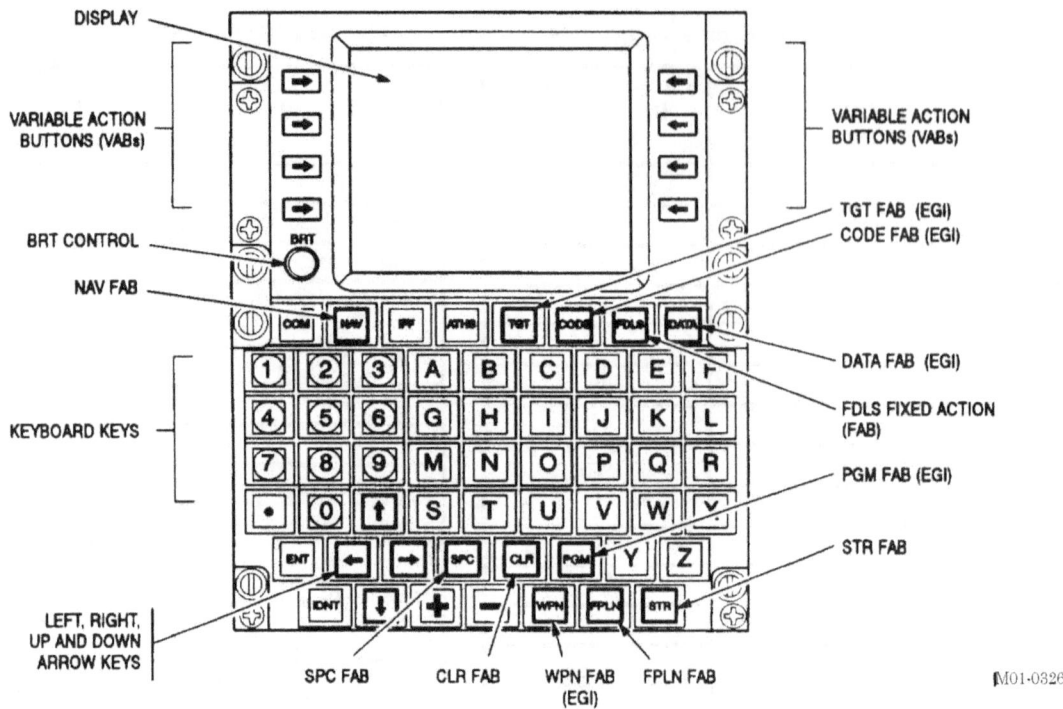

Figure 3-24.1. CDU IP-1552G, Used with AN/ASN-137 DNS

Table 3-22. IP-1552G CDU Control and Display Functions

Control/Display	Function
Display Screen	Display formats are organized as pages with 8 display lines; the bottom line is always used as the scratchpad for entering or editing data.
Keyboard	Alphanumeric keyboard is used for entering data.
Left/Right Arrows	Move the cursor one space left or right, as appropriate, per keystroke. When pressed constantly, will move the cursor at about 4 character positions per second.
Up/Down Arrows	Scroll through the waypoint dictionary pages, one page per keystroke. When constantly pressed, will scroll through all dictionary pages at about two pages per second.
BRT Control	Adjust display brightness.
COM	Not Used.
NAV	Displays NAV top level page.
IFF	Not Used.
ATHS	Not Used.
TGT	Access to Target List page.
CODE	Access to Laser Code page.

3-64.2   Change 3

corrections computed by the DNS computer and are applied to all subsequent Doppler radar velocities. The velocity bias (error), when present, will be most noticeable when the helicopter is at a hover or slow speeds. The reason is that at a hover the velocity error, although small is about the same amount as the actual helicopter velocity. When HBCM is selected, the CPG can manually start and stop the calibration. If a calibration NO-GO status is displayed at the conclusion, a recalibration may be restarted by the CPG. If a GO calibration is computed, the bias velocities will automatically be stored and applied continuously to all subsequent navigation computations.

Table 3-22. IP-1552G CDU Control and Display Functions

Control/Display	Function
Display Screen	Display formats are organized as pages with 8 display lines; the bottom line is always used as the scratch pad for entering or editing data.
Keyboard	Alphanumeric keyboard is used for entering data.
Left/Right Arrows	Move the cursor one space left or right, as appropriate, per keystroke. When pressed constantly, will move the cursor at about 4 character positions per second.
Up/Down Arrows	Scroll through the way point dictionary pages, one page per keystroke. When constantly pressed, will scroll through all dictionary pages at about two pages per second.
BRT Control	Adjust display brightness.
COM	Not Used.
NAV	Displays NAV top level page.
IFF	Not Used.
ATHS	Not Used.
TGT	Access to Target List page.
CODE	Access to Laser Code page.
FDLS	Access to Aircraft Fault Detection and Location System functions.
DATA	Access to Data Menu Top Level page.
SPC	Used only for entry of coordinate ID or FDLS/BST operations.
ENT	Not Used.
CLR	If pressed in response to an ERROR prompt in the scratch pad, will clear the error prompt and position the cursor at the first detected data entry error. If pressed a second time, will clear the entire scratch pad. If no ERROR prompt, will clear the entire scratch pad.
PGM	Access to Program Top Level page.
IDNT	Not Used.
WPN	Access to Weapon Control Top Level page.
FPLN	Access to Way point LIST page.
STR	Stores PPOS in next way point store location (31-40), overwriting existing data.

TM 1-1520-238-10

Table 3-22.1. NAV TOP LEVEL PAGE VARIABLE ACTION BUTTONS

Control/Display	Function
VAB 1	If **NAVSTAT 1** is displayed on line 6 of the **NAV** page, VAB 1 is disabled. Depressing the VAB 1 button with no data present in the scratchpad freezes the current displayed **PPOS** data and changes the **PPOS** legend to ⇒ **UPD** indicating the system is ready for a manual **PPOS** update. A second depression of VAB 1 with no scratchpad data present aborts the manual update. Depressing the VAB 1 button with data present in the scratchpad that satisfies the **PPOS**, Datum, or MAGVAR data entry rules, then the intended system parameter will be updated. Used for manually entering **PPOS** and Datum updates. Allows manual MAGVAR updates when the outlined arrow a is displayed in character position 1 on line 2.
VAB 2	Not Active.
VAB 3	Not Active.
VAB 4	When depressed, if the left-hand character field in line 7 displays the legend **UPD TO**, will complete a fly over type **PPOS** update; and display updated **PPOS** on line 1. When updated data is accepted or aborted, line 7 will return to its' previous legend of ⇒ **FLY TO**.
VAB 5	With valid **FLY TO** data symbol (WXX or TXX with valid previously entered coordinates) entered on the scratchpad, will change the active **FLY TO** to location entered in scratchpad. Invalid scratchpad coordinate data will result in "ERROR" message on scratchpad. Without data in scratchpad; will display active **FLY TO** coordinates in scratchpad with initial depression. Second depression will display the Associated data line (ID, Altitude and Datum) for the active **FLY TO** coordinates. Subsequent depressions will toggle between the **FLY TO** coordinates and the Associated Data. **FLY TO** coordinates and Associated Data displayed in the scratchpad are for review purposes only, and cannot be edited or transferred. If **NAVSTAT 1** is displayed on line 6 of the **NAV** page, VAB 5 is disabled. Depression of VAB 5 with other than **NAVSTAT 1** status while overflying a waypoint will freeze the **PPOS** display on line 1 and changes the **FLY TO** legend on line 7 to **UPD TO**. Reselection of VAB 5, or selection of any other page, will abort the update.
VAB 6	Presents the **ADMIN** sub page when depressed.
VAB 7	Not Active.
VAB 8	Not Active.

3.16.7 CDU Displays. The CDU display architecture is organized into page formats. Each page consists of 8 display lines. All lines are 22 characters wide. The bottom line (eighth) is used as a scratchpad for entering or editing data. Individual pages may display navigation data, display failure messages, or indicate selectable functions. The CDU displays are categorized into 8 top level page formats: **NAV, TGT, DATA, CODE, FDLS, PGM, WPN**, and **FLPN**. The CDU will automatically revert back to the **NAV** top level page from the **ADMIN** page after 30 seconds if there are no characters in the scratchpad and no other key presses are made.

0-64.4 Change 3

NAV Top Level Page. When the DNS is powered, the NAV top level page (fig 3-26.1) will display the following:

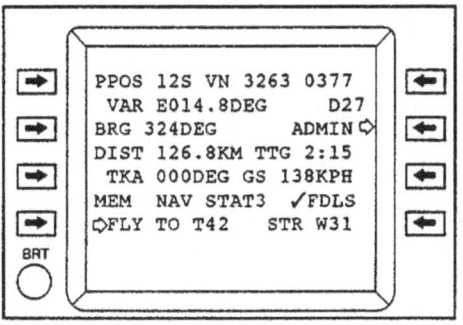

Figure 3-26.1. NAV Top Level Page (UTM Format)

Line 1. Displays the legend **PPOS** followed by the current **PPOS** coordinates. **PPOS** represents the computed present position in either **UTM** or **LAT/LONG** coordinates. **PPOS** changes to OUPD when an on-the move manual **PPOS** update is initiated.

Line 2. The left-hand character field displays the legend **VAR** followed by the **PPOS** MAGVAR in degrees. If the INS is NO-GO, an outlined arrow (O) will be displayed immediately to the left of **VAR** permitting manual **MAGVAR** updates. The right-hand character field displays the legend D followed by the **PPOS** Datum number.

Line 3. The left-hand character field displays the legend **BRG XXXDEG** reflecting the direct bearing to the destination specified in the **FLY TO** legend. The **BRG** legend is a read-only display. The right-hand character field displays the legend **ADMIN**. An outlined arrow (O) following **ADMIN** indicates the ability to select the **ADMIN** page for display.

Line 4. -The left-hand character field displays the legend **DIST XXX.XKM**. **DIST** is displayed in KM (kilometers) when the coordinate system in use is metric, and in NM (nautical miles) when the coordinate system in use is nautical. The display uses a floating decimal point allowing a distance accuracy of 10 meters (32 feet) when **DIST** is less than 100 KM units. The **DIST** numerals are updated at the rate of 1Hz regardless of aircraft speed. The right-hand character field displays

the legend **TTG X:XX** to the **FLY-TO** destination in Hours and Minutes (H:MM). When less than five (5) minutes from destination, these units change to Minutes and Seconds (M:SS). If **TTG** to the **FLY TO** exceeds 9 hours and 59 minutes, or the aircraft ground speed is less than 10 knots, the display is dashed. Both of the character fields are read-only and cannot be edited.

Line 5. Displays the legend **TKA XXXDEG** and **GS XXXIPH**. **TKA** denotes the computed Track Angle, in Degrees **(DEG)**, while **GS** reflects Ground Speed. **GS** is presented in Kilometers Per Hour **(KPH)** or Knots **(KTS)** dependent on display selected. The units for both **DIST** and **GS** are the same for a given selection: **KM** and **KPH**, or **NM** and **KTS**.

Line 6. Displays the System Annunciator Data which consists of three elements: A **DNS** Memory **(MEM)** element; a Navigation Status **(NAV STAT 1-3)** element; and a Check FDLS (V **FDLS**) element. These status messages may appear at any time on this line. A display on this line indicates that the **DNS** has either lost radar lock **(MEM)**; the navigation error exceeds certain parameters **(NAV STAT 1-3)**; or, the Continuous **FDLS** test has detected a fault in the system (/ **FDLS**). Navigation Status categories are as follows:

NAV STAT 1: Estimate of position error calculation is less than 50 meters. Actual position error may be substantially less. **PPOS** updates are disabled.

NAV STAT 2: Estimate of position error calculation at least 50 meters but less than 201 meters. Actual position error may be substantially less. All **PPOS** update operations are enabled.

NAV STAT 3: Estimate of position error calculation is greater than 200 meters. Actual position error may be substantially less. All **PPOS** update operations are enabled.

Line 7. The left-hand character field displays the legend ⇒ **FLY TO TXX** or ⇒ **FLY TO WXX**, depending upon the selected destination to which the system is navigating. The displayed value can be any of the 40 waypoint locations in the **FPLN** List; or any of the 40 target locations. ⇒ **UPD TO WXX** displayed indicates the destination to which the system is being updated. The right hand character field displays the legend **STR WXX** representing the waypoint number (from 31-40) to be used by the system for the next **PPOS** store.

Line 8. This is the scratchpad line for manual data entry.

TM 1-1520-238-10

b. **Administration (ADMIN) Page.** The **ADMIN** page (fig 3-26.2), is selected by depressing VAB 6 on the **NAV** Top Level Page. VAB functions are explained in Table 3-22.2. The **ADMIN** page contains the following:

Figure 3-26.2. ADMIN Page Top Level

Line 1. Displays the example legend ⇒ **12/24/94** denoting the current Month, Day, and Year., separated by slashes (/) in the left-hand character field. Update of this field is automatic through GPS data, and when updated, the O is blanked. If the system has no date available (GPS or Manual), this field is displayed as dashed (--/--/--). The current time example legend 15:42:18< is displayed in the right-hand character field representing Hours, Minutes, and Seconds, separated by a colon(:), in that order. If GPS time is not valid,(<: is displayed), manual update to the system is permitted.

Line 2. Displays the centered page title ADMIN.

Line 3. The left-hand character field displays an outlined arrow and the legend ⇒ **A XXXXXT** reflecting

the current aircraft altitude. This altitude value is always displayed in Feet only, and is referenced to MSL. Positive altitude above MSL is displayed with a blank preceding the altitude readout; negative altitude (below MSL) is displayed using a minus (-) sign preceding the altitude readout (**A-XXXXXFT**). The right-hand character field displays the legend **HGXX.XX<** denoting the current Barometric pressure setting.

Line 4. -The left-hand character field displays the legend **TKAE L XXX**.. **TKAE** represents the current Track Angle Error; **L** or **R** denotes Left or Right; and the **XXX** signifies the angle error in degrees. The right hand character field displays the legend **XTKE L XXX**.

XTKE represents the current Cross Track Error; **L or R** denotes Left or Right; and **XXX.X** signifies the distance in Kilometers or Nautical Miles, dependent upon units selected by the CPG, and displayed in line 7 **(DSPL:KM or DSPL:NM).**

Line 5. Displays the legend **MODE:LAND** or **MODE:WATER** in the left-hand character field. The right-hand character field displays the legend **HBCMO** ⇒ if the DNS system status is GO; otherwise this field is blanked.

Line 6. -Blank.

Line 7. The left-hand character field displays the legend **MODE:UTM** or **MODE:L/L** signifying coordinate data entry methodology. The right-hand character field displays the legend **DSPL:KM** or **DSPL:NM** signifying metric system or nautical system, respectively, in use.

Line 8. This is the scratchpad line for manual data entry.

3-64.6 Change 3

Table 3-22.2. ADMIN SUB-PAGE VARIABLE ACTION BUTTONS

Control/Display	Function
VAB 1	Used for manual entry of **MONTH DAY YEAR** update data (**MMDDYY**) when the outlined arrow is (0) displayed in character position 1 on line 1, and character string is present in the scratchpad. Valid entries are: 01-12 for **MM**; 01-31 for **DD**; and 00-99 for **YY**
VAB 2	Used for manual entry of Altitude update data. Positive altitude is assumed and does not require a "+" sign; Negative altitude (below MSL) requires a minus (-) sign preceding the altitude value. Altitude ranges are -900 feet MSL to 20000 MSL. Altitude is always displayed in feet.
VAB 3	Used for selecting MODE**:LAND or MODE :Water** navigation operations. System defaults to **MODE:LAND** at power up. Depressing VAB 3 toggles between two choices.
VAB 4	Selects the data presentation format to be used for **PPOS, CPOS, WAYPOINT**, and **TARGET** coordinate data entry. System defaults to **MODE:UTM** at power up. Depressing VAB 4 toggles between **MODE:UTM** AND **MODE:L/L**. Pages affected by pressing **VAB 4** are **NAV, OFFSET, WAYPOINT and TARGET**.
VAB 5	Used for manual entry **of HOUR MINUTE SECOND** update data (**HHMMSS**) when the outlined arrow is (0) displayed in character position 22 on line 1, and character string is present in the scratchpad. Valid entries are: 00-23 for **HH**; 00-59 for **MM**; and for SS.
VAB 6	Used for manual entry of **HG** (barometric pressure). Valid entries are: 27.92 to 31.92. Data is entered in scratchpad as **XXXX**; the decimal point entry is not required. This entry followed by an altitude entry within 60 seconds will cause the system to re-calibrate the ADS pressure sensor bias correction.
VAB 7	Used for selecting **HBCM** sub-page. Valid only if **HBCM** Ois present indicating the DNS system is "GO".
VAB 8	Used for selecting **DSPL:KM** or **DSPL:NM** signifying metric system or nautical system, respectively, in use for coordinate display. Depressing VAB 8 toggles between the two (2) choices.

TM 1-1520-238-10

c. **HBCM Page.** The **HBCM** page, (fig 3-26.3) is used to START and STOP the bias calibration of the DNS. VAB functions are explained in table 3-22.3. The display has the following:

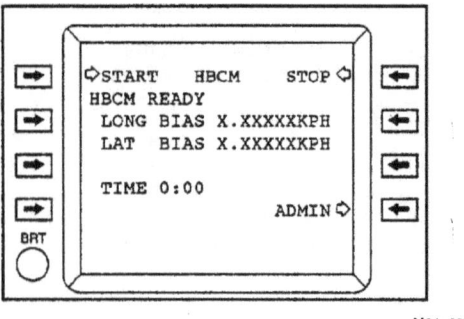

Figure 3-26.3. HBCM Page

Line 1. Displays the legend O **START** in the left-hand character field; the **HBCM** title in the center character field; and the legend **STOPO** in the right-hand character field.

Line 2. Displays the legend(s) **HBCM READY**;.

HBCM ACTIVE, HBCM GO, or **HBCM NO-GO** as appropriate in the left-hand character field.

Lines 3. & 4 If line 2 displays **HBCM READY** or **HBCM ACTIVE**, lines 3 & 4 will display the current DNS velocity biases. If line 2 displays **HBCM GO**, lines & 4 will display the new **DNS** velocity biases. If line 2 displays **HBCM NO-GO,** line 3 will display the **HBCM** fail condition. If a second fail condition exists, it will be displayed on line 4, otherwise line 4 is blanked. Possible **DNS** error messages are:

TIME BELOW 2 MIN BIAS EXCEEDS 0.3 KTS EXCESSIVE MEM Line 5. Blank.

Line 6. Displays the legend **TIME: XX:XX** in the left-hand character field indicating elapsed time in minutes/seconds for **DNS** computations of the bias calibration. The timer will stop while the **DNS** is in memory during the bias calibration. The legend **MEM** will be displayed following 'TIME: XX:XC' if the DNS velocity is invalid; otherwise the rest of line 6 is blanked.

Line 7. The right-hand character field displays the legend **ADMIN>O**.

Line 8. Blank.

Table 3-22.3. HBCM SUB-PAGE VARIABLE ACTION BUTTONS

Control/Display	Function
VAB 1	Used to initiate or re-start the calibration of the Hover Bias. Pressing VAB 1 causes the display on line 2 to change from **"HBCM READY'** to **"HBCM ACTIVE"** and starts the timer display located on line 6. Calibration time is dynamic and is displayed in minutes and seconds (M: SS).
VAB 2	Not Used
VAB 3	Not Used
VAB 4	Not Used
VAB 5	Used to stop the calibration of the Hover Bias. Pressing VAB 5 will cause the message **"HBCM GO"** or **"HBCM NO-GO"** to appear on line 2. This indicates a "within limits" or "out of limits" calibration condition.
VAB 6	Not Used
VAB 7	Not Used
VAB 8	Used to return to the **ADMIN** sub-page display.

3.64.8 Change 3

d. Offset Update (OFS UPDATE) Page. If **NAV-STAT 1**, the offset update operations are disabled and, the offset update page cannot be activated. If **NAV-STAT 2** or **NAVSTAT 3**, offset update is accomplished through actioning the Update/Store (UPDT/STR) rocker switch located on the ORT. A momentary depression of the UPDT side of the rocker switch results in the display of the **OFS UPDATE** page (figure 3-26.4) on the CDU. VAB functions are explained in table 3-22.4. The **OFS UPDATE** page is displayed as follows

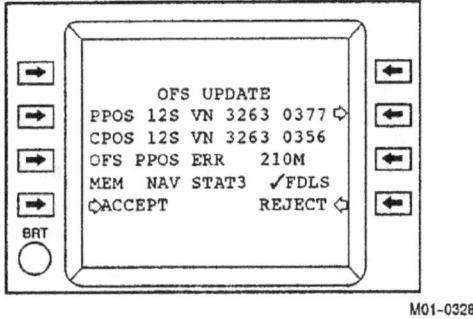

Figure 3-26.4. OFS UPDATE Page

Line 1. Blanked

Line 2. Displays the centered page title OFS UPDATE.

Line 3. Displays the legend PPOS XXX VN XXXX XXXX representing the current PPOS at the moment of activation of the OFS UPDATE page.

Line 4. Displays the legend CPOS XXS VN XXXX XXXX representing the computed, corrected aircraft present position based upon CPG range, LOS angles and aircraft bearing relative to the currently selected active target.

Line 5. Displays the legend OFS PPOS ERR X)OXXXXM. This denotes the radial distance between PPOS & CPOS. The value is used as a measure of PPOS error by the CPG to determine whether to accept or reject the update. The radial error is limited to 99999 meters on the display.

Line 6. Displays System Annunciator Data.

Line 7. The left-hand character field displays the legend **OACCEPT.** The right-hand character field displays the legend **REJECTO**.

Line 8. - Blanked.

Table 3-22.4. OFS UPDATE SUB-PAGE VARIABLE ACTION BUTTONS

Control/Display	Function
VABs 1, 2 & 3	Not Used.
VAB 4	Used to accept the offset update. Display returns to page previously displayed prior to selecting the offset update function.
VABs 5, 6 & 7	Not Used
VAB 8	Used to reject the offset update. Display returns to page previously displayed prior to selecting the offset update function.

e. FDLS page. Depression of the **FDLS FAB** displays the FDLS page on the CDU. The display on the CDU (Fig. 3-26.5) is intended to inform the operator that FDLS prompts/displays will not be provided on the CDU, and FDLS operations are to be conducted through the ORT and HMD. All VABs are inactive on the FDLS page display. Entry of any FDLS Menu two digit codes into the CDU keyboard will initiate the FDLS BIT. The CDU SPC key is used in conjunction with FDLS operations as follows:

TM 1-1520-238-10

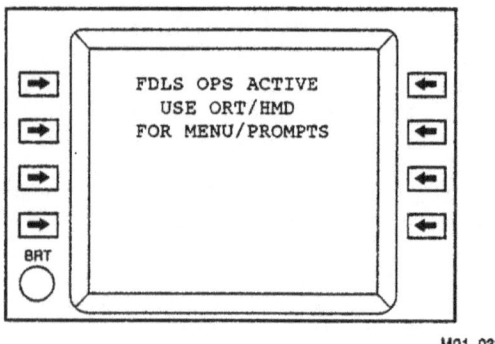

Figure 3-26.5. FDLS Page

(1) Upon initial selection of FDLS, with more than one FDLS NO-GO message displayed, depressing the SPC key allows scrolling among the continuous TEST NO-GO messages.

(2) With a FDLS Menu page displayed, depressing the SPC key allows scrolling among the FDLS Menu pages.

(3) At completion of a maintenance test, with more than one FDLS NO-GO message displayed, depressing the SPC key allows scrolling among the maintenance test NO-GO messages.

(4) With a FDLS Action/Acknowledge Prompt displayed, depressing the SPC key is interpreted by the system as an acknowledgement of the requested prompt.

(5) During an ongoing maintenance test without a FDLS Action/Acknowledge Prompt displayed, depressing the SPC key will abort the test.

3.16.8 FPLN CDU Displays. Depressing the **FPLN FAB** displays the Waypoint List (**WPT LIST**) page

a. WAYPOINT LIST (WPT LIST) Page.

NOTE

• All data and messages depicted in displays are typical representations.

• The Waypoint List and Target List are separately, but identically, structured. The following data applies to the operation of both. Where minor operating differences occur, each is described separately.

• Waypoints are numbered 1 through 40; Targets are numbered through 80.

(1) Each page provides access to a list of 40 Waypoints contained in the Waypoint List (Figures 3-26.6 and 3-26.7) or 40 targets contained in the Target List. The WPT LIST and TGT LIST are identically formatted except for the page title on line 2 of the display and numbering. Waypoints numbered 1-20 are displayed on the first page; waypoint numbered 21-40 are displayed on the second page. The first TGT LIST page will list numbers page 2 will list numbers 61-80. The second page of each list is accessed by depressing VAB 8 on page 1. Depressing VAB 8 on the second page will return the display to the first page.

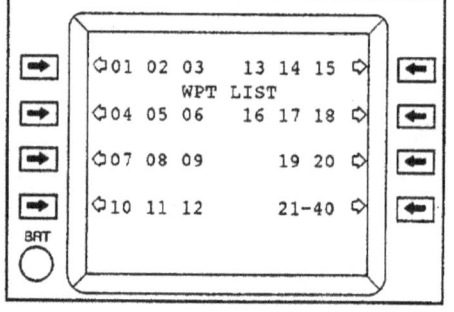

Figure 3-26.6. FPLN Top Level Page

3-64.10 Change 3

TM 1-1520-238-10

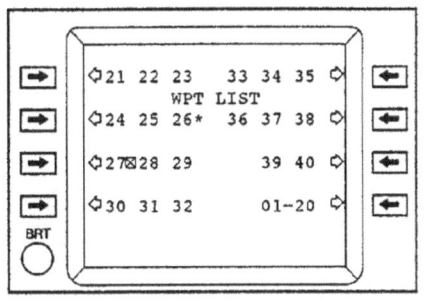

Figure 3-26.7. FPLN Second Level Page

(2) Three special symbols are associated with Active FLY-TOs and Active TGTs on these pages. An Asterisk (*) symbol displayed to the right of a TGT or WPT designator (TXX or WXX) indicates that location as the Active FLY-TO location. Similarly, an outlined cross (m) symbol indicates the location is the Active Target location. A pound (#) symbol displayed to the right of the designator indicates that point as being BOTH the Active FLY-TO and the Active TGT location.

(3) With either the TGT or WPT LIST page displayed, entry of a valid Coordinate Symbol (TXX or WXX with valid coordinates previously entered) into the scratchpad and depression of any of VABs 1-4 will cause the scratchpad entered data to become the Active FLY-TO location. Depression of any of VABs will cause the scratchpad entered data to become the Active TGT location. With an invalid coordinate symbol, the "ERROR" message is displayed in the scratchpad.

Line 1. Displays waypointS 001 02 03 in the Left-hand character field; waypoint 13 14 15> in the right-hand character field.

Line 2. Displays centered title **WPT LIST**.

Line 3. Displays waypoint 0 04 05 06 in the Left-hand character field; waypoint 16 17 180 in the right-hand character field.

Line 4. - Blanked.

Line 5. Displays waypoint 007 08 09 in the Left-hand character field; waypoint 19 20c> in the right-hand character field.

Line 6. Blanked.

Line 7. Displays waypoints 010 11 12 in the Left-hand character field; waypoints 21-400C in the right-hand character field.

Line 8. Scratchpad.

b. Waypoint Coordinate (WPT COORD) Page. Depressing the **FPLN FAB** displays the Waypoint List (**WPT LIST**) page. The selection and depression of a specific VAB (Table 3-22.5) will result in the display of a desired set of coordinates.

NOTE
All data and messages depicted in displays are typical representations.

The accessing of the sample waypoints and coordinates shown in figure 3-26.8 is the result of depressing the **FPLN FAB**, then **VAB 8**, then **VAB 2**. VAB functions for the displayed waypoint coordinate page are explained in table 3-22.6.

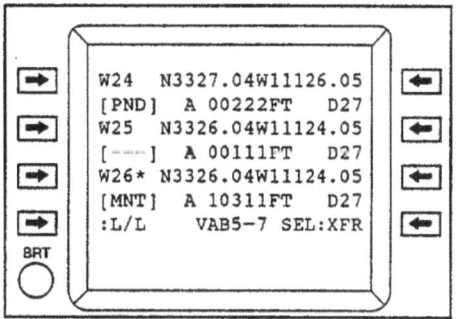

Figure 3-26.8. Selected Waypoint Coordinates

The Up/Down arrow keys can be used to scroll through the Waypoint Coordinate pages. An * is shown immediately after the storage location currently being used for **FLY TO** computations and it cannot be changed until some other location is chosen to **FLY TO**. The page has the following data organization:

Line 1. Displays the waypoint number (W24, as shown) and the coordinate location data of that waypoint in the **L/L** system.

Change 3 3-64.11

Line 2. Displays the associated data for waypoint 24, consisting of a three letter ID (PND, as shown) surrounded by brackets; Altitude (an A followed by up to five character spaces for numerals and the unit of measure FT); and the Datum number (D27, as shown).

Line 3. Displays the next waypoint number (W25) and coordinate location data for that waypoint.

Line 4. Displays the associated data for waypoint 25, consisting of a three letter ID surrounded by brackets; Altitude (an A followed by up to five character spaces for numerals and the unit of measure FT); and the Datum number (D27, as shown). If the characters within the brackets are dashed (---), it is an indication that no data has been entered for this waypoint.

Line 5. Displays the next waypoint number (W26) and coordinate location data for that waypoint.

Line 6. Displays the associated data for Waypoint 26, consisting of a three letter ID (MNT, as shown) surrounded by brackets; Altitude (an A followed by up to five character spaces for numerals and the unit of measure FT); and the Datum number (D27, as shown).

Line 7 The left-hand character field displays the data format selected for page display (either L/L or UTM).

The right-hand character field displays the legend VAB 5 7 SEL: followed by selection alternatives XFR, TGT, or R/B.

Line 8. Scratchpad for data input.

Table 3-22.5. FLPN PAGE 1 VAB OPERATIONS WITH SCRATCHPAD EMPTY

Control/Display	Function
VAB 1	Selects Waypoints 01-03 for display on Waypoint Coordinate Page
VAB 2	Selects Waypoints 04-06 for display on Waypoint Coordinate Page
VAB 3	Selects Waypoints 07-09 for display on Waypoint Coordinate Page
VAB 4	Selects Waypoints 10-12 for display on Waypoint Coordinate Page
VAB 5	Selects Waypoints 13-15 for display on Waypoint Coordinate Page
VAB 6	Selects Waypoints 16-18 for display on Waypoint Coordinate Page
VAB 7	Selects Waypoints 19-20 for display on Waypoint Coordinate Page
VAB 8	Displays page 2 of WPT LIST allowing access to Waypoints 21-40

Table 3-22.6. WAYPOINT COORDINATE PAGE VAB OPERATIONS

Control/Display	Function
VAB 1-3 (SCRATCHPAD EMPTY)	With valid data in the coordinate location, depression of VABs 1-3 will select the coordinates associated with the selected VAB as the Active FLY-TO location.
VAB 1-3 SCRATCHPAD CONTAINS DATA)	With Coordinate, ID, Altitude or Datum data in the scratchpad, depression of VABs 1-3 will cause the scratchpad data to be validated. If valid, the data item associated with the depressed VAB will change to the entered value. If data is invalid, an error message will display prompting operator to employ normal scratchpad editing rules. If the intended destination of the entry is either the Active Fly-To or Active Target, the prompt "NO MOD TO ACTIVE COORD" will appear in the scratchpad. This signifies that the system will not permit the Active Fly-To or Target coordinate to be modified.

Table 3-22.6. WAYPOINT COORDINATE PAGE VAB OPERATIONS - continued

Control/Display	Function
VAB 4	Toggles between **UTM** or **L/L** display formats.
VAB 5-7 (WITH DISPLAYED)	With valid data in the coordinate location, depression of any of these VABs will select the SEL:TGT coordinate associated with the depressed VAB as the Active Target location.
VAB 5-7 (WITH SEL:R/B DISPLAYED)	Displays the range and bearing (**Range XXXXXXBRGXXX**) to the selected coordinate on line 8.
VAB 5-7 (WITH SEL:XFR DISPLAYED	Copies the coordinate data associated with the depressed VAB into the scratchpad.
VAB 8	Depression of this VAB allows toggling between **VAB 5-7 SEL:XFR**, **VAB 5-7 SEL:TGT** and **VAB 5-7 SEL:R/B**. Upon system power-up, the system defaults to **VAB 5-7 SEL:TGT**. Following selection of **SEL:XFR** or **SEL:R/B**, the system defaults back to **SEL:TGT**.

3.16.9 Weapon Control (WPN CONTROL) Page. Depressing the **WPN FAB** allows the operator to access the **WPN CONTROL** page (figure 3-26.9). Variable Action Button operation is detailed in Table 3-22.7. When the **WPN CONTROL** page is selected, the **WPN CONTROL** display is formatted as follows:

Figure 3-26.9. Weapon Control Display Page

Line 1. - Blanked.

Line 2. Displays the centered page title **WPN CONTROL**.

Line 3. The left-hand character field displays the legend **RNG:MAN XXXXXXM** showing the current range source and current range value in meters. The current CPG range source may be any of the following sources: **AUTO, DFLT, LSR, MAN, NAV, NOAR or TGT** (See Table 3-22.8) Line 4. Blanked,

Line 5 Displays the legend **T(or W)XX DXXXXXX BRG XX.X** denoting Active Target number, distance to Active Target in meters (D000001 to D999999) and bearing (**BRG**) to Active Target in degrees and tenths of degrees (000.0 to 359.9) respectively.

Line 6. Displays System Annunciator Data.

Line 7. The left-hand character field displays the legend **TGT RPT: ON (or OFF)**. The right-hand character field displays **STR TXX** denoting the location for the next **TGT Store** function (71-80).

Line 8. Scratchpad for data input.

Change 3 3-64.13

Table 3-22.7. WEAPON CONTROL PAGE VAB OPERATIONS

Control/Display	Function
VAB 1	Inactive
VAB 2	With the scratchpad empty, or with a leading "A" or "0" entered in the first position in the scratchpad, depressing VAB 2 selects **AUTO** range as the CPG range source. With a scratchpad entry of a range (1 to 999,999 meters), depressing VAB 2 changes the CPG range source to manual and will update the CPG range source value to the entered value.
VAB 3	With a valid target symbol, **TXX or WXX**, and valid coordinates previously entered in the scratchpad, depressing VAB 3 will designate the scratchpad entry as the new Active Target. Invalid entry will result in the ERROR message being displayed in the scratchpad. Without data present in the scratchpad, depressing VAB 3 presents the Active Target coordinates in the scratchpad. With the Active Target coordinates in the scratchpad, a second depression of VAB 3 results in presentation of the Associated Data for the Target coordinates. Coordinate and Associated Data in the scratchpad are "Display Only"; and are not editable or transferable.
VAB 4	Depressing VAB 4 results in alternate presentation of the Target Report display (**TGT RPT:ON** or **TGT RPT:OFF**). With selection of **TGT RPT:ON**, the CPG's Video display will be as shown in figure 3-26.10. Upon power up, the system defaults to **BTGT RPT:OFF**.
VABs 5-8	Inactive (

Table 3-22.8. CPG RANGE SOURCE

Display	Range Source Definition
AUTO	Auto Range
DFLT	Default Range (Only displayed at system start-up or AUTO range selected and invalid CPG LOS or invalid radar altitude).
LSR	Laser Range
MAN	Manual Range
NAV	Nav Range
NOAR	No Auto range (only displayed with AUTO range selected and the CPG LOS elevation angle is approaching the horizon or the radar altitude is less than 5 meters).
TGT	Target range

TM 1-1520-238-10

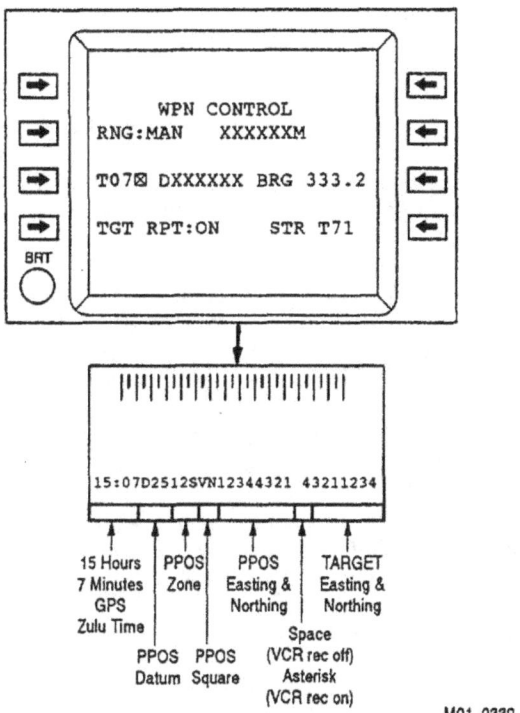

Figure 3-26.10. CPG Video Target Report Data

3.16.10 Missile Laser Codes (CODES) Page.
The Missile Laser Codes page (CODES) is accessed by depressing the CODES FAB. Each of the letter code positions displayed on lines 1, 3, 5, and 7 of the CDU display (figure 3-26.11) are accessible by depression of the associated VAB. Code entry is accomplished through the keyboard and scratchpad operation. Codes are entered into the scratchpad as numerical values only. The first character position will always be either a number or a number "2" only; character positions 2-4 will only utilize numbers 1-8; letters are not required to be entered.

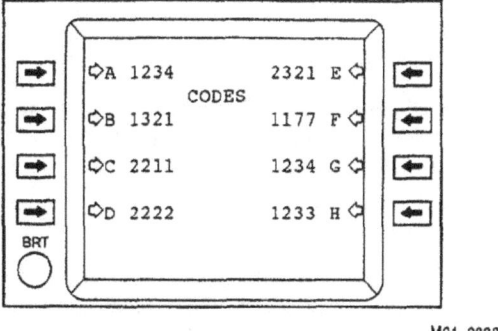

Figure 3-26.11. Missile Laser Codes Display

3.16.11 DATA MENU Page. Selection and depression of the **DATA FAB** will result in the display of the top level **DATA MENU** Page (figure 3-26.12). The **DATA MENU** page provides access to **HAVE QUICK (HQ)** radio functions and the following sub-pages: **NAV STAT, NAV SENSOR CONTROL, DTU, ZEROIZE,** and **GPS STAT**. Access to the menu sub-pages is accomplished through selection and depression of the associated VABs for the sub-page desired.

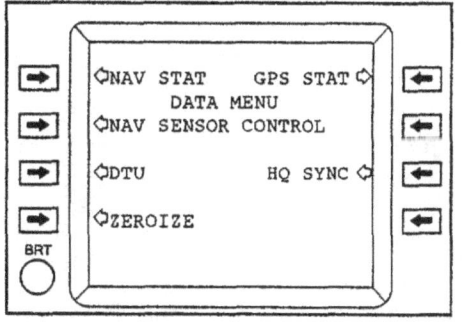

Figure 3-26.12. DATA Top Level Menu Page

Line 1. Left-hand character field displays the legend **ONAV STAT**. The right-hand character field displays the legend **GPS STATO**

Line 2. Displays the centered page title **DATA MENU**.

Line 3. The left-hand character field displays the legend **ONAV SENSOR CONTROL**.

Line 4. Blanked.

Change 3 3-64.15

TM 1-1520-238-10

Line 5 - Left-hand character field displays the legend **ODTU**.

Line 6. - Blanked.

Line 7. - The left-hand character field displays the legend **OZEROIZE**. The right-hand character field displays the legend **HQ:SYNCO**.

Line 8. - Blanked.

Navigation Status (NAV STATUS) Page. The **NAV STATUS** page is accessed by depressing VAB 1 on the **DATA** Menu page. The **NAV STATUS** page is displayed as shown in figure 3-26.13. All data displayed on the **NAV STATUS** page is read-only data and non editable. VABs 1-7 are inactive, and VAB 8 is used to return to the **DATA MENU** top level page.

Figure 3-26.13. NAV STATUS Page

Line 1. Left-hand character field displays the legend EPE (Estimated Position Error) followed by the error stated in Meters (XXXXM). The distance error display is limited to four (4) characters, or a maximum of 9999M. The right-hand character field displays the current INS mode of operation. Possible displays in this character field, dependent upon whether MODE: LAND or MODE: WATER is selected are:

INIT-----Initialization
LEVEL---Leveling
TEST----Internal Test
CGA-----Coarse Ground Align
FGA-----Fine Ground Align
ICGA----Interrupted Coarse Ground Align
IFGA----Interrupted Fine Ground Align

CIFA----Coarse Inflight Align
FIFA----Fine Inflight Align
CSA-----Coarse Sea Align
FSA-----Fine Sea Align

Line 2. - Displays the centered page title **NAV STATUS**.

Line 3. - The left-hand character field displays INS status information from the following list:

INS ATT?-----INS attitude data invalid
INS VEL?------INS airframe coord. velocity invalid
INS HDG?-----INS heading data invalid
INS NOGO----INS not operationally useable
INS GO----- INS aligned and operational

The right-hand character field displays GPS status information from the following list:

GPS NOGO-----GIPS not operationally useable
GPS BATT LO--GPS battery voltage low
GPS GO--------GPS is operational

Line 4. - Blanked.

Line 5 - The left-hand character field displays DNS status information of DNS GO or DNS NOGO. The right-hand character field displays HARS status information from the following list:

HARS CAL---HARS calibration mode
HARSATT?---HARS attitude data invalid
RARSVEL?---HARS velocity data invalid
HARSHDG?--HARS heading data may be invalid
HARS TST---HARS in internal test mode
HARSLAT?---HARS latitude may be invalid
HARSFAST---HARS in fast align mode
HARSNORM--HARS in normal align mode
HARS GO-----HARS aligned and operational
HARSNOGO---HARS is not operationally usable

Line 6. - Displays System Annunciator Data.
Line 7. - The right-hand character field displays the legend **DATAO.**
Line 8. - Blanked.

b. Navigation Sensor (NAV SENSOR CONTROL) Page. The **NAV SENSOR CONTROL** page is accessed by depressing VAB 2 on the **DATA** Menu page. The **NAV SENSOR CONTROL** page is displayed as shown in figure 3-26.14. VAB 7 is used to return to the **DATA MENU** top level page. VAB functions are explained in table 3-22.9.

3-64.16 Change 3

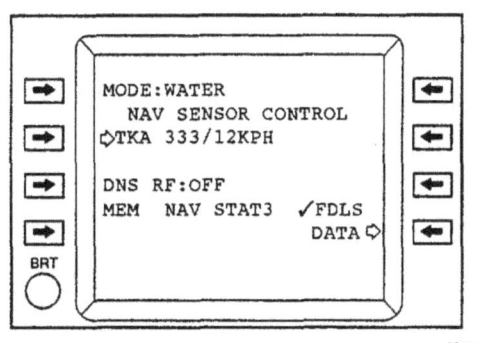

Figure 3-26.14. NAV SENSOR CONTROL Page

Line 1. - Left-hand character field displays the legend **MODE:WATER or MODE:LAND**.

Line 2. - Displays the centered page title NAV SENSOR CONTROL.

Line 3. The left-hand character field displays **OTKA XXX/YYKPH** only when **MODE: WATER** is displayed on line 1; otherwise, this field is blanked.. When **OTKA XXX/YYKPH** is displayed, the **XXX/YY** will appear as dashes until the Ground Track/Ground Speed entry is made. Subsequently, it will display system ground track and speed.

Line 4. - Blanked.

Line 5 - Left-hand character field displays the legend **DNS RF:OFF or DNS RF:ON.**

Line 6. - Displays System Annunciator Data.

Line 7. - The right-hand character field displays the legend **DATA**.

Line 8. - Scratchpad.

Table 3-22.9. NAV SENSOR CONTROL PAGE VAB OPERATIONS

Control/Display	Function
VAB 1	Toggles between **MODE: LAND** and **MODE: WATER**. The system defaults to **MODE: LAND** at power up. **NOTE** This entry is only useable for a water start (without GPS or EGI capability) If this operation is attempted (it is not recommended), the aircraft must be aligned with the moving platform to within ° the heading entry must be accurate to 5° and the speed accurate to within 2 knots. Additionally, when the aircraft departs the moving platform, aircraft inertial speed must be maintained and the DNS RF:OFF switched to ON. If manual speed is not within 2 knots of the DNS speed and if the DNS does not stay out of memory, then the possibility of the INS and HARS rejecting the DNS velocity is high. If this occurs (velocity vector error), an in flight restart is required.
VAB 2	With aircraft heading and **GRND SPD** data entered in the scratchpad, depressing the VAB will load the data into line 3 of the display. Entry limitation are 000 to 359 degrees and 00 to 99 KPH (or KTS) respectively. Both entries are required if this function is used to aid the INS. Entry is only effective during INS startup. If **MODE:LAND** is selected on line 1, this line is blanked and the function is inhibited.
VAB 3	Toggles between **DNS RF:ON** and **DNS RF:OFF**. System defaults to **DNS RF:ON** at power up. With "squat switch" activated and **MODE:WATER** selected, system automatically changes to **DNS RF:OFF**; this can be manually overridden by the CPG. When **DNS RF:OFF** is selected, an operator initiated BIT of the **DNS** will cause the legend to change to **DNS RF:ON** prior to BIT starting. This function is blanked until the system has initialized the DNS.
VABs 4 -7	Inactive.
VAB 8	Selects return **to DATA MENU** top level page.

c. Data Transfer Unit (DTU) Page. The **DTU** page is accessed by depressing VAB 3 on the **DATA** Menu page. The **DTU** page is displayed as shown in figure 3-26.15) VAB 8 is used to return to the DATA MENU top level
page. VAB functions are explained in table 3-22.10.

Line 1. - Left-hand character field displays the legend >**LOAD ALL** The right-hand character field displays the legend **LOAD PPOSO** except when the aircraft is airborne.

Line 2. - Displays the centered page title **DTU.**

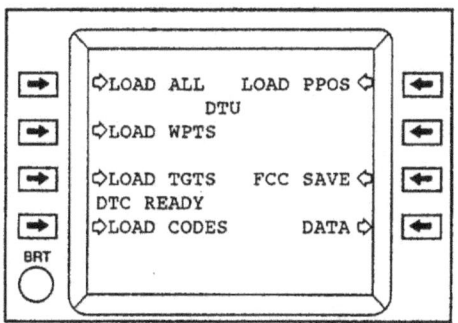

Figure 3-26.15. DATA TRANSFER UNIT Page

Line 3. - The left-hand character field displays the legend **OLOAD WPTS**. The right hand character field may be blank or may display the legend **DTC INITO**.

Line 4. - Blanked.

Line 5 - Left-hand character field displays the legend **OLOAD TGTS**. The right-hand character field displays the legend **FCC SAVEO**.

Line 6. - Displays the left justified **DTU LOAD/SAVE** prompts listed below. If any of the first three prompts are displayed, all DTU operations are inhibited.
DTU INOP
NO DTC
DTC FORMAT INVALID
DTC READY
TARGETS LOAD FAIL
WAYPOINTS LOAD FAIL
LASER CODES LOAD FAIL
PPOS LOAD FAIL
UPLOAD ALL FAIL
TARGETS LOAD COMPLETE
WPTS LOAD COMPLETE
CODES LOAD COMPLETE
PPOS LOAD COMPLETE
UPLOAD ALL COMPLETE
FCC SAVE-OK
FCC SAVE FAIL
The **LOAD FAIL** and **LOAD COMPLETE** prompts refer to initialization data that was loaded in the DTC using the Aviation Mission Planning System (AMPS).

Line 7. - The left-hand character field displays the legend **OLOAD CODES**. The right-hand character field displays the legend **DATAO**.

Line 8. - Scratchpad

TM 1-1520-238-10

Table 3-22.10. DTU PAGE VAB OPERATIONS

Control/Display	Function
VAB 1	Initiates LOAD of Waypoints, Targets, Laser Codes, and Present Position from cartridge. To preclude inadvertent load of PPOS, the LOAD ALL function will not load PPOS if the aircraft is airborne.
VAB 2	Loads WPTS from cartridge.
VAB 3	Loads TGTS from cartridge.
VAB 4	Loads CODES from cartridge.
VAB 5	Loads PPOS from cartridge unless aircraft is airborne.
VAB 6	Reformats DTC for normal use if **DTC INITO** is displayed; otherwise blanked.
VAB 7	Saves critical data in the FCC volatile memory (BST DATA, PPOS, Altitude, Altitude Corrections, MAGVAR, etc.) to non-volatile memory and saves the waypoints, targets, laser codes and PPOS to the DTC "SAVE" file.
VAB 8	Returns to the **DATA MENU** top level page.

d. ZEROIZE Page. The **ZEROIZE** page is accessed by depressing VAB 4 on the **DATA MENU** page. The ZEROIZE page is displayed as shown in figure VAB 8 is used to return to the **DATA MENU** top level page. If the **ZEROIZE** page is displayed for 5 seconds without an operator depression of VABs 1, 4, 7, or 8, the display automatically reverts back to the **DATA MENU** top level page.. VAB functions are explained in table 3-22.11.

Line 1. - Left-hand character field displays the legend >**CONFIRM ZEROIZE ALL.**

Line 2. - Displays the centered page title **ZEROIZE**

Line 3. - Blanked

Line 4. - Blanked.

Line 5 - The right-hand character field displays the legend **ABORT ZEROIZE 0** .

Line 6. -.Blanked

Line 7. - Left-hand character field displays the legend **ODNS ONLY**.

Line 8. - The right-hand character field displays the legend **DATAO.**

Figure 3-26.16. ZEROIZE Page

Change 3 3-64.19

TM 1-1520-238-10

Table 3-22.11. ZEROIZE PAGE VAB OPERATIONS

Control/Display	Function
VAB 1	Zeroizes the following avionics subsystems data: FCC Waypoint List, FCC Target List, FCC Laser Codes, FCC PPOS, DNS Ram Memory, GPS Y CODE Keys, and the DTU Cartridge. The CDU will then return to the page displayed before the zeroize action was initiated.
VABs 2-3	Inactive.
VAB 4	Zeroizes the DNS SDCC RAM Memory only.
VABs 5-6	Inactive.
VAB 7	Aborts the ZEROIZE action and returns the CDU to the page presented prior to the ZEROIZE action being initiated.
VAB 8	Returns to the DATA MENU Top Level page.

e **Global Positioning System Status (GPS STATUS) Page**. The **GPS STATUS** page is accessed by depressing VAB 5 on the **DATA MENU** page. The **GPS STATUS** page is displayed as shown in figure 3-26.17.

The functions on the GPS STATUS page are of a read only nature and are non-editable. VAB 8 is used to return to the **DATA MENU** top level page. All other VAB functions are inactive..

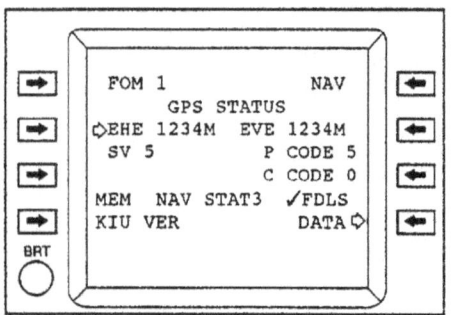

Figure 3-26.17. GPS STATUS Page Line 1. Left-hand character field displays the legend **FOM 1** (Figure Of Merit) and an associated value (1) reflecting GPS performance status. Performance is scaled from 1-9, with 1 indicating best GPS solution (1 = **FOM** of less than 25 meter error). Any value other than 1 to 9 indicates erroneous GPS performance.

Right-hand character field will display one of the legends **NAV**, or **INIT** or **TEST**, indicating GPS Receiver operational mode.

Line 2. Displays the centered page title **GPS STATUS**.

Line 3. Left-hand character field displays the legend **EHE XXXXM**, the GPS navigation solution Estimated **PPOS** Horizontal Error, in meters. **EHE** value is displayed in units of 0000 to 9999 meters. When 9999 is displayed, the actual GPS Horizontal Error may significantly exceed that value. The lower the number, the better the GPS Horizontal position performance.

Right-hand character field displays the legend **EVE XXXXM** the GPS navigation solution Estimated **PPOS** Vertical Error, in meters. EVE value is displayed in units of 0000 to 9999 meters. When 9999 is displayed, the actual GPS Vertical Error may significantly exceed that value. The lower the number, the better the GPS Vertical position performance.

Line 4. Left-hand character field displays the legend SV X, indicating the number of satellite vehicles the GPS receiver is using in its navigation solution. Displayed value can be 0 through 5.

Right-hand character field displays the legend **P CODE X**, indicating the number of **P CODE** (Precise Positioning Service PPS) satellites the GPS receiver is using in its navigation solution. Displayed value can be 0 through 5.

3-64.20 Change 3

Line 5. - Right-hand character field displays the legend C CODE X, indicating the number of C CODE (Standard Positioning Service - SPS) satellites the GPS receiver is using in its navigation solution. Displayed value can be 0 through 5.

Line 6. - Displays the System Annunciator Data.

Line 7. - Left-hand character field displays status prompts indicating operational status of the GPS Precise Positioning Service - PPS feature. Displayed prompts and their meaning are:

KIU VER	Keys in Unit Verified
KIU UNVER	Keys in Unit Unverified
KIU INCOR	Keys in Unit Incorrect
KEY PARITY ERR	Key Parity Error
INSUFF KEYS	Insufficient Keys

The right-hand character field displays the legend **DATAO**.

Line 8: Blanked.

Program Menu (PGM MENU) Page. The PGM MENU top level page is accessed by depressing the PGM FAB on the CDU. The PGM MENU page allows access to the maintenance-related functions of Boresight EGI (BST EGI), FCC CONFIG, AWS Harmonization (AWS HARM), FCC Memory READ (READ) and the Auxiliary Alphanumeric Display (AND).The PGM MENU top level page is shown in figure 3-26.18 and discussed below. VAB functions are explained in Table 3-22,12.

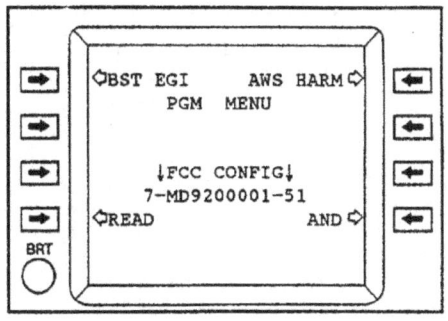

Figure 3-26.18. PGM MENU Page

Line 1. - Left-hand character field displays the legend **OBST EGI**. The right-hand character field displays the legend **AWS HARMO**.

Line 2. - Displays the centered page title PGM MENU.

Line 3. - Blanked.

Line 4. - Blanked.

Line 5. This line displays the centered legend **FCC CONFIGS**. The downward pointing arrows direct the operator's attention to the FCC software version data displayed on line 6.

Line 6. Displays the centered FCC software version.

Line 7. Left-hand character field displays the legend **OREAD**. The right-hand character field displays the legend **AND>**.

Line 8: Blanked.

Table 3-22.12. PGM MENU PAGE VAB OPERATIONS

Control/Display	Function
VAB 1	Presents the **BST EGI** page.
VABs 2-3	Inactive.
VAB 4	Presents the **READ** page.
VAB 5	Presents the **AWS HARM** page.
VABs 6-7	Inactive.
VAB 8	Presents the **AND** page.

Change 3   3-64.21

TM 1-1520-238-10

a. Boresight EGI (BST EGI) Page. The **BST EGI** page is accessed by depressing VAB 1 on the PGM MENU top level page. The BST EGI top level page is shown in figure 3-26.19 and discussed below. VAB functions are explained in Table 3-22.13.

Line 1. - Blanked.

Line 2. - Displays the centered page title **BST EGI**, followed by **-MR-**(milliradians).

Line 3. - Left-hand character field displays the legend **OAZ+XX..X**

Line 4. - Blanked.

Line 5. - Left-hand character field displays the legend **OEL+XLX**. Right-hand character field displays the legend **ROLL+XX.XCO**.

Line 6. - Blanked.

Line 7. - Left-hand character field displays the legend **OEGI RESET**. The right-hand character field displays the legend **PGMO**.

Line 8: Scratchpad.

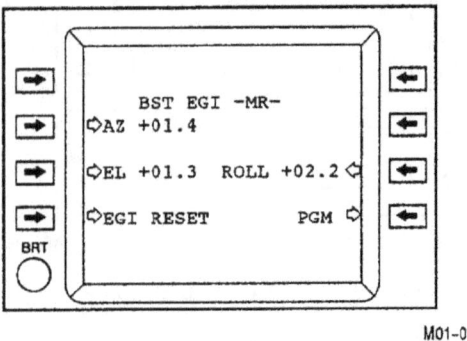

Figure 3-26.19. BST EGI Page

Table 3-22.13. BST EGI PAGE VAB OPERATIONS

Control/Display	Function
VAB 1	Inactive.
VAB 2	With the value for Azimuth boresight correction angle (Units in milliradians) in the scratchpad (E.g. -12.1), depressing VAB 2 enters the scratchpad value into the system. The Azimuth boresight correction angle entry range shall be +/99.9 mr. Azimuth boresight correction angle entry must be three digits (additional characters are ignored). It is not necessary to enter the decimal point; it will be automatically placed. A positive value is assumed if a +/sign is not entered.
VAB 3	With the value for Elevation boresight correction angle (Units in milliradians) in the scratchpad (E.g. -12.1), depressing VAB 3 enters the scratchpad value into the system. The Elevation boresight correction angle entry range shall be +/99.9 mr. Elevation boresight correction angle entry must be three digits (additional characters are ignored). It is not necessary to enter the decimal point; it will be automatically placed. A positive value is assumed if a +/sign is not entered.
VAB 4	Initiates an INS Reset action.
VABs 5-6	Inactive.

3-64.22 Change 3

Table 3-22.13. BST EGI PAGE VAB OPERATIONS - continued

Control/Display	Function
VAB 7	With the value for Roll boresight correction angle (Units in milliradians) in the scratchpad (E.g. -12.1), depressing VAB 7 enters the scratchpad value into the system. The Roll boresight correction angle entry range shall be +/99.9 mr. Roll boresight correction angle entry must be three digits (additional characters are ignored). It is not necessary to enter the decimal point; it will be automatically placed. A positive value is assumed if a +/sign is not entered.
VAB 8	Returns to the PGM MENU top level page.

b. READ Page. The **READ** page is accessed by depressing VAB 4 on the **PGM MENU** top level page.

With the **READ** page displayed and the OCTAL (or HEX) format selected, entering a six digit OCTAL (or four digit HEX) number into the scratchpad and then depressing any one of VABs 1-4 will result in the display of FCC memory data, in the format selected, at the memory location selected. The system defaults to OCTAL at power-up. The **READ** page is shown in figure and discussed below. VAB functions are explained in Table 3-22.14.

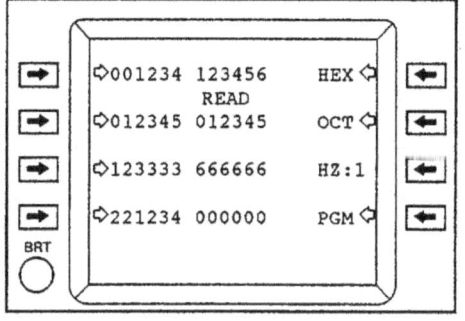

Figure 3-26.20. READ Page

Line 1. Left-hand character field displays the legend the **001234 123456** FCC memory read address 1 and the numerical data in that address. Right-hand character field displays the legend **HEXO**.

Line 2. Displays the centered page title **READ**.

Line 3. Left-hand character field displays the legend 012345 012345 FCC memory read address 2 and the numerical data in that address. Right-hand character field displays the legend OCTO.

Line 4. Blanked.

Line 5. Left-hand character field displays the legend 123333 666666 FCC memory read address 3 and the numerical data in that address. Right-hand character field displays the legend **HZ:1**, signifying the selectable data refresh rate (1, 2, or 5) for the **READ** page in Hertz..

Line 6. Blanked.

Line 7. Left-hand character field displays the legend **221234000000** FCC memory read address 4 and the numerical data in that address. The right-hand character field displays the legend **PGMO**.

Line 8: Scratchpad.

Table 3-22.14. READ PAGE VAB OPERATIONS

Control/Display	Function
VAB 1-4	With a recall address in the scratchpad, depressing any of VABs 1-4 causes memory recall data to be displayed on the line associated with the VAB depressed.
VAB 5	Depression changes the entry and memory recall data display to **HEX** format.
VAB 6	Depression changes the entry and memory recall data display to **OCTAL** format.
VAB 7	Allows selection of a data refresh rate for the **READ** page. Alternatives are 1, 2, or 5 Hz. System defaults to 1Hz at power up.
VAB 8	Returns to the **PGM MENU** top level page.

TM 1-1520-238-10

c. **Area Weapon Subsystem (AWS HARMONIZATION) Page**. The **AWS HARMONIZATION** page is accessed by depressing VAB 5 on the **PGM MENU** top level page. The **AWS HARMONIZATION** page is shown in figure 3-26.21 and discussed below. VAB functions are explained in Table 3-22.15.

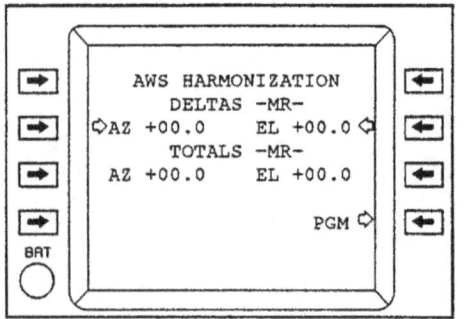

Figure 3-26.21. AWS HARMONIZATION Page

Line 1. Displays the centered page title **AWS HARMONIZATION.**

Line 2. Displays the centered legend **DELTAS -MR-**.

Line 3. Left-hand character field displays the legend **OAZ+XXX,** indicating the last azimuth correction entered. Right-hand character field displays the legend E**L+XLX,O**, indicating the last elevation correction entered..

Line 4. Displays the centered legend **TOTALS -MR-**.

Line 5. Left-hand character field displays the legend **AZ+XLX**, signifying the total correction value for azimuth. This value is limited to +/20.0. Right-hand character field displays the legend **EL+X.XX** signifying the total correction value for elevation. This value is limited to +/20.0.

Line 6. Blanked.

Line 7. The right-hand character field displays the legend **PGMO**.

Line 8: Scratchpad.

TABLE 3-22.15. AWS HARMONIZAT'ION PAGE VAB OPERATIONS

Control/Display	Function
VABs 1, 3, 4, 5 & 7	Inactive.
VAB 2	Allows for entry of additional azimuth bias corrections. Azimuth correction angle entry must be three digits (additional characters are ignored). It is not necessary to enter the decimal point; it will be automatically placed. A positive value is assumed if a +/- sign is not entered.
VAB 6	Allows for entry of additional elevation bias corrections. Elevation correction angle entry must be three digits (additional characters are ignored). It is not necessary to enter the decimal point; it will be automatically placed. A positive value is assumed if a +/- sign is not entered.
VAB 8	Returns to the **PGM MENU** top level page.

d. **Alphanumeric Display (AND) Page**. The **AND** page is accessed by depressing VAB 8 on the **PGM MENU** top level page. All data presented on the **AND** page is Read-Only. The **AND** page is shown in figure and discussed below. VAB functions are inactive. To return to **PGM MENU** top level page, depress the **PGM FAB.**

3-64.24 Change 3

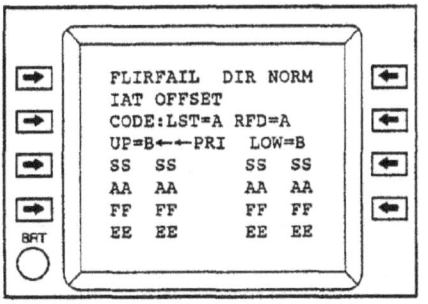

Figure 3-26.22. ALPHANUMERIC DISPLAY Page Line 1. Left-hand character field displays the **AND** Sight Status Field. Right-hand character field displays the **AND** Weapon Status Field.

Line 2. Displays the **AND TADS** Status Field.

Line 3. Displays the **AND LST/RFD** Status Field.

Line 4. Displays the **AND** Missile Enhancement Display Field.

Lines 5 8. Displays the AND Missile status displays (read vertically):

Cols. 3 & 4 show the LH Outbd pylon missile status

Cols. 7 & 8 show the LH Inbd pylon missile status

Cols. 15 & 16 show the RH Inbd pylon missile status

Cols. 19 & 20 show the RH Outbd pylon missile status

3.16.13. Pilot High Action Display (HAD) The Pilot HAD Sight Status and Weapon Status fields have been modified to allow display of Distance-To-Go and TimeTo-Go (figure 3-26.23) to the current **FLY-TO** destination and the active fly-to symbol.. The new Sight St tus and Weapon Status Field prompts replace the existing, lowest priority displays. The new displays are out-prioritized by all other HAD displays.

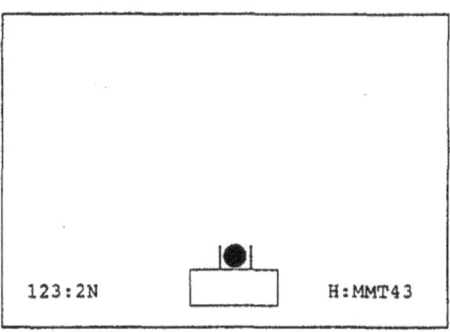

Figure 3-26.23. PILOT HIGH ACTION DISPLAY (HAD)

a. Sight Status Field. The Sight Status Field is modified to provide Distance-To-Go from **PPOS** to the selected **FLY-To** location. Character space 1 and 2 are blank. Character spaces 3 to 8 provide the DistanceTo-Go, with a colon in space 6 (providing 100 meter resolution to the pilot) and a **K** or **N** (**Kilometers or Nautical** miles) in space 8 . Units are selectable by the CPG using the CDU. All numerals in this display are refreshed at 1Hz.

b. Weapon Status Field. The Weapon Status Field provides, as its three lowest priorities prompts, **HSI CUE?,** Time-To-Go (**TTG**) from **PPOS** to the selected **FLY-TO** and the active fly_ to symbol. The **HSI CUE?** is the highest priority and is the same indication to the pilot as the advisory message **DNS/HSI NAV CUES** ?? is to the CPG. It is removed from the pilot weapon status display when the CPG clears the Advisory message. The **TTG** display is the lowest priority and is in hours and minutes (H: MM) if the **TTG** is greater than 5 minutes. This display changes to minutes and seconds when **TTG** is less than 5 minutes. The first character space will be H (Hours) or M (minutes). The second character space is a colon (:). The third and fourth character spaces will be either MM (minute values 00 to 59) or SS second values 00 to 59). The fifth, sixth and seventh character spaces will display the fly to symbol (W01 through T80) The eighth character space is blank. All characters relating to **TTG** bocome dashed when the **TTG** is greater than 9 hours and 59 minutes or the ground speed is less than 10 KTS.

c. Trim Ball. The trim ball will flash when the HARS has been free inertial for 20 seconds or more, and aircraft ground speed is less than 10 knots.

3.16.14 Co-Pilot Gunner High Action Display (HAD)
The CPG HAD (figure 3-26.24) has been modified to provide an Active Target Number in the Sight Status Field and an ORT STORE prompt in the Weapon Status Field. This information will be displayed unless other Sight Status and Weapon Status messages out prioritize it.

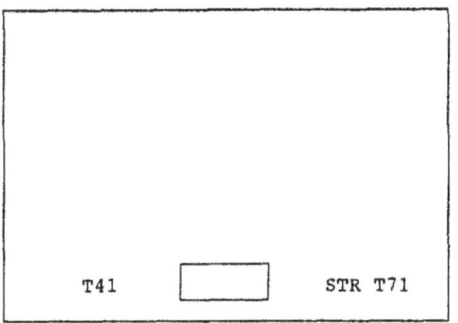

Figure 3-26.24. CPG HIGH ACTION DISPLAY (HAD)

a. Sight Status Field. The Sight Status Field is modified to provide an indication of the Active Target Number. Character spaces 1 thru 3 are blank. Character space 4 will be a **T or W** indicating the Active Target is in either the Target or Waypoint List. Character space 5 and 6 will be a number from 01-80 indicating the location within the **TGT/WPT** List. Character spaces 7 and 8 are blank.

b. Weapon Status Field. The Weapon Status Field is modified to provide the **ORT STORE** prompt. The **ORT STORE** prompt is formatted as follows: **STR TXX. XX** is a number from 71-80 to designate which location in the **TGT List** has received the new coordinate data. The **ORT STORE** prompt will be displayed for a duration of 2 seconds and will then be removed. **STR** will be located in character spaces 1-3. Character space 4 is blank. Character space 5 will be a **T**. Character spaces 6 and 7 will contain the numbers from 71 to Character space 8 is blank.

3-64.26 Change 3

3.16 15 Horizontal Situation Indicator (HSI).

The HSI, (fig 3-26.25), on the pilot instrument panel is an electromechanical indicator that presents position information in relation to various navigational inputs.

The HSI interfaces with the (HARS), the (ADF), and the doppler navigation set. The instrument displays consist of a fixed aircraft symbol, a compass card, two bearing-to-station pointers with back-course markers, a course bar, a (**KM**) indicator, a (**HDG**) knob and marker, a course set (**CRS**) knob, a COURSE digital readout, a to-from arrow, a **NAV** flag, and a compass HDG flag. Operating power for the HSI is taken from the 115 vac No. 1 essential bus through a circuit breaker marked **HSI** on the pilot center circuit breaker panel. Controls and indicators for the horizontal situation indicator are described in table 3-22.16.

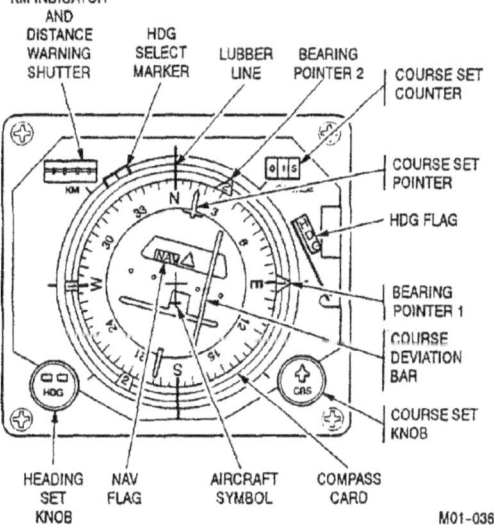

Figure 3-26.25. Horizontal Situation Indicator (Typical)

Table 3-22.16. HSI Controls and Indicators

Control/Indicator	Function
Compass Card	The compass card is a 360-degree scale that turns to display heading data obtained from the HARS. The aircraft headings are read at the upper lubber line.
Bearing pointer No.1	The pointer operates in conjunction with the Doppler. It indicates relative bearing to the active FLY-TO-location. The No. 1 bearing pointer "parks" at the 3 o'clock position when the Doppler "bearing-to-destination" signal is invalid.
Bearing pointer No. 2	This pointer operates in conjunction with the ADF receiver. The pointer is read against the compass card and indicates the magnetic bearing to the ADF station (nondirectional beacon).
Course deviation bar	This bar indicates the lateral deviation from the desired navigation course. When the helicopter is flying the desired navigation course, the course bar will be aligned with the course set pointer and will be centered on the fixed aircraft symbol.
CRS knob	(CRS) knob and the course set counter operate in conjunction with the course pointer and allow the pilot to select any of 360 courses. Once set, the course pointer will turn with the compass card and will be centered on the upper lubber line when the helicopter is flying the selected course, providing there is no wind to blow the helicopter off course.
KM indicator	The digital distance display in (KM) to the FLY-TO-location.
HDG knob	(HDG) knob operates in conjunction with the heading select marker and allows the pilot to select any one of 360 headings. Seven full turns of the knob produce a 360-degrees turn of the marker. To-from arrow Works with VOR. Not applicable.
NAV flag	The NAV flag, on the HSI course carriage, turns with the compass card. The flag retracts from view when a reliable course deviation signal is available from the doppler.
HDG flag	The HDG flag retracts when a reliable heading signal is available from the HARS. During HARS alignment, retraction of the heading flag indicates that HARS is aligned (ready-to-fly).
Distance shutter	The distance shutter (upper left corner) retracts when the "the distance-to-destination" signal from the Doppler is valid.

Table 3-22. IP-1652G CDU Control and Display Functions - continued

Control/Display	Function
FDLS	Access to Aircraft Fault Detection and Location System functions.
DATA	Access to Data Menu Top Level page.
SPC	Used only for entry of coordinate ID or FDLS/BST operations.
ENT	Not Used.
CLR	If pressed in response to an ERROR prompt in the scratchpad, will clear the error prompt and position the cursor at the first detected data entry error. If pressed a second time, will clear the entire scratchpad. If no ERROR prompt, will clear the entire scratchpad.
PGM	Access to Program Top Level page.
IDNT	Not Used.
WPN	Access to Weapon Control Top Level page.
FPLN	Access to Waypoint LIST page.
STR	Stores PPOS in next waypoint store location (31-40), overwriting existing data.

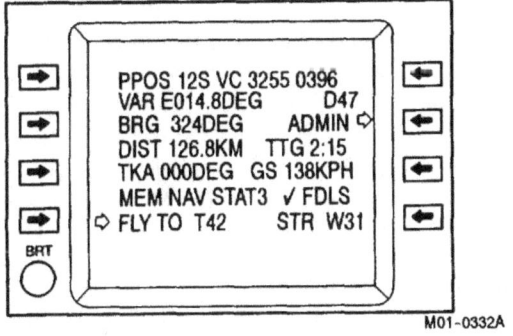

Figure 3-25. NAV Top Level Page (UTM Format)

b. Three distinct techniques are used for navigation mode startup under varying conditions. These are Stationary, Airborne (with/without GPS) and Moving Platform (Sea, with/without GPS) alignments.

1. Stationary Start. This method assumes a normal aircraft startup while the aircraft is parked and the LAND mode is used. EGI alignment is automatic. HARS mode switch is placed to NORM. This method is not dependent upon GPS availability, although GPS availability does improve INU alignment time and heading accuracy. This is the most often used startup method.

NOTE

Do not attempt airborne or sea start with HARS mode switch in **NORM**. HARS will assume the aircraft is stationary with switch in this position.

2. Automatic Airborne Start (with GPS). This method assumes an aircraft power interrupt or a desired in-flight navigation restart while the aircraft is airborne. This method can be accomplished in either LAND or WATER mode. The EGI will automatically use GPS data to aid in startup. The HARS mode switch is placed to FAST after the heading tape on the video is visible. When the HARS Heading Flag is retracted from the HSI, the HARS mode switch is placed to **OPERATE**. The DASE may then be engaged.

Change 4   3-64.3

3. **Automatic Sea Start (with GPS).** A sea start assumes the aircraft is moving relative to the earths' surface, the squat switch indicates ground, and WATER mode is selected soon after power is applied. The EGI will automatically use INU data to aid in startup. The HARS mode switch is placed to FAST after the heading tape on the video is visible. When the HARS Heading Flag is retracted from the HSI, the HARS mode switch is placed to **OPERATE**. The DASE may then be engaged.

NOTE

If the following operations are attempted (it is NOT recommended) Heading entry must be accurate to within 5 degrees of the actual aircraft heading and the speed entry accurate to within 2 knots. For Sea starts, when the aircraft departs the moving platform, aircraft inertial speed must be maintained and the DNS RF: **OFF** switched to **ON**. If manually entered speed is not within 2 knots of the DNS speed and DNS does not stay out of memory, then the possibility of the INS and HARS rejecting the DNS velocity is high. If this occurs (Velocity vector error), a forced inflight restart is required.

4. **Manual Airborne or Sea Start (without GPS or INU).** Manual entry of inertial speed and heading is required. When the entries are completed, the heading tape will be displayed and the HARS mode switch is placed to **FAST**. When the HARS Heading Flag is retracted from the HSI, the HARS mode switch is placed to **OPERATE**. The DASE may then be engaged. For Sea Starts the pilot, after liftoff, must maintain speed and constant heading with the ship (same as entered speed and heading); move off the ship over water; then the CPG will toggle the CDU DNS RF switch from **OFF** to **ON**.

5. **Forced Inflight Restart.** Place the HARS mode switch to **FAST** to power up HARS and DNS. When HARS heading flag is retracted, place HARS mode switch to **OPR**. HARS and EGI will use DNS velocity to complete alignment.

3.18.7 CDU Displays. The CDU display architecture is organized into page formats. Each page consists of 8 display lines. All lines are 22 characters wide. The bottom line (eighth) is used as a scratchpad for entering or editing data. Individual pages may display navigation data, display failure messages, or indicate selectable functions. The CDU displays are categorized into 8 top level page formats: **NAV, TGT, DATA, CODE, FDLS, PGM, WPN,** and **FLPN**. The CDU will automatically revert back to the **NAV** top level page from the **ADMIN** page after 30 seconds if there are no characters in the scratchpad and no other key presses are made.

a. **NAV Top Level Page.** The **NAV** top level page (fig 3-25.1) will display the following:

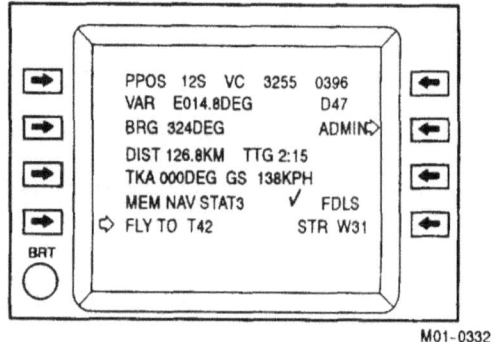

Figure 3-25.1. NAV Top Level Page (UTM Format)

Line 1. - Displays the legend **PPOS** followed by the current **PPOS** coordinates. **PPOS** represents the computed present position in either UTM or LAT/LONG coordinates. **PPOS** changes to **OUPD** when an on-the-move manual **PPOS** update is initiated.

Line 2. - The left-hand character field displays the legend **VAR** followed by the **PPOS** MAGVAR in degrees. If the INS is NO-GO, an outlined arrow (◊) will be displayed immediately to the left of **VAR** permitting manual MAGVAR updates. The right-hand character field displays the legend **D** followed by the **PPOS** Datum number.

Line 3. - The left-hand character field displays the legend **BRG XXXDEG** reflecting the direct bearing to the destination specified in the **FLY TO** legend. The **BRG** legend is a read-only display. The right-hand character field displays the legend **ADMIN**. An outlined arrow (◊) following **ADMIN** indicates the ability to select the **ADMIN** page for display.

Line 4. -The left-hand character field displays the legend **DIST XXX.XKM**. **DIST** is displayed in KM (kilometers) when the coordinate system in use is metric, and in NM (nautical miles) when the coordinate system in use is nautical. The display uses a floating decimal point allowing a distance accuracy of 10 meters (32 feet) when **DIST** is less than 100 KM units. The **DIST** numerals are updated at the rate of 1Hz regardless of aircraft speed. The right-hand character field displays the legend **TTG X:XX** to the **FLY-TO** destination in Hours and Minutes (H:MM). When less than five (5) minutes from destination, these units change to Minutes and Seconds (M:SS). If **TTG** to the **FLY TO** exceeds 9 hours and 59 minutes, or the aircraft groundspeed is less than 10 knots, the display is dashed. Both of the character fields are read-only and cannot be edited.

Line 5. - Displays the legend **TKA XXXDEG** and **GS XXXKPH**. **TKA** denotes the computed Track Angle, in Degrees **(DEG)**, while **GS** reflects Ground Speed. **GS** is presented in Kilometers Per Hour **(KPH)** or Knots **(KTS)** dependent on display selected. The units for both **DIST** and **GS** are the same for a given selection: **KM** and **KPH**, or **NM** and **KTS**.

Line 6. - Displays the System Annunciator Data which consists of three elements: A DNS Memory **(MEM)** element; a Navigation Status **(NAV STAT 1-3)** element; and a Check FDLS (✓ FDLS) element. These status messages may appear at any time on this line. A display on this line indicates that the DNS has either lost radar lock **(MEM)**; the navigation error exceeds certain parameters **(NAV STAT 1-3)**; or, the Continuous FDLS test has detected a fault in the system (✓ **FDLS**). Navigation Status categories are as follows:

NAV STAT 1: Estimate of position error calculation is less than 50 meters. Actual position error may be substantially less. **PPOS** updates are disabled.

NAV STAT 2: Estimate of position error calculation at least 50 meters but less than 201 meters. Actual position error may be substantially less. All **PPOS** update operations are enabled.

NAV STAT 3: Estimate of position error calculation is greater than 200 meters. Actual position error may be substantially less. All **PPOS** update operations are enabled.

Line 7. - The left-hand character field displays the legend ⇨] **FLY TO TXX** or ⇨] **FLY TO WXX,** depending upon the selected destination to which the system is navigating. The displayed value can be any of the 40 waypoint locations in the FPLN List; or any of the 40 target locations. ⇨] **UPD TO WXX** displayed indicates the destination to which the system is being updated. The right-hand character field displays the legend **STR WXX** representing the waypoint number (from 31-40) to be used by the system for the next **PPOS** store.

Line 8. - This is the scratchpad line for manual data entry.

Access to **NAV** system pages is accomplished using Variable Action Buttons (VABs) located on either side of the CDU display screen. Refer to table 3-22.1 for their respective function. There are three (3) **NAV** system pages: **ADMIN,** Hover Bias Calibration Mode **(HBCM),** and Offset Update **(OFS UPDT).**

Table 3-22.1. NAV TOP LEVEL PAGE VARIABLE ACTION BUTTONS

Control/Display	Function
VAB 1	If **NAVSTAT 1** is displayed on line 6 of the **NAV** page, VAB 1 is disabled. Depressing the VAB 1 button with no data present in the scratchpad freezes the current displayed PPOS data and changes the **PPOS** legend to ⇨ **UPD** indicating the system is ready for a manual **PPOS** update. A second depression of VAB 1 with no scratchpad data present aborts the manual update. Depressing the VAB 1 button with data present in the scratchpad that satisfies the **PPOS,** Datum, or MAGVAR data entry rules will cause the intended system parameter to be updated. Used for manually entering **PPOS** and Datum updates. Allows manual MAGVAR updates when the outlined arrow ⇨ is displayed in character position 1 on line 2.
VAB 2	Not Active.
VAB 3	Not Active.

Table 3-22.1. NAV TOP LEVEL PAGE VARIABLE ACTION BUTTONS - continued

Control/Display	Function
VAB 4	When depressed, if the left-hand character field in line 7 displays the legend **UPD TO**, will complete a flyover type **PPOS** update; and display updated **PPOS** on line 1. When updated data is accepted or aborted, line 7 will return to its' previous legend of ⇨ **FLY To**. With valid **FLY** TO data symbol **(WXX or TXX** with valid previously entered coordinates) entered on the scratchpad, will change the active **FLY TO** to location entered in scratchpad. Invalid scratchpad coordinate data will result in **"ERROR"** message on scratchpad. Without data in scratchpad; will display active **FLY** TO coordinates in scratchpad with initial depression. Second depression will display the Associated data line (ID, Altitude and Datum) for the active **FLY** To coordinates. Subsequent depressions will toggle between the **FLY** TO coordinates and the Associated Data. **FLY TO** coordinates and Associated Data displayed in the scratchpad are for review purposes only, and cannot be edited or transferred.
VAB 5	If **NAVSTAT 1** is displayed on line 6 of the **NAV** page, VAB 5 is disabled. Depression of VAB 5 with other than **NAVSTAT 1** status while overflying a waypoint will freeze the **PPOS** display on line 1 and changes the **FLY TO** legend on line 7 to **UPD TO**. Reselection of VAB 5, or selection of any other page, will abort the update.
VAB 6	Presents the **ADMIN** sub page when depressed.
VAB 7	Not Active.
VAB 8	Not Active.

b. Administration (ADMIN) Page. The **ADMIN** page (fig 3-25.2), is selected by depressing VAB 6 on the **NAV** Top Level Page. VAB functions are explained in Table 3-22.2. The **ADMIN** page contains the following:

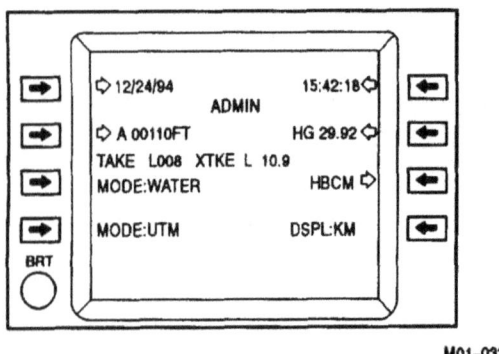

Figure 3-25.2. ADMIN Page - Top Level

Line 1. - Displays the example legend ⇨ **12 /24/94** denoting the current Month, Day, and Year., separated by slashes (/) in the left-hand character field. Update of this field is automatic through GPS data, and when updated, the ⇨ is blanked. If the system has no date available (GPS or Manual), this field is displayed as dashed (--l--/--). The current time example legend 15:42:18 ⇨ is displayed in the right-hand character field representing Hours, Minutes, and Seconds, separated by a colon(:), in that order. If GPS time is not valid,(⇨ is displayed), manual update to the system is permitted.

Line 2. - Displays the centered page title ADMIN.

Line 3. - The left-hand character field displays an outlined arrow and the legend ⇨ **A XXXXXFT** reflecting the current aircraft altitude. This altitude value is always displayed in Feet only, and is referenced to MSL. Positive altitude above MSL is displayed with a blank preceding the altitude readout; negative altitude (below MSL) is displayed using a minus (-) sign preceding the altitude readout **(A-XXXXXFT)**. The right-hand character field displays the legend **HGXX.XX** ⇨ denoting the current Barometric pressure setting.

Line 4. -The left-hand character field displays the legend **TKAE L XXX.**. **TKAE** represents the current Track Angle Error; **L** or **R** denotes Left or Right; and the **XXX** signifies the angle error in degrees. The right-hand character field displays the legend **XTKELXX.X**. **XTKE** represents the current Cross Track Error; **L** or **R** denotes Left or Right; and **XXX.X** signifies the distance in Kilometers or Nautical Miles, dependent upon units selected by the the CPG, and displayed in line 7 **(DSPL:KM or DSPL:NM)**.

Line 5. - Displays the legend **MODE:LAND** or **MODEWATER** in the left-hand character field. The right-hand character field displays the legend **HBCM** ⇨ if the DNS system status is **GO**; otherwise this field is blanked.

Line 6. -Blank.

Line 7. - The left-hand character field displays the legend **MODE:UTM** or **MODE:L/L** signifying coordinate data entry methodology. The right-hand character field displays the legend **DSPL:KM** or **DSPL:NM** signifying metric system or nautical system, respectively, in use.

Line 8. - This is the scratchpad line for manual data entry.

Table 3-22.2. ADMIN SUB-PAGE VARIABLE ACTION BUTTONS

Control/Display	Function
VAB 1	Used for manual entry of **MONTH DAY YEAR** update data **(MMDDYY)** when the outlined arrow is (⇨) displayed in character position 1 on line 1, and character string is present in the scratchpad. Valid entries are: 01-12 for **MM**; 01-31 for **DD**; and 00-99 for **YY**.
VAB 2	Used for manual entry ofAltitude update data. Positive altitude is assumed and does not require a "+" sign; Negative altitude (below MSL) requires a minus (-) sign preceding the altitude value. Altitude ranges are -900 feet MSL to 20000 MSL. Altitude is always displayed in feet.
VAB 3	Used for selecting **MODE:LAND** or **MODE:Water** navigation operations. System defaults **to MODE:LAND** at power up. Depressing VAB 3 toggles between two choices.
VAB 4	Selects the data presentation format to be used for **PPOS, CPOS, WAYPOINT,** and **TARGET** coordinate data entry. System defaults **to MODE:UTM** at power up. Depressing VAB 4 toggles between **MODE:UTM** AND **MODE:L/L**. Pages affected by pressing **VAB 4** are **NAV, OFFSET, WAYPOINT** and **TARGET**.
VAB 6	Used for manual entry of **HOUR MINUTE SECOND** update data **(HHMMSS)** when the outlined arrow is (⇨) displayed in character position 22 on line 1, and character string is present in the scratchpad. Valid entries are: 00-23 for **HH**; **00-59** for **MM**; and 00-59 for **SS.**
VAB 6	Used for manual entry of **HG** (barometric pressure). Valid entries are: 27.92 to 31.92. Data is entered in scratchpad as **XXXX**; the decimal point entry is not required. This entry followed by an altitude entry within .60 seconds will cause the system to re-calibrate the ADS pressure sensor bias correction.
VAB 7	Used for selecting **HBCM** sub-page. Valid only if **HBCM** ⇨ is present indicating the DNS system is "GO".
VAB 8	Used for selecting **DSPL:KM** or **DSPL:NM** signifying metric system or nautical system, respectively, in use for coordinate display. Depressing VAB 8 toggles between the two (2) choices.

c. **HBCM Page.** The **HBCM** page, (fig 3-25.3) is used to **START** and **STOP** the velocity bias calibration of the DNS. The HBCM calibrates the DNS system for small velocity errors that may be present in the doppler receiver/ transmitter subsystem. The velocity bias corrections are computed by the DNS computer and are applied to all subsequent doppler radar velocities. When **HBCM** is selected, the CPG can manually start and stop the calibration. If a NO-GO condition is detected, a recalibration can be initiated by pressing **START** again. This will zero the timer and allow for additional bias velocities calculations. If TIME message is displayed, re-initiate the calibration. If the BIAS message is displayed, re-initiate calibration one time. If the BIAS message is displayed again, write up the system. If the excessive memory message is displayed, attempt to re-locate the aircraft over a surface area that is a non-reflective surface (short grass/coarse surface area) and re-initiate the calibration. If a GO calibration is computed, the bias velocities will automatically be stored and applied continuously to all subsequent DNS velocities. VAB functions are explained in table 3-22.3. The display has the following:

Line 1. - Displays the legend ⇨ **START** in the left-hand character field; the **HBCM** title in the center character field, and the legend STOP ⇨ in the right-hand character field.

Line 2. - Displays the legend(s) **HBCM READY;. HBCM ACTIVE, HBCM GO,** or **HBCM NO-GO** as appropriate in the left-hand character field.

Lines 3. & 4 - If line 2 displays **HBCM READY** or **HBCM ACTIVE,** lines 3 & 4 will display the current DNS velocity biases. If line 2 displays **HBCM** GO, lines 3 & 4 will display the new DNS velocity biases. If line 2 displays **HBCM NO-GO,** line 3 will display the HBCM fail condition. If a second fail condition exists, it will be displayed on line 4, otherwise line 4 is blanked. Possible DNS error messages are:

TIME BELOW 2 MIN

BIAS EXCEEDS 0.3 KTS

EXCESSIVE MEM

Line 5. - Blank.

Line 6. - Displays the legend **TIME: XX:XX** in the left-hand character field indicating elapsed time in minutes/seconds for DNS computations of the bias calibration. The timer will stop while the DNS is in memory during the bias calibration. The legend **MEM** will be displayed following **"TIME: XX:XX"** if the DNS velocity is invalid; otherwise the rest of line 6 is blanked.

Line 7. - The right-hand character field displays the legend **ADMIN**⇨.

Line 8. - Blank.

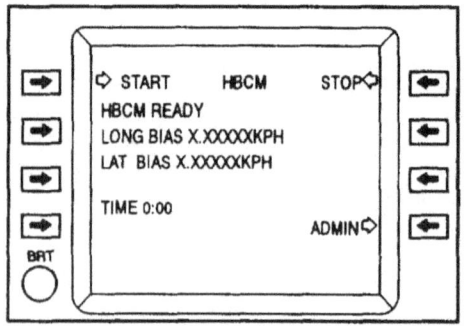

Figure 3-25.3. HBCM Page

M01-0330

Table 3-22.3. HBCM SUB-PAGE VARIABLE ACTION BUTTONS

Control/Display	Function
VAB 1	Used to initiate or re-start the calibration of the Hover Bias. Pressing VAB 1 causes the display on line 2 to change from **"HBCM READY"** to **"HBCM ACTIVE"** and starts the timer display located on line 6. Calibration time is dynamic and is displayed in minutes and seconds (M:SS).
VAB 2	Not Used
VAB 3	Not Used

Table 3-22.3. HBCM SUB-PAGE VARIABLE ACTION BUTTONS - continued

Control/Display	Function
VAB 4	Not Used
VAB 5	Used to stop the calibration of the Hover Bias. Pressing VAB 5 will cause the message **"HBCM GO"** or **"HBCM NO-GO"** to appear on line 2. This indicates a "within limits" or "out of limits" calibration condition.
VAB 6	Not Used
VAB 7	Not Used
VAB 8	Used to return to the **ADMIN** sub-page display.

d. Offset Update (OFS UPDATE) Page. If **NAV-STAT 1**, the offset update operations are disabled and the offset update page cannot be activated. If **NAV-STAT 2** or **NAVSTAT 3**, offset update is accomplished through actioning the Update/Store **(UPDT/STR)** rocker switch located on the ORT. A momentary depression of the **UPDT** side of the rocker switch results in the display of the **OFS UPDATE** page (figure 3-25.4) on the CDU. VAB functions are explained in table **3-22.4**. The **OFS UPDATE** page is displayed as follows:

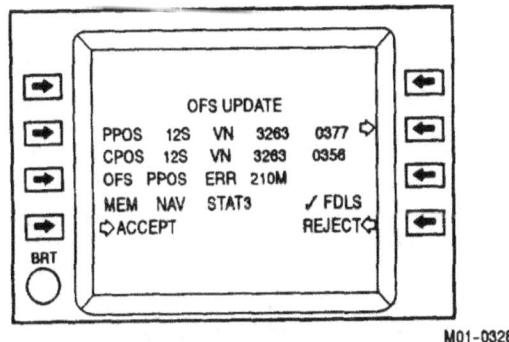

Figure 3-25.4. OFS UPDATE Page

Line 1. - Blanked

Line 2. - Displays the centered page title OFS UPDATE.

Line 3. - Displays the legend PPOS XXX VN XXXX XXXX representing the current PPOS at the moment of activation of the OFS UPDATE page.

Line 4. - Displays the legend CPOS XXS VN XXXX XXXX representing the computed, corrected aircraft present position based upon CPG range, LOS angles and aircraft bearing relative to the currently selected active target.

Line 5. - Displays the legend OFS PPOS ERR XXXXXM. This denotes the radial distance between PPOS & CPOS. The value is used as a measure of PPOS error by the CPG to determine whether to accept or reject the update. The radial error is limited to 99999 meters on the display.

Line 6. - Displays System Annunciator Data.

Line 7. - The left-hand character field displays the legend ◇**ACCEPT**. The right-hand character field displays the legend **REJECT**◇.

Line 8. - Blanked.

Change 4   3-64.9

Table 3-22.4. OFS UPDATE SUB-PAGE VARIABLE ACTION BUTTONS

Control/Display	Function
VABs 1, 2, & 3	Not Used.
VAB 4	Used to accept the offset update. Display returns to page previously displayed prior to selecting the offset update function.
VABs 5, 6 & 7	Not Used
VAB 8	Used to reject the offset update. Display returns to page previously displayed prior to selecting the offset update function.

3.16.8 FDLS page. Depression of the **FDLS** FAB displays the FDLS page on the CDU. The display on the CDU (Fig. 3-25.5) is intended to inform the operator that FDLS prompts/displays will not be provided on the CDU. All VABs are inactive on the FDLS page display. FDLS prompts will be provided through the ORT and HMD. FD/LS operations are conducted through the CDU keyboard and entry of any FDLS Menu two digit code will initiate the FDLS BIT

NOTE

- The Waypoint List and Target List are separately, but identically, structured. The following data applies to the operation of both. Where minor operating differences occur, each is described separately.

- Waypoints are numbered 1 through 40; Targets are numbered 41 through 80.

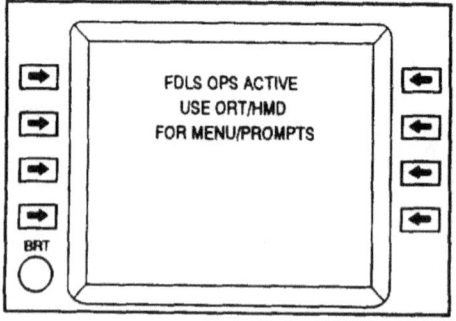

Figure 3-26.6. FDLS Page

3.16.9 FPLN CDU Displays. Depressing the **FPLN** FAB displays the Waypoint List **(WPT LIST)** page.

a. **Waypoint List (WPT LIST) Page.**

1. Each page provides access to a list of 40 Waypoints contained in the Waypoint List (Figures 3-26.6 and 3-25.7) or 40 targets contained in the Target List. The **WPT LIST** and **TGT LIST** are identically formatted except for the page title on line 2 of the display and numbering. Waypoints numbered 1-20 are displayed on the first page; waypoints numbered 21-40 are displayed on the second page. The first **TGT LIST** page will list numbers 41-60; page 2 will list numbers 61-80. The second page of each list is accessed by depressing VAB 8 on page 1. Depressing VAB 8 on the second page will return the display to the first page.

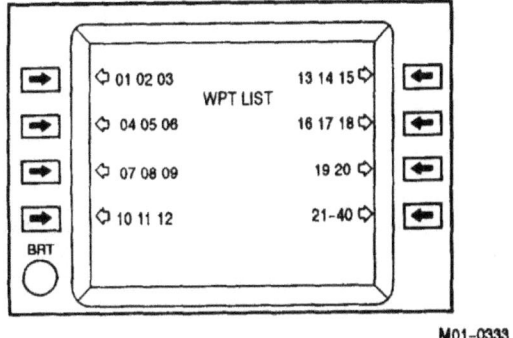

Figure 3-25.6. FPLN Top Level Page

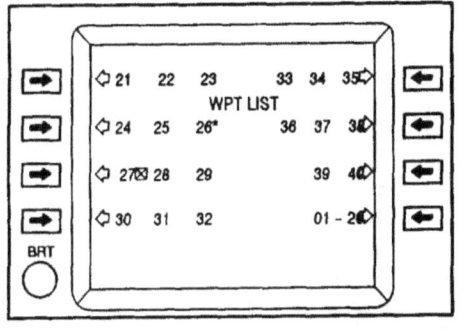

Figure 3-25.7. FPLN Second Level Page

2. Three special symbols are associated with Active **FLY-TOs** and Active **TGTs** on these pages. An Asterisk (*) symbol displayed to the right of a **TGT** or **WPT** designator **(TXX or WXX)** indicates that location as the Active **FLY-TO** location. Similarly, an outlined cross (⊠) symbol indicates the location is the Active Target location. A pound (#) symbol displayed to the right of the designator indicates that point as being BOTH the Active FLY-TO and the Active **TGT** location.

3. With either the TGT or WPT LIST page displayed, entry of a valid Coordinate Symbol **(TXX or WXX** with valid coordinates previously entered) into the scratchpad and depression of any of VABs 1-4 will cause the scratchpad entered data to become the Active **FLY-TO** location. Depression of any of VABs 5-8 will cause the scratchpad entered data to become the Active **TGT** location. With an invalid coordinate symbol, the **"ERROR"** message is displayed in the scratchpad.

Line 1. - Displays waypoints ◊01 02 03 in the Left-hand character field; waypoints 13 14 15◊ in the right-hand character field.

Line 2. - Displays centered title **WPT LIST**.

Line 3. - Displays waypoints ◊04 05 06 in the Left-hand character field; waypoints 16 17 18◊ in the right-hand character field.

Line 4. - Blanked.

Line 5. - Displays waypoints ◊07 08 09 in the Left-hand character field; waypoints 19 20◊ in the right-hand character field.

Line 6. - Blanked.

Line 7. - Displays waypoints ◊10 11 12 in the Left-hand character field; waypoints 21-40◊ in the right-hand character field.

Line 8. - Scratchpad.

b. **Waypoint Coordinate (WPT COORD) Page.** Depressing the **FPLN FAR** displays the Waypoint List **(WPT LIST)** page. The selection and depression of a specific VAB (Table 3-22.5) will result in the display of a desired set of coordinates.

NOTE

All data and messages depicted in displays are typical representations.

The accessing of the sample waypoints and coordinates shown in figure 3-25.8 is the result of depressing the **FPLN FAR**, then **VAB 8**, then **VAB 2**, VAB functions for the displayed waypoint coordinate page are explained in table 3-22.6.

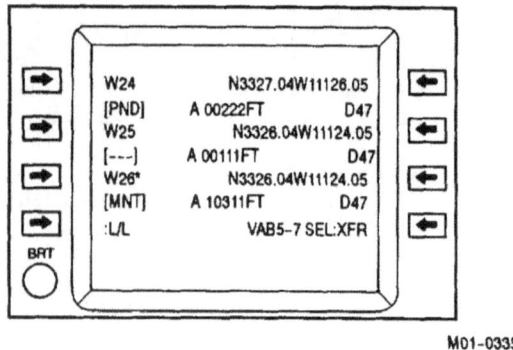

Figure 3-25.8. Selected Waypoint Coordinates

The Up/Down arrow keys can be used to scroll through the Waypoint Coordinate pages. An * is shown immediately after the storage location currently being used for **FLY TO** computations and it cannot be changed until some other location is chosen to **FLY TO**. The page has the following data organization:

Line 1. - Displays the waypoint number (W24, as shown) and the coordinate location data of that waypoint in the **L/L** system.

Line 2. - Displays the associated data for waypoint 24, consisting of a three letter ID (**PND**, as shown) surrounded by brackets; Altitude (an **A** followed by up to five character spaces for numerals and the unit of measure - **FT**); and the Datum number (**D47**, as shown).

Line 3. - Displays the next waypoint number (**W25**) and coordinate location data for that waypoint.

Line 4. - Displays the associated data for waypoint 25, consisting of a three letter ID surrounded by brackets; Altitude (an **A** followed by up to five character spaces for numerals and the unit of measure - **FT**); and the Datum number (**D47**, as shown). If the characters within the brackets are dashed (---), it is an indication that no data has been entered for this waypoint.

Line 5. - Displays the next waypoint number (**W26**) and coordinate location data for that waypoint.

Line 6. - Displays the associated data for Waypoint 26, consisting of a three letter ID (**MNT**, as shown) surrounded by brackets; Altitude (an **A** followed by up to five character spaces for numerals and the unit of measure - **FT**); and the Datum number (**D47**, as shown).

Line 7 - The left-hand character field displays the data format selected for page display (either **L/L** or **UTM**). The right-hand character field displays the legend **VAB 5 - 7 SEL:** followed by selection alternatives **XFR**, **TGT**, or **R/B**.

Line 8. - Scratchpad for data input.

Table 3-22.5. FLPN PAGE 1 VAB OPERATIONS WITH SCRATCHPAD EMPTY

Control/Display	Function
VAB 1	Selects Waypoints 01-03 for display on Waypoint Coordinate Page
VAB 2	Selects Waypoints 04-06 for display on Waypoint Coordinate Page
VAB 3	Selects Waypoints 07-09 for display on Waypoint Coordinate Page
VAB 4	Selects Waypoints 10-12 for display on Waypoint Coordinate Page
VAB 5	Selects Waypoints 13-15 for display on Waypoint Coordinate Page
VAB 6	Selects Waypoints 16-18 for display on Waypoint Coordinate Page
VAB 7	Selects Waypoints 19-20 for display on Waypoint Coordinate Page
VAB 8	Displays page 2 of WPT LIST allowing access to Waypoints 21-40

Table 3-22.6. WAYPOINT COORDINATE PAGE VAB OPERATIONS

Control/Display	Function
VAB 1-3 (SCRATCHPAD EMPTY)	With valid data in the coordinate location, depression of VABs 1-3 will select the coordinates associated with the selected VAB as the Active **FLY-TO** location.
VAB 1-3 (SCRATCHPAD CONTAINS DATA)	With Coordinate, ID, Altitude or Datum data in the scratchpad, depression of VABs 1-3 will cause the scratchpad data to be validated. If valid, the data item associated with the depressed VAB will change to the entered value. If data is invalid, an error message will display prompting operator to employ normal scratchpad editing rules. If the intended destination of the entry is either the Active Fly-To or Active Target, the prompt **"NO MOD TO ACTIVE COORD"** will appear in the scratchpad. This signifies that the system will not permit the Active Fly-To or Target coordinate to be modified.
VAB 4	Toggles between **UTM** or **L/L** display formats.
VAB 5-7 (WITH SEL:TGT DISPLAYED)	With valid data in the coordinate location, depression of any of these VABs will select the coordinate associated with the depressed VAB as the Active Target location.
VAB 5-7 (WITH SEL:R/B DISPLAYED)	Displays the range and bearing **(RangeXXXXXXBRGXXX.X)** to the selected coordinate on line 8.
VAB 5-7 (WITH SEL:XFR DISPLAYED	Copies the coordinate data associated with the depressed VAB into the scratchpad.
VAB 8	Depression of this VAB allows toggling between **VAB 5-7 SEL:XFR**, **VAB 5-7 SEL:TGT** and **VAB 5-7 SEL:R/B**. Upon system power-up, the system defaults to **VAB 5-7 SEL:TGT**. Following selection of **SEL:XFR** or **SEL:R/B**, the system defaults back to SEL:TGT.

3.16.10 Weapon Control (WPN CONTROL) Page. Depressing the WPN FAB allows the operator to access the **WPN CONTROL** page (figure 3-25.9). Variable Action Button operation is detailed in Table 3-22.7. When the **WPN CONTROL** page is selected, the **WPN CONTROL** display is formatted as follows:

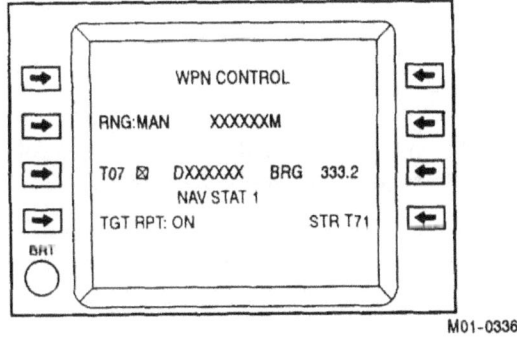

Figure 3-25.9. Weapon Control Display Page

Line 1. - Blanked.

Line 2. - Displays the centered page title **WPN CONTROL**.

Line 3. - The left-hand character field displays the legend **RNG:MAN XXXXXXM** showing the current range source and current range value in meters. The current CPG range source may be any of the following sources: **AUTO, DFLT, LSR, MAN, NAV, NOAR** or **TGT** (See Table 3-22.8)

Line 4. - Blanked.

Line 5 - Displays the legend **T(or W)XX DXXXXXX BRG XXX.X** denoting Active Target number, distance to Active Target in meters (D000001 to D999999) and bearing **(BRG)** to Active Target in degrees and tenths of degrees (000.0 to 359.9) respectively.

Line 6. - Displays System Annunciator Data.

Line 7. - The left-hand character field displays the legend **TGT RPT:** ON **(or OFF)**. The right-hand character field displays **STR TXX** denoting the location for the next **TGT Store** function (71-80).

Line 8. - Scratchpad for data input.

Table 3-22.7. WEAPON CONTROL PAGE VAB OPERATIONS

Control/Display	Function
VAB 1	Inactive
VAB 2	With the scratchpad empty, or with a leading "A" or "0" entered in the first position in the scratchpad, depressing VAB 2 selects **AUTO** range as the CPG range source. With a scratchpad entry of a range (1 to 999,999 meters), depressing VAB 2 changes the CPG range source to manual and will update the CPG range source value to the entered value.
VAB 3	With a valid target symbol, **TXX** or **WXX,** and valid coordinates previously entered in the scratchpad, depressing VAB 3 will designate the scratchpad entry as the new Active Target. Invalid entry will result in the **ERROR** message being displayed in the scratchpad. Without data present in the scratchpad, depressing VAB 3 presents the Active Target coordinates in the scratchpad. With the Active Target coordinates in the scratchpad, a second depression of VAB 3 results in presentation of the Associated Data for the Target coordinates. Coordinate and Associated Data in the scratchpad are "Display Only", and are not editable or transferable.
VAB 4	Depressing VAB 4 results in alternate presentation of the Target Report display **(TGT RPT:ON or TGT RPT:OFF).** With selection of **TGT RPT:ON,** the CPG's Video display will be as shown in figure 3-25.10. Upon power up, the system defaults to **TGT RPT:OFF.**
VABs 5-8	Inactive.

Table 3-22.8. CPG RANGE SOURCE

Display	Range Source Definition
AUTO	Auto Range
DFLT	Default Range (Only displayed at system start-up or AUTO range selected and invalid CPG LOS or invalid radar altitude).
LSR	Laser Range
MAN	Manual Range
NAV	Nav Range
NOAR	No Auto range (only displayed with AUTO range selected and the CPG LOS elevation angle is approaching the horizon or the radar altitude is less than 5 meters).
TGT	Target range

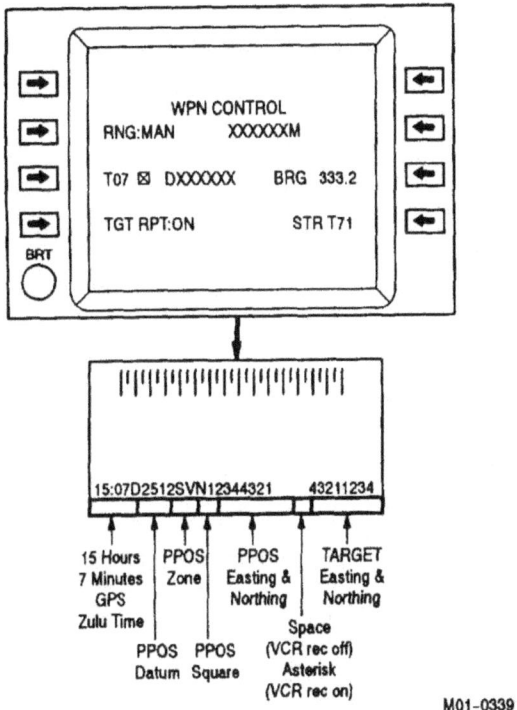

Figure 3-25.10. CPG Video Target Report Data

3.16.11 Missile Laser Codes (CODES) Page. The Missile Laser Codes page (CODES) is accessed by depressing the CODES FAB. Each of the letter code positions displayed on lines 1, 3, 5, and 7 of the CDU display (figure 3-25.11) are accessible by depression of the associated VAB. Code entry is accomplished through the keyboard and scratchpad operation. Codes are entered into the scratchpad as numerical values only. The first character position will always be either a number "1" or a number "2" only; character positions 2-4 will only utilize numbers 1-8; letters are not required to be entered.

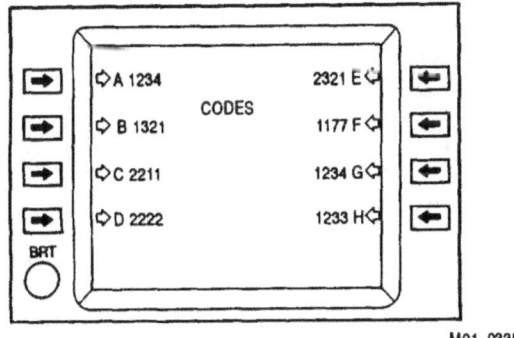

Figure 3-25.11. Missile Laser Codes Display

3.16.12 DATA MENU Page. Selection and depression of the **DATA** FAB will result in the display of the top level **DATA MENU** Page (figure 3-25.12). The **DATA MENU** page provides access to HAVE QUICK **(HQ)** radio functions and the following sub-pages: **NAV STAT, NAV SENSOR CONTROL, DTU, ZEROIZE,** and **GPS STAT.** Access to the menu sub-pages is accomplished through selection and depression of the associated VABs for the sub-page desired. VAB functions are explained in table 3-22.9.

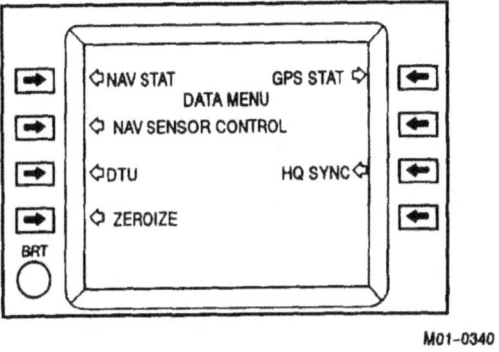

Figure 3-25.12. DATA Top Level Menu Page

Line 1. - Left-hand character field displays the legend ⇔**NAV STAT**. The right-hand character field displays the legend **GPS STAT**⇔.

Line 2. - Displays the centered page title **DATA MENU**.

Line 3. - The left-hand character field displays the legend ⇔**NAV SENSOR CONTROL**.

Line 4. - Blanked.

Line 5 - Left-hand character field displays the legend ⇔**DTU**.

Line 6. - Blanked.

Line 7. - The left-hand character field displays the legend ⇔ ZEROIZE. The right-hand character field displays the legend **HQ:SYNC**⇔.

Line 8. - Blanked.

TM 1-1520-238-10

Table 3-22.9. DATA PAGE VAB OPERATIONS

Control/Display	Function
VAB 1	Selectes NAV STATUS Page.
VAB 2	Selects NAV SENSOR CONTROL Page.
VAB 3	Selects DTU CONTROL Page.
VAB 4	Selects ZEROIZE Page.
VAB 5	Selects GPS STATUS Page.
VAB 6	Inactive
VAB 7	Energizes HQ Sync relay for 5 seconds. Supplies GPS UTC signal to ARC-164
VAB 8	Inactive

a. Navigation Status (NAV STATUS) Page. The NAV STATUS page is accessed by depressing **VAB 1** on the **DATA** Menu page. The **NAV STATUS** page is displayed as shown in figure 3-25.13. All data displayed on the **NAV STATUS** page is read-only data and non-editable. VABs 1-7 are inactive, and VAB 8 is used to return to the **DATA MENU** top level page.

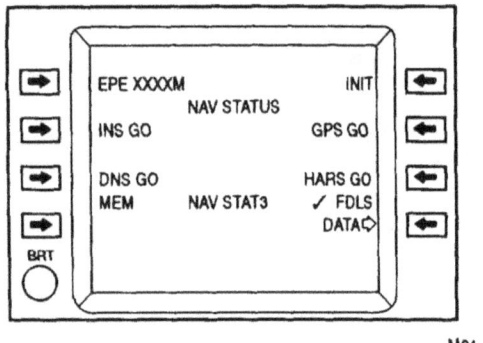

Figure 3-25.13. NAV STATUS Page

Line 1. - Left-hand character field displays the legend **EPE** (Estimated Position Error) followed by the error stated in Meters **(XXXXM)**. The distance error display is limited to four (4) characters, or a maximum of 9999M. The right-hand character field displays the current INS mode of operation. Possible displays in this character field, dependent upon whether **MODE:LAND** or **MODE:WATER** is selected are:

INIT - - - - Initialization

LEVEL - - -Leveling

TEST - - - - Internal Test

CGA- - - - - Coarse Ground Align

FGA - - - - - Fine Ground Align

ICGA - - - - Interrupted Coarse Ground Align

IFGA - - - - Interrupted Fine Ground Align

CIFA - - - - Coarse Inflight Align

FIFA - - - - Fine Inflight Align

CSA- - - - - Coarse Sea Align

FSA - - - - - Fine Sea Align

Line 2. - Displays the centered page title **NAV STATUS.**

Line 3. - The left-hand character field displays INS status information from the following list:

INS ATT? - - - - - INS attitude data invalid

INS VEL? - - - - - -INS airframe coord. velocity invalid

INS HDG? - - - - - **INS** heading data invalid

INS NOGO ----INS not operationally useable

INS GO - - - - - INS aligned and operational

The right-hand character field displays GPS status information from the following list:

3-64.16   Change 4

GPS NOGO ----- GPS not operationally useable

GPS BATT LO -- GPS battery voltage low

GPS GO -------- GPS is operational

Line 4. - Blanked.

Line 5 - The left-hand character field displays DNS status information of **DNS GO** or **DNS NOGO**. The right-hand character field displays **HARS** status information from the following list:

HARS CAL --- HARS calibration mode

HARSATT? - - - HARS attitude data invalid

HARSVEL? - - - HARS velocity data invalid

HABSHDG? - - HARS heading data may be invalid

HARS TST --- HARS in internal test mode

HARSLAT? --- HARS latitude may be invalid

HARSFAST - - - HARS in fast align mode

HARSNORM - - BARS in normal align mode

HARS GO ----- HARS aligned and operational

HARSNOGO --- HARS is not operationally usable

Line 6. - Displays System Annunciator Data.

Line 7. - The right-hand character field displays the legend **DATA◇**.

Line 8. - Blanked.

b. Navigation Sensor (NAV SENSOR CONTROL) Page. The **NAV SENSOR CONTROL** page is accessed by depressing VAB 2 on the **DATA** Menu page. The **NAV SENSOR CONTROL** page is displayed as shown in figure 3-25.14. VAB 7 is used to return to the **DATA MENU** top level page. VAB functions are explained in table 3-22.10.

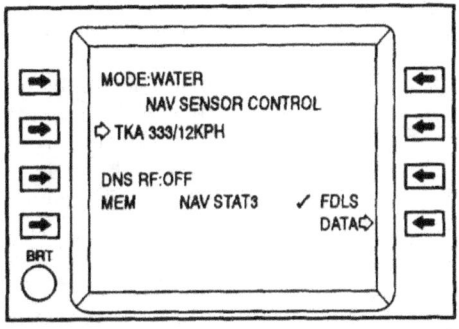

Figure 3-25.14. NAV SENSOR CONTROL Page

Line 1. - Left-hand character field displays the legend **MODEWATER** or **MODE:LAND**.

Line 2. - Displays the centered page title **NAV SENSOR CONTROL**.

Line 3 - The left-hand character field displays ◇**TKA XXX/YYKPH** only when **MODE:WATER** is displayed on line 1; otherwise, this field is blanked.. When ◇**TKA XXX/YYRPH** is displayed, the **XXX/YY** will appear as dashes until the Ground Track/Ground Speed entry is made. Subsequently, it will display system ground track and speed.

Line 4. - Blanked.

Line 5 - Left-hand character field displays the legend **DNS RF:OFF** or **DNS RF:ON**.

Line 6. - Displays System Annunciator Data.

Line 7. - The right-hand character field displays the legend **DATA**.

Line 8. - Scratchpad.

Table 3-22.10. NAV SENSOR CONTROL PAGE VAB OPERATIONS

Control/Display	Function
WAB 1	Toggles between **MODE&AND** and **MODE:WATER**. The system defaults to **MODE:LAND** at power up. **NOTE** This entry is only useable for Airborne or Sea Starts without GPS or EGI capability.
VAB 2	With aircraft heading and **GRND SPD** data entered in the scratchpad, depressing the VAB will load the data into line 3 of the display. Entry limitation are 000 to 359 degrees and 00 to 99 KPH (or KTS) respectively. Both entries are required if this function is used to aid the INS. Entry is only effective during INS startup. If **MODE:LAND** is selected on line 1, this line is blanked and the function is inhibited.
VAB 3	Toggles between **DNS RF:ON** and **DNS RF:OFF**. System defaults to **DNS RF:ON** at power up. With "squat switch" activated and **MODE:WATER** selected, system automatically changes to **DNS RF:OFF**; this can be manually overridden by the CPG. When **DNS RF:OFF** is selected, an operator initiated BIT of the DNS will cause the legend to change to **DNS RF:ON** prior to BIT starting. This function is blanked until the system has initialized the DNS.
VABs 4-7	Inactive.
VAB 8	Selects return to **DATA MENU** top level page.

c. **Data Transfer Unit (DTU) Page.** The **DTU** page is accessed by depressing VAB 3 on the **DATA** Menu page. The **DTU** page is displayed as shown in figure 3-25.15) VAB 8 is used to return to the **DATA MENU** top level page. VAB functions are explained in table 3-22.11.

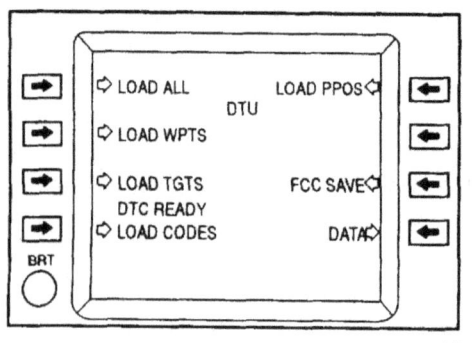

Figure 3-25.15. DATA TRANSFER UNIT Page

Line 1. - Left-hand character field displays the legend ◊**LOAD** ALL The right-hand character field displays the legend **LOAD PPOS**◊ except when the aircraft is airborne.

Line 2. - Displays the centered page title **DTU**.

Line 3. - The left-hand character field displays the legend ◊**LOAD** WPTS. The right-hand character field may be blank or may display the legend **DTC INIT**◊.

Line 4. - Blanked.

Line 5 - Left-hand character field displays the legend ◊**LOAD TGTS**. The right-hand character field displays the legend **FCC SAVE**◊.

Line 6. - Displays the left justified **DTU LOAD/SAVE** prompts listed below. If any of the first three prompts are displayed, all DTU operations are inhibited.

DTU INOP

NO DTC

DTC FORMAT INVALID

DTC READY

TARGETS LOAD FAIL

WAYPOINTS LOAD FAIL

LASER CODES LOAD FAIL

PPOS LOAD FAIL

UPLOAD ALL FAIL

TARGETS LOAD COMPLETE

WPTS LOAD COMPLETE

CODES LOAD COMPLETE

PPOS LOAD COMPLETE

UPLOAD ALL COMPLETE

FCC SAVE-OK

FCC SAVE FAIL

The **LOAD FAIL** and **LOAD COMPLETE** prompts refer to initialization data that was loaded in the DTC using the Aviation Mission Planning System (AMPS).

Line 7. - The left-hand character field displays the legend ◊**LOAD** CODES. The right-hand character field displays the legend **DATA**◊. -

Line 8. - Scratchpad

Table 3-22.11. DTU PAGE VAB OPERATIONS

Control/Display	Function
VAB 1	Initiates LOAD of Waypoints, Targets, Laser Codes, and Present Position from cartridge. To preclude inadvertent load of PPOS, the LOAD ALL function will not load PPOS if the aircraft is airborne.
VAB 2	Loads WPTS from cartridge.
VAB 3	Loads TGTS from cartridge.
VAB 4	Loads CODES from cartridge.
VAB 5	Loads PPOS from cartridge unless aircraft is airborne.
VAB 6	Reformats DTC for normal use if **DTC INIT**◊ is displayed; otherwise blanked.
VAB 7	Saves critical data in the FCC volatile memory (BST DATA, PPOS, Altitude, Altitude Corrections, MAGVAR, etc.) to non-volatile memory and saves the waypoints, targets, laser codes and PPOS to the DTC "SAVE" file.
VAB 8	Returns to the **DATA MENU** top level page.

d. ZEROIZE Page. The **ZEROIZE** page is accessed by depressing VAB 4 on the **DATA MENU** page. The **ZEROIZE** page is displayed as shown in figure 3-25.16. VAB 8 is used to return to the **DATA MENU** top level page. If the **ZEROIZE** page is displayed for 5 seconds without an operator depression of VABs 1,4,7, or 8, the display automatically reverts back to the **DATA MENU** top level page.. VAB functions are explained in table 3-22.12.

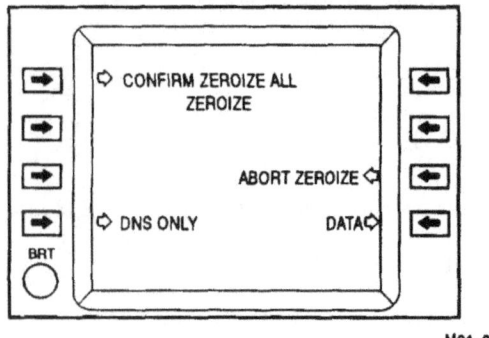

Figure 3-25.16. ZEROIZE Page

Line 1. - Left-hand character field displays the legend ◇**CONFIRM** ZEROIZE ALL.

Line 2. - Displays the centered page title **ZEROIZE**

Line 3. - Blanked

Line 4. - Blanked.

Line 5 - The right-hand character field displays the legend **ABORT** ZEROIZE ◇.

Line 6. - Blanked

Line 7. - Left-hand character field displays the legend ◇**DNS** ONLY.

Line 8. - The right-hand character field displays the legend **DATA**◇.

Table 3-22.12. ZEROIZE PAGE VAB OPERATIONS

Control/ Display	Function
VAB 1	Zeroizes the following avionics subsystems data: FCC Waypoint List, FCC Target List, FCC Laser Codes, FCC PPOS, DNS Ram Memory, GPS Y CODE Keys, and the DTU Cartridge. The CDU will then return to the page displayed before the zeroize action was initiated.
VABs 2-3	Inactive.
VAB 4	Zeroizes the DNS SDCC RAM Memory only.
VABs 5-6	Inactive.
VAB 7	Aborts the **ZEROIZE** action and returns the CDU to the page presented prior to the **ZEROIZE** action being initiated.
VAB 8	Returns to the **DATA MENU** Top Level page.

e. **Global Positioning System Status (GPS STATUS) Page.** The **GPS STATUS** page is accessed by depressing VAB 5 on the **DATA MENU** page. The **GPS STATUS** page is displayed as shown in figure 3-25.17. The functions on the **GPS STATUS** page are of a read-only nature and are non-editable. VAB 8 is used to return to the **DATA MENU** top level page. All other VAB functions are inactive.

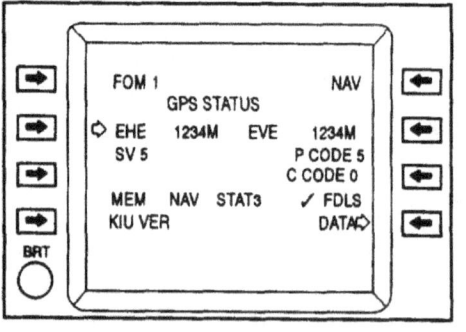

Figure 3-25.17. GPS STATUS Page

Line 1. - Left-hand character field displays the legend **FOM** 1 (Figure Of Merit) and an associated value **(1)** reflecting GPS performance status. Performance is scaled from 1-9, with 1 indicating best GPS solution (1 = **FOM** of less than 25 meter error). Any value other than 1 to 9 indicates erroneous GPS performance.

Right-hand character field will display one of the legends **NAV,** or **INIT** or **TEST,** indicating GPS Receiver operational mode.

Line 2. - Displays the centered page title **GPS STATUS.**

Line 3. - Left-hand character field displays the legend, **EHE** XXXXM, the GPS navigation solution Estimated **PPOS** Horizontal Error, in meters. **EHE** value is displayed in units of 0000 to 9999 meters. When 9999 is displayed, the actual GPS Horizontal Error may significantly exceed that value. The lower the number, the better the GPS Horizontal position performance.

Right-hand character field displays the legend **EVE** XXXXM, the GPS navigation solution Estimated **PPOS** Vertical Error, in meters. **EVE** value is displayed in units of 0000 to 9999 meters. When 9999 is displayed, the actual GPS Vertical Error may significantly exceed that value. The lower the number, the better the GPS Vertical position performance.

Line 4. - Left-hand character field displays the legend **SV X,** indicating the number of satellite vehicles the GPS receiver is using in its navigation solution. Displayed value can be 0 through 5.

Right-hand character field displays the legend **P CODE X,** indicating the number of **P CODE** (Precise Positioning Service - PPS) satellites the GPS receiver is using in its navigation solution. Displayed value can be 0 through 5.

Line 5. - Right-hand character field displays the legend **C CODE** X, indicating the number of C **CODE** (Standard Positioning Service - SPS) satellites the GPS receiver is using in its navigation solution. Displayed value can be 0 through 5.

Line 6. - Displays the System Annunciator Data.

Line 7. - Left-hand character field displays status prompts indicating operational status of the GPS Precise Positioning Service - PPS feature. Displayed prompts and their meaning are:

KIU VER	Keys in Unit Verified
KIU UNVER	Keys in Unit Unverified
KIU INCOR	Keys in Unit Incorrect
KEY PARITY ERR	Key Parity Error
INSUFF KEYS	Insufficient Keys

The right-hand character field displays the legend **DATA◊.**

Line 8: Blanked.

3.16.13 Program Menu (PGM MENU) Page. The **PGM MENU** top level page is accessed by depressing the **PGM** FAB on the CDU. The **PGM MENU** page allows access to the maintenance-related functions of Boresight EGI **(BST EGI)**, FCC CONFIG, AWS Harmonization **(AWS HARM)**, FCC Memory READ **(READ)** and the Auxiliary Alphanumeric Display **(AND)**. The PGM MENU top level page is shown in figure 3-25.18 and discussed below. VAB functions are explained in Table 3-22.13

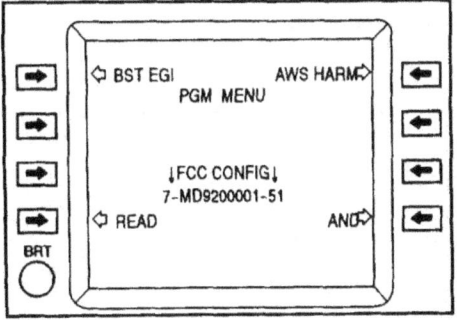

Figure 3-25.18. PGM MENU Page

Line 1. - Left-hand character field displays the legend **◊BST EGI.** The right-hand character field displays the legend **AWS HARM◊.**

Line 2. - Displays the centered page title **PGM MENU.**

Line 3. - Blanked.

Line 4. - Blanked.

Line 5. - This line displays the centered legend **FCC CONFIG**. The downward pointing arrows direct the operator's attention to the FCC software version data displayed on line 6.

Line 6. - Displays the centered FCC software version.

Line 7. - Left-hand character field displays the legend **◊READ.** The right-hand character field displays the legend **AND◊.**

Line 8: Blanked.

Table 3-22.13. PGM MENU PAGE VAB OPERATIONS

Control/Display	Function
VAB 1	Presents the **BST EGI** page.
VABs 2-3	Inactive.
VAB 4	Presents the **READ** page.
VAB 5	Presents the **AWS HARM** page.
VABs 6-7	Inactive.
VAB 8	Presents the **AND** page.

a. Boresight EGI (BST EGI) Page. The **BST EGI** page is accessed by depressing VAB 1 on the **PGM MENU** top level page. The **BST EGI** top level page is shown in figure 3-25.19 and discussed below. VAB functions are explained in Table 3-22.14.

Line 1. - Blanked.

Line 2. - Displays the centered page title **BST EGI**, followed by -MR-(milliradiana).

Line 3. - Left-hand character field displays the legend ⇔**AZ+XX.X**.

Line 4. - Blanked.

Line 5. - Left-hand character field displays the legend ⇔**EL+XX.X**. Right-hand character field displays the legend **ROLL+XX.X** ⇔

Line 6. - Blanked.

Line 7. - Left-hand character field displays the legend ⇔**EGI RESET**. The right-hand character field displays the legend **PGM**⇔.

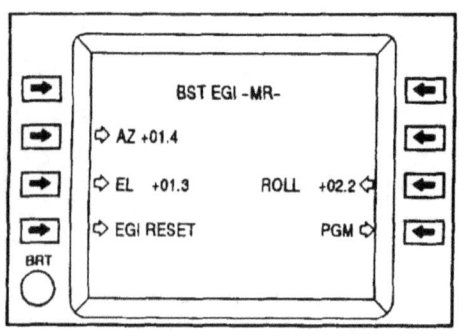

Table 3-22.14. BST EGI PAGE VAB OPERATIONS

Control/Display	Function
VAB 1	Inactive.
VAB 2	With the value for Azimuth boresight correction angle (Units in milliradians) in the scratchpad (E.g. -12.1). depressing VAB 2 enters the scratchpad value into the system. The Azimuth boresight correction angle entry range shall be +/- 99.9 mr. Azimuth boresight correction angle entry must be three digits (additional characters are ignored). It is not necessary to enter the decimal point; it will be automatically placed. A positive value is assumed if a +/- sign is not entered.

Table 3-22.14. BST EGI PAGE VAB OPERATIONS - continued

VAB 3	With the value for Elevation boresight correction angle (Units in milliradians) in the scratchpad (E.g. -12.1), depressing VAB 3 enters the scratchpad value into the system. The Elevation boresight correction angle entry range shall be +/- 99.9 mr. Elevation boresight correction angle entry must be three digits (additional characters are ignored). It is not necessary to enter the decimal point; it will be automatically placed. A positive value is assumed if a +I- sign is not entered.
VAB 4	Initiates an INS Reset action.
VABs 5-6	Inactive.
VAB 8	With the value for Roll boresight correction angle (Units in milliradians) in the scratchpad (E.g. -12.1), depressing VAB 7 enters the scratchpad value into the system. The Roll boresight correction angle entry range shall be +/- 99.9 mr. Roll boresight correction angle entry must be three digits (additional characters are ignored). It is not necessary to enter the decimal point; it will be automatically placed. A positive value is assumed if a +/- sign is not entered.
VAB 8	Returns to the **PGM MENU** top level page.

b. READ Page. The **READ** page is accessed by depressing VAB 4 on the **PGM MENU** top level page. With the **READ** page displayed and the OCTAL (or HEX) format selected, entering a six digit OCTAL (or four digit HEX) number into the scratchpad and then depressing any one of VABs 1-4 will result in the display of FCC memory data, in the format selected, at the memory location selected. The system defaults to OCTAL at power-up. The **READ** page is shown in figure 3-25.20 and discussed below. VAB functions are explained in Table 3-22.15.

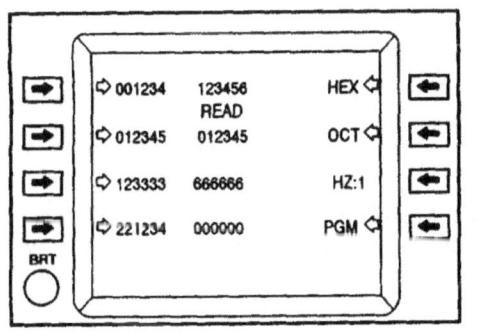

Figure 3-25.20. READ Page

Line 1. - Left-hand character field displays the legend
⇨ 661234 123456, the FCC memory read address 1 and the numerical data in that address. Right-hand character field displays the legend **HEX**⇨.

Line 2. - Displays the centered page title **READ**.

Line 3. - Left-hand character field displays the legend
⇨ **012346 012346**, the FCC memory read address 2 and the numerical data in that address. Right-hand character field displays the legend **OCT**⇨.

Line 4. - Blanked.

Line 5. - Left-hand character field displays the legend
⇨ 123333 666666, the FCC memory read address 3 and the numerical data in that address. Right-hand character field displays the legend **HZ:l**, signifying the selectable data refresh rate (1,2, or 5) for the **READ** page in Hertz..

Line 6. - Blanked.

Line 7. - Left-hand character field displays the legend
⇨ 221234 000000, the FCC memory read address 4 and the numerical data in that address. The right-hand character field displays the legend **PGM**⇨.

Line 8: Scratchpad.

Table 3-22.15. READ PAGE VAB OPERATIONS

Control/Display	Function
VAB 1-4	With a recall address in the scratchpad, depressing any of VABs 1-4 causes memory recall data to be displayed on the line associated with the VAB depressed.
VAB 5	Depression changes the entry and memory recall data display to **HEX** format.
VAB 6	Depression changes the entry and memory recall data display to **OCTAL** format.
VAB 7	Allows selection of a data refresh rate for the **READ** page. Alternatives are 1, 2, or 5 Hz. System defaults to 1Hz at power up.
VAB 8	Returns to the **PGM MENU** top level page.

c. Area Weapon Subsystem (AWS HARMONIZATION) Page. The AWS HARMONIZATION page is accessed by depressing VAB 5 on the **PGM MENU** top level page. The **AWS HARMONIZATION** page is shown in figure 3-25.21 and discussed below. VAB functions are explained in Table 3-22.16 .

Line 1. - Displays the centered page title **AWS HARMONIZATION**.

Line 2. - Displays the centered legend **DELTAS -MR-**.

Line 3. - Left-hand character field displays the legend ◊**AZ+XX.X**, indicating the last azimuth correction entered. Right-hand character field displays the legend EL+XX.X◊, indicating the last elevation correction entered..

Line 4. - Displays the centered legend **TOTALS -MR-**.

Line 5. - Left-hand character field displays the legend **AZ+XX.X**, signifying the total correction value for azimuth. This value is limited to +/- 20.0. Right-hand character field displays the legend **EL+XX.X**, signifying the total correction value for elevation. This value is limited to +/- 20.0.

Line 6. - Blanked.

Line 7. - The right-hand character field displays the legend **PGM◊**.

Line 8: Scratchpad.

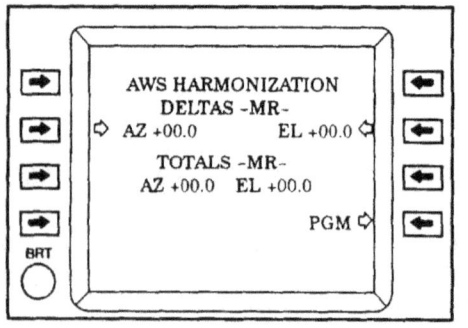

Figure 3-25.21. AWS HARMONIZATION Page

Table 3-22.16. AWS HARMONIZATION PAGE VAB OPERATIONS

Control/Display	Function
VABs 1,3,4,5 & 7	Inactive.
VAB 2	Allows for entry of additional azimuth bias corrections. Azimuth correction angle entry must be three digits (additional characters are ignored). It is not necessary to enter the decimal point; it will be automatically placed. A positive value is assumed if a +/- sign is not entered.
VAB 6	Allows for entry of additional elevation bias corrections. Elevation correction angle entry must be three digits (additional characters are ignored). It is not necessary to enter the decimal point; it will be automatically placed. A positive value is assumed if a +/- sign is not entered.
VAB 8	Returns to the **PGM MENU** top level page.

d. Alphanumeric Display (AND) Page. The **AND** page is accessed by depressing VAB 8 on the **PGM MENU** top level page. All data presented on the **AND** page is Read-Only. The AND page is shown in figure 3-25.22 and discussed below. VAB functions are inactive. To return to PGM MENU top level page, depress the PGM FAB.

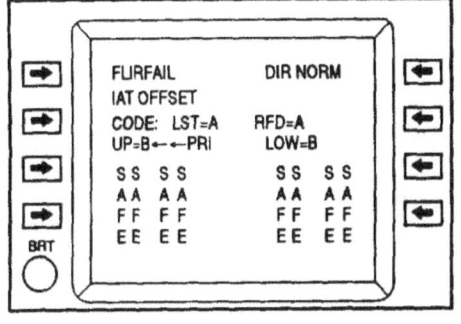

Figure 3-25.22. ALPHANUMERIC DISPLAY Page

Line 1. - Left-hand character field displays the **AND** Sight Status Field. Right-hand character field displays the **AND** Weapon Status Field.

Line 2. - Displays the **AND** TADS Status Field.

Line 3. - Displays the **AND** LST/RFD Status Field.

Line 4. - Displays the **AND** Missile Enhancement Display Field.

Lines 5 - 8. - Displays the **AND** Missile status displays (read vertically):

Cols. 3 & 4 show the LH Outbd pylon missile status

Cols. 7 & 8 show the LH Inbd pylon missile status

Cols. 15 & 16 show the RH Inbd pylon missile status

Cols. 19 & 20 show the RH Outbd pylon missile status

System Advisory Messages. System Advisory messages are displayed on the scratchpad line (line 8) of the CDU. Advisory messages may occur at any time on any page. When a message is displayed, it will remain displayed until depression of the CLR FAB. No other data may be entered while an advisory message is displayed, but an advisory message will overwrite existing data on the scratchpad line. Once the advisory message is cleared from the scratchpad, previously entered data in the scratchpad will be recalled automatically. A listing of the system advisory messages and their meanings is presented in table 3-22.17. If the FCC detects a fault (on power up) that occurred during the last FCC shutdown sequence, an extended, or full page advisory message (fig. 3-25.23) will be displayed on the CDU indicating the detected fault. Up to two faults from the Extended Advisory Message Table (table 3-22.18) can be displayed on lines 3 and 4 of the CDU display. The extended message(s) will only be displayed if the FCC has enough functionality to perform the task at the time it is required. Depressing any key on the CDU will cause the FCC to attempt to clear the fault(s) and proceed with normal operation. Information presented on the display should be recorded for maintenance action prior to clearing the fault condition.

Table 3-22.17. System Advisory Messages

ADVISORY MESSAGE	MEANING
BST DATA LOAD FAIL	FCC EEPROM TO RAM LOAD CHECKSUM ERROR
INIT DATA LOAD FAIL	FCC EEPROM TO RAM LOAD CHECKSUM ERROR
FCC SAVE COMPLETE	FCC RAM TO EEPROM SUCCESSFUL
FCC SAVE FAIL	FCC RAM TO EEPROM SAVE CHECKSUM ERROR
DTC FORMAT INVALID	DTC FILE FORMAT INVALID (DTC IS NOT USEABLE)
WPTS AUTOLOAD FAIL	INVALID WPT CHECKSUM (WPT SAVE DATA INVALID)
TGTS AUTOLOAD FAIL	INVALID TGT CHECKSUM (TGT SAVE DATA INVALID)
CODES AUTOLOAD FAIL	INVALID CODES CHECKSUM (CODES SAVE DATA INVALID)
PPOS AUTOLOAD FAIL	INVALID PPOS CHECKSUM (PPOS SAVE DATA INVALID)
DTC BATTERY LOW	DTC BATTERY INDICATES LOW VOLTAGE
DNS/HSI NAV CUES ??	DNS FLY-TO DATA VALIDATION (WRAP) ERROR

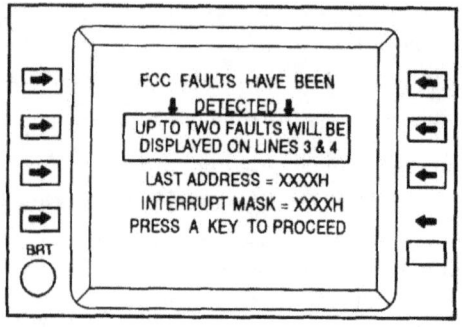

Figure 3-25.23. EXTENDED ADVISORY MESSAGE

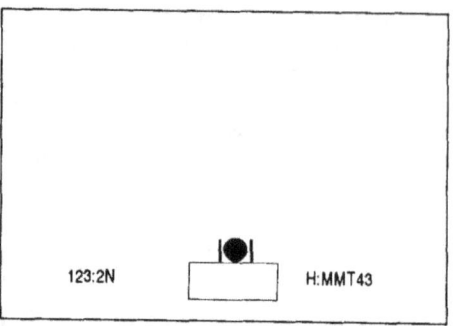

Figure 3-25.24. PILOT HIGH ACTION DISPLAY (HAD)

Table 3-22.18. EXTENDED ADVISORY MESSAGES

CPU Fault
ROM Fault
RAM Fault
50 Hz Interrupt Fault
Interrupt Mask Fault
Program Lost Fault
RAM Overwrite Fault
Watchdog Timeout Fault

3.16.14 Pilot High Action Display (HAD) The Pilot HAD Sight Status and Weapon Status fields have been modified to allow display of Distance-To-Go and Time-To-Go (figure 3-25.24) to the current **FLY-TO** destination and the active fly-to symbol.. The new Sight Status and Weapon Status Field prompts replace the existing, lowest priority displays. The new displays are out-prioritized by all other HAD displays.

a. Sight Status Field. The Sight Status Field is modified to provide Distance-To-Go from **PPOS** to the selected **FLY-To** location. Character space 1 and 2 are blank. Character spaces 3 to 8 provide the Distance-To-Go, with a colon in space 6 (providing 100 meter resolution to the pilot) and a **K** or **N** (Kilometers or Nautical miles) in space 8 . Units are selectable by the CPG using the CDU. All numerals in this display are refreshed at 1Hz.

b. Weapon Status Field. The Weapon Status Field provides, as its three lowest priorities prompts, **HSI CUE?**, Time-To-Go **(TTG)** from **PPOS** to the selected **FLY-TO** and the active fly . to symbol. The **HSI CUE?** is the highest priority and is the same indication to the pilot as the advisory message **DNS/HSI NAV CUES ??** is to the CPG. It is removed from the pilot weapon status display when the CPG clears the Advisory message. The **TTG** display is the lowest priority and is in hours and minutes (H:MM) if the **TTG** is greater than 5 minutes. This display changes to minutes and seconds (M:SS) when **TTG** is less than 5 minutes. The first character space will be H (Hours) or M (minutes). The second character space is a colon (:). The third and fourth character spaces will be either MM (minute values 00 to 59) or SS second values 00 to 59). The fifth, sixth and seventh character spaces will display the fly-to symbol (W01 through T80) The eighth character space is blank. All characters relating to **TTG** become dashed when the **TTG** is greater than 9 hours and 59 minutes or the ground speed is less than 10 KTS.

c. Trim Ball. The trim ball will flash when the HARS has been free inertial for 20 seconds or more, and aircraft groundspeed is less than 10 knots.

3.16.15 Co-Pilot Gunner High Action Display (HAD)

The CPG HAD (figure 3-25.25) has been modified to provide an Active Target Number in the Sight Status Field and an **ORT STORE** prompt in the Weapon Status Field. This information will be displayed unless other Sight Status and Weapon Status messages outprioritize it.

Figure 3-25.25. CPG HIGH ACTION DISPLAY (HAD)

a. **Sight Status Field.** The Sight Status Field is modified to provide an indication of the Active Target Number. Character spaces 1 thru 3 are blank. Character space 4 will be a **T** or **W** indicating the Active Target is in either the Target or Waypoint List. Character space 5 and 6 will be a number from 01–80 indicating the location within the **TGT/WPT** List. Character spaces 7 and 8 are blank.

b. **Weapon Status Field.** The Weapon Status Field is modified to provide the **ORT STORE** prompt. The **ORT STORE** prompt is formatted as follows: **STR TXX. XX** is a number from 71–80 to designate which location in the **TGT List** has received the new coordinate data. The **ORT STORE** prompt will be displayed for a duration of 2 seconds and will then be removed. **STR** will be located in character spaces 1–3. Character space 4 is blank. Character space 5 will be a **T**. Character spaces 6 and 7 will contain the numbers from 71 to 80. Character space 8 is blank.

NOTE

During an electrical system malfunction and operating on EMERG BATT power, the HSI/RMI will not provide adequate indications to the station.

3.16.16 Horizontal Situation Indicator (HSI).

The HSI, (fig 3-25.26), on the pilot instrument panel is an electromechanical indicator that presents position information in relation to various navigational inputs. The HSI interfaces with the (HARS), the (ADF), and the doppler navigation set. The instrument displays consist of a fixed aircraft symbol, a compass card, two bearing-to-station pointers with back-course markers, a course bar, a (**KM**) indicator, a (**HDG**) knob and marker, a course set (**CRS**) knob, a COURSE digital readout, a to-from arrow, a **NAV** flag, and a compass HDG flag. Operating power for the HSI is taken from the 115 vac No. 1 essential bus through a circuit breaker marked **HSI** on the pilot center circuit breaker panel. Controls and indicators for the horizontal situation indicator are described in table 3-22.19.

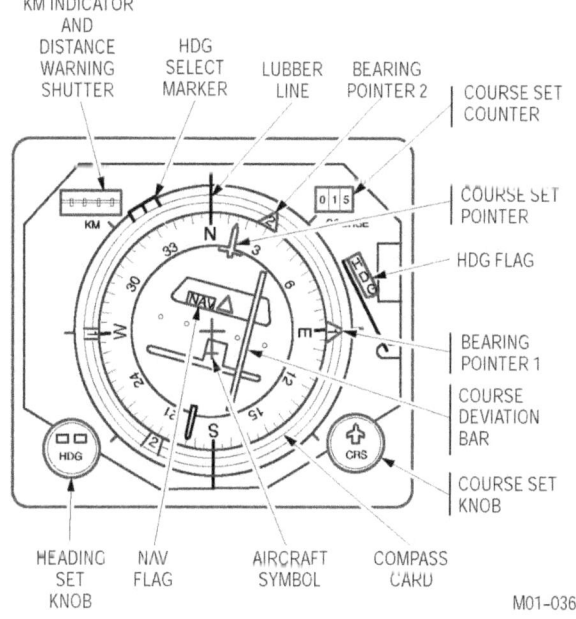

Figure 3-25.26. Horizontal Situation Indicator (Typical)

Table 3-22.19. HSI Controls and Indicators

Control/Indicator	Function
Compass Card	The compass card is a 360-degree scale that turns to display heading data obtained from the HARS. The aircraft headings are read at the upper lubber line.
Bearing pointer No. 1	The pointer operates in conjunction with the doppler. It indicates relative bearing to the active FLY-TO-location. The No. 1 bearing pointer "parks" at the 3 o'clock position when the doppler "bearing-to-destination" signal is invalid.
Bearing pointer No. 2	This pointer operates in conjunction with the ADF receiver. The pointer is read against the compass card and indicates the magnetic bearing to the ADF station (nondirectional beacon).
Course deviation bar	This bar indicates the lateral deviation from the desired navigation course. When the helicopter is flying the desired navigation course, the course bar will be aligned with the course set pointer and will be centered on the fixed aircraft symbol.
CRS knob	(CRS) knob and the course set counter operate in conjunction with the course pointer and allow the pilot to select any of 360 courses. Once set, the course pointer will turn with the compass card and will be centered on the upper lubber line when the helicopter is flying the selected course, providing there is no wind to blow the helicopter off course.
KM indicator	The digital distance display in (KM) to the FLY-TO-location.
HDG knob	(HDG) knob operates in conjunction with the heading select marker and allows the pilot to select any one of 360 headings. Seven full turns of the knob produce a 360-degrees turn of the marker.
To-from arrow	Works with VOR Not applicable.
NAV flag	The NAV flag, on the HSI course carriage, turns with the compass card. The flag retracts from view when a reliable course deviation signal is available from the doppler.
HDG flag	The HDG flag retracts when a reliable heading signal is available from the HARS. During HARS alignment, retraction of the heading flag indicates that HARS is aligned (ready-to-fly).
Distance shutter	The distance shutter (upper left corner) retracts when the ''the distance-to-destination'' signal from the doppler is valid.

Section IV. TRANSPONDER AND RADAR

3.17 TRANSPONDER RT-1295/APX-100(V)1(IFF) AND RT-1557/APX-100(V)(IFF).

The transponder set (fig 3-26) provides automatic radar identification of the helicopter to all suitably equipped challenging aircraft as well as surface or ground facilities within the operating range of the system. The RT-1296/APX-100(V) and the RT-1557/APX-1000 receives, decodes, and responds to the characteristic interrogation of operational **MODE 1,2,3/A, C,** and **4**. Specially coded identification of position (IP) and emergency signals may be transmitted to interrogation stations when conditions warrant. The transceiver can be operated in any one of four master modes, each of which may be selected by the operator at the control panel. Five coding modes are available to the operator. The first three modes may be used independently or in combination. **MODE 1** provides 32 possible code combinations, any one of which may be selected in flight. **MODE 2** provides 4096 possible code combinations, but only one is available and is normally preset before takeoff. **MODE 3/A** provides 4096 possible codes, any one of which **maybe** selected in flight. **MODE 4** is an external computer mode and can be selected to provide any one of many classified operational codes for security identification. Power to operate the IFF system is provided from the dc emergency bus through the IFF circuit breaker on the pilot overhead circuit breaker panel.

3.17.1 Antenna. Two blade antennas (fig 3-1) are used to receive interrogating signals and to transmit reply signals. The upper IFF antenna is installed on top of the fuselage area aft of the canopy, and the lower IFF antenna is installed on the bottom of the fuselage as an integral part of the UHF-AM antenna. Some helicopters have the upper IFF antenna installed on the work platform forward of the main rotor mast and the lower IFF antenna installed on the bottom of the fuselage aft of the tail boom jack pad.

3.17.2 Controls and Functions. All operating and mode code select switches for transceiver operation are on identical control panels (fig 3-26) for the RT-1296/APX-100(V) and the RT-1557/APX-100(V). Control and indicator functions of the RT-1296/APX-100(V) and the RT-1557/APX-100(V) transponder are described in table 3-23.

TM 1-1520-238-10

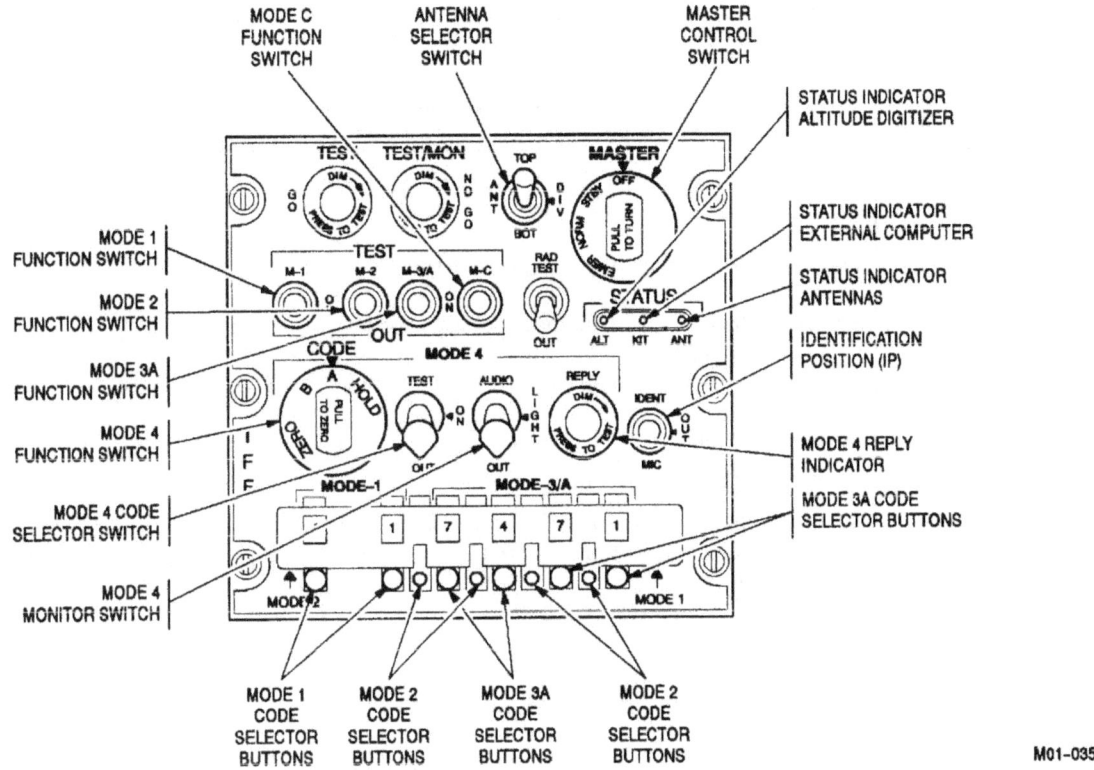

Figure 3-26. Control Panel RT-1296/APX-100(V) 1(IFF) and RT-1557/APX-100(V) 1(IFF)

Table 3-23. RT-1296/APX-100(V) 1(IFF) and RT-1557/APX-100(V) 1(IFF) Control and Indicator

Control/Indicator	Function
TEST GO	Indicates successful Built In Test (BIT).
TEST/MON NO GO	Indicates unit malfunction.
ANT	Selects antenna(s) to be used.
MASTER/OFF/ STBY/NORM/ EMER	Selects operating condition.
M-1 TEST/ON/OUT	Determines whether Mode 1 is on, off, or in BIT operation.
M-2 TEST/ON/OUT	Determines whether Mode 2 is on, off, or in BIT operation.
M-3A TEST/ON/OUT	Determines whether Mode 3A is on, off, or in BIT operation.
M-C TEST/ON/OUT	Determines whether Mode C is on, off, or in BIT operation.

3-67

Table 3-23. RT-1296/APX-100(V)1(IFF) and RT-1557/APX-100(V)1(IFF) Control and Indicator Functions - continued

Control/Indicator	Function
MODE 4 CODE Selector	Selects condition of code changer in remote computer.
A	Selects mode 4 code setting for present period.
B	Selects mode 4 code setting for succeeding period.
HOLD	Retains mode 4 code setting when power is removed from transponder.
MODE 4 TEST/ON/OUT	Determines whether Mode 4 is on, off, or in BIT operation.
MODE 4 AUDIO/ LIGHT/OUT	Enables or disables audio and visual Mode 4 indicators. Visual indicators signify valid replies. Audible signal indicates the interrogation and reply computers are set to opposite codes.
RAD TEST/OUT	Enables TEST mode.
STATUS ALT	Indicates that BIT or MON failure is due to altitude digitizer.
STATUS KIT	Indicates that BIT or MON failure is due to external computer.
STATUS ANT	Indicates that BIT or MON failure is due to high voltage standing wave ratio (VSWR) in antenna.
MODE 4 REPLY	Indicates that a Mode 4 reply is generated.
IDENT/OUT/MIC	Controls transmission of I/P pulse.
MODE 1 Selector buttons	Selects Mode 1 code to be transmitted.
MODE 2 Selector buttons	Selects Mode 2 code to be transmitted.
MODE 3/A Selector buttons	Selects Mode 3/A code to be transmitted.

3.17.3 Operation.

CAUTION

Due to possible operating problems of the (IFF), usage of the UHF radio on frequency ranges 341.325 thru 345.325 is prohibited while operating the IFF in MODE 4.

a. **Starting Procedures.** If the **MODE 2** code has not been set previously, loosen two screws which hold the **MODE 2** numeral cover, and slide the cover upward to expose numerals of **MODE 2** code switches. Set these switches to code assigned to helicopter. Slide cover down and tighten screws.

1. **MASTER** control switch - **STBY.** Allow 2 minutes for warm up.

2. **MODES 1** and/or 3A code selector buttons - Press and release until desired code shows.

3. **TEST. TEST/MON,** and **REPLY** indicators - **PRESS-TO-TEST.** "If **MODE 1** is to be used, check as follows:

   a. **MASTER** control switch - **NORM.**

   b. **M-1** switch - Hold at **TEST.** Observe that no indicators are on.

   c. **M-1** switch - Return to **ON.** If **MODE 2, 3/A** or **C** are to be used, check as follows:

      (1) M-2, M-3/A, and M-C switches - Repeat steps (3b) and (3c).

   NOTE

   Do not make any checks (1) near a radar site; (2) with **MASTER** control switch in **EMER,** or (3) with **M-3/A** codes 7600 or 7700 without first obtaining authorization from the interrogating stations.

4. **MASTER** control switch - **NORM.** If **MODE 4** is to be used, check as follows:

   NOTE

   This procedure utilizes bit to perform a self-test on the AN/APX-100 IFF. Because bit cannot test the KIT-1A/TSEC, this test will only verify the operational readiness of the IFF in **MODES 1, 2, 3/A** and **C.**

   a. **MODE 4 CODE** selector switch - position A.

   b. **MODE 4 AUDIO/LIGHT/OUT** switch - **OUT.**

   c. **MODE 4 TEST/ON/OUT** switch - **TEST.** Observe that **TEST GO** indicator lights and **MODE 4 REPLY** indicator does not light.

   d. **MODE 4 TEST/ON/OUT** switch - **ON.**

5. **ANT** switch - **BOT.**

6. Repeat steps (3b) through (3c(1), but observe that **TEST GO** indicator lights.

7. **ANT** selector switch - **TOP.**

8. Repeat steps (3b) through (3c(1), but observe that **TEST GO** indicator lights.

9. **ANT** selector switch - **DIV.**

10. Repeat steps (3b) through (3c(1), but observe that **TEST GO** indicate; lights.

NOTE

When possible, request permission from interrogating station to activate radar test mode.

11. **RAD TEST** switch - **RAD TEST** and hold.

12. **RAD TEST** switch - Return to **OUT.**

13. Verify from interrogating station that **TEST MODE** reply was received.

3.17.4 Normal Operation. Completion of the starting procedure leaves the APX-100(V) in operation. The following steps may be required, depending upon mission.

| CAUTION |

Due to possible operating problems of the (IFF), usage of the UHF radio on frequency ranges 341.325 thru 36325 is prohibited while operating the IFF in MODE 4.

a. Normal Mode.

   1. **MODE 4 CODE** selector switch - **A** or **B** as required.

      a. If **MODE 4 code** retention is desired, after landing and before turning the **MASTER** control switch **OFF** or helicopter power off, place the **MODE 4 CODE** selector switch to **HOLD** and release. Allow approximately 15 seconds for the **HOLD** function to operate before turning the **MASTER** control switch **OFF** or helicopter power off.

      b. If **MODE 4** code retention is not desired, either place the **MODE 4 CODE** selector switch to **ZERO** or with the helicopter on the ground turn the **MASTER** control switch **OFF.** For both methods zeroizing is immediate.

2. Mode M-1, M-2, M-3/A, M-C, or MODE 4 switches - Select desired mode.

3. IP Switch - IDENT, when required, to transmit identification of position pulses or set IP switch to MIC to transmit IP pulse only when microphone press-to-talk switch is actuated. (IP pulses will be for 15 to 30-second duration when activated.)

b. Emergency Mode. During an emergency or distress condition, the APX-100(V) may be used to transmit specially coded emergency signals on MODE 1, 2, 3/4 and 4 to all interrogating stations. Those emergency signals will be transmitted as long as the MASTER control switch on the control panel remains in EMER.

1. MASTER control switch - EMER.

c. Stopping Procedure. Refer to paragraph 3.17.4 a. before stopping transponder.

1. MASTER control switch - OFF.

3.17.5 Transponder Computer KIT-1A/TSEC and KIT-1C/TSEC. The transponder computer operates in conjunction with MODE 4. A caution light on the pilot and CPG caution/warning panel, marked IFF, will go on when a malfunction occurs that will prevent a reply when interrogated or the KIT-1A or KIT-1C computers have failed. MODE 4 operation is selected by placing the MODE 4 switch ON. Placing the MODE 4 switch to OUT disables MODE 4. As with the other modes of the transponder, the MASTER control switch must be placed in the NORM position to provide power and permit functioning of the selected modes. MODE 4 CODE switch is placarded ZERO, B, A, and HOLD. The switch must be lifted over a detent when rotated to ZERO. Position A selects the MODE 4 CODE for one period, and position B selects the MODE 4 CODE for another period. The switch is spring-loaded to return from HOLD to the A position. Both A and B codes are either mechanically inserted into the KIT-1A or electronically inserted into the KIT-1C. A KYK-13 or similar keying device is used to load the KIT-1C. The KIT-1C must have operable batteries or have been turned on, as previously described, before it will accept and hold the codes. The KIT-1A does not require power because it is mechanically keyed. Once keyed the KIT-IA and KIT-1C (with operable batteries) will retain the codes regardless of MASTER control switch position or helicopter power as long as the helicopter is on the ground. Once the helicopter has become airborne (determined by the squat switch) the MODE 4 codes will automatically zeroize anytime the MASTER control switch or helicopter power is turned off or interrupted. The codes may be retained by using the HOLD function. After landing place the MODE 4 CODE selector switch to HOLD and release before turning the MASTER control switch OFF or helicopter power off. Allow approximately 15 seconds for the HOLD function to operate before turning the MASTER control switch OFF or helicopter power off. The HOLD function must be selected with the helicopter on the ground (determined by the squat switch). Selecting the HOLD function while the helicopter is airborne has no effect whatsoever. MODE 4 codes can be zeroized anytime the MASTER control switch is not in the OFF position and the helicopter is powered by placing the MODE 4 CODE switch in the ZERO position. Zeroizing is immediate. The transponder computer KIT-1A/TSEC and KIT-1C/TSEC operation is classified.

3.18 ALTIMETER AN/APN-209(V).

The radar altimeter system (fig 3-27) provides instantaneous indication of actual terrain clearance height. Altitude, in feet, is displayed on a radar altimeter indicator on the instrument panel in front of the pilot. The radar altimeter indicator contains a pointer that indicates altitude on a linear scale from 0 to 200 feet (10 feet per unit) and on a second-linear scale from 200 to 1500 feet (100 feet per unit). An **on/OFF/LO** altitude bug set knob, on the lower left corner of the indicator, combines functions to serve as low level warning bug set control and **on/OFF** power switch. The system is turned on by turning the **LO** control knob, marked **SET**, clockwise from **OFF**. Continued clockwise turning of the control knob will permit the pilot to select any desired low-altitude limit as indicated by the **LO** altitude bug. Whenever the altitude pointer exceeds low-altitude set limit, the **LO** altitude warning light will go on. Turning the **PUSH-TO-TEST HI SET** control on the lower right corner of the indicator positions the high altitude bug. Whenever the pointer exceeds the **HI** altitude set limit, the high altitude warning light will come on. Pressing the **PUSH-TO-TEST HI SET** control provides a testing feature of the system at any time and altitude. When the **PUSH-TO-TEST HI SET** control knob is pressed, a reading between 900 feet and 1100 feet on the indicator, and digital display will be displayed. The **OFF** flag removed from view indicates satisfactory system operation. Releasing the **PUSH-TO-TEST HI SET** control knob restores the system to normal operation. Loss of system power will be indicated by the indicator pointer moving behind the dial mask and the **OFF** flag reappearing in the center of the instrument. If the system should become unreliable, the flag will appear and the indicator point will go behind the dial mask to prevent the pilot from obtaining erroneous readings. Flight operations above 1500 feet do not require that the system be turned off. The pointer will go behind the dial mask but the transmitter will be operating. Power to operate the AN/APN-209 is supplied from the emergency dc bus through the **RDR ALT** circuit breaker on the pilot center circuit breaker panel.

3.18.1 Antenna. The radar altimeter antenna (fig 3-1) are flush-mounted in a fairing on the bottom of the helicopter. The aft antenna is the transmitting antenna and the forward is the receiving antenna.

3.18.2 Controls, Indicators, and Functions. Control of the radar altimeter set is provided by the **LO SET OFF** knob on the front of the height indicator. The knob marked **HI SET** also controls the **PUSH TO TEST**. Control and indicator functions of the AN/APN-209 altimeter set are described in table 3-24.

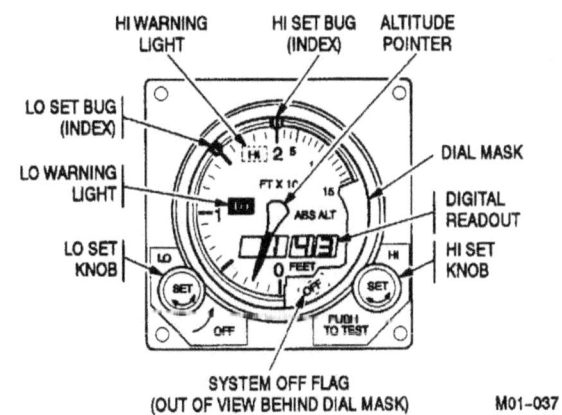

Figure 3-27. Altimeter AN/APN-209(V)

Table 3-24. AN/APN-209(V) Altimeter Control and Indicator Functions

Control/Indicator	Function
LO SET knob	Power control turned counterclockwise to OFF; clockwise to on.
LO SET bug	Sets altitude trip point of LO warning light.
HI SET bug	Sets altitude trip point of HI warning light.
HI SET knob	Pushing knob actuates BIT system to self-test altimeter.
Altitude pointer	Provides an analog indication of absolute altitude from zero to 1500 feet.
Digital readout	Provides a direct-reading four-digit indication of absolute from zero to 1500 feet.
LO warning light	Lights whenever dial pointer goes below L altitude bug setting.
HI warning light	Lights whenever dial pointer goes above H altitude bug setting.
OFF flag	Moves into view whenever altimeter loses track while power is applied.

3.18.3 Operation.

  a. Starting Procedure.

    1. **LO SET** knob - On.

    2. **LO** set bug - Set to 80 feet.

    3. **HI** set bug - Set to 800 feet.

  b. Track Operation. After about two minutes of warmup, the altimeter will go into track mode with these indications:

    1. **OFF** flag - Not in view.

    2. Altitude pointer -0 to +5 feet.

    3. Digital readout -0 to +3 feet.

    4. **LO** warning light - Will light.

    5. **HI** warning light - Will be off.

  c. Self Test. Press and hold HI SET Knob. The altimeter will indicate a track condition as follows:

    1. **OFF** flag - Not in view.

    2. Altitude pointer -1000 ± 100 feet.

    3. Digital readout -1000 ± 100 feet.

    4. **LO** warning light - Will be off.

    5. **HI** warning light - Will light.

    6. **HI SET** knob - Release. The altimeter will return to indications in step b, Track Operation.

3.18.4 Stopping Procedure.

    1. **LO SET** knob - **OFF**.

CHAPTER 4
MISSION EQUIPMENT

Section I. MISSION AVIONICS

4.1 INFRARED COUNTERMEASURES SET AN/ALQ-144A, 144(V)3. Data moved to Section III, paragraph 4.342A.

4.1.1 Infrared Countermeasures Set Control Panel. Data moved to Section III, para. 4.34A.1.

4.2 RADAR COUNTERMEASURES SET AN/ALQ-136. Data moved to Section III, paragraph 4.34B.

4.2.1 Radar Countermeasures Set Control Panel. Data moved to Section III, para. 4.34B.1.

4.3 FLIGHT SYMBOLOGY.

The flight symbology has four modes; cruise, transition, hover, and bob-up. The symbology modes (fig 4-1) are selectable using the flight symbology mode switch on the cyclic. The symbols are defined in table 4-1.

WARNING

When conducting sea operations the velocity vector in the pilots symbology will indicate the speed of the aircraft relative to the earth and not to the ship.

The pilot flight symbology, particularly the velocity vector, is always driven by inertial velocity. The indicated velocity is relative to actual motion over ground or water. When attempting to move or hover near or over a moving platform (such as a ship), the relative velocity between the aircraft and the moving platform is different than the true inertial velocity indicated by the velocity vector. If hovering over a ship moving at 10 knots through the water, the velocity vector will indicate 10 knots, not 0 knots, which is the relative velocity between the aircraft and the ship. For this situation the pilot must use <u>visual cues</u> to derive the actual relative motion with respect to the moving platform. The velocity vector <u>must not</u> be used as an indiator of relative motion between the aircraft and moving platform, since it will be inaccurate to the degree of the platforms' velocity through the water.

TM 1-1520-238-10

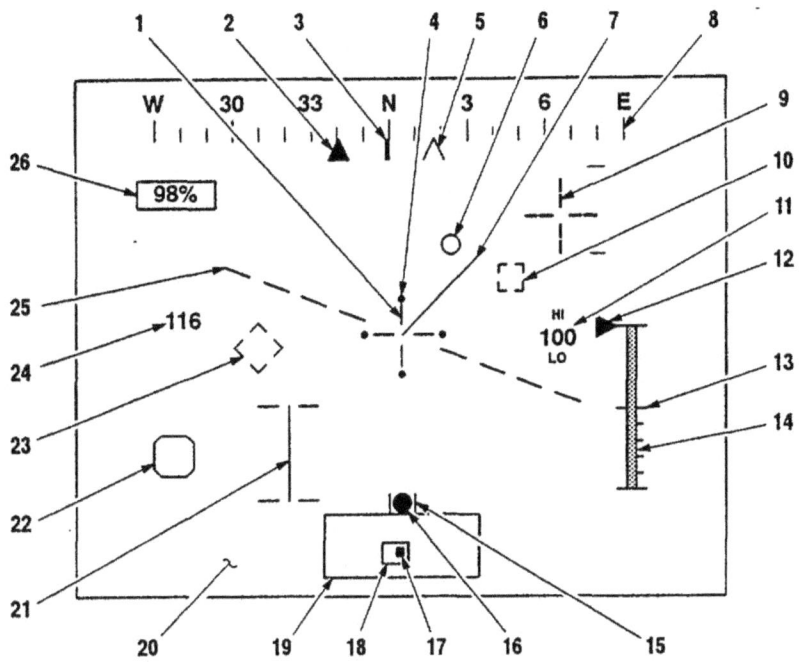

Figure 4-1. Flight Symbology Modes

Table 4-1. Flight Symbology Definitions

Fig. 4-1 Ref No	Symbol Name	Description
1	LOS Reticle	Represents the line-of-sight of the crew member selected sight. The reticule will flash whenever the selected sight LOS is invalid or has failed. The reticule will also flash whenever the "ACTIONED" weapon is in a NO-GO state. The High Action Display will prompt the crewmember for the appropriate condition.
2	Alternate Sensor Bearing	Indicates to the crewmember the other crewmember sensor relative bearing with respect to helicopter center line.
3	Lubber Line	Index indicates helicopter magnetic heading.
4	Cueing Dots	Indicates cued direction for target acquisition. All four dots present and flashing indicate IHADSS boresight is required.
5	Command Heading	Indicates heading to fly to next navigation waypoint designated by the Doppler navigation system if bob-up mode of the flight symbology is not selected. Indicates fixed heading reference when bob-up mode is selected.

Change 3   4-3

Table 4-1. Flight Symbology Definitions - continued

Fig. 4-1 Ref No	Symbol Name	Description
6	Acceleration Cue	A vectorial representation of the helicopter longitudinal and lateral acceleration; the display origin is normally the end of the velocity vector. In the hover and bob-up modes, when the velocity vector exceeds its maximum scale, the display origin changes to the center of the LOS reticule. The Acceleration Cue will flash when the HARS inertial platform has gone into free inertial mode, usually as a result of the LDNS going into memory.
7	Velocity Vector	A vectorial representation of the helicopter longitudinal and lateral ground velocities; in hover mode, maximum scale is 6 knots ground speed; in transition mode 60 knots ground speed. The velocity vector will flash when the HARS inertial platform has gone into the free inertial mode if MUX switch is in SEC.
8	Heading Scale	Helicopter magnetic heading scale.
9	Cued LOS Reticule	A virtual reticule indicating the cued LOS to the appropriate crew member. Used with the cueing dots.
10	Missile Constraints	Indicates the required orientation to align the helicopter into constraints for Hellfire missile engagements. When all constraints for the mode of engagement are satisfied, the box will go from 'dashed' to 'solid'.
11	Radar Altitude	A digital display of radar altitude. Displays in 1-foot increments to 50 feet and in 10-foot increments above 50 feet.
12	Rate of Climb	An analog display of rate of climb moving along the left side of the vertical scale. Tick marks designate 500 and 1000 fpm rates of climb or descent.
13	Radar Altitude Vertical Scale	A vertical scale for the analog display of radar altitude. Tick marks designate 10-foot increments from 0 to 50-feet, and 50-foot increments from 50 to 200 feet. The scale is blanked when radar altitude is greater than 200 feet.
14	Radar Altitude Vertical Tape	An analog display of radar altitude moving within the vertical scale; also blanked above 200 feet.
15	Skid/Slip Lubber Lines	Represent the limits for 'ball centered' flight.
16	Skid/Slip Ball	Indicates the amount of skid or slip the helicopter is experiencing.
17	Cued LOS DOT	Indicates the cued LOS location within the field of regard. The cued LOS DOT will flash when the HARS inertial platform has gone into the free inertial mode.
18	Field of View	Represents the instantaneous FOV of the crewmember sensor within the field of regard.
19	Sensor Field of Regard	Represents the total gimbal limits possible for the respective crew member sensor.
20	High Action Display	Refer to paragraph 4.28.

Table 4-1. Flight Symbology Definitions - continued

Fig. 4-1 Ref No.	Symbol Name	Description
21	Rocket Steering Cursor	Indicates the required orientation to align the helicopter into constraints for 2.75 inch FFAR rocket engagements. During fixed or flight stow rocket delivery, a broken I beam will appear.
22	Hover Position Box	Displays helicopter relative position when bob-up mode is selected and represents an 8 foot square. Maximum displacement is 48 feet laterally or longitudinally. The Hover Position Box drifts while in a stationary hover. A drift of 6 feet the first minute is possible, and as much as 23 feet after 5 minutes when using EGI with GPS keyed tracking 4 or more satellites. If these conditions are not met, a drift of 21 feet per minute is possible.
23	Head Tracker	Indicates the pilot head position relative to the center line of the helicopter. This is a virtual symbol whose range of display is 30 degrees vertically and 40 degrees horizontally about the nose of the helicopter. When CPG selects PLT/GND ORIDE and SIGHT SEL NVS, this symbol indicates his head position.
24	Airspeed	A digital display of true airspeed when the ADSS is turned on or not failed. If the ADSS is OFF or failed, display is ground speed in knots from the doppler navigation system. Range is 0 to 200, omnidirectional.
25	Horizon Line	Indicates pitch and roll attitude of the helicopter.
26	Engine Torque	Indicates the engine torque output by the engines. The magnitude is the larger of the two engine torque values. If a greater than 12% torque split occurs between engines, the displayed torque value will flash. At an engine torque value of 98% or higher, a box around the torque value flashes to indicate an impending engine torque limit. Symbolic torque value maximum is 120%.

4.4 VIDEO DISPLAY UNIT (VDU).

The VDU (fig 4-2) is located in the pilot vertical instrument panel (fig 2-9). The basic configuration has a red night filter which is stored in a bracket on top of the unit. The -3 configuration VDU has a gray night filter which is stored in a protective pouch on top of the unit. The red and gray filters are used on evening and night flights to provide a display which does not affect the pilot night vision. The VDU has the capability of displaying the video from either the pilot or CPG selected sensors independent of the IHADSS. This permits the pilot to have a simultaneous display of the PNVS video on the HMD and CPG video on the VDU. In the event of pilot HDU failure, a limited night terrain flight capability is available by selecting **PLT** on the VDU and placing the PNVS in the NW FXD position. The VDU receives 28 vdc power from the No. 2 essential dc bus and is protected by the **VDU** circuit breaker located on the pilot overhead circuit breaker panel. The VDU is not monitored by FD/LS. The function of each switch is described in table 4-2.

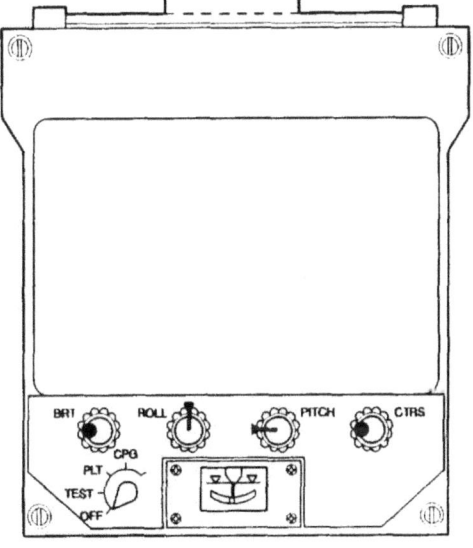

Figure 4-2. Video Display Unit

Table 4-2. Video Display Unit Control/Indicator Functions

Switch/Control	Position	Function
	OFF	Removes all power to the VDU.
	TEST	Displays a vertically oriented test pattern.
	PLT	Displays pilot selected video.
	CPG	Displays CPG selected video.
BRT		Adjusts display brightness.
ROLL		Adjusts roll trim on the symbolic horizon displayed in the transition and cruise modes of flight symbology.
PITCH		Adjusts the pitch trim on the symbolic horizon displayed in the transition and cruise modes of flight symbology.
CTRS		Adjusts display contrast in PLT and CPG modes.

4.5 VIDEO RECORDER SUBSYSTEM (VRS).

The VFtS consists of an airborne video recorder, located in the left side of the aft avionics bay (fig 2-2), and the video recorder control panel, placarded **RECORDER** (fig 4-3), located in the CPG left console (fig 2-12). The VRS has the capability of recording up to 72 minutes of either pilot or CPG selected video. Capability exists for playback onboard the helicopter for real time damage assessment and reconnaissance. The VRS receives 115 vac from the No. 2 essential ac bus through the **RCDR** circuit breaker on the CPG No.1 circuit breaker panel. The VRS is not monitored by FD/LS. The video tape recorded onboard the helicopter requires a special video playback unit to view the imagery off the helicopter. The video recorder control panel control/indicator functions are discribed in table 4-3.

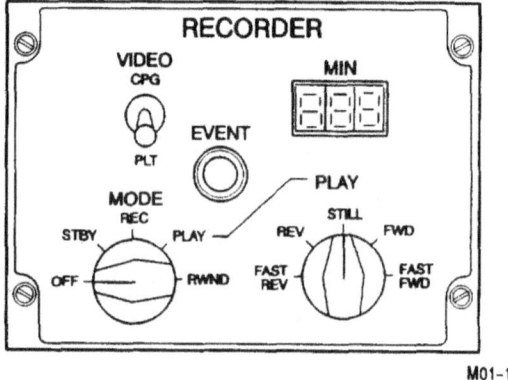

Figure 43. Video Recorder Subsystem Control Panel

Table 43. Video Recorder Control Panel/Indicator Functions

Control/Indicator	Position	Function
MODE	OFF	Removes all power to the VRS; commands the recorder to unthread the video tape cassette.
	STBY	Enables power to the VRS and commands the recorder to thread the video tape cassette.
	REC	Places the recorder in the record mode. Actual recording starts or stops when the VID RCD switch on the ORT RHG is pressed.
	PLAY	Places the recorder in the play mode and activates the PLAY rotary switch for further commands. Automatically commands the symbol generator to display the recorder video on the CPG displays.
	RWND	Commands the recorder to rewind the video tape cassette; once commanded, the recorder will completely rewind the video tape prior to following any further commands.
PLAY		The PLAY switch is active only when the MODE switch is in the PLAY position.
	FAST REV	Commands the recorder to fast reverse.
	REV	Plays back the recorded video in reverse on the CPG displays at normal speed.
	STILL	Freezes the video image.
	FWD	Plays back the recorded video in forward on the CPG displays at normal speed.
	FAST FWD	Fast forwards the recorder.
WENT		A pushbutton switch which when pressed will place an event mark (EMK) and audio tone on the video tape.
VIDEO	CPG	Selects the CPG displayed video to be sent to the recorder when recording.
	PLT	Selects the pilot displayed video to be sent to the recorder when recording.
MIN		Tape used indicator; the value displayed is NOT in minutes or feet. It is a relative tape used indicator. At the beginning of the tape the indicator will read 00.0.

4.5.1 Video Recorder Operation. The VRS is turned on by placing the **MODE** switch in the STBY position. When this occurs, the tape recorder will thread the video cassette. Placing the **MODE** switch in the **REC** position will arm the **VID RCD** pushbutton on the ORT RHG. Actuating the **VID RCD** pushbutton will turn on or off the recording of video. When recording video, whether pilot or CPG, the prompt **RECORDER** 45 will be displayed or the prompt RCDR ON 49A / 51 will be displayed initially for eight seconds and for eight seconds once every minute in the sight status block of the CPG HAD and AND. To playback recorded video place the **MODE** switch to **PLAY** and select the desired playback method on the **PLAY** switch: **STILL, REV, FAST REV, FWD** or **FAST FWD**. If the **MODE** switch is placed in the **RWND** position, the digital display, **MIN,** will automatically reset to 00.0 and the video tape recorder will start to rewind the cassette. Although the rewinding of the tape may be aborted by placing the **MODE** switch in any other position,

DO NOT use **RWND** position unless the intent is to completely rewind the tape because all elapsed counter references will no longer be valid. If an event mark is encountered during playback the video recorder will automatically stop and freeze the image. To proceed past the event mark, either forward or backward, place the **MODE** switch in the **STBY** position then back to the **PLAY** position. The **RCDR OFF** 49A / 51 will be displayed for eight seconds in the sight status block of the CPG HAD and AND after the **VID RCD** pushbutton is activated to stop recording.

4.6 DATA ENTRY KEYBOARD (DEK).

NOTE

The DEK is functional on 45 / 49A aircraft. It is functional on 51 aircraft as a backup to the CDU when the BBC is in control.

The DEK placarded **DATA ENTRY** (fig 4-4), is located in the CPG left console (fig 2-12) and receives power from the No. 1 essential bus and No. 3 essential dc bus; protection is provided by the **FC AC** and **FC DC** circuit breakers located on the pilot overhead circuit breaker panel. The DEK is continuously monitored by FD/LS; the on command test (test 15-UTIL) will fault isolate to the LRU. Control and functions are described in table 4-4.

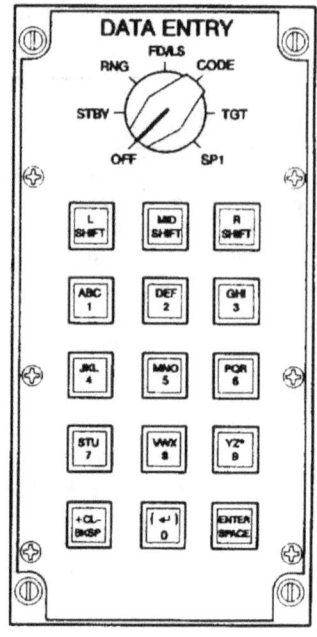

Figure 44. Data Entry Keyboard

Table 4-4. Data Entry Keyboard Control Functions

Control	Position	Function
DATA ENTRY	OFF	Deactivates the keyboard.
	STBY	Power applied to keyboard and self test initiated, all inputs from the keyboard are ignored.
	RNG	Input of CPG range, manual or automatic, in meters, aircraft to target. Automatic range calculations are commanded by the input of 0 range. Maximum manual range entry: 31,999.
	FD/LS	Input to the fault detection/location system.
	CODE	Input of standard NATO/TRI-Service laser code data.
	TGT	Input of coordinate data for the waypoint/ targeting subsystem.
	SP1	Input of data to the fire control computer.

4.6.1 DEK Operation.

a. Key in a Letter: Press the shift key corresponding to the relative location of the desired letter. Then press the key with the desired letter.

b. Key in a Number: Press the desired key.

c. Enter Keyed-In Data: Press any shift key. Then press the **ENTER/SPACE** key.

d. DEK Special Characters:

(1) **BKSP:** Moves the cursor one position to the left.

(2) **CL:** Clears the entire current input line and moves the cursor to the leftmost position.

(3) **(↵):** Advance to the first position of the next input line.

(4) **SPACE:** Skip the next character position or perform automatic character inserts.

e. DEK Saving Functions.

Data entered through the DEK will be automatically saved ▬45▬. Data entered through the DEK, in flight storage of waypoint/targets and last PPOS will be automatically saved if a take off and landing was conducted (two squat switch activations) and the HARS switch was placed to OFF (standard engine shutdown procedures for pilot) ▬49A▬.

For ground only operations, DEK entries can be manually saved by selecting **SP1,** page 1 and entering an E. The message DATA SAVED will be displayed in the upper left corner of the display verifying that the data was saved. Manual saving of data using the same procedure as described for ground operations can be conducted by the flight crew at their option during flight operations ▬49A▬.

Section II. ARMAMENT

4.7 AUTHORIZED ARMAMENT CONFIGURATIONS.

For authorized configurations, refer to figure 7-18 **701** or figure 7A-18 **701C**.

4.8 AREA WEAPON SYSTEM 30MM, M-230E1.

WARNING

In the event of IHADSS failure with gun selected, the gun will be commanded to the fixed forward position. Once in this position, the gun can still be fired without having to re-action the gun.

Prior to initiating AWS FD/LS check (IBIT), ensure the pilot ground override (PLT/GND ORIDE) switch is in the OFF position. Failure to perform this action may result in uncommanded gun turret slewing or uncommanded gun firing during a AWS FD/LS (IBIT) manual abort.

CAUTION

If ECP 1206 (Mechanical Gun Stops) is installed, but not ECP 1251 (Electronic Gun Stops), the M230E1 AWS may be used with the following restrictions:

- If external tanks are installed, limit AWS azimuth travel to ± 45 degrees.
- If external tanks are not installed, limit AWS azimuth travel to ± 70 degrees.

If both ECPs are installed, the AWS is not restricted when external tanks are mounted.

If neither ECP is installed, the AWS is restricted from use when external tanks are mounted.

The M-230E1,30mm gun (fig 4-5) is a single barrel, externally powered, chain drive weapon using M788/789 or ADEN/DEFA type ammunition. The 30mm gun is mounted in a hydraulically driven turret capable of slewing the gun 100° **15** or 86° **49A** / **51** left or right of the helicopter centerline and up 11° to 60° down. In the event of loss of hydraulics, the turret will lock in the current azimuth position and the gun will return to the elevation stow position of 11° up. The rate of fire is set for 600 to 650 rounds per minute. The maximum capacity of the linkless storage subsystem is 1200 rounds. The gun duty cycle is as follows: six 50-round bursts with 5 seconds between bursts followed by a lo-minute cooling period. For bursts limiter settings other than 50, the duty cycle can be generalized as no more than 300 rounds fired within 60 seconds before allowing the gun to cool for ten minutes, after which the cycle may be repeated. The FCC limits the fire control solution to a maximum range of 4000 meters. The maximum range of the 30mm gun is approximately 4000 meters.

4.8.1 AWS Dynamic Harmonization 49A / 51. Dynamic harmonization is the capability of the flight crew to apply an in flight boresight correction procedure. Dynamic harmonization corrects for AWS system variations and firing characteristics for different aircraft and AWS combinations. The procedure involves live-fire of the 30mm gun to fine tune the system in addition to the CBHK static boresight.

Performance of dynamic harmonization is not a requirement but a crews option. This procedure is not intended to replace the CBHK static boresight or proper AWS maintenance.

Any modification of boresight correctors must be entered in the aircraft logbook. Correctors remain in the FCC until changed via the boresight procedures.

TM 1-1520-238-10

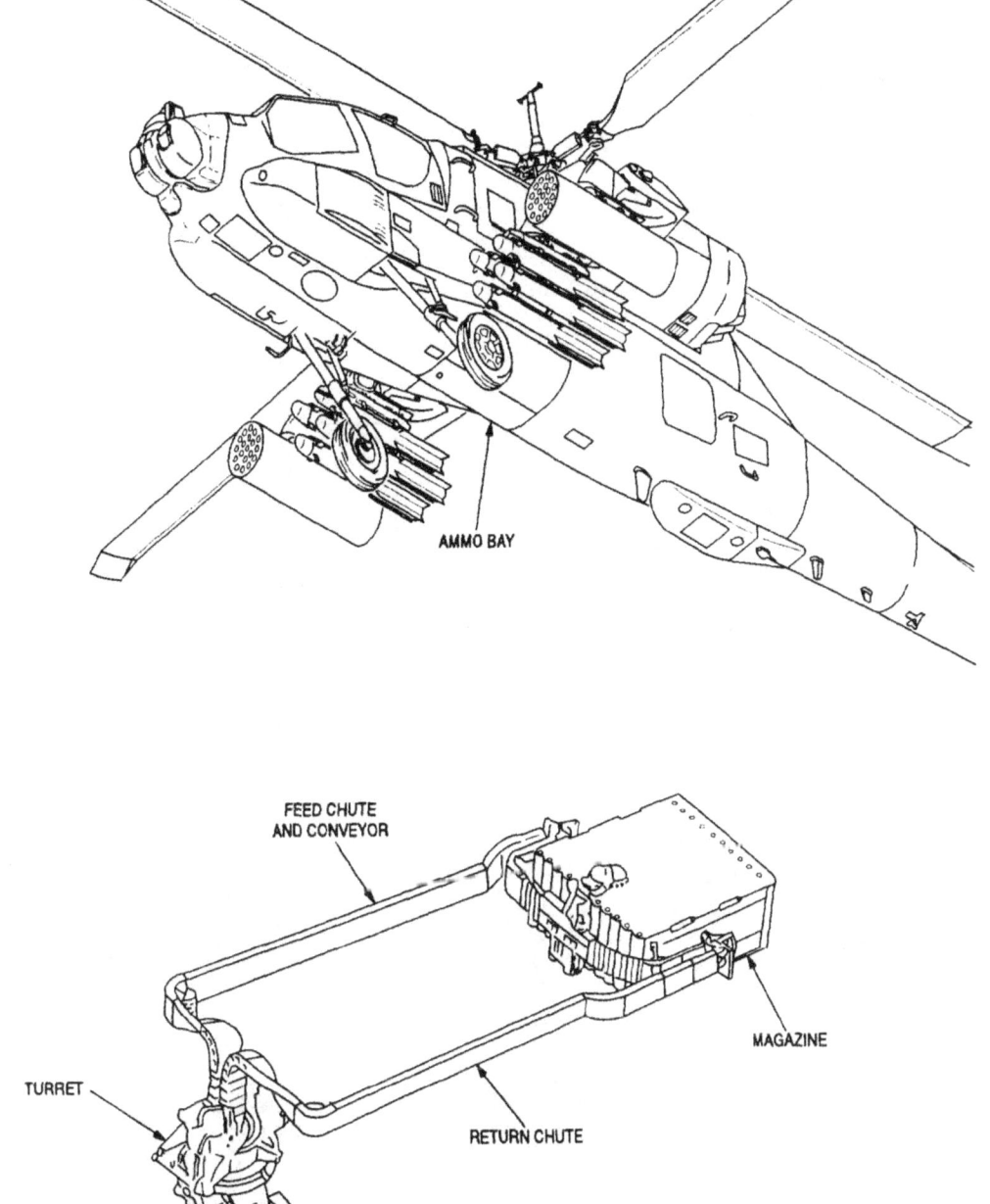

Figure 4-5. Area Weapon System 30mm, M-230E1

4.9 AERIAL ROCKET CONTROL SYSTEM (ARCS), 2.75 INCH.

CAUTION

2.75 inch Hydra-70 Rocket configurations (Fuse/Warhead/Motor), are limited to the following flight conditions:

Firing of M439/M261/Mark 66 and M442/M257/Mark 66 at ranges less than 1000 meters and/or airspeeds greater than 90 knots is not authorized.

The ARCS (fig 4-6) is a light antipersonnel assault weapon. The ARCS consists of a rocket control panel located in the pilots station and four station directors located in each of the four wing station pylons. The ARCS permits the pilot to select the desired type of 2.75 inch folding fin aerial rocket (FFAR) warhead, fuze, quantity, and range. The lightweight nineteen-tube launchers can be mounted on any of the four wing stations. There are three modes of firing rockets: pilot, CPG, and cooperative (precision) mode. The FCC limits the fire control solution to a maximum range of 6000 meters (MK-40) and 7500 meters (MK-66).

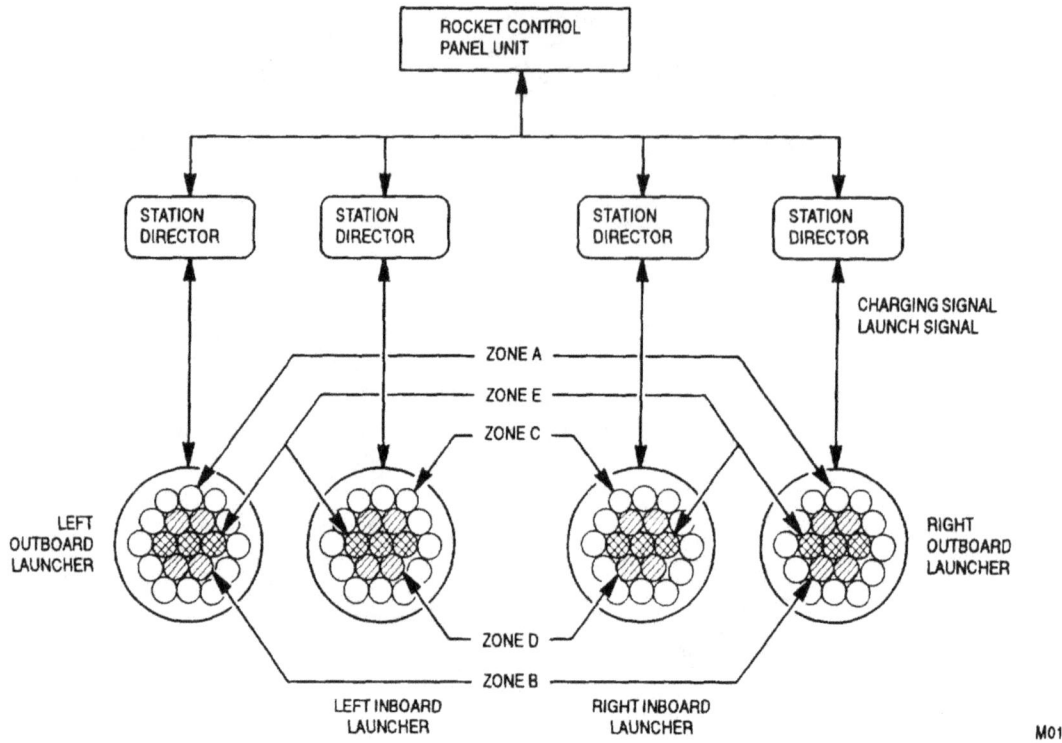

Figure 4-6. Aerial Rocket (2.75 inch) Delivery System

4.10 POINT TARGET WEAPONS SYSTEM.

The point target weapons system (fig 4-7), commonly called Hellfire, is the primary armament on the helicopter for the destruction of tanks and other hard material targets. It provides the capability of firing missiles on the ground and airborne, at speeds from hover to the maximum level flight speed. The Hellfire currently has the tri-service type laser seeker. This gives the flight crew the capability of two types of launches: lock-on-before-launch (LOBL) and lock-on-after-launch (LOAL). The type depends only on when the laser designator is fired; before or after launch of the missile. The LOAL type of launch can specify three prelaunch programmed trajectories. The missiles can be launched in two types of modes: normal, sometimes referred to as rapid; and ripple. During normal mode only the priority channel missiles can be fired. During ripple mode priority and alternate channel, missiles are fired alternately.

The CPG has the capability to launch missiles in all modes and types of engagements: **RF ORIDE, NORM, RIPL, MANL, LOAL, or LOBL.** The pilot has the capability to launch missiles in the following modes and types of engagements: **RF ORIDE, NORM, LOAL, or LOBL.** If the missile system is in any mode other than **NORM,** including **STBY,** when the pilot "ACTIONS" missiles, the missile system will automatically mode to **NORM.**

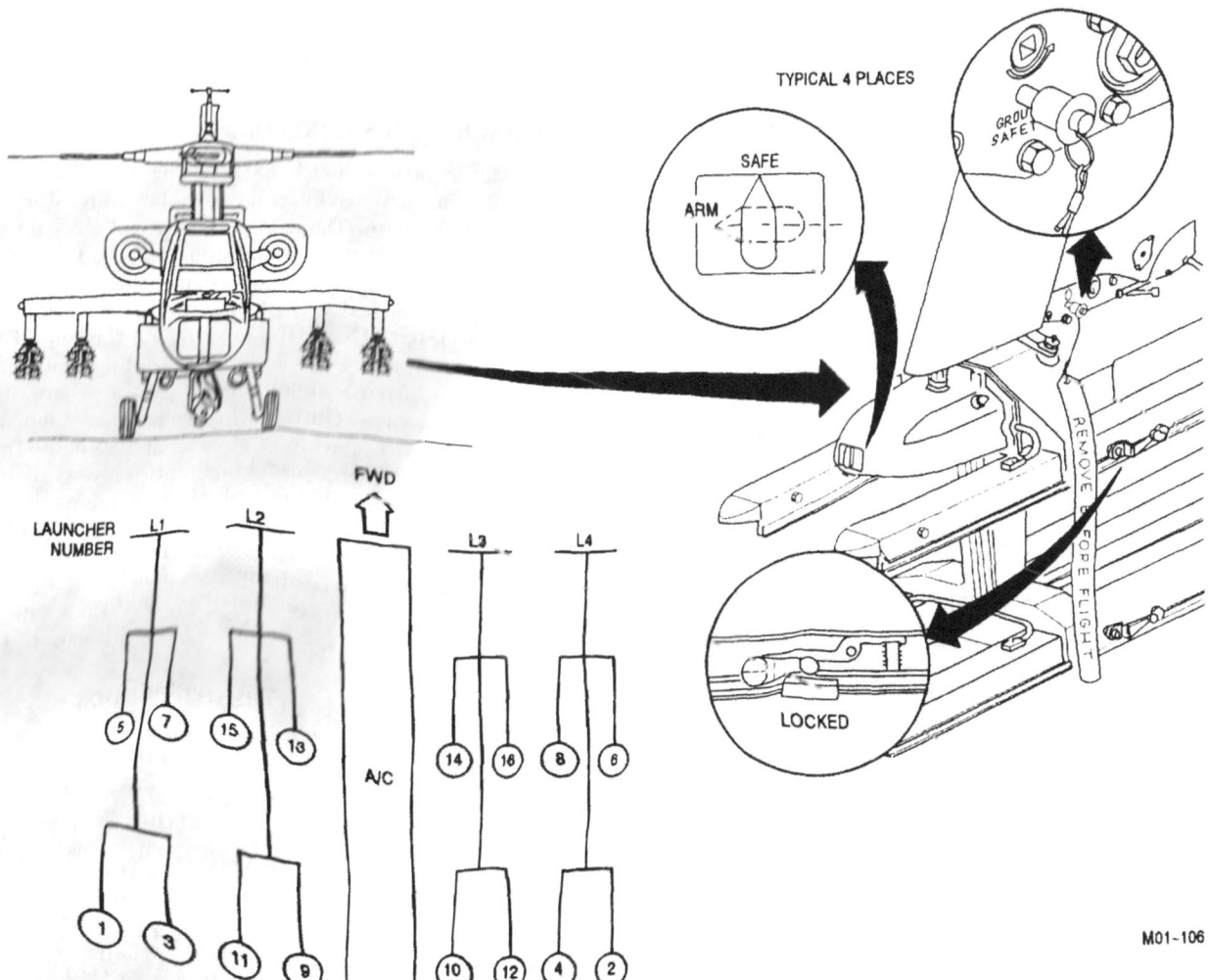

Figure 4-7. Point Target (HELLFIRE) Preferred Missile Firing Order

The missile system prompts presented to the CPG are shown in either the sight status or weapons status section of the high action display. Autonomous designation missile launch prompts are shown in the weapons status section; remote designation missile launch prompts are shown in the sight status section. Autonomous designation is defined as when the missile code of the priority channel and the code of the LRF/D are the same and the selected sight is TADS plus the laser power is turned on. Remote designation is defined as when the missile code of the priority channel and the code of the LRF/D are not the same.

Time of flight is displayed for all missile launches and reflects the time of flight for the missile closest to the target. Temperature compensation is incorporated when calculating missile time of flight -49A / -51. Ripple fire engagements have individual time of flight counters displayed, one for the autonomous missile, and one for the remotely designated missile, regardless of the order of their launch.

With -45 software the pilot cannot launch missiles for autonomous designation; consequently only one set of prompts is displayed in the weapons status section of the high action display -45. Time of flight is displayed to the pilot for any missile that he has launched. With -49 software the pilot can conduct autonomous launches. Pilot prompts and TOF displays remain the same as described above. Steering commands for the missile constraints box will be driven by the TADS LOS during LOAL DIR. The missile constraint steering box will be driven by the NAV LOS regardless of autonomous or remote moding during LOAL HI and LOAL LO launches -49A / -51.

The sight status section will reflect the codes of the upper and lower channels, UP=A LO=B; the code of the priority channel will flash. Authorized engagement ranges for the Hellfire System are from 500 meters to 8000 meters.

4.11 FIRE CONTROL COMPUTER (FCC).

The FCC is the primary bus controller. The FCC controls all data transmissions on the multiplex bus during normal operations. The FCC processes and computes data for all fire control capabilities on board the helicopter (fig 4-8). The FCC continuously executes internal built-in-tests and in the event of a failure signals the BBC to assume control of the bus. The FCC, BBC and FAB MRTUs are turned on with the switch labeled FCC/MUX ON (fig 4-14). Flight crew personnel shall ensure that this switch is in the ON position at all times, except for when applying power to the aircraft systems when the aircraft has been heat soaked in direct sunlight for more than 1 hour and the ambient temperature exceeds 100 degrees Fahrenheit. With these conditions present, the FCC/MUX switch shall be placed in the OFF position for at least two minutes after power has been applied with the FAB fans running in the STBY position to remove the superheated air from within the critical LRUs. Otherwise the switch is ON during any ground or flight operations. The FCC receives 115 vac power from the No. 1 essential bus and 28 vdc power from the No. 3 essential dc bus and is protected by the FCC AC and FCC DC circuit breakers located on the CPG circuit breaker panel. To determine the -49A FCC software code, place the DEK to SPI and enter C. The software code is displayed to the right of the dash. To determine the -51 FCC software code, press the CDU PGM FAB. The software version is displayed on line 6 of the CDU display.

4.12 BACKUP BUS CONTROLLER (BBC).

The BBC is part of the CPG MRTU (fig 4-8). The BBC monitors the primary bus controller for faults during normal operations. The BBC automatically assumes control of the multiplex bus when it senses a failure of the FCC. The CPG can manually select the BBC as the primary bus controller by placing the MUX switch on the CPG FIRE CONTROL panel (fig 4-14) to the SEC position. When the BBC is the primary bus controller several fire control operational capabilities are not available. They are: Gun and Rocket ballistic solutions; Fault Detection/Location System; and Waypoint/Targeting. All other fire control capabilities are available and function identically as under FCC bus control. The BBC receives 28 vdc power from the No. 3 Essential DC bus and is protected by the MUX/CPG circuit breaker located on the CPG circuit breaker panel. The status of the BBC is continuously monitored by FD/LS; the on command test (test OS-MUX) will fault isolate to the LRU.

4.13 MULTIPLEX BUS SUBSYSTEM (MUX).

The MUX (fig 4-8) consists of remote terminal (RT) units and a redundant data bus. The RTs are used to interface the various subsystems on board the helicopter. Three helicopter subsystems, the digital automatic stabilization equipment (DASE), the remote Hellfire electronics (RHE) unit, and the symbol generator also function as RTs. The RHE provides multiplex bus interface for the missile system electronics. The DASE provides multiplex bus interface for itself, ADSS, transmission and APU FD/LS, BUCS LVDTs, and the heading attitude reference system. The symbol generator receives only data from the bus to use in generation of symbols on the video displays. With the EGI modification, additional RTs are connected to the MUX bus. The DTU and EGI are new LRUs on the aircraft, and they are connected directly to the MUX bus. The CDU and DNS SDCC are already installed on the aircraft, but are now connected to the MUX bus via data bus switches (DBS) that are normally de-energized. The DBSs energize when the BBC takes control of the MUX bus, either manually via the **MUX PRI/SEC** switch on the CPG FCP or automatically when the FCC indicates a failure to the BBC. The data bus consists of two identical bus networks separated down each side of the helicopter and interconnected in such a manner that loss of data communications is minimized or avoided in the event of battle damage. The multiplex bus system receives 115 vac power from the No. 1 essential bus and 28 vdc power from the No. 3 essential dc bus. The MRTUs are protected by their own circuit breakers: the DASE by the **ASE/AC** and **ASE/DC** circuit breakers located on the pilot overhead circuit breaker panel; the symbol generator by the **SYM GEN** circuit breaker located on the pilot overhead circuit breaker panel; the remote Hellfire electronics unit by the **MSL/ DC ELEC** circuit breaker located on the CPG circuit breaker panel. The remaining MRTU circuit breakers are located on the CPG circuit breaker panel and are labeled **MUX/ L PYL OUTBD, MUX/L PYL INBD, MUX/ R PYL OUTBD, MUX/R PYL INBD, MUX/FAB L and MUX/ FAB R.** With the EGI modification, the DTU and EGI have dedicated circuit breakers on the pilots forward circuit breaker panel. The CDU circuit breaker is located on the CPG circuit breaker panel, and the DNS circuit breaker is located on the pilots forward circuit breaker panel. The status of the multiplex bus subsystem is continuously monitored by FD/LS; the on command test (test OS-MUX) will fault isolate to the specific LRU **45** or LRU and LRU bus failure **49A** / **51**. With -45 software the RHE and symbol generator MRTU functions are tested as part of each subsystem overall on command test (tests 07-MSL; 13-SYMG) **45**. With -49A or -51 software the RHE MRTU functions are tested as part of the 08 MUX test **49A** / **51**.

TM 1-1520-238-10

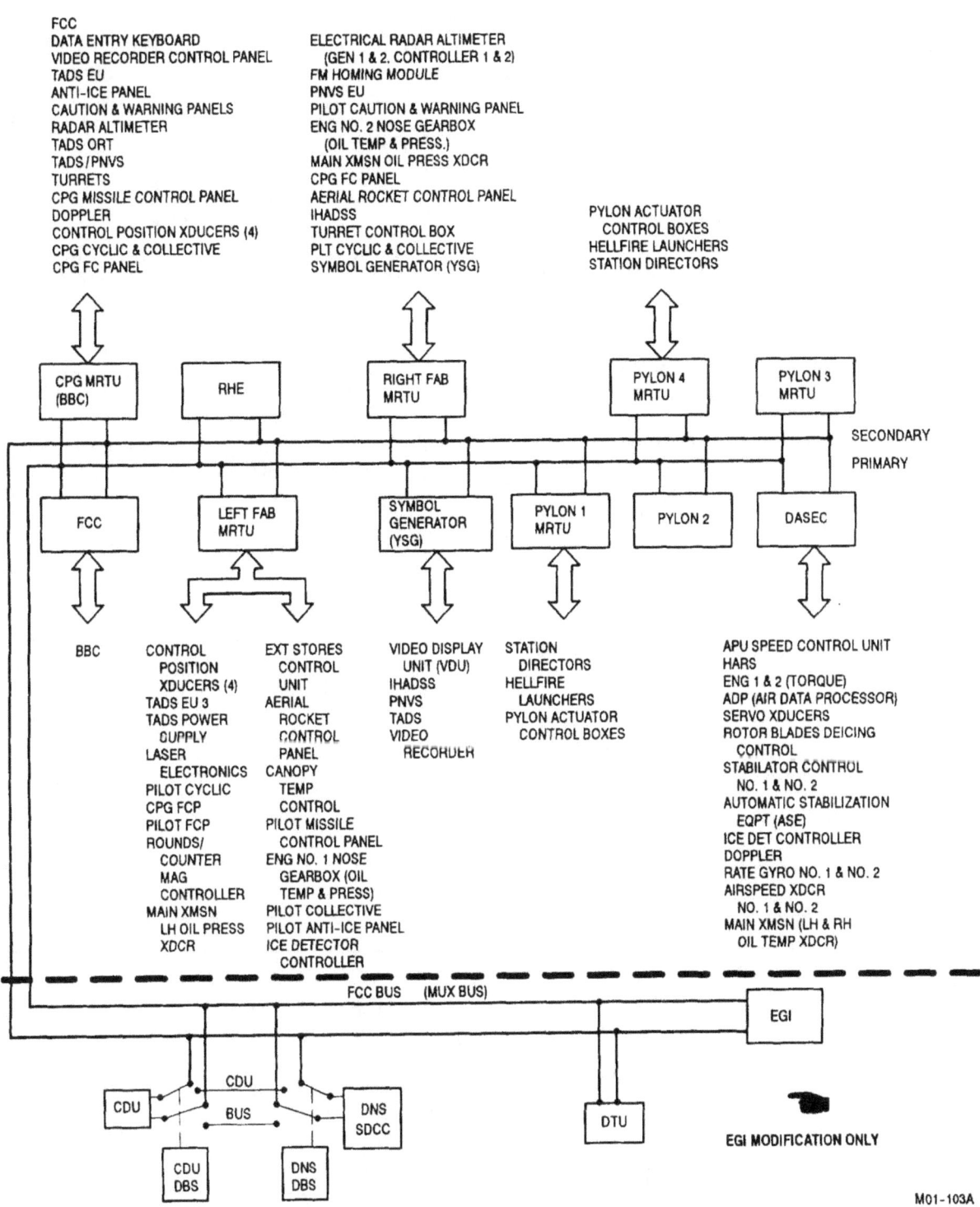

Figure 4-8. Multiplex Bus (1553)

4.14 SYMBOL GENERATOR.

The symbol generator receives data from the bus controller and creates the symbols on the video image. The symbol generator also serves as the video switching unit for the helicopter. The symbol generator receives incoming video from the TADS, PNVS and video recorder and routes them to crew stations as requested by the crew members. The symbol generator will display video and symbology or symbology only on any of the displays; IHADSS, TADS ORT, or VDU. In the event of a symbol generator failure the IHADSS DEU will display the PNVS FLIR 2 video images without symbology on both the pilot and CPG HDUs. The TADS FLIR 2 video image will be displayed on the TADS ORT. The only symbology on the TADS FLIR 2 will be the TADS reticle and the IAT tracking gates when the IAT is engaged. The pilot VDU will not have an image for display. The symbol generator receives 115 vac power from the No. 1 Essential Bus and is protected by the **SYM GEN** circuit breaker on the pilot overhead circuit breaker panel. The symbol generator is continuously monitored by FD/LS; the on command test (test 13-SYMG) will fault isolate to the LRU.

4.15 FAULT DETECTION/LOCATION SYSTEM (FD/LS).

The FD/LS is contained within the fire control computer (FCC) program. The continuous monitor functions whenever power is applied to the FCC. If the BBC is the primary bus controller, no FD/LS capability exists. The function of the FD/LS is to detect faults in the subsystems monitored and announce the NO-GO conditions to the flight crew or FCC repairman. The FD/LS identifies the NO-GO subsystem component and its location within the helicopter. Form FCC software, the CDU is used to access the **FD/LS** and/or **BST** functions. For ▮45▮ and ▮49A▮ FCC software, the DEK is used to access the FD/LS and/or BST functions. There are two modes of the FCC operation: continuous monitor; and on-command initiated. Most weapon, sensor, and fire control subsystems have a continuous check on their status and report their status to FD/LS without an on-command initiated test. If the FD/LS detects a NO-GO condition during the continuous monitor of any of the weapons, TADS or PNVS, the results are displayed on the pilot and CPG caution/warning panels. If the exact status information regarding the NO-GO condition for ▮45▮ and ▮49A▮ software is desired, the DEK switch must be set to **FD/LS**. For ▮51▮ software the CDU FDLS FAB must be pressed. The FD/LS will automatically display the status of all NO-GO subsystem components. An asterisk will be displayed to the left of the fault message that initiated the FD/LS annunciator ▮49A▮ / ▮51▮. The status list may be paged by pressing the **SPACE** key on the DEK for ▮45▮ and ▮49A▮ software or the **SPC** key on the CDU for ▮51▮ software. The FD/LS MENU will be displayed when pressing any other key. The list is completed when the message **ANY KEY FOR FD/LS MENUS** appears on the display.

WARNING

Prior to initiating AWS FD/LS check (IBIT), ensure the pilot ground override (PLT/GND ORIDE) switch is in the OFF position. Failure to perform this action may result in uncommanded gun turret slewing or uncommanded gun firing during a AWS FD/LS (IBIT) manual abort.

4.15.1 FD/LS On-Command Tests. Most command initiated tests are fully automatic and do not require any intervention between initiation and completion. There are certain tests which do require some interaction between the initiator and the fire control system. Interaction may require either switch actuation or evaluating a specific parameter or function and the decision regarding its acceptability. The interaction is fully prompted for all switch actions or evaluations. The on-command initiated tests are displayed as a menu of the specific tests. There are two pages of the test menu. They are:

01-ADS	05-HARS	09-PNVS
02-DASE	06-IHDS	10-PYLN
03-DICE	07-MSL	11-RKT
04-GUN	08-MUX	12-STAB

13-SYMG	17-APU	33-CDU
14-TADS	18-GEN	34-DNS
15-UTIL	19-TRAN	35-DTU
16-ETE	32-TAGA	36-EGI

These tests are initiated by entering the corresponding two-digit number for the test; tests 32 - 36 are fo ▮51▮ only. Any required intervention is fully prompted and the results of the test are displayed on the TADS ORT, VDU and helmet mounted displays. The results may be paged by pressing the **SPC** key on the CDU ▮51▮ ; or the **SPACE** key on the DEK for ▮45▮ and ▮49A▮; this should be done until the message ANY KEY FOR **FD/LS** MENUS is displayed. The FD/LS status or on-command initiated test may be accomplished without turning the TADS on. The TADS heads out display can be used independently of the TADS for FD/LS by turning on the symbol generator to display the FD/LS on-command tests. The FD/LS tests may request that the operator acknowledge that an action has been completed. The method used to acknowledge (ACK) is to press the **SPACE** key on the DEK 45 and 49A 49A software. For 51 software the CDU **SPC** key must be pressed.

4.16 AIR DATA SENSOR SUBSYSTEM (ADSS).

The ADSS consists of an omni-directional airspeed sensor (OAS) and air data processor (ADP). The OAS (fig 2-2) measures the airspeed and temperature. The air data processor senses static pressure and with the data from the OAS calculates the air mass data required for fire control. The ADSS is interfaced with the multiplex bus through the DASE MRTU (fig 4-8). The ADSS is turned on with the switch labeled **ADSS** on the **AUX/ANTI-ICE** control panel in the CPG left console (fig 2-12). If the ADSS fails or is turned **OFF,** the displayed airspeed will revert to ground speed in knots as computed by the doppler navigation system. The ADSS receives 115 vac power from the No. 1 essential bus and 23 vdc from the No. 3 essential dc bus and is protected by the **AIR DATA AC** and **AIR DATA DC** circuit breakers on the pilot overhead circuit breaker panel. The status of the ADSS is continuously monitored by the Fault Detection/Location System (FD/LS); the on command test (test 01-ADS) will fault isolate to the specific line replaceable unit (LRU).

4.17 WAYPOINT/TARGETING. -45 / -49A

NOTE

If 51 software is installed go to paragraph 3.16 for Waypoint/Targeting system description and operation.

Waypoint/targeting functions are performed solely by the CPG. The pilot cannot actively participate. Four functions can be implemented. These are: preflight, storing, position update, and cueing.

4.17.1 Waypoint/Targeting Data Entry. The waypoint/targeting function requires certain data. The data is input using both the **TGT** and **SPl** positions on the DEK rotary switch. The data entered under **SPl is** displayed as a menu with two pages. They are:

TIME:	HH: MM: SS
PPOS:	11S QG 5551 5221
ALT:	+AAAA' HG: 29.92IN.

-45

SPH:CL6 MV:	EO13.8	DEG
DOPP BIAS:	L0.0	DEG
HEADING (MAG):	000.0	DEG

-49A

SPH:CL6 MV:	EO13.8	DEG
HEADING (MAG):	000.0	DEG

a. Data Entry. To enter data input the first character of the desired parameter. The cursor will jump to the first digit position of the data, then input the full data required. On completion the data will be automatically entered.

NOTE

The LAT/LONG conversion process within the FCC contains non-linear transformation errors in accordance with the table below:

Latitude	Possible Error
0 to ± 34°	Less than 2 meters
± 35° to ± 44°	Less than 50 meters
± 45° to ± 54°	Less than 1100 meters
± 55° to ± 63°	Less than 2000 meters
± 64° and up	More than 2000 meters

(1) FCC Present Position Data. The FCC present position (PPOS) data are used only for the initial alignment of the HARS when the helicopter is on the ground. The present position shall be entered prior to a normal or fast alignment of the HARS. It is not required for a stored heading alignment or inflight restart. The accuracy of the present position input should be within 500 - 750 meters, straight line distance, from the actual location. Inflight this display will periodically update reflecting the doppler present position. The doppler display updates at a substantially higher rate than the PPOS display, consequently there may not be any correlation of the doppler and PPOS display in forward flight. This is normal and does not indicate a problem condition. If at any time ?? appear following the PPOS data, as shown in the example, the FCC is rejecting the doppler navigation data as being in error. This may be an indication of doppler malfunction or that the doppler has not been updated recently. The displayed PPOS data, in this case, will then reflect the best computed present position based on HARS velocity data.

(2) Altitude and Altimeter Data. The altitude and altimeter setting data are to be used in the following manner: Altitude above mean sea level shall be entered when the aircraft is on the ground. The value is determined from referring to navigational charts or by tactical map study. The Fire Control System will compute the corresponding altimeter setting. The altimeter setting shall be entered when the aircraft is in flight. The altitude display may be observed in flight; if an error exists in the dynamic altitude display when compared to the aircraft barometric altimeter, the altimeter setting should be entered. The altimeter setting can normally be obtained from a local air traffic control agency. In the event of a failure of the air data sensor, question marks, ??.??, will be displayed in the altimeter setting display. The fire control system will attempt to derive the aircraft altitude above mean sea level by adding radar altitude to the last valid altitude value displayed. The altitude value should then be periodically updated in flight. The altitude value computed from radar altitude is accomplished based on standard atmosphere.

(3) lime Display Functions. The TIME display functions only when power is applied to the fire control computer. The TIME display is not part of the backup bus controller software. The TIME display will increment from whatever time value is entered; on initial power up the TIME value is zero. If a TIME value of 00:00:00 is input, the TIME display will function as an elapsed time clock. The TIME display is a twenty-four hour clock with an accuracy of plus or minus 1 second while running.

Change 4   4-17

b. Conventional Uses:

NOTE

- If an erroneous magnetic variation is entered into **SPI**, the aircraft heading scale and the HSI will also indicate that amount of error.

- Entries of magnetic headings greater than 360° or a magnetic variation entered greater than 179.9° will be ignored. The message **ERROR** will be displayed on the upper left of the display field prompting the operator to re-enter valid data ▮49A▮.

(1) **Magnetic variation.** Use **E** for East and **W** for West, as appropriate.

(2) **Grid Convergence** ▮15▮. Use **E** for East and **W** for West, as appropriate. Data may be obtained from G-M angle diagram on UTM maps if applicable.

(3) **Spheroid.** Use the same codes utilized by the doppler.

(4) **Latitude.** Use - for North and **S** for South, as appropriate.

4.17.2 Waypoint/Targeting Coordinate Data Storing. Up to ten sets of coordinate data may be stored in the FCC at anytime. Two methods can be used to store coordinate data; these are either the data entry keyboard, or by use of the store (**ST**) position of the **UPDT/ST** switch on the ORT LHG.

The coordinate data are displayed as four pages of a coordinate data menu. The various pages may be paged by using the **SPACE** key on the DEK. An example page is shown below:

```
                              TRGT
0  11S QG51285172A +0820'
1  NO DATA
2  NO DATA
```

a. Store Coordinate Data Using DEK. To enter data input the appropriate coordinate data address. The cursor will jump to the first position for data entry, i.e., grid zone. Either input the specific grid zone or use the **SPACE** key for automatic entry if the coordinate data grid zone is the same as the aircrafts current present position of the 100 KM ID; either input the data or use the **SPACE** key followed by the eight or six-digit coordinates. Pressing the **SPACE** key will automatically enter the **A** for the altitude. If the altitude of the coordinate data is above sea level directly input the numbers; if necessary a leading zero is required. The plus sign will be automatically entered; if coordinate data is below sea level the minus sign must be input prior to the numeric value.

If a particular storage location has no data the message **NO DATA** will be shown. If it is desired to clear a specific storage location, key in the letter C followed by the desired coordinate data storage location address. The location will be cleared and the message **NO DATA** will be displayed.

b. Storing Coordinate Data Using UPDT/ST Switch.

1. Waypoint/target index - Desired storage location.

2. Enter range, aircraft to waypoint/target.

NOTE

Range to geographic feature or object must be input to the FCC. This may be accomplished by three means: 1. Manual range input through the keyboard (determined from a map or by estimation); 2. Automatically calculated range based on LOS angles from selected sight; 3. Laser range input by firing the laser after placing reticle on feature or object (This is the most accurate range data).

3. Place reticle of selected sight on the geographic feature or object.

4. **UPDT/ST - ST** (Waypoint/target coordinate data will be automatically calculated and placed in the location specified by the index.)

4.17.3 Waypoint/Targeting Position Update.

NOTE

This subsystem of the waypoint/targeting calculates the present position of the helicopter based on previously stored coordinate data. The update present position is stored within the FCC and NOT automatically sent to the doppler navigation system.

a. Position Update.

1. Waypoint/target indexer - index of location of previously stored coordinate data.

2. Enter range, helicopter to coordinate data. See explanation under storing coordinate data [para 4.17.1 (2)].

3. Place reticle of selected sight on the geographic feature or object of previously entered coordinate data.

4. **UPDT/ST - UPDT.** (The aircraft present position is calculated automatically, based on the previously entered coordinate data.)

b. **Recall Updated Hellcopter Present Position.**

1. Data entry keyboard - **SP1.**

2. Input - **U** (The updated present position data is automatically displayed.)

c. **Update Doppler.**

1. Recall updated aircraft present position immediately after accomplishing update function.

2. Insert the FCC displayed present position into the doppler using normal procedures.

d. **Example of Updated Present Position Data.**

1. Input - **U.**

2. Display After Entering: UPDATE

   A/C 11S QG56035160

   DELEAS = +0065

   DELNOR = -0012

3. Which decodes as follows:

   Grid Zone - 11S

   Coordinates - QG56035160

   Doppler easting error - +65 meters

   Doppler northing error - 12 meters

If the coordinates updated from had not been previously stored, an attempt to recall the updated present position will result in a **NO DATA** message on the display.

4.17.4 Waypoint/Targeting Cueing.

Once waypoint/targeting data has been stored, sightline cueing or TADS slaving to the stored coordinates can be accomplished as follows:

a. **Sightline Cueing or TADS Slaving.**

1. Waypoint/target indexer - Index of desired coordinate data.

2. Acquisition Select - **TGT** or **NAV.**

3. Slave button - Press.

   a. If the selected sight was TADS, TADS will be slaved so as to place its LOS reticle over the geographic feature or object on the ground.

   b. If the selected sight was other than TADS, the CPG will be cueing in the normal manner so as to cue the CPG to place the reticle of his selected sight over the geographic feature or object.

b. **Recall Range and Bearing, Hellcopter Present Position to Stored Coordinate Data.**

1. Data entry keyboard - **TGT.**

2. Input - **Rn.** (Where n = coordinate data storage address.)

c. **Example.**

1. Input - R3.

2. After entering, display shows - **3R 15990 B042.4** degrees.

   Which decodes as follows:

   a. Waypoint/target storage (index) No.-3.

   b. Range, helicopter to coordinate data - **15990** meters.

   c. Bearing, helicopter to coordinate data - **042.4** degrees.

   NOTE

- If no previously stored coordinate data exists in the index location selected, a message **NO DATA** will be displayed.

- The range and bearing function can be accomplished at any time after storing waypoint/target data.

- Maximum possible displayed range for this function is 32000 meters.

4.18 INTEGRATED HELMET AND DISPLAY SIGHT SUBSYSTEM (IHADSS).

WARNING

In the event of IHADSS failure with gun selected, the gun will be commanded to the fixed forward position. Once in this position, the gun can still be fired without having to re-action the gun.

The IHADSS (fig 4-9) consists of the crewmember helmet; a helmet display unit (HDU), sensor survey units (SSU), a display adjust panel (DAP), and a boresight reticle unit (BRU) located in each crew station; and the sight electronics unit (SEU) and display electronics unit (DEU) located in the right forward avionics bay. Each helmet has two cable connections in the cockpit; they are: a TEMPEST ICS cable and a cable connecting the helmet electronics to the IHADSS. The sensor survey units, in conjunction with the sight electronics unit, determine the crewmembers line-of-sight (LOS). The weapons and sensor turrets can be directed by either crew member LOS. The helmet display unit (HDU) consists of a CRT with optical elements which project the selected symbology and sensor imagery onto a combining lens. The HDU is attached to the right side of the helmet during normal use. the attached HDU is rotated in front of the right eye for viewing of the display or can be rotated vertically away from the eye when not in immediate use. When not attached to the helmet the HDU shall be stored in the holster located on the right console inside kick panel. The display electronics unit provides power and video signals to each crewmember HDU through the display adjust panel located in each crew station. The display adjust panel is used to adjust the image size, centering and electronic focus as it appears on the combining lens. These adjustments are normally required only after replacement of the DAP or HDU. The boresight reticle unit in each cockpit is used to boresight the crewmember helmet. The IHADSS receives 115 vac power from the No. 1 essential bus and the SEU is protected by the **IHADSS** circuit breaker located on the pilot overhead circuit breaker panel; the DEU is protected by the **IHADSS** circuit breaker located on the CPG circuit breaker panel. The IHADSS is continuously monitored by FD/LS; the on command test (test 06-IHDS) will fault isolate to specific LRU including the crewmember helmet.

4.18.1 IHADSS Operation. Two actions shall be accomplished by each crewmember each time the IHADSS is turned on. The HDU CRT shall be adjusted, and the crewmember helmet, boresighted. Adjustment of the HDU is accomplished by selecting the gray scale and adjusting the brightness and contrast to establish a displayed 10 shades of gray image on the HDU. Boresighting the helmet is accomplished by turning on the BRU, aligning the LOS reticle projected on the combining lens with the BRU alignment reticle, and actuating the **BRSIT HMD** switch. The requirement to boresight the IHADSS is announced to the crewmember in the sight status block of the high action display and by the flashing of all four cueing dots on the symbology projected on the combining lens. The IHADSS can be boresighted in flight or on the ground. Display adjust panel adjustments are made usually following a maintenance action that replaced either the HDU or the DAP. The procedure requires assistance from the crew chief or FCC repairman. To accomplish: the crewmember attaches the HDU to the helmet in the normal manner and rotates the HDU in front of his eye, the gray scale is selected and viewed, adjustments are first made for image centering so that the gray scale appears centered **(HORIZONTAL CTRG, VERTICAL CTRG)** (fig 4-9), next the gray scale size is adjusted so that the four borders of the image are just contained on the edges of the combining lens **(HORIZONTAL SIZE, VERTICAL SIZE),** and the corners are not visible, the last step is to adjust the electronic focus, **(FOCUS),** so that the gray scale image appears crisp and sharp.

TM1-1520-238-10

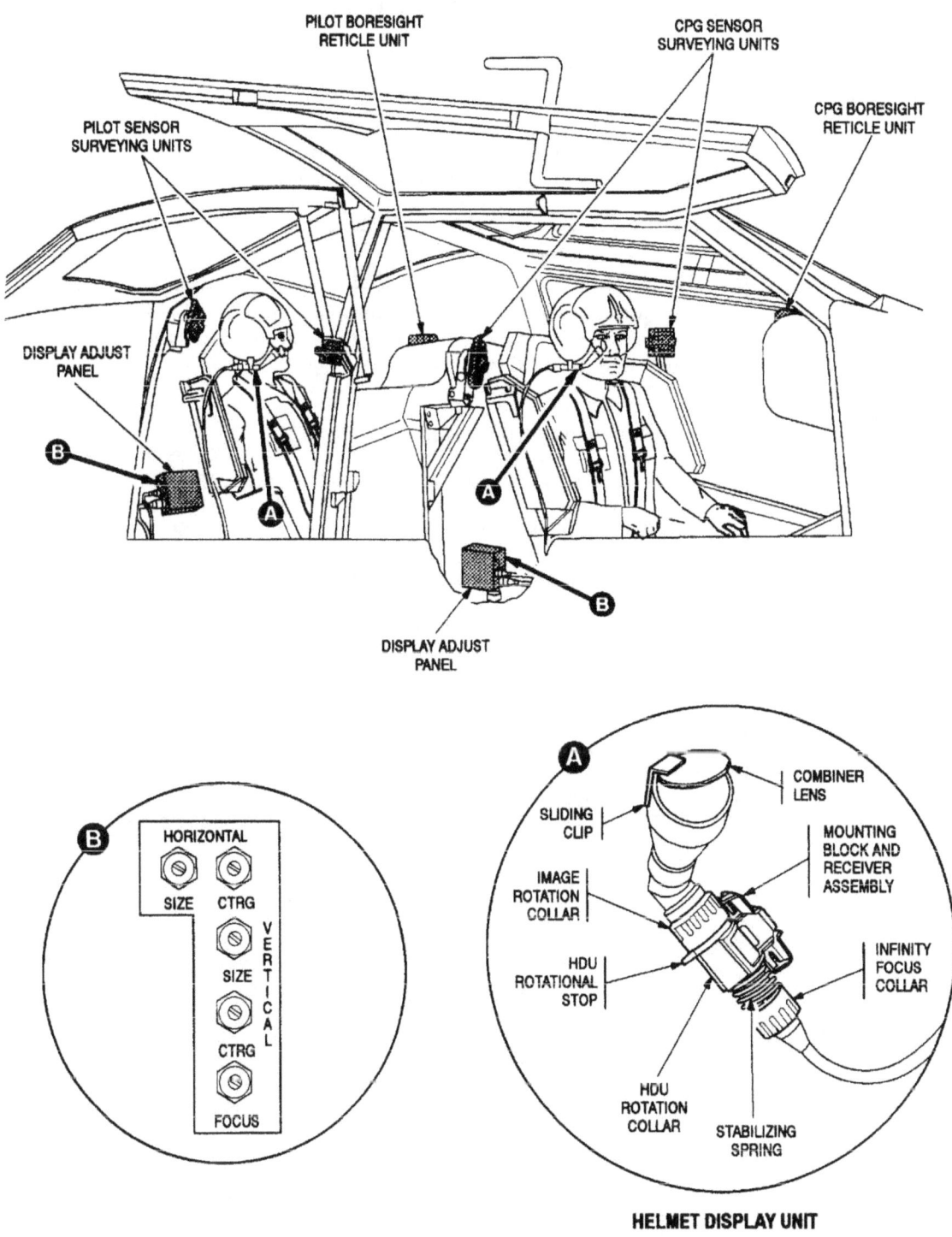

Figure 4-9. Integrated Helmet and Display Sight Subsystem

4-21

TM 1-1520-238-10

4.19 CYCLIC AND COLLECTIVE MISSION EQUIPMENT SWITCHES.

The cyclic and collective have several switches used for the mission equipment (fig 2-26).

4.19.1 Cyclic Switches. Cyclic stick mission equipment switch position and functions are described in table 4-5.

Table 45. Cyclic Stick Switoh Functions

Control/ Switch	Position	Function
Weapon Action Switch (WAS)		A five position momentary switch which actions the selected weapon and removes the electrical trigger interlock. The actioned weapon maybe deselected by reactioning the same weapon or by selecting another weapon.
	G	Actions the 30mm Gun (12:00 position on the WAS).
	R	Actions the Aerial Rocket system (9:00 position on the WAS).
	M	Actions the Hellfire missile system (3:00 position on the WAS).
	C	Fires the chaff, if Chaff system armed (6:00 position on the WAS).
Flight Symbology Mode Switch		A three position momentary switch which permits the selection of the different flight symbology modes. Cruise and Transition, alternately, in the forward position and Hover and Bob-up in the aft position.
Trigger		A two position guarded trigger which enables weapons firing. The first detent is for normal selected weapons usage; the second will override the selected weapons performance inhibits but not the safety inhibits of the selected weapon.

4.19.2 Collective Switches. Collective stick mission equipment switch position and functions are described in table 4-6.

Table 46. Collective Stick Switch Functions

Control/Switch	Position	Function
RF OVRD		Immediately modes the missile system for an RF/IR missile.
NVS		This two position switch is not active unless the crewmember has selected the NVS position on the sight select switch on the fire control panel.
	PNVS	Pilot: Select the PNVS as the sensor to be used by the pilot. PNVS video with flight symbology will be displayed on the pilot HMD.
		CPG: If the PLT/GND ORIDE switch is OFF, the position is inactive. If the PLT/GND ORIDE switch is in ORIDE, this position allows takeover of control of the PNVS from the pilot by placing the NVS select switch to PNVS and coupling it to the CPG helmet LOS. PNVS video with flight symbology will be displayed on the CPG HMD.
	TADS	Pilot: Permits the pilot to take control of the TADS from the CPG and automatically place the TADS in WFOV FLIR with symbology. This would be done in the event of a PNVS failure and is the PNVS backup mode.
		CPG: Automatically modes the TADS for WFOV FLIR with flight symbology and couples the TADS to the CPG helmet LOS. This mode is accomplished independent of the PNVS and would be used by the CPG or IP to provide separate monitor for obstacle or terrain avoidance.
PLRT/BRSIT HMD	PLRT	Alternates the FLIR polarity; back and forth; black hot or white hot.
	BRSIT HMD	Activates the IHADSS sight electronics unit to compute the boresight bias.

4.20 PILOT ARMAMENT CONTROL PANELS.

The pilot Armament control panels are the Aerial Rocket Control Panel, the Missile Control Panel, and the Fire Control Panel.

4.20.1 Pilot Aerial Rocket Control Panel (ARCS).
The pilot ARCS panel, placarded **ROCKETS** (fig 4-10), is located in the pilot left console (fig 2-11). The ARCS receives power from the No. 3 essential dc bus and is protected by the **RKT ELEX** circuit breaker on the pilot overhead circuit breaker panel. The ARCS is continuously monitored by FD/LS; the on command test (test 11-RKT) will fault isolate to the LRU. The function of each control is described in table 4-7.

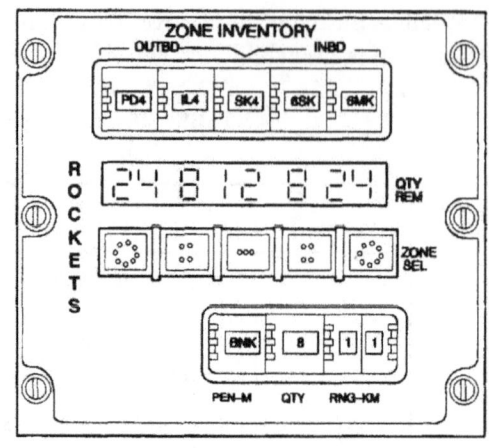

Figure 4-10. Pilot Aerial Rocket Control Panel

Table 4-7. Pilot Aerial Rocket Control Panel/Indicator Functions

Control/Indicator	Switch	Function
Zone Inventory		Indicates the type of 2.75 FFAR round loaded in the zone.
	PD4	High explosive warhead with PD fuzing with a MK 40 motor.
	RC4	High explosive warhead with remote set fuzing with a MK40 motor.
	DP4	High explosive dual purpose warhead with PD fuzing with a MK 40 motor.
	WP4	White phosphorus warhead with PD fuzing with a MK 40 motor.
	IL4	Illumination warhead with remote set fuzing with a MK 40 motor.
	SK4	Smoke warhead with remote set fuzing with a MK 40 motor.
	6PD	High explosive warhead with PD fuzing with a MK 66 motor.
	6RC	High explosive warhead with remote set fuzing with a MK 66 motor.
	6IL	Illumination warhead with remote set fuzing with a MK 66 motor.
	6SK	Smoke warhead with remote set fuzing with a MK 66 motor.
	6MP	Multi-purpose submunitions warhead with remote set fuzing with a MK 66 motor.
	Blank	Not implemented.
QTY REM		Indicates the functional rounds remaining in inventory by zone location.

TM 1-1520-238-10

Table 4-7. Pilot Aerial Rocket Control Panel/Indicator Functions - continued

Control/Indicator	Switch	Function
ZONE SEL		Pressing the appropriate ZONE SEL switch arms the selected zone. When any one zone is selected all other zones displaying the same zone inventory will be automatically armed and visually indicated by illumination of the appropriate ZONE SEL switches. Any currently selected (armed) zones which are different from the ZONE SEL switch pressed will automatically be deselected (de-armed).
PEN-M		Permits the pilot to select tree height of burst detonation (5 meter increments), superquick detonation, or bunker penetration prior to detonation.
	45	
	40	
	35	Canopy height in meters.
	30	
	25	Enables rocket to penetrate canopy of
	20	foliage or building and
	15	detonate at a more
	10	effective point.
	BNK	Bunker penetration - To defeat bunkers (logs and dirt) up to 3 meters thick.
	SQ	Superquick - detonates when fuze makes contact with any object.
QTY		Selects the quantity of rockets to be fired.
	ALL	Sustained or continuous fire of all rockets in zones selected.
	24	24 rockets to be fired.
	12	12 rockets to be fired.
	8	8 rockets to be fired.
	4	4 rockets to be fired.
	2	2 rockets to be fired.
	1	1 rocket to be fired.
	-	No rockets to be fired

Change 3 4-25

TM 1-1520-238-10

Table 4-7. Pilot Aerial Rocket Control Panel/Indicator Functions - continued

Control/Indicator	Switch	Function
RNG-KM		**NOTE** The RNG-KM thumbwheel selectors on the rocket panel are the pilot manual range entry for ALL armament the pilot may select. Automatic range calculations are based on the pilot LOS and radar altitude; a flat earth model. Left Thumbwheel selector
	Numerals	Selects multiples of 1000 meters.
	A	Signals for automatic range calculation to be executed. During rocket engagements time to function fuzing data is passed by the FCC to the ARCS. Right Thumbwheel selector
	Numerals	Selects multiples of 100 meters.

4.20.2 Pilot Missile Control Panel. The pilot missile control panel, placarded **MSL** (fig 4-11), is located in the pilot left console (fig 2-11). The function of each control is described in table 4-8.

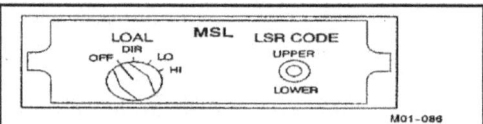

Figure 4-11. Pilot Missile Control Panel

Table 4-8. Pilot Missile Control Panel Control/Indicator Functions

Control/Indicator	Switch	Function
LOAL	OFF	Disables Lock-On- After-Launch mode and defaults to Lock-On- Before-Launch (LOBL) missile launch mode.
	DIR	Direct fire LOAL launch trajectory.
	LO, HI	Indirect fire LOAL launch trajectories.
LSR CODE		This is the CHAN SEL switch and used to command the missile subsystem to accept and validate the mode, code and quantity of missiles selected. The selection of UPPER or LOWER tells the missiles subsystem which coded missiles to establish as the priority channel, by default, the other coded missiles are the alternate channel.

4-26 Change 3

4.20.3 Pilot Fire Control Panel. The pilot fire control panel, placarded **FIRE CONTROL (fig 4-12)**, is located in the lower left quadrant of the pilot vertical instrument panel (fig 2-9) and receives power from the No. 1 essential bus and the No. 3 essential dc bus; protection is provided by the **FC AC** and **FC DC** circuit breakers located on the pilot overhead circuit breaker panel. The function of each control is described in table 4-9.

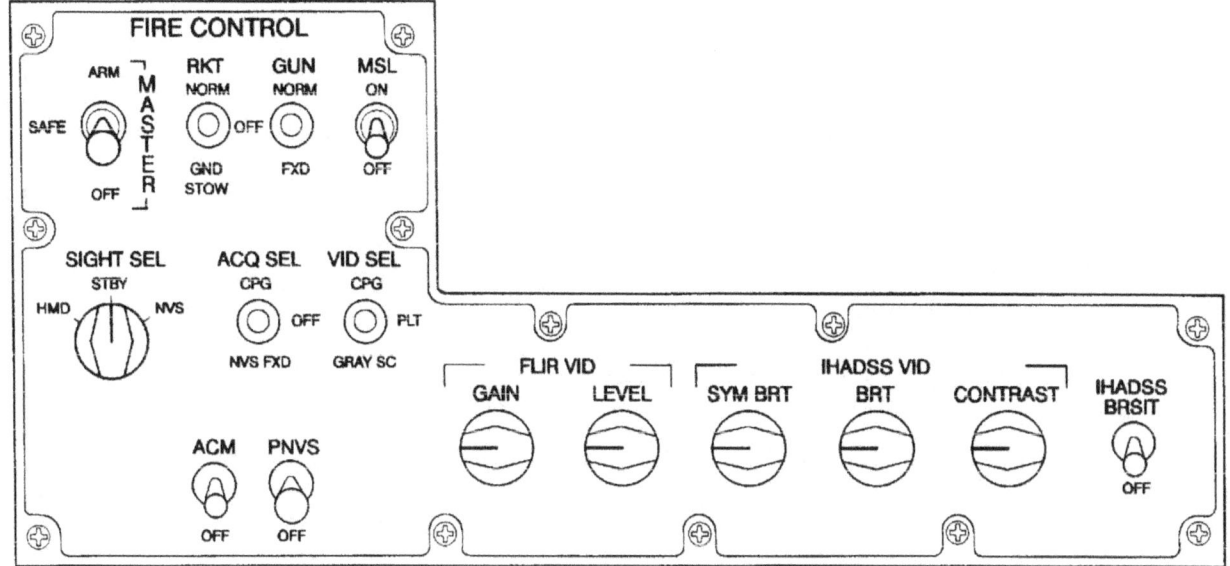

Figure 4-12. Pilot Fire Control Panel

Table 4-9. Pilot Fire Control Panel/Indicator Functions

Control/Indicator	Position	Function
		WARNING
		With CPG PLT/GND switch in ORIDE and pilots MASTER switch OFF, pilots SAFE/ARM lights will not indicate a SAFE or ARMED condition of the CPG ARM/SAFE switch. Further, with CPG SIGHT SELECT switch in NVS, HAD is not present to display weapons status messages.
MASTER	OFF	Disables all weapon control and power to weapons.
	SAFE	Enables SAFE power to the pilot weapons select switches and provides safe power to the CPG ARM switch. A green SAFE light indicates a valid safe condition.
	ARM	Enables ARM power to all weapon firing circuits and provides ARM power to the CPG ARM switch. An amber ARM light indicates a valid arm condition.

Table 4-9. Pilot Fire Control Panel/Indicator Functions - continued

Control/Indicator	Position	Function
RKT	OFF	Disables SAFE/ARM power to the Aerial Rocket Control Subsystem (ARCS).
	NRML	Enables pilot SAFE/ARM power to the ARCS and permits pylon articulation during pilot initiated rocket firing.
	GND STOW	Places the wing stores in the ground stow position; parallel with the ground on level terrain.
GUN	OFF	Disables pilot SAFE/ARM power to the M-230E1, 30mm gun subsystem.
	NRML	Enables pilot SAFE/ARM power to the gun subsystem and permits slaving of the gun turret to the pilot LOS when actioned.
	FXD	Enables pilot SAFE/ARM power to the gun subsystem and permits positioning of the gun turret to the fixed forward position when actioned.
MSL	OFF	Disables pilot SAFE/ARM power to the Hellfire missile subsystem.
	ON	Enables pilot SAFE/ARM power to the Hellfire missile subsystem.
SIGHT SEL	STBY	Deselects all lines-of-sight and stows sensor turret. IHADSS LOS active for boresight only.
	HMD	Selects the helmet mounted display (HMD) as the pilot reference LOS and activates the display electronics for display of flight symbology.
	NVS	Activates the NVS select switch on the collective switchbox for sensor selection. The HDU will display the selected night vision sensor FLIR video with flight symbology.
ACQ SEL	OFF	No cueing symbology displayed to the pilot.
	CPG	Pilot receives cueing symbology to the CPG selected sight LOS.
	NVS FXD	If the pilot has selected an NVS sensor, the sensor turret will be slaved to the fixed forward position.
VID SEL	CPG	Permits the pilot to observe the CPG selected video on his HMD.
	PLT	Permits the pilot to observe his selected sensor video and symbology on the HMD.
	GRAY SC	Displays to the pilot on his HMD a 10 shades of gray scale. To be used in calibration of the HMD brightness and contrast.
ACM		Auto Control Module. The switch is used to maintain an optimum gain and level setting of the selected night vision sensor during varying thermal scene content or when switching FLIR polarity.

Table 4-9. Pilot Fire Control Panel/Indicator Functions - continued

Control/Indicator	Position	Function
PNVS	OFF	Disables all power to the pilot night vision sensor.
	ON	Enables power to the pilot night vision sensor and begins cooldown of the FLIR.
FLIR VID	GAIN	A rheostat adjustment varying the thermal image gain, analogous to display contrast.
	LEVEL	A rheostat adjustment varying the thermal image level, analogous to display brightness.
IHADSS VID	SYM BRT	A rheostat adjustment varying symbol brightness from bright white to black.
	BRT	A rheostat adjustment varying image brightness on the IHADSS HDU. Used for initially setting brightness with the gray scale.
	CONTRAST	A rheostat adjustment varying image contrast on the IHADSS HDU. Used for initially setting contrast with the gray scale.
IHADSS BRSIT		Enables power to the boresight reticle unit and places the pilot IHADSS in boresight mode.
	OFF	Disables power to the boresight reticle unit.

4.21 CPG ARMAMENT CONTROL PANELS.

The CPG Armament control panels are the Missile Control panel and the Fire control panel.

4.21.1 CPG Missile Control Panel. The CPG missile control panel, placarded **MSL (fig 4-13), is located** in the left console (fig 2-12) and receives its power as part of the overall missile system. Controls and functions are described in table 4-10.

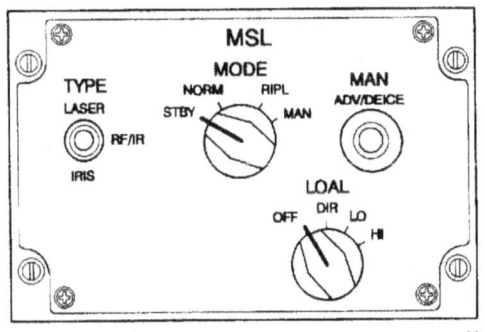

Figure 4-13. CPG Missile Control Panel

Table 4-10. CPG Missile Control Panel Functions

Control	Position	Function
TYPE	LASER	Specifies that laser seeker missiles are to be used for firing.
	RF/IR	Growth position for advance missile concepts; currently inactive.
	IRIS	Growth position for advanced missile concepts; currently inactive.
MODE	STBY	Places the missile system in standby mode.
	NORM	Places the missile system in normal mode in which only priority channel missiles can be fired.
	RIPL	Places the missile system in ripple mode in which priority and alternate channel missiles can be fired alternately.
	MAN	Permits manual advance of missile selection instead of using auto replenishment.
MAN ADV/ DEICE		Used with the MAN position of missile mode to actually advance missile selection manually. This DEICE mode is used to manually initiate the separation of the ice protection dome on the HELLFIRE missile. Separation will occur for the priority missile of the priority channel.
LOAL		This switch and related positions function identically to the pilot LOAL switch.

4.21.2 CPG Fire Control Panel.

CAUTION

With the PLT/GND ORIDE switch in the ORIDE position and the CPG SIGHT SELECT switch set to NVS, the high action display is disabled. The AWS can be actioned without any message in weapon control or weapon status and can result in damage during takeoff or landing if the 30 mm gun is actioned and depressed inadvertently.

The CPG fire control panel, placarded **FIRE CONTROL** (fig 4-14), is located in the left vertical instrument panel (fig 2-10) and receives power from the No. 1 essential bus and No. 3 essential dc buss; protection is provided by the **FC AC** and **FC DC** circuit breakers located on the pilot overhead circuit breaker panel. The CPG **FIRE CONTROL** panel is continuously monitored by FD/LS; the on command test (test 15-UTIL) will fault isolate to the LRU. Controls and functions are described in table 4-11.

TM 1-1520-238-10

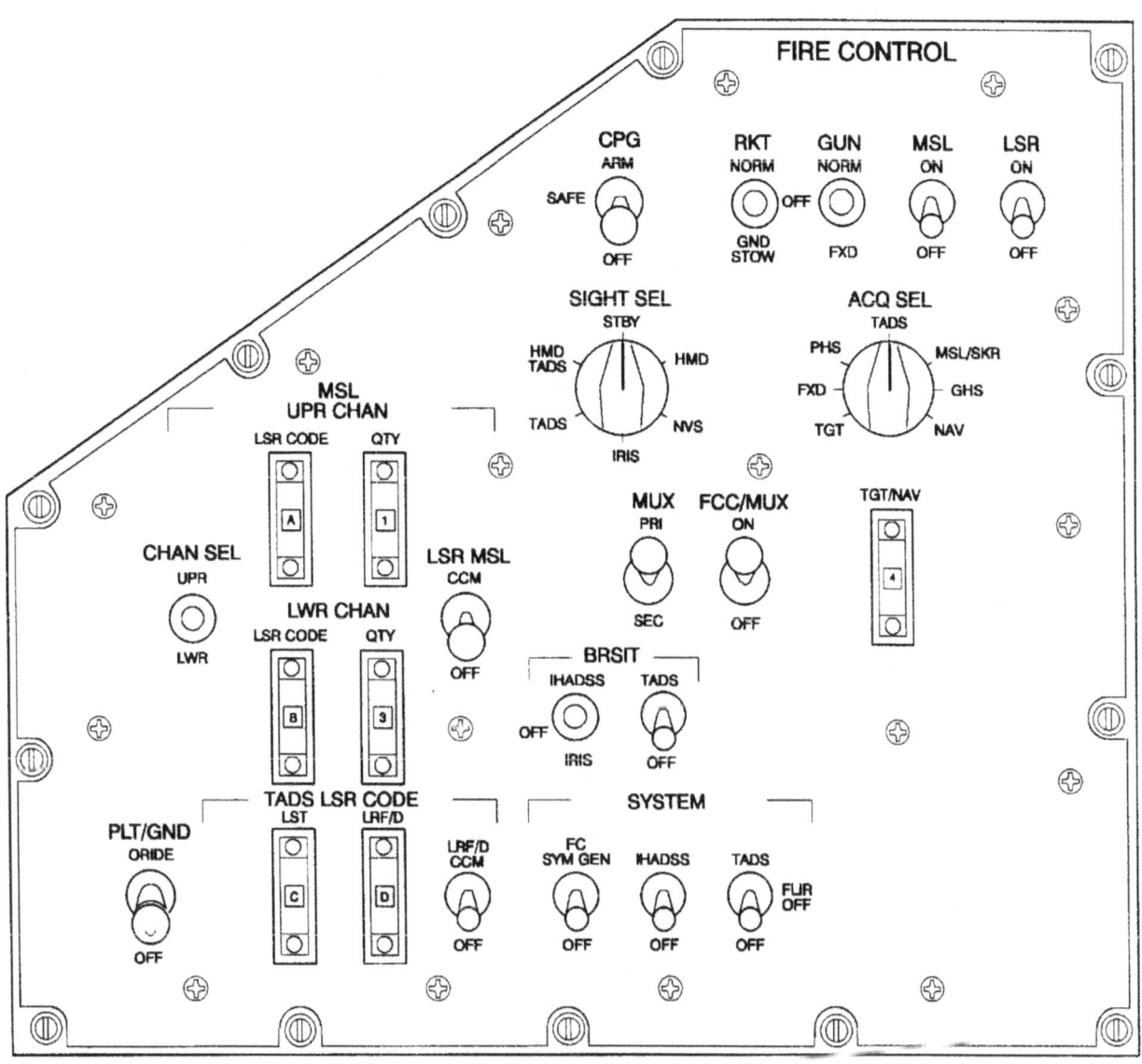

Figure 4-14. CPG Fire Control Panel

4-31

Table 4-11. CPG Fire Control Panel Functions

Control	Position	Function
PLT/GND	OFF	Squat switch inhibits arm/safe power while on the ground. Airborne MASTER ARM switch controls arm/safe power.
	ORIDE	Squat switch overridden in all modes and CPG ARM switch controls arm/ safe power for the aircraft.
CPG ARM	OFF	Disables all weapon control/firing circuits of the CPG.
	SAFE	Enables safe power to the CPG weapons select switches only if the MASTER ARM/SAFE switch is in arm or safe or if PLT GND ORIDE is in override. A green SAFE light indicates a valid safe condition.
	ARM	Enables ARM power to the CPG firing circuits only if the MASTER ARM/SAFE switch is in arm or if the PLT/ GND ORIDE is in override. An amber ARM light indicates a valid arm condition.
RKT		All positions identical to the pilot.
GUN		All positions identical to the pilot.
MSL		All positions identical to the pilot.
LSR		The laser select switch provides power to the laser electronics and laser transceiver units.
		The laser cannot be fired unless a valid arm condition exists for the CPG and aircraft.
SIGHT SEL		Permits reference sight selection by the CPG.
	STBY	Deselects all lines-of-sight and stows sensors.
	HMD/ TADS	Selects the helmet mounted display as the reference LOS and permits slaving of the TADS to the helmet LOS.
	TADS	Selects TADS as the reference LOS.
	IRIS	Selects the IRIS missile as the reference LOS; growth position, not currently active.
	NVS	Allows the CPG to view TADS FLIR wide field of view with flight symbology. The HMD helmet configuration is the only valid configuration for utilization of this position.
	HMD	Selects the helmet mounted display as the reference LOS. The HDU is active with CPG weapons symbology only. TADS may be slaved to the helmet, however imagery from the selected TADS sensor is not displayed.

Table 4-11. CPG Fire Control Panel Functions - continued

Control	Position	Function
ACQ SEL		Enables cueing or slaving to selected acquisition LOS when the slave pushbutton is actuated. Only the TADS may be slaved to a selected acquisition LOS, all others receive cueing.
	FXD	Defines the fixed forward position.
	TGT	Defines the selected target as the acquisition source.
	NAV (-49)	Same as TGT position.
	NAV (-51)	Defines FLY-TO as selected acquisition source.
	GHS	CPG helmet LOS is defined as the acquisition source.
	MSL/SKR	The Hellfire missile seeker angles define the acquisition LOS.
	TADS	The TADS LOS is defined as the acquisition LOS. This mode is not active if the sight selected is TADS.
	PHS	The pilot helmet LOS is defined as the acquisition LOS.
CHAN SEL		Functions the same as the pilot channel select switch.
MSL UPR/ LWR CHAN		Labels used on the CPG fire control panel for set-up of two missile channels. Once a set-up has been made, the channel select switch is actuated either to upper or lower establishing that code and quantity set-up as the priority channel; by default, the other code and quantity set-up becomes the alternate channel.
	LSR CODE	Specifies the alphabetic laser code designator assigned to that channel set-up.
	QTY	Specifies the number of missiles per channel that are to be coded and brought to a ready state at one time. A maximum of three missiles is permitted per channel at one time.
LSR MSL CCM		The LSR MSL CCM switch location lower center of the CPG HELLFIRE fire control panel, provides HELLFIRE weapons system hardening in the true or up position.
		The LSR MSL CCM switch position is set during the sequence of operations in which the code is loaded to a specific missile location.
		In the absence of enemy countermeasures, the CCM switch shall remain in the false or down position.
TGT/NAV	-49A	Numeric value in the window specifies the storage location of selected target coordinate data. The range is 0-9.
	-51	The switch is inoperative.

Change 4 4-33

Table 4-11. CPG Fire Control Panel Functions - continued

Control	Position	Function
BRSIT		Enables various sight or missile boresight modes.
	TADS	Places the TADS in internal boresight mode.
	OFF	Disables TADS internal boresight and returns to normal control.
	IHADSS	Enables IHADSS boresight for the CPG and turns on the boresight reticle unit.
	OFF	Disables all switch functions.
	IRIS	Enables boresight of IRIS missiles; growth position, currently inactive.
MUX		Specifies which bus controller will control the multiplex bus.
	PRI	Specifies the fire control computer.
	SEC	Specifies the backup bus controller. The normal position is PRI. In the event of failure of the FCC, switching to the BBC will occur automatically.

CAUTION

FCC/MUX switch shall be left ON.

Control	Position	Function
FCC/MUX	ON	Enables power to the FCC.
	OFF	Disables power to the FCC.
TADS LSR CODE	LST	This index specifies the alphabetic laser code designation to be assigned to the laser spot tracker.
	LRF/D	This index specifies the alphabetic laser code designation to be assigned to the LRF/D.
	LRF/D CCM	Enables the first pulse range circuit. CCM Off enables last pulse logic. If erratic laser range readouts occur, enable the LRF/D CCM.
FC SYM GEN		Turns on the fire control symbol generator. Required for symbology generation and video switching on board the aircraft.
IHADSS		Turns on the IHADSS and activates the IHADSS display electronics unit.
TADS		Turns on the TADS and initiates TADS FLIR cooldown.
	FLIR OFF	Powers TADS but does not cool down the FLIR; not normally used.
	OFF	Disables all power to the TADS.

4.22 EXTERNAL STORES SUBSYSTEM (ESS).

The ESS consists of an external stores controller and up to four pylon assemblies. The external stores controller commands the pylons to the required elevation angles for the various fire control modes. The modes are: ground stow; flight stow; and FCC control.

The ground stow mode articulates the pylons so that the wing stores are parallel with the ground over level terrain. The ground stow mode is automatically commanded on landing when the squat switch goes to ground mode, or in flight whenever the rocket select switch, **RKT**, (fig 4-14) is placed in the **GND STOW** position. The **GND STOW** position would normally be selected immediately prior to landing on uneven terrain or when making a slope landing. The flight stow mode articulates the pylons to a single fixed position so that the wing stores present minimum flat plate drag area in forward flight. The flight stow mode is automatically commanded on takeoff whenever the squat switch goes air mode. The **PLT/GND ORIDE** switch will not override the squat switch for these functions. In flight, the pylons remain in the flight stow mode until missiles or rockets are actioned, which then places the pylons under FCC control. Under FCC control the pylons can be commanded through a range of +4.9 to -15 degrees. If the forward flight airspeed as sensed by the ADSS is 100 KTAS or greater, the pylons will remain in the flight stow position under FCC control and will not articulate. The external stores controller receives 115 vac power from the No. 1 essential bus and 28 vdc from the No. 2 essential dc bus. The external stores controller is monitored continuously by FD/LS; the on command test (test 10-PYLN) will fault isolate to the LRU.

4.22.1 Pylon Assemblies. The pylon assembly contains the pylon MRTU, the pylon aerial rocket control subsystem station director, the ejector rack, and the hydraulic actuator for articulation (fig 2-2). The pylon MRTU is used to interface the missile launcher and station director with the multiplex bus. The ejector rack contains the attaching lugs for the wing store, the winch assembly for lifting the wing store, and the ballistic ejector for stores jettison. The pylon assembly connects to electrical and hydraulic power when attached to the helicopter wing.

a. Wing Stores Jettison. Each pylon is equipped with an electrically operated ballistic ejector to jettison the attached wing store. Selectable wing stores jettison is activated by lifting the guard and pressing the spring loaded switch for the desired store (fig 2-11). An emergency stores jettison switch is located on each collective switchbox (fig 2-26) which when activated will jettison all wing stations simultaneously. The wings stores jettison receives 28 vdc power from the No. 2 essential dc bus and the emergency dc bus; the selectable stores jettison circuit is protected by the **MISSION JETT** circuit breaker on the pilot overhead circuit breaker panel; the emergency stores jettison circuit is protected by the **JETT** circuit breaker on the pilot overhead circuit breaker panel. The two stores jettison circuits are completely independent.

TM 1-1520-238-10

4.23 OPTICAL RELAY TUBE (ORT) CONTROLS AND DISPLAYS.

The ORT (fig 4-15) is located between the CPG left and right vertical instrument panels (fig 2-8) and receives power as part of the overall TADS system. The ORT is continuously monitored as part of the overall TADS FD/LS; the on command test (test 14-TADS) will fault isolate to the LRU. Controls and functions are described in table 4-12.

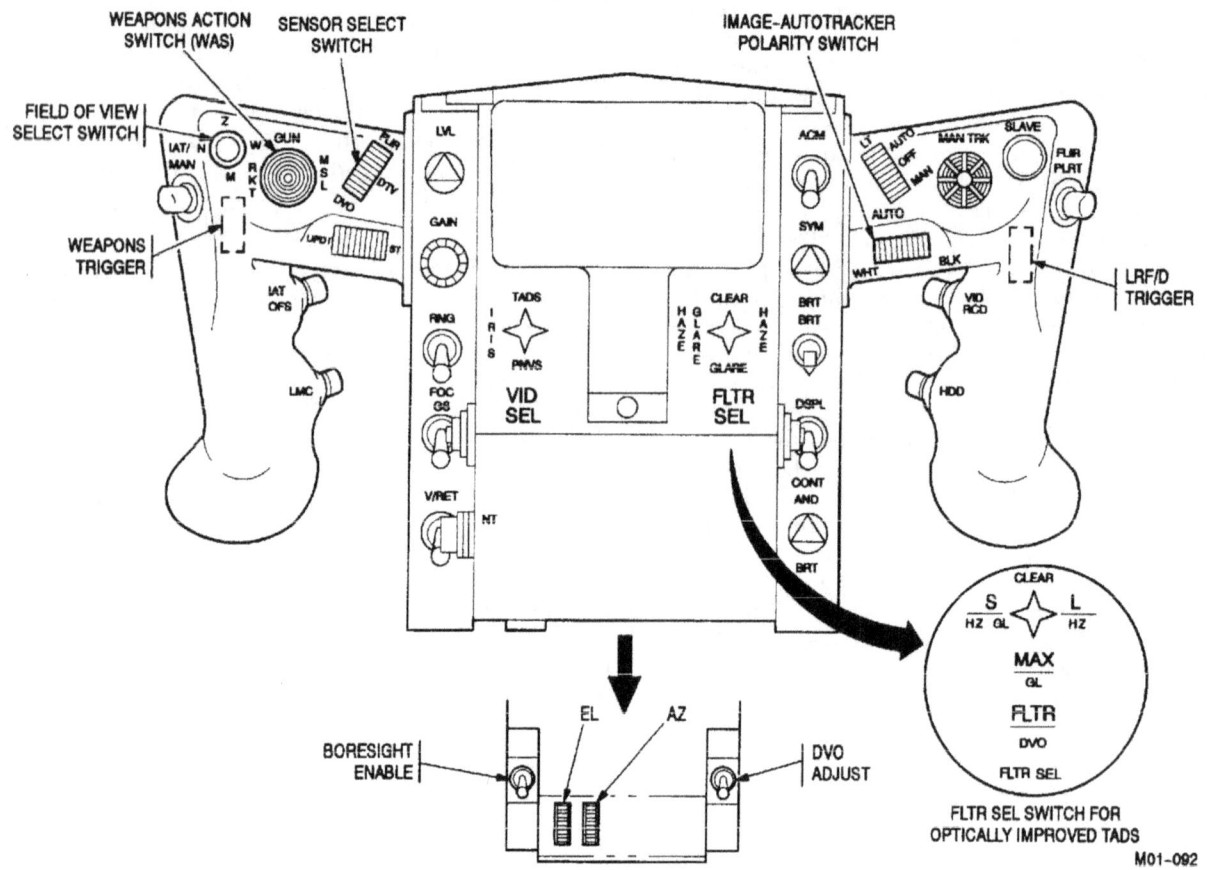

Figure 4-15. ORT and Hand Controls

TM 1-1520-238-10

Table 412. ORT and Hand Controls/Display Functions

Control/Display	Position	Function
Left Hand Grip		
IAT/MAN		Engages or disengages the image autotracker. On disengagement the TADS turret is placed in manual track.
Field of view select		Selects the field of view desired.
	W	Wide field of view.
	M	Medium field of view.
	N	Narrow field of view.
	Z	Zoom or underscan field of view.
WAS		Weapons action switch. Actions the selected armament and removes the electrical trigger interlock.
	RKT	Actions the Aerial Rocket Control System; pylons will articulate to commanded quadrant elevation and displays the rocker steering indicator. The ORT WAS in the RKT position defines the cooperative engagement mode.
	GUN	Actions the M-230E1 gun system and slaves the gun turret to the selected LOS.
	MSL	Actions the Hellfire missile system and displays constraints symbology.
Sensor select		Selects the desired sensor to TADS.
	DVO	Selects the direct view optics.
	DTV	Selects the day television.
	FLIR	Selects the FLIR.
UPDT/ST		Actuation signals the FCC to execute the update or store functions of waypoint/targeting.
IAT OFS		Engages offset track mode of the image auto tracker.
LMC		Turns on the linear motion compensation and rate integrator tracking aids.
Weapons trigger		Two-position guarded trigger which enables weapons firing. The first detent is for normal selected weapons usage; the second will override the selected weapons performance inhibits, but not the safety inhibits of the selected weapons.

4-37

Table 4-12. ORT and Hand Controls/Display Functions - continued

Control/Display	Position	Function
Left Side of ORT		
LVL		Adjusts the FLIR level.
GAIN		Adjusts the FLIR gain.
RNG FOC		Adjusts the narrow FOV focus in DTV or FLIR. Minimum distance that maybe focused in narrow field of view is 500 meters.
GS		Activates the gray scale for the CPG. On some TADS systems, one or more video displays may be in gray scale after power up on completion of FD/LS.
VID SEL		Selects the video source to be displayed to the CPG.
	TADS	Displays the video from the CPG selected sight.
	PNVS	Displays the video from the pilot selected sight.
	IRIS	Displays the video from the IRIS missile; growth position, currently will mode the ORT displays for 525 line video format.
NT		Actuates the heads down display red night filter.
V/RET		This function has been deleted. The switch is inactive.
Right Side of ORT		
ACM		This switch is utilized to maintain optimum gain and level settings as a result of varying thermal scene content or when switching FLIR polarities.
SYM BRT		Adjusts the symbology brightness. Range is: bright white through black.
DSPL BRT		Adjusts heads up or heads down display brightness. When an IHADSS mode is selected on the SIGHT SEL switch, this control adjusts HDU brightness.
DSPL CONT		Adjusts heads up or heads down display contrast. When an IHADSS mode is selected on the SIGHT SEL switch, this control adjusts HDU contrast.
AND BRT		Adjusts alphanumeric display brightness.
FLTR SEL		Permits selection of various optical filters when sensor select is DVO. This switch will change the various DVO filters. In the optically improved (OI) TADS, filter selection can also be made when sensor select is FLIR. The FLIR filters are used as countermeasures protection.
When sensor select is DVO	CLEAR/CLEAR	Selects a clear filter for DVO.

Table 4-12. ORT and Hand Controls/Display Functions - continued

Control/Display	Position	Function
Right Side of ORT (cont)		
	HAZE GLARE/ HZ-GZ	Selects a haze/glare filter for DVO.
	HAZE/HZ	Selects a haze filter for DVO.
	GLARE/GL	Selects a glare filter for DVO.
When sensor select is FLIR	CLEAR	Selects a clear filter for FLIR.
	S	Selects laser protection filter for FLIR against short wavelength laser.
	L	Selects laser protection filter for FLIR against long wavelength laser.
	MAX	Selects maximum (short and long wavelength) laser protection for FLIR.
Right Hand Grip		
LT		Laser spot tracker engagement switch.
	AUTO	Arms the laser spot tracker and places the TADS turret into a four bar scan about the point of engagement.
	OFF	Disarms the laser spot tracker.
	MAN	Arms the laser spot tracker and leaves the turret in manual track mode.
Image auto-tracker polarity		Selects the contrast polarity for the image auto-tracker.
	WHT	Selects white on black contrast polarity.
	AUTO	Permits the tracker to automatically select the tracker polarity.
	BLK	Selects black on white contrast polarity.
MAN TKR		Thumbforce controller that controls slewing the turret during manual track mode. Slew rates are proportional to the sensor and FOV selected.
SLAVE		Actuates the slave latch to slave or cue the selected sight to an acquisition LOS.

TM 1-1520-238-10

Table 4-12. ORT and Hand Controls/Display Functions - continued

Control/Display	Position	Function
Right Hand Grip (cont)		
FLIR PLRT		Allows switching of FLIR polarities: black or white hot.
VID RCD		Starts or stops the video recorder when the VRS is in REC or PLAY mode.
HDD		Turns off heads out display and turns on heads down display.
LRF/D trigger		Two-position guarded trigger. The first detent will fire a ranging burst and stop. If continuous range or designation is desired, press trigger to second detent.
Boresight Enable	Up	Enables the IAT during TADS internal Boresighting or enables the AZ and EL pots during TADS outfront boresight.
Center		Disables all functions.
	Down	Enables DVO boresight adjust switch.
DVO adjust		Three position, center maintain switch. Used to slew the DVO boresight adjustment in a helical motion.

4.24 TARGET ACQUISITION DESIGNATION SIGHT (TADS) AN/ASQ-170.

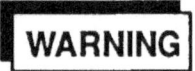

WARNING

If the pilot commands TADS away from the CPG, the CPGs sight defaults to IHADSS. The gun can still be fired without re-actioning, once the sight has changed.

The TADS provides the CPG with day and night target acquisition by means of a direct view optical (DVO) telescope, a day television (DTV), and a forward looking infrared (FLIR) sensor system. The sensors may be used singly, or in combination, depending on the tactical, weather, and visibility conditions. Target tracking may be accomplished manually; automatically, using the image auto-tracker (IAT); or by using the laser spot tracker (LT). The laser spot tracker facilitates target handoff from another laser designator. Linear motion compensation (LMC) aids in the tracking of moving targets either manually or automatically. The image autotracker has the capability to offset track one target, while automatically tracking another target. The TADS receives 115 vac power from the No. 1 essential ac bus and 28 vdc power from the No. 1 essential dc bus and is protected by the TADS AC and TADS DC circuit breakers located on the CPG circuit breaker panel. The TADS is continuously monitored by FD/LS; on-command test (test 14-TADS) will fault isolate to the LRU.

The Optically Improved (OI) configuration TADS contains special optical filters which provide laser threat protection in the FLIR and Direct View Optics (DVO) modes of operation.

4.24.1 OI Configuration. The OI configuration Night Sensor Assembly (NSA) contains selectable filters which will block different laser wavelength bands. These filters prevent lasers which operate in these bands from jamming the FLIR video (causing it to "bloom") should the lasers be directed at the TADS. The OI configuration ORT **FLTR SEL** switch is used to select one of four possible filter types: (a) the CLEAR filter is used for normal operation and provides no laser threat protection; (b) the S (for short) filter provides protection against lasers operating in the short wavelength band; (c) the L (for long) filter provides protection against lasers operating in the long wavelength band; and (d) the MAX filter provides protection against lasers operating in both the S (short) and L (long) wavelength band.

a. OI Special Filter Coatings. Special filter coatings are applicable to Direct View Optics lenses contained in the OI configuration Day Sensor Assembly. These filter coatings provide the CPG with eye protection against certain laser threats should the CPG happen to view those laser threats through the DVO. Because the filters are coatings applied to fixed optics, they are always in place and require no action on the part of the CPG to activate.

4.24.2 TADS Equipment Data. TADS equipment data is contained in figure 4-16 and table 4-13.

Table 4-13. TADS Equipment Data

Parameter	Function
Normal Azimuth Operating Range	+120 degrees/ -120 degrees
Azimuth Stow Position	-180 degrees
Elevation Operating Range	+30 degrees/ -60 degrees
Maximum Slew Rate (WFOV FLIR)	60 degrees per second
Fields of View (diagonal measurement)	
DVO	
Wide	18.0 degrees
Narrow	4.0 degrees
DTV	
Wide	4.0 degrees
Narrow	0.9 degree
Zoom	0.45 degree
FLIR	
Wide	50.0 degrees (identical to PNVS)
Medium	10.0 degrees
Narrow	3.1 degrees
Zoom	1.6 degrees

4.24.3 Operation Data.

a. Drift Null. The TADS system has the capability to null the turret servo drift by either of two methods. Servo drift can be identified as happening when ever the TADS turret is under manual control, the CPG is not making any inputs with the thumbforce controller and the reticle appears to drift off the aim point. The aim point should remain within the narrow FOV of the DTV for 30-seconds. If it does not, or a lesser amount of drift is desirable, adjust as follows:

(1) Manual Servo Drift Null. The manual servo drift null procedure can be accomplished when the helicopter is on the ground or during flight. First, select **DTV or FLIR** narrow FOV as the sensor; then aim the reticle at an easily observed object. Place the Boresight Enable switch in **UP** position and wait for at least 5 seconds; then use the **AZ** and **EL** thumbwheels to null the turret drift. When the servo drift is nulled, place the Boresight Enable switch in **CENTER** position and the procedure is complete.

(2) Auto Drift Null. Auto drift null is active from the time the TADS is first turned on. The only time it is inactive is when the manual servo drift null procedure has been accomplished. If the manual procedure was accomplished, the auto drift null can be reactivated by placing the Boresight Enable switch in **UP** position for 2 seconds and then placing it back to **CENTER** position. Auto drift is not as precise as the manual procedure so some residual servo drift will always be present when it is active.

WARNING

The narrow and zoom FOV TADS FLIR imagery has inaccuracies for lasing or weapons direction following TADS internal boresight. TADS outfront boresight validation and adjustment (if necessary) shall be performed prior to using the TADS FLIR imagery for laser or weapons operations.

b. Boresight. Boresight procedures consist of cue update, internal boresight, outfront boresight, and manual boresight.

(1) Cue Update. Performance of cue update ensures that the TADS turret is in proper position relative to the boresight assembly prior to firing the laser for internal boresight. If cue update is adjusted, outfront boresight must be performed following the cue update. Cue update procedure is contained in paragraph 4.31.7.a.

(2) Internal Boresight. Performance of internal boresight aligns DTV and FLIR to laser LOS and DVO to DTV LOS. Internal boresight shall be performed as part of preflight procedures prior to any firing of laser or weapons when TADS is used as imaging sensor. Internal boresight can also be performed in flight to ensure boresight accuracy without requiring. outfront boresight as a follow-up. Internal boresight procedure is contained in paragraph 4.31.7.b.

(3) Outfront Boresight. Performance of outfront boresight ensures FLIR LOS is in coincidence with laser LOS. Outfront boresight shall be checked as part of preflight procedures prior to any firing of laser or weapons when TADS FLIR is used as the imaging sensor. Outfront boresight must be performed after a cue update adjustment. Target requirements for the outfront boresight procedure are as follows:

(a) A target a minimum of 0.5 km away from the helicopter is required. Target must be clearly visible and trackable through both FLIR and TV sensor NFOV. Target must have the same center as viewed in both FLIR and TV sensors.

(b) Outfront boresight procedure is contained in paragraph 4.31.7 c.

(4) Manual Boresight. The manual boresight procedure is used only to recapture or center the FLIR laser spot. It is not an acceptable boresight procedure for normal flight operations. Manual boresight procedure is contained in paragraph 4.31.7 d.

4.25 PILOT NIGHT VISION SENSOR (PNVS) AN/AAQ-11.

WARNING

IFF transmission on the upper antenna may cause the PNVS to slew toward the fixed forward position and then back to LOS. If this occurs, turn the Transponder off or use the lower antenna only during night NOE PNVS operation.

The PNVS is used by the pilot for externally aided vision at night or during adverse weather. The PNVS consists of a stabilized FLIR contained in a rotating turret mounted above the TADS (fig 4-16). Refer to table 4-14 for PNVS equipment data. When selected, the turret is slaved to the crewmember helmet LOS. This is accomplished using the IHADSS which also presents the FLIR image and symbology video to the crewmember on the HMD. The PNVS image and symbology can be displayed on any of the displays in the helicopter through use of the VID SEL switch. Control of the PNVS turret is governed by the SIGHT SEL switch. Normal operation calls for the pilot to have priority control of the PNVS turret; however, in the event the pilot becomes incapacitated, the CPG may take control of the PNVS through use of the PLT/GND ORIDE switch, the SIGHT SEL switch, and the collective NVS switch. Four degraded modes of operation exist when using the PNVS. In two of the degraded modes corrective action by the fire control system occurs automatically. The remaining two require pilot corrective action. In all degraded modes there will be a loss of some operational capability, as described in paragraphs thru 4.25.5.

TM 1-1520-236-10

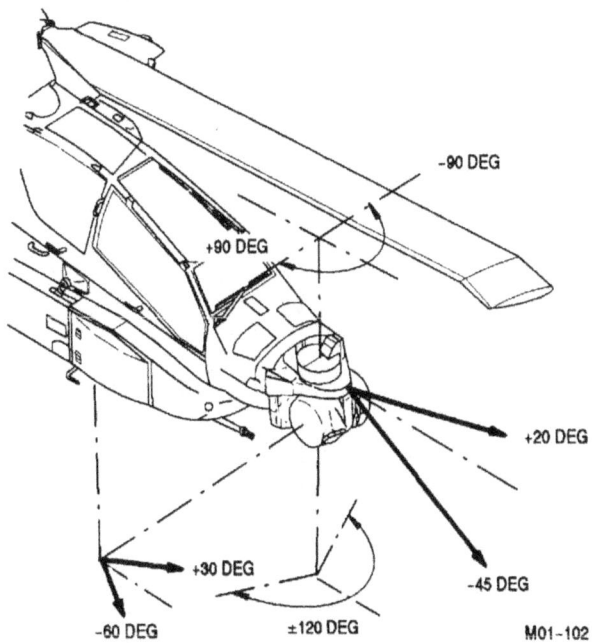

Figure 4-16. **TADS/PNVS** Equipment Data

Table 4-14. PNVS Equipment Data

Parameter	Function
Normal Azimuth Operating Range	+90 degrees/ -90 degrees
Azimuth Stow Position	-118 degrees. or greater
Elevation Operating Range	+20 degrees/ -45 degrees
Field of View	30 degrees vertical/ 40 degrees horizontal
Maximum Slew Rate	120 degrees per second

4.25.1 PNVS Electrical Power. The PNVS receives 115 vac power from the No. 1 essential ac bus and 28 vdc power from the No. 3 essential dc bus through the **PNVS AC** and **PNVS DC** circuit breakers on the pilot overhead circuit breaker panel. The PNVS is continuously monitored by FD/LS which will illuminate the PNVS caution light segment for a PNVS malfunction. The on command DD/LS test (test 09-PNVS) will fault isolate to the LRU.

4.25.2 TADS Electronic Unit Failure. In the event of a TADS Electronic Unit (TADS computer) failure, the PNVS will be placed in the direct mode automatically by the fire control system. The PNVS turret will be directly driven by the IHADSS sight electronics unit.

PNVS turret motion maybe limited to 75 degrees left and 75 degrees right in azimuth; turret motion may be erratic beyond the 75 degree azimuth position. The message PNVS...DIRECT will be displayed in the sight status block of the pilot's high action display. No corrective action is required by or available to the pilot.

4.25.3 Symbol Generator Failure. In the event of a symbol generator failure, the fire control system will automatically command the IHADSS display electronics unit to display **PNVS FLIR 2** on the pilot/CPG HDU. The PNVS will continue to function normally; however, the video displayed to the pilot will not have symbology. VDU video will blank and the recorder will be inoperative. Additionally, the CPG will receive TADS FLIR 2 video on the ORT. No corrective action is required by or available to the pilot.

WARNING

If night or simulated night NOE, reaction to the following malfunctions must be immediate, exit of the NOE environment maybe required.

4.25.4 PNVS FLIR or Turret Failure. In the event of a failure of the PNVS FLIR or turret assembly, the pilot must place the NVS switch on the collective switchbox (figs 2-11 and 2-12) in the **TADS** position. TADS will be slaved to the pilot helmet LOS and moded to WFOV FLIR. Normally, this action will provide useable TADS video to the pilot. However, in some cases, the TADS FLIR video may be scrambled following loss of the PNVS FLIR video. If this should occur, reselect **PNVS** on the pilot's collective NVS switch and then select **TADS**. The out-of-synch video will be corrected. The slew rates of the TADS turret are noticeably slower than those of the PNVS and some gain and level adjustment may be required to obtain optimum image quality.

4.25.5 IHADSS Failure. In the event of an IHADSS failure, either inability to command the PNVS turret or loss of video on the HDU, the pilot must first exit the NOE environment. and then set the VDU control switch to **PLT** (fig 4-2) and the **ACQ SEL** switch to NVS **FXD**. This configuration will allow the pilot to fly a fixed panel mounted display. Terrain flight capability will be directly dependent on pilot proficiency.

4.26 WEAPONS SYMBOLOGY.

The weapons symbology is displayed to the CPG on the video from the selected sight. The symbology is shown in figure 4-17. Refer to table 4-15 for symbol definition.

4-43

TM 1-1520-238-10

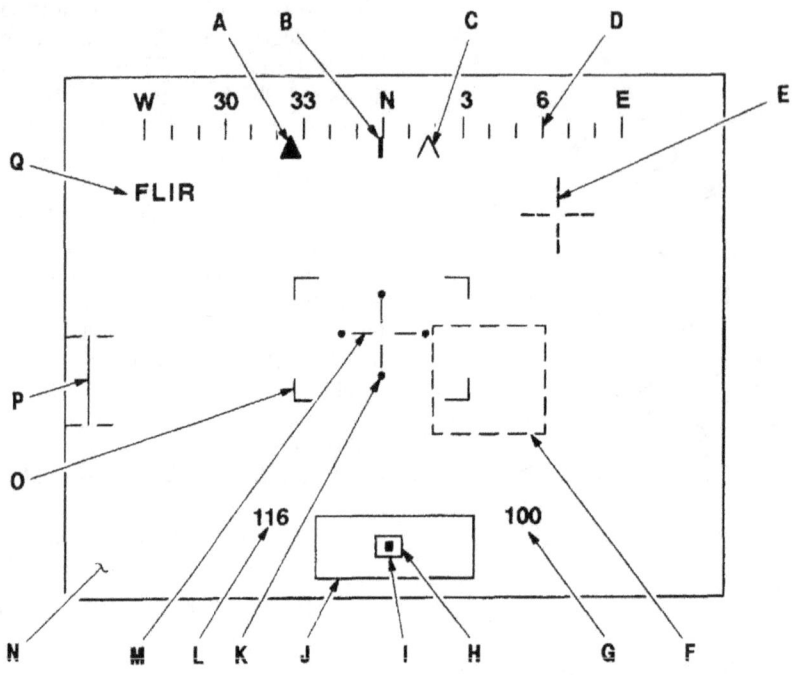

Figure 4-17. Weapons Symbology Modes

Table 4-15. Weapons Symbology Definitions

Fig. 4-17 Ref No.	Symbol Name	Description
M	LOS Reticle	Represents the line-of-sight of the crewmember selected sight. The reticle will flash whenever the selected sight LOS is invalid or has failed. The reticle will also flash whenever the "ACTIONED" weapon is in a NO-GO state. The High Action Display will prompt the crewmember for the appropriate condition.
A	Alternate Sensor Bearing	Indicates to the crewmember the other crewmember sensor relative bearing with respect to helicopter center line.
B	Lubber Line	Index indicates helicopter magnetic heading.
K	Cueing Dots	Indicates cued direction for target acquisition. All four dots present and flashing indicate IHADSS boresight is required.

4-44

TM 1-1520-238-10

Table 4-15. Weapons Symbology Definitions - continued

Fig. 4-17 Ref No.	Symbol Name	Description
C	CPG LOS	Indicates relative bearing of CPG selected LOS.
D	Heading Scale	Helicopter magnetic heading scale.
E	Cued LOS Reticle	A virtual reticle indicating the cued LOS to the appropriate crew member. Used with the cueing dots.
F	Missile Constraints	Indicates the required orientation to align the helicopter into constraints for Hellfire missile engagements. When all constraints for the mode of engagement are satisfied, the box will go from 'dashed' to 'solid'.
G	Radar Altitude	A digital display of radar altitude. Displays in 1-foot increments to 50 feet and in 10-foot increments above 50 feet.
Q	Selected Sensor	Displays the name of the TADS selected sensor. Also when FD/LS detects a fault in the continuous monitor mode, the message 'FD/LS' will alternately flash with the sensor name.
O	TADS FOV Gates	Indicate the amount of the currently displayed imagery of TADS that will be displayed in the next narrower FOV.
I	Cued LOS DOT	Indicates the cued LOS location within the field of regard. The cued LOS DOT will flash when the HARS inertial platform has gone into the free inertial mode.
H	Field of View	Represents the instantaneous FOV of the crewmember sensor within the field of regard.
J	Sensor Field of Regard	Represents the total gimbal limits possible for the respective crewmember sensor
N	High Action Display	Refer to paragraph 4.28.

4-45

TM 1-1520-238-10

Table 4-15. Weapons Symbology Definitions - oontinued

Fig. 4-17 Ref No.	Symbol Name	Description
P	Rocket Steering Cursor	Indicates the required orientation to align the helicopter into constraints for 2.75 inch FFAR rocket engagements. During fixed or flight stow rocket delivery, a broken I beam will appear.
L	Airspeed	A digital display of true airspeed when the ADSS is turned on or not failed. If the ADSS is OFF or failed, display is ground speed in knots from the doppler navigation system. Range is 0 to 200, omnidirectional.

4..27 FIRE CONTROL SYSTEM MESSAGES.

The fire control system displays to the crewmembers messages and prompts for management of the subsystems on board the helicopter. The messages and prompts are displayed in the high action display (HAD) or the alphanumerical display (AND) contained in the TADS ORT.

4-46

4.28 HIGH ACTION DISPLAY (HAD).

CAUTION

With the PLT/GND switch in the ORIDE position and the CPG SIGHT SELECT switch set to NVS, the HAD is disabled. The 30mm gun can be actioned without any messages in weapon control or weapon status. This can result in damage during takeoff or landing if the 30mm gun is actioned and depressed inadvertently.

The HAD (fig 4-18) provides information to the pilot and CPG for use in operation of the mission equipment. The HAD is located along the bottom of the displayed symbology and is split into two fields by the gimbal limits box. The two fields on either side of the gimbal limits display are further subdivided for a total of four message fields: sight status, range and range source on the left, weapon status, and opposite crew station weapon control. Tables 4-16, sight status, 4-17, range and range source, 4-18, weapons status, and 4-19, opposite crew station weapon control list the messages and location.

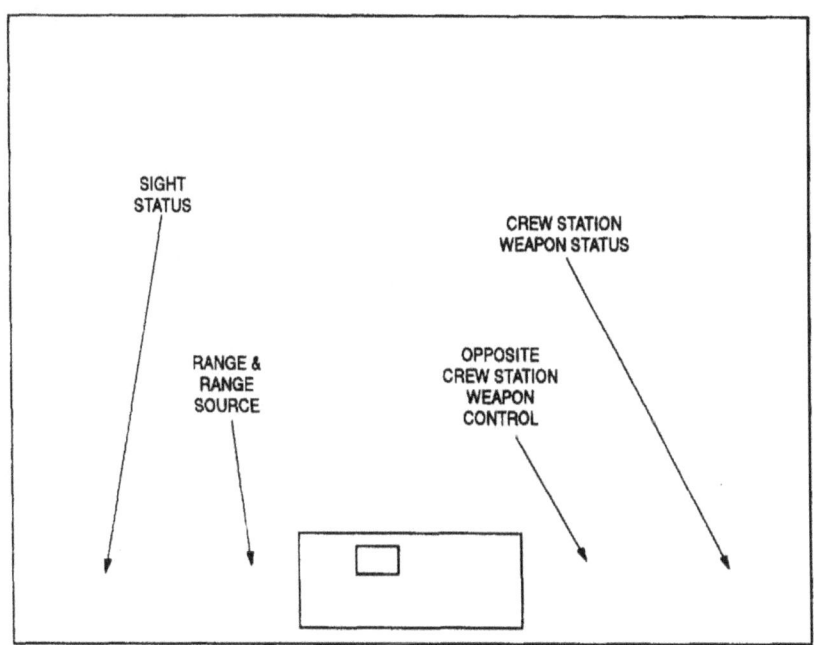

Figure 4-18. High Action Display

Table 4-16. High Action Display Sight Status

Crewmember	Message	Description
P/CPG	IHADFAIL	The IHADSS has been detected as NO-GO for that crew station by FD/LS. The IHADSS LOS defaults to fixed forward for the affected crew station.
P/CPG	BORESIGHT REQUIRED	IHADSS boresight required.
P/CPG	IHADSLOS INVALID	The IHADSS SEU has computed an invalid LOS for that crew station. This can be caused by the crewmembers head outside of the head motion box or blockage of the line-of-sight between the SSU and the helmet. The IHADSS LOS is frozen at the last valid computation until the LOS is computed as valid.
P	PNVSFAIL	FD/LS has detected the PNVS turret or electronics as NO-GO.
P	PNVS NOT COOLED	The PNVS FLIR has not cooled down for optimum performance; cool down time should not exceed 20 minutes.
P	PNVS DIRECT	FD/LS has detected the TADS Electronics Unit (TADS computer) as NO-GO. The PNVS turret is being directed by the IHADSS SEU. The PNVS turret azimuth range may be limited to 75 degrees left and right.
P/CP	LIMITS	The PNVS or TADS is at a gimbal limit.
P/CPG	FORWARD	The PNVS or TADS has been slaved to fixed forward through use of the ACQ SEL switch.
P/CPG	PNVSBST?	The FCC has detected an alteration of internal values affecting the PNVS alignment. The PNVS will function normally but without alignment correction.
P/CPG	TADSBST?	The FCC has detected an alteration of internal values affecting TADS boresight harmonization. The TADS will function normally but without CBHK correction.
P	INVALID CPGLOS'	ACQ SEL switch is in the CPG position and there is no valid CPG LOS.
P	UP=? LO=?	Pilot has actioned the missile system. If a ? is present, no missiles are coded and ready. When the missiles are coded and ready or tracking, the code letter will be displayed. The code of the priority channel will flash.
CPG	TADSFAIL	FD/LS has detected the TADS system as NO-GO.

Table 4-16. High Action Display Sight Status - continued

Crewmember	Message	Description
CPG	FLIR NOT . . . COOLED	The TADS FLIR has not cooled down sufficiently for optimum performance; cool down should not exceed 20 minutes.
CPG	FLIR OFF	TADS FLIR is turned off, but FLIR has been selected as the sensor.
CPG	MSL LNCH	A missile has been launched for a remote designator.
CPG	SIM LNCH	A training missile has been launched for a simulated remote designator.
CPG	TOF = TT `45`	The time-of-flight remaining for the missile launched for the remote designator.
CPG	TOF=TTT `49A` / `51`	The time-of-flight remaining for the missile launched for the remote designator.
CPG	FIRE . . . MISSILES	A subsequent missile may be fired for the remote designator.
CPG	LASE 1 . . . TARGET LASE 2 . . . TARGET LASE 3 . . . TARGET LASE 4 . . . TARGET THRU LASE 16 . . . TARGET	The remote designator must have his laser on for missile terminal guidance. The crewmember may use this message to cue the remote designator when to turn on the laser.
CPG	EOT	End of tape. The video tape cassette is at the end; no tape remains.
CPG	BOT	Beginning of tape. The VRS has completely rewound the video tape cassette.
CPG	RECORDER `45`	The video recorder has been selected to record.
CPG	RCDR ON `-49A` / `51`	The video recorder has been selected to record.
CPG	RCDR FAIL	The video recorder has been selected and has been detected as NO-GO.
CPG	RCDR OFF `-49A` / `51`	The video recorder has been deselected and has stopped recording.
CPG	INTERNAL . . . BORESITE	The TADS internal boresight has been selected by having the BRSIT TADS/OFF switch placed in the TADS position.
CPG	OUT-FRNT . . . BORESITE	The TADS out-front boresight has been selected by having the LAT tracking when the BRSIT TADS/OFF switch is placed in the TADS position.
CPG	NAV?	A Northing or Easting distance differential greater than 6100 meters exists between the doppler and the FCC and the doppler data is now considered bad or invalid.

Change 4   4-49

Table 4-16. High Action Display Sight Status - continued

Crewmember	Message	Description
CPG	CUE UPDT . . . BORESITE	The TADS cue update boresight has been selected by having the IAT in manual and the SLAVE latched Low when the BRSIT TADS/OFF switch is placed in the TADS position.
CPG	INVALID . . . PLT LOS	ACQ SEL switch is in the PLT position and there is no valid PLT LOS.
CPG	TGT = X	Indicates selected address in the target NAV indexer (0-9). If X is replaced by a ?, no data is stored in target location.
CPG	FCC LOAD —49A	Power up, checksum test has failed.

NOTE

The FCC LOAD message will be automatically cleared when the aircraft is airborne. The message clearing does not indicate that appropriate corrections have been entered. The following data must be verified and corrected as appropriate. Any or all of the data may be in error: CBHK boresight parameters, AWS harmonization values, aircraft data (PPOS ALT, SPH and MV), laser codes, and waypoint/target latitude/longitude and altitudes.

Table 4-17. High Action Display Range and Range Source

Crewmember	Message	Description
NOTE		
The Fire Control Computer (FCC) is controlling the MUX whenever the symbol located between Kilometers and tenths of Kilometers is a colon (:). The Backup Bus Controller (BBC) in control is indicated by a period (.) instead of a colon.		
P	AX . X	The range being used is an automatic solution based on the selected sight LOS and radar altitude. The letter A is set in the leftmost RNG-KM thumbwheel on the ARCS panel. Range displayed is in kilometers.
CPG	AX . X	The range being used is an automatic solution based on selected sight LOS and radar altitude. The value 0 is entered through the DEK RNG data entry to command the automatic range solution. Range displayed is in kilometers.
P	MX . X	A manual range has been entered using the RNG-KM thumbwheels on the ARCS panel. Range displayed is in kilometers.
CPG	MX . X	A manual range has been entered through the DEK RNG data entry. Range displayed is in kilometers even if range entered was in meters.

Table 4-17. High Action Display Range and Range Source - continued

Crewmember	Message	Description
CPG	NX.X	The fire control system has computed the range to the target or waypoint indicated in the TGT/NAV indexer. This range will increase or decrease dynamically with helicopter ground track. Commanded by placing the ACQ SEL switch in the TGT or NAV position. The displayed index storage location must have valid coordinate data. Range displayed is in kilometers.
CPG	XXXX	Laser range is being used by the CPG. The displayed laser range will increase or decrease for about 4 seconds after stopping the laser. The increase or decrease in range is based on the helicopter ground track relationship with the object lased. Laser range is in meters.
CPG	XXXX	First detent, 3 pulse ranging burst.
CPG	*XXXX ▇49A▇ / ▇51▇	Second detent, continous ranging.
CPG	*XXXX ▇49A▇ / ▇51▇	Second detent, multiple target laser returns (asterisk is flashing).
CPG	XX.X	CPG range display for any range source above if the range value is 10.0 KM or more.
P	3.0 ▇45▇ / ▇49A▇ 1.5 ▇51▇	This is the pilots default range the fire control system will use for PLT range in the event no other range is available.
	0.5 ▇49A▇	AWS only.
CPG	2.0 ▇45▇ / ▇49A▇ 3.0 ▇51▇	This is the CPG default range the fire control system will use for CPG range in the event no other range is available.
	0.5 ▇49A▇	AWS only.
P/CPG	10.0 ▇45▇ / ▇49A▇	AUTO RNG default range. Indicates selected LOS is looking higher than 1° below horizontal or altitude is less than 33 feet AGL.

Table 4-18. High Action Display Weapons Status

Crewmember	Message	Description
P/CPG	WEAPON?	The weapons trigger has been pulled without "actioning" a weapon.
P/CPG	RKT FAIL	The rocket subsystem has been checked and is in a NO-GO state.
P	CPG MSL	The CPG has actioned the missile system and has outprioritized the rocket system action.

Table 4-18. High Action Display Weapons Status - continued

Crewmember	Message	Description
CPG	PLT MSL	The pilot has actioned the missile system and has outprioritized the CPG's rocket system action.
P/CPG	ROCKETS	The rocket system has been actioned with neither crewmember RKT select switch in GND STOW and airspeed less than 100 KTAS as sensed by the ADSS.
P/CPG	RKT-G-S	The rocket system has been actioned, but one crewmember RKT select switch in GND STOW and airspeed less than 100 KTAS as sensed by the ADSS.
P/CPG	RKT-F-S	The rocket system has been actioned and the airspeed is 100 KTAS or greater as sensed by the ADSS.
P/CPG	TOF=TT	Rockets have been launched. Message indicates time of flight to rocket impact, in seconds, decrementing.
P/CPG	GUN FAIL	The gun system has been actioned but has been detected as NO-GO.
P/CPG	LIMITS	The gun system has been actioned but the weapon has reached an azimuth or elevation limit.
P/CPG	RNDS DDDD	The gun system has been actioned and the DDDD indicates the number of rounds remaining and will count down one unit for each firing PULSE.
CPG	MSL FAIL	The missile system has been actioned but has been deleted as NO-GO.
P/CPG	INVALID... COMMAND	The remote Hellfire electronics has detected an improper command from a crewmember to the Hellfire missile system. This message is displayed for 2 seconds.
P/CPG	BIT IN... PROGRESS	The RHE is performing BIT.
P/CPG	HANGFIRE	The missile fire signal was sent but umbilical separation did not occur at the predicted time. This message is displayed for 6 seconds, during which time all missiles on that side of the helicopter become "not available" and are inhibited from firing.
P/CPG	GUNBST?	The FCC has detected an alteration of internal values affecting gun boresight harmonization. The gun will function normally but without boresight corrections.
P/CPG	MSL LNCH	The missile system is functioning properly and the weapons trigger has been pulled. This message is displayed for 2 seconds.
CPG	TOF=TT TOF=TTT	Time of flight remaining in seconds to missile impact, decrementing for autonomous launches. If more than one missile is in flight, the display will show TOF for the first launched missile, then subsequently launched missiles.

Table 4-18. High Action Display Weapons Status - continued

Crewmember	Message	Description
P/CPG	FIRE ... MISSILES	The minimum launch separation time has elapsed in a rapid fire missile engagement. The prompt is displayed whenever more than one missile is present on the priority channel after launch of a missile. This message is displayed for 2 seconds.
P/CPG	SIM LNCH	Displayed if training missiles are being used in place of MSL LNCH. This message is displayed for 2 seconds.
P/CPG	RF ORIDE	Growth position for RF/IR missile seekers.
P/CPG	2 CHANLS...TRACKING	Missiles selected on each channel have locked onto and are tracking laser energy of two different codes.
P/CPG	PRI CHAN...TRACKING	Missiles selected on the priority channel have locked onto and are tracking laser energy.
P/CPG	ALT CHAN...TRACKING	Missiles selected on the alternate channel have locked onto and are tracking laser energy.
		If no channels are tracking, the weapon status section will display the delivery mode and the firing delivery messages which are "HI", "LO", "DIR", or "LOBL". The fire delivery messages are "NORM", "STBY", "MANL", or "RIPL" (i.e., "LOBL NORM").
P/CPG	LASE 1 ... TARGET LASE 2 ... TARGET LASE 3 ... TARGET LASE 4 ... TARGET THRU LASE 16 ... TARGET	The laser designator must be lasing the target for missile terminal guidance. This message is displayed during the last 8 seconds of the calculated missile time of flight in place of the TOF-TT message.
P/CPG	MSL SEL?	Missiles have been actioned, but no missiles have been selected.
P/CPG	ZONE?	Rockets have been actioned, but no zone has been selected for firing.
P/CPG	SIGHT?	A weapon system has been actioned but the SIGHT SEL switch is in STBY.
P/CPG	PYLNBST?	The FCC has detected an alteration of internal values affecting pylon boresight harmonization. The pylons, rockets and missile subsystems will function normally but without pylon boresight corrections.
P	HSI CUE 51	DNS FLY-TO data validation (wrap) error

TM 1-1520-238-10

Table 4-19. High Action Display Opposite Crew Station Weapon Control

Crewmember	Message	Description
P	CGUN	The CPG has actioned the area weapon system.
P	CRKT	CPG has actioned the rocket system.
P	CMSL	CPG has actioned the missile system.
CPG	PGUN	Pilot has actioned the area weapon system.
CPG	PRKT	Pilot has actioned the rocket system.
CPG	PMSL	Pilot has actioned the missile system.

4.29 ALPHA/NUMERIC DISPLAY (AND).

The AND provides information to the CPG for use in the operation of the mission equipment. The AND is located below the heads-down-display in the ORT. The display is continuously visible to the CPG whenever he views heads down. The display fields are outlined in Figure 4-19. In the event of an AND failure, the Alpha/Numeric Display information may be symbolically duplicated on the TADS displays by placing the DEK rotary switch in the SP1 position and keying-in an 'I'. The AND information will then be duplicated on the selected TADS display in the same orientation as on the AND. Table 4-20 lists the message and visual location.

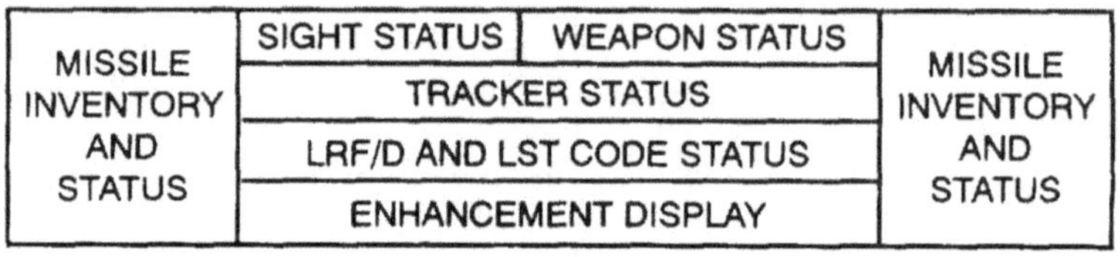

MISSILE INVENTORY AND STATUS	SIGHT STATUS	WEAPON STATUS	MISSILE INVENTORY AND STATUS
	TRACKER STATUS		
	LRF/D AND LST CODE STATUS		
	ENHANCEMENT DISPLAY		

Figure 4-19. Alpha/Numeric Display (AND)

Table 4-20. Alpha/Numeric Display Message Location and Description

Message	Description
SIGHT STATUS SECTION	
RECORDER -45	The ORT VID RCD switch has been pressed and the video recorder is recording.
RCDR ON -49A / -51	The ORT VID RCD switch has been pressed and the video recorder is recording.
RCDRFAIL	The ORT VID RCD switch has been pressed and the video recorder has malfunctioned.
RCDR OFF -49A / -51	The ORT VID RCD switch has been pressed and the video recorder has stopped recording.
NON TADS	PNVS is the sight in use.
TADS FAIL	
TV FAIL	
FLIR FAIL	
IRISFAIL	
TADS...BORESITE	The boresight mode is in effect.

4-54 Change 4

Table 4-20. Alpha/Numeric Display Message Location and Description - continued

Message	Description
IRIS.. .BORESIGHT	
LIMITS	The TADS is at an azimuth or elevation gimbal limit.
PILOT...CONTROLS...TADS	
IMPEND...LASER...INHIBIT	
LASER...INHIBIT	
INVALID...PILOTLOS	
SLAVE PL	The acquisition select switch is in PLT.
SLAVE TG	The acquisition select switch is in TGT.
SLAVE NV	The acquisition select switch is in NAV.
INVALID...TARGTLOS	There are no valid UTM coordinates in the FCC for the target number selected on the thumbwheel switch.
INVALID...NAV LOS	
SYMBOL G	The symbol generator has been detected as NO-GO.
KBRDFAIL	The data entry keyboard (DEK) has been detected as NO-GO.
TADS TEMP	The TADS has been selected and the turret internal temperature has been detected as overheating.
TADS NOT. . READY	The TADS has been selected and the turret internal temperature has not reached operating temperature.
LRFD . . COOLANT	The TADS LRF/D power is on and the transceiver coolant level has been detected as low. The LRF/D temperature is above maximum limit.
ENRGY LO	The TADS LRF/D power is on and the output energy level has dropped below a specified level.
WEAPON STATUS SECTION	
WEAPON?	A weapons trigger has been pulled without a weapon selection.
GUN FAIL	
TADS FAIL	The selected video is not operational.
RNDSDDDD	
RKT FAIL	The rocket subsystem has been checked and is in a NO-GO state; message blinks at a 2 Hz rate.
RKTINV? -49A / 51	Rockets have been actioned but the FCC has no rocket inventory. Re-inventory can be accomplished by cycling pilots SAFE/ARM switch to OFF, then back to SAFE/ARM.

TM 1-1520-238-10

Table 4-20. Alpha/Numeric Display Message Location and Description - continued

Message	Description
ROCKETS	The rocket system has been actioned with neither crewmember RKT select switch in GND STOW-and airspeed less than 100 KTAS as sensed by the ADSS.
TOF=TT	The WAS is in the rocket position. The trigger has been pulled. The message indicates the time of flight to impact in seconds, decrementing.
MSL FAIL	
INVALID...COMMAND	
NO TYPE	The type of missile selected on the missile control panel is not present on the launchers.
BIT IN...PROGRESS	
PLT MSL	The pilot has control of the missile system.
HANGFIRE	
MSL SEL?	Missiles have been actioned but no missiles have been selected.
ZONE?	Rockets have been actioned but no zone has been selected for firing.
MSL LNCH	
FIRE... MISSILES	
SIM LNCH	A training missile is being utilized.
TOF = TT	The WAS is in the MSL position. The trigger has been pulled. Time of flight in seconds to impact, decrementing.
RF ORIDE	
2 CHANLS...TRACKING	
PRI CHAN...TRACKING	
ALT CHAN...TRACKING	

NOTE

If no channels are tracking, the weapon status section will display the delivery mode and the firing delivery mode. The delivery messages are "HI", "LO", "DIR', or "LOBL". The fire delivery messages are "NORM", "STBY", "MANL", or "RIPL" (i.e., "LOBLNORM").

TRACKER STATUS SECTION

TADS...STOWED	The TADS is either stowed or in an internal boresight mode.
TADS...FORWARD	
LST FAILED	
LST TRACKING	LST tracking remotely designated laser energy.

4-56 Change 1

Table 4-20. Alpha/Numeric Display Message Location and Description - continued

Message	Description
LST SEARCH	Manual search.
LST AUTO SEARCH	LST scanning in a programmed search pattern.
IAT FAILED	
IAT BREAK- LOCK	Image auto tracker has broken track.
IAT OFFSET	IAT offset tracking engaged.
IAT TRACKING	
IAT B/W	IAT polarity is black on white.
IAT W/B	
IAT AUTO	
ENERATOR	The symbol generator has been detected as NO-GO.
RKT-G-S	
RKT-F-S	
LIMITS	
SIGHT?	
LASE 1 . . . TARGET LASE2. . . TARGET LASE3. . . TARGET LASE 4 . . . TARGET THRU LASE 16 . . . TARGET	
LST AND LRF/D CODES STATUS SECTION	
	NOTE If neither system is operable, the section is blanked.
CODE	
LST=C RFD=D	Both systems are operational and the code selected by the FCP pushbutton is displayed for each.
RFD = D	The LST is not operational but the RFD is operational.
LST = C	LRF/D is not operational but the LST is operational and code selected is "C".
LSR CCM	The LRF/D has detected a multiple target return while ranging or designating. The proper action is to engage laser countermeasures devices.

Table 4-20. Alpha/Numeric Display Message Location and Description - continued

Message	Description
ENHANCEMENT DISPLAY SECTION	
FIXD IMPACT RTCL	Rocket Action - Gun Action. Rocket fixed or gun fixed.
UP=?PRI LOW=?	No missiles ready.
UP = A<< PRI LOW = B	Priority channel is upper. Upper code is "A" with missile(s) ready. Lower channel missiles are ready.
UP . ? pRI >> LOW= H	No missiles ready on upper channel. Priority channel is lower and lower code is "H" with missile(s) ready.
MISSILE INVENTORY AND STATUS SECTION	

NOTE

The following displays will be presented based on the position of the TYPE switch on the missile panel and the types of missiles loaded.

L	Laser missile.
I	IRIS missile.
R	RF/IR missile.
A through H	Code of laser missile in place of the "L" indication.

NOTE

The following characters will be displayed based on the status of a missile. The single status characters are displayed below the inventory type or code characters. The multiple character status indicators utilize both positions.

s	Selected.
R (steady)	Ready.
R (flashing)	LOAL priority missile, next missile to be launched.
T (steady)	Missile seeker in track mode.
T (flashing)	LOBL priority missile, next missile to be launched.
M U	Missile is unlatched on launcher.
M F	Missile has failed BIT or has been detected as failed subsequent to BIT.
S F	Missile launch station detected as failed.

Table 4-20. Alpha/Numeric Display Message Location and Description - continued

Message	Description
T F	Pylon MRTU cannot communicate with selected missile. Once detected for one missile on a particular launcher, all missiles present on that launcher will reflect this status This status may be cleared by moding the missile system to STBY and then moding again to desired mode. The remote Hellfire electronics will attempt to communicate with the missiles on that launcher.
N A	Missile status has been determined to be: Not Available, Low Coolant or Hangfire in process. Missiles that are: Selected, Ready, or Tracking on the same side of the aircraft as a missile that is hangfiring will reflect a 'Not Available' status for six seconds (0:06) from onset of hangfire. After six seconds the missiles will revert to their previous status.
M A	Missile launch sequence has aborted or missile has misfired.

WARNING

This is a valid indication of misfired ordnance. Appropriate safety and ordnance disposal procedures shall be accomplished.

M H	Missile is in the process of hangfiring or has hangfired.

WARNING

This is a valid indication of malfunctioning ordnance. Appropriate emergency procedures shall be accomplished.

Launcher Status

Message	Description
FF AA II LL	Launcher has failed BIT or the serial/digital data link between the pylon MRTU and the launcher has failed.
SS AA FF EE	Launcher ARM/SAFE switch is in the SAFE position and may be armed by taking the aircraft ARM/SAFE to ARM and back to SAFE.

TM 1-1520-238-10

4.30 CBHK DATA VALIDATION.

NOTE

- If the fire control computer battery fails, boresight correctors will not be maintained ▪45▪.

- Non-volatile parameter storage is accomplished by means of Electrically-Eraseable Programmable Read-Only Memory (EEPROM) ▪49A▪ / ▪51▪.

The fire control computer (FCC) continually monitors certain critical internal values derived from the Captive Boresight Harmonization Kit (CBHK) boresight correctors. These critical values are used in the correction or adjustment of line-of-sight and weapons aiming. In the event of any alteration of these internal FCC values, the crewmembers will be notified of the affected sight or weapons subsystem by a message in the HAD and FD/LS continuous monitoring.

The appropriate message in the HAD will be displayed whenever the affected sight or weapon is selected. The FD/LS messages will be announced as part of the continuous tests described in paragraph 4.15, Fault Detection/Location System. The effect on the selected sight or weapon will be that no correction to the line-of-sight or weapon aiming based on the CBHK correctors will be made. The sight or weapon will function normally but without apparent boresight correction or alignment.

Validate the CBHK boresight correction values for the affected subsystem prior to using the subsystem.

To check the FCC boresight corrector numbers, proceed as follows:

1. Place the DEK mode select switch to **FD/LS** ▪45▪ / ▪49A▪ or press CDU **FD/LS FAB** ▪51▪. Enter the letter **B**. The boresight menu page will be displayed as follows:

TADS	AL-20	VF-21	ED-22
GUN	AL-23	VF-24	ED-25
PYLN	AL-26	VF-27	ED-28
PNVS	AL-29	VF-30	ED-31

The edit column (ED) is located to the right of the align (AL) and verify (VF) columns and is the only column the pilot will reference. The alignment, verification and edit numbers are established by armament personnel using the CBHK during aircraft systems boresight.

2. To check the boresight numbers in the FCC against the boresight numbers on the boresight page in the aircraft logbook, the aircraft must be on the ground with the FCC operational. To check the TADS, GUN, PYLN, or PNVS boresight numbers, enter **22, 25, 28** or **31** respectively. The selected system boresight edit numbers will be displayed in lieu of the menu page. For example, to check the PNVS boresight numbers, enter **31**. The boresight menu will blank and the PNVS edit page will be displayed as follows:

PNVS	CORRECTORS	-MR-
EL +09.3	AZ -15.6	

3. If the FCC numbers displayed match the logbook numbers, proceed to another system by entering a **B**, which recalls the boresight menu, and enter the next appropriate number for the desired system. The PYLN (ED-28) data is displayed as one pylon number per page across four pages. To move from one pylon to the next sequentially, press the DEK **SPACE** key ▪45▪ / ▪49A▪ or press CDU **SPC KEY** ▪51▪ and the next pylon in sequence (1-4 repeated) will be displayed.

4. If the FCC numbers displayed are not the same as the logbook numbers, the pilot must change the values to match the logbook values. The protocol used to access the edit numbers assigns an integer (1, 2, 3, 4, . .) to data locations on the page following the rule: left to right, top to bottom. The top left data location is addressed by entering a **1**. The next data location to the right is addressed by entering a **2**, etc. When a data location is addressed by the appropriate integer entry, the flashing cursor will move from the top left corner to the first digit in the data location. For example, using the PNVS page from above, if the elevation (EL) numbers were correct but the azimuth (AZ) numbers were not the same as the logbook values, the pilot would access the azimuth location by entering a **2**. The flashing cursor would relocate beneath the 1 in -15.6.

5. Type in the logbook values. If the plus (+) or minus (-) sign is not correct, use the **DEK BKSP** key ▪45▪ / ▪49A▪ or CDU left arrow key ▪51▪ to move the cursor beneath the sign and make the appropriate correction to match the logbook values.

4-60 Change 4

4.31 ARMAMENT PREFLIGHT PROCEDURES.

NOTE
Checks herein are only applicable if the armament is installed and are in addition to those listed in Chapter 8. Except for safety, Chapter 4 does not duplicate Chapter 8 checks.

4.31.1 FLIR Operational Check.

1. **SIGHT SEL** switch **NVS**, check turret function.

2. Adjust gain and level for optimum image.

3. Verify capability to select the various modes of flight symbology.

4. **FLIR PLRT** pushbutton Check polarity reversal.

5. Registration Check.

    a. Align crewmembers LOS forward to the 12 o'clock position ± 5°.

    b. Select a reference object approximately 90 feet in front of the aircraft.

    c. Check registration (alignment differential) between the thermal image and reference object in azimuth.

    d. The allowable registration error is 1 foot at 90 feet. The center open position of the LOS reticle is equivalent to 1 foot at 90 feet.

    e. If registration is out of tolerance refer to TM 9-4935-476-13.

6. Alternate sensor Check.

4.31.2 IHADSS Boresight - Pilot.

OFFSET boresight is not authorized.

1. **SIGHT SEL** switch **HMD**.
2. **VID SEL** switch **GRAY SC**.
3. **BRT** and **CONTRAS**T control Adjust.
4. **VID SEL** switch **PLT**.
5. **SYM BRT** control Adjust.
6. **IHADSS BRSIT** switch On.
7. **INST** light control Adjust BRU intensity.
8. Align HMD reticle with BRU.
9. **BRSIT HMD** switch Actuate then release.
10. **IHADSS BRSIT** switch **OFF**.
11. **INST LT** control As desired.

4.31.3 IHADSS Boresight CPG.

OFFSET Boresight not authorized.

1. **SIGHT SEL** switch **HMD** or **HMD TADS**.
2. **GS** switch - Press.
3. Adjust brightness and contrast.
4. **VID SEL** switch **TADS**.
5. **SYM BRT** control Adjust.
6. **IHADSS BRSIT** switch On.
7. **INTR LT** control As Desired.
8. Align HMD reticle with BRU.
9. **BRSIT HMD** switch Actuate, then release.
10. **IEADSS BRSIT** switch **OFF**.
11. **INTR LT** control As Desired.

4.31.4 TADS Operational Checks - CPG.

> **CAUTION**
>
> The FLIR zoom FOV shall not be activated for more than 2 minutes of continuous operation to prevent excessive raster retention. Permanent damage may occur if it is engaged for more than 5 minutes continuously.

1. **TADS** switch - **TADS or FLIR OFF** - (Announce to pilot).

2. **GS** switch - Press.

3. Heads-up and heads-down displays. Adjust brightness and contrast as desired.

4. **VID SEL** switch - **TADS**.

5. **SIGHT SEL** switch - **TADS**.

6. Sensor select switch - **TV**.

7. **SLAVE** pushbutton - Press. Turret should return to manual track.

8. Thumbforce controller - Exercise turret.

9. Field-of-view select switch - Evaluate various FOVs.

10. **IAT/MAN** pushbutton - Engage image auto tracker and check for proper function.

11. Sensor select switch - **DVO**.

12. Field-of-view select switch - Evaluate various FOVs. If scene is not vertical in both FOVs, perform Pechan Alignment, paragraph 4.31.5.

13. Sensor select switch - **FLIR**.

14. Adjust gain and level for optimum image, then engage ACM, as desired.

15. Field-of-view select switch - Evaluate various FOVs.

16. **IAT/MAN** pushbutton - Engage image auto tracker and check for proper performance.

17. Polarity reversal - Check, leave in white hot.

18. Sensor select switch - **TV** or **FLIR**.

19. Drift null - If Auto Drift Null does not reduce drift to desired level, perform Manual Servo Drift Null, paragraph 4.31.6.

4.31.5 TADS Pechan Alignment.

NOTE

This procedure aligns the voltages used to erect the Pechan (DVO) lens. It should be used in order to make the scene appear upright when viewing through the DVO. Should alignment be unsuccessful, refer to TM 9-1270-476-20.

1. Sensor select switch - **DVO**.

2. Field-of-view select switch - **N**.

3. TADS turret - Manual control.

4. BRSIT ENABLE switch - UP.

5. Left thumbwheel control - Adjust as required to align horizontal crosshair with horizon.

6. BRSIT ENABLE switch - CENTER.

7. Field-of-view select switch - **W**.

8. Recheck alignment of crosshair and horizon.

    a. If satisfactorily aligned procedure is complete. Select desired operating mode.

    b. If not satisfactorily aligned:

    (1) BRSIT ENABLE switch - UP.

    (2) Adjust right thumbwheel control as required to align horizontal crosshair with horizon.

    (3) BRSIT ENABLE switch - CENTER.

    (4) Field-of-view select switch - **N**.

    (5) Reverify alignment of NFOV. If satisfactory, procedure is complete. If unsatisfactory, repeat steps 4 thru 8 above as required. If still unsatisfactory, see note above.

4.31.6 TADS Manual Servo Drift Null.

1. **SLAVE** switch - Press and verify dotted crosshair is present on selected display (HOD or HDD).

2. **SIGHT SEL** switch - **TADS**.

3. Verify TADS turret slews to fixed forward position by observing solid crosshair coming to center of display and cueing LOS centering on cueing dot.

4. **SLAVE** Switch - Press.

5. Sensor select switch - **TV** or **FLIR**. For greatest accuracy, use **TV**.

6. Field-of-view select switch - **N**.

7. Aim reticle at an easily observed object.

8. BRSIT ENABLE switch - UP, wait 5 seconds then adjust left and right thumbwheel controls for drift null.

9. BRSIT ENABLE switch - CENTER.

4.31.7 TADS Boresight.

WARNING

Ensure proper laser safety procedures are followed.

NOTE

Boresighting of the DTV and FLIR narrow and zoom fields-of-view occurs independently; the actual order is not critical. However, DTV must be boresighted prior to boresighting the DVO.

a. CUE Update Procedure.

NOTE

The CUE update procedure should be accomplished whenever the TADS FLIR is reported as being unable to be internally boresighted. Once the CUE update procedure has been accomplished and the TADS FLIR still cannot be boresighted, continue with troubleshooting as specified in TM 9-1270-476-20.

1. **SIGHT SEL** switch - **TADS**.

2. Sensor select switch - **TV**.

3. Field-of-view select switch - **W**.

4. **BRSIT** switch - **TADS**.

5. **SLAVE** switch - Actuate. Verify internal boresight message replaced by cue update message.

6. If reticle does not appear centered on the black cross, execute the following:

   a. BRSIT ENABLE switch - UP.

   b. Use thumbforce controller to position DTV reticle in proximity to the black cross.

   c. BRSIT ENABLE switch - CENTER.

   d. **SLAVE** switch - Actuate. Verify cue update message replaced by internal boresight message.

   e. **BRSIT TADS** switch - **OFF**. Procedure complete, continue as desired.

b. TADS Internal Boresight.

1. DTV

   a. **PLT/GND ORIDE** switch - As required.

   b. **CPG ARM/SAFE** switch - **ARM**.

   c. **SIGHT SEL** switch - **TADS**.

   d. **BRSIT** switch - **TADS**. Confirm **INTERNAL BORESIGHT** message is present.

   e. **LSR** switch - **ON**.

   f. Sensor select switch - **TV**.

   g. Field-of-view select switch - **N**.

   h. Tracker polarity white over black.

   i. Laser trigger - Press and hold.

   j. BRSIT ENABLE switch - UP. Observe tracking gates capture laser spot. Continue to fire laser until tracking gates disappear. If spot cannot be captured by tracking gates, perform manual boresight adjustment, paragraph 4.31.7 d.

   k. BRSIT ENABLE switch - CENTER.

l. Laser trigger - Release.

CAUTION

The DTV zoom FOV shall not be activated for more than 2 minutes of continuous operation to prevent excessive raster retention. Permanent damage may occur if it is engaged for more than 5 minutes continuously.

　　m. Field-of-view select switch - **Z**.

　　n. Repeat steps i. thru l.

2. FLIR.

　　a. Sensor select switch - **FLIR**.

　　b. Field-of-view select switch - **N**.

　　c. FLIR - Adjust level fully counterclockwise.

　　d. FLIR - Adjust gain mid-range.

　　e. Laser trigger - Press and hold.

　　f. Observe laser spot, and optimize FLIR. If laser spot is not visible, perform CUE Update Procedure, paragraph 4.31.7 a.

　　g. BRSIT ENABLE switch - UP. Observe tracking gates capture laser spot. Continue to fire the laser until tracking gates disappear. If spot cannot be captured by tracking gates, perform manual boresight adjustment, paragraph 4.31.7 d.

　　h. BRSIT ENABLE switch - CENTER.

　　i. Laser trigger - Release.

CAUTION

The FLIR zoom FOV shall not be activated for more than 2 minutes of continuous operation to prevent excessive raster retention. Permanent damage may occur if it is engaged for more than 5 minutes continuously.

　　j. Field-of-view select switch - **Z**.

　　k. Repeat steps e. thru i.

　　l. LSR switch - **OFF**.

　　m. CPG ARM/SAFE switch - **SAFE**.

　　n. PLT/GND ORIDE switch - **OFF**.

3. DVO

　　a. Sensor select switch - **DVO**.

　　b. Field-of-view select switch - **N**.

　　c. Observe position of DVO crosshairs; if coincident with DTV reticle, go to step g, below.

　　d. BRSIT ENABLE switch - DOWN.

　　e. DVO BRSIT - Adjust DVO crosshairs into coincidence with DTV reticle.

　　f. BRSIT ENABLE switch - CENTER.

　　g. **BRSIT TADS - OFF**

　　h. **ACQ SEL** switch - **FXD**.

　　i. **SLAVE** switch - Press, TADS returns to the fixed forward position.

c. TADS Outfront Boresight.

WARNING

TADS Outfront Boresight validation and adjustment (if necessary) shall be performed prior to using the TADS FLIR imagery for laser or weapons operations, after performing a cue update, a TADS component change or if the helicopter experiences an abnormal electrical shutdown.

1. Position helicopter on the ground over identified location for this procedure.

2. If on an approved laser firing range, fire laser to obtain range. Otherwise, enter manually the range from helicopter to the outfront boresight target.

NOTE

If the target is between 0.5 km and 1.5 km, the range entered on the DEK must be within 10 meters. If the target is between 1.5 km and 5 km, the range entered on the DEK must be within 50 meters. If the target is greater than 5 km, the range entered on the DEK must be within 1 km.

3. Observe the outfront boresight target in NFOV FLIR, adjust gain and level for optimum autotracker image. Engage LAT.

4. Sensor select switch - **TV,** observe light source.

5. If not precisely centered on the TADS reticle, perform the following:

   a. **BRSIT TADS** switch - **TADS.** Confirm message, **OUTFRONT BORESIGHT** is present.

   b. BRSIT ENABLE switch - UP.

   c. Adjust right and left thumbwheel controls to center the target on the reticle.

   d. BRSIT ENABLE switch - CENTER.

6. **IAT** switch - Off.

7. **BRSIT TADS** switch - **OFF**

d. TADS Manual Boresight Adjust.

NOTE

The manual boresight procedure is used only to recapture or center the laser spot. It is not an acceptable boresight procedure for normal flight operations.

1. **BRSIT TADS** switch - **TADS.** Confirm message **INTERNAL BORESIGHT** is present.

2. **LSR** select switch - **ON.**

3. Sensor select switch - **FLIR** or **TV** as required.

4. **FLIR PLRT** pushbutton - Press to select white hot.

5. Field-of-view select switch - **N.**

6. **GAIN** - Midrange (if using FLIR).

7. **LVL** - Fully counterclockwise (if using FLIR).

WARNING

Ensure proper laser safety procedures are followed.

8. Laser trigger - Press.

NOTE

If the laser spot is too weak or not visible, perform Cue update.

9. Observe laser spot and optimize FLIR.

10. BRSIT ENABLE switch - UP for 1 second, then move to center position. After 5 seconds, the left and right thumbwheel controls will be active; adjust laser spot in elevation and azimuth.

11. Laser trigger - Release.

12. **LSR** switch - **OFF.**

4.32 RAPID REARMING.

1. **TAIL WHEEL** switch - **LOCK.**

2. **PARK BRAKE** - Set.

3. **HARS** switch - **NORM.**

4. Weapons select switches - On.

5. **PLT/GND ORIDE** switch - **ORIDE.**

6. **CPG ARM/SAFE** switch - **ARM.**

7. **ROCKETS** control panel - Arm.

8. **ANTI-COL** switch - **OFF.**

9. Stray current check - Perform.

10. Weapons select switches - **OFF**

11. **CPG ARM/SAFE** switch - **OFF**

12. **PLT/GND ORIDE** switch - **OFF.**

13. Armament and pylon safety pins - Installed.

14. Rearming - Monitor.

15. **ANTI-COL** switch - As Desired.

16. Armament and pylon safety pins - Removed.

4.33 ARMAMENT INFLIGHT PROCEDURES.

4.33.1 30mm Gun System.

WARNING

If 300 or more rounds have been fired in the preceding ten minutes, and a stoppage occurs, personnel must remain clear of the aircraft for 30 minutes. Aircraft crewmembers should remain in the aircraft and continue positive gun control.

1. **CPG ARM/SAFE** switches - **ARM**.

2. **GUN** select switch - As desired.

3. Crewmember desiring to fire - Establish range to target and track target with selected sight.

4. WAS switch - **GUN**.

5. Weapons trigger - Press, continue to fire as required.

6. WAS - Deselect **GUN**.

4.33.2 AWS Harmonization Procedures 49A / 51.

1. Locate target approximately 1000 meters (± 50) from aircraft position.

2. Hover aircraft 100 FT (± 20) above target altitude and maintain heading (± 5°) azimuth.

3. Sensor select switch - **TV**

4. Field-of-view select switch - **N** or **Z**.

5. **LSR** select switch - **ON**.

6. Laser trigger - Press. Ensure the displayed range to target is accurate to ± 10 meters.

7. Laser trigger - Press. Confirm range.

NOTE

Harmonization burst shall be accomplished in the narrowest field of view which the CPG can observe the rounds impact (wide/narrow).

8. Field-of-view select switch - **W** or **N**.

NOTE

A video recording may be used to verify impact area.

9. Fire one or two 10 round bursts. Note centroid of impact area.

10. Determine the appropriate sector from the appropriate field of view Corrector Guides (figs 4-19.1 and 4-19.2) for harmonization offset values.

NOTE

• A fraction of the offset values in a sector can be used to improve accuracy

• Go to step 12 51 to enter modified correctors.

11. For 49A enter modified correctors:

  a. DEK switch - **SP1**.

  b. Enter **G** on the DEK.

  c. The screen will display harmonization corrector Deltas and Totals in the following format:
  AWS HARMONIZATION
  DELTAS -MR-
  AZ=+00.0   EL=+00.0
  TOTALS -MR-
  AZ=+00.0   EL=+00.0

  d. On the DEK enter 1 to access azimuth or 2 to access elevation corrector value(s) to be changed.

  e. Enter corrector Delta value(s) from the Corrector Guides (figs 4-19.1 and 4-19.2).

  f. Repeat steps 1 through 11 while reducing field of view as appropriate to DTV-NFOV until center of impact is verified to be within zoom FOV gates. When FOV gates are achieved, proceed to step g.

  g. DEK **ENTER** button - Press. This will complete the modification.

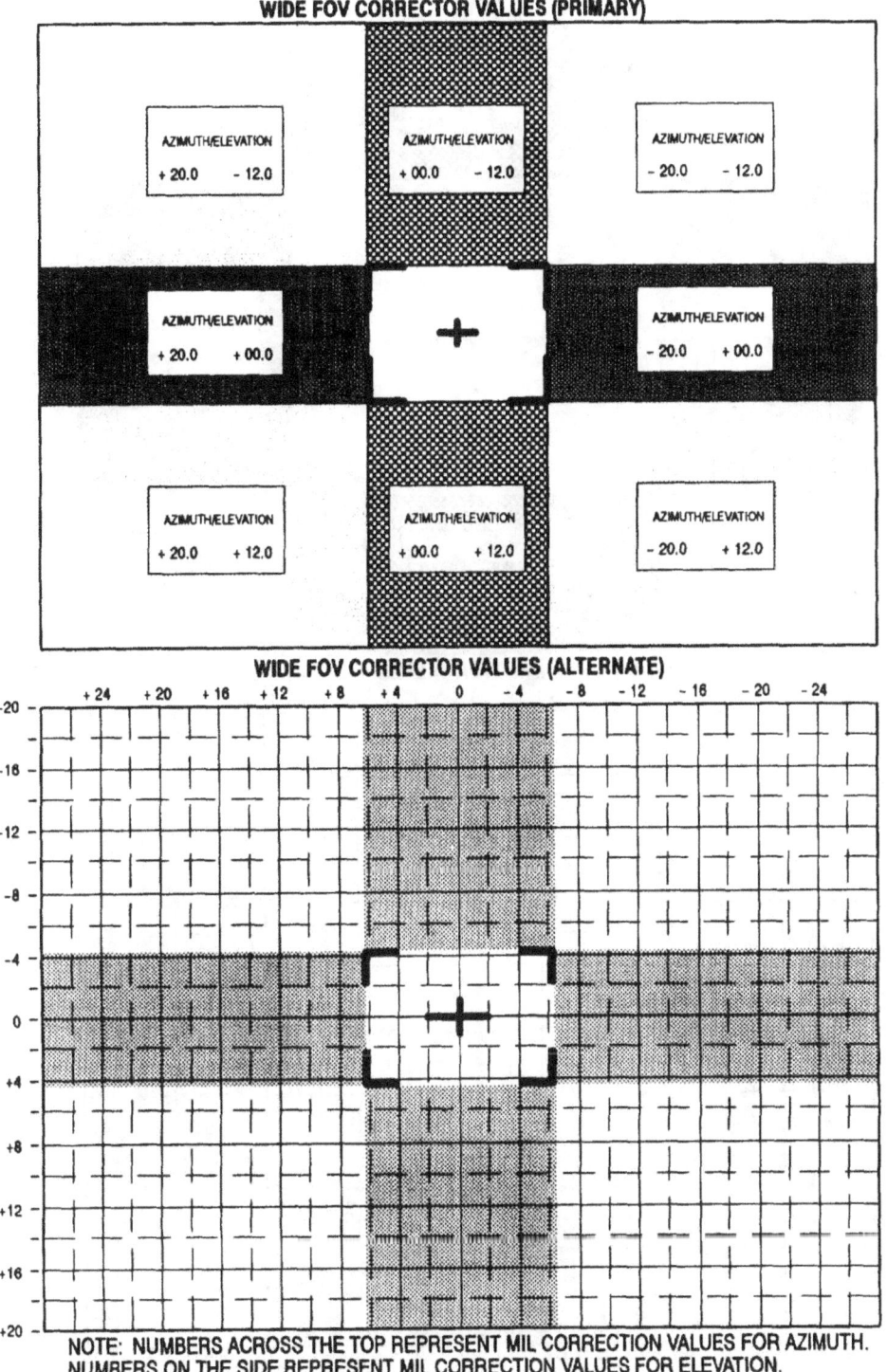

Figure 4-19.1. AWS Harmonization Wide FOV Corrector Guide 49A / 51

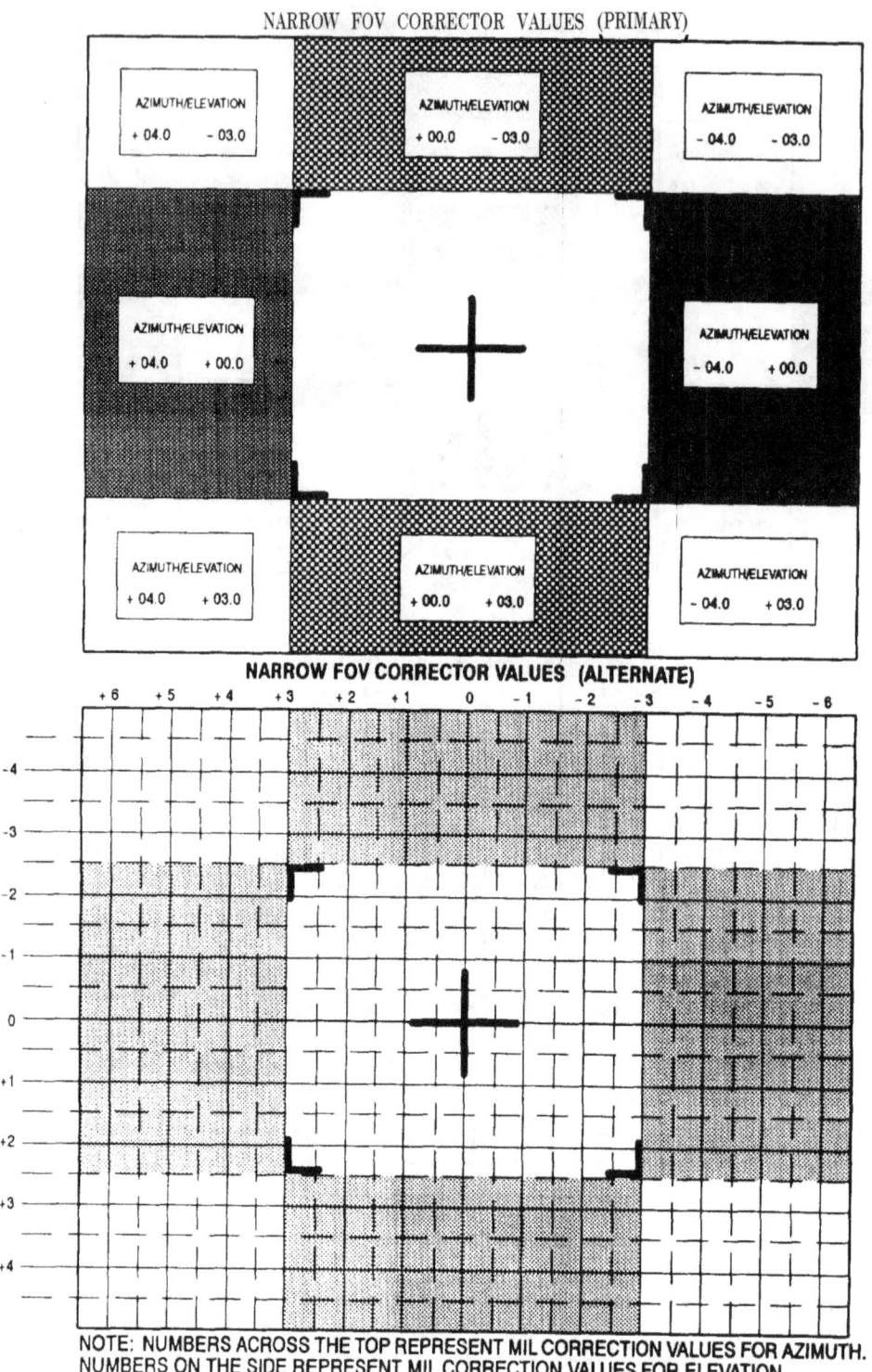

Figure 4-19.2. AWS Harmonization Narrow FOV Corrector Guide

12. For ▮51▮ enter modified correctors:

    a. CDU - Select **PGM**.

    b. CDU - Select **AWS HARM**.

    c. Enter appropriate correction data **(AZ or EL)** on CDU scratchpad.

    d. Depress VAB 2 **(AZ)** or VAB 6 **(EL)**, as appropriate, to enter correction factors into FCC.

    e. The CDU screen will display harmonization corrector Deltas and Totals in the following format:

    ```
    AWS HARMONIZATION
        DELTAS -MR-
    AZ=+00.0        EL=+00.0
        TOTALS -MR-
    AZ=+00.0        EL=+00.0
    ```

    f. Repeat steps 1 through 10 and 12 while reducing field of view as appropriate to DTV-NFOV until center of impact is verified to be within zoom FOV gates. When FOV gates are achieved, proceed to step g.

    g. Press **PGM** VAB to return to top level page.

4.33.3 Aerial Rocket Control System.

NOTE

If one or more pylons have failed, rocket firing from the failed pylon is disabled without inhibiting rocket firing from the remaining pylons.

a. CPG Cooperative Engagement Mode.

1. **CPG ARM/SAFE** switch - **ARM**.

2. **RKT** select switch - As desired.

3. Desired target - Acquire and track.

4. CPG ORT WAS - **RKT**.

5. Laser trigger - Press. Establish range of helicopter to target.

6. On completion of engagement: ORT WAS - Deselect **RKT**

b. Pilot Cooperative Engagement Mode.

1. **MASTER ARM/SAFE** switch - **ARM**.

2. **RKT** select switch - As desired.

3. **ROCKETS** panel - Set **QTY, PEN-M** and **ZONE SEL**.

4. WAS - **RKT**.

5. Align rocket steering symbology.

6. Weapons trigger - Press until selected quantity has been fired or target neutralized.

7. WAS - Deselect **RKT**.

c. Pilot or CPG Only Mode.

1. **CPG ARM/SAFE** switches - **ARM**.

2. **RKT** select switch - As desired.

3. **ROCKETS** control panel - Set **QTY, PEN-M** and **ZONE SEL**.

4. Crewmember desiring to fire - Track target with HMD, maintain LOS reticle on target. Establish range.

5. WAS - **RKT**.

6. Align rocket steering symbology.

7. Weapons trigger - Press until selected quantity has been fired or target neutralized.

8. WAS - Deselect **RKT**.

4.33.4 Point Target Weapons System.

WARNING

When firing Hellfire missile AGM-114A at temperatures below -10 °C, severe ice fog can be formed by the missile exhaust plume. The ice fog can obscure crew visibility and also interfere with autonomous designation.

1. **MSL** switch - **ON**, observe **BIT IN PROGRESS** message on high action display and AND. When message disappears any failed BIT status will be displayed on AND.

NOTE

- If BIT failure indicates MSL failure, and that missile is required to complete the mission, another BIT cycle may be initiated. If additional BIT indicates a go, then missile is considered functional and may be launched. If missile fails second Built-In-Test (BIT) then missiles shall be rejected.

- BIT may be overridden at any time by placing the **MSL MODE** switch in any mode other than **STBY**.

a. Types of Engagements.

1. **Normal Mode - LOBL.**

    a. **UPR/LWR CHAN CODE** - As desired.

    b. **UPR/LWR CHAN QTY** - As desired.

    c. **MSL MODE - NORM.**

    d. **CHAN SEL** switch - Establish priority channel.

    e. Observe AND for missile selection, coding and ready status.

    f. **CPG ARM/SAFE** switch - **ARM**.

    g. Lase target or call for remote designator.

    h. Observe AND or HIGH ACTION display for proper missile track status.

    i. CPG WAS - **MSL**.

    j. Pilot establish helicopter in constraints.

    k. Weapons trigger - Press and release.

NOTE

If 'in constraints' criteria are not met, the 2nd detent of the weapons trigger may be used to override constraints inhibits and fire missile.

   l. If autonomously designating, continue lasing until missile impact.

2. **Normal Mode - LOAL.**

    a. **UPR/LWR CHAN CODE** - As desired.

    b. **UPR/LWR CHAN QTY** - As desired.

    c. **MSL MODE - NORM.**

    d. **CHAN SEL** switch - Establish priority channel.

    e. Observe AND or HIGH ACTION display for missile selection, coding and ready status.

    f. **LOAL** select switch - As desired.

    g. **CPG ARM/SAFE - ARM**.

    h. CPG WAS - **MSL**.

    i. Pilot establish helicopter in constraints.

    j. Weapons trigger - Press and release.

    k. Lase target or call for remote designator in adequate time for terminal guidance.

3. **Ripple Mode - LOBL.**

    a. **UPR/LWR CHAN CODE** - As desired.

    b. **UPR/LWR CHAN QTY** - As desired.

    c. **MSL MODE - RIPL.**

    d. **CHAN SEL** switch - Establish priority channel.

    e. Observe AND for missile selection, coding and ready status.

f. **CPG ARM/SAFE - ARM.**

g. Lase target or call for remote designator.

h. Observe AND or HIGH ACTION display for missile system messages.

i. **CPG WAS - MSL.**

j. Pilot establish the helicopter in constraints.

k. Weapons trigger - Press and release.

l. Pilot establish helicopter in constraints.

m. Repeat steps j. thru l. until desired number of missiles have been fired.

4. **Ripple Mode - LOAL.**

   a. **UPR/LWR CHAN CODE** - As desired.

   b. **UPR/LWR CHAN QTY** - As desired.

   c. **MSL MODE - RIPL.**

   d. **CHAN SEL** switch - Establish priority channel.

   e. Observe AND or HIGH ACTION display for missile selection, coding, ready status.

   f. **LOAL** select switch - As desired.

   g. **CPG ARM/SAFE - ARM.**

   h. CPG WAS - **MSL.**

   i. Pilot establish helicopter in constraints.

   j. Weapons trigger - Press and release.

   k. Pilot establish helicopter in constraints for alternate channel.

l. Weapons trigger - Press and release.

m. Lase target or call for terminal guidance.

n. Execute steps i. thru m. until desired number of missiles have been fired.

b. **Missile System Shutdown Procedures.**

  1. **MSL MODE - STBY.**

  2. **LOAL** select switch - **OFF.**

  3. **CHAN SEL** - Actuate either direction.

  4. **MSL** switch - **OFF.**

4.34 ARMAMENT POSTFLIGHT PROCEDURES.

If armament system has been used, check as follows:

1. Right FAB - Open. Check carrier drive for rounds.

2. Gun chute assembly - Check for rounds.

3. Bolt status indicator - FEED (green range).

4. Transfer door - Open. Check bolt is to rear, no rounds in transfer assembly, and chamber clear.

5. Transfer door - Secure.

6. AWS - General condition and security. Check for leaks and proper piston index groove indication.

7.

7. Wing stores pylon - Check for unexpended ordnance.

Section III. ACTIVE AND PASSIVE DEFENSE EQUIPMENT

4.34A INFRARED COUNTERMEASURES SET AN/ALQ-144A, -144(V)3.

WARNING

Do not continuously look at the infrared countermeasure transmitter (fig 2-2) during operation, or for a period of over 1 minute from a distance of less than 3 feet. Skin exposure to countermeasure radiation for longer than 10 seconds at a distance less than 4 inches shall be avoided.

The infrared countermeasures set (IR jammer) provides infrared countermeasure capability. It transmits radiation modulated mechanically at high and low frequencies using an electrically heated source. The IR jammer consists of a control panel (fig 4-20), on the right side of the pilot instrument panel, and a transmitter, located on the fairing immediately aft of the main rotor (fig 2-2). A built-in test monitors operation and alerts the pilot when a malfunction occurs. If the IR jammer malfunctions, the **IR JAM** segment on the pilot caution/warning panel illuminates. The IR jammer is not monitored by FD/LS. The IR transmitter receives 28 vdc from the No. 1 essential dc bus through the **IR JAMMER** circuit breaker in the electical power center when the **IR JAM/PWR** circuit breaker is closed. The IR jammer cooling fan receives 28 vdc from the No. 2 essential dc bus through the **IR JAM/XMTR** circuit breaker. The **IR JAM/PWR** and **IR JAM/XMTR** circuit breakers are located on the pilot overhead circuit breaker panel.

4.34A.1 Infrared Countermeasures Set Control Panel. Control of the IR jammer is provided by the IRCM (infrared countermeasures) part of the **IRCM/RDRCM** control panel (fig 4-20) located on the right side of the pilot instrument panel. The control panel receives 28 vdc from the No. 2 essential dc bus through the **PEN AIDS CONTR** circuit breaker on the pilot overhead circuit breaker panel. When the **ON-OFF** switch on the **IRCM** panel is set to **ON**, the power distribution and control circuits are activated. The source begins to heat, and the servo motor and drive circuits are energized, turning on the high and low speed modulators. A signal is applied to stabilize subsystem operations before energizing the built-in test function. After a one minute warmup period, the stabilizing signal is removed and the subsystem operates normally. Setting the **ON-OFF** switch to **OFF** causes the power distribution and control circuits to de-energize the source and initiates a cool-down period. During the one minute cool-down period, the servo motor drive circuits remain in operation, applying power to the motors that cause the modulators to continue turning and the **IR JAM** segment on the pilot caution/warning panel to illuminate. After the cool-down period, the power distribution and control circuits de-energize, all subsystem voltages are removed and the **IR JAM** caution light segment is extinguished.

4.34B RADAR COUNTERMEASURES SET AN/ALQ-136.

WARNING

Do not stand within 3 feet of the front of the transmit antenna when the equipment is turned on. High frequency electromagnetic radiation can cause internel burns without causing any sensation of heat. If you feel the slightest warming effect while near the transmit antenna, move away quickly.

The radar countermeasures set (radar jammer) provides the helicopter with protection against ground based fire control radars. When operating, the radar jammer transmits modulating signals at radar frequencies, causing range and angle measurement errors to the radar. The radar jammer consists of a control panel (fig 4-20), located on the right side of the pilot instrument panel, a receiver transmitter, and transmit and receive antennas. A built-in test monitors operation and alerts the pilot when a malfunction occurs. If the radar jammer malfunctions, the **RDR JAM** segment on the pilot caution/warning panel illuminates. The radar jammer is not monitored by FD/LS. The receiver/transmitter receives 28 vdc from the No. 2 essential dc bus through the **RDR JAMMER** circuit breaker in the electrical power center when the **RDR JAM** circuit breaker is closed. The radar jammer set cooling fan receives 115 vac from the No. 1 essential ac bus through the **RDR JAM AC** circuit breaker. The **RDR JAM AC** and **RDR JAM DC** circuit breakers are located on the pilot overhead circuit breaker panel.

4.34B.1 Radar Countermeasures Set Control Panel. Control of the radar jammer is provided by the RDRCM (radar countermeasures) part of the **IRCM/RDRCM** control panel (fig 4-20) located on the right side of the pilot instrument panel. The control panel receives 23 vdc from the No. 2 essential dc bus through the **PEN AIDS CONTR** circuit breaker on the pilot overhead circuit breaker panel. When the control switch is set to **STBY**, the receiver/transmitter is placed in the warm-up mode (3 minute time period) and the **RDR JAM** segment on the pilot caution/warning panel will illuminate if it senses a failure. When the control switch is set to **ON**, the radar jammer automatically provides jamming power to the transmit antenna upon receipt of threat radar energy at the receive antenna. Setting the control switch to **OFF** removes power to the radar jammer and the **RDR JAM** caution/warning light is extinguished.

4.35 COUNTERMEASURES CONTROL PANELS.

The countermeasures control panels (fig 4-20), located on the right side of the pilot instrument panel, consist of the radar/infrared countermeasures panel, the chaff dispenser panel and the radar signal detector control panel. The flare panel is not used.

4.36 DISPENSER KIT M-130.

The general purpose dispenser M-130 (fig 4-20) consists of a dispenser control panel located on the right of the pilot instrument panel (fig 2-9), a dispenser assembly, a payload module assembly, and an electronic module to dispense M-1 Chaff. It provides effective survival countermeasures against radar guided weapons systems. The dispenser subsystem has the capability of dispensing 30 chaff cartridges. The dispenser subsystem receives 28 vdc from the Emergency dc Bus through the **CHAFF** circuit breaker on the pilot overhead circuit breaker panel. The dispenser subsystem is not monitored by FD/LS. The function of each control/indicator is described in table 4-21.

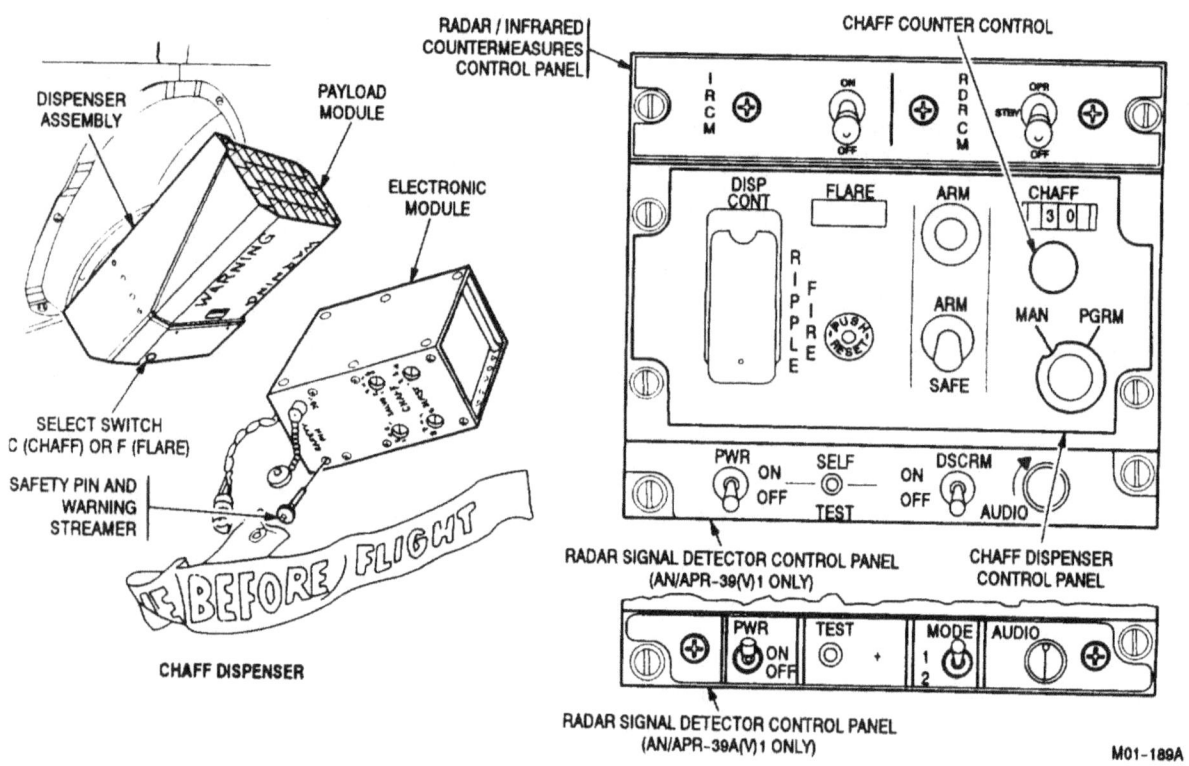

Figure 4-20. Chaff Dispenser and Countermeasures Control Panels

Table 4-21. Dispenser Control Panel Control/Indicator Function

Control/Indicator	Position	Function
CHAFF counter		Displays the number of chaff cartridges remaining in the payload module.
CHAFF counter knob		Adjusts counter to correspond to number of chaff cartridges loaded in payload module.
ARM light		Indicates that the ARM/SAFE switch is at ARM, safety flag pin is removed, and payload module is armed.
ARM/SAFE switch	ARM	Applies electrical power through safety flag switch to cyclic WAS for firing.
	SAFE	Removes power from dispenser subsystem.
MODE selector		Selects type of chaff release.
	MAN	Dispenses one chaff cartridge each time the WAS chaff fire position is toggled.
	PGRM	Allows setting electronic module controls before flight to automatically dispense chaff according to predetermined number of chaff cartridges per burst and number of salvos.

4.36.1 Dispenser Assembly. The dispenser assembly (fig 4-20) contains the C-F (chaff-flare) selector switch, chaff counter control, and a housing containing the sequencer assembly. The sequencer assembly receives 28 vdc through the WAS and furnishes pulses to each of the 30 contacts of the breech assembly, in sequential order 1 through 30, thus firing each of the chaff cartridges

4.36.2 Remote Safety Switch. The remote safety switch is located on the tailboom just forward of the dispenser assembly. This switch safes the dispenser by grounding the chaff dispenser contacts and opening the firing circuit when the safety pin is installed in the switch.

4.36.3 Payload Module Assembly. The payload module assembly (fig 4-20) consists of the payload module and the retaining plate. The payload module has 30 chambers which will accept chaff. The chaff cartridges are loaded, one per chamber, and are held in place by the retaining plate. The payload module assembly is loaded into the dispenser assembly.

4.36.4 Electronic Module. The electronic module (fig 4-20) provides signals to fire the chaff and to change the chaff counter accordingly. It contains a programmer and a cable assembly which includes a 28 volt supply receptacle. The programmer consists of a programming circuit which allows the setting of chaff burst quantity, burst interval, salvo quantity, and salvo interval. The safety switch on the electronic module is not used. The electronic module is located in the right aft avionics bay.

4.36.5 Electronic Module Controls. The switches on the electronic module (fig 4-20) are used to program the chaff dispenser for predetermined release of chaff cartridges. The function of each switch is described in table 4-22.

TM 1-1520-238-10

Table 4-22. Electronic Module Switch Functions

Switch	Function
SALVO COUNT	Programs the number of salvos: 1, 2, 4, 8, or C (continuous). A salvo is the same as one burst.
SALVO INTERVAL	Programs the time in seconds between salvos: 1, 2, 3, 4, 5, 8, or R (random timing, e.g. 2, 5,3,4,3, 2).
BURST COUNT	Programs the number of chaff cartridges that are fired in one burst: 1, 2, 3, 4, 6, or 8.
BURST INTERVAL	Programs the time interval in tenths of a second between individual chaff cartridge firings within a burst: 0.1, 0.2, 0.3, or 0.4.

4.36.6 Kit Safety Procedures. The following special safety procedures shall be followed:

WARNING

Avoid exposure to high concentrations of chaff; this can cause temporary irritation to eyes and throat.

a. Chaff and Impulse cartridges. Chaff and impulse cartridges shall be kept away from all fires and excessively high temperatures.

b. Impulse cartridges. Impulse cartridges must be handled with extreme care. Each cartridge generates extremely high gas pressure and temperature when fired.

c. Safety pin. The safety pin shall be installed in the remote safety switch whenever the helicopter is parked. Remove the safety pin prior to takeoff.

4.36.7 Kit Operation.

CAUTION

Operation is totally independent of aircraft ARM/SAFE power.

NOTE

When power is removed from the dispenser assembly, the firing order resets to position No. 1 in the payload module. If payload module is partially expended, the payload module may be rotated 180° prior to power up to ensure an unexpended chamber is in the No. 1 position.

1. Chaff counter - Set for number of chaff cartridges in payload module.

2. Mode switch - **MAN**.

CAUTION

Mode switch should always be in the MAN position prior to setting the ARM/SAFE switch to ARM to prevent immediate salvo of chaff.

3. **ARM/SAFE** switch - **ARM**; **ARM** indicator light on.

4. Mode switch - **PGRM** -if desired or required.

5. Cyclic WAS - C (6:00 position) as desired to fire chaff.

6. Stopping Procedure - **ARM/SAFE** switch - **SAFE**.

4.37 RADAR WARNING SYSTEM AN/APR-39(V)1.

The radar warning (RW) system provides visual and aural warnings of radar reception in bands generally associated with hostile fire control radar. Each radial strobe displayed on the RW display represents a line of bearing to an active radar transmission. When a radar signal represents a threat an audio signal is sent to the RW control panel and an audio alarm is sounded in the pilots helmet. The audio alarm frequency represents the relative strength of the intercepted radar signals. Also during radar signal threats, the RW display missile alert (**MA**) lamp is illuminated. The **MA** lamp flashes to represent the relative strength of the intercepted radar signals. The operating controls and indicators of the RW control panel (fig 4-20) are described in table 4-23.

4-71

Table 4-23. AN/APR-39(V)1 Radar Warning System (CRT) Control/Indicator Functions

Control/Indicator	Position	Function
PWR	ON	Turns set on. Fully operational after 1-minute warmup.
	OFF	Turns set off.
DSCRM		Selects mode of operation.
	ON	Activates discriminator circuit.
	OFF	Deactivates discriminator circuit.
SELF-TEST		When pressed, initiates self-test check (except for antenna and receiver).
AUDIO		Adjusts volume to the intercommunications system.

4.37.1 AN/APR-39(V)1 Operating Procedures. The procedure for turning on and off the RW equipment is as follows:

CAUTION

To prevent damage to the receiver detector crystals, assure that the radar warning set antennas are at least 60 meters from active ground radar antennas or 6 meters from active airborne radar antennas. Allow an extra margin for new, unusual, or high power emitters.

1. Equipment on. **PWR** switch - **ON**. Allow 1-minute for warm-up.

2. **BRIL** and filter controls - Adjust as desired.

3. **AUDIO** control - Adjust volume as desired.

4. **DSCRM** switch - Set for mission requirement.

5. Equipment off. **PWR** switch - **OFF**.

4.37.2 AN/APR-39(V)1 Radar Warning Display Controls. The operating controls of the RW display, located on the right side of the pilot instrument panel (fig 2-9 and 4-21), are described in table 4-24. The display shows a line-of-bearing radial stroke for each processed signal.

Table 4-24. AN/APR-39(V) 1 Radar Warning Display Control/Indicator Functions

Control/Indicator	Function
BRIL Control	Adjust CRT display brightness.
MA Indicator	Missile Alert. Flashes on and off to indicate time correlation between missile guidance and associated tracking radar.
Filter Control	Varies density of red polarized CRT faceplate filter (used for day or night operation) by moving tang right or left.

4.37.3 AN/APR-39(V)1 Operation Modes. The RW set may be operated in either the discriminator off or discriminator on mode. The RW set receives 28 vdc from the emergency dc bus through the **RDR WARN** circuit breaker on the pilot overhead circuit breaker panel. The RW set is not monitored by FD/LS.

NOTE

Display strobe lengths indicate only relative signal amplitude. Since many variables can affect the atmospheric attenuation of the signals, strobe length should not be considered as a direct interpretation of the distance to the emitter.

a. Discriminator Off Mode. When operating in the discriminator off mode, the **DSCRM** switch is placed **OFF**. In this mode all high band received signals with amplitude greater than a predetermined threshold level are displayed on the CRT and an audio signal, representative of the combined amplitudes and pulse repetition frequencies (PRF), is present at the headset. The displays indicate the total radar environment in which the helicopter is operating. Each radial strobe on the CRT is a line of bearing to an active emitter. When a SAM radar complex becomes a threat to the helicopter (low band signals correlated with high band signals), the alarm audio is superimposed on the PRF audio signal and the **MA** light and associated strobe start flashing. Lengths of strobes and audio levels depend on the relative strength of the intercepted signals. A typical display when operating in the discriminator off mode is shown in figure 4-21.

NOTE

In this mode, received low band signals which are not correlated with a high band intercept will cause the **MA** light to flash and an alarm audio will sound.

b. Discriminator On Mode. When operating in the discriminator on mode, the **DSCRM** switch is set to **ON**. In this mode, signals are processed to determine their conformance to certain threat associated criteria as follows:

(1) The signal level must be greater than the minimum threshold level.

(2) Pulse width must be less than the maximum pulse width.

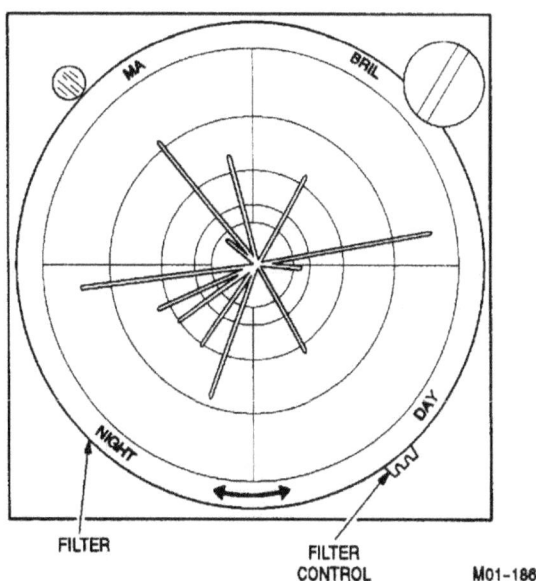

Figure 4-21. Radar Warning Discriminator Off Mode Display

(3) PRF must be greater than the minimum pulses per second.

(4) The pulse train must exist with not less than the minimum pulse train persistence.

The CRT display is divided into eight sectors. Strobes are displayed only in those sectors in which signals meeting all threat criteria are present. This reduces display clutter by eliminating low-level and wide pulse width signals and by selective sector display. Intercepts which meet these requirements are displayed as described in paragraph 4.37.3 a.

NOTE

In this mode, uncorrelated low band signals will not give any indications.

A typical display when operating in the discriminator **ON** mode is shown in figure 4-22. Conditions are the same as for figure 4-21, but it is assumed that one or more threats have been identified in the 225 to 270 degree sector.

4-73

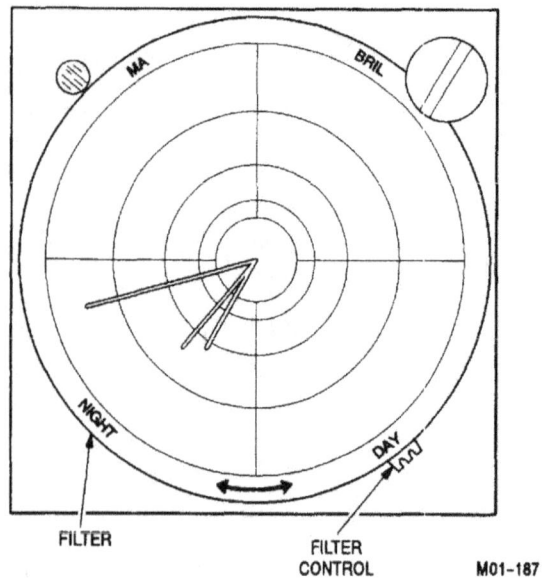

Figure 4-22. Radar Warning Discriminator On Mode Display

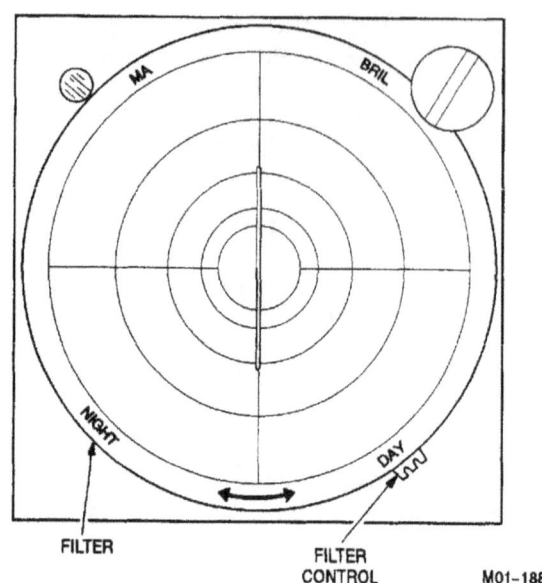

Figure 4-23. Radar Warning Self-Test Mode Display

4.37.4 AN/APR-39(V)1 Self-Test Operation. The self-test checks all RW set circuits except antennas, high-pass filters and detectors in the high band receiver, bandpass filter and detector in the low-band receiver, analysis signal commutator, and high and low-band blanking circuits. The self-test procedure is done before operation and when any malfunction is detected. A self-test display is shown in figure 4-23.

  a. Self-test procedures are as follows:

   1. **PWR** switch - **ON.** Panel lights illuminate.

   2. **DSCRM** switch - **OFF.** Wait 1-minute for warmup. Monitor display CRT and audio.

   3. **SELF-TEST** button - Press and hold.

     a. FWD and AFT strobes appear, extending to about the third circle on the display graticule, and a 2.5 kHz (approximately) PRF audio is present immediately.

     b. Within about 6 seconds, alarm audio is present and the **MA** light starts flashing.

     c. Display **BRIL** control - Turn cw and ccw. Strobes brighten (cw) and dim (ccw) as control is turned. Set **BRIL** control for desired brightness level.

     d. **AUDIO** control - Turn between max ccw and max cw. Audio will not be audible at max ccw and clearly audible at max cw.

     e. **SELF-TEST** pushbutton - Release. All indications cease.

   4. **DSCRM** switch - **ON.**

   5. **SELF-TEST** button - Press and hold.

     a. Within about 4 seconds a FWD or AFT strobe (either may appear first) and a 1.2 kHz (approximately) PRF audio will be present.

NOTE

Occasionally, during the period between pressing **SELF-TEST** and appearance of the first strobe, a distorted dot on the display and intermittent audio will be present. This is not a fault isolation.

     b. Within about 6 seconds, the other strobe will appear and the PRF audio frequency will double.

     c. Several seconds later alarm audio will be present and the MA light starts flashing.

     d. **SELF-TEST** button - Release. All indications cease.

   6. **PWR** switch - As desired.

4.38 RADAR WARNING SYSTEM AN/APR-39A(V)1.

The radar warning (RW) system provides visual and aural warnings of radar reception. This is done by receiving, processing, and displaying potential threats in the radio frequency (RF) environment. Potential threats are displayed as symbols on the RW display and announced over the intercommunication system. The emitters that it displays are derived from the emitter identification data (EID) contained in the user data module (UDM) that is inserted in the top of the digital processor. The UDM contains the electronic warfare threat data that goes to make up the specific library for a specific mission or geographical location. When a match of the electronic warfare data occurs the processor generates the appropriate threat symbology and synthetic audio. The operating controls and indicators of the RW control panel (fig 4-20) are described in table 4-25. When installed, the AN/AVR-2A(V)1 laser detecting set provides early warning against laser threats. In the event of simultaneous warnings, one from the radar warning and one from the laser detecting set, the laser warning signals have priority. Refer to paragraph 4.39.

Table 4-25. AN/APR-39A(V)1 Radar Warning System Control/Indicator Functions

Control/Indicator	Position	Function
PWR	ON	Turns set on. Listen for synthetic voice message - **APR 39 POWER UP** Plus (+) sign will appear and stabilize in center of CRT.
	OFF	Turns set off.
TEST		When momentarily depressed initiates self-test confidence check except for antennas and antenna receiver cabling.
MODE	1	Selects normal voice message format.
	2	Selects terse/abbreviated voice message format.
AUDIO		Adjusts volume to the intercommunications system.

4.38.1 AN/APR-39(V)1 Operating Procedures. The procedure for turning on and off the RW equipment is as follows:

CAUTION

- To prevent damage to the receiver detector crystals, assure that the radar warning set antennas are at least 60 yards from active ground radar antennas or 6 yards from active airborne radar antennas. Allow an extra margin for new, unusual, or high power emitters.

- Excessive indicator display brightness may damage the CRT. Set BRIL control for readable display.

1. Equipment on. **PWR** switch - **ON**. Listen for synthetic voice message - **APR 39 POWER UP**. Plus (+) sign will appear and stabilize in center of CRT.

2. **BRIL** - Adjust as desired for best indicator display of + symbol.

3. **MODE** switch - Select for desired synthetic voice message format.

4. **TEST** - Press pushbutton to start system self-test.

5. **AUDIO** control - Adjust volume as desired.

6. Equipment off. **PWR** switch - **OFF**. (Pull **PWR** switch handle out to disengage lock.)

4.38.2 AN/APR-39A(V)1 Radar Warning Display Controls. The operating controls of the RW display, located on the right side of the pilot instrument panel (fig 2-9 and 4-24), are described in table 4-26. The display shows the + symbol.

4.38.3 ANIAPR-39A(V)1 Operation Modes. The RW set may be operated in either MODE 1 or MODE 2. The RW set receives 28 vdc from the emergency dc bus through the RDR WARN circuit breaker on the pilot overhead circuit breaker panel. The RW set is not monitored by FD/LS.

 a. MODE 1 Operation. By selecting **MODE 1** normal synthetic voice messages will be heard when an emitter has been processed (ie. RW will announce; SA, **SA-8 TWELVE O'CLOCK TRACKING**). Selection of this mode does not have any effect on emitters received, processed, or displayed. It only affects synthetic voice audio.

 b. MODE 2 Operation. By selecting **MODE 2** terse or abbreviated synthetic voice messages will be heard (ie. RW will announce; **MISSILE, MISSILE TWELVE O'CLOCK TRACKING**).

Symbol generation and position relative to the center of the RW display shows threat lethality. It does not show or represent any lethality of range but of condition/mode of the emitter. Highest priority threats (most lethal) are shown nearest the center. Each symbol defines a generic threat type. The definition of what the symbols mean is classified. The complete set of symbols and definitions are contained in TM 11-5841-294-30-2.

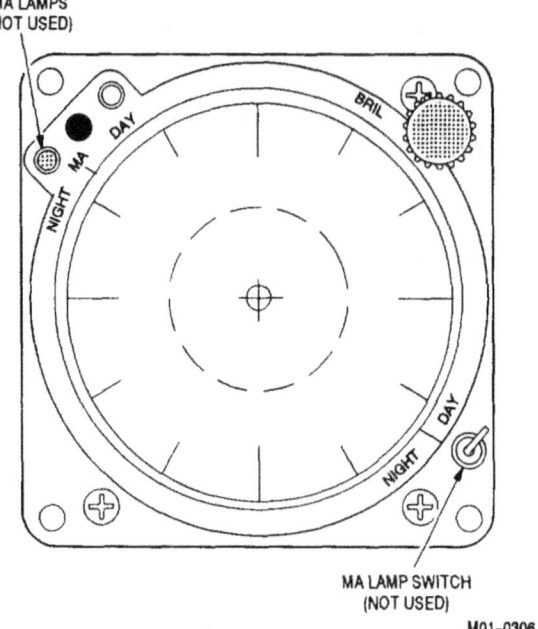

Figure 4-24. Radar Warning Display

4.38.4 AN/APR-39A(V)1 Self-Test Operation. Provides a GO/NO-GO test of system functions. Self-test can be run at anytime. The complete self-test runs in less than 30 seconds. When installed, the AN/AVR-2A(V)1 laser detecting set is tested simultaneously when the AN/APR-39A(V)1 radar warning system is tested.

 a. Self-test procedures are as follows:

  1. **PWR** switch - ON. Listen for synthetic voice message - **APR 39 POWER UP**.

  2. **MODE** switch - 1 or 2 position.

   MODE 1 provides long count for self-test.
   MODE 2 provides short count for self-test,

Table 4-26. AN/APR-39A(V)1 Radar Warning Display (CRT) Control/Indicator Functions

Control/Indicator	Function
BRIL Control	Adjust CRT display brightness.
MA Indicator	Not Used.
NIGHT/DAY Switch	Not Used.

3. **TEST** pushbutton - Press

   a. **AUDIO** control - Adjust as desired during long or short count.

   b. Listen for synthetic voice long or short count. **SELF-TEST SET VOLUME 1, 2, 3, 4, 5, 6, 7, 8, 9, 10, 11, 12.**

   c. The RW display will show specific numbers, operational flight program (OFP) at the twelve o'clock position and the emitter identification data (EID) at the six o'clock position (fig 4-25).

   NOTE

   OFP and EID numbers should be correct for mission or geographical location.

   d. The RW display will then show a RW receivers check. For a *good* self-test the RW display will show two triangles with slashes at the six and twelve o'clock positions (fig 4-26). Snowflakes will appear at the two, four, eight, and ten o'clock positions. All will flash if the laser detecting set is *not* installed and OFP is lower than 23.9. Otherwise, snowflakes will be steady. This is a normal indication and does not affect RW system performance. If the laser detecting set is installed a *good* self-test will display a steady snowflake in each quadrant. Faulty quadrants are displayed with a flashing snowflake.

   e. Listen for synthetic voice message at end of self-test. A *good* self-test ends with the message: **APR-39 OPERATIONAL.**

   f. A *bad* self-test ends with the message: **APR-39 FAILURE.**

   g. After completing system self-test, verify that the + symbol is shown at the center of the RW display. The + symbol shall be displayed anytime the system is **ON**.

4. Equipment off. **PWR** switch - **OFF**. Pull **PWR** switch handle out to disengage lock.

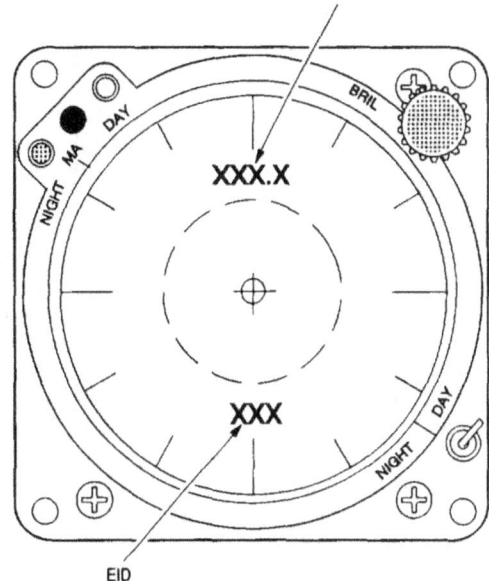

Figure 4-25. Radar Warning OFP and EID Display

Figure 4-26. Radar Warning Receiver Check Display

4.39 AN/AVR-2 (V)1 LASER DETECTING SET

4.39.1 AN/AWMA(V)1 Laser Detecting Set (LDS) System Description. The AN/AVR-2A(V) Laser Detecting Set (LDS) is a passive electronic warfare system that detects, locates, and identifies hostile laser-aided weapon threats fired from both airborne and ground-based platforms. The LDS is a frequency extension of the Radar Warning (RW) system and interfaces with the RW receivers and processor to function as an integrated Radar Laser Warning Receiver (RLWR). The system detects optical radiation illuminating the helicopter, processes this laser data into laser threat messages, and sends these messages to the RW digital processor. The digital processor processes these inputs to provide for both visual and aural threat indications for the system. The LDS can also be used with both the RW and the Air-to-Ground Engagement Systems (AGES) to provide an engagement simulation system, in the operational training mode. The system is composed of five components: four laser sensor units and an Interface Unit Comparator (IFU). The four sensor units are strategically located around the helicopter with two mounted forward, facing forward and two mounted aft, facing aft. Each sensor unit provides a 100° Field-Of-View (FOV) and +/- 45° of coverage in elevation. This configuration provides for 360° coverage in azimuth and t/-45° in elevation about the helicopter with substantial overlap. Each sensor unit contains four separate laser detectors. They are located under a special optical window and supply coverage of three different spectral regions: Electra-Optical (EO) bands I, II, and III. Two detectors are employed in the band III region, the band IIIA and band IIIB detectors, to provide the required band III detection coverage.

4.39.2 System Operation. The sensor units perform the actual laser detection function for the system and contain the necessary electronics to process detected laser signals. If a valid laser signal is detected, a threat message containing the laser type (band I, II, or III) is sent to the IFU for processing. Each sensor unit contains optical and electrical Built-In Test (BIT) electronics to perform a self-test upon command from the IFU. When a self-test command is received, the sensor unit disables detection of all externally generated signals and performs a self-test. When the self-test is completed, the appropriate pass or fail message is sent to the IFU for processing and normal operation is resumed. The IFU is located in the LH aft avionics bay. It is mounted just forward of the RW digital processor with which it directly interfaces. The IFU provides the control and timing necessary for the interface with the sensor units. It also provides the interface with the RW system. The LDS was designed to operate in conjunction with the RW system, therefore, being an integral part of the RW system. The IFU provides the majority of the wiring interface between the RW system and the associated helicopter systems. If the IFU is removed from the helicopter, a jumper box must be installed in the system or an alternate connector configuration employed to permit the RW system to operate. The LDS employs a removable User Data Module (UDM) which is mounted in the face of the IFU. The UDM contains the classified operational software required for tactical operation of the system. This software gets downloaded into volatile memory within the sensor units during system power-up and initialization, and the sensor units then become classified. When system power is removed, the sensor units zeroize the classified software and become unclassified components. The removal of system power and the UDM for the IFU effectively declassifies the system. The LDS has the capability to operate in two modes, training and tactical. In the training mode, the system operates with AGES in the Multiple Integrated Laser Engagement System (MILES) to provide the crewmembers with a realistic combat tactical training system that closely simulates the effect of weapon engagements.

a. Training. During training operation, the LDS operates as a detecting system in a MILES environment and the operating software within the LDS does not recognize .904 micron gallium arsenide (GaAs) MILES laser hits as actual laser threats.

b. Tactical. During tactical operation, the LDS detects, identifies, and characterizes three different types of optical signals. Each sensor unit provides laser threat detection in three different spectral bands; band I, band II, and band III. When a sensor unit detects optical, coherent radiation within its FOV, it provides band and pulse characteristics as laser threat data to the IFU. The IFU further processes this threat data, thus comparing received signal characteristics with stored parameters. It then determines the existence of a laser threat, threat type, and Angle-Of-Arrival (AOA) (quadrant resolution only). This threat data is sent as laser threat messages to the RW digital processor for manipulation to provide visual threat indications on the RW display and aural voice threat messages over the helicopter ICS. Both the visual and aural threat indications provide threat type and relative position information to the crewmembers.

c. Operating Procedures. Refer to para. 4.38.1.

CHAPTER 5
OPERATING LIMITS AND RESTRICTIONS

Section I. GENERAL

5.1 PURPOSE.

This chapter identifies, or refers to, all important operating limits and restrictions that shall be observed during ground and flight operations.

5.2 GENERAL.

The operating limitations set forth in this chapter are the direct results of design analysis, test, and operating experiences. Compliance with these limits will allow the pilot to safely perform the assigned missions and to derive maximum utility from the aircraft.

NOTE
See current Interim Statement of Airworthiness Qualification for additional limitations/restrictions.

5.3 EXCEEDED OPERATIONAL LIMITS.

Any time an operational limit is exceeded, an appropriate entry shall be made on DA Form 2408-13-1. Entry shall state what limit or limits were exceeded, range, time beyond limits, and any additional data that would aid maintenance personnel in the maintenance action that may be required.

5.4 MINIMUM CREW REQUIREMENTS.

The minimum crew requirements for flight is two, pilot and CPG. A technical observer may be authorized to occupy the CPG station during ground maintenance on a case by case basis at the discretion of the commander.

Section II. SYSTEM LIMITS

5.5 INSTRUMENT OPERATING RANGES AND MARKINGS.

5.5.1 Instrument Marking Color Code. Operating limitations and ranges (fig 5-1) are illustrated by colors on the engine, flight, and utility system instruments. RED markings indicate the limit above or below which continued operation is likely to cause damage or shorten component life. GREEN markings indicate the safe or normal range of operation. YELLOW (light or dark) markings indicate the range when special attention should be given to that operation covered by the instrument. Operation is permissible in the yellow range, but may be time limited or cautionary. Scales with green-coded, yellow-coded, or red-coded segments above green-coded segments operate in this manner; the segment will light in normal progression and remain on as the received signal level increases. Those segments will go off in normal progression as the received signal level decreases. Scales with red-colored and/or yellow-coded segments below green-coded segments operate in this manner; when the received signal level is zero or bottom scale, the segments will light in normal progression and will remain on. When the first segment above the red or yellow range goes on, all red-coded or yellow-coded segments will go off. These segments will remain off until the received signal level indicates a reading at or within the red or yellow range. At that time all red-coded or yellow-coded segments will go on and the scale display will either go on or off in normal progression, depending upon the received signal level. Blue colored segments indicate that power is on.

5.5.2 Rotor Limitations. It is not abnormal to observe a % RPM 1 and 2 speed split during autorotation descent when the engines are fully decoupled from the transmission. A speed increase from 100% reference to 104% is possible. Refer to figure 5-1 for limitations.

    a. Rotor Start and Stop Limits. Maximum wind velocity for rotor start or stop is 45 knots.

    b. Rotor Speed Limitations. Refer to figure 5-1 for rotor limitations.

TM 1-1520-238-10

701 ENGINE TURBINE GAS TEMPERATURE (TGT °C)

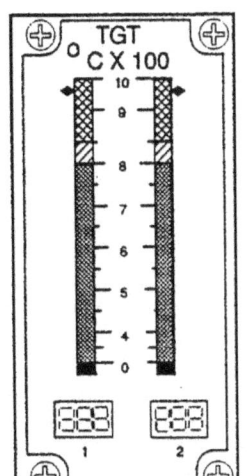

NOTE:
LIMITS BASED ON INDICATED TGT

Color		LIMITS	
RED	▨	950	MAXIMUM
RED	▨	917 – 950	TRANSIENT (12 SECONDS)
RED	▨	867 – 917	SINGLE ENGINE CONTINGENCY (2.5 MINUTE LIMIT)
RED	▨	867	AUTOMATIC DUAL ENGINE TGT LIMITING
YELLOW	▨	852	MAXIMUM DURING START
YELLOW	▨	805 – 867	IRP (30 MINUTES)
GREEN	▨	805	MCP
GREEN	▨	0 – 805	NORMAL OPERATION
BLUE	■		INSTRUMENT POWER ON

701C ENGINE TURBINE GAS TEMPERATURE (TGT °C)

Color			LIMITS	
RED	▨		950	MAXIMUM
RED	▨	☞	901 – 950	TRANSIENT (12 SECONDS)
RED	▨	☞	875 – 901	SINGLE ENGINE CONTINGENCY (2.5 MINUTE LIMIT)
RED	▨		867	AUTOMATIC DUAL ENGINE TGT LIMITING
YELLOW	▨		852	MAXIMUM DURING START
YELLOW	▨	☞	852 – 875	IRP (10 MINUTES)
YELLOW	▨		805 – 852	IRP (30 MINUTES)
GREEN	▨		805	MCP
GREEN	▨		0 – 805	NORMAL OPERATION
BLUE	■			INSTRUMENT POWER ON

LEGEND

▨	RED
▨	YELLOW
▨	GREEN
■	BLUE

ENGINE GAS GENERATOR SPEED (N_G) (RPM%)

Color			LIMITS	
RED	▨	(UPPER)	102 – 105	TRANSIENT 12 SECOND MAXIMUM
YELLOW	▨		99 – 102	30 MINUTE LIMIT
GREEN	▨		63 – 99	NORMAL OPERATION
RED	▨	(LOWER)	63	MINIMUM–ENGINE OUT WARNING LIGHT SET AT THIS VALUE
BLUE	■			INSTRUMENT POWER ON
RED	⊗			BEGINNING OF A RED RANGE (FROM A NORMAL OPERATING REFERENCE)
YELLOW	▷			BEGINNING OF A YELLOW RANGE (FROM A NORMAL OPERATING RANGE)

M01-131-1A

Figure 5-1. Instrument Markings (Sheet 1 of 6)

TM 1-1520-238-10

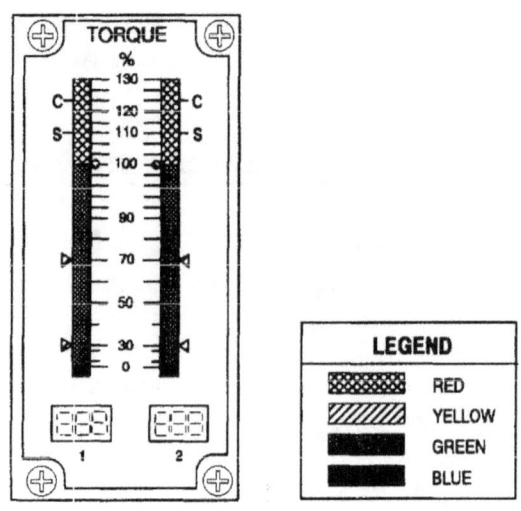

TORQUE (%)

RED	⊗		BEGINNING OF RED RANGE (FROM A NORMAL OPERATING REFERENCE)
RED	▦	122 - 125	SINGLE ENGINE TRANSIENT (6 SECOND LIMIT)
WHITE	C	122	SINGLE ENGINE CONTINGENCY POWER (2.5 MINUTE LIMIT)
RED	▦	100 - 115	DUAL ENGINE TRANSIENT (6 SECOND LIMIT)
WHITE	S	110	SINGLE ENGINE MAXIMUM CONTINUOUS POWER
GREEN	▪	0 - 100	DUAL ENGINE NORMAL OPERATION
YELLOW	▷▷	70	DO NOT EXCEED WITH LESS THAN 90% Nr
YELLOW	▷▷	30	DO NOT EXCEED WITH LESS THAN 50% Nr
BLUE	▬		INSTRUMENT POWER ON

M01-131-2

Figure 5-1. Instrument Markings (Sheet 2 of 6)

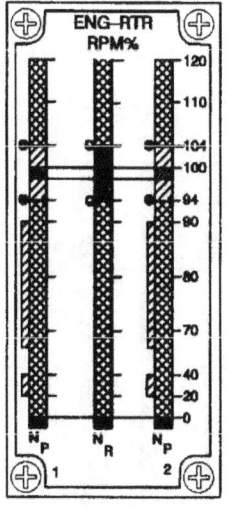

ENGINE POWER TURBINE SPEED (Np) (RPM%)

LIMITS

RED	▨	(UPPER)	110	MAXIMUM
RED	▨		104 – 110	TRANSIENT OPERATION
YELLOW	▨	(UPPER)	100 – 104	NORMAL OPERATION
GREEN	■	(UPPER)	98 – 100	NORMAL OPERATION
YELLOW	▨	(LOWER)	94 – 98	TRANSIENT OPERATION ONLY
RED	●	(LOWER)	94	ENGINE OUT WARNING LIGHT SET AT THIS VALUE

ROTOR SPEED (Nr) (RPM%)

LIMITS

RED	▨		110	MAXIMUM
RED	▨		104 – 110	TRANSIENT OPERATION
GREEN	■		94 – 104	NORMAL OPERATION
RED	▨	(LOWER)	0 – 94	TRANSIENT POWER OFF
BLUE	■			INSTRUMENT POWER ON
RED	●			BEGINNING OF RED RANGE (FROM A NORMAL OPERATING REFERENCE)

Figure 5-1. Instrument Markings (Sheet 3 of 6)

TM 1-1520-238-10

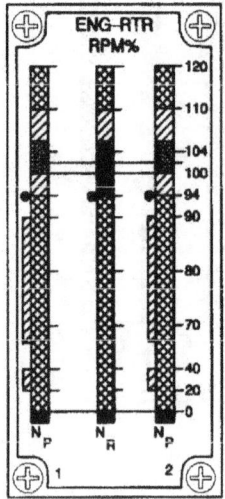

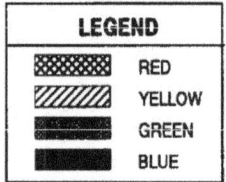

ENGINE POWER TURBINE SPEED (Np) (RPM%)

LIMITS

RED	▓	(UPPER)	110	MAXIMUM
YELLOW	▨	(UPPER)	104 - 110	TRANSIENT OPERATION
GREEN	▪	(UPPER)	98 - 104	NORMAL OPERATION
YELLOW	▨	(LOWER)	94 - 98	TRANSIENT OPERATION
RED	▓	(LOWER)	94	ENGINE OUT WARNING LIGHT SET AT THIS VALUE

ROTOR SPEED (Nr) (RPM%)

LIMITS

RED	▓		110	MAXIMUM
YELLOW	▨		104 - 110	TRANSIENT OPERATION
GREEN	▪		94 - 104	NORMAL OPERATION
RED	▓	(LOWER)	0 - 94	TRANSIENT POWER OFF
BLUE	▪			INSTRUMENT POWER ON
RED	⊛			BEGINNING OF RED RANGE (FROM A NORMAL OPERATING REFERENCE)

M01-131-7

Figure 5-1. Instrument Markings (Sheet 4 of 6)

ENGINE LUBRICATION OIL PRESSURE (PSI)

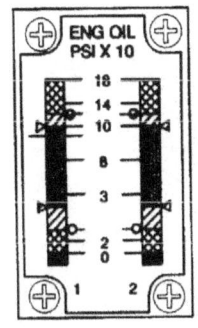

		LIMITS	
RED	▧	120	MAXIMUM
YELLOW	▨	100 – 120	5 MINUTE LIMIT
GREEN	■	28 – 100	NORMAL OPERATION
YELLOW	▨	22 – 28	IDLE OPERATION
RED	▧	22	MINIMUM
BLUE	■		INSTRUMENT POWER ON
RED	●		BEGINNING OF RED RANGE (FROM A NORMAL OPERATING REFERENCE)
YELLOW	▷	28 AND 100	BEGINNING OF A YELLOW RANGE (FROM A NORMAL OPERATING RANGE)

LEGEND

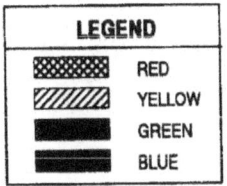

▧	RED
▨	YELLOW
■	GREEN
■	BLUE

FUEL QUANTITY (LB)

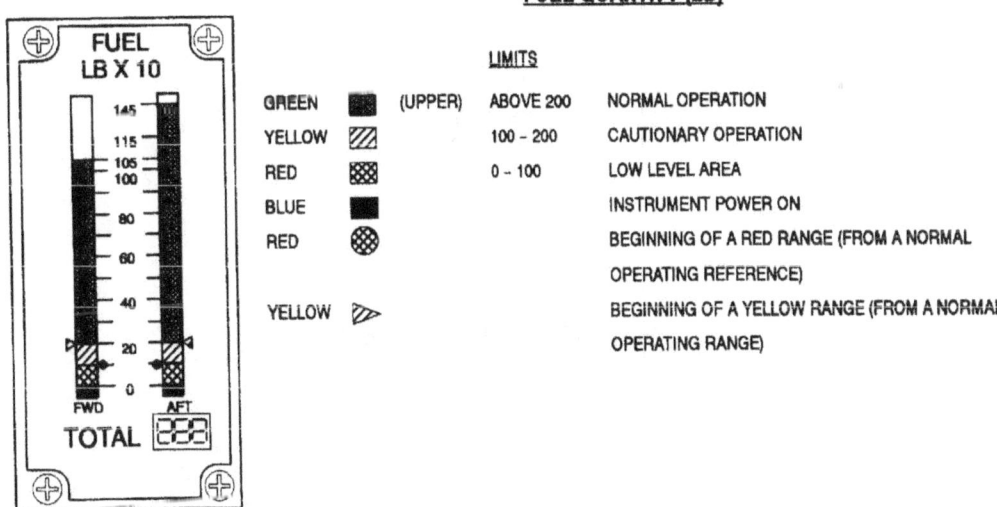

			LIMITS	
GREEN	■	(UPPER)	ABOVE 200	NORMAL OPERATION
YELLOW	▨		100 – 200	CAUTIONARY OPERATION
RED	▧		0 – 100	LOW LEVEL AREA
BLUE	■			INSTRUMENT POWER ON
RED	●			BEGINNING OF A RED RANGE (FROM A NORMAL OPERATING REFERENCE)
YELLOW	▷			BEGINNING OF A YELLOW RANGE (FROM A NORMAL OPERATING RANGE)

Figure 5-1. Instrument Markings (Sheet 5 of 6)

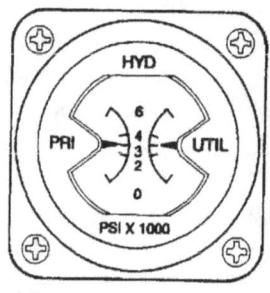

DUAL HYDRAULIC PRESSURE INDICATOR

3250 PSI MAXIMUM OPERATING PRESSURE
3000 PSI NORMAL OPERATING PRESSURE

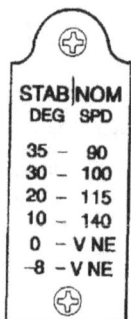

**(STABILATOR POSITION INDICATOR)
PLACARD**

Figure 5-1. Instrument Markings (Sheet 6 of 6)

Section III. POWER LIMITS

5.6 ENGINE POWER LIMITATIONS.

The absolute limitations, regardless of atmospheric conditions, are shown in figure 5-1. For variation in power available with temperature and pressure altitude, refer to the charts in Chapter 7 **701** or Chapter 7A **701C**.

NOTE

The N_p and N_r triple tachometer for the -701 **701** and -701C **701C** engines have different markings. However, both engines should be operated within the same normal operation limit of 104% N_p and a maximum N_r limit of 110%.

5.7 ENGINE START LIMITS.

Refer to figure 5-1 for limitations.

5.8 ENGINE STARTER LIMITATIONS.

The pneumatic starter is capable of making the number of consecutive start cycles listed below, when exposed to the environmental conditions specified, with an interval of at least 60 seconds between the completion of one cycle and the beginning of the next cycle. A starting cycle is the interval from start initiation and acceleration of the output drive shaft, from zero rpm, to starter dropout. The 60-second delay between start attempts applies when the first attempt is aborted for any reason and it applies regardless of the duration of the first attempt. If motoring is required for an emergency, the 60-second delay does not apply.

   a. At ambient temperatures of 16 °C (61 °F) and below, two consecutive start cycles may be made, followed by a 3-minute rest period, followed by two additional consecutive start cycles. A 30-minute rest period is then required before any additional starts.

   b. At temperatures above 16 °C (61 °F), two consecutive start cycles may be made. A 30-minute rest period is then required before any additional start cycles.

   c. Dual engine starts are prohibited.

5.9 ENGINE TEMPERATURE LIMITATIONS.

Refer to figure 5-1 for limitations.

5.10 PNEUMATIC SOURCE INLET LIMITS.

The minimum ground air source (pneumatic) required to start the helicopter engines is 40 psig and 30 ppm. The maximum ground-air source to be applied to the helicopter is 50 psig.

5.11 ENGINE OVERSPEED CHECK LIMITATIONS.

Engine overspeed check in flight is prohibited. Only maintenance test flight pilots are authorized to perform an overspeed check.

5.12 APU OPERATIONAL LIMITS.

CAUTION

Avoid prolonged operation at 94% - 96% N_R with the APU running. The APU clutch will oscillate from engaged to disengaged. This creates high loads on the clutch and shall be avoided.

   a. APU operation is prohibited during normal flight. After a fault or aborted start, wait 30 seconds after compressor has stopped before attempting another start. After two consecutive start attempts, wait 20 minutes before third start attempt. No more than three start attempts are permitted in one hour.

CAUTION

Do not operate the APU for more than five minutes at a main transmission oil temperature of 120 degrees C (248 degrees F). Shut down APU to prevent damaging accessory gearbox components.

   b. During prolonged ground operations greater than 30 minutes the on-command test 19 TRAN shall be periodically executed and the XMSN 1 and XMSN 2 temperatures observed. If the temperature exceeds 130 °C (266 °F), the APU shall be secured and the transmission fluid allowed to cool for 30 minutes prior to resuming APU ground operations; or transmission fluid may be cooled by operating an engine with rotor turning. There is **NO** requirement to remove transmission side panels during extended APU ground operations. However, the transmission fluid will not get as hot under high ambient temperature conditions if the side panels are removed.

Section IV. LOADING LIMITS

5.13 CENTER OF GRAVITY LIMITS.

Center of gravity limits for the helicopter to which this manual applies and instructions for computation of the center of gravity are contained in Chapter 6.

5.14 WEIGHT LIMITATIONS.

The maximum gross weight of the helicopter is 21,000 pounds.

5.15 TURBULENCE.

Intentional flight in extreme turbulence is prohibited.

Section V. AIRSPEED LIMITS MAXIMUM AND MINIMUM

5.16 AIRSPEED OPERATING LIMITS.

See figure 5-2 to determine the never exceed velocity (V_{NE}) as a function of weight, altitude, and temperature. Additional airspeed limits are:

a. Maximum airspeed during autorotation is 145 KTAS.

b. Maximum airspeed with one engine inoperative shall not exceed the greater of

   (1) 67% of V_{NE} determined from figure 5-2 sheet 1 using the GROSS WEIGHT line.

   (2) The speed for minimum power determined from the cruise charts in Chapter 7 **701** or Chapter 7A **701C** using the MAX END/MAX R/C lines.

c. The **NOM SPD** values depicted on the stabilator position **(STAB POS)** indicator placard (fig 5-1) shall be observed as maximum indicated airspeeds during manual stabilator operations.

d. Maximum rearward/sideward flight speed is 45 KTAS for all gross weights.

e. Maximum airspeed for searchlight extension is 90 KIAS.

f. Maximum airspeed with symetrically loaded external fuel tanks (2 or 4) installed is 130 KLAS.

g. Jettison of external armament stores is not authorized except for emergency conditions, and then only from unaccelerated flight during:

   (1) Maximum airspeed for stores jettison is 120 KIAS, or

   (2) Level flight - Hover to 40 KIAS (minimize side slip), or

   (3) Level flight - 40 to 120 KIAS (ball centered), or

   (4) Descents (0 fpm to full autorotation) 80 to 100 KIAS (ball centered).

h. Jettison of external fuel tanks is not authorized except for emergency conditions, and then only from airspeeds less than 100 KIAS. Jettison from level flight if possible, and if not, jettison at an airspeed which minimizes the rate of descent at the time of jettison.

5.16.1 Airspeed Operating Limits Chart. Referring to figure 5-2 sheet 1, note that free air temperature lines and pressure altitude scale are provided in the upper grid, and gross weight lines and true airspeed scale on the lower grid. Using the observed free air temperature and altitude obtained from the aircraft instruments and the calculated aircraft weight, enter the chart as directed in the chart example. Determine maximum true airspeed at the left side of the lower grid. To determine the maximum indicated airspeed (pilots gauge), refer to figure 5-2 sheet 2 and enter as directed in the chart example with the true airspeed and density altitude determined from figure 5-2 sheet 1.

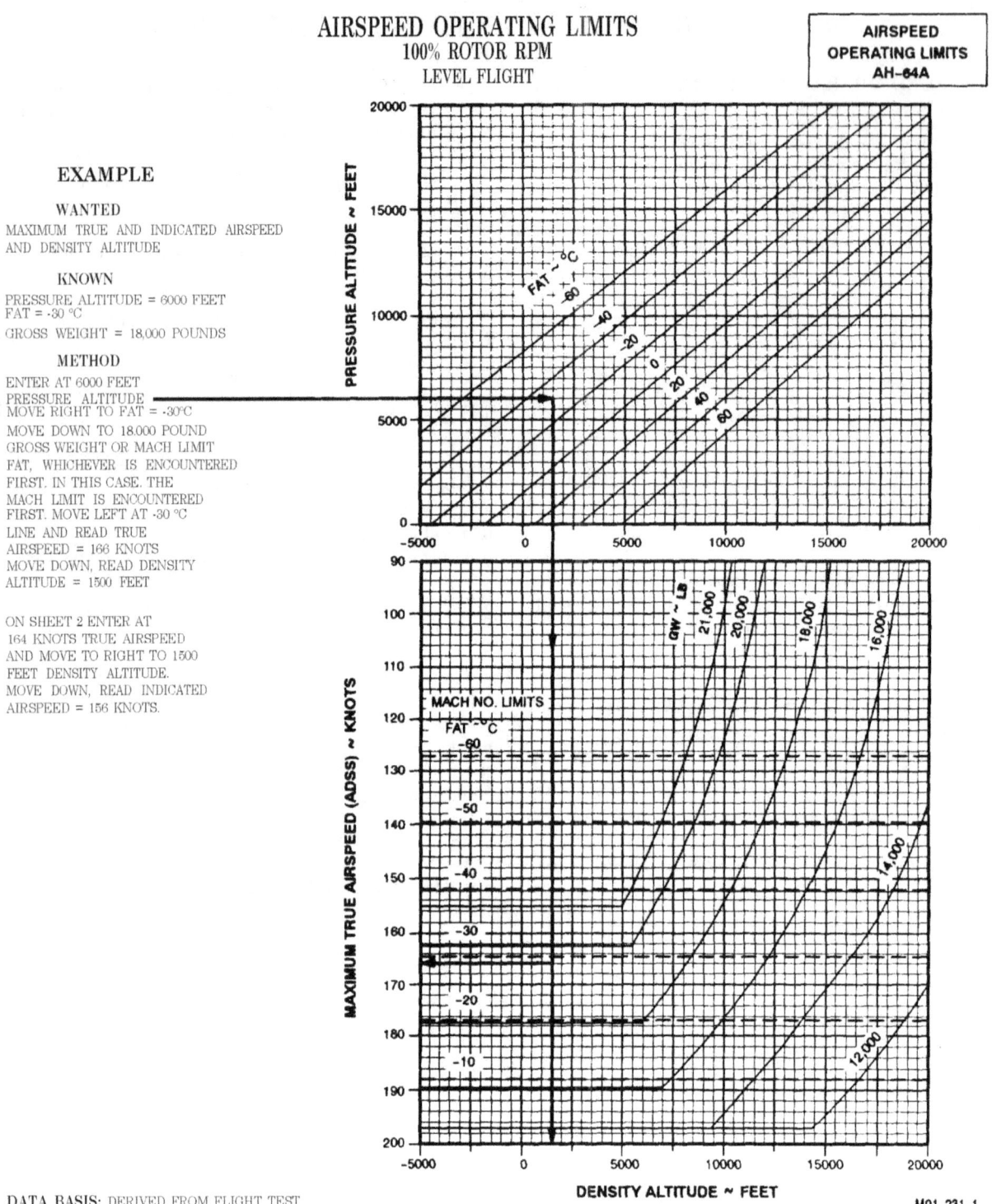

Figure 5-2. Airspeed Operating Limits Chart (Sheet 1 of 2)

AIRSPEED CONVERSION

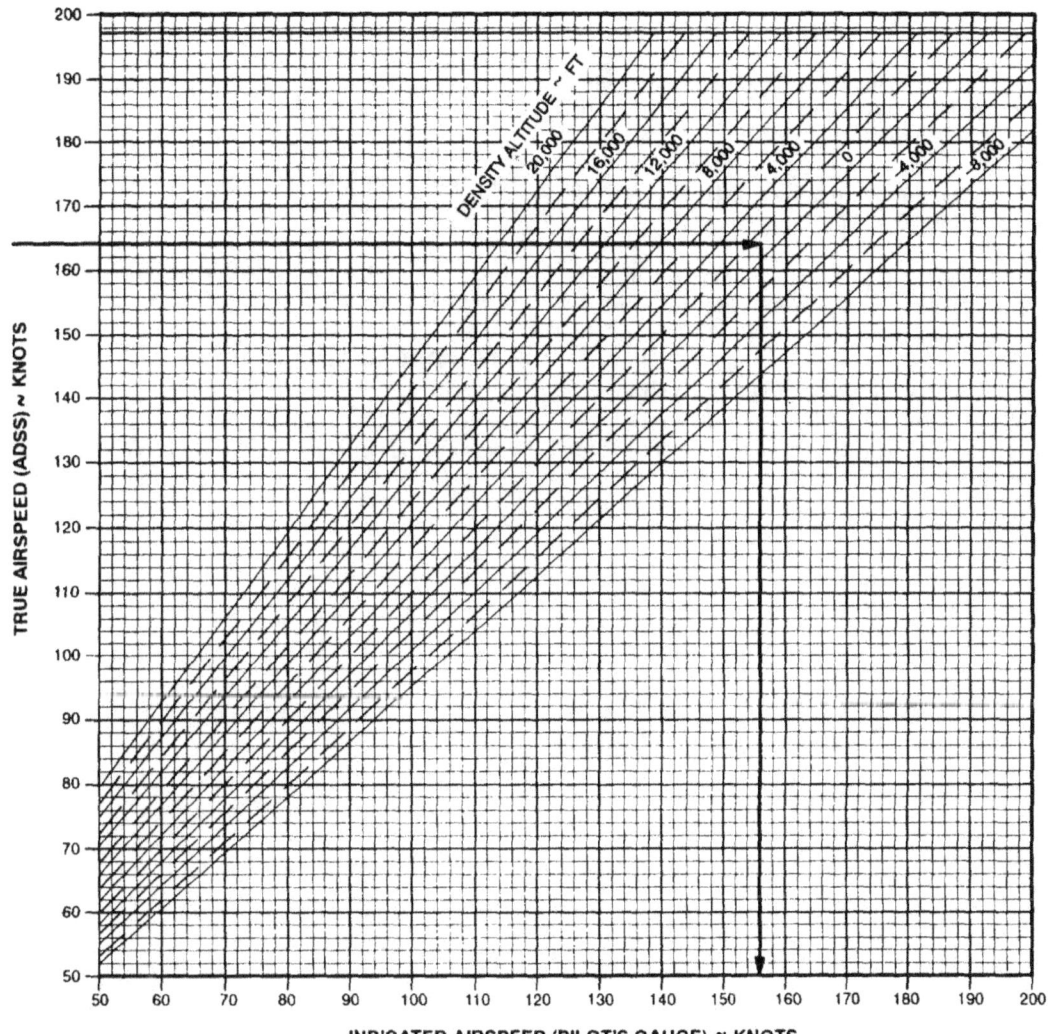

DATA BASIS: DERIVED FROM FLIGHT TEST

Figure 5-2. Airspeed Operating Limits Chart (Sheet 2 of 2)

Section VI. MANEUVERING LIMITS

5.17 MANEUVERING LIMITS.

The AH-64A helicopter is subject to the maneuvering restrictions shown in figure 5-3.

a. Avoid large, abrupt pedal inputs in arresting right hovering/low speed yawing turns greater than 60°/sec. This is to avoid excessive tail rotor drive system loads. Avoid rapid, abrupt pedal inputs when any installed fuel tank(s) contain fuel. This is to avoid excessive torquing of pylon structure.

b. Intentional maneuvers beyond attitudes of ± 30° in pitch or ± 60° in roll are prohibited.

c. Flight, hovering flight, and ground taxiing with the canopy enclosure open is prohibited, except for smoke/fume elimination.

d. The helicopter shall be limited to a maximum of 2.0g's when any external tank(s) contain fuel. There are no limitations on normal load factor, except for figure 5-3, when all external fuel tank(s) are empty.

e. With external fuel tanks (2 or 4) symmetrically installed, the following restrictions apply:

(1) Normal load factor of 2 g's shall not be exceeded.

(2) Maneuvers are limited to those required to takeoff, climb to optimum altitude, heading/course corrections, obstacle avoidance, descend, and land.

(3) Two hundred thirty (230) gallon external fuel tanks shall be in the flight stowed position [four degrees (4°) nose-up with respect to waterline (WL)].

(4) Rapid and step-shaped pedal inputs in excess of 1/2 inch shall be avoided.

5.18 LANDING LIMITS.

Do not complete a landing on terrain which produces a pitch attitude change from a hover greater than 7° nose up or 12° nose down; or a roll attitude greater than 10°.

TM 1-1520-238-10

FLIGHT ENVELOPE

FLIGHT ENVELOPE
AH-64A

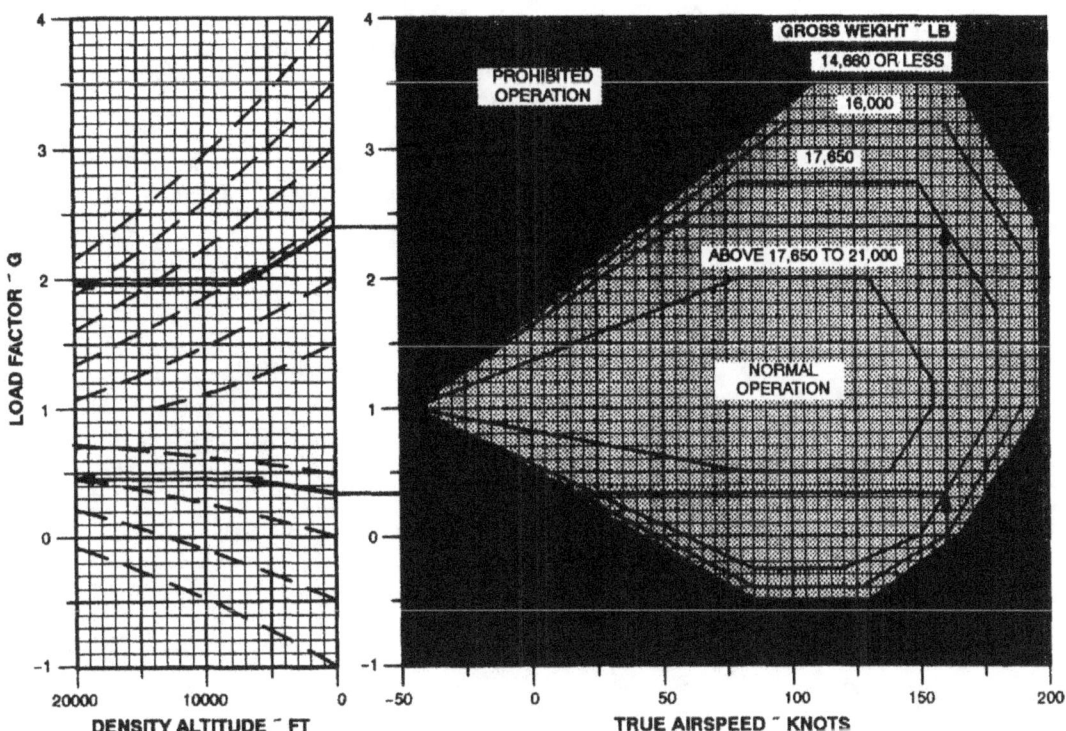

EXAMPLE

WANTED
MAXIMUM AND MINIMUM LOAD FACTOR

KNOWN
GROSS WEIGHT = 17,650 POUNDS
DENSITY ALTITUDE = 7000 FEET
AIRSPEED = 160 KTAS

METHOD
ENTER AT V = 160 KT. MOVE UP TO UPPER AND LOWER ENVELOPE BOUNDARIES FOR
GROSS WEIGHT = 17,650 POUNDS.
MOVE LEFT TO DENSITY ALTITUDE OF ZERO FEET.
SLIDE TO LEFT ALONG DASHED LINES TO 7000 FEET DENSITY ALTITUDE.
MOVE LEFT TO LOAD FACTOR SCALE, READ MAX G = 1.96, MIN G = 0.46.

DATA BASIS: DERIVED FROM FLIGHT TEST

M01-232

Figure 5-3. Flight Envelope Chart

Section VII. ENVIRONMENTAL RESTRICTIONS

5.19 ENVIRONMENTAL RESTRICTIONS.

> **CAUTION**
>
> **Intentional flight into known or forecast moderating icing is prohibited.**

a. The AH-64A helicopter is equipped with de-icing and anti-icing equipment for flights into light icing conditions.

b. This aircraft is qualified for operation in instrument meteorological conditions.

c. Deleted.

d. For operation in adverse enviormental conditions, reference Chapter 8, Section V.

Section VIII. OTHER LIMITS

5.20 WING STORES CONFIGURATION.

For authorized wing stores configurations refer to figure 7-18 **701** or figure 7A-18 **701C**.

5.21 ALTERATION OF CBHK VALUES.

Only maintenance qualified personnel are authorized to alter values from current CBHK values, unless performing the AWS dynamic harmonization procedure. Aviators may only verify and correct CBHK values, other than the gun, to the current CBHK values as recorded in the aircraft logbook. When aviators modify gun values while performing dynamic harmonization, the new values will be entered on DA form 2408-13-1 for maintenance review and transcription.

5.22 TRIM AND FORCE FEEL.

Use of the Trim and Force Feel Release switch in the FORCE TRIM OFF position with the helicopter on the ground is not authorized. Inflight operation with the Trim and Force Feel Release switch in FORCE TRIM OFF position is authorized when briefed and the FORCE TRIM OFF selection is acknowledged by both crewmembers.

CHAPTER 6
WEIGHT/BALANCE AND LOADING

Section I. GENERAL

6.1 INTRODUCTION.

This chapter provides information required for helicopter loading and computing weight and balance. This chapter contains sufficient instructions and data so that an aviator, knowing the basic weight and moment of the helicopter, can compute any combination of weight and balance using the prescribed Army charts and forms.

6.2 HELICOPTER CLASS.

The Army AH-64A helicopter is in Class 2. Additional directives governing weight and balance of Class 2 aircraft forms and records are contained in AR 95-1.

6.3 HELICOPTER COMPARTMENTS AND STATIONS.

The helicopter has many compartments (fig 6-1). Most of them contain electronic or other equipment. The boundaries of all compartments and a listing of the equipment in each compartment are provided in the helicopter records Chart A - Basic Weight Checklist, DD Form 365-1. The compartments of primary concern to the pilot, when loading, are the CPG and pilot crew stations, left aft storage bay, and survival kit bay. These compartments may contain personal items or extra equipment not accounted for in basic weight. (Refer to paragraph 6.18). Any additional items must be entered on the Weight and Balance Form F (DD 365-4).

6.3.1 CPG Crew Station. The CPG station extends from station 35.5 to station 115.0. The CPG nominal centroid is at station 82.2, but the seat can be adjusted so the CPG centroid can vary from station 81.9 to 82.8. This produces a small moment variation and should be ignored.

6.3.2 Pilot Crew Station. The pilot station extends from station 115.0 to station 168.0. The pilot nominal centroid is at station 143.3, but the seat can be adjusted so the pilot centroid can vary from station 142.8 to 143.8. This produces a small moment change and should be ignored.

6.3.3 Left Aft Storage Bay. The aft storage bay is on the left side from fuselage station 280.0 to 310.0 and can be loaded to 15 pounds per square foot with a capacity of 60 pounds. The floor area is approximately 4.16 square feet and the volume is approximately 21 cubic feet. Floor tiedown fittings are in place to accommodate the flyaway kit.

6.3.4 Survival Kit Bay. The survival kit bay is reached from either side. From fuselage station 310.0 to 340.0, it can be loaded to 15 pounds per square foot with a capacity of 100 pounds. A single concentrated load of 45 pounds with a load density of 45 pounds per square foot may be carried. The floor area is approximately 7.3 square feet and the volume is approximately 17 cubic feet. Floor tiedown fittings are in place to accommodate the survival kit.

NOTE

When loading these two bays, check for exceeding the aft cg limit.

TM 1-1520-238-10

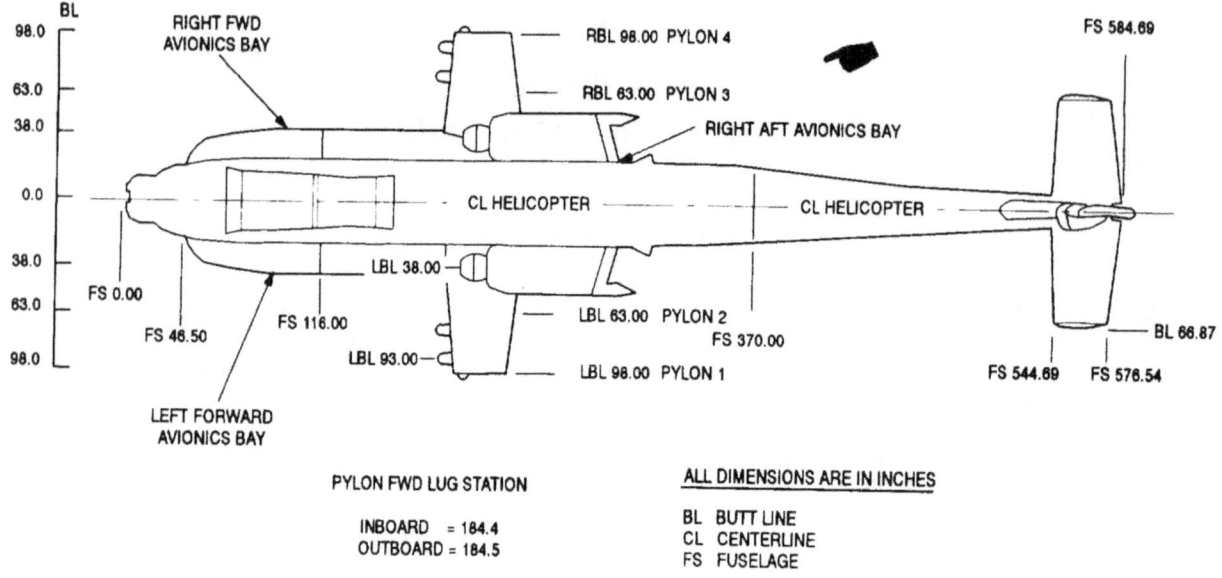

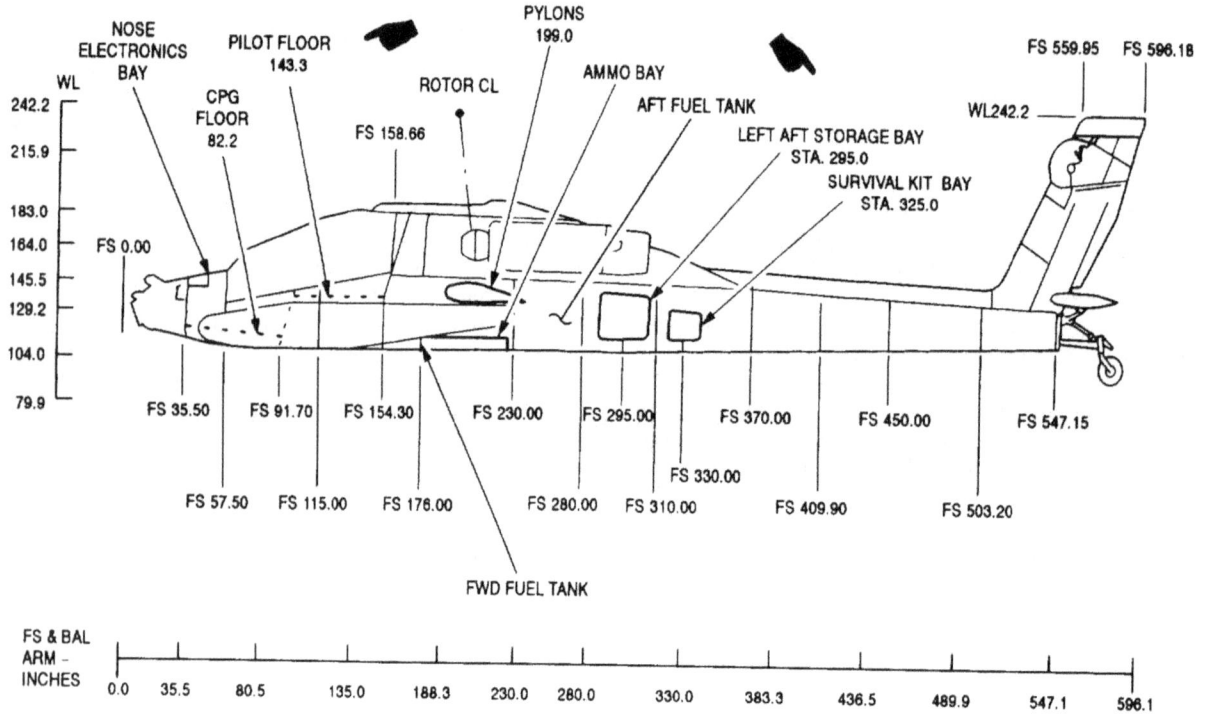

Figure 6-1. Station Diagram

6-2 Change 4

Section II. WEIGHT AND BALANCE

6.4 WEIGHT AND BALANCE.

This section contains information needed to compute the weight and balance for an individual helicopter by using the prescribed standard charts and forms.

6.5 WEIGHT DEFINITIONS.

6.5.1 Basic Weight. The normal basic weight of this helicopter includes wing pylons, all fixed operating equipment, 30mm gun, all oil, and trapped fuel. It is only necessary to add the variables or expendable to these items for the missions.

NOTE

The basic weight of the helicopter will vary with mission requirements and structural modifications such as addition or removal of wing pylons, peculiar kits, 30mm gun, turret, ammo handling system, etc. A continuing record of an individual helicopter's basic weight is maintained on Chart C - Basic Weight and Balance Record, DD Form 365-3.

6.5.2 Operating Weight. The operating weight of the helicopter is the basic weight plus those variables which remain substantially constant for a particular mission. These items include crew, baggage, rocket launchers, Hellfire launchers, and any emergency or extra equipment that may be required.

6.5.3 Gross Weight. The gross weight is the total weight of the helicopter and its contents.

6.6 BALANCE Definitions.

6.6.1 Reference Datum. The reference datum is an imaginary vertical plane from which all horizontal distances are measured (in inches) for balance purposes.

6.6.2 Arm. For balance purposes, the term ARM is the horizontal distance (in inches) from the reference datum to the center of gravity of a given item. For special cases, ARM can be determined form figure 6-1. For the AH-64A helicopter, ARM and fuselage station (FS) are the same.

6.6.3 Moment. Moment is the weight of an item multiplied by its arm. For the AH-64A helicopter, moment divided by 100 (moment/100) is used to simplify calculations by reducing the number of digits.

NOTE

Throughout this chapter, moment/100 figures have been rounded off to the nearest whole number. When moments from other sources are being used, they must be divided by 100 and rounded off.

6.6.4 Average Arm. Average arm is the arm obtained by adding the weights and the moments of a number of items and dividing the total moment by the total weight.

6.6.5 Basic Moment. Basic moment is the sum of the moments of all items making up the basic weight with respect to the helicopter reference datum.

6.6.6 Center of Gravity (CG). Center of gravity is the point about which the helicopter would balance if suspended. Distance from the reference datum is found by dividing the total moment by the gross weight of the helicopter.

6.6.7 CG Limits. The cg limits are the extremes of movement to which the helicopter cg can travel without endangering controllability or structural integrity. The cg of the loaded helicopter must remain within these limits at takeoff, throughout flight, and during landing.

6.7 LOADING DATA.

The loading data in this chapter is intended to provide information necessary to work loading problems for the helicopter. From this data, weight and moment/100 are obtained for all variable load items and are added to the current basic weight and moment/100 from Chart C (DD Form 365-3) to determine the gross weight and moment/100 using Form F (DD Form 365-4). The effect on helicopter cg of expending the fuel and armament in a logical sequence may be checked by subtracting the weight and moment/100 of each item from the takeoff gross weight and moment/100; then, checking the new moment (or helicopter cg) with the cg limits chart. This check should be made to determine if the cg will remain within limits during the entire flight. Refer to paragraph 6.10 for helicopter cg management.

6.8 CHART C - BASIC WEIGHT AND BALANCE RECORD, DD FORM 365-3.

Chart C is a continuous history of the basic weight and moment resulting from structural and equipment changes in service. At all times, the last weight and moment/100 entries are considered the current weight and balance status of the basic helicopter.

6.9 WEIGHT AND BALANCE CLEARANCE FORM F, DD FORM 3654.

Form F is the summary of the actual disposition of load in the helicopter. It records the balance status of the helicopter step-by-step. It serves as a work sheet on which to record weight and balance calculations and any corrections that must be made to ensure that the helicopter will be within weight and cg limits throughout the mission. There are two versions of this form: Transport and Tactical. Each was designed to provide for the respective loading arrangement of these two types of aircraft. The general use and fulfillment of either version is the same.

6.10 CENTER OF GRAVITY MANAGEMENT.

This paragraph contains fuel loading/management methods that can be used to maintain cg limits in flight and during the expending of external stores for some helicopter configurations. The table of expendable (table 6-1) provides a guide for quick definition of intermediate flight cg as stores/fuel are expended at various gross weights and at forward and aft cg limits. Table 6-1 eliminate calculation of intermediate cg when the helicopter is well within limits. When flight limits are doubtful or when operation is close to cg limits, a detailed calculation must be made to determine any cg limit violation. When the storage bays are used for miscellaneous equipment it is possible to cause an aft cg condition.

6.10.1 Fuel Loading. The helicopter takeoff cg can be moved by loading either tank with more fuel than the other. Example: to move the cg forward, fill the forward tank (1014 pounds of JP-4) and reduce fuel load in the aft tank depending on cg shift required. For some missions, it maybe necessary to reduce the stowed weight.

6.10.2 Fuel Management. The following example presents normal cg/fuel management where each fuel tank supplies an engine. This procedure prevents drastic helicopter cg shifts as fuel is expended. Refer to Chapter 2 for fuel system details.

EXAMPLE:

1. Using table 6-1, refer to FUEL EXPENDED 500 lb EACH line. Note that fuel expended at 15,000 gross weight at forward cg limit produces a helicopter forward cg shift of -0.13 inch; and at aft cg limits, +0.29 inch. When the helicopter cg is at the combined fuel cg (202.8 inches) expending fuel will produce a zero shift. This is true only when the forward tank fuel remains in the lower portion of the L-shape tank. Refer to paragraph 6.12.

2. When filling the forward tank into the upper portion of the L-shape, the combined fuel expended cg moves aft to 212.4 inches which means that the helicopter cg will always shift forward during fuel burnoff in this area.

3. Refer to the 1000 lb fuel expended each tank line in table 6-1. Note that fuel expended at 15,000 gross weight at aft cg limit is +0.41 inches. This is correct when the total 2,000 pounds of fuel is expended, but this includes a helicopter cg shift forward during initial fuel burn-off due to the forward L-shaped tank.

4. For the full fuel example based on a conservative gross weight of 15,000 pounds and cg of 201.0 inches, 208 pounds of aft fuel must be transferred so that all fuel can be expended equally. A helicopter cg shift of -1.49 to -1.68 inches is produced depending on when the transfer occurs. When working with full fuel, it is advisable to calculate several intermediate points and determine when it's best to transfer fuel.

Table 6-1. Helicopter CG Movement When Loaded Items Are Expended

		When Aircraft CG Near Fwd CG Limit			When Aircraft CG Near Aft CG Limit		
Aircraft Gross Weight		13000	15000	17650	13000	15000	17650
Expended Items	Qty						
Ammo	200	+0.32	+0.27	+0.23	+0.39	+0.34	+0.27
Ammo	600	+0.19	+0.16	+0.14	+0.41	+0.35	+0.25
Ammo	1200	-0.08	-0.08	-.0.06	+0.37	+0.31	+0.18
H-F Missiles	8	+0.66	+0.57	+0.48	+1.05	+0.89	+0.68
H-F Missiles	16	+1.14	+1.21	+1.01	+2.25	+1.88	+1.44
Rockets 27.1 lbs ea.	38	+1.00	+0.86	+0.72	+1.51	+1.28	+0.99
Rockets 27.1 lbs ea.	76	+2.18	+1.85	+1.52	+3.21	+2.75	+2.11
Chaff	30	-0.23	-0.20	-0.17	-0.22	-0.19	-0.16
*Fuel	500 lbs per tank	-0.15	-0.13	-0.11	-0.35	+0.29	+0.15
*Fuel	1000 lbs per tank		-.048	-0.40		+0.41	+016
*Fuel	Fill Tank		-2.31	-1.91		-1.18	-1.21

Note: A plus (+) means aircraft CG moves aft and negative (-) means forward CG movement.

*The above fuel values represent the total fuel on board such that the 500 lb, 1000 lb, or full in each tank is the starting point and the expended fuel is 500 lb, 1000 lb, or full leaving zero fuel in each tank.

Section III. FUEL AND OIL

6.11 OIL WEIGHT AND MOMENT.

For weight and balance purposes, oil is considered a part of aircraft basic weight.

6.12 FUEL WEIGHT AND MOMENT.

When the actual or planned fuel loading (pounds or gallons) and type is known, the total fuel weight and moment/100 can be determined from the fuel moment Tables 6-2 and 6-3. The tables present data for JP-4, JP-5, and JP-8 based on the approximate density for these fuels at 15 degrees centrigrade. The following information is provided to show the general range of fuel density to be expected. Density of the fuel will vary depending on fuel temperature. Density will decrease as fuel temperature rises and increase as fuel temperature decreases at the rate of approximately 0.1 lb/gal for each 15 degree centrigrade temperature change. Density may also vary between lots of the same fuel at the same temperature by as much as 0.5 lb/gal. The full tank usable fuel weight presented in the tables for density closest to that of the fuel being used may be used for mission planning. The aircraft fuel gage system was designed for use with JP-4, but does tend to compensate for other fuels and provide acceptable readings. When possible the weight of fuel onboard should be determined by direct reference to aircraft fuel gages.

6.12.1 Fuel Moments. The forward fuel moment calculations are complicated by the L-shape of the tank. Consider the tank being filled from empty to 132.8 gallons; the fuel cg remains constant at 150.6 inches. From 132.8 gallons to full (156 gallons), the cg of the total fuel moves aft linearly to 153.7 inches at capacity. The aft fuel cg is constant at 255 inches.

6.13 AUXILIARY FUEL TANKS.

WARNING

A slight increase in the risk of post-crash fire exists if a mishap occurs after tanks are pressurized. Crashworthiness of the fuel system is reduced by external fuel tanks, which are designed for <u>FERRY MISSION ONLY</u>. External fuel tank installation is prohibited for use in tactical missions.

The auxiliary fuel tanks are installed on the wing pylons in sets of two or four and are for extending the helicopter ferry range. Plumbing from the fuselage to the tank is provided with each tank. Each tank has a capacity of approximately 230 gallons. Table 6-4 lists the weight and moment/100 of each fuel tank, wing plumbing and total fuel for JP-4, JP-5, and JP-8. Note that the data are given for one tank, wing plumbing, and indicated fuel so that any combination can be determined. The table can be used for inboard and outboard locations because the small moment/100 differences can be ignored. Remember to add the tank, and specific fuel together for one location, then multiply by 2 or 4 depending on the number of tanks carried. Add wing plumbing for 2 or 4 tanks as appropriate.

Table 6-2. Fuel Loading - Forward Tank

JP-4 Density = 6.5 lb/gal
JP-5 Density = 6.8 lb/gal
JP-8 Density = 6.7 lb/gal

Weight (lb)	Moment/100	U.S. Gallons JP-4	U.S. Gallons JP-5	U.S. Gallons JP-8
50	75	7.7	7.4	7.5
100	151	15.4	14.7	14.9
150	226	23.1	22.1	22.4
200	301	30.8	29.4	29.9
250	377	38.5	36.8	37.3
300	452	46.2	44.1	44.8
350	527	53.8	51.5	52.2
400	602	61.5	58.8	59.7
450	678	69.2	66.2	67.2
500	753	76.9	73.5	74.6
550	828	84.6	80.9	82.1
600	904	92.3	88.2	89.6
650	979	100.0	95.6	97.0
700	1054	107.7	102.9	104.5
750	1130	115.4	110.3	111.9
800	1205	123.1	117.6	119.4
850	1281	130.8	125.0	126.9
900	1365*	138.5	132.4	134.3
950	1450*	146.2	139.7	141.8
1000	1535*	153.8	147.1	149.3
1012	1555	155.7 (full)	-	-
1043	1603	-	-	155.7 (full)
1059	1627	-	155.7 (full)	-

NOTE

*Moment varies slightly with fuel density. Moments listed are based on JP-4 @ 6.5 lb/gal. These values may be used for JP-5 or JP-8 as the variation with fuel density is small.

Table 6-3. Fuel Loading - Aft Tank

JP-4 Density = 6.5 lb/gal
JP-5 Density = 6.8 lb/gal
JP-8 Density = 6.7 lb/gal

Weight (lb)	Moment/100	U.S. Gallons JP-4	U.S. Gallons JP-5	U.S. Gallons JP-8
50	127	7.7	7.4	7.5
100	255	15.4	14.7	14.9
150	382	23.1	22.1	22.4
200	510	30.8	29.4	29.9
250	637	38.5	36.8	37.3
300	765	46.2	44.1	44.8
350	892	53.8	51.5	52.2
400	1020	61.5	58.8	59.7
450	1147	69.2	66.2	67.2
500	1275	76.9	73.5	74.6
550	1402	84.6	80.9	82.1
600	1530	92.3	88.2	89.6
650	1657	100.0	95.6	97.0
700	1785	107.7	102.9	104.5
750	1912	115.4	110.3	111.9
800	2040	123.1	117.6	119.4
850	2167	130.8	125.0	126.9
900	2295	138.5	132.4	134.3
950	2422	146.2	139.7	141.8
1000	2550	153.8	147.1	149.3
1050	2677	161.5	154.4	156.7
1100	2805	169.2	161.8	164.2
1150	2932	176.9	169.1	171.6
1200	3060	184.6	176.5	179.1
1250	3187	192.3	183.8	186.6
1300	3315	200.0	191.2	194.0
1350	3442	207.7	198.5	201.5
1400	3570	215.4	205.9	209.0
1430	3646	220.0 (full)	-	-
1474	3758	-	-	220.0 (full)
1496	3815	-	220.0 (full)	-

Table 6-4. Auxiliary Fuel Tanks

JP-4 Density = 6.5 lb/gal
JP-5 Density = 6.8 lb/gal
JP-8 Density = 6.7 lb/gal

Auxiliary Tank

Item	Weight Each (lb)	Moment/100	US. Gallons
Auxiliary Fuel Tank	140.0	268	—
Wing Tank Plumbing			
2 Tanks	16.0	35	—
4 Tanks	20.0	43	—
JP-4 Fuel (Capacity)	1495.0	2906	230.0
JP-5 Fuel (Capacity)	1564.0	3040	230.0
JP-8 Fuel (Capacity)	1541.0	2996	230.0

Note 1: Weight and moment are shown for one tank with fuel; and must be doubled for two inboard or multiplied by 4 when inboard and outboard stations are used.

Note 2: Add wing tank plumbing for 2 tanks or 4 tanks as appropriate.

Section IV. PERSONNEL

6.14 GENERAL.

Personnel provisions consist of the pilot and CPG located in the cockpit.

6.15 PERSONNEL WEIGHT.

When aircraft are operated at critical gross weights, the exact weight of each individual occupant plus equipment should be used. If weighting facilities are not available, or if the tactical situation dictates, a crewmember with no equipment, compute weight according to each individuals estimate.

6.16 PERSONNEL MOMENT.

Always try to use exact weight of each crewmember including all equipment and any personal items stored in the crew station. If weighing facilities are not available, then use the best estimate available. Table 6-5 presents the crew moments with the seat in its nominal location.

Table 6-5. Crew Moments

Crewmember Weight Including Equipment (lb)	CPG Station		Pilot Station	
	Nominal Arm (in.)	Moment (in.-lb/100)	Nominal Arm (in.)	Moment (in.-lb/100)
100	82.2	82	143.3	143
110	82.2	90	143.3	158
120	82.2	99	143.3	172
130	82.2	107	143.3	186
140	82.2	115	143.3	201
150	82.2	123	143.3	215
160	82.2	132	143.3	229
170	82.2	140	143.3	244
180	82.2	148	143.3	258
190	82.2	156	143.3	272
200	82.2	164	143.3	287
210	82.2	173	143.3	301
220	82.2	181	143.3	315
230	82.2	189	143.3	330
240	82.2	197	143.3	344
250	82.2	206	143.3	358

Section V. MISSION EQUIPMENT

6.17 MISSION EQUIPMENT.

The AH-64A helicopter mission equipment includes Hellfire missiles and launchers, 2.75-inch rockets and launchers, 30mm ammunition, chaff cartridges, and CBR filters/blowers. External fuel tanks are described in paragraph 6.13. All electronic mission equipment is part of basic weight and may be found in Chart A (Form DD 365-1).

6.17.1 Hellfire Launchers. Figure 6-2 presents the M-272 Hellfire launchers, in pairs, weight, and moment/100. When four launchers are required, double the numbers.

6.17.2 Hellfire Missiles. The present training, dummy, and laser seeker missiles weigh the same. Table 6-6 lists the weight and moment/100 of each missile accumulated to a capacity of four missiles per launcher. When a pair of inboard launchers are filled to capacity, double the (missile No. 4) weight and moment/100. When four launchers are used, multiply the numbers by four.

6.17.3 Rocket Launchers (2.75 inch). Figure 6-2 presents the M-261 2.75-inch rocket launchers, in pairs, weight, and moment/100. Double the weight and moment/100 when four launchers are used.

6.17.4 Rockets (2.75-inch). Present training rockets are designated H519 with an M156 warhead, M423 fuse, and a MK 40 MOD 3 rocket motor. Dummy and training rockets are the same weight. Table 6-7 lists the 2.75-inch rocket accumulated weight and moment/100 to a capacity of 19 per launcher. Select the correct number of rockets, then double the weight and moment/100 for a pair of launchers loaded evenly. Table 6-8 lists an MPSM warhead with a MK 66 rocket motor. The tables are presented so that any combination or mix can be easily determined for a launcher.

6.17.5 Ammunition (30mm). Table 6-9 presents the M-788 or M-789 30mm linkless aluminum case ammunition accumulated weight, and moment/100 to a capacity of 1200 rounds. Approximately 90 rounds are in the right chute and the remainder is in the magazine. A note at the bottom of the table allows conversion of the table numbers to ADEN or DEFA ammunition.

6.17.6 Chaff Cartridges. The helicopter survivability equipment is a kit. It is added to the helicopter as dictated by mission requirements. The electronic equipment, controls, and chaff supports are part of the basic weight and are listed in Chart A (DD 365-1). Table 6-10 lists the chaff dispenser empty and the chaff dispenser (full) with 30 chaff(M-1) cartridges. These values are to be used on Form F (DD 365-4) when chaff is on board.

6.17.7 Chemical, Biological, and Radiological (CBR) Filters/Blowers. The CBR filters/blowers are used by the crewmembers as dictated by mission requirements. The CBR mounting bracket, located on the seats left armored wing, is listed on Chart A (DD 365-1) and when installed shall be accounted for on Chart C (DD 365-3) as part of the basic weight. When the filters/blowers are aboard, the values listed in table 6-11 are to be used to record the weights and moments of the filters/blowers as a part of the helicopter loading on Form F (DD 365-4).

TM 1-1520-238-10

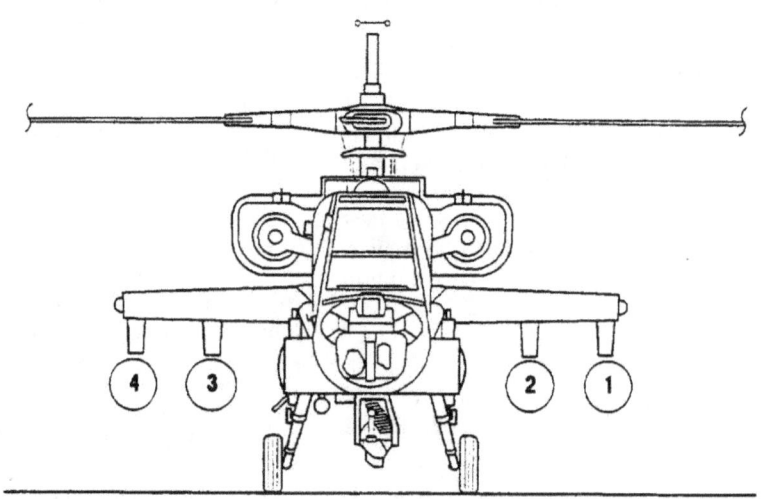

PYLONS AND LAUNCHERS

ITEM	STATIONS	WEIGHT of PAIR (lb)	MOMENT (in.-lb/100)
M-272 HELLFIRE LAUNCHERS	2 & 3 or 1 & 4	278.2	528
M-261 ROCKET LAUNCHERS	2 & 3 or 1 & 4	173.6	344

NOTE

FWD HOOKS ARE AT F.S. 184.5
(BOTH INBOARD & OUTBOARD PYLONS)

Figure 6-2. External Stores and Stations

6-12   Change 2

Table 6-6. Hellfire Missile Loading

		Inboard Station 2, 3 or Outboard Station 1,4	
Item	Qty	Accum Weight (lb)	Moment (in.-lb/100)
Missile	1	99.9	191
Missile	2	199.8	381
Missile	3	299.7	573
Missile	4	399.6	762

Table 6-7. Rocket (2.75) Loading for H519 or Dummy

		Inboard Station 2, 3 or Outboard Station 1, 4	
Item	Qty	Accum Weight (lb)	Moment (in.-lb/100)
H519 Rocket	1	20.6	41
Rocket	2	41.2	81
Rocket	3	61.8	122
Rocket	4	82.4	162
Rocket	5	103.0	203
Rocket	6	123.6	243
Rocket	7	144.2	284
Rocket	8	164.8	324
Rocket	9	185.4	365
Rocket	10	206.0	406
Rocket	11	226.6	446
Rocket	12	247.2	487
Rocket	13	267.8	527
Rocket	14	288.4	568
Rocket	15	309.0	608
Rocket	16	329.6	649
Rocket	17	350.2	690
Rocket	18	370.8	730
Rocket	19	301.4	771

Table 6-8. Rocket (2.75) Loading for MPSM Warhead with MK66 Motor

Item	Qty	Inboard Station 2,3 or Outboard Station 1,4	
		Accum Weight (lb)	Moment (in.-lb/100)
Rocket	1	27.1	51
Rocket	2	54.2	102
Rocket	3	81.3	154
Rocket	4	108.4	205
Rocket	5	135.5	257
Rocket	6	162.6	308
Rocket	7	189.7	359
Rocket	8	216.8	411
Rocket	9	243.9	462
Rocket	10	271.0	513
Rocket	11	298.1	565
Rocket	12	325.2	616
Rocket	13	352.3	667
Rocket	14	379.4	719
Rocket	15	406.5	770
Rocket	16	433.6	821
Rocket	17	460.7	873
Rocket	18	487.8	924
Rocket	19	514.9	975

Table 6-9. Ammunition Loading for M-788 or M-789 30mm Rounds (Aluminum Cartridges)

Number of Rounds	Weight (lb)	Moment (in.-lb/100)
50	38.5	42
100	77.0	110
150	115.5	189
200	154.0	269
250	192.5	346
300	231.0	427
350	269.5	504
400	308.0	585
450	346.5	662
500	385.0	744
550	423.5	819
600	462.0	905
650	500.5	988
700	539.0	1063
750	577.5	1146
800	616.0	1221
850	654.5	1304
900	693.0	1379
950	731.5	1462
1000	770.0	1537
1050	808.5	1620
1100	847.0	1696
1150	885.5	1778
1200	924.0	1868

Note: When ADEN (brass cartridges) are used multiply weight and moment by 1.354. When DEFA (steel cartridges) are used multiply by 1.343.

Table 6-10. Chaff Dispenser and Cartridges

Item	Weight (lb)	Moment (in.-lb/100)
Chaff Dispenser (Empty)+ Payload Module	9.0	44
30 Chaff Cartridges	10.0	49
Total Chaff Dispenser + 30 Cartridges (Full)	19.0	93

Table 6-11. CBR Filters/Blowers

Item	Weight (lb)	CPG Moment (in.-lb/100)	Pilot Moment (in.-lb/100)
CBR Filters/Blowers	4.50	3.7	6.5

Section VI. CARGO LOADING

6.18 CARGO LOADING.

CAUTION

To prevent damage to the helicopter, all cargo must be securely tied down.

There are two aft storage bays in the helicopter. The left aft storage bay contains the flyaway equipment kit which consists of:

a. Tiedown and mooring kit.

b. Main rotor tiedown assembly.

c. Main rotor blade tiedown pole assembly.

d. Safety pins and stowage pouch.

e. Protective covers kit.

NOTE

All *calculated* moments must be divided by 100 before being entered on Form F (DD 365-4).

The survival kit bay may carry a survival equipment kit and/or personal equipment. The flyaway equipment kit and the survival equipment kit are basic weight items and are listed on Chart A (DD 365-1). Personal items or extra equipment that has not been identified as basic weight must be entered on Form F (DD 365-4). Table 6-12 lists various weights and moments for the storage bay and survival bay equipment.

6.19 EXTRA CARGO.

All extra cargo should be weighed so that exact weight and moments are used for the weight and balance computations. If weighing facilities are not available, weight should be estimated in terms of probable maximum weight to reduce the possibility of exceeding the aft cg limits.

Table 6-12. Storage Bay and Survival Kit Bay Equipment Weights and Moments

Left Aft Storage Bay		Survival Kit Bay	
Accum Wt. (lb)	Moment (in.-lb/100)	Accum Wt. (lb)	Moment (in.-lb/100)
5	14	5	16
10	29	15	47
15	43	30	95
20	57	45	142
30	86	50	164
40	114	65	213
45	128	80	262
50	143	90	279
*60	171	95	320
		*100	337
*Max. Load 60 lb. @ 15 lb/ft^2		*Max. Load 100 lb @ 15 lb/ft^2	

Section VII. CENTER OF GRAVITY

6.20 CENTER OF GRAVITY.

This section contains information needed to determine whether the helicopter loading (gross weight and moment combination) will fall within the helicopter center of gravity limits.

6.21 CENTER OF GRAVITY LIMITS CHART.

a. The normal forward cg limit is at fuselage station 201.0 inches to 19,400 pounds and a straight taper from 201.0 to 202.2 inches from 19,400 to 21,000 pounds. The normal aft cg limit is at fuselage station 207.0 inches to 14,660 pounds and a straight taper from 207.0 to 203.3 inches from 14,660 to 21,000 pounds.

b. The normal center of gravity limits chart is shown in figure 6-3. All flight cg must remain within these limits. This chart is used in conjunction with Chart F (DD 365F) as follows:

1. Load the helicopter to takeoff condition and determine takeoff cg (Form F, ref 12 and ref 13).

2. Check cg limits using the chart (fig 6-3). If cg limits are exceeded, then the loading must be revised. Refer to paragraph 6.10 for guidance.

3. After the takeoff cg limits are satisfied, determine estimated landing weight and cg (Form F, ref 15 and ref 16).

4. Check cg limits using the chart (fig 6-3). If cg limits are exceeded, 'the loading must be revised. Refer to paragraph 6.10 for guidance.

5. When either takeoff or landing cg is close to the cg limits, further analysis is required to determine if intermediate flight conditions will exceed limits. Refer to paragraph 6.10 for guidance.

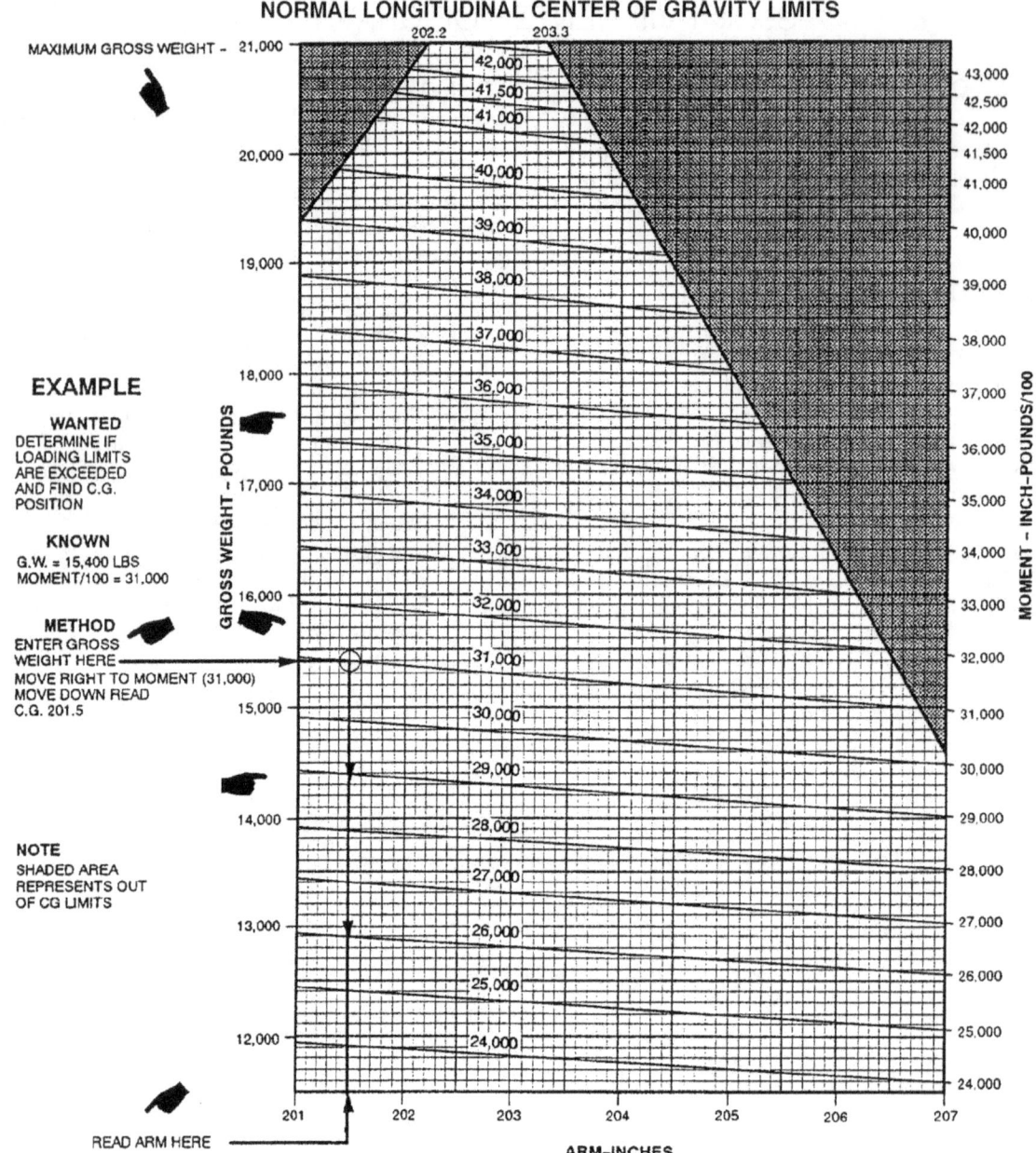

Figure 6-3. Center of Gravity Limits

TM 1-1520-238-10

CHAPTER 7
PERFORMANCE DATA FOR AH-64A HELICOPTERS EQUIPPED WITH T-700-GE-701 ENGINES

Section I. INTRODUCTION

NOTE

This chapter contains performance data for helicopters equipped with T-700-GE-701 engines. Performance data for helicopters equipped with T-700-GE-701C engines is contained in Chapter 7A. Users are authorized to remove the chapter that is not applicable and are not required to carry both chapters on the helicopter.

7.1 PERFORMANCE DATA.

The purpose of this chapter is to provide the best available performance data for the AH-64A helicopter equipped with -701 engines. Regular use of this information will allow maximum safe use of the helicopter. Although maximum performance is not always required, regular use of the information in this chapter is recommended for the following reasons:

   a. Knowledge of performance margins will allow better decisions when unexpected conditions or alternate missions are encountered.

   b. Situations requiring maximum performance will be more readily recognized.

   c. Familiarity with the data will allow performance to be computed more easily and quickly.

   d. Experience will be gained in accurately estimating the effects of conditions for which data is not presented.

NOTE

The information is primarily intended for mission planning and is most useful when planning operations in unfamiliar areas or at extreme conditions. The data may also be used in flight, to establish unit or area standing operating procedures, and to inform ground commanders of performance/risk tradeoffs.

The data presented covers the maximum range of conditions and performance that can reasonably be expected. In each area of performance, the effects of altitude, temperature, gross weight and other parameters relating to that phase of flight are presented. In addition to the presented data, judgment and experience will be necessary to accurately determine performance under a given set of circumstances. The conditions for the data are listed under the title of each chart. The effects of different conditions are discussed in the text accompanying each phase of performance. Where practical, data is presented at conservative conditions. However, NO GENERAL CONSERVATISM HAS BEEN APPLIED.

7.2 OPERATIONAL LIMITS.

CAUTION

Exceeding operational limits can cause permanent damage to critical components. Overlimit operation can decrease performance, cause immediate failure or failure on a subsequent flight.

Applicable limits are shown on the charts as bold lines with a description. Performance generally deteriorates rapidly beyond limits. If limits are exceeded, minimize the amount and time. Enter the maximum value and time beyond limits on DA Form 2408-13 to ensure proper maintenance action is taken.

TM 1-1520-238-10

7.3 USE OF PERFORMANCE CHARTS.

7.3.1 Chart Explanation. The first page of each section describes the chart or charts in that section, and explains how each chart is used.

7.3.2 Reading the Charts. The primary use of each chart is given in an example. The use of a straight edge (ruler or page edge) and a hard fine-point pencil is recommended to avoid cumulative errors. The majority of the charts provide a standard pattern for use as follows: Enter first variable on top left scale, move right to second variable, deflect down at right angles to third variable, deflect left at right angles to fourth variable, and deflect down, etc, until final variable is read out on final scale. In addition to the primary use, other uses of each chart are explained in the text accompanying each set of performance charts. Abbreviations and symbols used in the charts are listed in Appendix B.

NOTE
An example of an auxiliary use of the performance charts follows: Although the hover chart is primarily arranged to find torque required to hover, maximum wheel height for hover can also be found by entering torque available as torque required. In general, any single variable can be found if all others are known. Also, the tradeoffs between two variables can be found. For example, at a given pressure altitude and wheel height you can find the maximum gross weight capability as free air temperature changes.

7.3.3 Data Basis. The type of data used is indicated at the bottom of each performance chart under DATA BASIS. The applicable report and date are also given. The data provided generally is based on one of the following categories.

    a. Flight Test Data. Data obtained by flight test of the aircraft at precisely known conditions using sensitive calibrated instruments.

    b. Calculated Data. Data based on test, but not on flight test of the complete aircraft.

    c. Estimated Data. Data based on estimates using aerodynamic theory or other means but not verified by flight test.

7.4 PERFORMANCE SPECIFIC CONDITIONS.

The data presented is accurate only for specific conditions listed under the title of each chart. Variables for which data is not presented, but which may affect that phase of performance, are discussed in the text. Where data is available or reasonable estimates can be made, the amount that each variable affects performance is given.

7.5 PERFORMANCE GENERAL CONDITIONS.

In addition to the specific conditions, the following general conditions are applicable to the performance data:

7.5.1 Rigging. All airframe and engine controls are assumed to be rigged within allowable tolerances.

7.5.2 Pilot Technique. Normal pilot technique is assumed. Control movements should be smooth and continuous.

7.5.3 Aircraft Variation. Variations in performance between individual helicopters are known to exist; they are considered small, however, and cannot be individually accounted for.

7.5.4 Instrument Variation. The data shown in the performance charts does not account for instrument inaccuracies or malfunctions.

7.5.5 Configurations. Except as otherwise noted, all data is for the primary mission configuration consisting of the basic helicopter plus a pylon and a fully loaded Hellfire missile launcher on each inboard stores station, no pylons or stores on outboard stations.

7.5.6 Types of Fuel. All flight performance data is based on JP-4 fuel. The change in fuel flow and torque availability, when using approved alternate fuels (table 2-8), is insignificant.

7.6 PERFORMANCE DISCREPANCIES.

Regular use of this chapter will also allow monitoring instruments and other helicopter systems for malfunction, by comparing actual performance with planned performance. Knowledge will also be gained concerning the affects of variables for which data is not provided, thereby increasing the accuracy of performance predictions.

7.7 TEMPERATURE CONVERSION.

A temperature conversion chart (fig 7-1) is included in this section for the purpose of converting Fahrenheit temperatures to Celsius.

TM 1-1520-238-10

7.8 ABBREVIATIONS.

Appendix B is a list of abbreviations and symbols used on the charts in this chapter. For units of measure, the same abbreviation applies to either the singular or plural form of the unit.

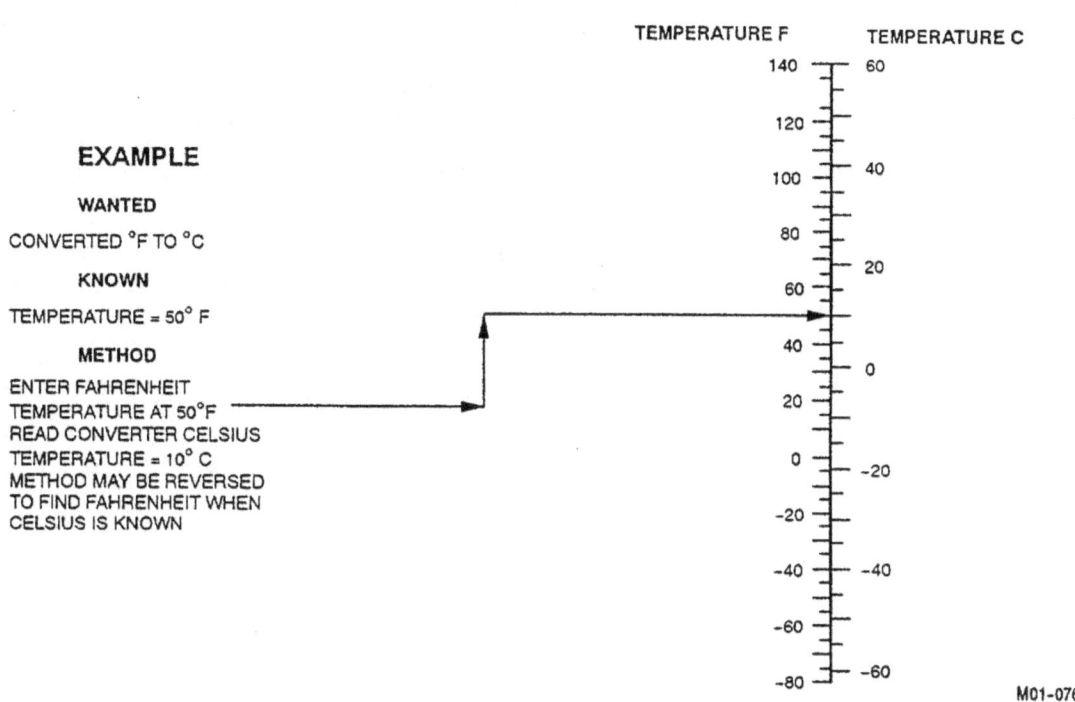

Figure 7-1. Temperature Conversion Chart

Change 3   7-3

Section II. MAXIMUM TORQUE AVAILABLE

7.9 DESCRIPTION.

The maximum torque available chart shows the maximum specification torque available per engine for 30 minute operation (fig 7-2 sheet 1) at various conditions of pressure altitude and free air temperature. Both single and dual engine operation limits are as shown.

The maximum torque available chart for 2.5 minute operation (fig 7-2 sheet 2) shows the maximum specification torque available when one engine is inoperative; only single engine operation limits are shown.

The torque factor charts (figs 7-3 and 7-4) provide an accurate indication of available power for the engines installed in each individual aircraft.

7.10 USE OF CHARTS.

The primary use of the maximum torque available charts (fig 7-2 sheets 1 and 2) is illustrated by the example. To determine the maximum specification torque available, 30 minute limit, it is necessary to know pressure altitude and free air temperature. Enter the left side of the chart (fig 7-2 sheet 1) at the known temperature and move right to the known pressure altitude, and then move down and read maximum specification torque available, 30 minute limit. This is torque per engine. For dual engine operation, if the torque per engine exceeds the two-engine limit, the maximum torque available must be reduced to the two engine limit.

For one engine inoperative, enter the left side of the 2.5 minute limit chart (fig 7-2 sheet 2) at the known temperature and move right to the known pressure altitude, and then move down and read the maximum specification torque available for one engine. If the torque exceeds the one engine limit, maximum torque available must be reduced to the one engine limit.

7.11 CONDITIONS.

The maximum torque available charts (fig 7-2 sheets 1 and 2) are based on 100% rotor rpm, zero airspeed, JP-4 fuel and **ENG INLET** anti-ice switch **OFF**. With **ENG INLET** anti-ice switch **ON**, available torque is reduced by as much as 16%. For example, if the value from the 30-minute limit chart is 90%, with anti-ice **ON**, torque available would be 90-16 = 74%.

7.12 TORQUE FACTOR METHOD.

The torque factor method provides an accurate indication of available power by incorporating ambient temperature effects on degraded engine performance. The torque factor method provides the procedure to determine the maximum dual or single engine torque available for the engines installed in each individual aircraft. The specification power is defined for a newly delivered low time engine. The aircraft HIT log form for each engine provide the engine and aircraft torque factors which are obtained from the maximum power check and recorded to be used in calculating maximum torque available.

7.12.1 Torque Factor Terms. The following terms are used when determining the maximum torque available for an individual aircraft:

   a. Torque Ratio (TR). The ratio of torque available to specification torque at the desired ambient temperature.

   b. Engine Torque Factor (ETF). The ratio of an individual engine torque available to specification torque at reference temperature of 35 °C. The ETF is allowed to range from 0.85 to 1.0.

   c. Aircraft Torque Factor (ATF). The ratio of an individual aircrafts power available to specification power at a reference temperature of 35 °C. The ATF is the average of the ETFs of both engines and its value is allowed to range from 0.9 to 1.0.

7.12.2 Torque Factor Procedure. The use of the ATF or ETF to obtain the TR from figure 7-3 for ambient temperatures between -15 °C and 35 °C is shown by the example. The ATF and ETF values for an individual aircraft are found on the engine HIT log. The TR always equals 1.0 for ambient temperatures of -15 °C and below, and the TR equals the ATF or ETF for temperatures of 35 °C and above.

When the TR equals 1.0 the torque available may be read directly from the specification torque available scales When the TR is less than 1.0, the actual torque available is determined by multiplying the specification torque available by the TR (example for TR = 0.98: 90% TRQ X 0.98 = 88.2% TRQ). The torque conversion chart (fig 7-4) is provided to convert specification data to actual torque available. The single and dual engine transmission limits are shown and should not be exceeded.

MAXIMUM TORQUE AVAILABLE
30-MIN LIMIT 100% N_P, ANTI-ICE OFF
ZERO AIRSPEED

MAXIMUM TORQUE AVAILABLE/IRP
AH-64A
T700-GE-701

EXAMPLE

WANTED

SPECIFICATION TORQUE AVAILABLE
30-MIN. LIMIT.

KNOWN

FAT = +20 °C.
PRESSURE ALTITUTE = 4000 FT.

METHOD

ENTER AT KNOWN FAT = + 20 °C.
MOVE RIGHT TO PRESSURE ALTITUDE = 4000 FT. THEN MOVE DOWN TO READ 97.5% TORQUE AVAILABLE PER ENGINE. THIS DOES NOT EXCEED 2-ENGINE RED LINE. FOR DUAL ENGINE OPERATION, TORQUE IS LIMITED TO 100% PER ENGINE.

CONTINUOUS TORQUE LIMIT 2-ENGINE

CONTINUOUS TORQUE LIMIT 1-ENGINE

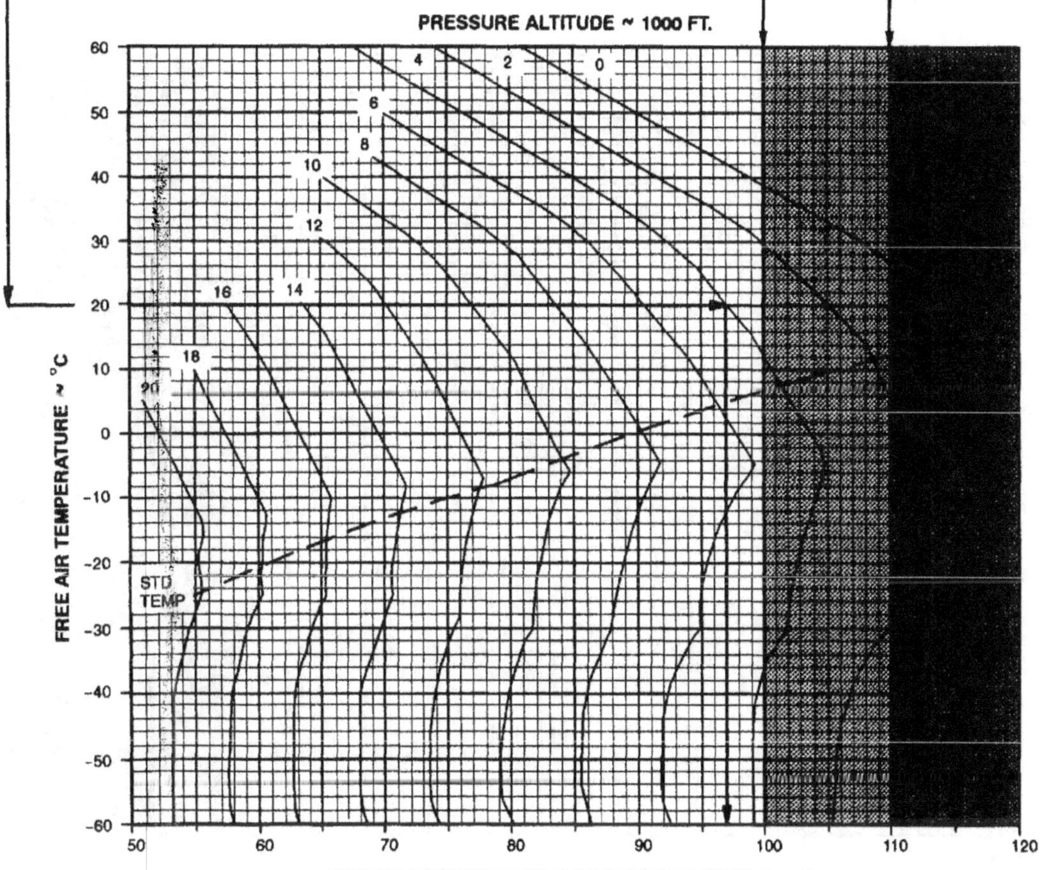

M01-001

Figure 7-2. Maximum Torque Available Chart (Sheet 1 of 2) 701

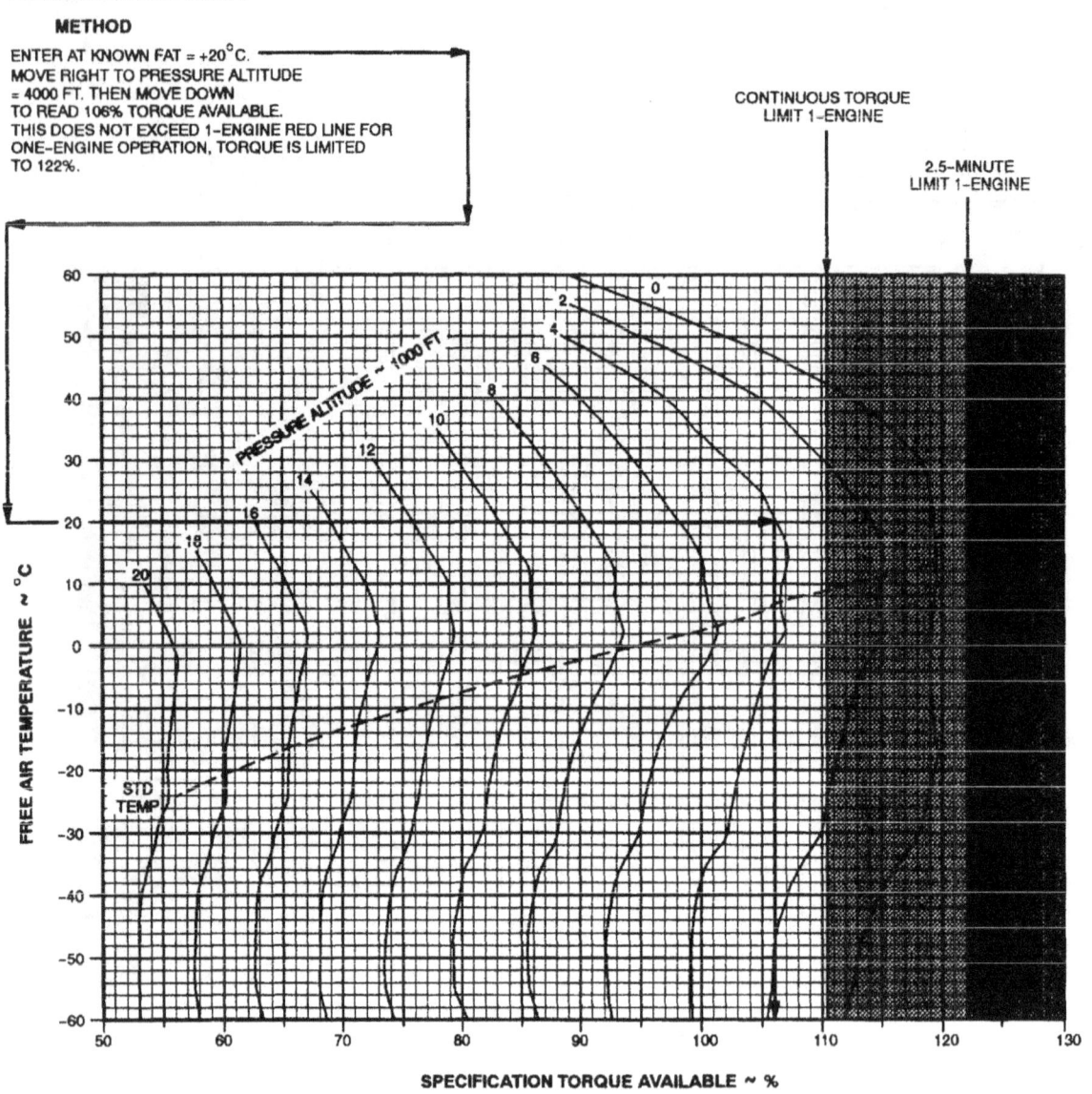

Figure 7-2. Maximum Torque Available Chart (Sheet 2 of 2)

TORQUE FACTOR
T700-GE-701 ENGINE, 100% RPM

TORQUE FACTOR
AH-64A
T700-GE-701

EXAMPLE

WANTED
TORQUE RATIO AND MAXIMUM TORQUE AVAILABLE 30-MIN LIMIT

KNOWN
ATF = .95
PRESSURE ALTITUDE = 4000 FT
FAT = +20° C

METHOD
TO OBTAIN TORQUE RATIO

1. ENTER TORQUE FACTOR CHART AT KNOWN FAT
2. MOVE RIGHT TO THE ATF VALUE
3. MOVE DOWN, READ TORQUE RATIO = .967

TO CALCULATE MAXIMUM TORQUE AVAILABLE:

4. ENTER MAXIMUM TORQUE AVAILABLE CHART 30 MIN LIMIT AT KNOWN FAT
5. MOVE RIGHT TO KNOWN PRESSURE ALTITUDE
6. MOVE DOWN, READ SPECIFICATION TORQUE = 97.5%

TO OBTAIN ACTUAL TORQUE VALUE AVAILABLE FROM THE TORQUE CONVERSION CHART:

7. ENTER TORQUE CONVERSION CHART AT % TORQUE OBTAINED FROM 30 MIN LIMIT CHART
8. MOVE UP TO TORQUE RATIO OBTAINED FROM TORQUE FACTOR CHART
9. MOVE LEFT, READ MAXIMUM TORQUE AVAILABLE = 94.3%

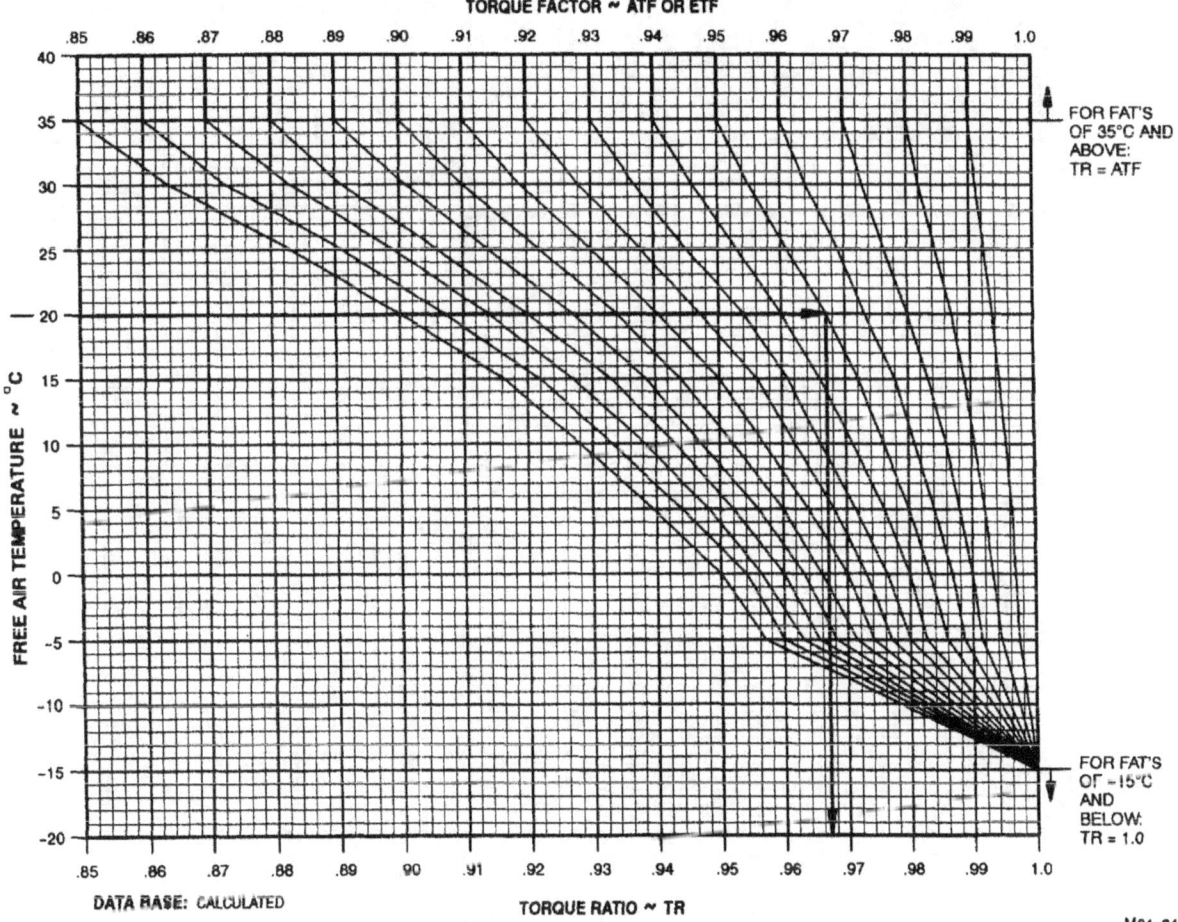

Figure 7-3. Torque Factor Chart

TORQUE CONVERSION

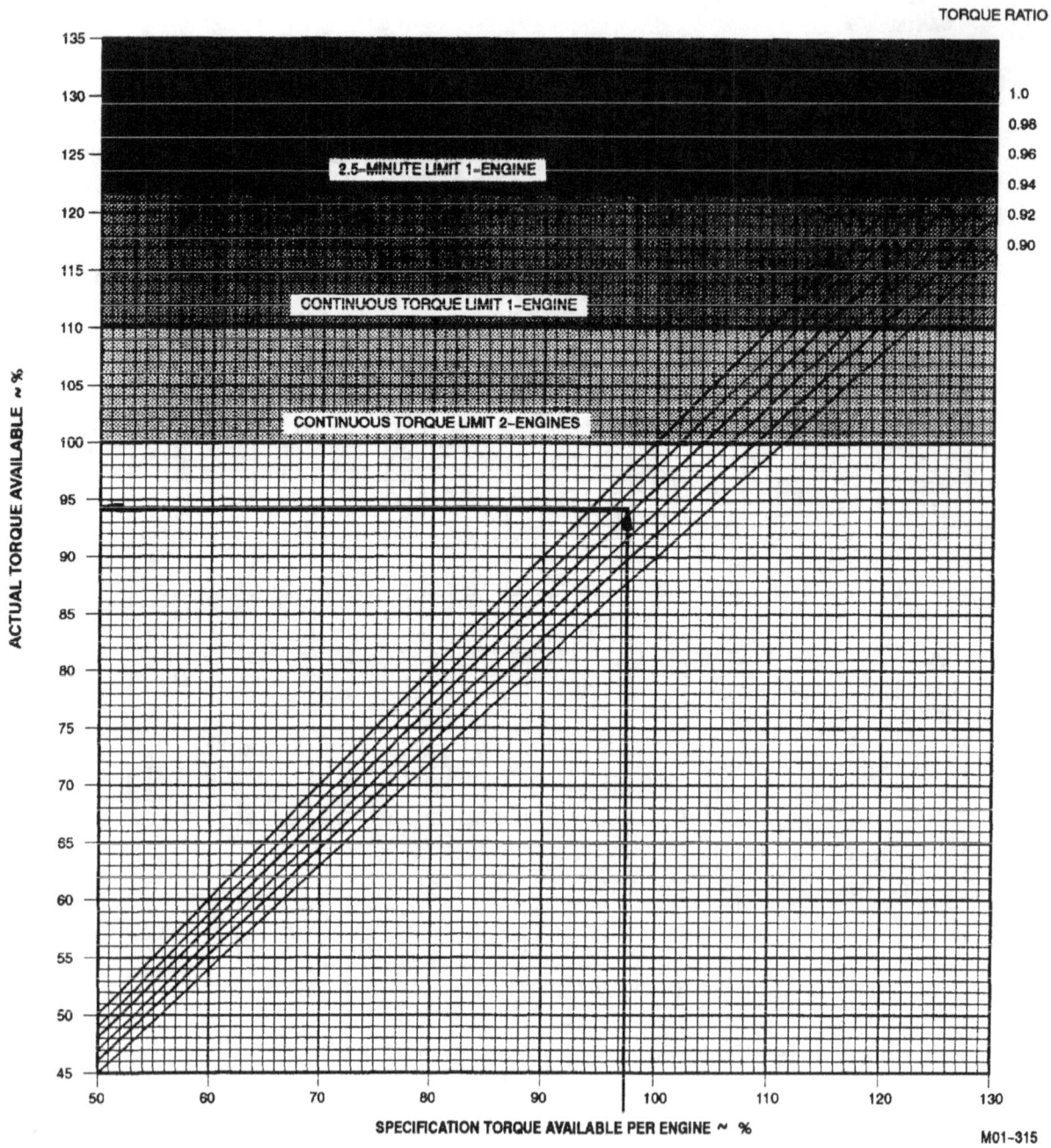

Figure 7-4. Torque Conversion Chart 701

Section III. HOVER CEILING

7.13 DESCRIPTION.

The hover ceiling chart (fig 7-5) presents the maximum gross weight for hover at various conditions of pressure altitude, free air temperature, and wheel height, using maximum torque available, 30 minute limit.

7.14 USE OF CHART.

The primary use of the chart is illustrated by the example. To determine the maximum gross weight for hover, it is necessary to know the pressure altitude, free air temperature, and desired wheel height. Enter the chart at the pressure altitude, move right to FAT, move down to the desired wheel height, and then move left and read maximum gross weight.

7.15 CONDITIONS.

The hover ceiling chart is based on maximum torque available, 30 minute limit, 100% rotor rpm, ATF = 1.0, **ENG INLET ANTI-ICE** switch **OFF** and rotor **BLADE** de-ice switch off. For **ENG INLET ANTI-ICE** switch **ON**, use dashed lines. Applicable configuration is all external stores except four external fuel tanks. For the four external tank configuration, reduce the maximum gross weight for hover, as calculated from the hover ceiling chart, by 11 pounds for each 1000 pounds of gross weight. See the examples below.

HOVER CEILING

EXAMPLE I

WANTED

MAXIMUM GROSS WEIGHT FOR HOVER AT 10-FOOT WHEEL HEIGHT, ENGINE INLET ANTI-ICE OFF.

KNOWN

PRESSURE ALTITUDE = 10,000 FEET.
FAT = -10°C.
WHEEL HEIGHT = 10 FEET.

METHOD

ENTER PRESSURE ALTITUDE SCALE AT 10,000 FT.
MOVE RIGHT TO -10°C FAT, SOLID LINE.
MOVE DOWN TO 10 FEET WHEEL HEIGHT.
MOVE LEFT TO READ GROSS WEIGHT TO HOVER = 16,400 POUNDS.
WITH 4 EXTERNAL TANKS INSTALLED HOVER GW = 16,400 -11 (16,400/1000)
= 16,220 POUNDS.

EXAMPLE II

WANTED

MAXIMUM GROSS WEIGHT FOR HOVER AT 10-FOOT WHEEL HEIGHT, ENGINE INLET ANTI-ICE ON.

KNOWN

PRESSURE ALTITUDE = 10,000 FEET.
FAT = -10° C.
WHEEL HEIGHT = 10 FEET.

METHOD

ENTER PRESSURE ALTITUDE SCALE AT 10,000 FT.
MOVE RIGHT TO -10°C FAT, DASHED LINE.
MOVE DOWN TO 10 FEET WHEEL HEIGHT.
MOVE LEFT TO READ GROSS WEIGHT TO HOVER = 15,200 POUNDS.
WITH 4 EXTERNAL TANKS INSTALLED HOVER GW = 15,200 -11 (15,200/1000)
= 15,033 POUNDS.

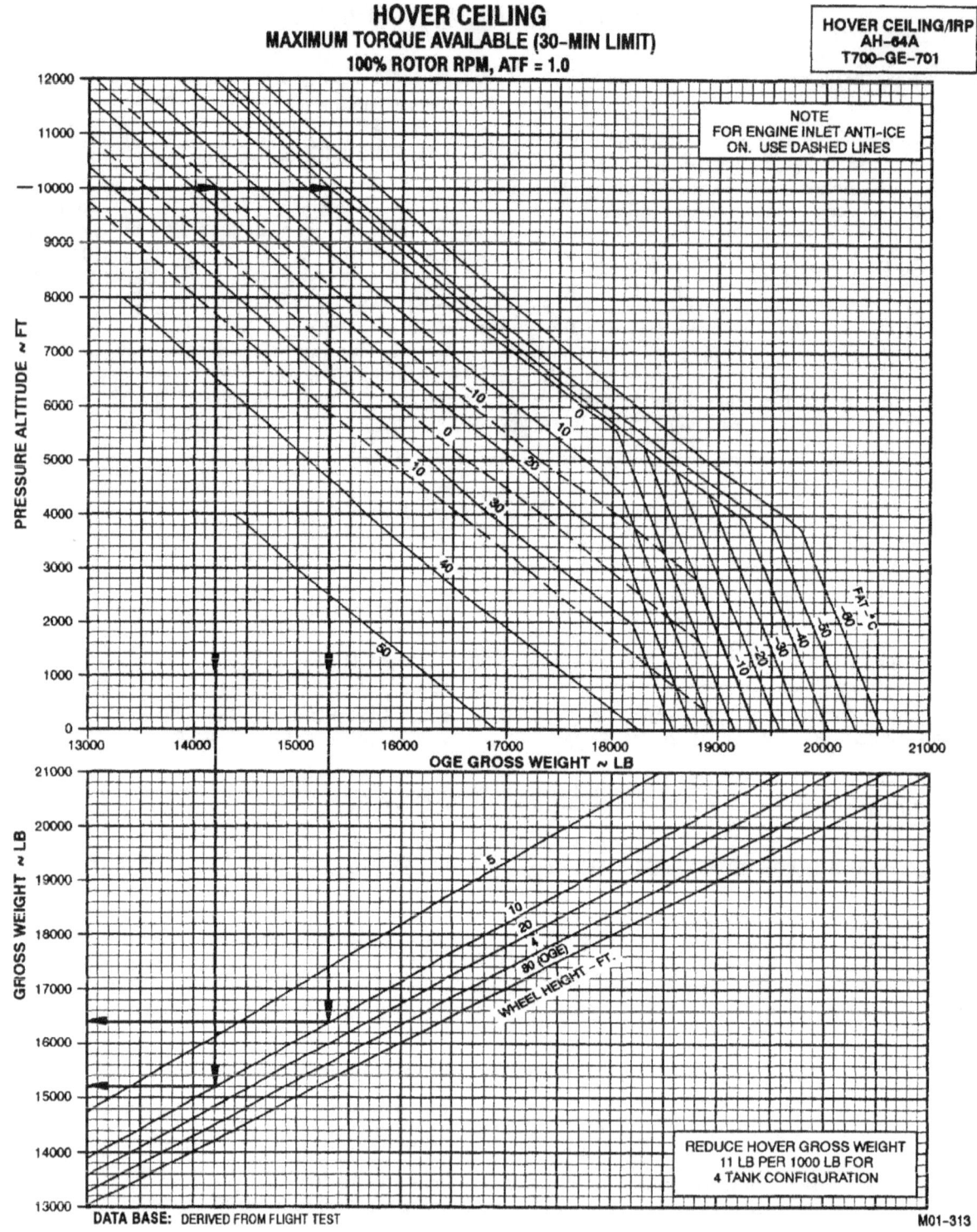

Figure 7-5. Hover Ceiling Chart 701

Section IV. HOVER LIMITS

7.16 DESCRIPTION.

The hover chart (fig 7-6) shows the torque required to hover at various conditions of pressure altitude, free air temperature, gross weight, wheel height, and with external tanks or without external tanks.

7.17 USE OF CHART.

The primary use of the chart is illustrated by the example. To determine the torque required to hover, it is necessary to know the pressure altitude, free air temperature, gross weight, and desired wheel height. Enter the upper right grid at the known pressure altitude, move right to the temperature, move down to the gross weight, move left to the desired wheel height, and then move down and read the torque required to hover.

In addition to its primary use, the hover chart maybe used to predict the maximum hover height. To determine maximum hover height, it is necessary to know pressure altitude, free air temperature, gross weight, and maximum torque available. Enter the known pressure altitude, move right to the temperature, move down to the gross weight, then move left to intersection with maximum torque available and read wheel height. This wheel height is the maximum hover height.

The hover chart may also be used to determine the maximum gross weight for hover at a given wheel height, pressure altitude, and temperature condition. Enter at the known pressure altitude, move right to the temperature, then draw a line down to the bottom of the lower grid. Now enter lower grid at maximum torque available, move up to wheel height, and then move right to intersect the previously drawn line and read gross weight. This is maximum gross weight at which the helicopter will hover.

7.18 CONDITIONS.

The hover chart is based on calm wind, level surface, 100% rotor rpm and rotor **BLADE** de-ice switch off. With rotor **BLADE** de-ice switch **ON,** torque required will increase 1.4%

TM 1-1520-238-10

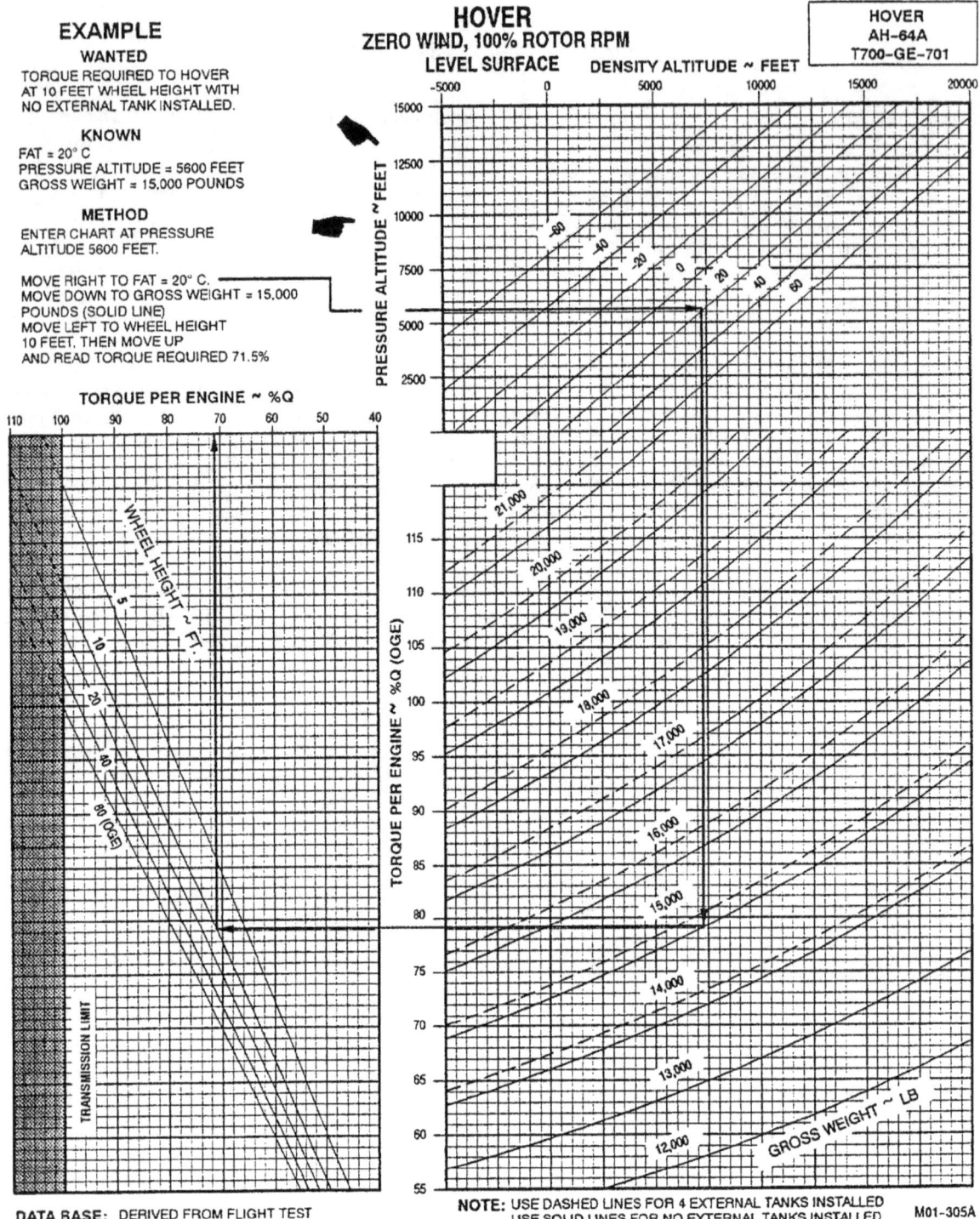

Figure 7-6. Hover Chart Mill

7-12 Change 3

Section V. CRUISE

7.19 DESCRIPTION.

The cruise charts (figs **7-7** thru **7-17**) present the level-flight torque required and total fuel flow at various conditions of airspeed, pressure altitude, free air temperature, and gross weight. Cruise charts are provided for pressure altitudes from sea level to 16,000 feet in 2000-foot increments. Free air temperatures range from -50° to +60° in 10 °C increments. In addition to basic cruise information, the charts show speed for maximum range, maximum endurance, and maximum rate of climb. Change in torque with change in frontal area information is presented in the upper left corner of each chart.

7.20 USE OF CHARTS.

The primary uses of the charts are illustrated by the examples. To use the charts, it is usually necessary to know the planned pressure altitude, estimated free air temperature, planned cruise speed, IAS, and gross weight. First, select the proper chart on the basis of pressure altitude and FAT. Enter the chart at the cruise airspeed, IAS, move right and read TAS, move left to the gross weight, move down and read torque required, and then move up and read associated fuel flow. Maximum performance conditions are determined by entering the chart where the maximum range line or the maximum endurance and rate-of-climb line intersects the gross weight line; then read airspeed, fuel flow, and torque required. Normally, sufficient accuracy can be obtained by selecting the chart nearest the planned cruising altitude and FAT or, more conservatively, by selecting the chart with the next higher altitude and FAT. If greater accuracy is required, interpolation between altitudes and/or temperatures is permissible. To be conservative, use the gross weight at the beginning of the cruise flight. For greater accuracy on long flights, however, it is preferable to determine cruise information for several flight segments to allow for the decreasing gross weight.

7.20.1 Airspeed. True and indicated airspeeds are presented at opposite sides of each chart. On any chart, indicated airspeed can be directly converted to true airspeed (or vice versa) by reading directly across the chart without regard for the other chart information. The applicable MACH No. or gross weight maximum permissible airspeed limits (V_{NE}) determined from figure 5-2 appear on the appropriate charts.

7.20.2 Torque. Since pressure altitude and temperature are fixed for each chart, torque required varies according to gross weight and airspeed. The torque required and the torque limits shown on these charts are for dual engine operation. The torque available shown on these charts are maximum continuous torque available and maximum torque available, 30 minute limit, where less than the maximum torque-two-engine transmission limit. These torque lines are the minimum torque available for ATF = 1 at the engine limits specified in Chapter 5. Higher torque than that represented by these lines maybe used if it is available without exceeding the limitations presented in Chapter 5. The limit torque line shown on these charts is for dual engine transmission limit and is defined as 100% torque. An increase or decrease in torque required because of drag area change is calculated by adding or subtracting the change in torque from the torque change (ΔQ) curve on the chart, and then reading the new total fuel flow.

7.20.3 Fuel Flow. Fuel flow scales are provided opposite the torque scales. On any chart, torque maybe converted directly to fuel flow without regard to other chart information. Sea level ground fuel flow at flat pitch and 100% N_p is approximately 550 pounds per hour.

7.20.4 Maximum Range. The maximum range lines indicate the combinations of gross weight and airspeed that will produce the greatest flight range per pound of fuel under zero wind conditions.

7.20.5 Maximum Endurance and Rate of Climb. The maximum endurance and rate of climb lines indicate the combinations of gross weight and airspeed that will produce the maximum endurance and the maximum rate of climb. The torque required for level flight at this condition is a minimum, providing a minimum fuel flow (maximum endurance) and a maximum torque change available for climb (maximum rate of climb).

7.20.6 Change In Frontal Area. Since the cruise information is given for the primary mission configuration, adjustments to torque should be made when operating with alternative wing-stores configurations. To determine the change in torque, first obtain the appropriate multiplying factor from the drag chart (fig 7-18), then enter the cruise chart at the planned cruise speed TAS, move right to the broken ΔQ line, and move up and read ΔQ. Multiply ΔQ by the multiplying factor to obtain change in torque, then add or subtract change in torque from torque required for the primary mission configuration. Enter the cruise chart at resulting torque required, move up, and read fuel flow. If the resulting torque required exceeds the governing torque limit, the torque required must be reduced to the limit. The resulting reduction in airspeed may be found by subtracting the change in torque from the limit torque; then enter the cruise chart at the reduced torque, and move up to the gross weight. Move left or right to read TAS or IAS. To determine the airspeed for maximum range for alternative wing stores configuration, reduce the value from the cruise chart by 2 knots for each 5 square feet increase in drag area, ΔF, or increase maximum range airspeed 2 knots for each 5 square feet reduction in drag area. For example, for 16 Hellfire configuration $\Delta F = 9.6$ square feet, from figure 7-18. Therefore, maximum range airspeed would be reduced by 2/5 x 9.6= 3.84 knots, or approximately 4 knots.

7.20.7 Additional Uses. The low-speed end of the cruise chart (below 40 knots) is primarily to familiarize you with the low-speed power requirements of the helicopter. It shows the power margin available for climb or acceleration during maneuvers, such as NOE flight. At zero airspeed, the torque represents the torque required to hover out of ground effect. In general, mission planning for low-speed flight should be based on hover out of ground effect.

7.21 CONDITIONS.

The cruise charts are based on 100% rotor rpm, ATF or ETF = 1.0, **ENG INLET ANTI-ICE** switch **OFF**, JP-4 fuel and dual-engine operation. Engine inlet anti-ice and rotor blade de-ice effect are as follows:

a. With ENG **INLET ANTI-ICE** switch **ON**, fuel flow will increase approximately 60 pounds per hour, maximum torque available could be reduced by as much as 16%, and maximum continuous torque available could be reduced by as much as 17%.

b. With rotor **BLADE** de-ice switch **ON**, fuel flow will increase approximately 30 pounds per hour, and torque required will increase 1.4%.

For example, with **ENG INLET ANTI-ICE** and rotor **BLADE** de-ice off, torque required, from cruise chart is 50%, and maximum continuous torque is 92%. With **ENG INLET ANTI-ICE** and rotor **BLADE** de-ice **ON**, torque required will be 50 + 1.4= 51.4%, and maximum continuous torque will be approximately 92 - 17 = 75%.

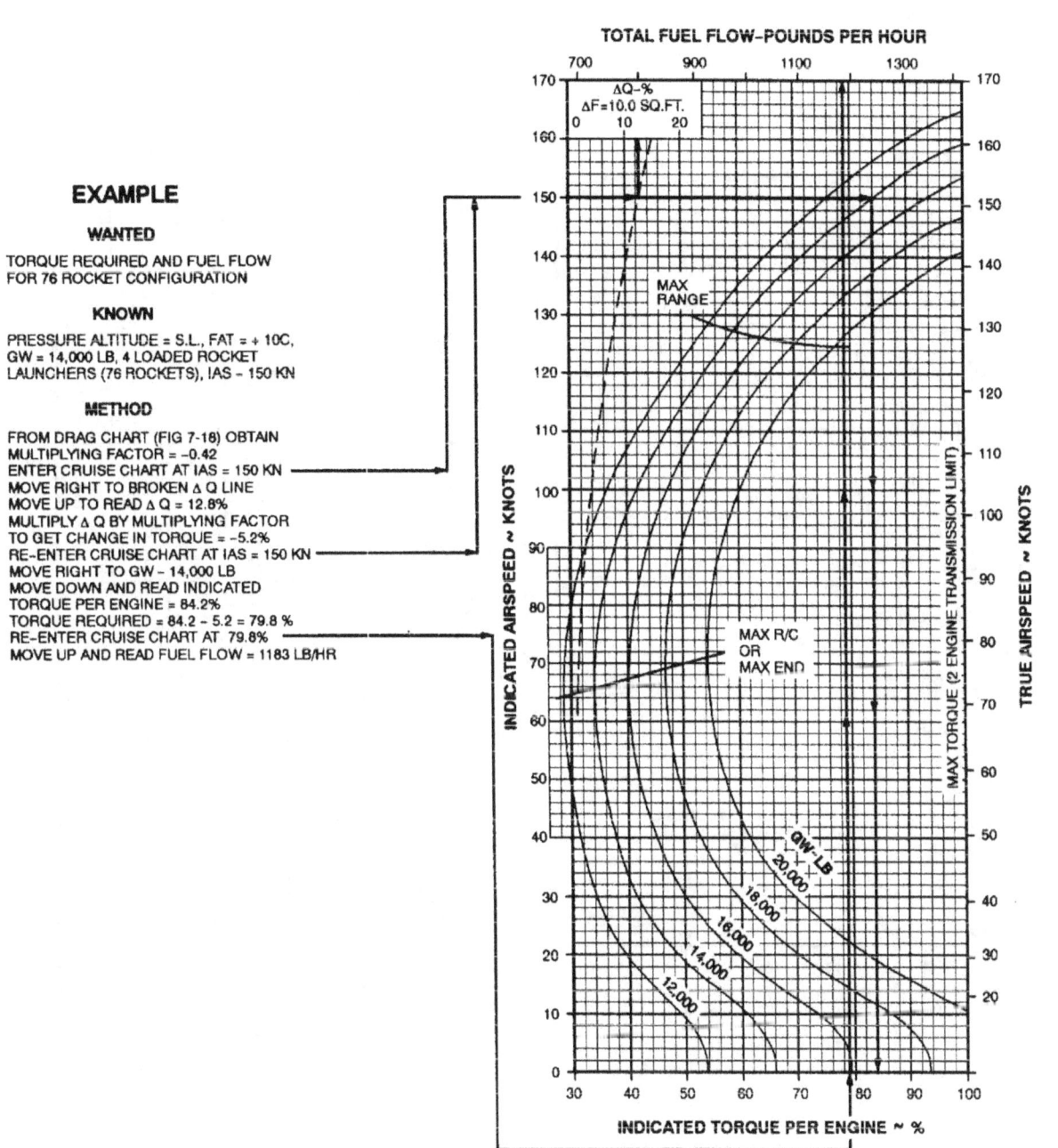

Figure 7-7. Cruise Chart, Example

CRUISE
PRESSURE ALTITUDE - SEA LEVEL
100% RPM, 8 HELLFIRE CONFIGURATION, JP-4 FUEL

CRUISE
AH-64A
T700-GE-701

EXAMPLE I

WANTED
TORQUE REQUIRED, AIRSPEED, AND FUEL FLOW FOR MAXIMUM RANGE.

KNOWN
PRESSURE ALTITUDE = SL, FAT = +10°C, AND GROSS WEIGHT = 14,000 POUNDS.

METHOD
AT THE INTERSECTION OF THE MAXIMUM RANGE LINE AND THE 14,000 POUND LINE
MOVE LEFT, READ IAS = 127 KN.
MOVE RIGHT, READ TAS = 130 KN.
MOVE UP, READ TOTAL FUEL FLOW = 970 LB/HR.
MOVE DOWN, READ INDICATED TORQUE/ENGINE = 59%.

EXAMPLE II

WANTED
TORQUE REQUIRED, AIRSPEED, AND FUEL FLOW FOR MAXIMUM ENDURANCE.

KNOWN
PRESSURE ALTITUDE = SL, FAT = +10°C, AND GROSS WEIGHT = 14,000 POUNDS.

METHOD
AT THE INTERSECTION OF THE MAXIMUM ENDURANCE AND THE 14,000 POUND LINE
MOVE LEFT, READ IAS = 66KN.
MOVE RIGHT, READ TAS = 73 KN.
MOVE UP, READ TOTAL FUEL FLOW = 717 LB/HR.
MOVE DOWN, READ INDICATED TORQUE/ENGINE = 34%

EXAMPLE III
(INTERPOLATION NOT ILLUSTRATED)

WANTED
MAXMIUM ENDURANCE, AIRSPEED, FUEL FLOW, AND TORQUE.

KNOWN
PRESSURE ALTITUDE = 1000 FT, FAT = +15°C, AND GROSS WEIGHT = 14,000 POUNDS.

METHOD
READ AIRSPEED, TORQUE, AND FUEL FLOW FOR EACH ADJACENT ALTITUDE AND FAT, THEN INTERPOLATE BETWEEN FAT AND ALTITUDE AS FOLLOWS:

ALTITUDE	SEALEVEL		2000 FEET		SOLUTION 1000 FEET
FAT	20	10	20	10	15
TORQUE	34	34	35	35	34
FUEL FLOW	720	717	700	700	709
IAS	65	66	63	64	64
TAS	73	73	73	73	73

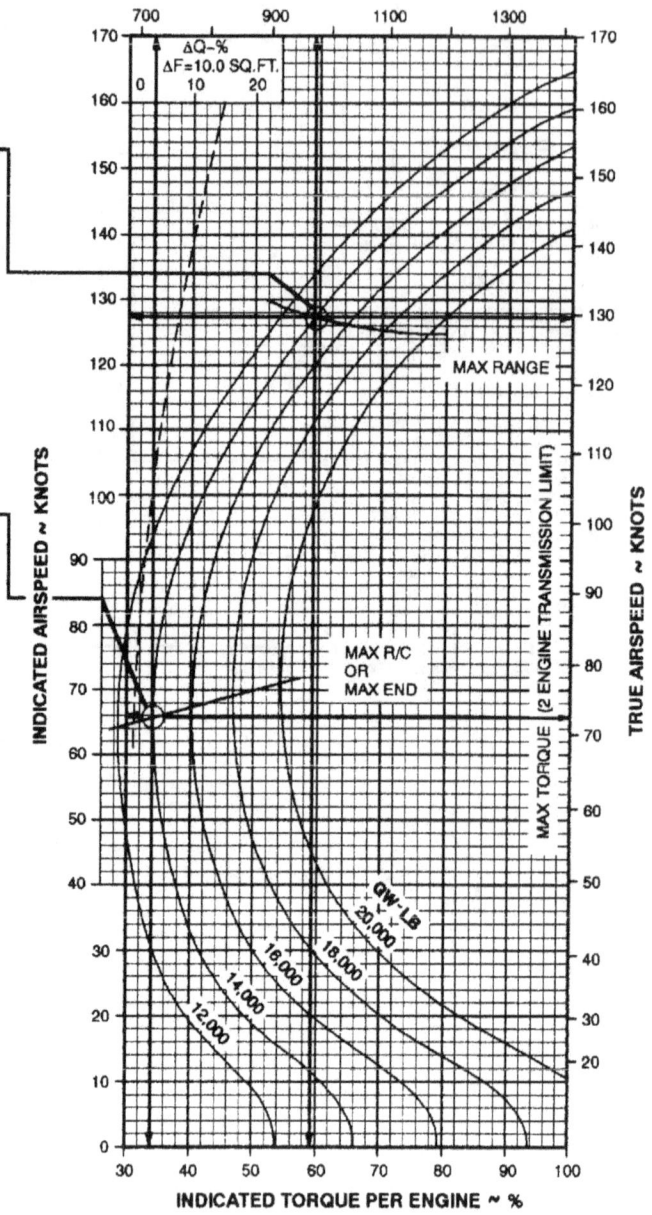

Figure 7-8. Cruise Chart, Example, Sea Level, +10°C

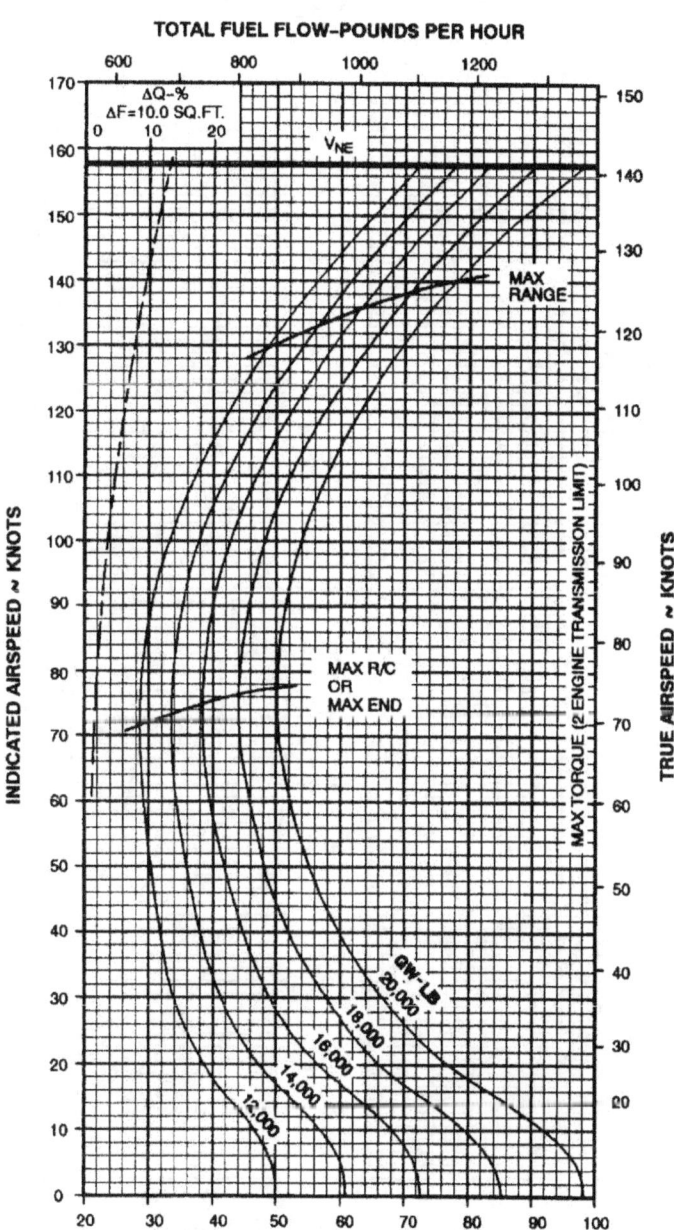

Figure 7-9. Cruise Chart, Sea Level -50 °C (Sheet 1 of 7)

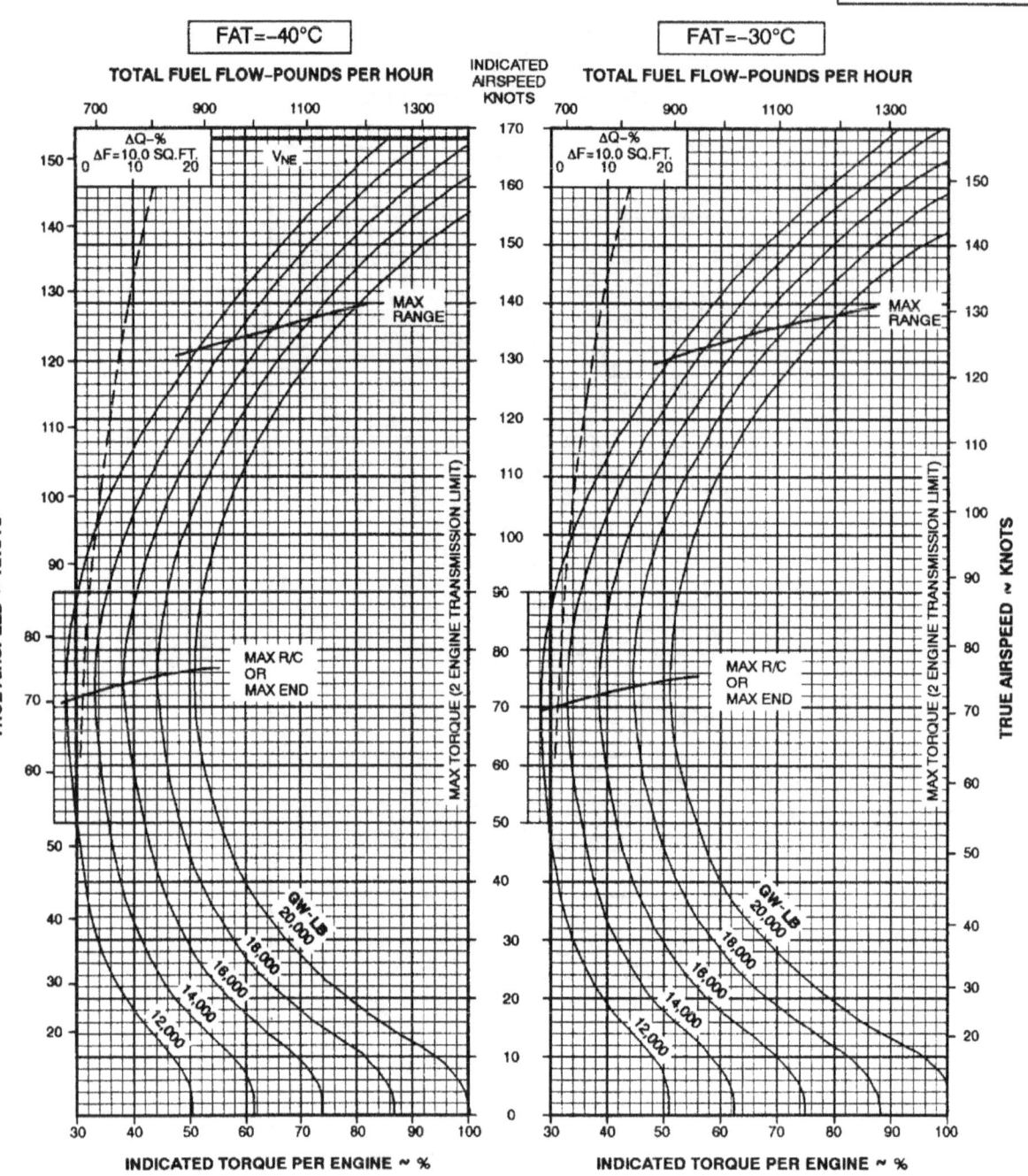

Figure 7-9. Cruise Chart, Sea Level, -40° and -30°C (Sheet 2 of 7)

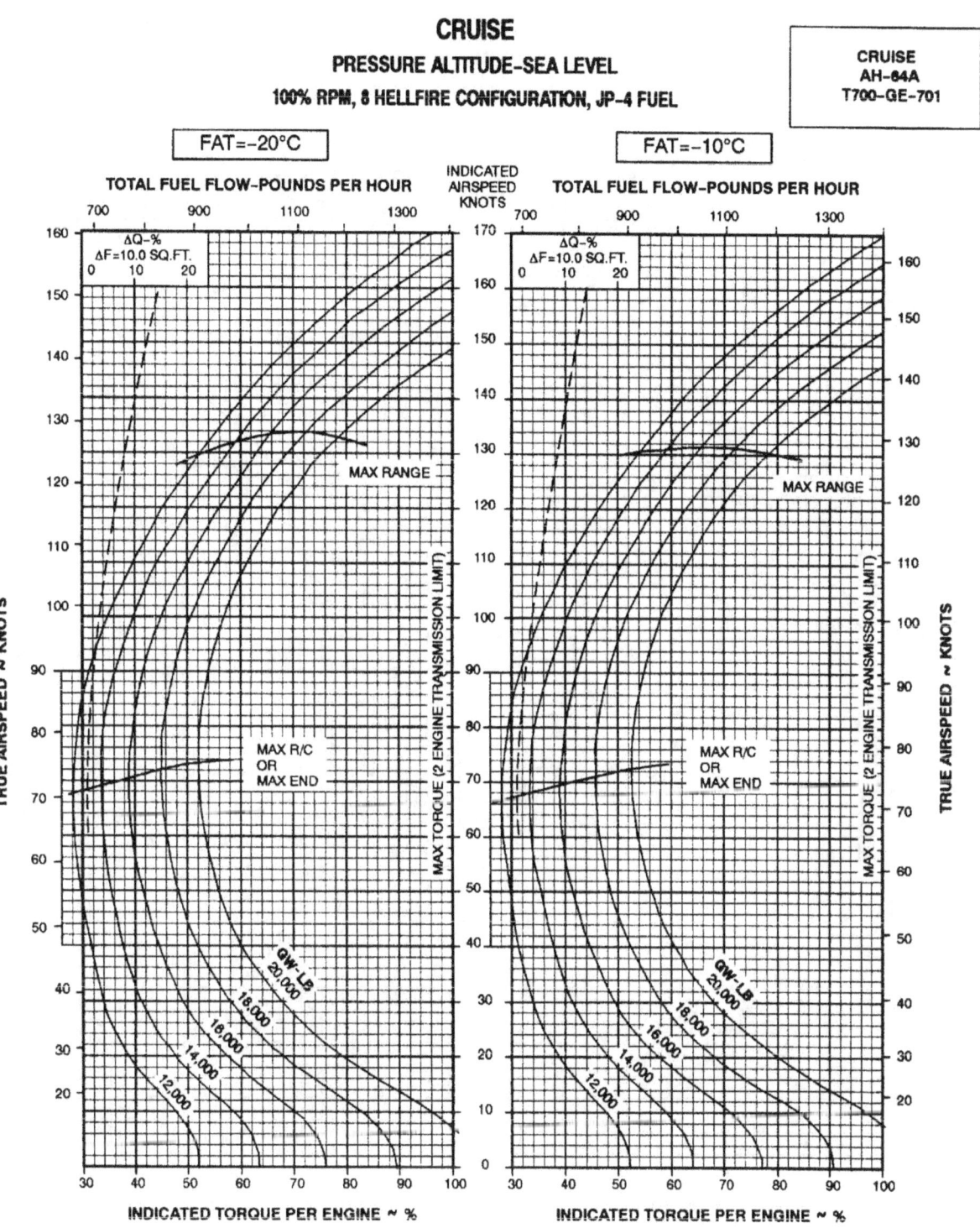

Figure 7-9. Cruise Chart, Sea Level, -20° and -10°C (Sheet 3 of 7)

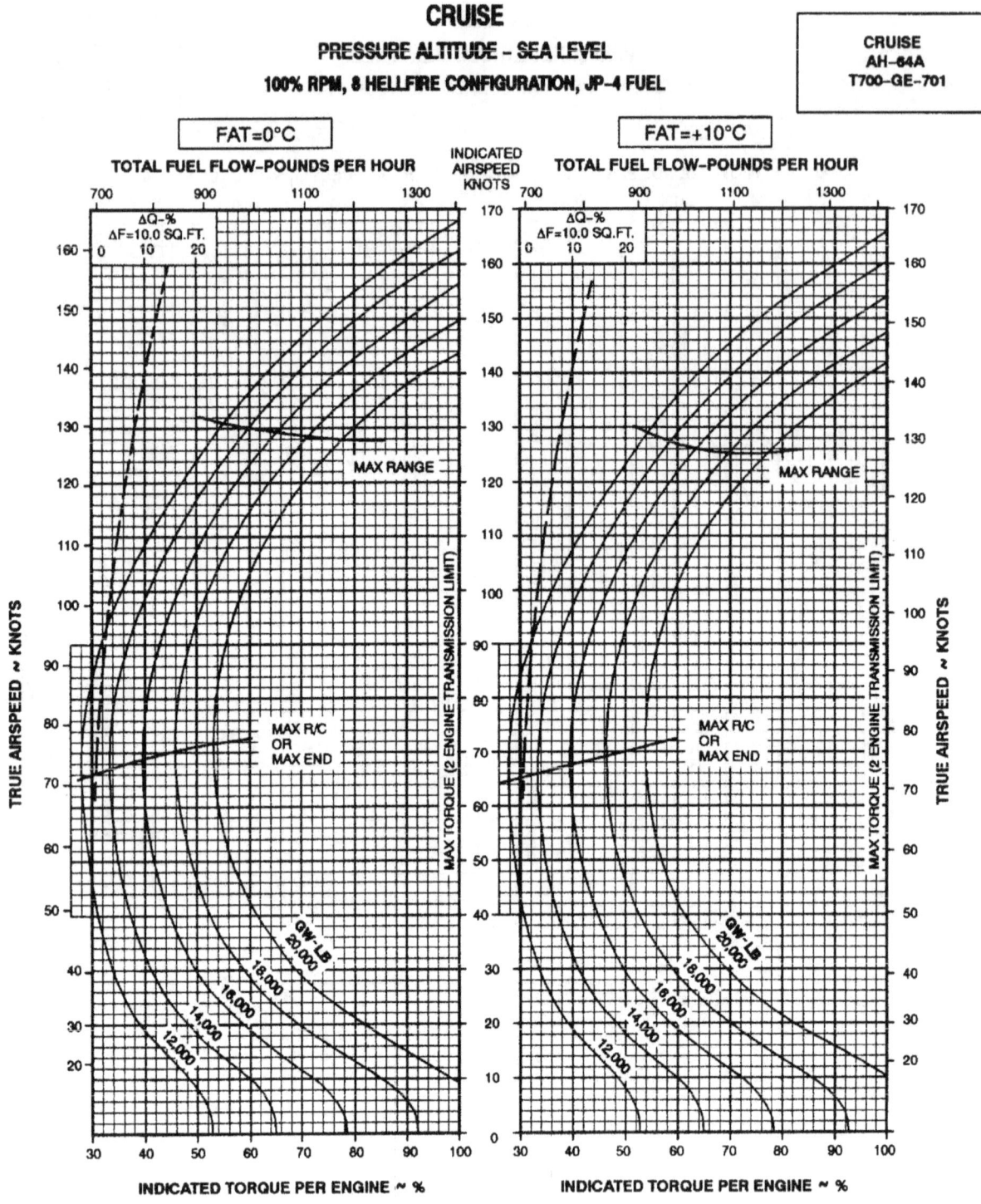

Figure 7-9. Cruise Chart, Sea Level, 0° and +10°C (Sheet 4 of 7)

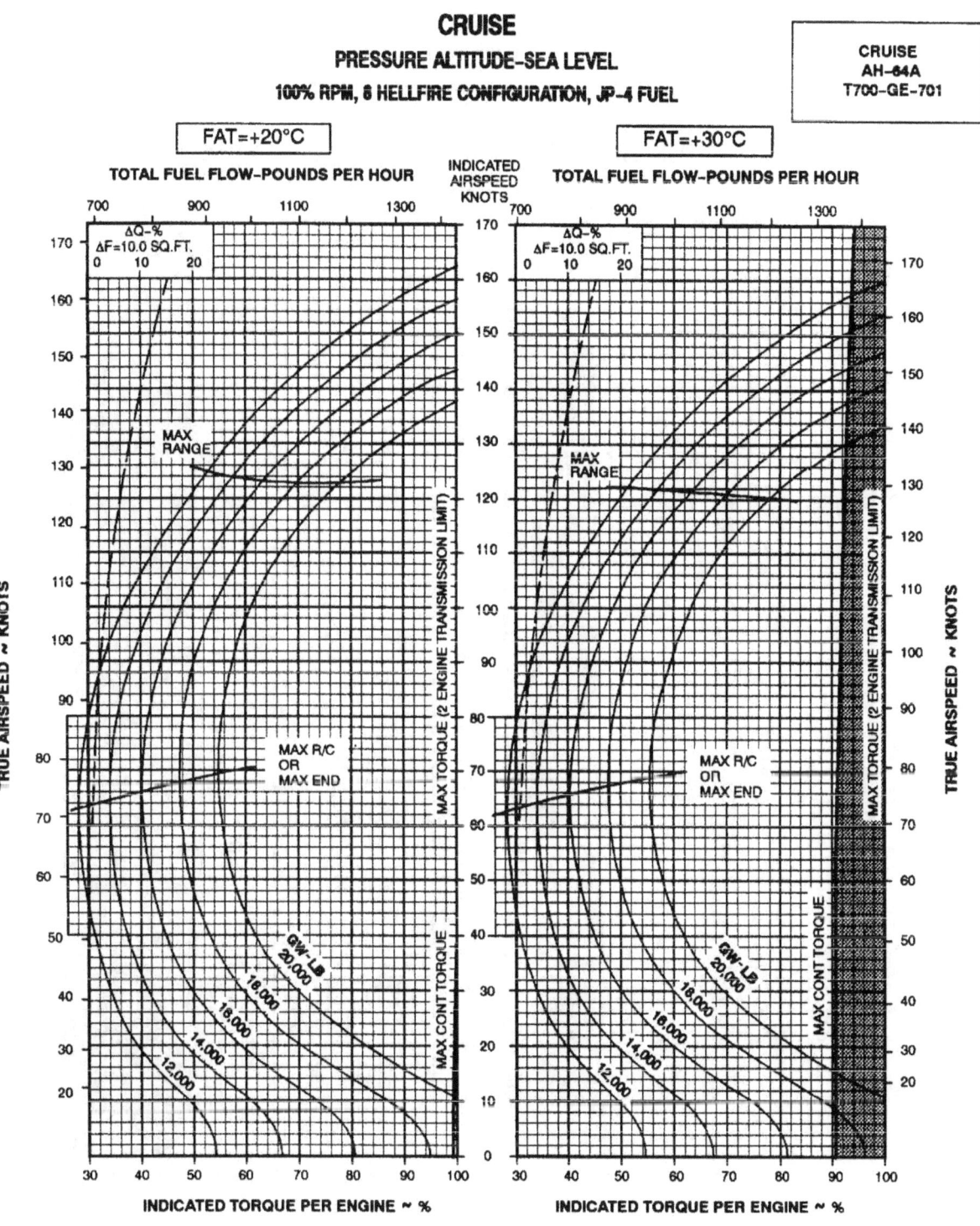

Figure 7-9. Cruise Chart, Sea Level, +20° and +30°C (Sheet 5 of 7)

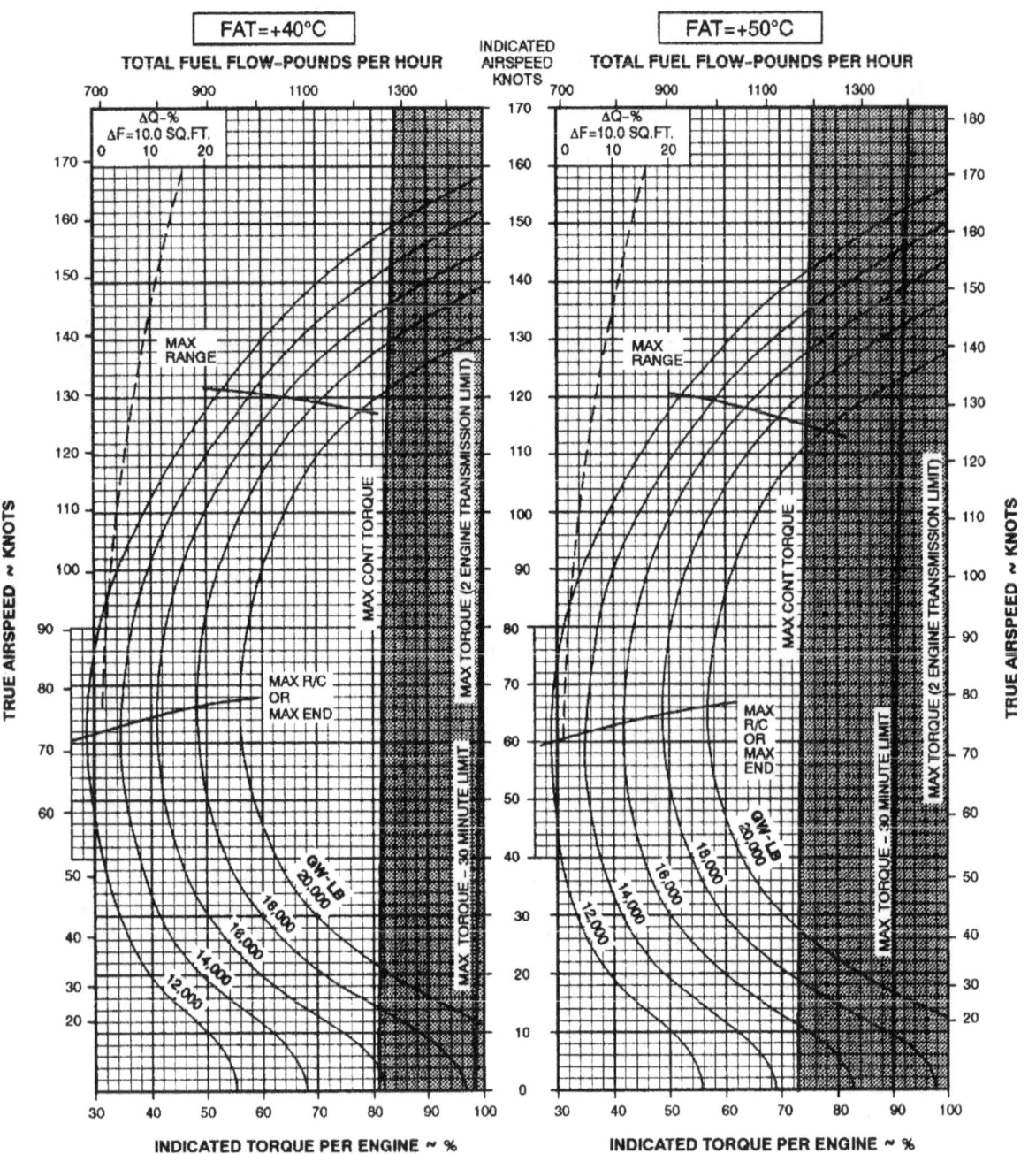

Figure 7-9. Cruise Chart, Sea Level, +40° and +50°C (Sheet 6 of 7) 701

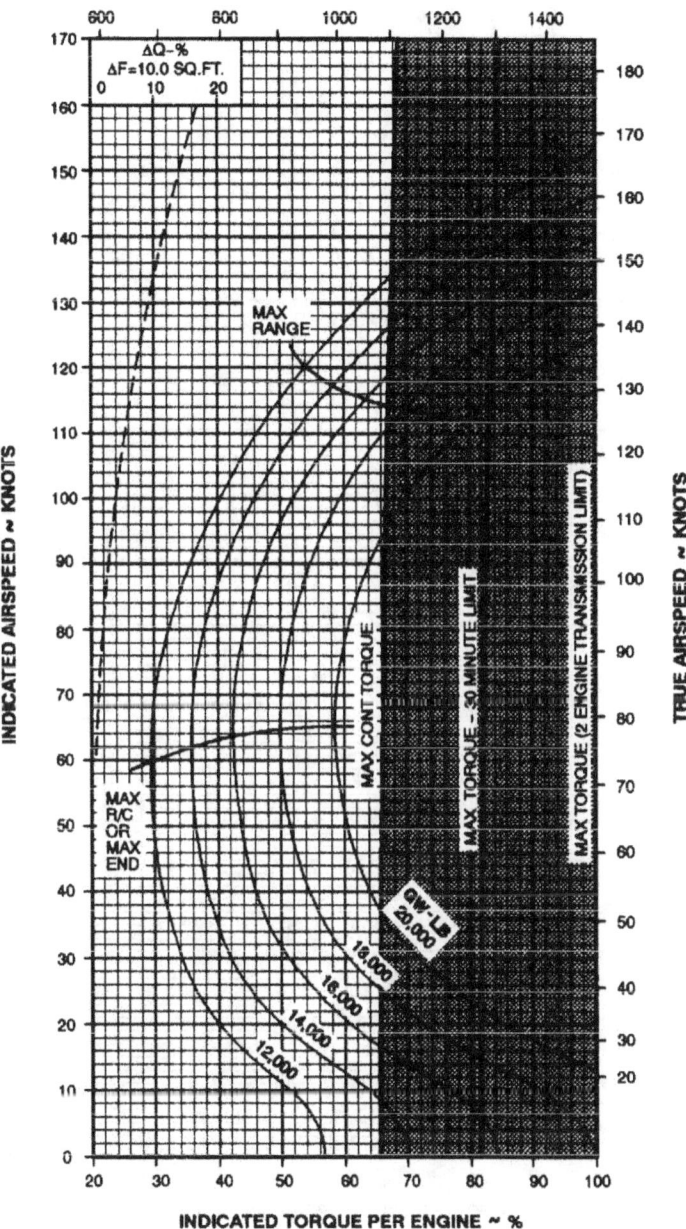

Figure 7-9. Cruise Chart, Sea Level, +60°C (Sheet 7 of 7)

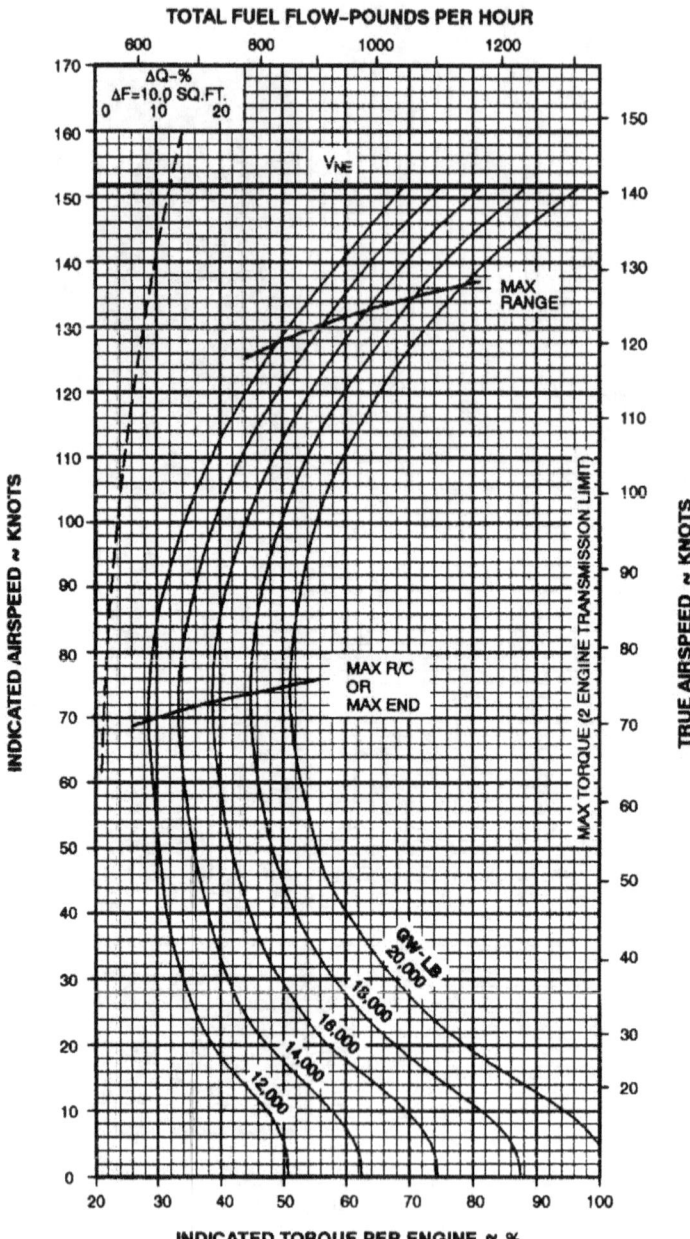

Figure 7-10. Cruise Chart, 2,000 Feet, -50°C (Sheet 1 of 7)

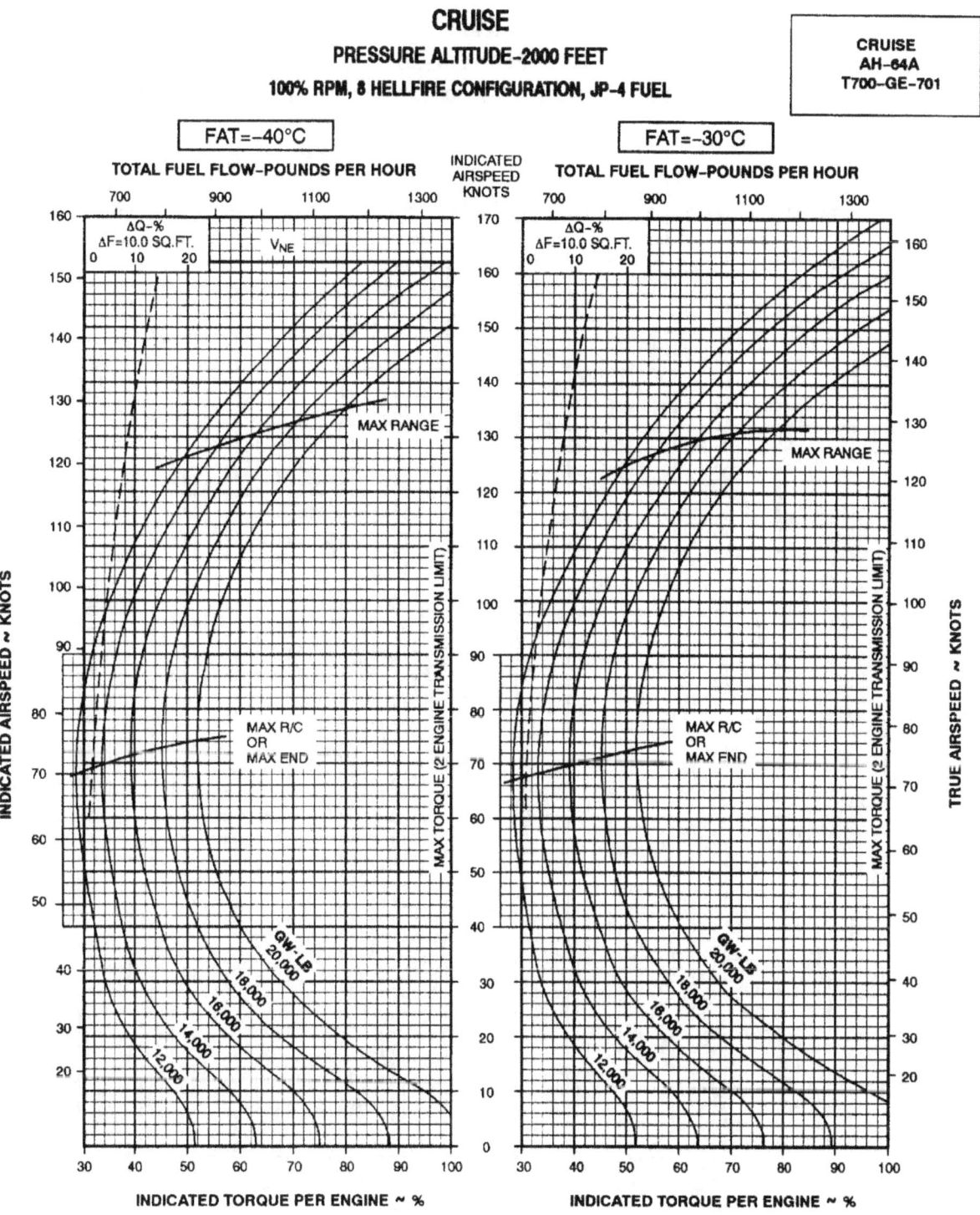

Figure 7-10. Cruise Chart, 2,000 Feet, -40° and -30°C (Sheet 2 of 7)

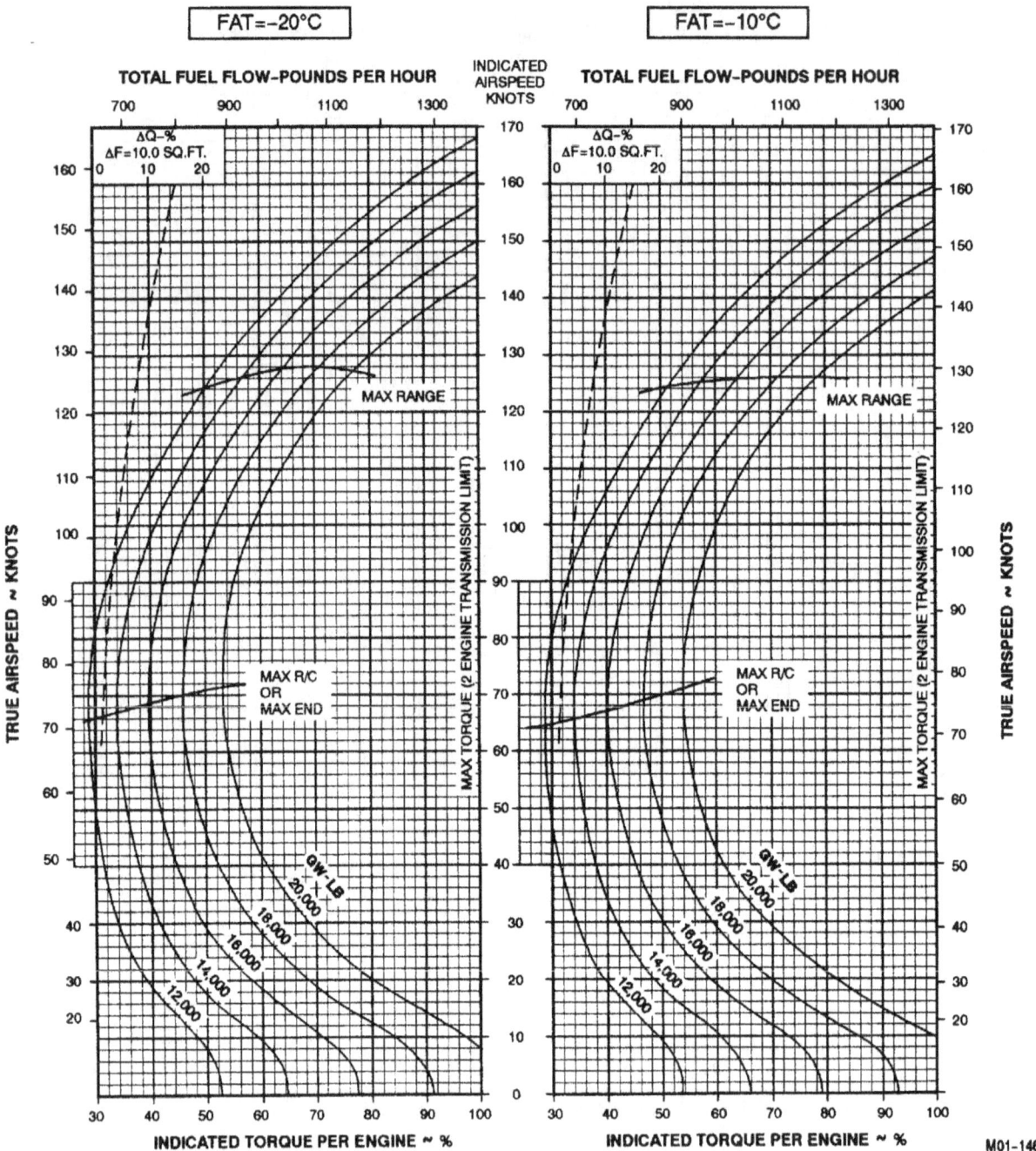

Figure 7-10. Cruise Chart, 2,000 Feet, -20° and -10°C (Sheet 3 of 7)

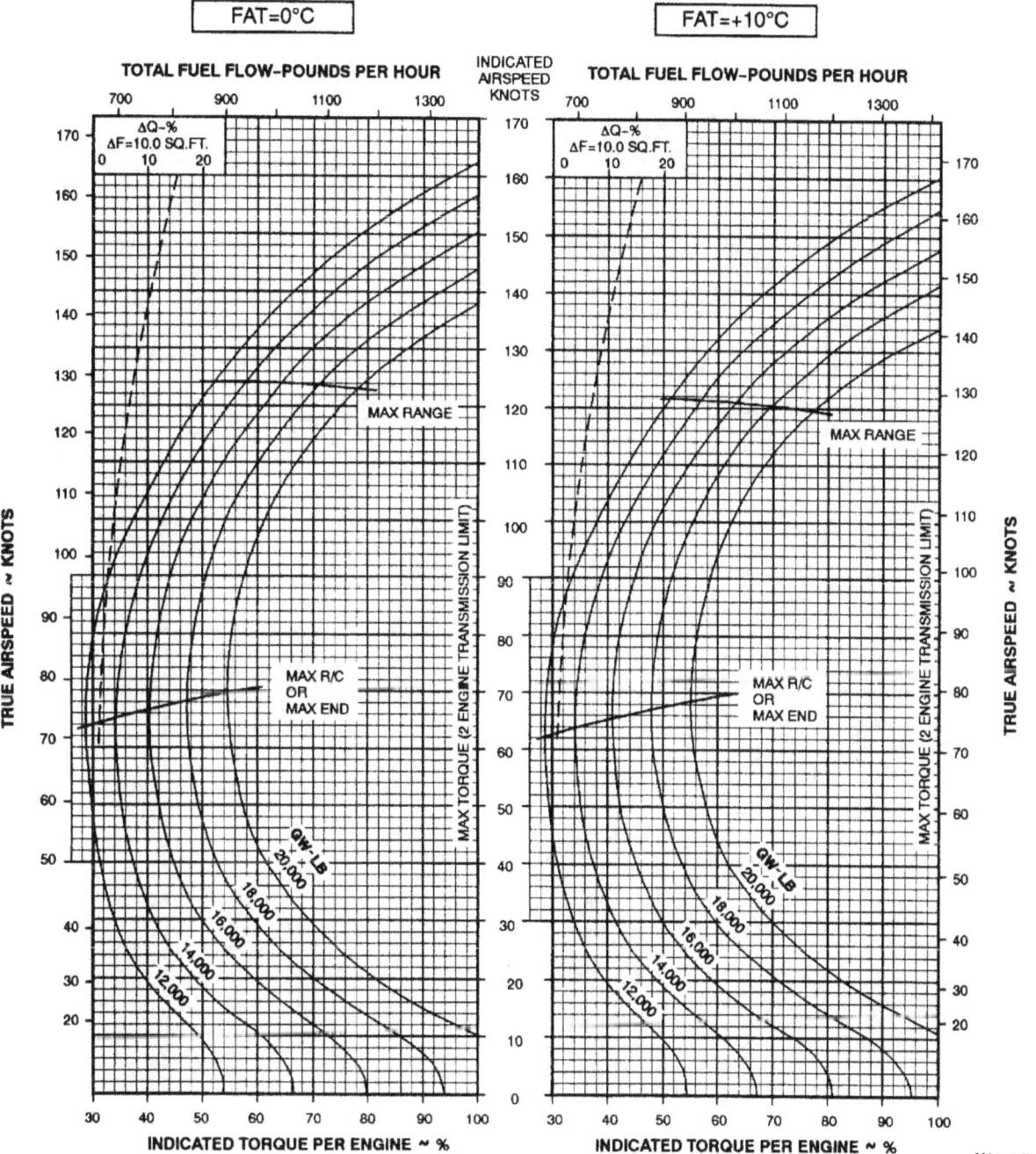

Figure 7-10. Cruise Chart, 2,000 Feet, 0° and +10°C (Sheet 4 of 7)

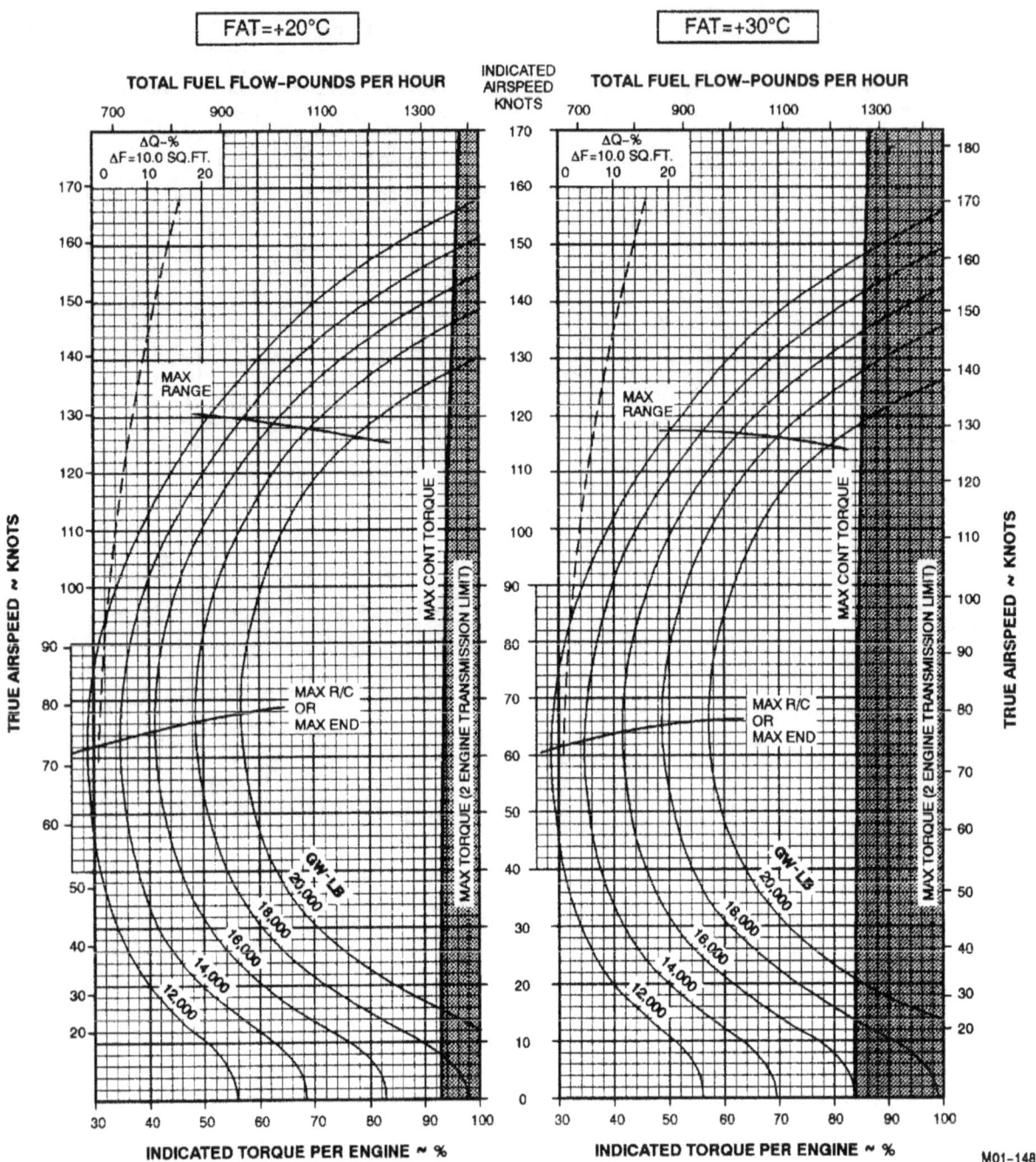

Figure 7-10. Cruise Chart, 2,000 Feet, +20° and +30°C (Sheet 5 of 7) 701

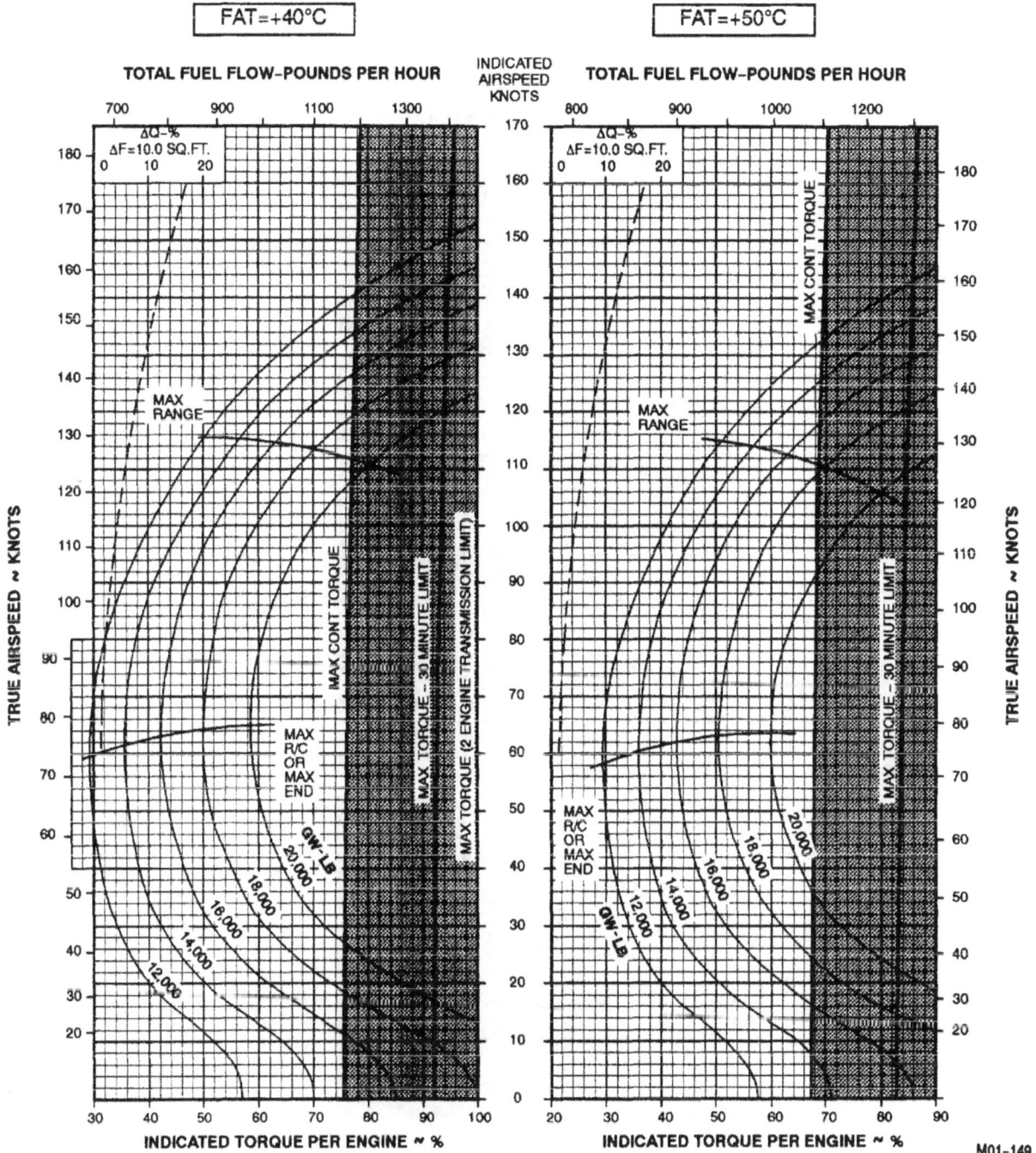

Figure 7-10. Cruise Chart, 2,000 Feet, +40° and +50°C (Sheet 6 of 7)

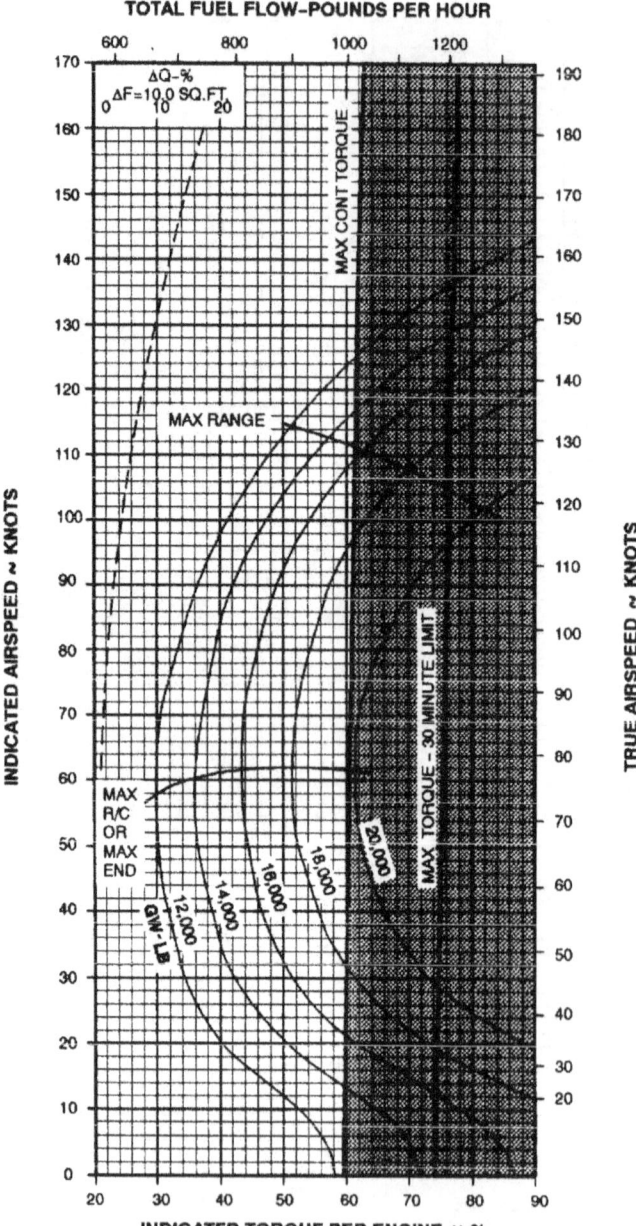

Figure 7-10. Cruise Chart, 2,000 Feet, +60°C (Sheet 7 of 7)

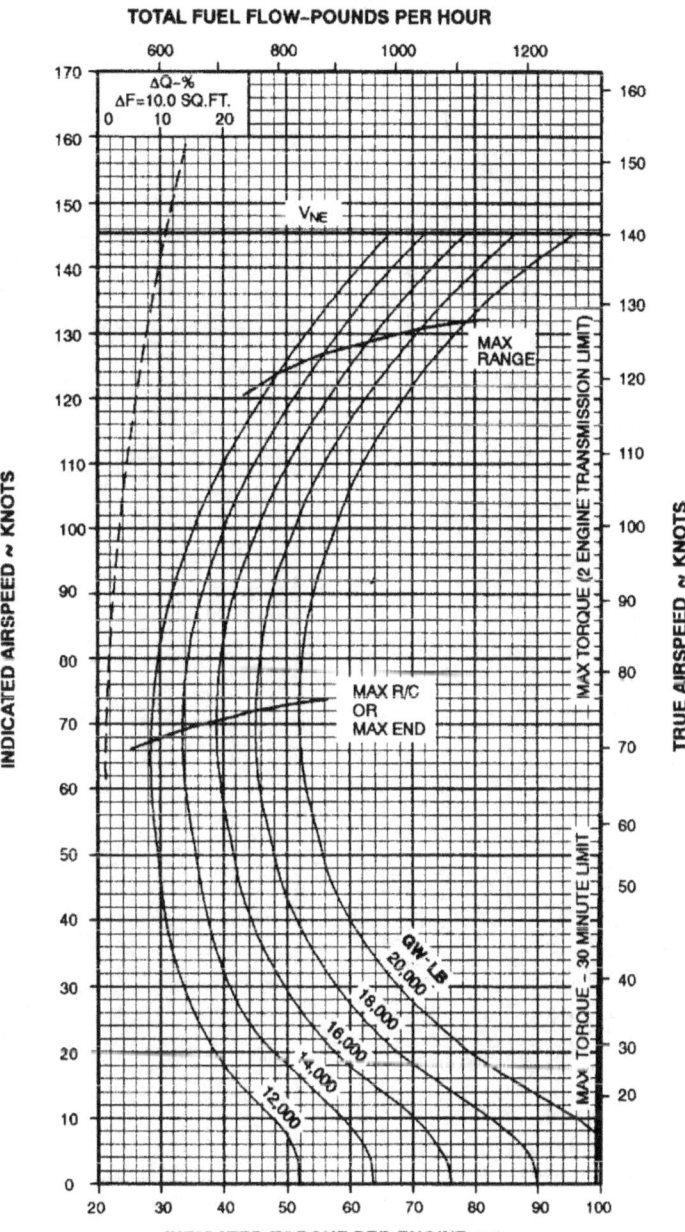

Figure 7-11. Cruise Chart, 4,000 Feet, -50°C (Sheet 1 of 7)

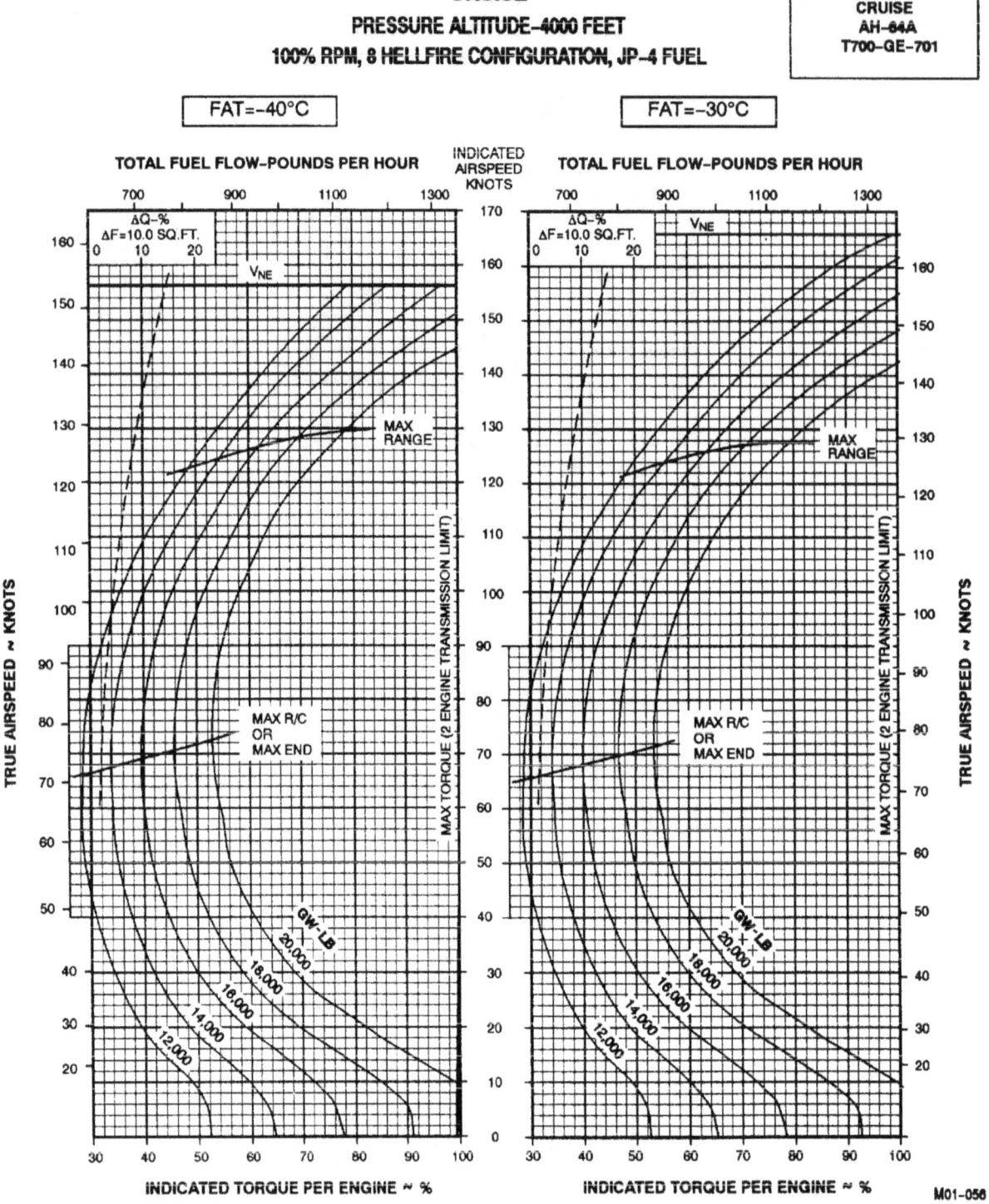

Figure 7-11. Cruise Chart, 4,000 Feet, -40° and -30°C (Sheet 2 of 7)

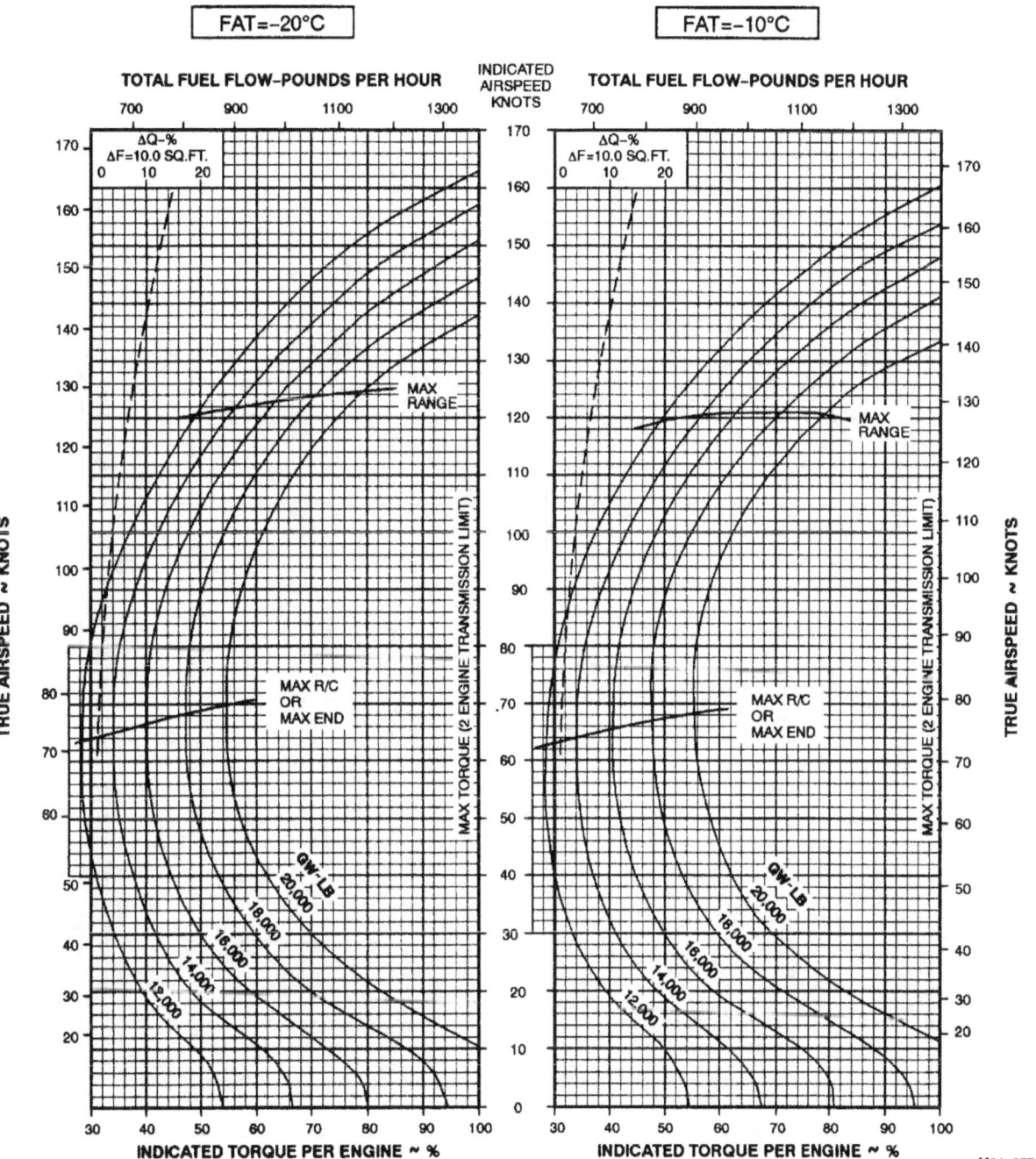

Figure 7-11. Cruise Chart, 4,000 Feet, -20° and -10° C (Sheet 3 of 7) 701

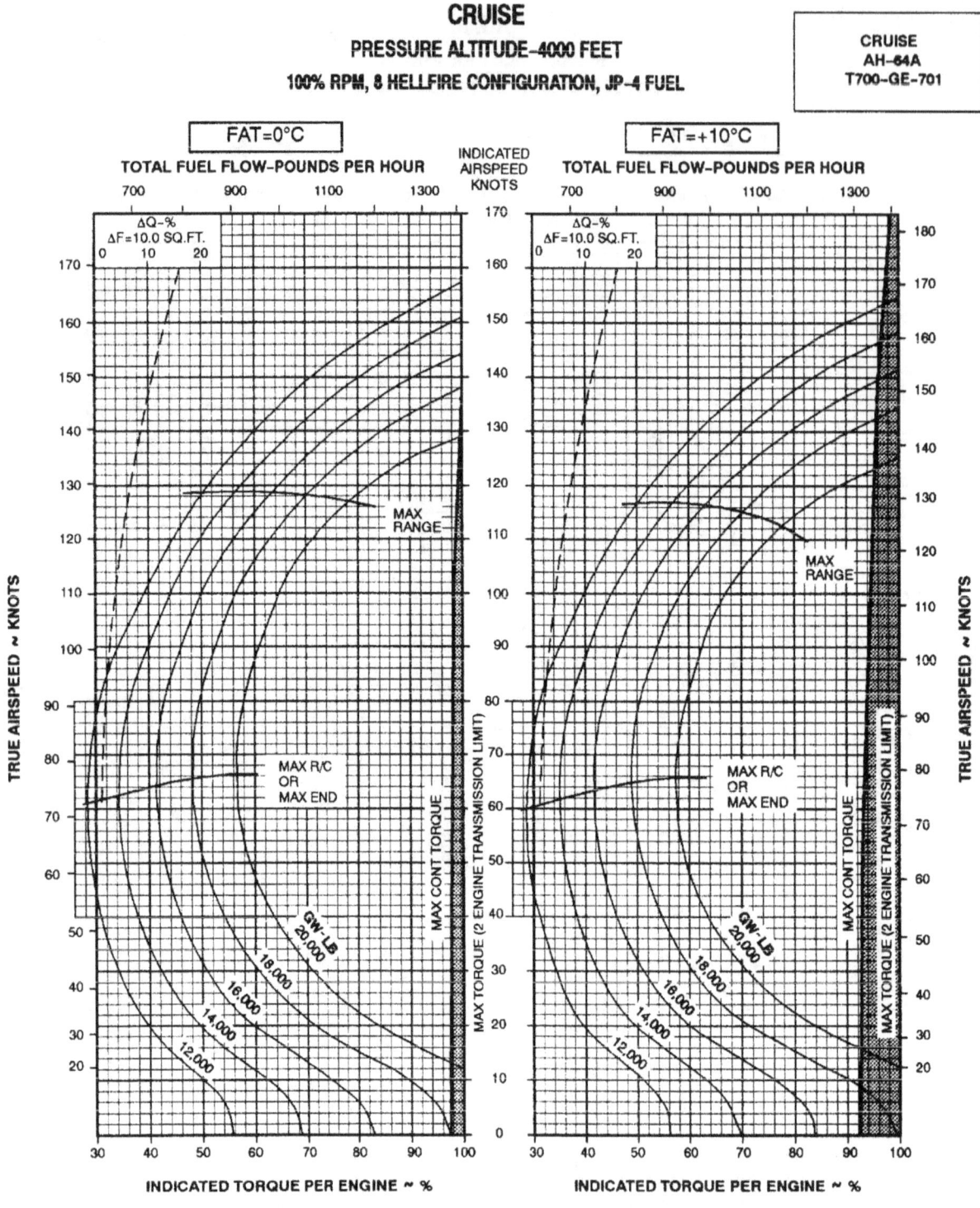

Figure 7-11. Cruise Chart, 4,000 Feet, 0° and +10°C (Sheet 4 of 7)

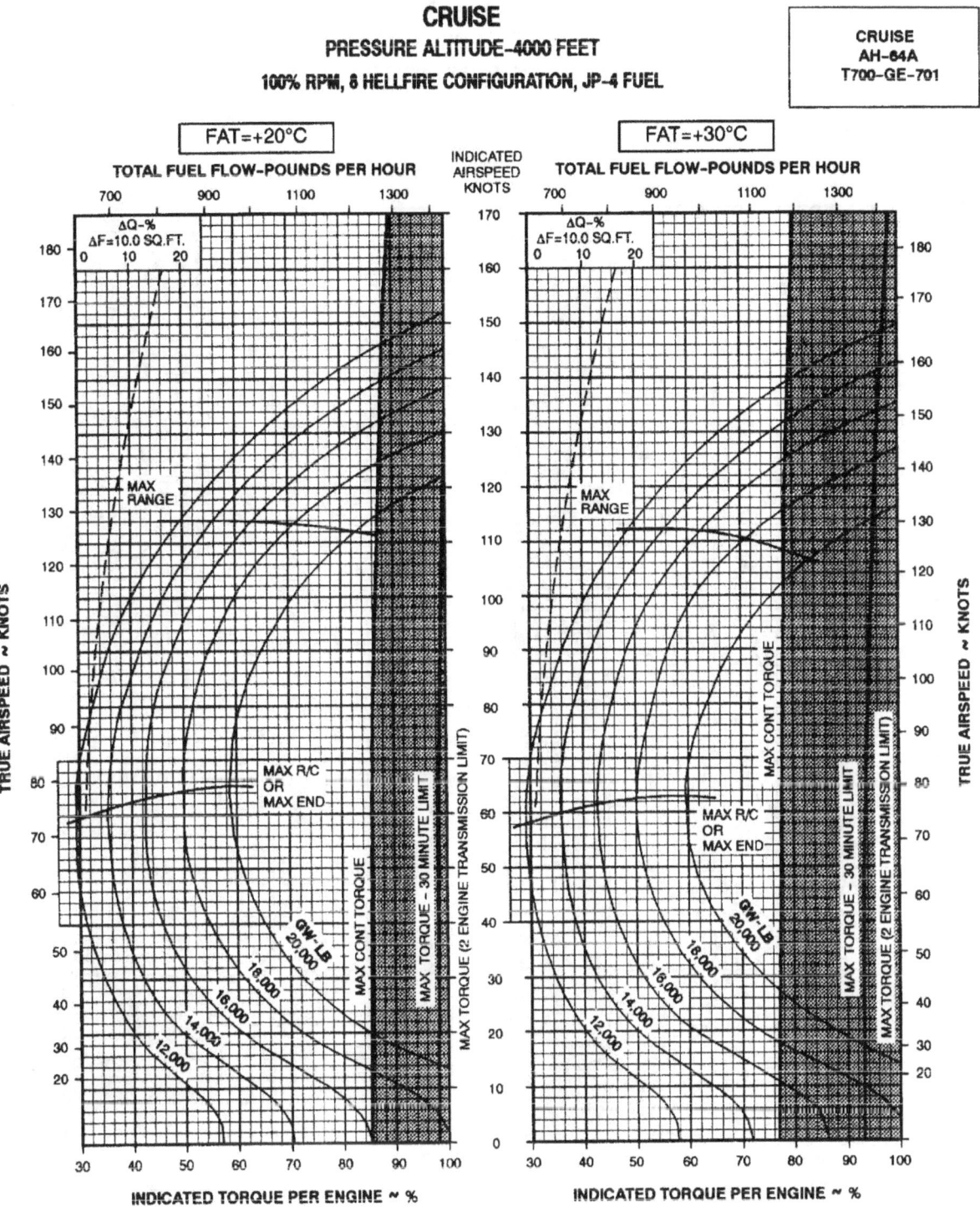

Figure 7-11. Cruise Chart, 4,000 Feet, +20° and +30°C (Sheet 5 of 7)

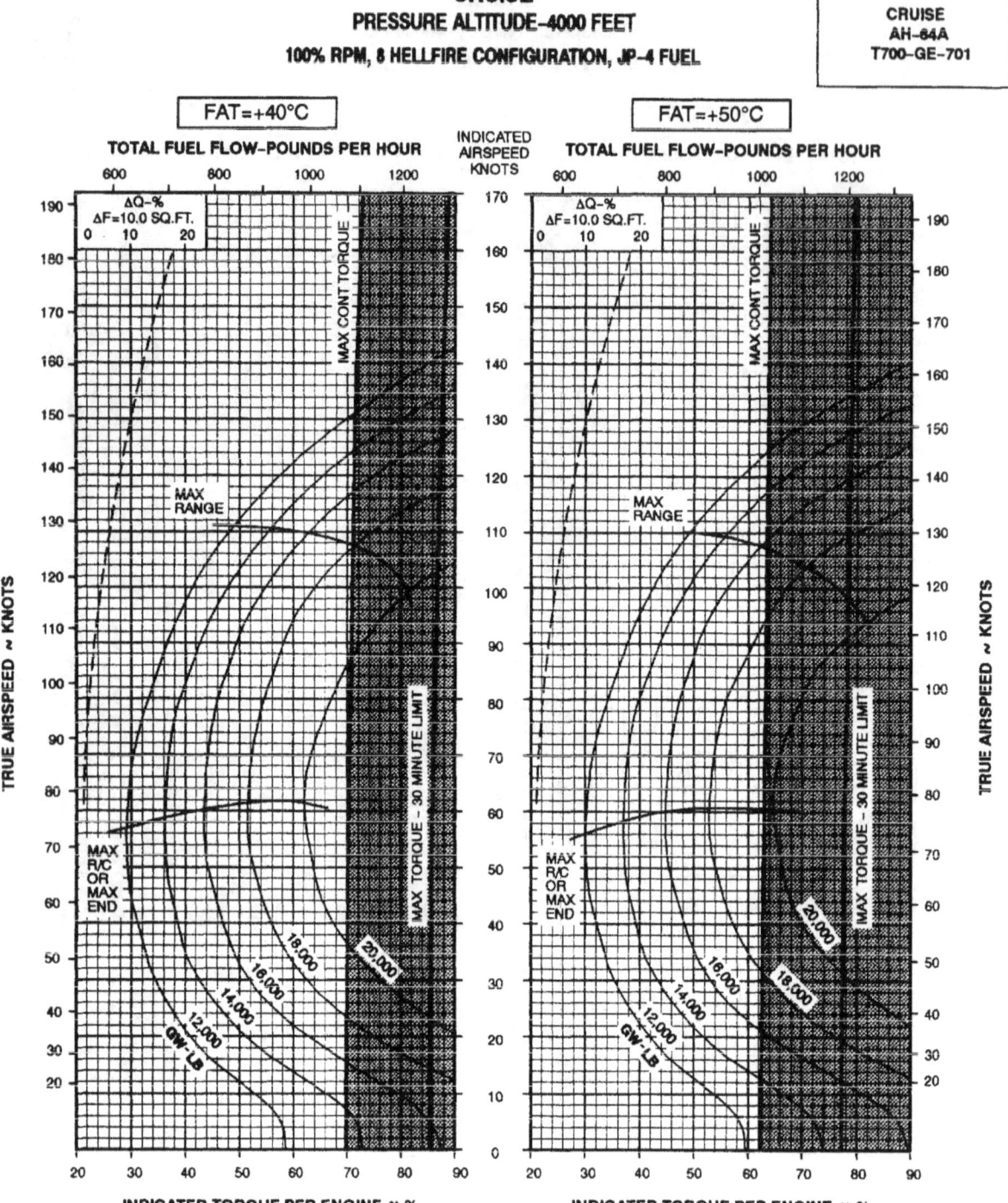

Figure 7-11. Cruise Chart, 4,000 Feet, +20° and +30°C (Sheet 5 of 7)

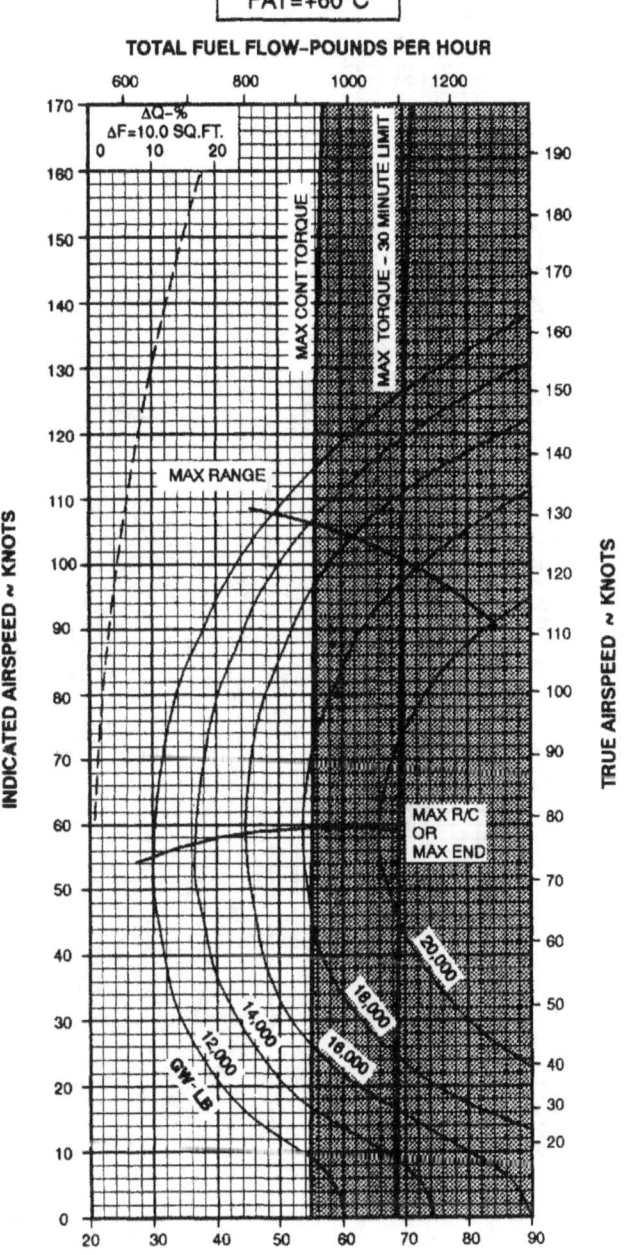

Figure 7-11. Cruise Chart, 4,000 Feet, +40° and +50°C (Sheet 6 of 7) 701

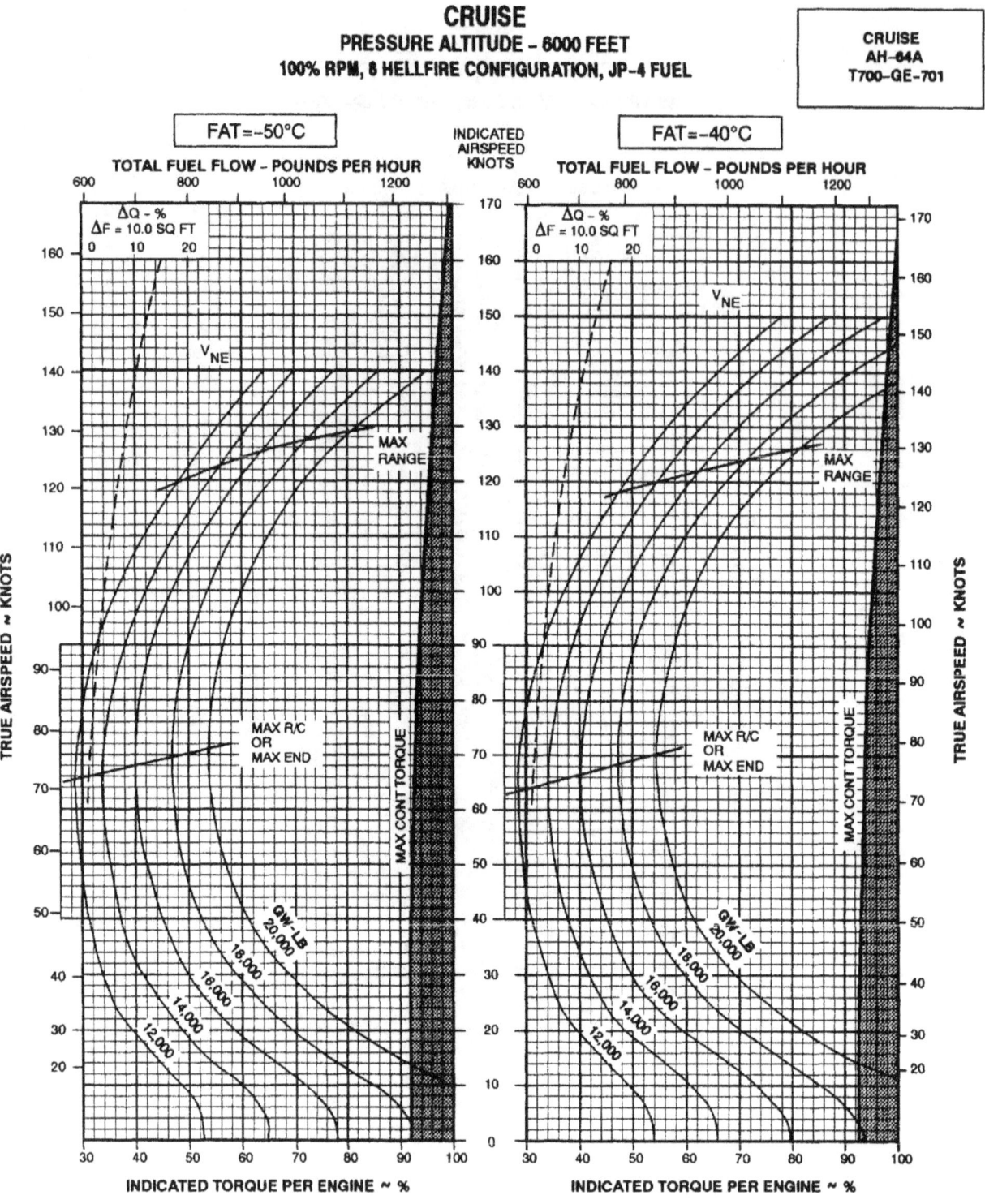

Figure 7-11. Cruise Chart, 4,000 Feet, +60°C (Sheet 7 of 7)

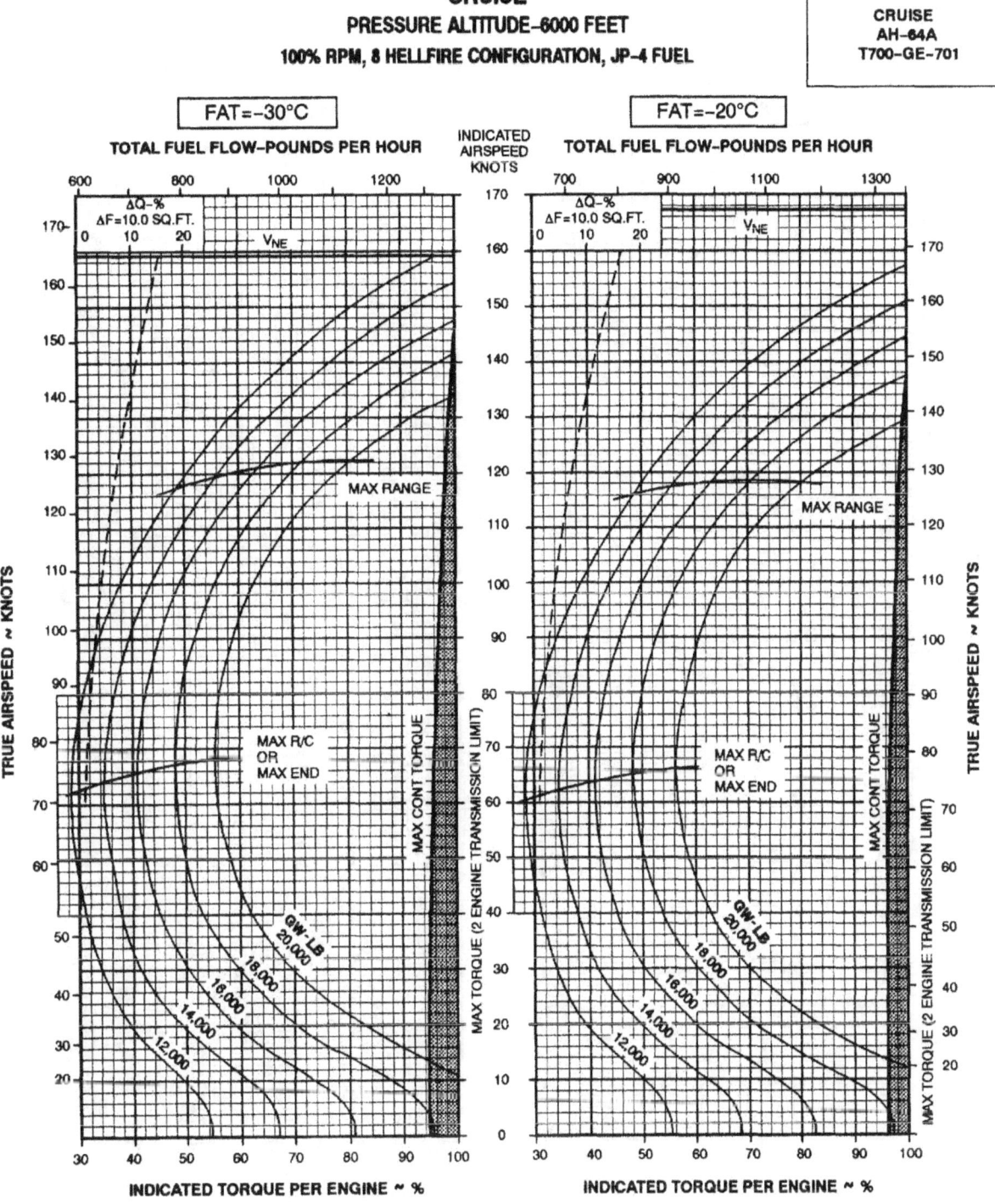

Figure 7-12. Cruise Chart, 6,000 Feet, -30° and -20 °C (Sheet 2 0f 5)

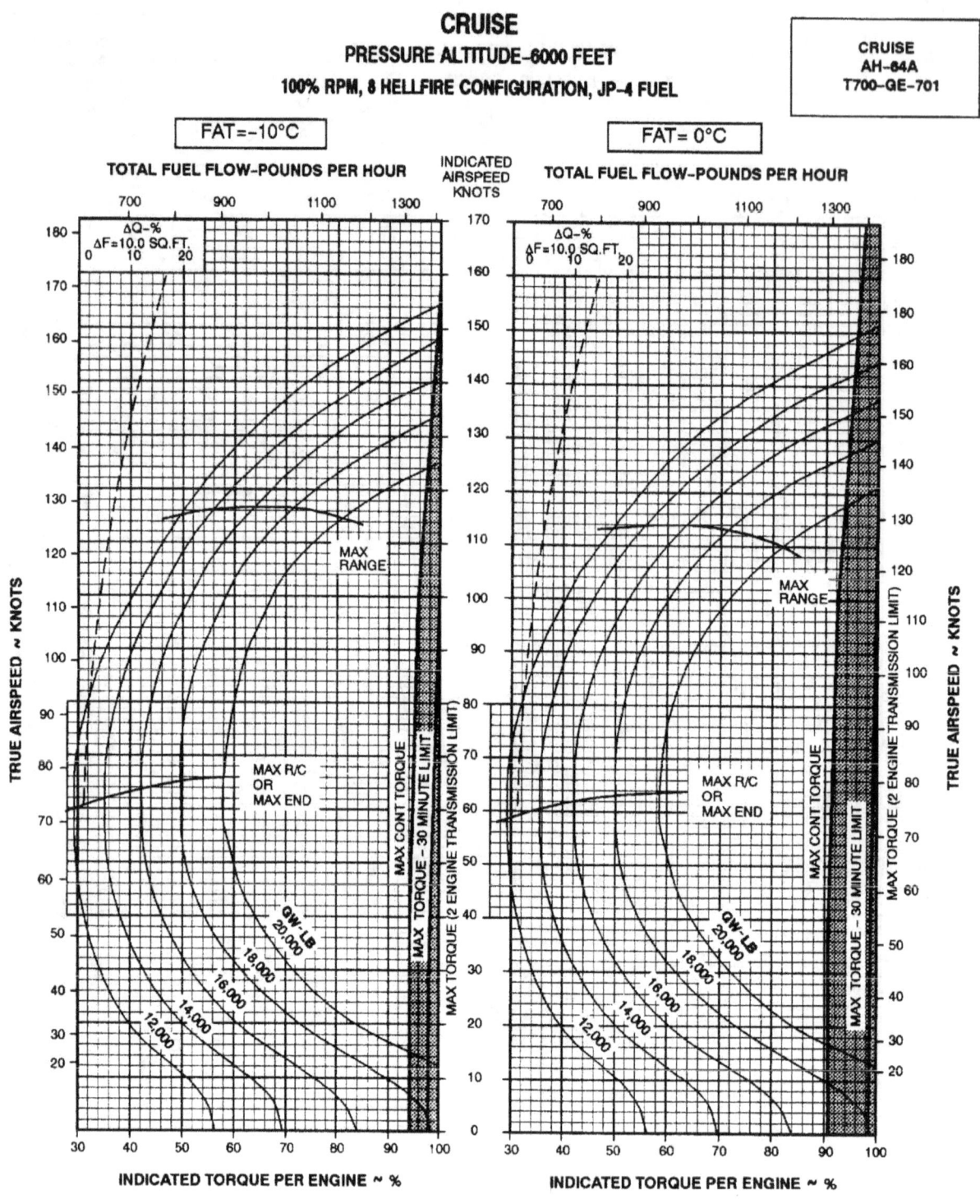

Figure 7-12. Cruise Chart, 6,000 Feet, -10° and 0°C (Sheet 3 of 5)

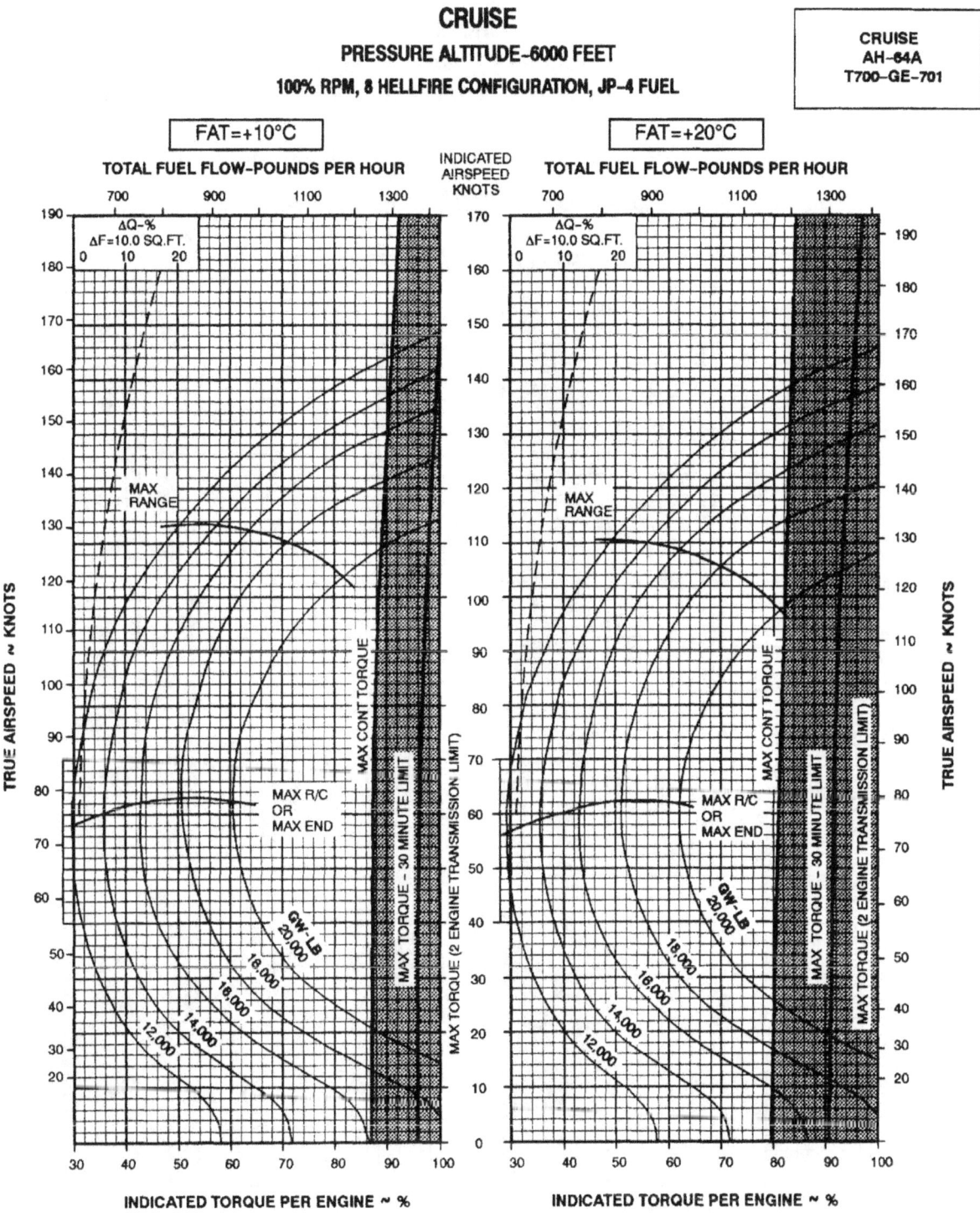

Figure 7-12. Cruise Chart, 6,000 Feet, +10° and +20 °C (Sheet 4 of 5)

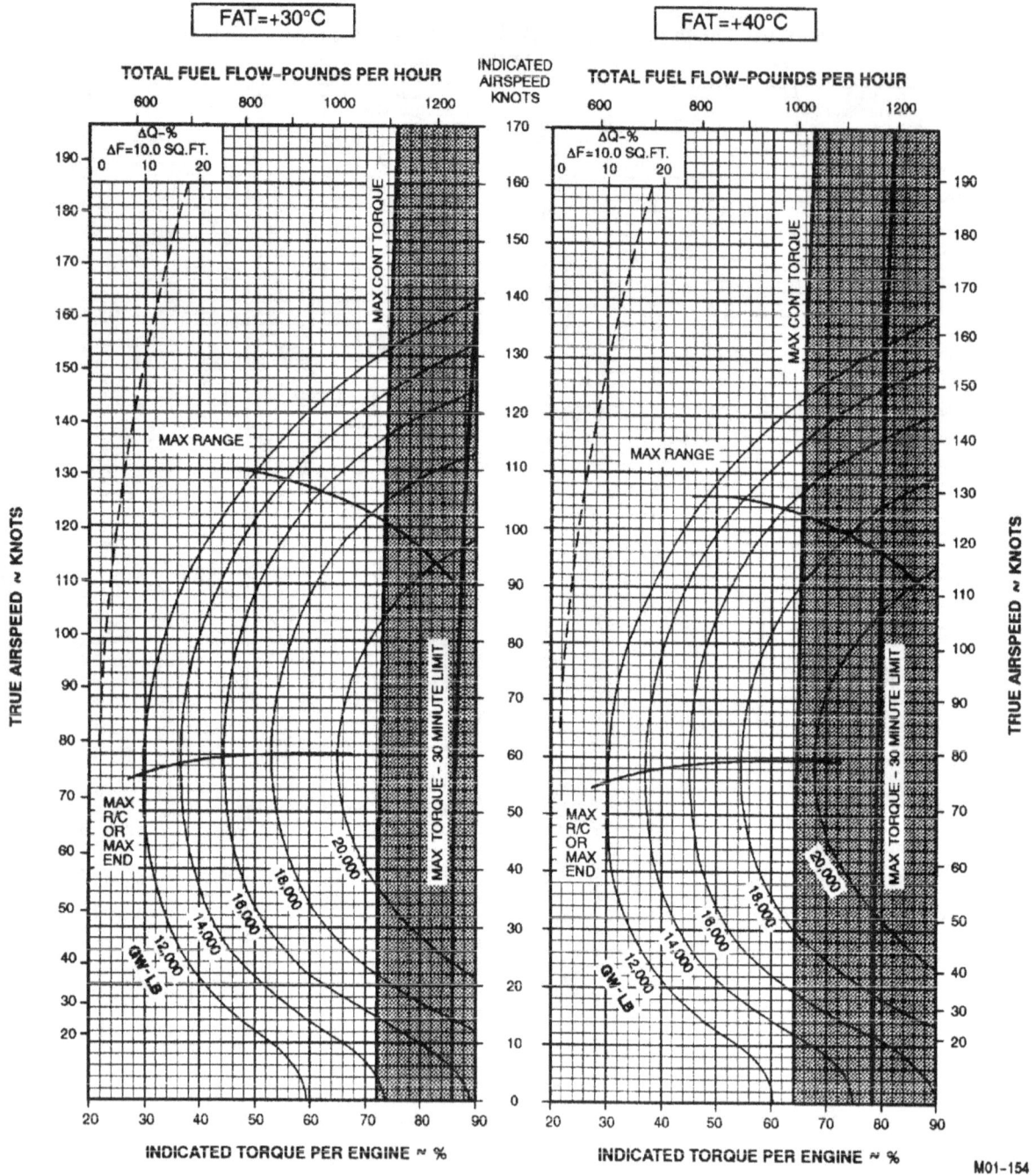

Figure 7-12. Cruise Chart, 6,000 Feet, +30° and +40°C (Sheet 5 of 5)

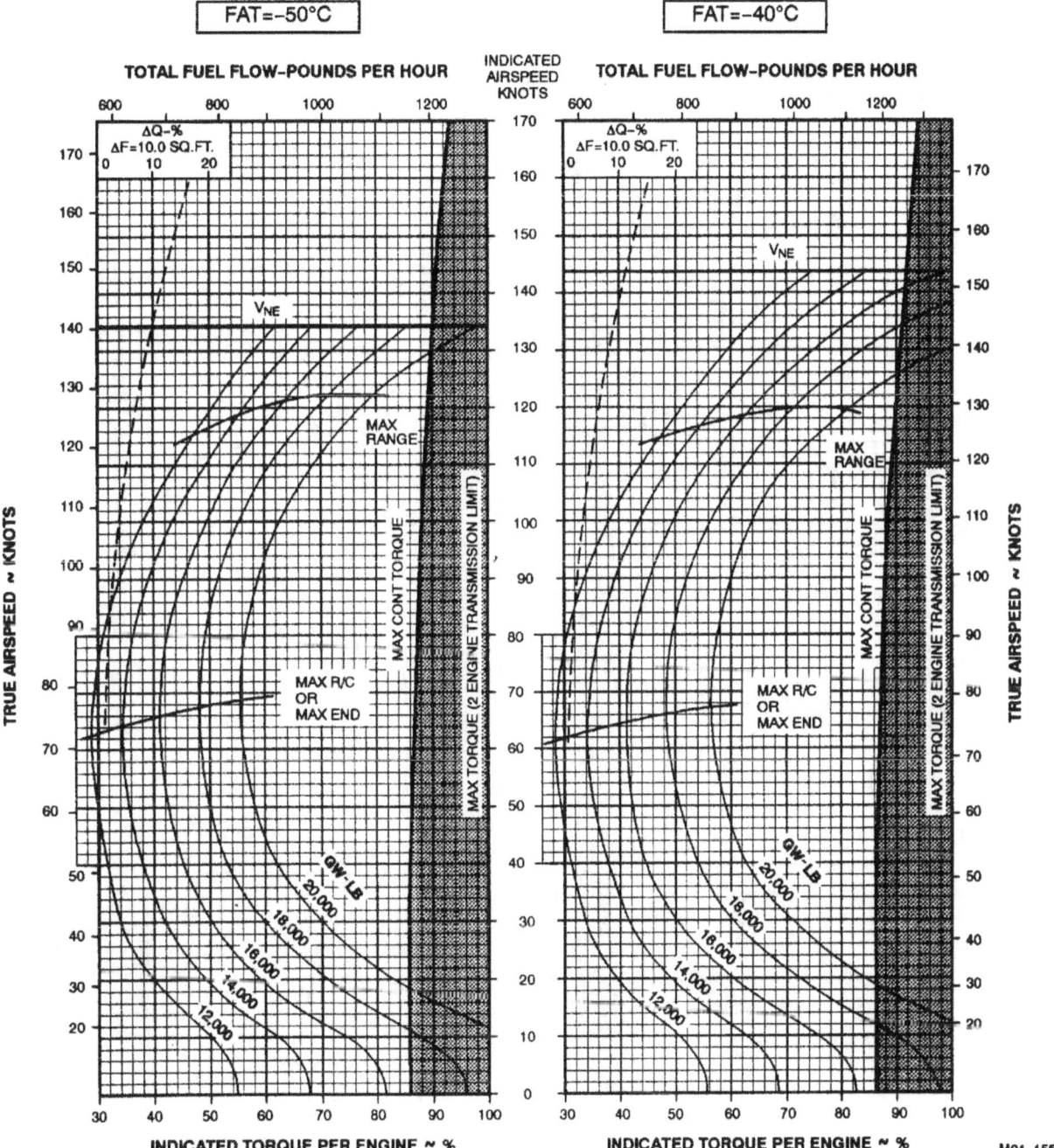

Figure 7-13. Cruise Chart, 8,000 Feet, -50° and -40°C (Sheet 1 of 5)

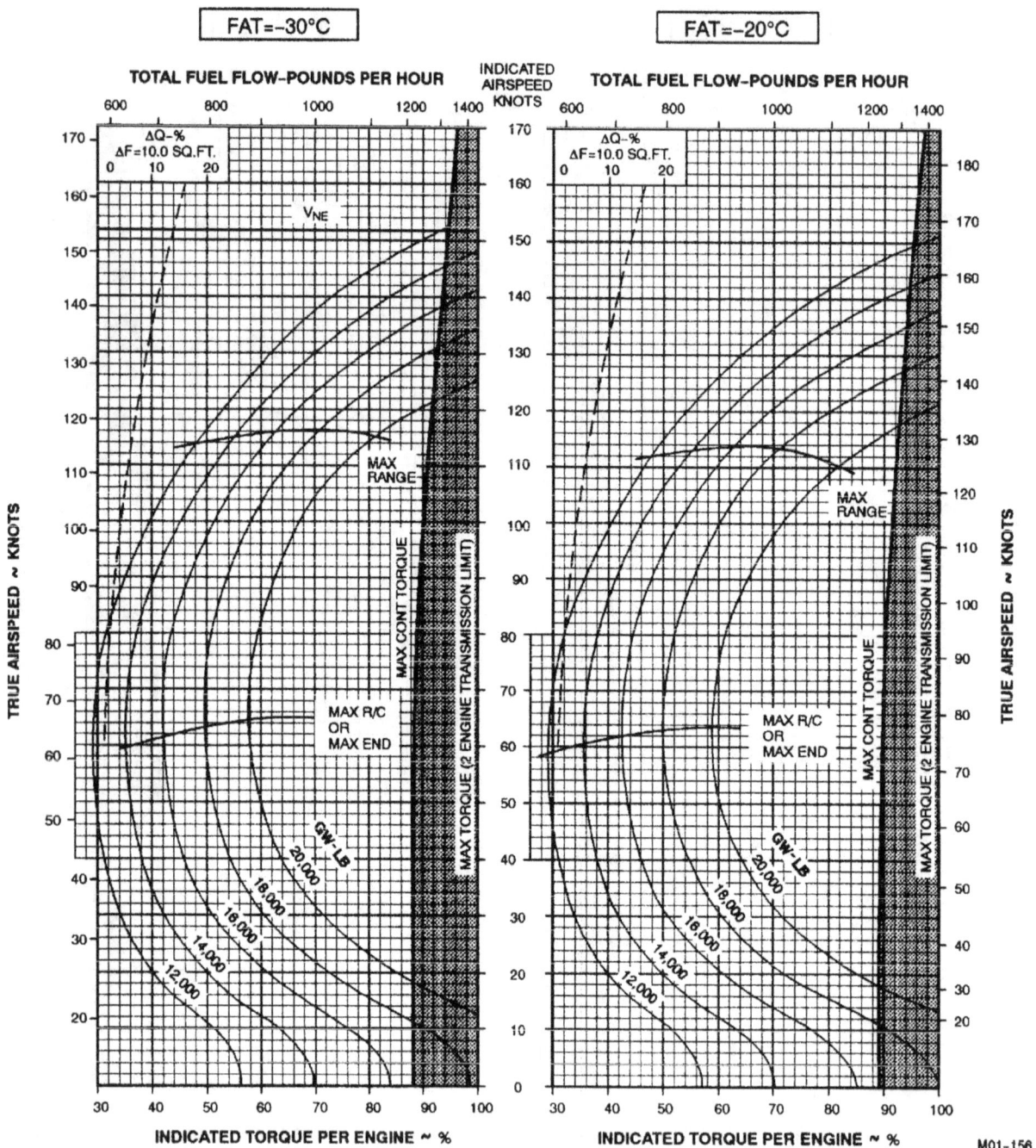

Figure 7-13. Cruise Chart, 8,000 Feet, -30° and -20°C (Sheet 2 of 5) 701

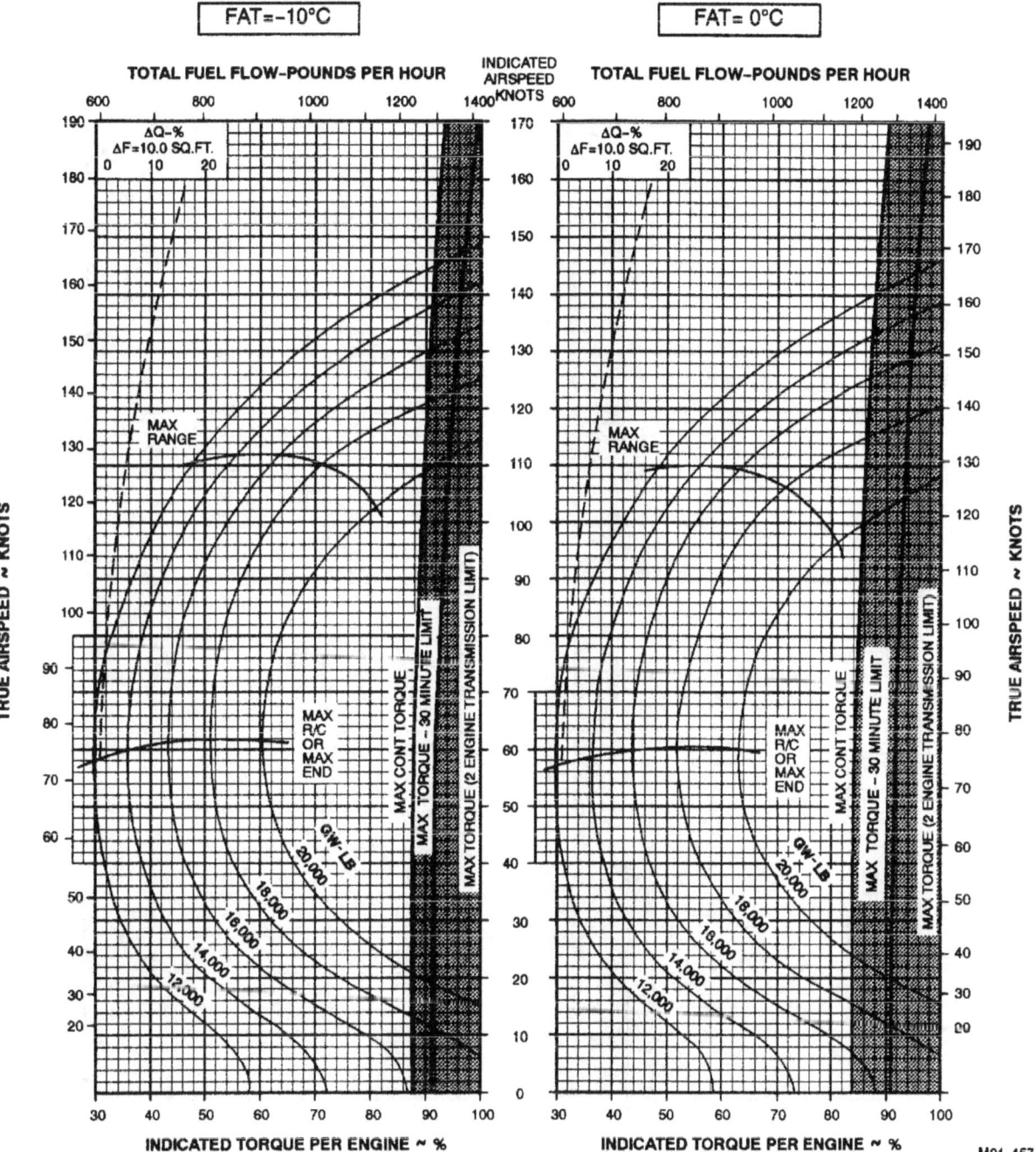

Figure 7-13. Cruise Chart, 8,000 Feet, -10° and 0°C (Sheet 3 of 5) 701

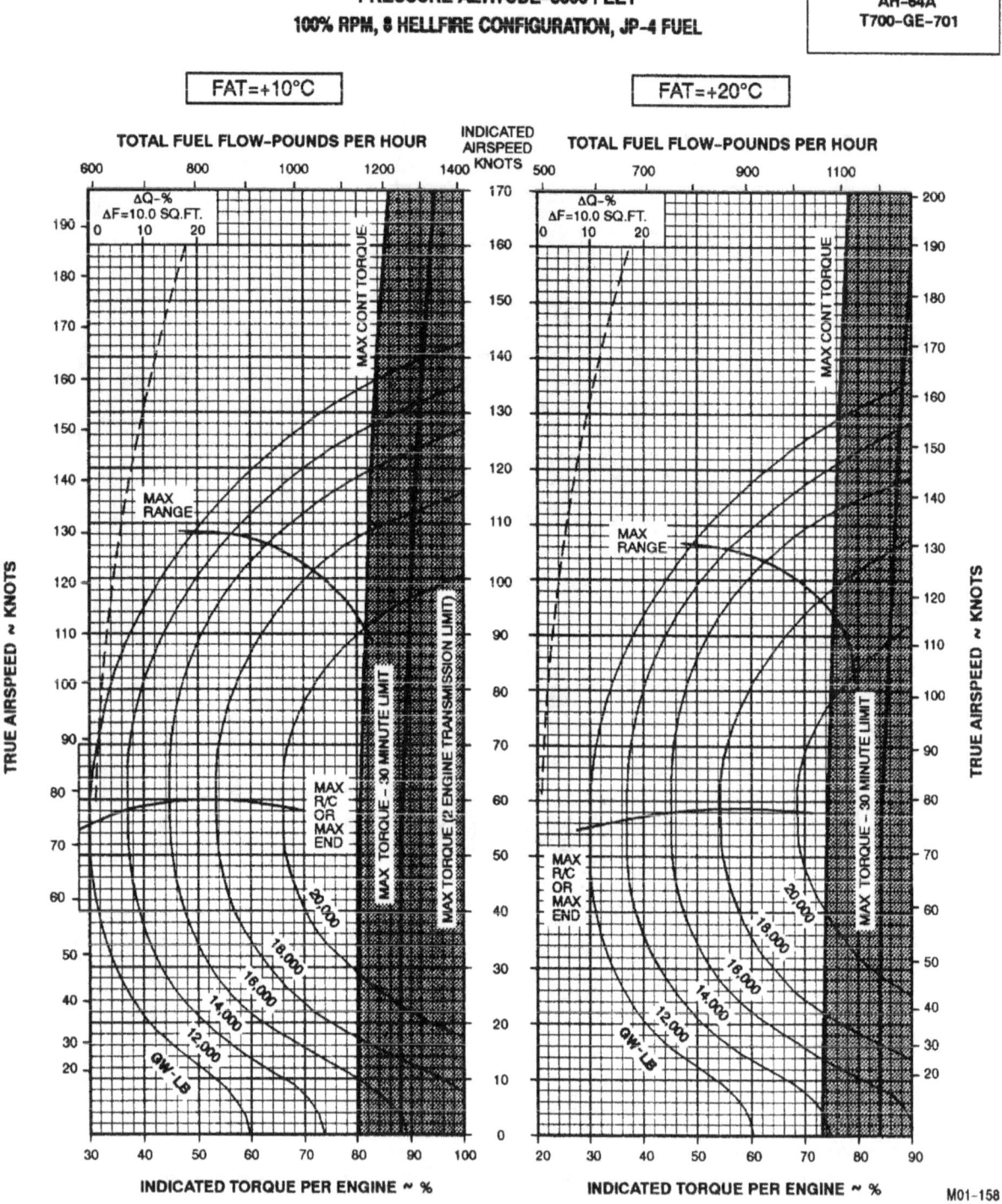

Figure 7-13. Cruise Chart, 8,000 Feet, +10° and +20°C (Sheet 4 of 5)

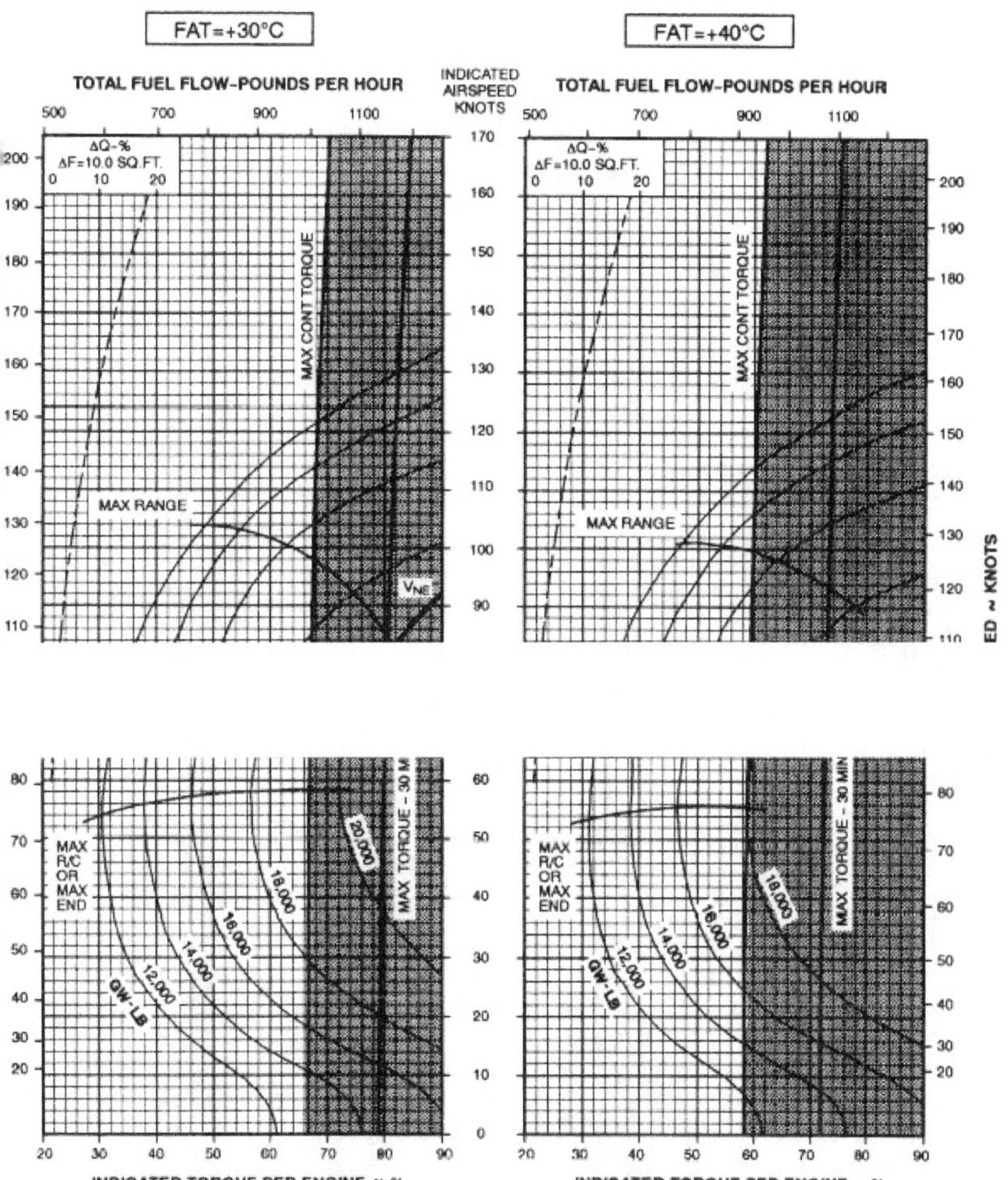

Figure 7-13. Cruise Chart, 8,000 Feet, +30° and +40°C (Sheet 5 of 5)

TM 1-1520-238-10

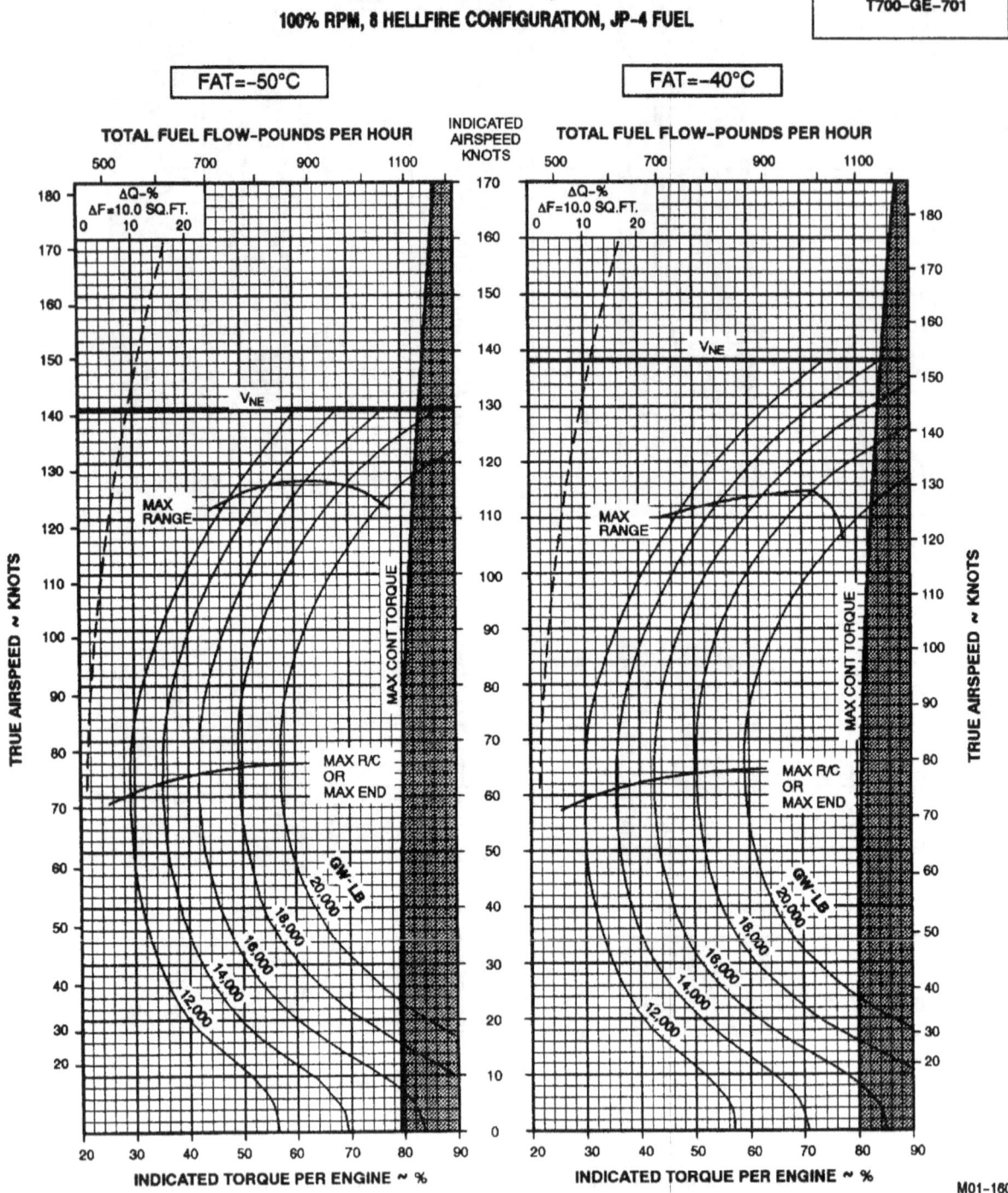

Figure 7-14. Cruise Chart, 10,000 Feet, -50° and -40°C (Sheet 1 of 6)

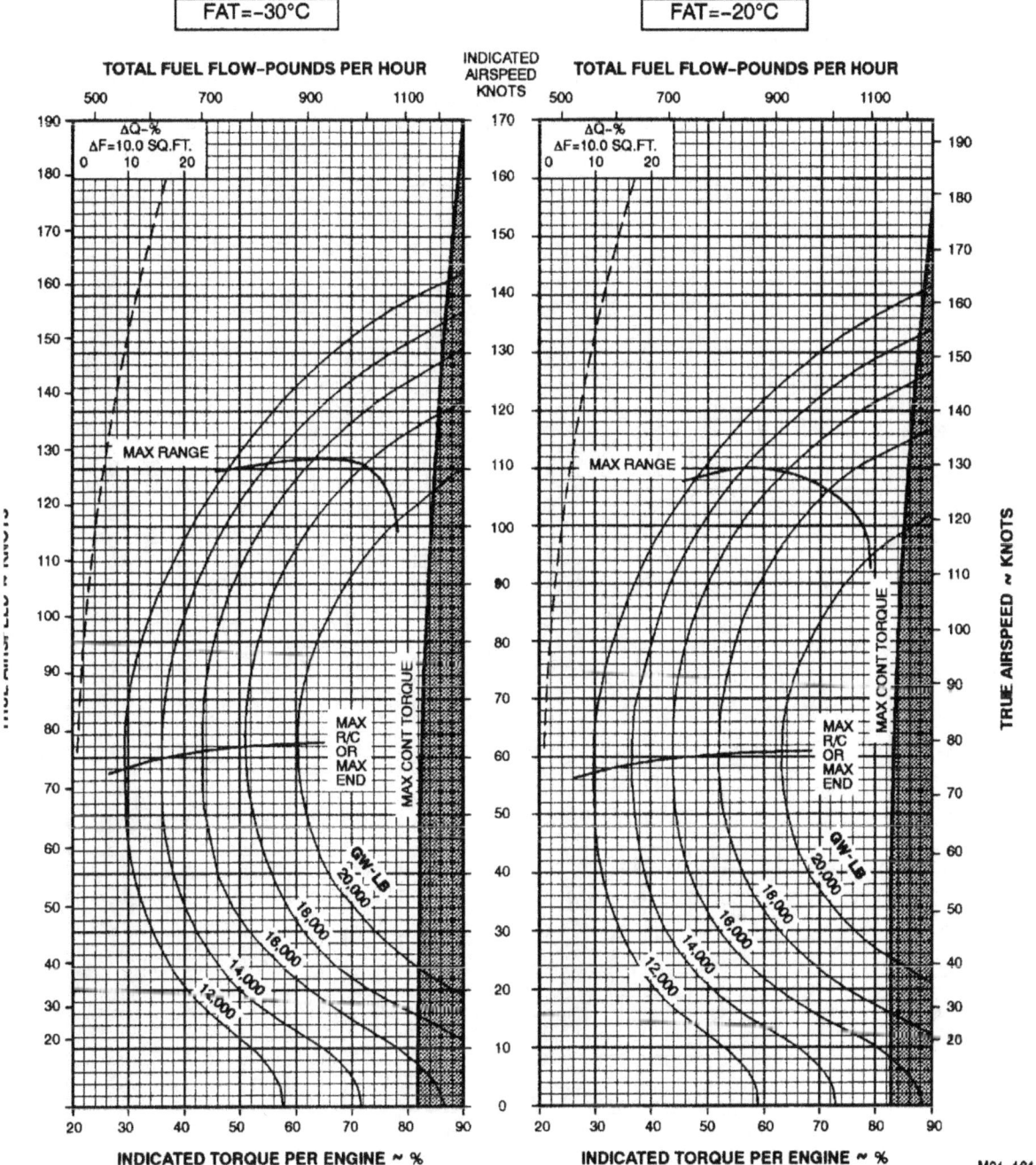

Figure 7-14. Cruise Chart, 10,000 Feet, -30° and -200 °C (Sheet 2 of 6) 701

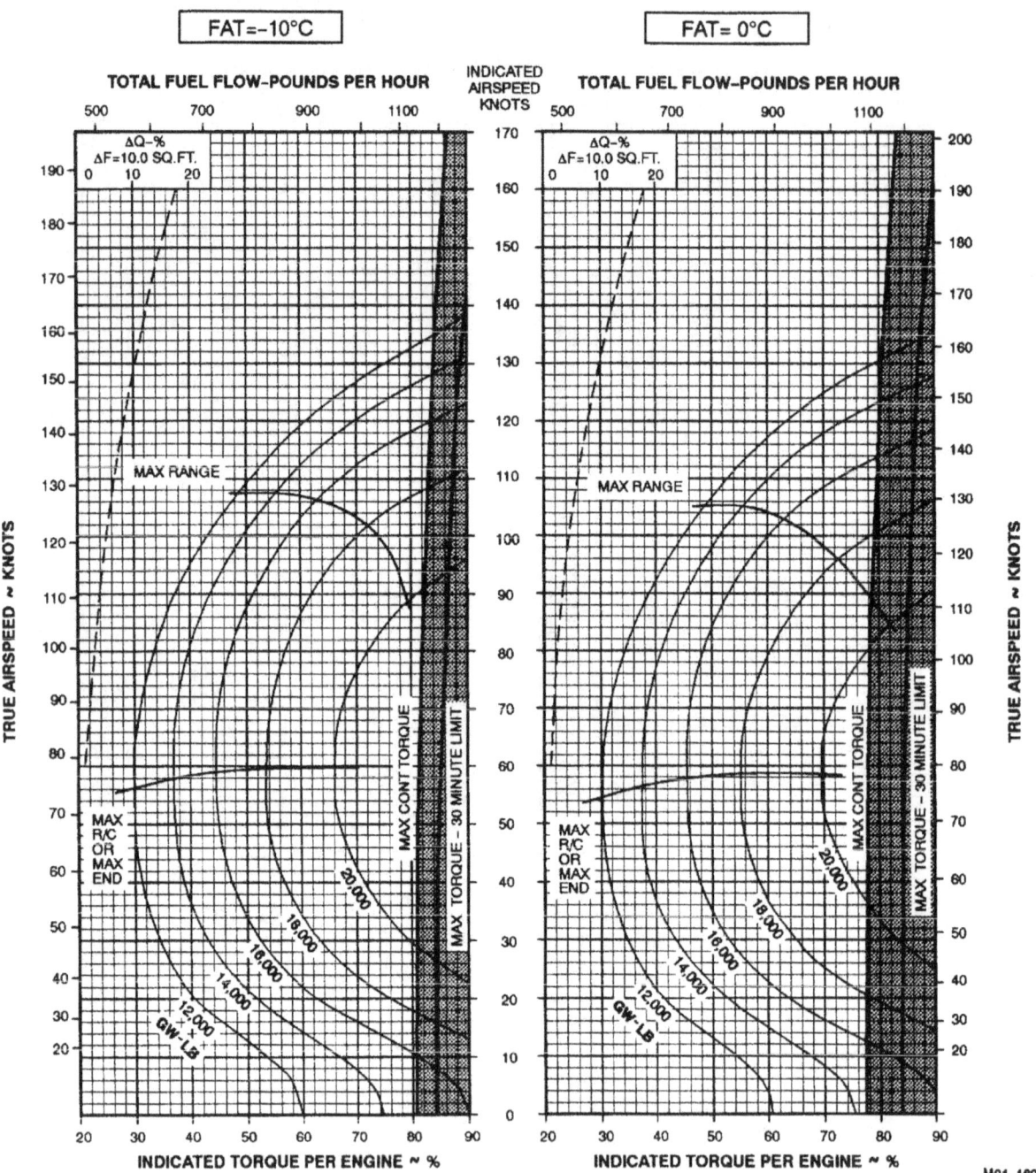

Figure 7-14. Cruise Chart, 10,000 Feet, -10° and 0 °C (Sheet 3 of 6) 701

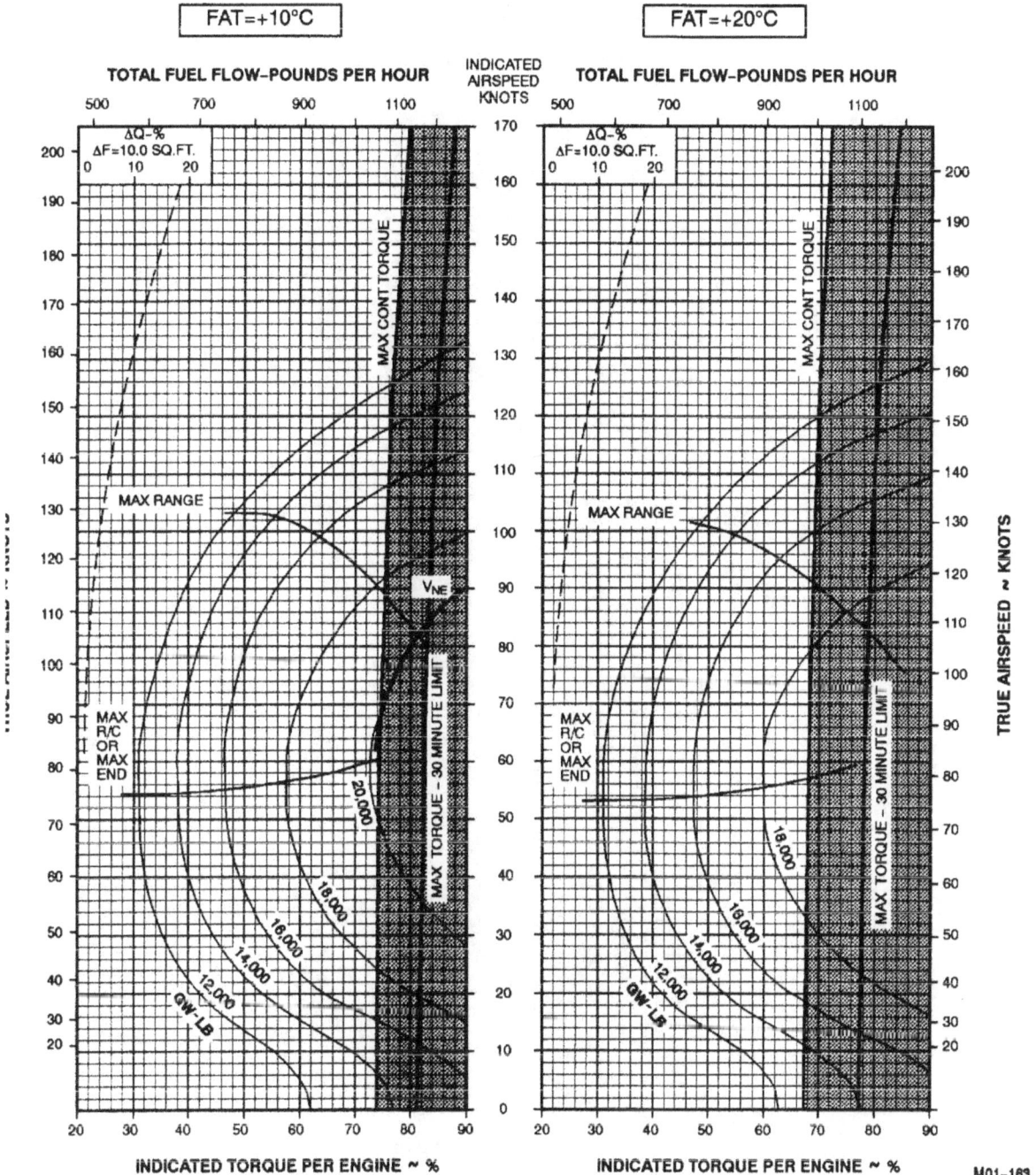

Figure 7-14. Cruise Chart, 10,000 Feet, +10° and +20 °C (Sheet 4 of 6) 701

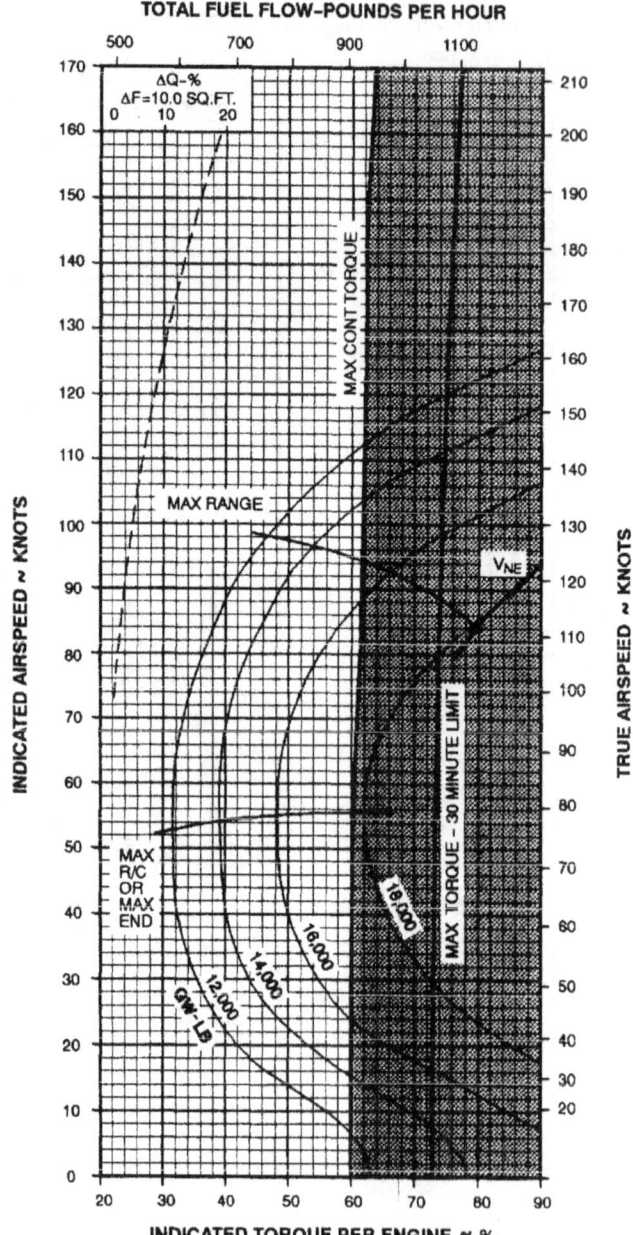

Figure 7-14. Cruise Chart, 10,000 Feet, +30 °C (Sheet 5 of 6)

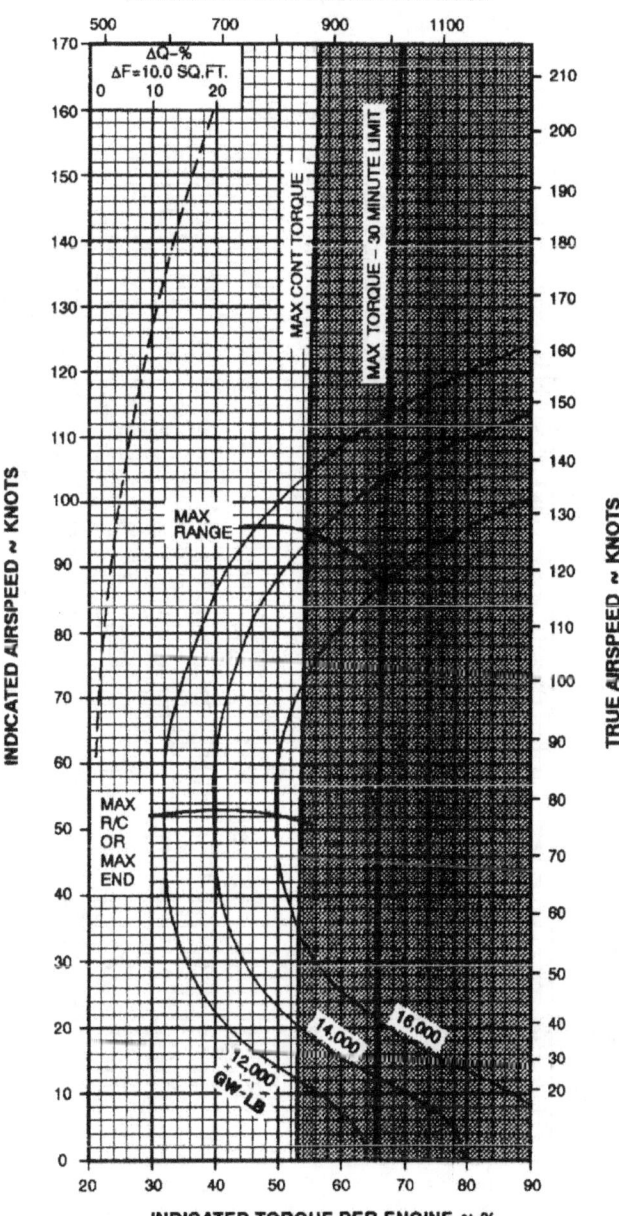

Figure 7-14. Cruise Chart, 10,000 Feet, +40 °C (Sheet 6 of 60)

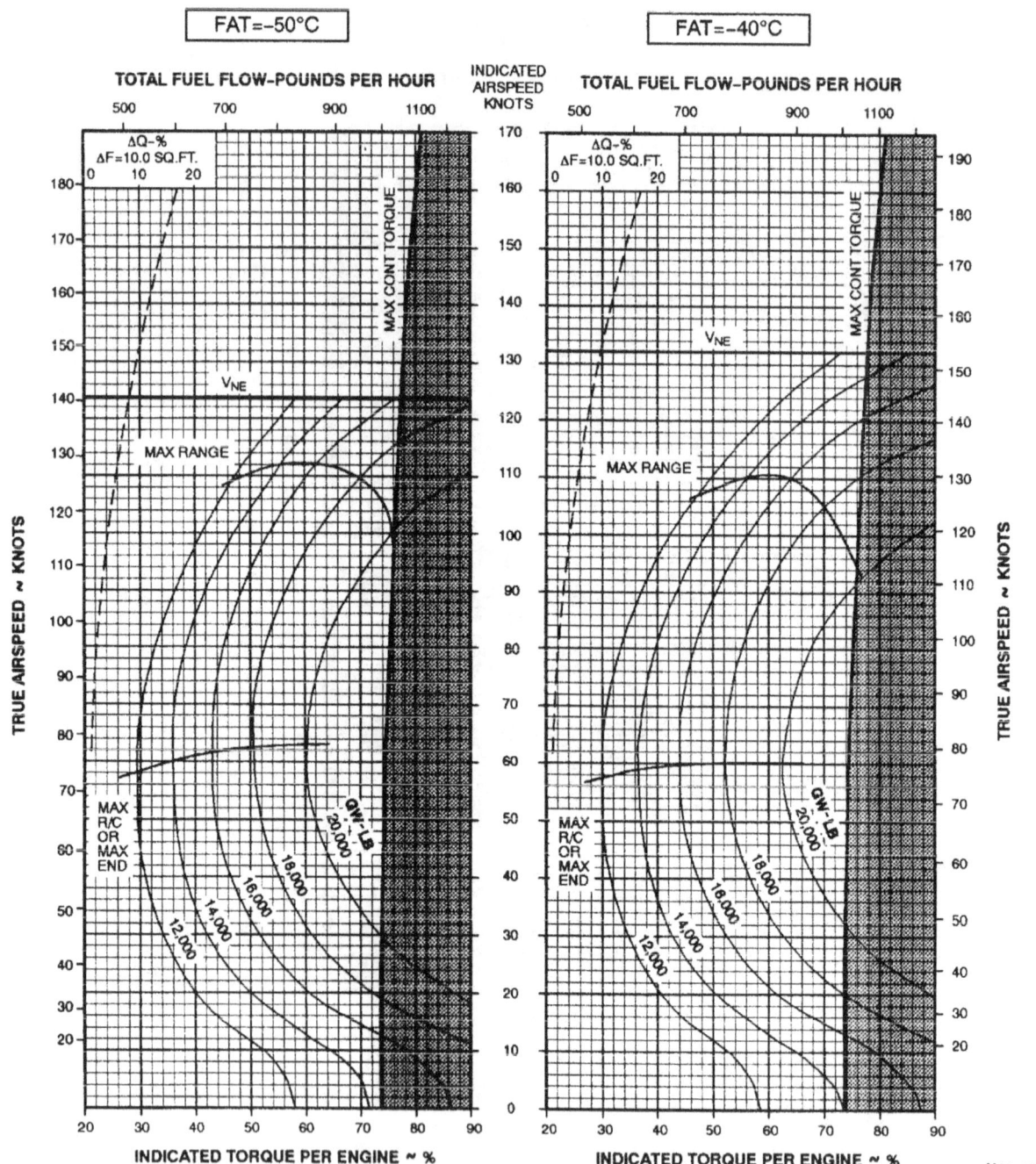

Figure 7-15. Cruise Chart, 12,000 Feet, -50° and -40 °C (Sheet 1 of 5)

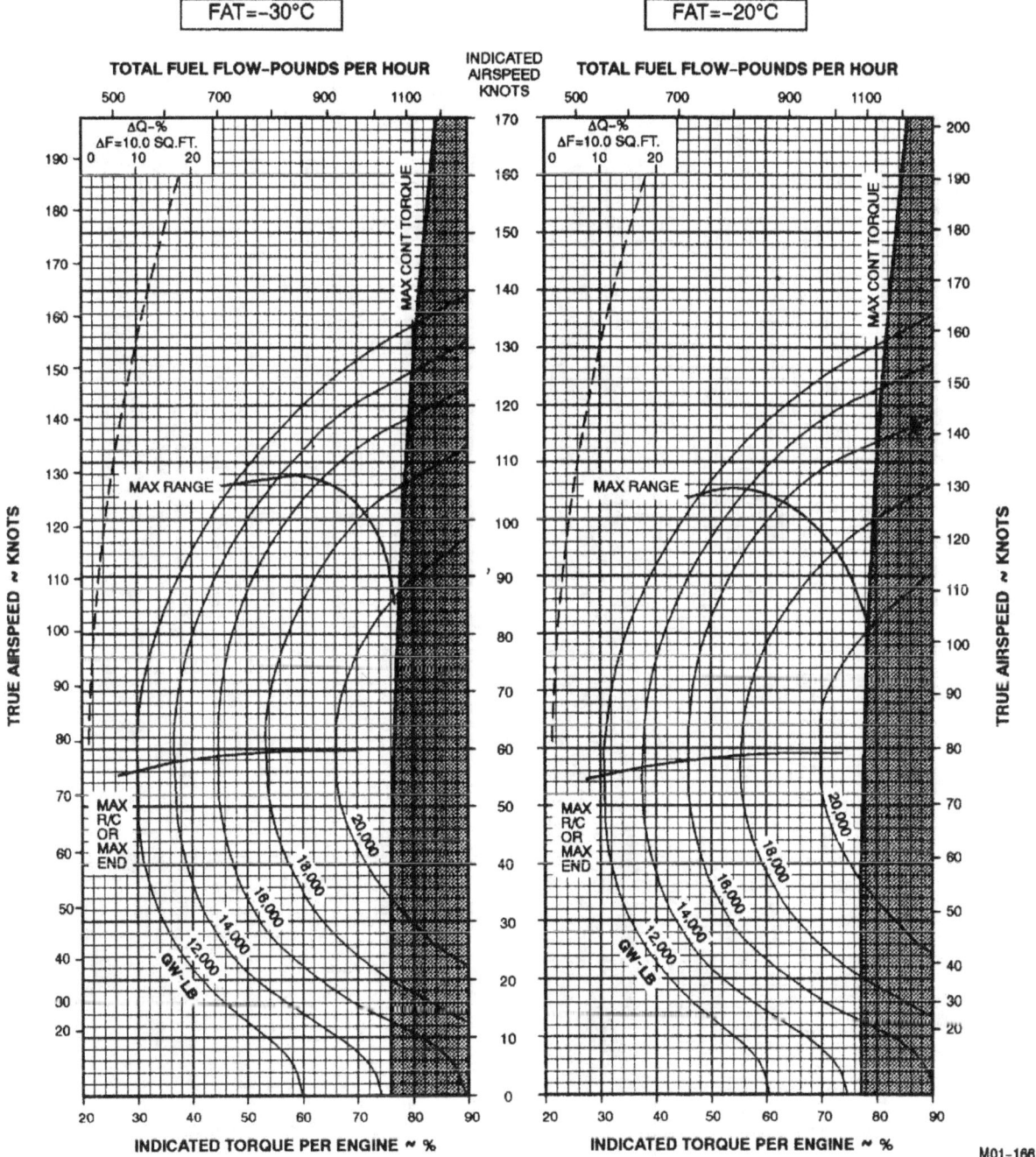

Figure 7-15. Cruise Chart, 12,000 Feet, -30° and -20 °C (Sheet 2 of 5)

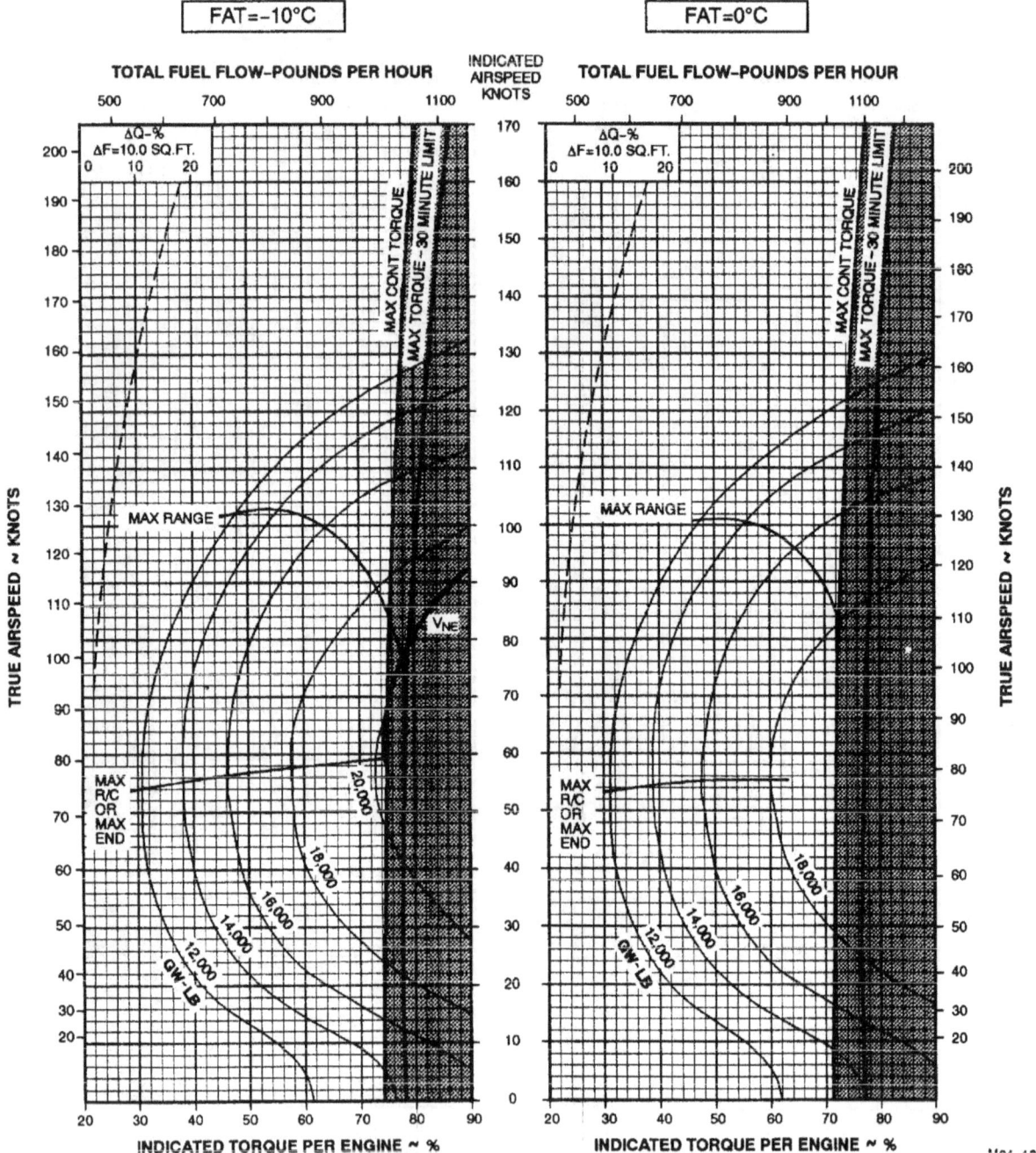

Figure 7-15. Cruise Chart, 12,000 Feet, -10° and 0 °C (Sheet 3 of 5)

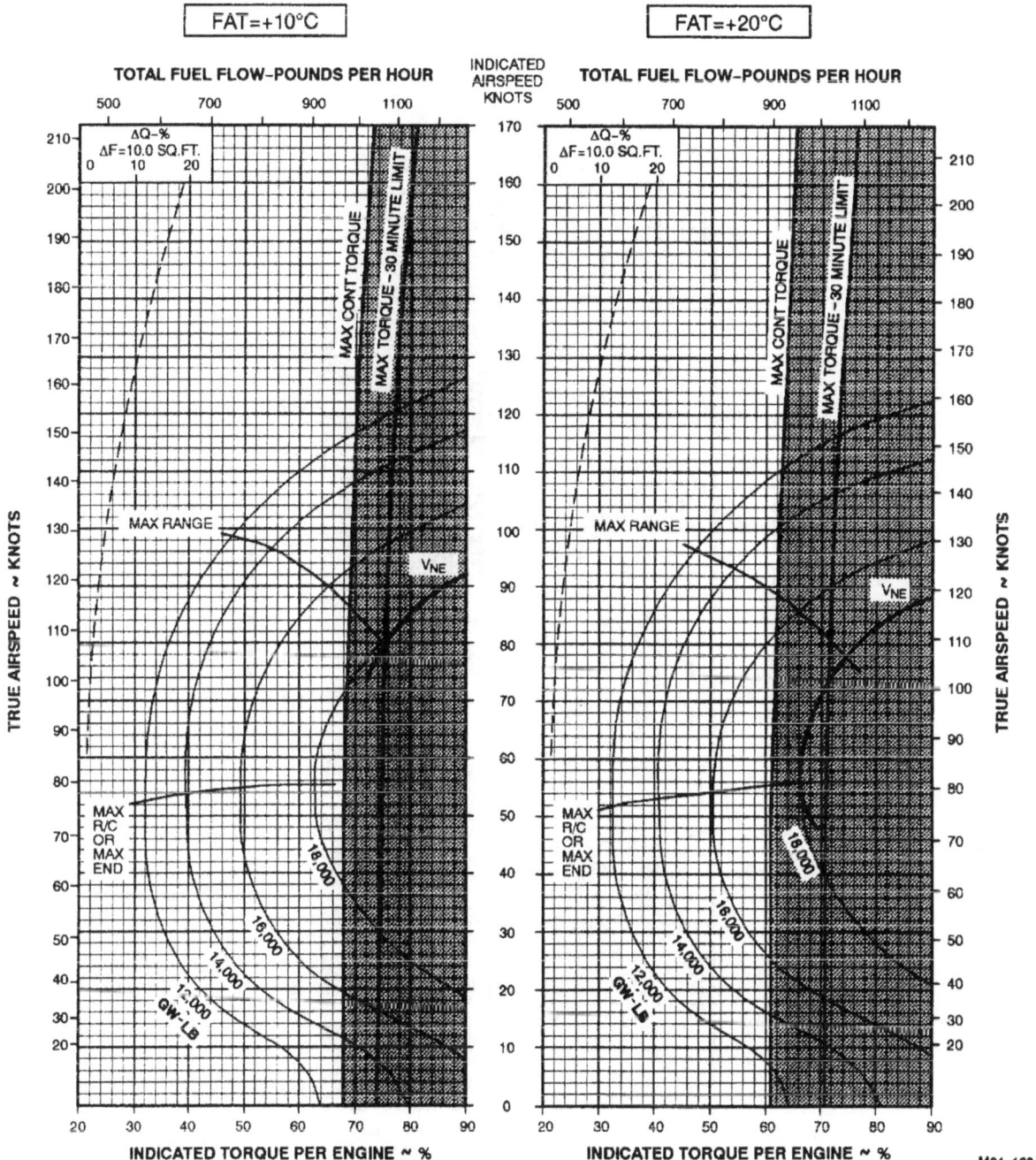

Figure 7-15. Cruise Chart, 12,000 Feet, +10° and +20 °C (Sheet 4 of 5)

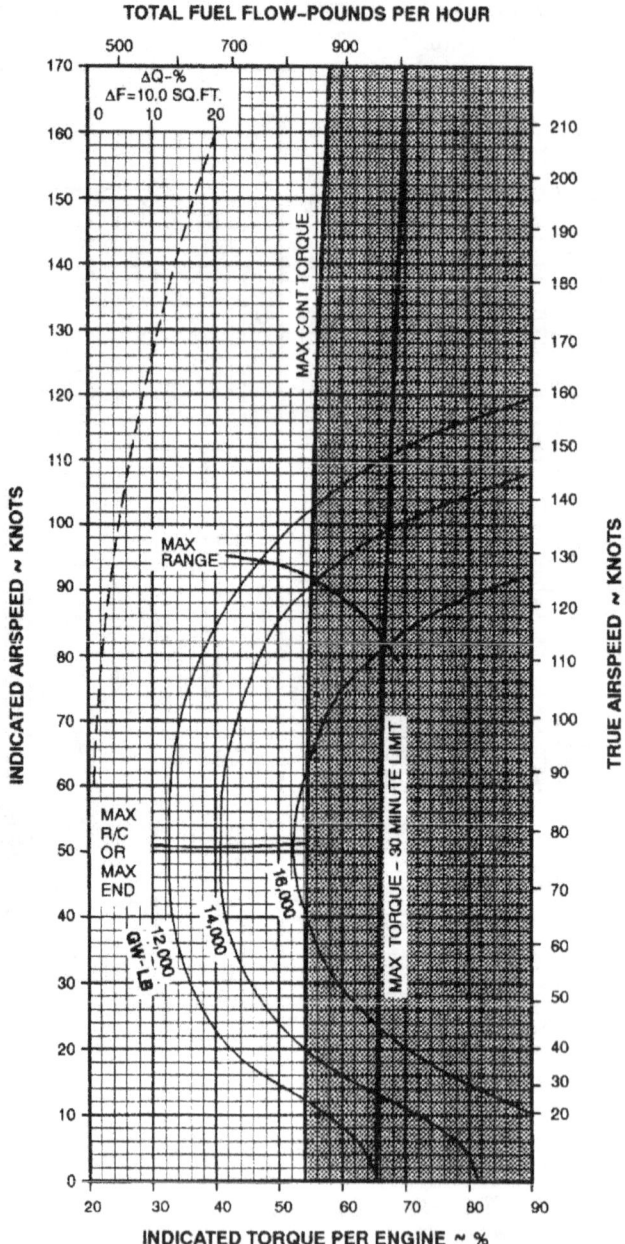

Figure 7-15. Cruise Chart, 12,000 Feet, +30 °C (Sheet 5 of 5)

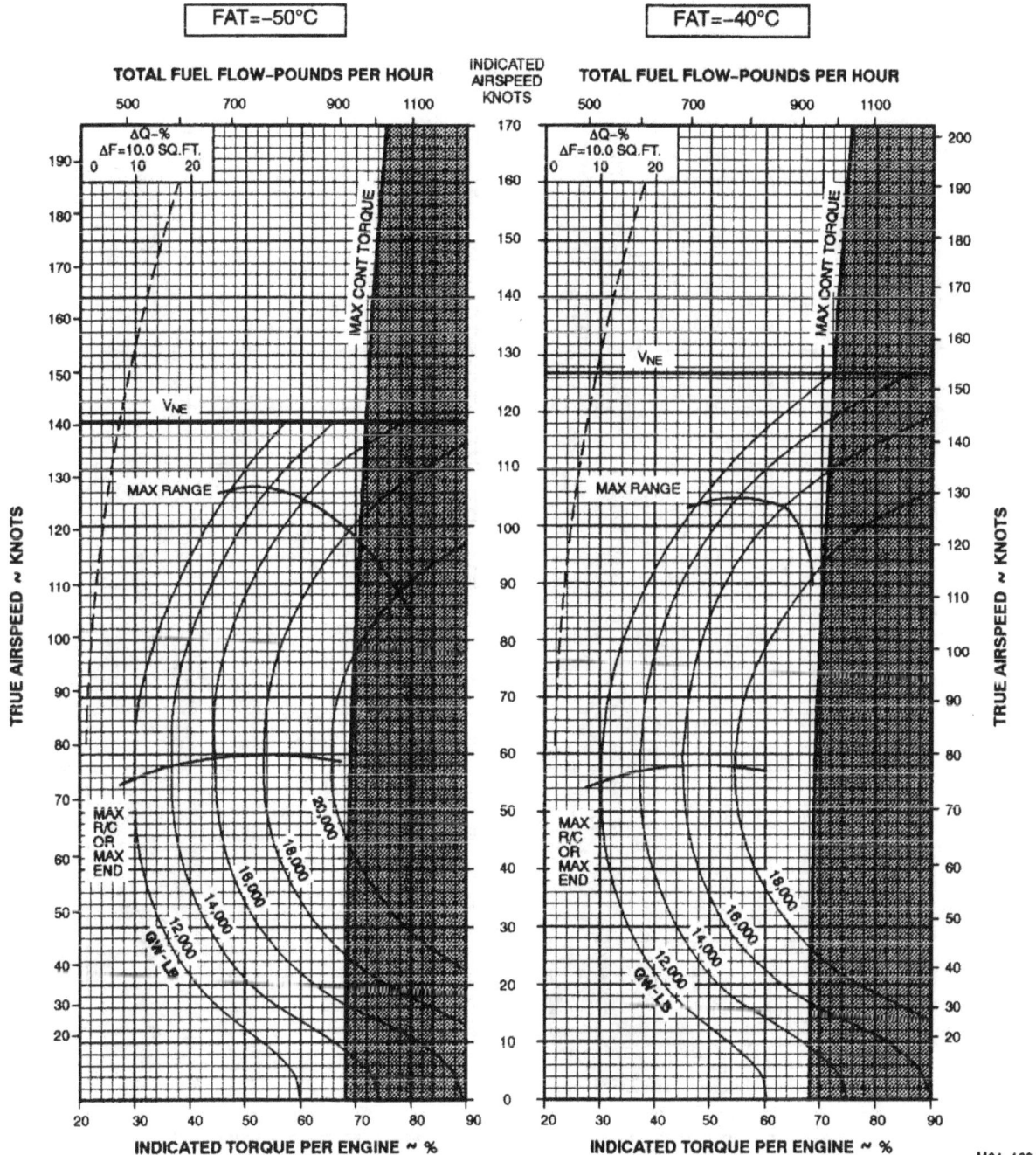

Figure 7-16. Cruise Chart, 14,000 Feet, -50 and -40 °C (Sheet 1 of 5)

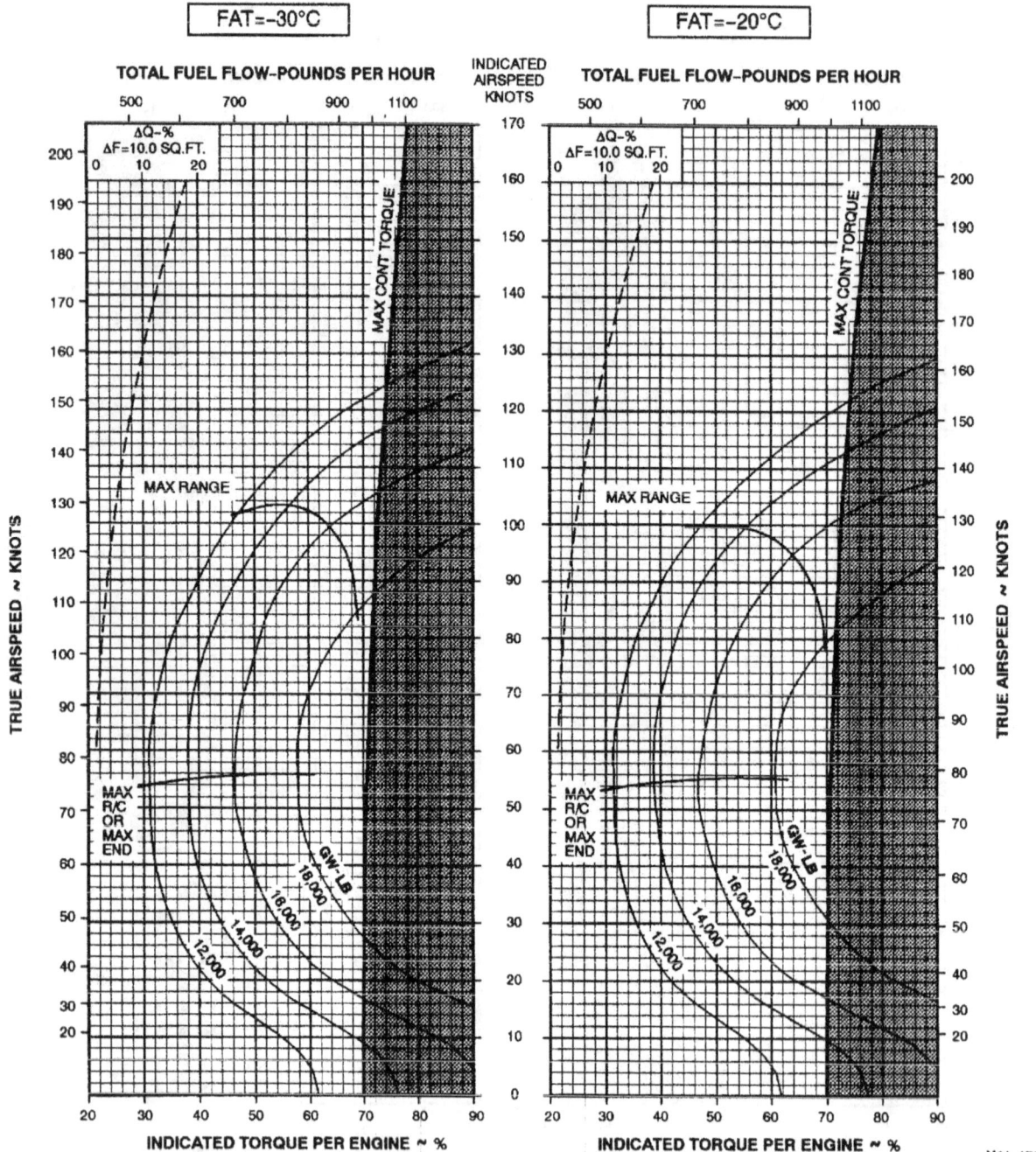

Figure 7-16. Cruise Chart, 14,000 Feet, -30° and -20 °C (Sheet 2 of 5)

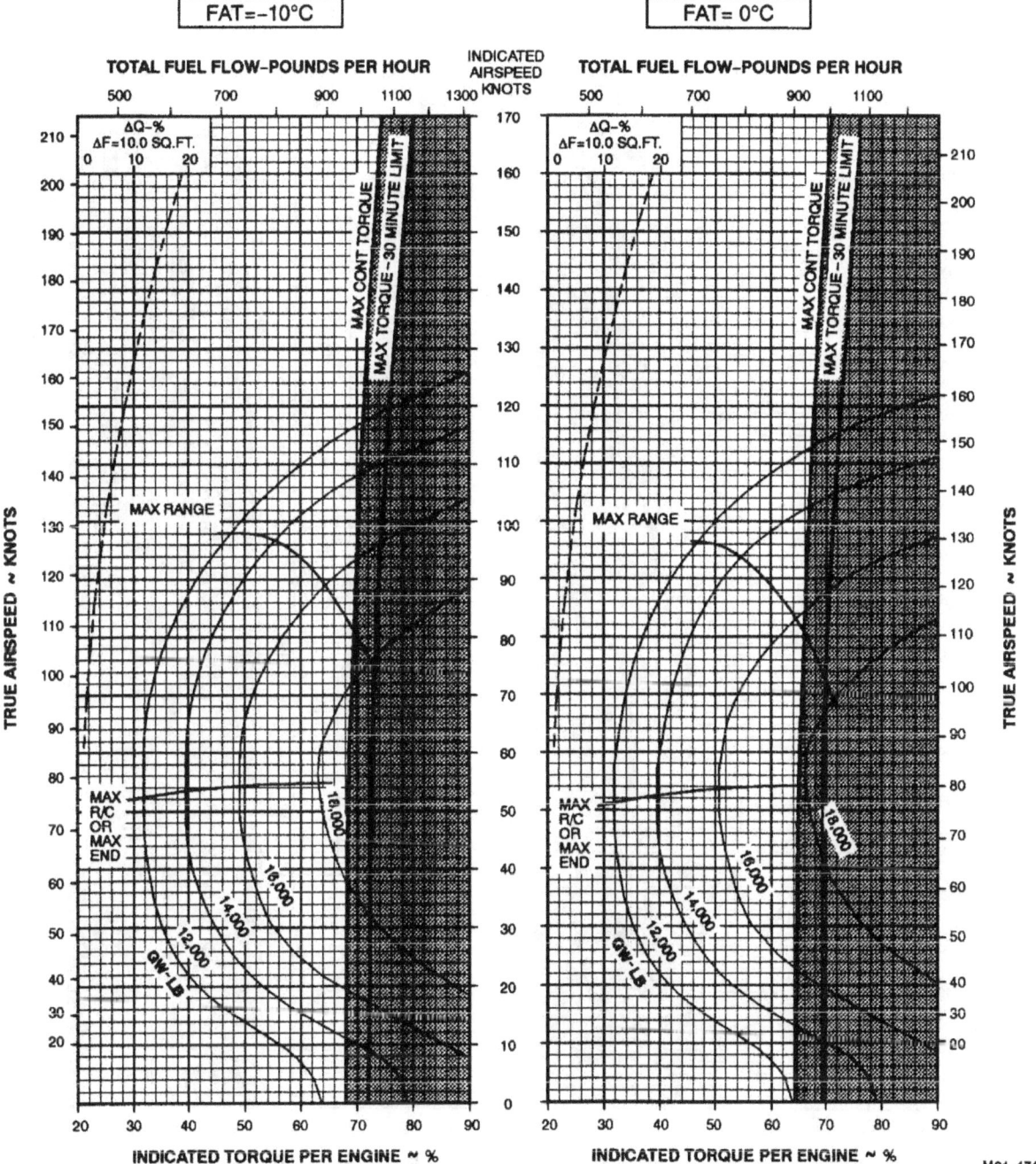

Figure 7-16. Cruise Chart, 14,000 Feet, -10° and 0 °C (Sheet 3 of 5)

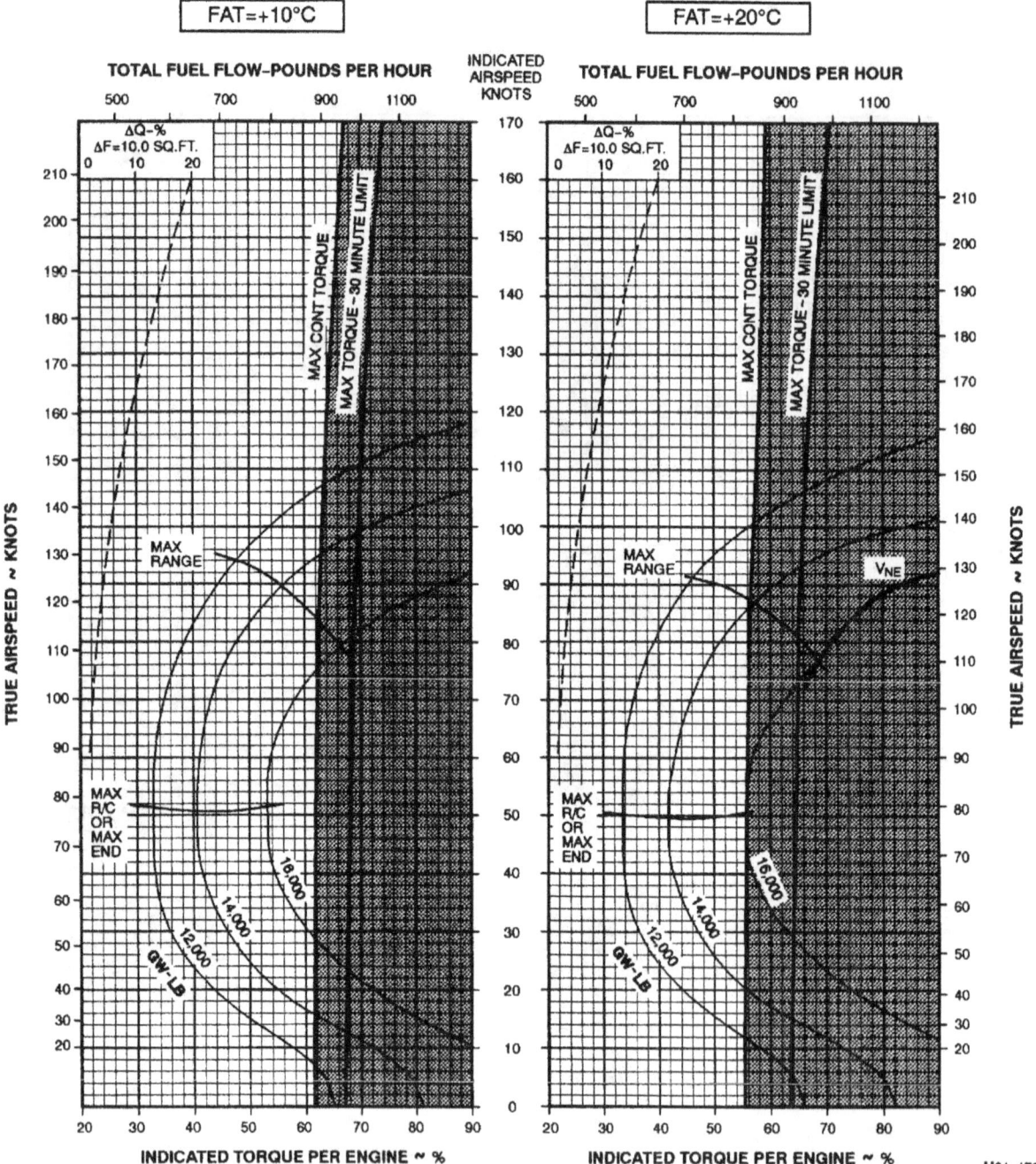

Figure 7-16. Cruise Chart, 14,000 Feet, +10° and +20 °C (Sheet 4 of 5) 701

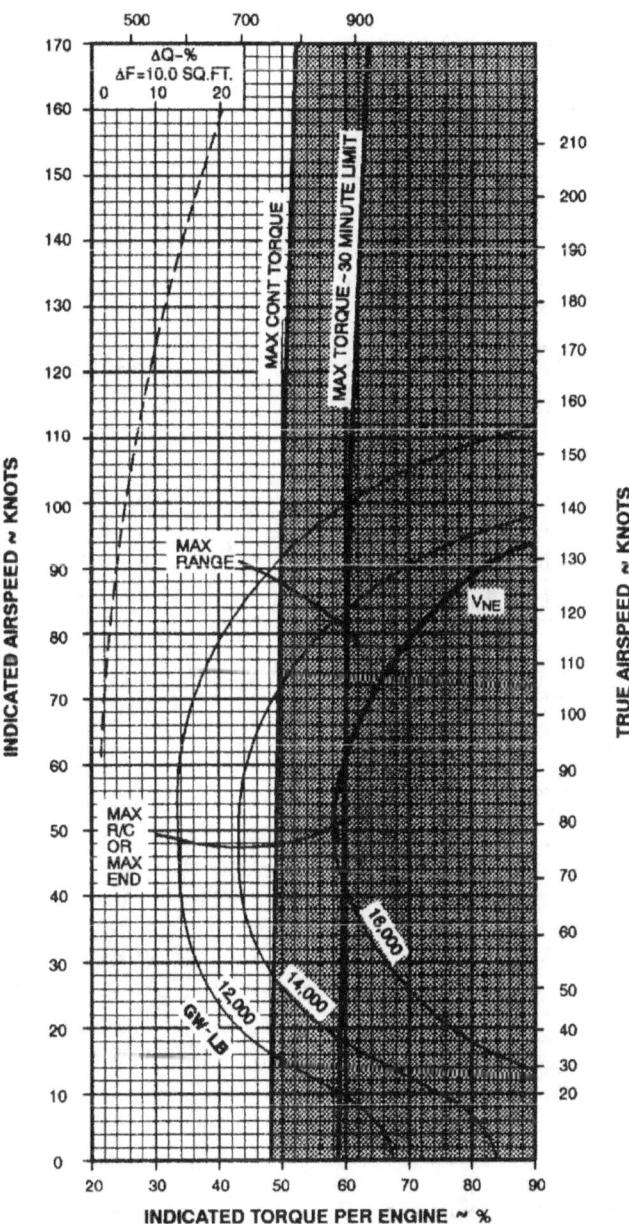

Figure 7-16. Cruise Chart, 14,000 Feet, +30 °C (Sheet 5 of 5)

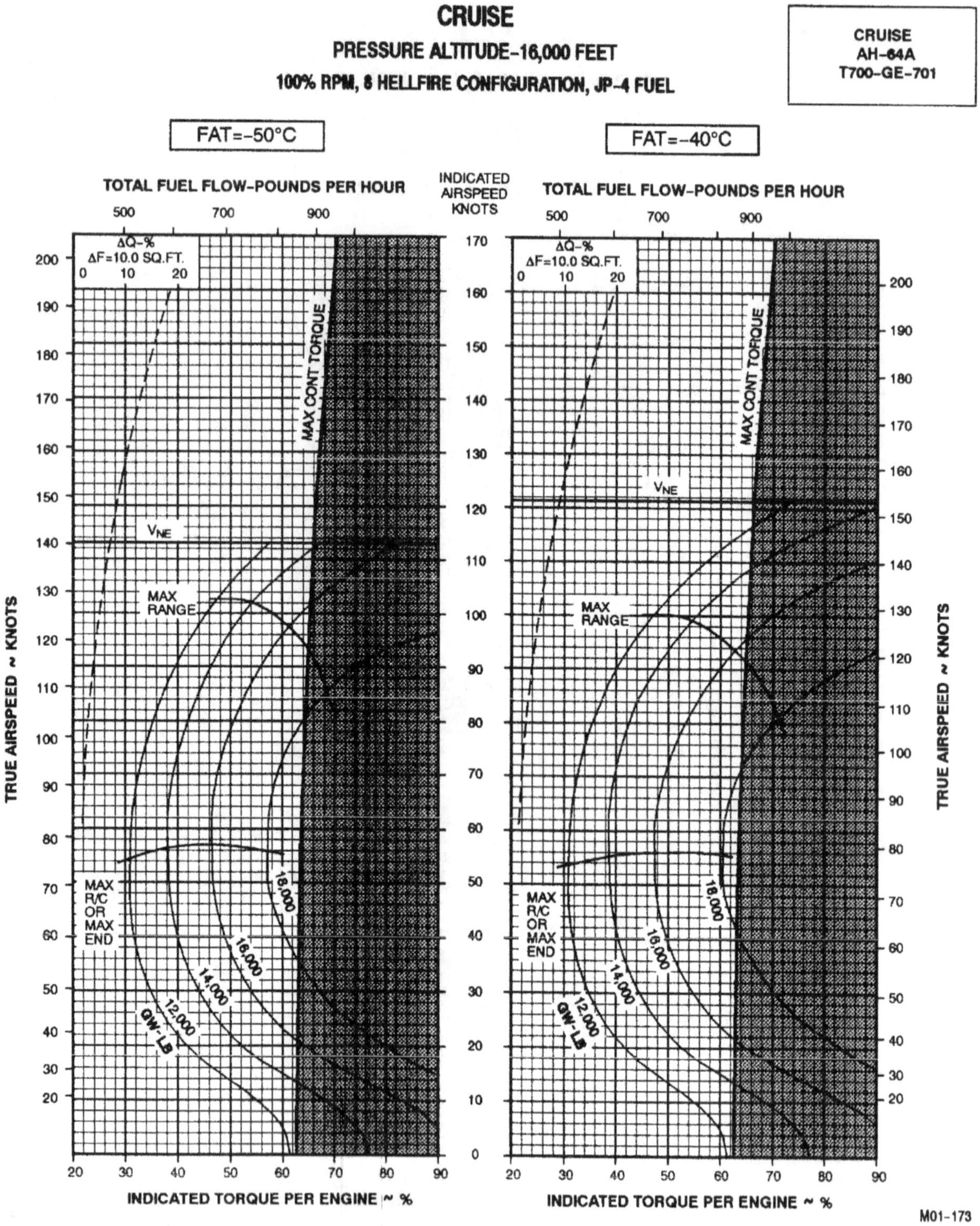

Figure 7-17. Cruise Chart, 16,000 Feet, -50° and -40 °C (Sheet 1 of 5)

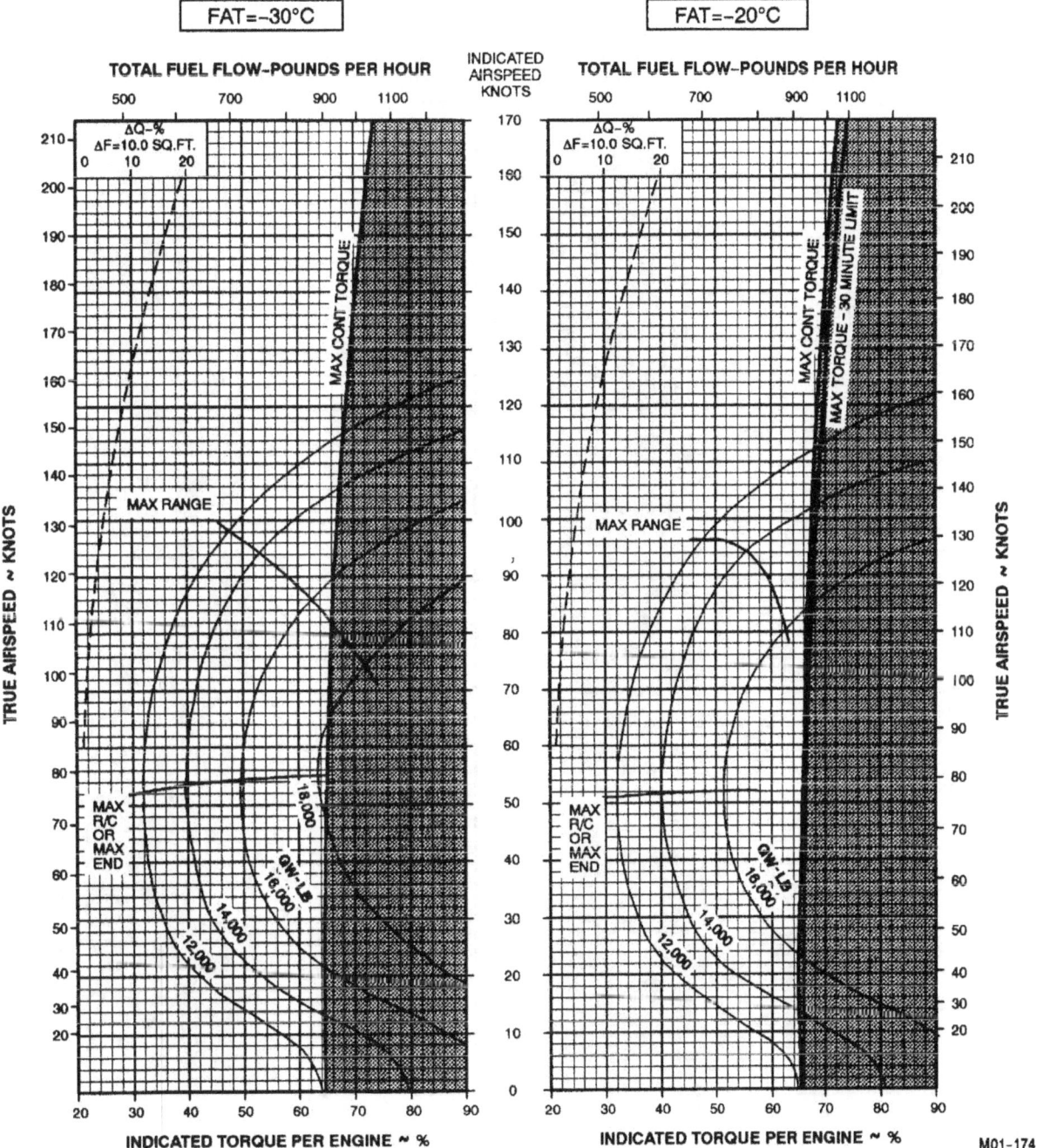

Figure 7-17. Cruise Chart, 16,000 Feet, -30° and -20 °C (Sheet 2 of 5)

TM 1-1520-238-10

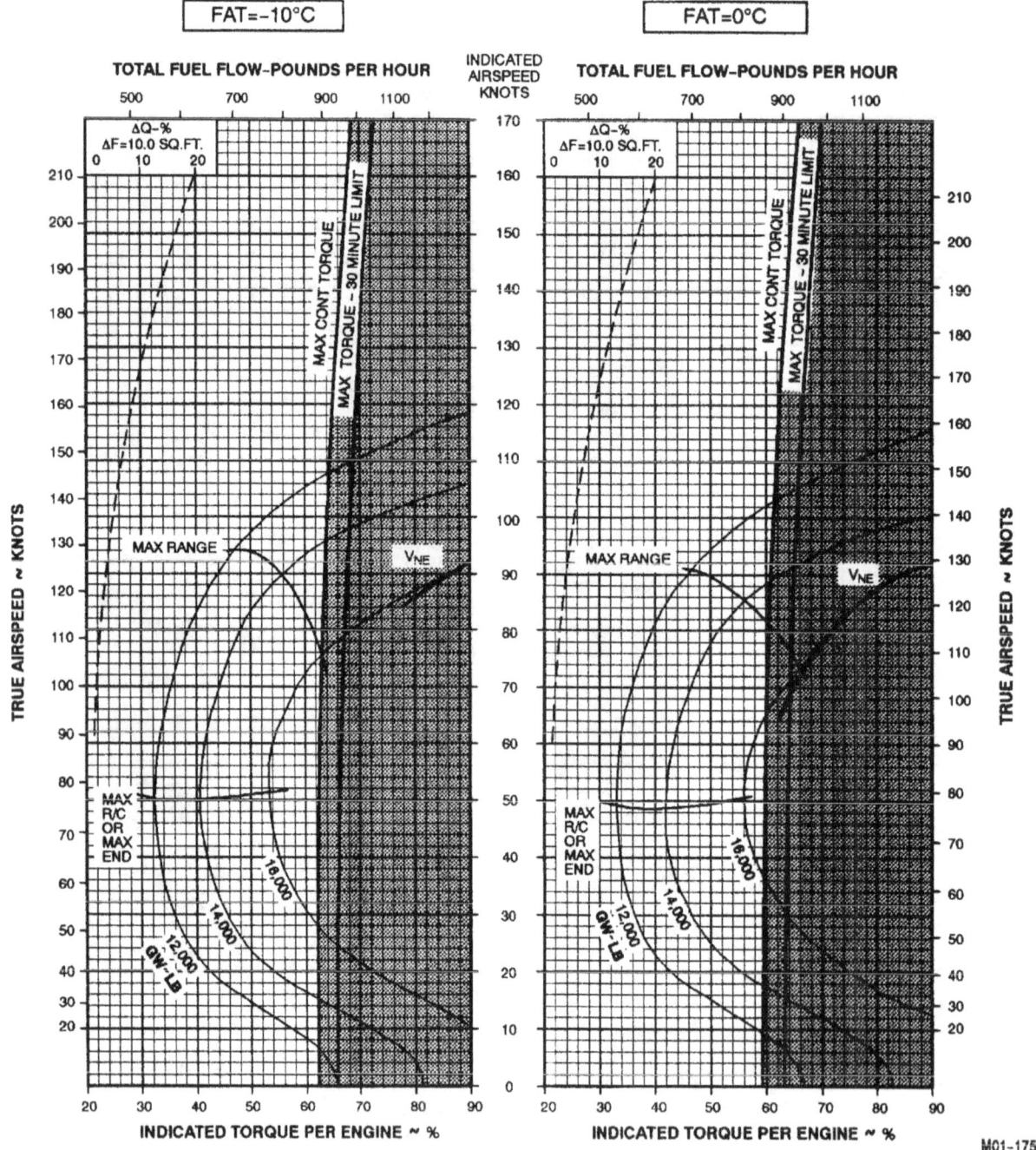

Figure 7-17. Cruise Chart, 16,000 Feet, -10° and 0 °C (Sheet 3 of 5)

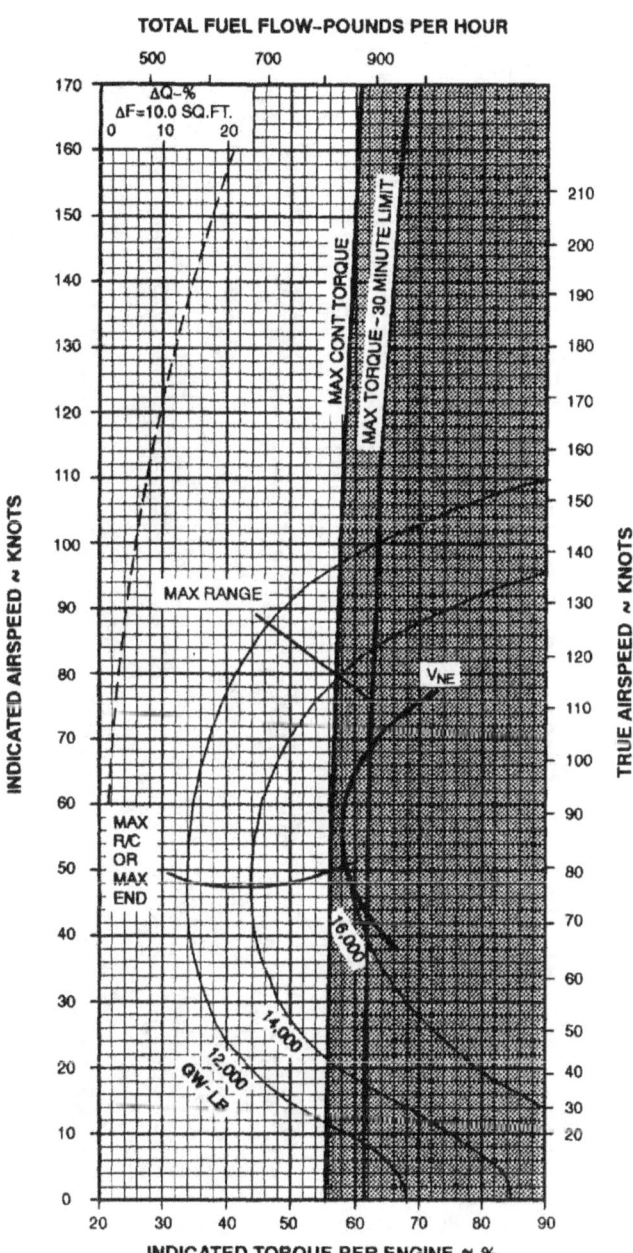

Figure 7-17. Cruise Chart, 16,000 Feet, +10 °C (Sheet 4 of 5) 701

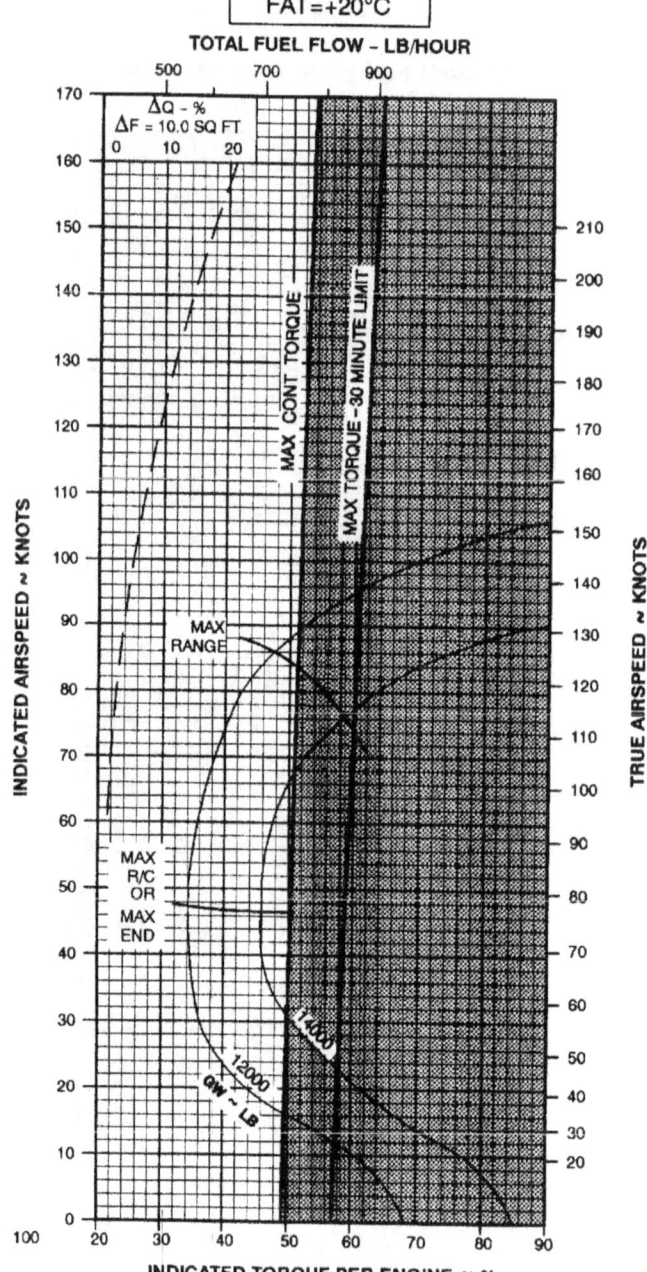

Figure 7-17. Cruise Chart, 16,000 Feet, +20 °C (Sheet 5 of 5) 701

Section VI. DRAG

7.22 DESCRIPTION.

The drag chart (fig 7-18) shows the change in frontal area for each wing-stores combination that can be installed on the helicopter. The baseline configuration does not include the Wire Strike Protection System (WSPS).

7.23 USE OF CHART.

To determine the change in frontal area (AF) and the associated multiplying factor, it is necessary to know what combination. of stores is installed. Enter the chart at the top, move down to the illustration that matches the desired combination, and then move right and read AF and the multiplying factor. Use the multiplying factor and data in Section VI, Cruise to determine the resuiting change in torque.

7.24 CONDITIONS.

The drag chart is based on the primary mission configuration having zero change in frontal area. If the WSPS is installed, increase the frontal area for all configurations in the drag table (including the primary mission configuration) by 1.9 sq ft and increase the multiplying factor by 0.19.

TM 1-1520-238-10

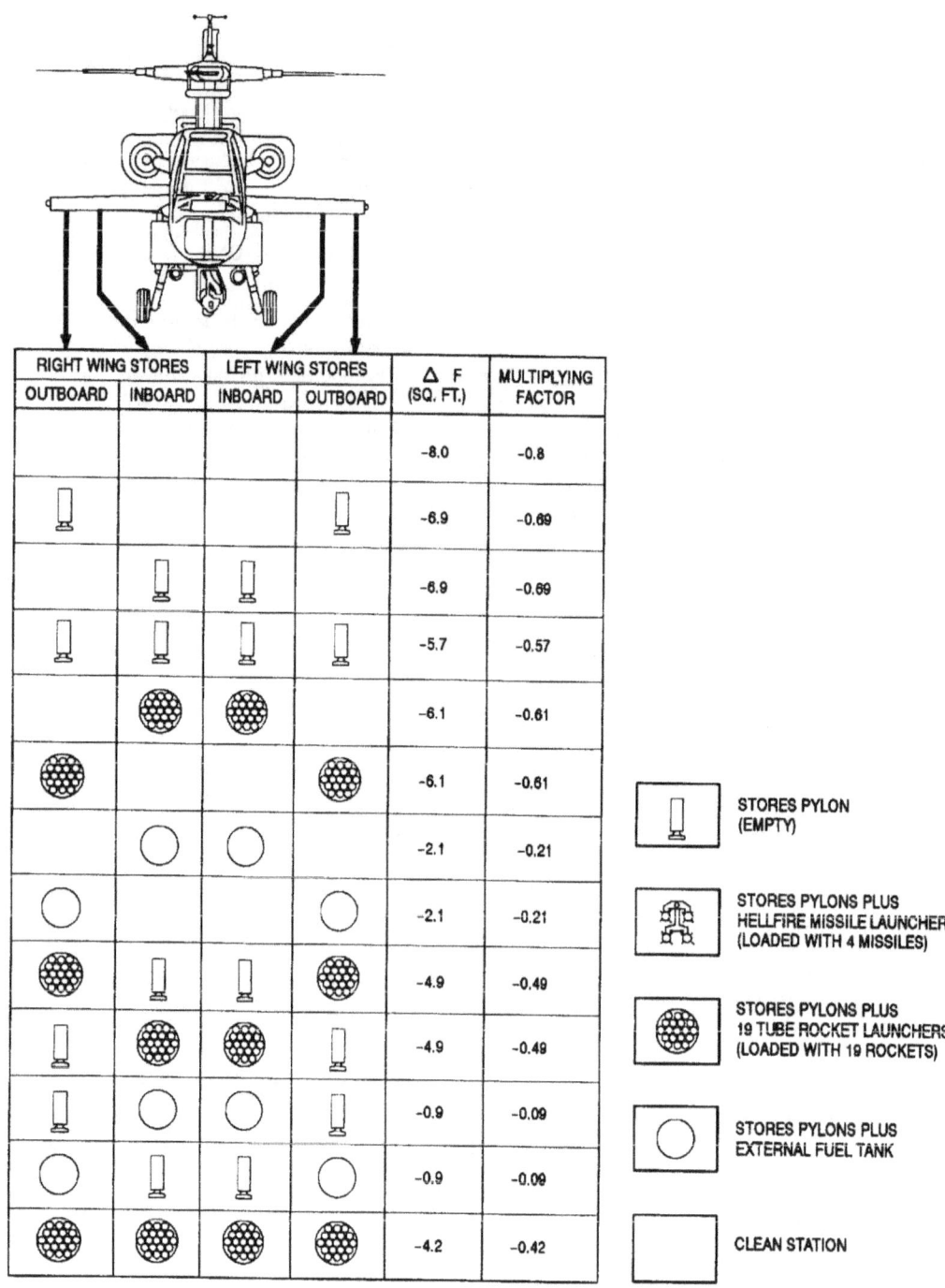

Figure 7-18. Drag Chart and Authorized Armament Configurations (Sheet 1 of 2)

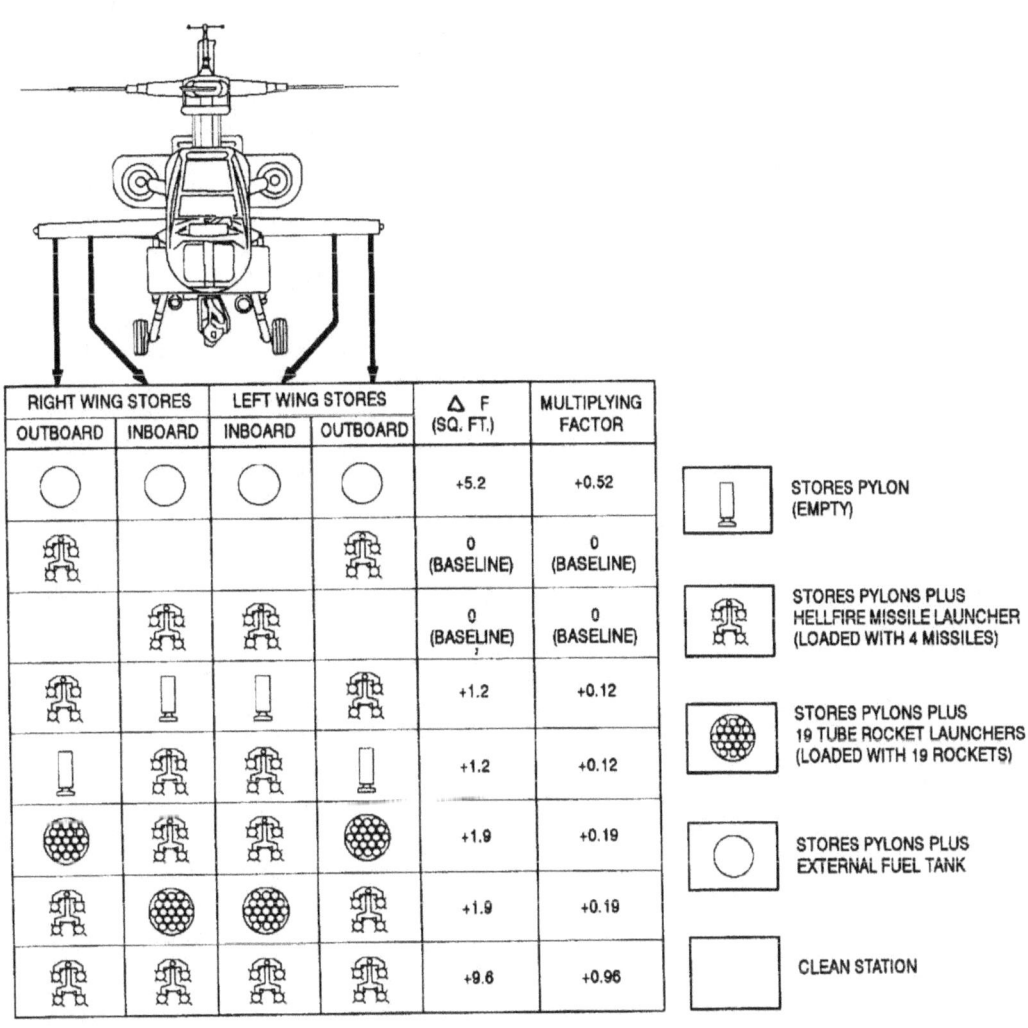

Figure 7-18. Drag Chart and Authorized Armament Configurations (Sheet 2 of 2)

Section VII. CLIMB-DESCENT

7.25 DESCRIPTION

The climb descent chart (fig 7-19) shows the change in torque (above and below the torque required for level flight under the same configuration, gross weight, and atmospheric conditions) to obtain a desired rate of climb or descent.

7.26 USE OF CHART.

The primary use of the chart is illustrated by the example. To determine the change in torque, it is necessary to know the gross weight and the desired rate of climb or descent. Enter the chart at the desired rate of climb or descent, move right to the known gross weight, and then move down and read the torque change, This torque change must be added to (for climb) or subtracted from (for descent) the torque required for level flight (obtained from the appropriate cruise chart) to obtain a total climb or descent torque.

By entering the chart with a known torque change, and moving up to the known gross weight and then left, the corresponding rate of climb or descent can also be obtained.

7.27 CONDITIONS.

The climb-descent chart is based on 100% rotor rpm.

CLIMB - DESCENT

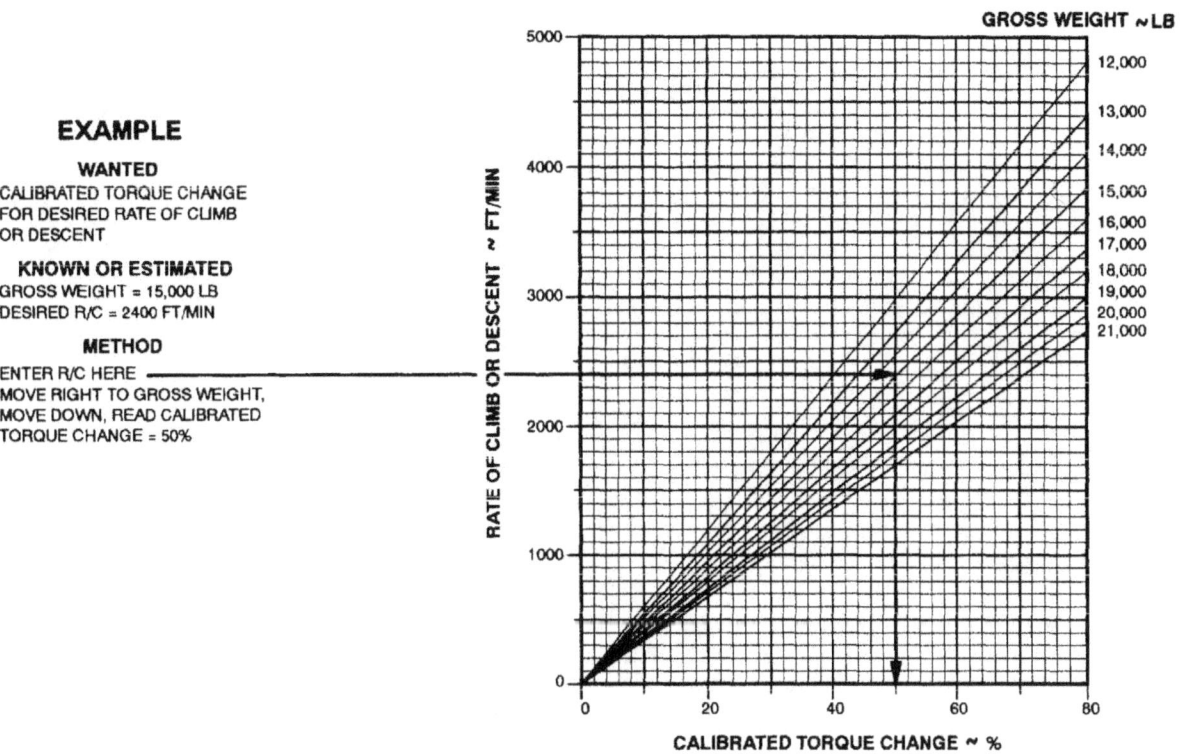

Figure 7-19. Climb-Descent Chart 701

CHAPTER 7A
PERFORMANCE DATA FOR AH-64A HELICOPTERS EQUIPPED WITH T-700-GE-701C ENGINES

Section I. INTRODUCTION

NOTE

This chapter contains performance data for helicopters equipped with T-700-GE-701C **701C** engines. Performance data for helicopters equipped with T-700-GE-701 **701** engines is contained in Chapter 7. Users are authorized to remove the chapter that is not applicable and are not required to carry both chapters on the helicopter.

7A.1. PERFORMANCE DATA.

The purpose of this chapter is to provide the best available performance data for the AH-64A helicopter equipped with -701C engines. Regular use of this information will allow maximum safe use of the helicopter. Although maximum performance is not always required, regular use of the information in this chapter is recommended for the following reasons:

    a. Knowledge of performance margins will allow better decisions when unexpected conditions or alternate missions are encountered.

    b. Situations requiring maximum performance will be more readily recognized.

    c. Familiarity with the data will allow performance to be computed more easily and quickly.

    d. Experience will be gained in accurately estimating the effects of conditions for which data is not presented.

NOTE

The information is primarily intended for mission planning and is most useful when planning operations in unfamiliar areas or at extreme conditions. The data may also be used in flight, to establish unit or area standing operating procedures, and to inform ground commanders of performance/risk tradeoffs.

The data presented covers the maximum range of conditions and performance that can reasonably be expected. In each area of performance, the effects of altitude, temperature, gross weight and other parameters relating to that phase of flight are presented. In addition to the presented data, judgment and experience will be necessary to accurately determine performance under a given set of circumstances. The conditions for the data are listed under the title of each chart. The effects of different conditions are discussed in the text accompanying each phase of performance. Where practical, data is presented at conservative conditions. However, NO GENERAL CONSERVATISM HAS BEEN APPLIED.

7A.2. OPERATIONAL LIMITS.

CAUTION

Exceeding operational limits can cause permanent damage to critical components. Overlimit operation can decrease performance, cause immediate failure or failure on a subsequent flight.

Applicable limits are shown on the charts as bold lines with a description. Performance generally deteriorates rapidly beyond limits. If limits are exceeded, minimize the amount and time. Enter the maximum value and time beyond limits on DA Form 2408-13 to ensure proper maintenance action is taken.

TM 1-1520-238-10

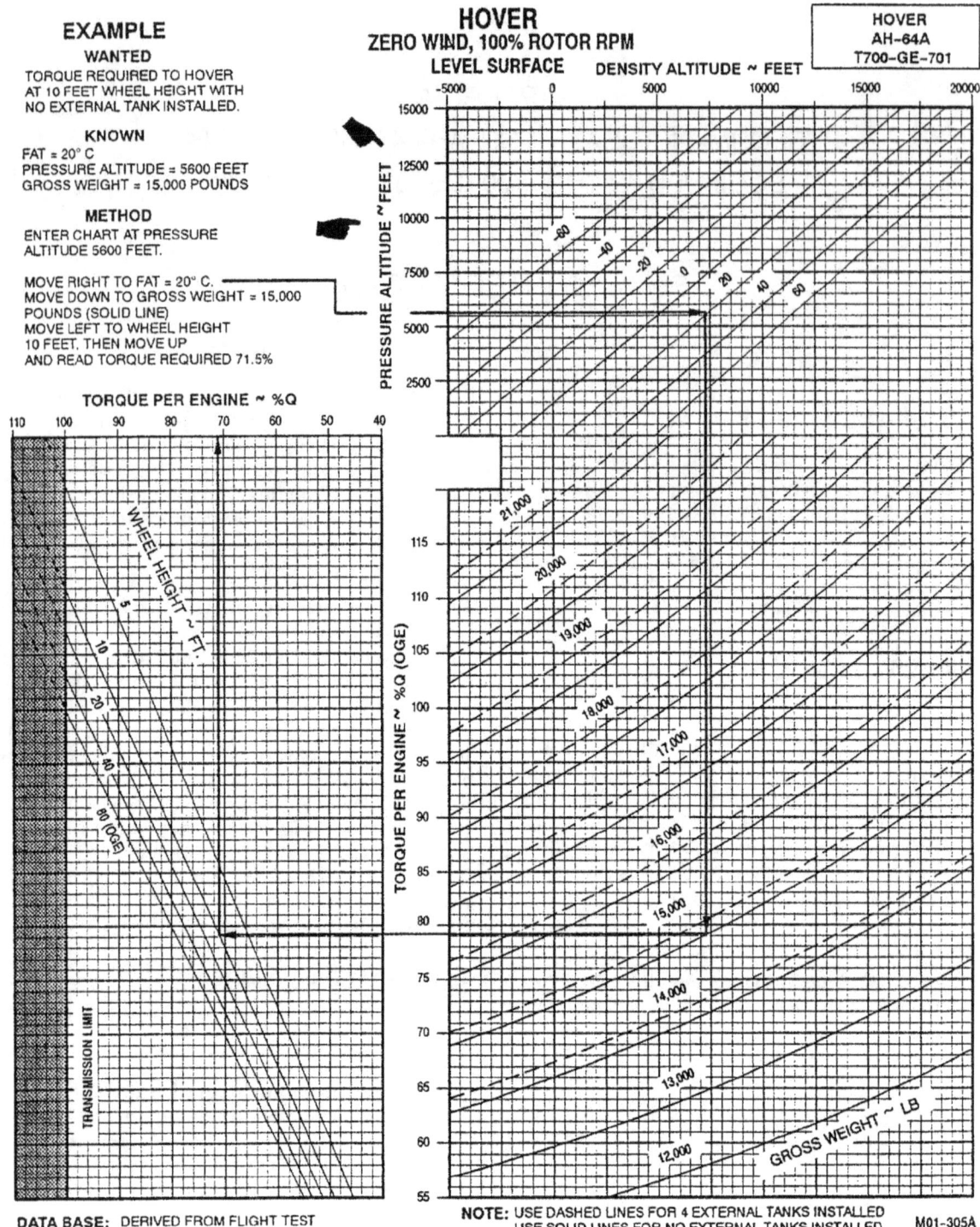

Figure 7-6. Hover Chart Mill

7-12 Change 3

TM 1-1520-238-10

7A.8. ABBREVIATIONS.

Appendix B is a list of abbreviations and symbols used on the charts in this chapter. For units of measure, the same abbreviation applies to either the singular or plural form of the unit.

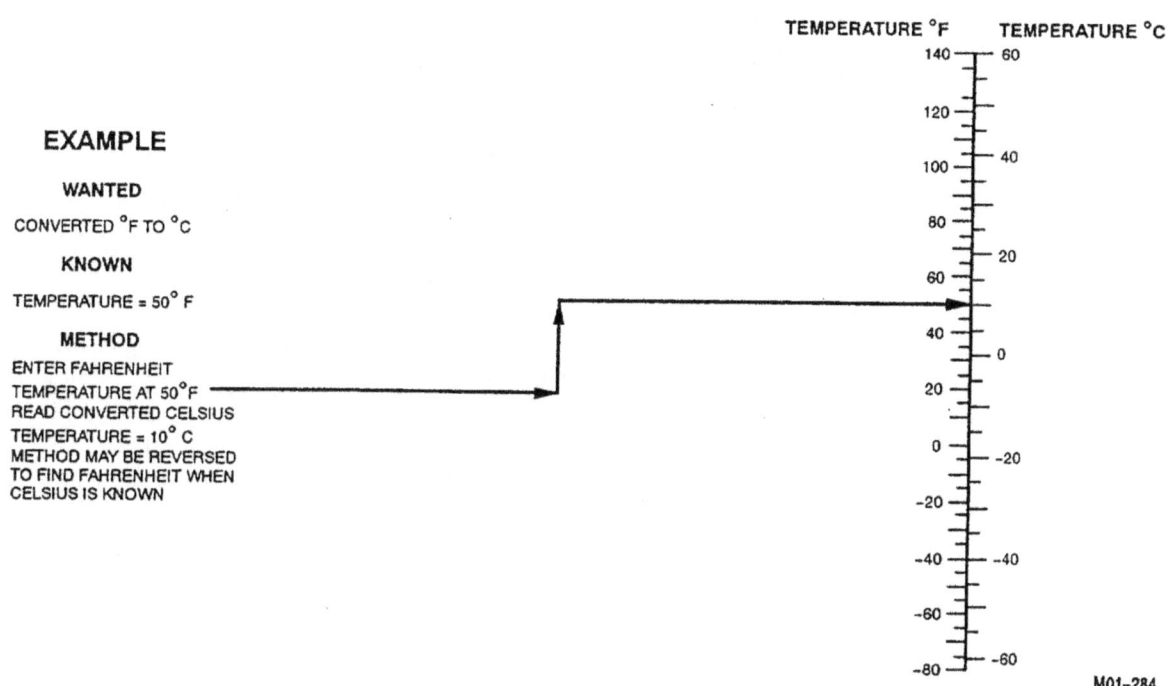

Figure 7A-1. Temperature Conversion Chart

Change 3   7A-3

Section II. MAXIMUM TORQUE AVAILABLE

7A.9. DESCRIPTION.

The maximum torque available charts shows the maximum specification torque available per engine for 30 minute operation (fig 7A-2 sheet 1) and 10 minute operation (fig 7A-2 sheet 2) at various conditions of pressure altitude and free air temperature. Both single and dual engine operation limits are shown.

The maximum torque available for 2.5 minute operation (fig 7A-2 sheet 3) shows the maximum specification torque available when one engine is inoperative; only single engine operation limits are shown.

The torque factor charts (figs 7A-3 and 7A-4) provide an accurate indication of available power for the engines installed in each individual aircraft.

7A.10. USE OF CHARTS.

The primary use of the maximum torque available charts (fig 7A-2 sheets 1, 2, and 3) is illustrated by the example. To determine the maximum specification torque available, it is necessary to know pressure altitude and free air temperature. Enter the left side of either the 30 minute or the 10 minute chart (fig 7A-2 sheets 1 or 2) at the known temperature and move right to the known pressure altitude, and then move down and read the maximum specification torque available. This is torque per engine. For dual engine operation, if the torque per engine exceeds the two engine limit, the maximum torque available must be reduced to the two engine limit.

For one engine inoperative, enter the left side of the 2.5 minute limit chart (fig 7A-2 sheet 3) at the known temperature and move right to the known pressure altitude, and then move down and read the maximum specification torque available for one engine. If the torque exceeds the one engine limit, maximum torque available must be reduced to the one engine limit.

7A.11. CONDITIONS.

The maximum torque available charts (fig 7A-2 sheets 1, 2, and 3) are based on 100% rotor rpm, zero airspeed, JP-4 fuel and **ENG INLET** anti-ice switch **OFF**. With **ENG INLET** anti-ice switch **ON**, available torque is reduced by as much as 19.3% for 30 minute operation and 18.6% for 10 minute operation. For example, if value from the chart is 90%, with anti-ice **ON**, torque available would be 90 - 19.3 = 70.7%, 30 minute limit.

7A.12. TORQUE FACTOR METHOD.

The torque factor method provides an accurate indication of available power by incorporating ambient temperature effects on degraded engine performance. The torque factor method provides the procedure to determine the maximum dual or single engine torque available for the engines installed in each individual aircraft. The specification power is defined for a newly delivered low time engine. The aircraft HIT log form for each engine provide the engine and aircraft torque factors which are obtained from the maximum power check and recorded to be used in calculating maximum torque available.

7A.12.1 Torque Factor Terms. The following terms are used when determining the maximum torque available for an individual aircraft:

a. Torque Ratio (TR). The ratio of torque available to specification torque at the desired ambient temperature.

b. Engine Torque Factor (ETF). The ratio of an individual engine torque available to specification torque at reference temperature of 35 °C. The ETF is allowed to range from 0.85 to 1.0.

c. Aircraft Torque Factor (ATF). The ratio of an individual aircrafts power available to specification power at a reference temperature of 35 °C. The ATF is the average of the ETFs of both engines and its value is allowed to range from 0.9 to 1.0.

7A.12.2 Torque Factor Procedure. The use of the ATF or ETF to obtain the TR from figure 7A-3 for ambient temperatures between -15 °C and 35 °C is shown by the example. The ATF and ETF values for an individual aircraft are found on the engine HIT log. The TR always equals 1.0 for ambient temperatures of -15 °C and below, and the TR equals the ATF or ETF for temperatures of 35 °C and above.

When the TR equals 1.0 the torque available may be read directly from the specification torque available scales. When the TR is less than 1.0, the actual torque available is determined by multiplying the specification torque available by the TR (example for TR = 0.98: 90% TRQ X 0.98 = 88.2% TRQ). The torque conversion chart (fig 7A-4) is provided to convert specification data to actual torque available. The single and dual engine transmission limits are shown and should not be exceeded.

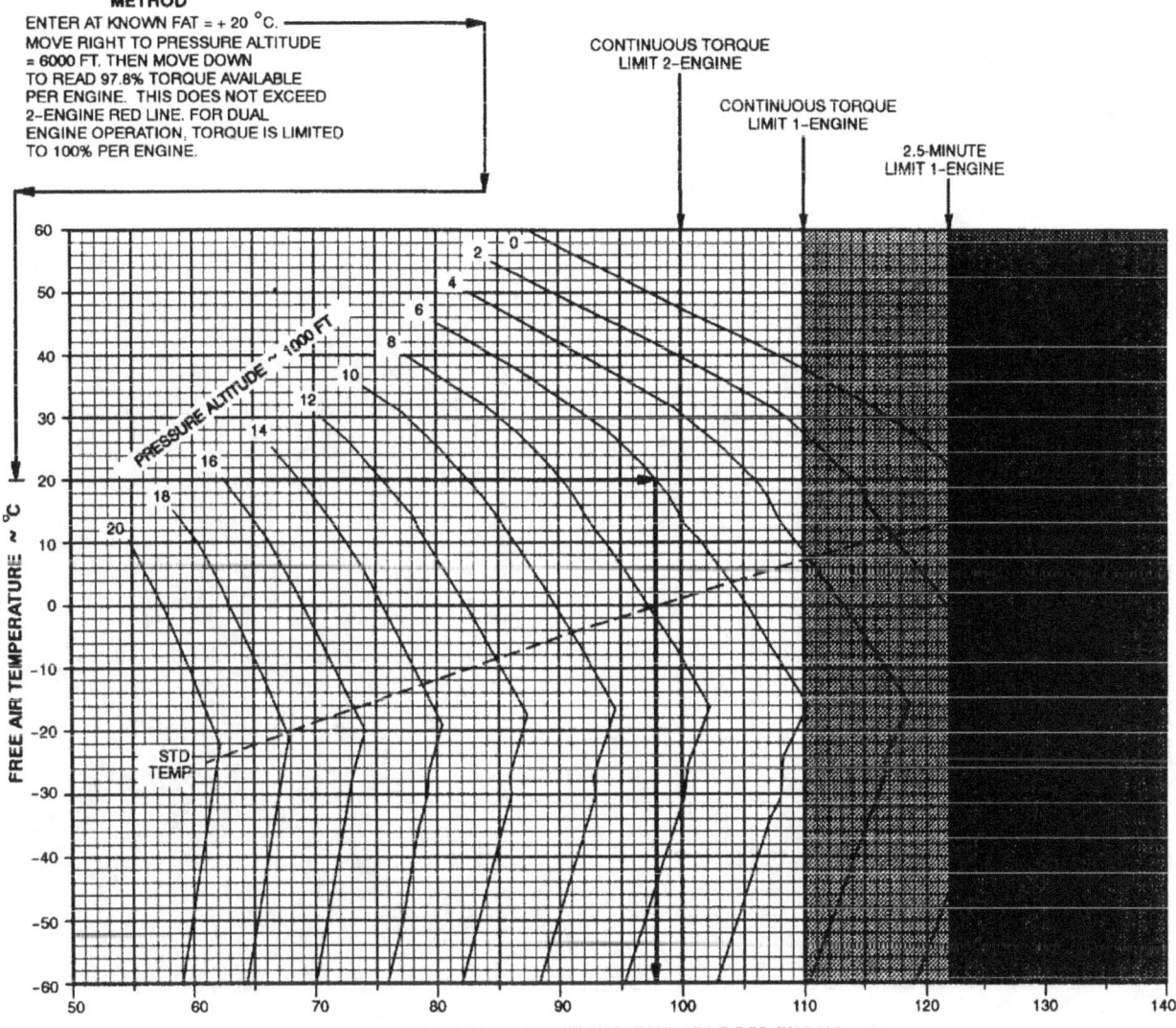

Figure 7A-2. Maximum Torque Available Chart (Sheet 1 of 3)

TM 1-1520-238-10

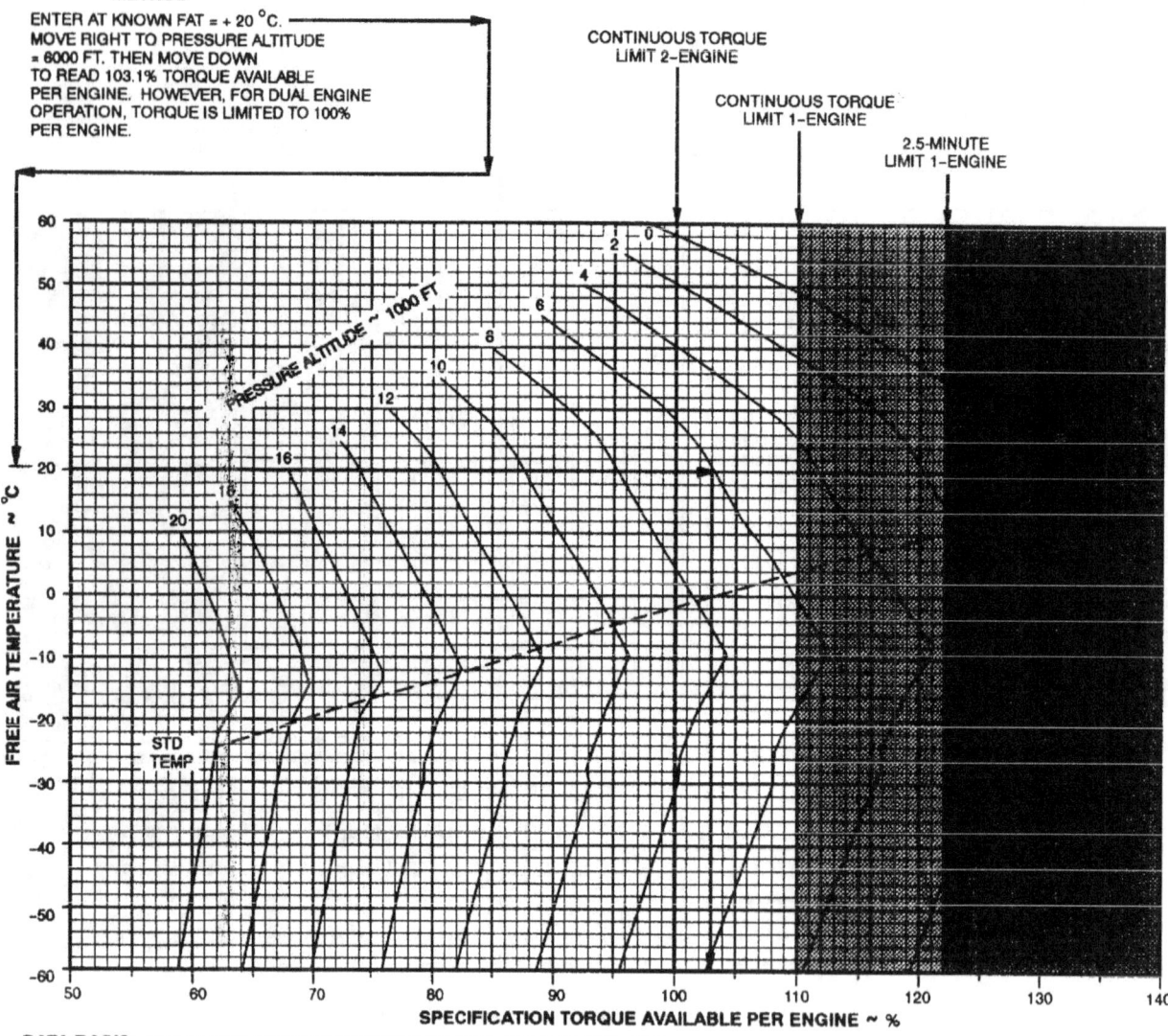

Figure 7A-2. Maximum Torque Available Chart (Sheet 2 of 3) 701C

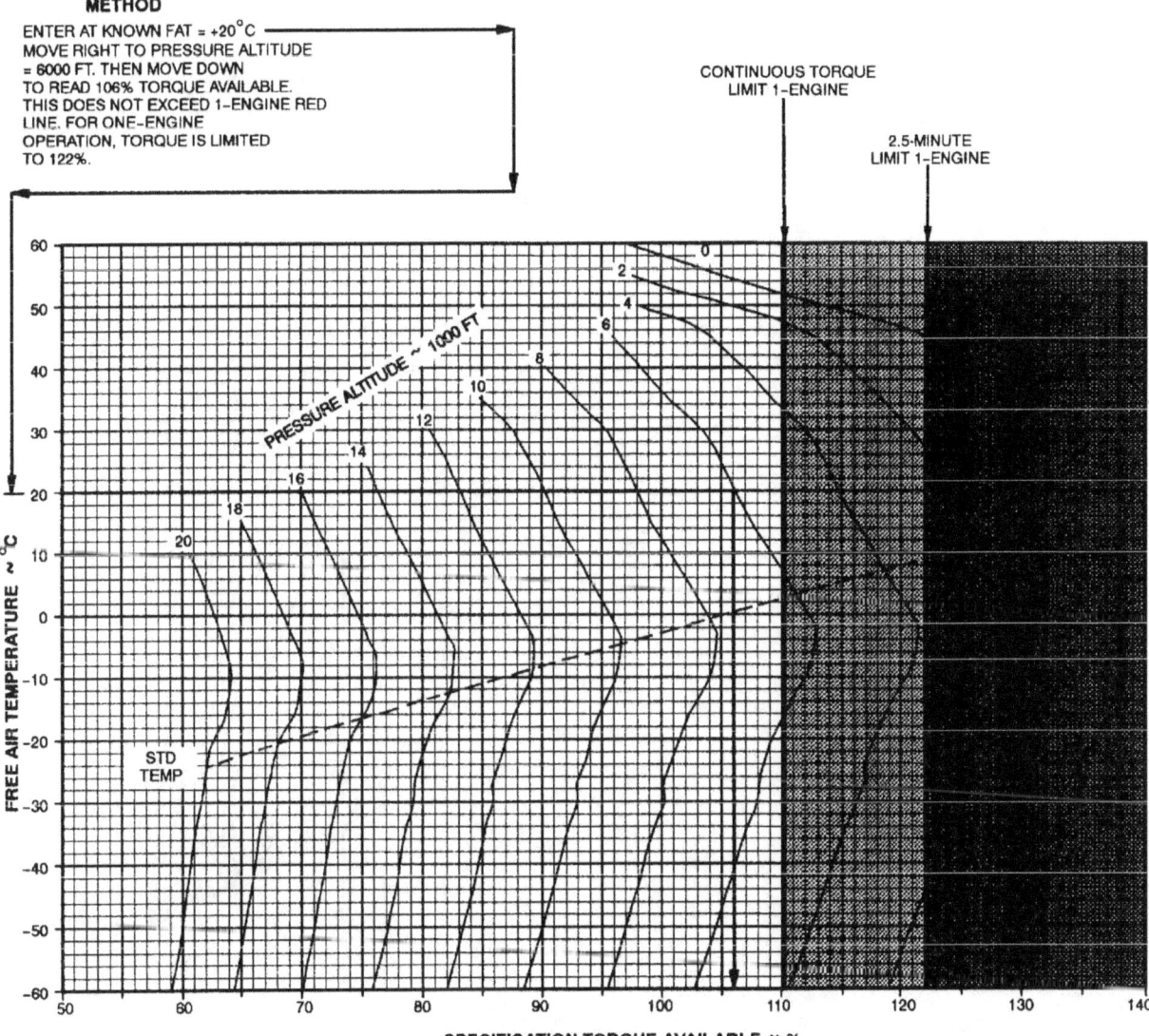

Figure 7A-2. Maximum Torque Available Chart (Sheet 3 of 3)

TM 1-1520-238-10

TORQUE FACTOR
T700-GE-701C ENGINE, 100% RPM

TORQUE FACTOR
AH-64A
T700-GE-701C

EXAMPLE

WANTED
TORQUE RATIO AND MAXIMUM TORQUE AVAILABLE
30-MIN LIMIT

KNOWN
ATF = .95
PRESSURE ALTITUDE = 6000 FT
FAT = +20° C

METHOD
TO OBTAIN TORQUE RATIO

1. ENTER TORQUE FACTOR CHART AT KNOWN FAT
2. MOVE RIGHT TO THE ATF VALUE
3. MOVE DOWN, READ TORQUE RATIO = .967

TO CALCULATE MAXIMUM TORQUE AVAILABLE:

4. ENTER MAXIMUM TORQUE AVAILABLE CHART 30 MIN LIMIT AT KNOWN FAT
5. MOVE RIGHT TO KNOWN PRESSURE ALTITUDE
6. MOVE DOWN, READ SPECIFICATION TORQUE = 97.8%

TO OBTAIN ACTUAL TORQUE VALUE AVAILABLE FROM THE TORQUE CONVERSION CHART:

7. ENTER TORQUE CONVERSION CHART AT % TORQUE OBTAINED FROM 30 MIN LIMIT CHART
8. MOVE UP TO TORQUE RATIO OBTAINED FROM TORQUE FACTOR CHART
9. MOVE LEFT, READ MAXIMUM TORQUE AVAILABLE = 94.6%

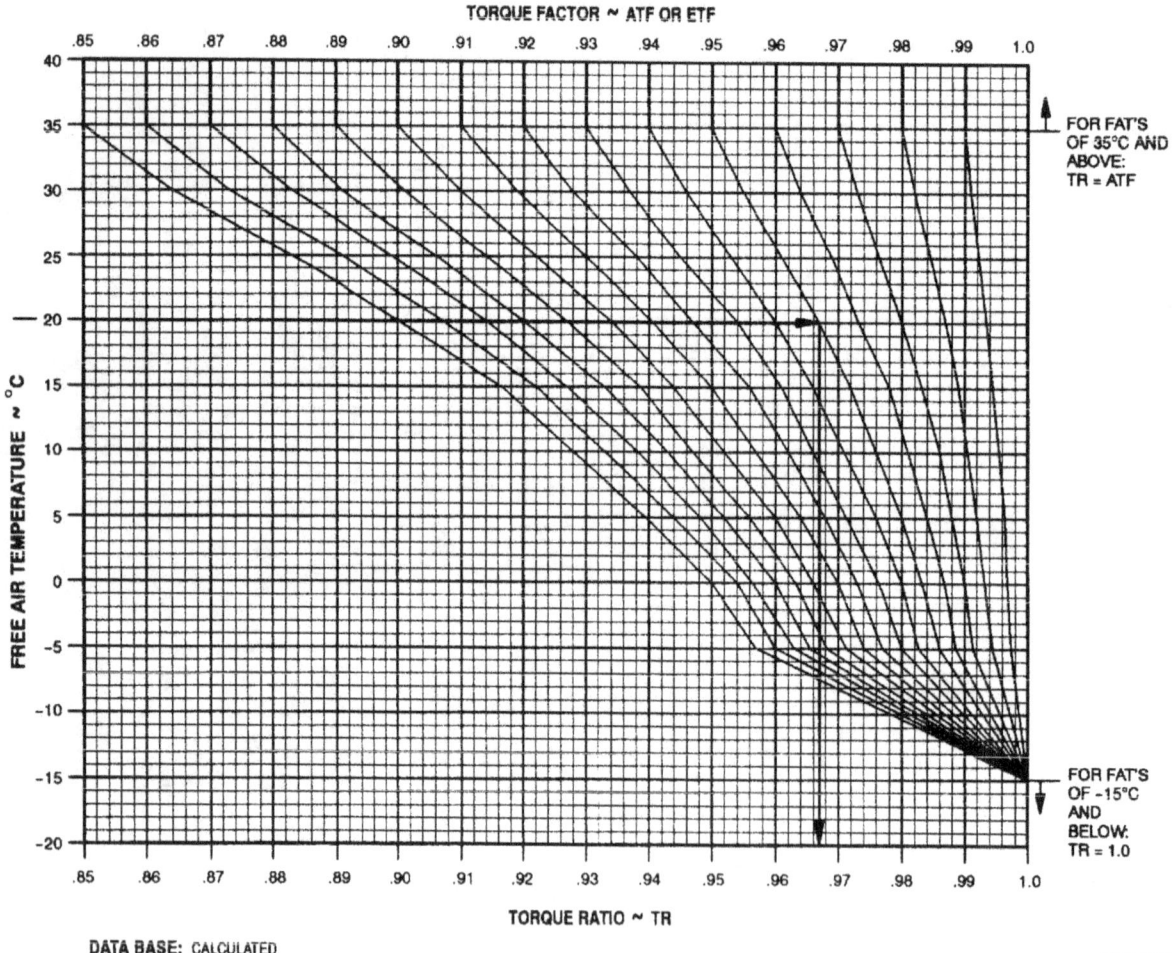

Figure 7A-3. Torque Factor Chart 701C

7A-8

TM 1-1520-238-10

TORQUE CONVERSION

TORQUE CONVERSION
AH-64A
T700-GE-701C

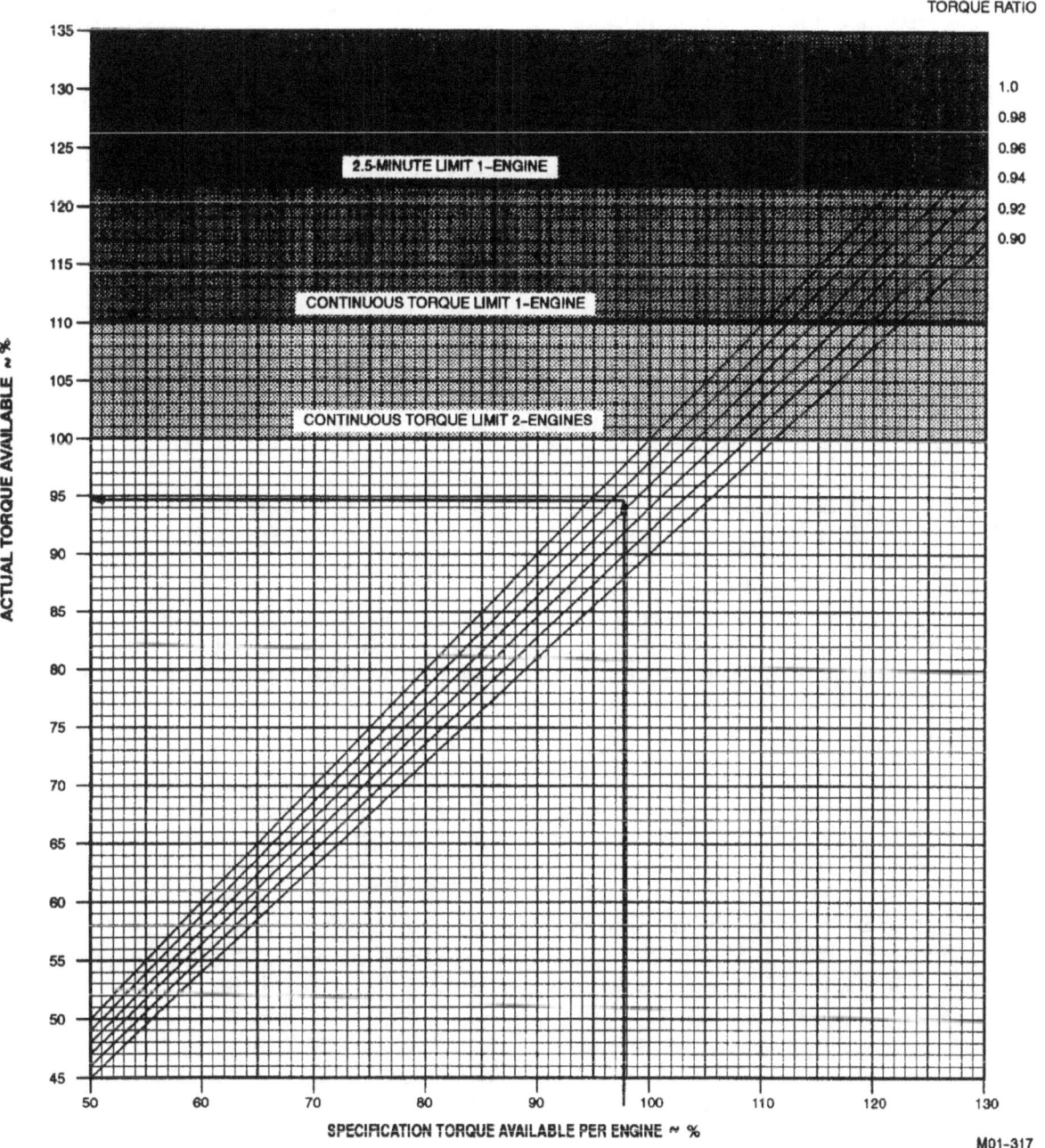

Figure 7A-4. Torque Conversion Chart 701C

Section III. HOVER CEILING

7A.13. DESCRIPTION.

The hover ceiling charts (fig 7A-5 sheets 1 and 2) presents the maximum gross weight for hovering at various conditions of pressure altitude, free air temperature, and wheel height, using maximum torque available, 30 minute limit or maximum torque available, 10 minute limit.

7A.14. USE OF CHARTS.

The primary use of the hover ceiling charts is illustrated by the examples. To determine the maximum gross weight for hover, it is necessary to know the pressure altitude, free air temperature, and desired wheel height. Enter the appropriate power available chart at the pressure altitude, move right to FAT, move down to the desired wheel height and then move left and read maximum gross weight.

7A.15. CONDITIONS.

The hover ceiling charts are based on maximum torque available, 30 minute limit and 10 minute limit; 100% rotor rpm, ATF = 1.0, **ENG INLET** anti-ice switch **OFF**; and rotor **BLADE** de-ice switch off. For **ENG INLET ANTI-ICE** switch **ON**, use dashed lines. Applicable configuration is all external stores except four external fuel tanks. For the four external tank configuration, reduce the maximum gross weight for hover, as calculated from the hover ceiling charts, by 11 pounds for each 1000 pounds of gross weight. See the examples below.

HOVER CEILING

EXAMPLE 1 (SHEET 1)

WANTED

MAXIMUM GROSS WEIGHT FOR HOVER AT 10-FOOT WHEEL HEIGHT, 30-MINUTE LIMIT TORQUE AVAILABLE, FOR ENGINE INLET ANTI-ICE OFF AND ON

KNOWN

PRESSURE ALTITUDE = 10,000 FEET
FAT = -10°C
WHEEL HEIGHT = 10 FEET

METHOD

ENTER PRESSURE ALTITUDE SCALE AT 10,000 FT
MOVE RIGHT TO -10°C FAT, SOLID LINE FOR ANTI-ICE OFF, DASHED LINE FOR ANTI-ICE ON
MOVE DOWN TO 10 FEET WHEEL HEIGHT
MOVE LEFT TO READ GROSS WEIGHT FOR HOVER:
ANTI-ICE OFF, HOVER GW = 17,520 LB
ANTI-ICE ON, HOVER GW = 15,850 LB
WITH 4 EXT TANKS INSTALLED
ANTI-ICE OFF
HOVER GW = 17,520-11 (17,520/1000)
= 17,327 LB

EXAMPLE II (SHEET 2)

WANTED

MAXIMUM GROSS WEIGHT FOR HOVER AT 10-FOOT WHEEL HEIGHT, 10-MINUTE LIMIT TORQUE AVAILABLE, FOR ENGINE INLET ANTI-ICE OFF AND ON

KNOWN

PRESSURE ALTITUDE = 10,000 FEET
FAT = -10°C
WHEEL HEIGHT = 10 FEET

METHOD

ENTER PRESSURE ALTITUDE SCALE AT 10,000 FT
MOVE RIGHT TO -10°C FAT, SOLID LINE FOR ANTI-ICE OFF, DASHED LINE FOR ANTI-ICE ON
MOVE DOWN TO 10 FEET WHEEL HEIGHT
MOVE LEFT TO READ GROSS WEIGHT FOR HOVER:
ANTI-ICE OFF, HOVER GW = 18,000 LB
ANTI-ICE ON, HOVER GW = 16,400 LB
WITH 4 EXTERNAL TANKS INSTALLED
ANTI-ICE OFF
HOVER GW = 18,000-11 (18,000/1000)
= 17,802 LB

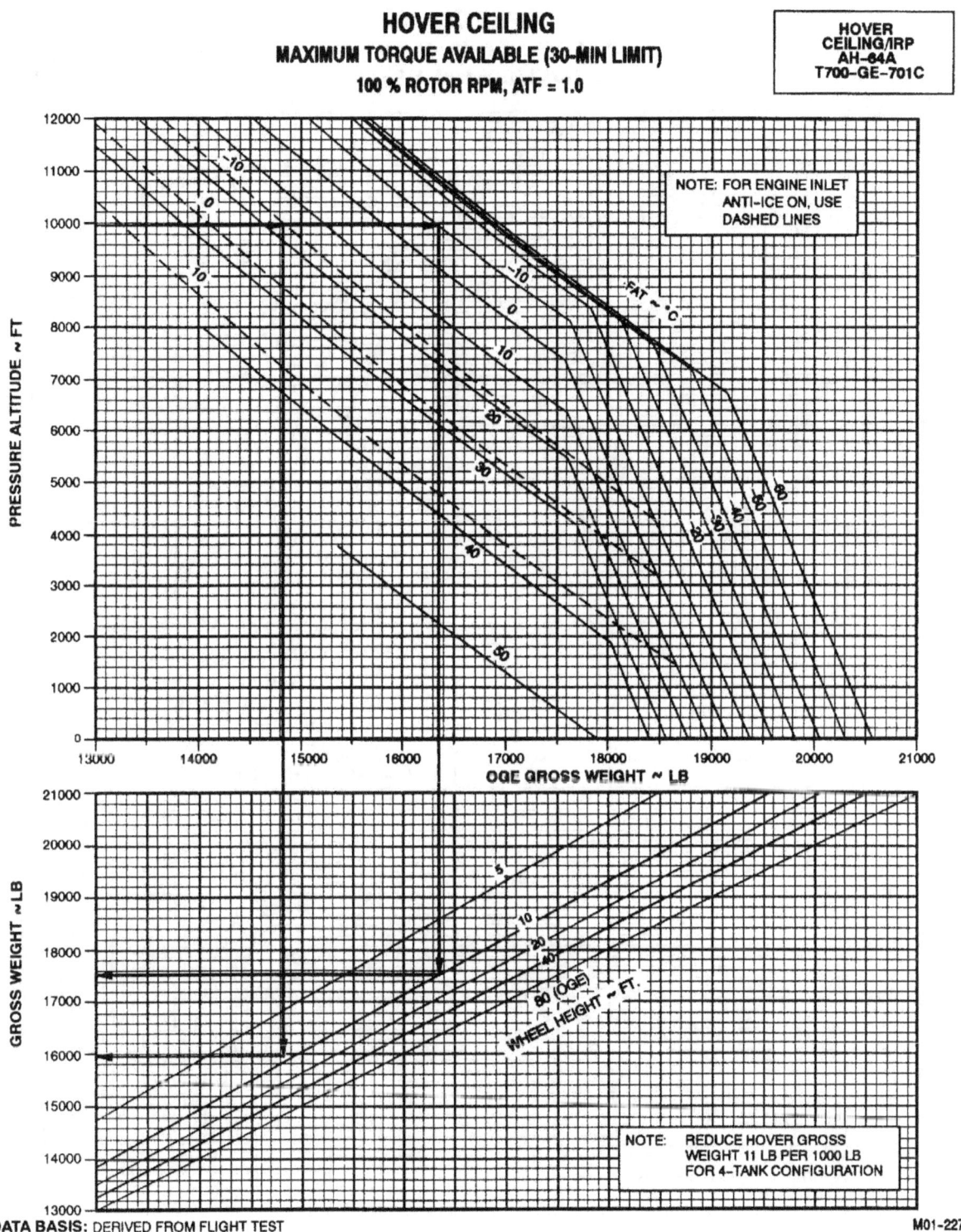

Figure 7A-5. Hover Ceiling Chart (Sheet 1 of 2)

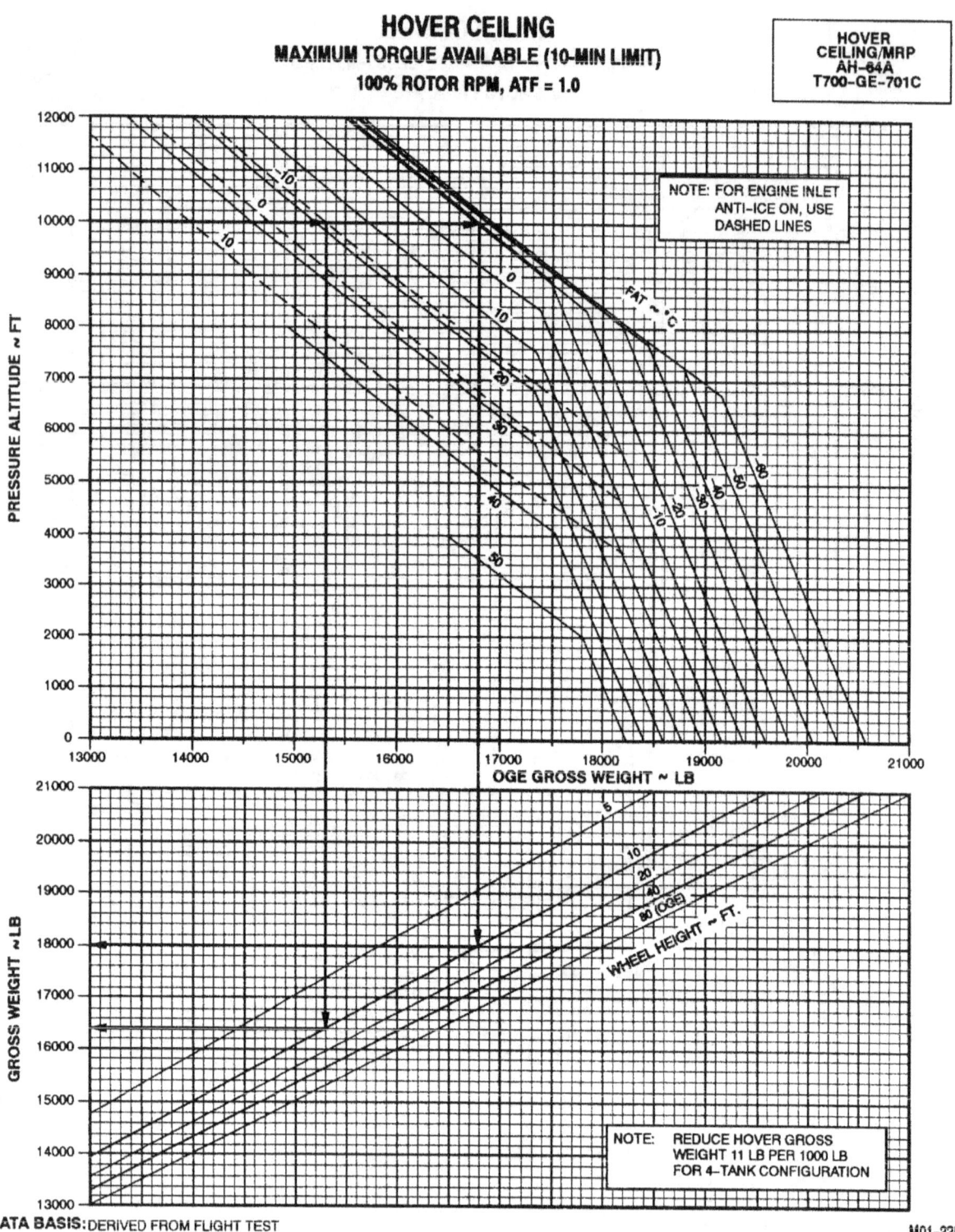

Figure 7A-5. Hover Ceiling Chart (Sheet 2 of 2)

Section IV. HOVER LIMITS

7A.16. DESCRIPTION.

The hover chart (fig 7A-6) shows the torque required to hover at various conditions of pressure altitude, free air temperature, gross weight, wheel height, and with external tanks or without external tanks.

7A.17. USE OF CHART.

The primary use of the chart is illustrated by the example. To determine the torque required to hover, it is necessary to know the pressure altitude, free air temperature, gross weight, and desired wheel height. Enter the upper right grid at the known pressure altitude, move right to the temperature, move down to the gross weight, move left to the desired wheel height, and then move up and read the torque required to hover.

In addition to its primary use, the hover chart maybe used to predict the maximum hover height. To determine maximum hover height, it is necessary to know pressure altitude, free air temperature, gross weight, and maximum torque available. Enter the known pressure altitude, move right to the temperature, move down to the gross weight, then move left to intersection with maximum torque available and read wheel height. This wheel height is the maximum hover height.

The hover chart may also be used to determine the maximum gross weight for hover at a given wheel height, pressure altitude, and temperature condition. Enter at the known pressure altitude, move right to the temperature, then draw a line down to the bottom of the lower grid. Now enter upper left grid at maximum torque available, move down to wheel height, and then move right to intersect the previously drawn line and read gross weight. This is maximum gross weight at which the helicopter will hover.

7A.18. CONDITIONS.

The hover chart is based on calm wind, level surface, 100% rotor rpm and rotor **BLADE** de-ice switch off. With rotor **BLADE** de-ice switch **ON,** torque required will increase 1.4%.

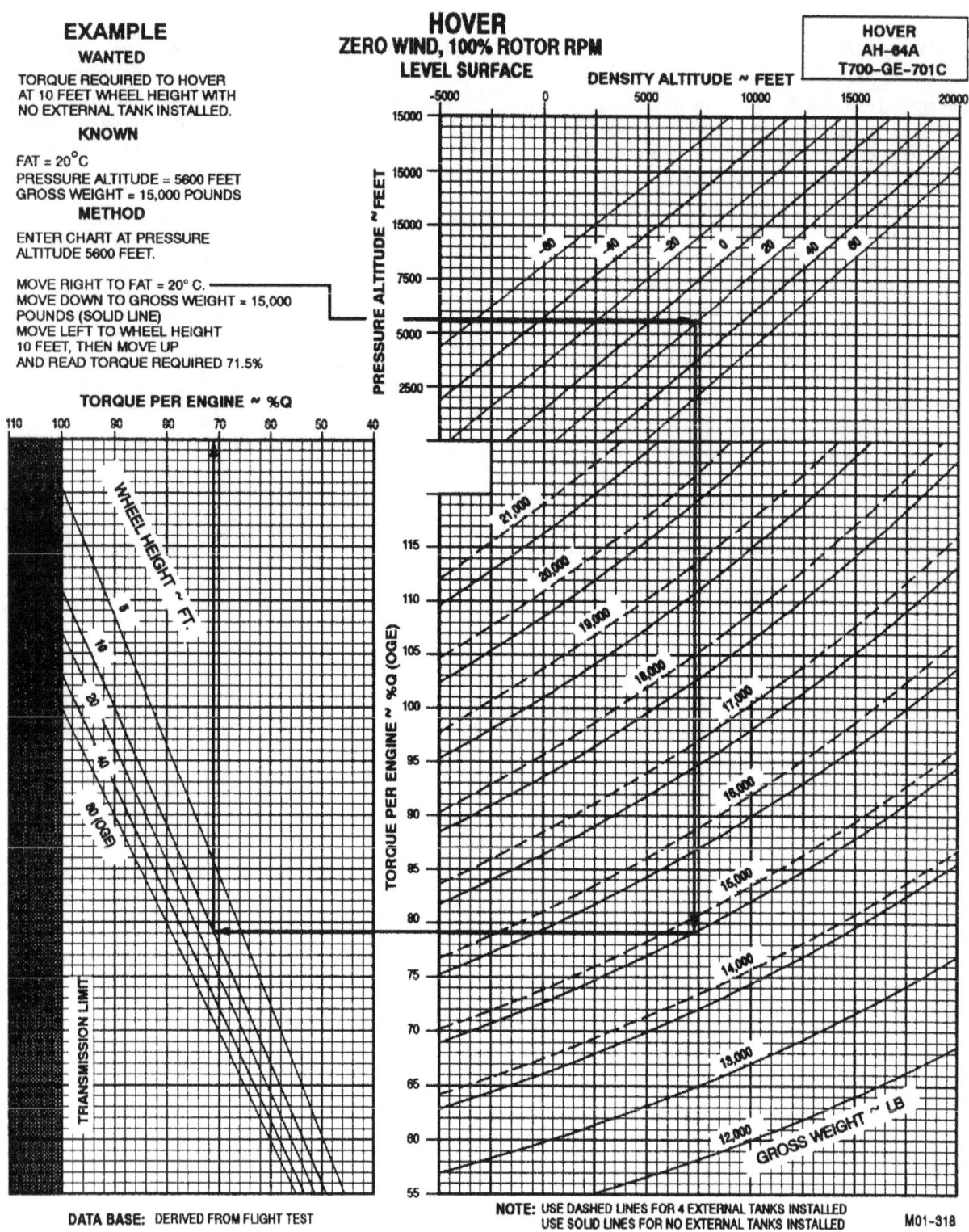

Figure 7A-6. Hover Chart 701C

Section V. CRUISE

7A.19. DESCRIPTION.

The cruise charts (figs 7A-7 thru 7A-17) present the level-flight torque required and total fuel flow at various conditions of airspeed, pressure altitude, free air temperature, and gross weight. Cruise charts are provided for pressure altitudes from sea level to 16,000 feet in 2000-foot increments. Free air temperatures range from -50° to +60 °C in 10 °C increments. In addition to basic cruise information, the charts show speed for maximum range, maximum endurance, and maximum rate of climb. Change in torque with change in frontal area information is presented in the upper left corner of each chart.

7A.20. USE OF CHARTS.

The primary uses of the charts are illustrated by the examples. To use the charts, it is usually necessary to know the planned pressure altitude, estimated free air temperature, planned cruise speed, IAS, and gross weight. First, select the proper chart on the basis of pressure altitude and FAT. Enter the chart at the cruise airspeed, IAS, move right and read TAS, move left to the gross weight, move down and read torque required, and then move up and read associated fuel flow. Maximum performance conditions are determined by entering the chart where the maximum range line or the maximum endurance and rate-of-climb line intersects the gross weight line; then read airspeed, fuel flow, and torque required. Normally, sufficient accuracy can be obtained by selecting the chart nearest the planned cruising altitude and FAT or, more conservatively, by selecting the chart with the next higher altitude and FAT. If greater accuracy is required, interpolation between altitudes and/or temperatures is permissible. To be conservative, use the gross weight at the beginning of the cruise flight. For greater accuracy on long flights, however, it is preferable to determine cruise information for several flight segments to allow for the decreasing gross weight.

7A.20.1 Airspeed.
True and indicated airspeeds are presented at opposite sides of each chart. On any chart, indicated airspeed can be directly converted to true airspeed (or vice versa) by reading directly across the chart without regard for the other chart information. The applicable MACH No. or gross weight maximum permissible airspeed limits (V_{NE}) determined from figure 5-2 appear on the appropriate charts.

7A.20.2 Torque.
Since pressure altitude and temperature are fixed for each chart, torque required varies according to gross weight and airspeed. The torque required and the torque limits shown on these charts are for dual-engine operation. The torque available shown on these charts are maximum continuous torque available, and maximum torque available, 30 minute limit, where less than the two-engine transmission limit. These torque lines are the minimum torque available for ATF = 1 at the engine limits specified in Chapter 5. Higher torque than that represented by these lines may be used if it is available without exceeding the limitations presented in Chapter 5. The limit torque line shown on these charts as is for dual engine transmission limit and is defined as 100% torque. An increase or decrease in torque required because of drag area change is calculated by adding or subtracting the change in torque from the torque change (ΔQ) curve on the chart, and then reading the new total fuel flow.

7A.20.3 Fuel Flow.
Fuel flow scales are provided opposite the torque scales. On any chart, torque may be converted directly to fuel flow without regard to other chart information. Sea level ground fuel flow at flat pitch and 100% Np is approximately 555 pounds per hour.

7A.20.4 Maximum Range.
The maximum range lines indicate the combinations of gross weight and airspeed that will produce the greatest flight range per pound of fuel under zero wind conditions.

7A.20.5 Maximum Endurance and Rate of Climb.
The maximum endurance and rate of climb lines indicate the combinations of gross weight and airspeed that will produce the maximum endurance and the maximum rate of climb. The torque required for level flight at this condition is a minimum, providing a minimum fuel flow (maximum endurance) and a maximum torque change available for climb (maximum rate of climb).

7A.20.6 Change in Frontal Area.
Since the cruise information is given for the primary mission configuration, adjustments to torque should be made when operating with alternative wing-stores configurations. To determine the change in torque, first obtain the appropriate multiplying factor from the drag chart (fig 7A-18), then enter the cruise chart at the planned cruise speed TAS, move right to the broken ΔQ line, and move up and read ΔQ. Multiply ΔQ by the multiplying factor to obtain change in torque, then add or

subtract change in torque from torque required for the primary mission configuration. Enter the cruise chart at resulting torque required, move up, and read fuel flow. If the resulting torque required exceeds the governing torque limit, the torque required must be reduced to the limit. The resulting reduction in airspeed may be found by subtracting the change in torque from the limit torque; then enter the cruise chart at the reduced torque, and move up to the gross weight. Move left or right to read TAS or LAS. To determine the airspeed for maximum range for alternative wing stores configuration, reduce the value from the cruise chart by 2 knots for each 5 square feet increase in drag area, ΔF, or increase maximum range airspeed 2 knots for each 5 square feet reduction in drag area. For example, for 16 Hellfire configuration $\Delta F = 9.6$ square feet, from figure 7A-18. Therefore, maximum range airspeed would be reduced by 2/5 x 9.6 = 3.84 knots, or approximately 4 knots.

7A.20.7 Additional Uses. The low-speed end of the cruise chart (below 40 knots) is primarily to familiarize you with the low-speed power requirements of the helicopter. It shows the power margin available for climb or acceleration during maneuvers, such as NOE flight. At zero airspeed, the torque represents the torque required to hover out of ground effect. In general, mission planning for low-speed flight should be based on hover out of ground effect.

7A.21 . CONDITIONS.

The cruise charts are based on 100% rotor rpm, ATF or ETF = 1.0, **ENG INLET ANTI-ICE** switch **OFF,** JP-4 fuel, and dual-engine operation. Engine inlet anti-ice and rotor blade de-ice effect are as follows:

a. With **ENG INLET ANTI-ICE** switch **ON,** fuel flow will increase approximately 85 pounds per hour; maximum continuous torque available could be reduced by as much as 24.5%; maximum torque available, 30 minute limit could be reduced by as much as 20%; and maximum torque available, 10 minute limit could be reduced by as much as 18.6%.

b. With rotor **BLADE** de-ice switch ON, fuel flow will increase approximately 30 pounds per hour, and torque required will increase 1.4%.

For example, with **ENG INLET ANTI-ICE** and rotor **BLADE** de-ice off, torque required, from cruise chart is 50%, and maximum continuous torque is 92%. With **ENG INLET ANTI-ICE** and rotor **BLADE** de-ice switches **ON,** torque required will be 50 + 1.4 = 51.4%, and maximum continuous torque will be approximately 92 -24.5 = 67.5%.

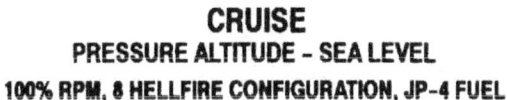

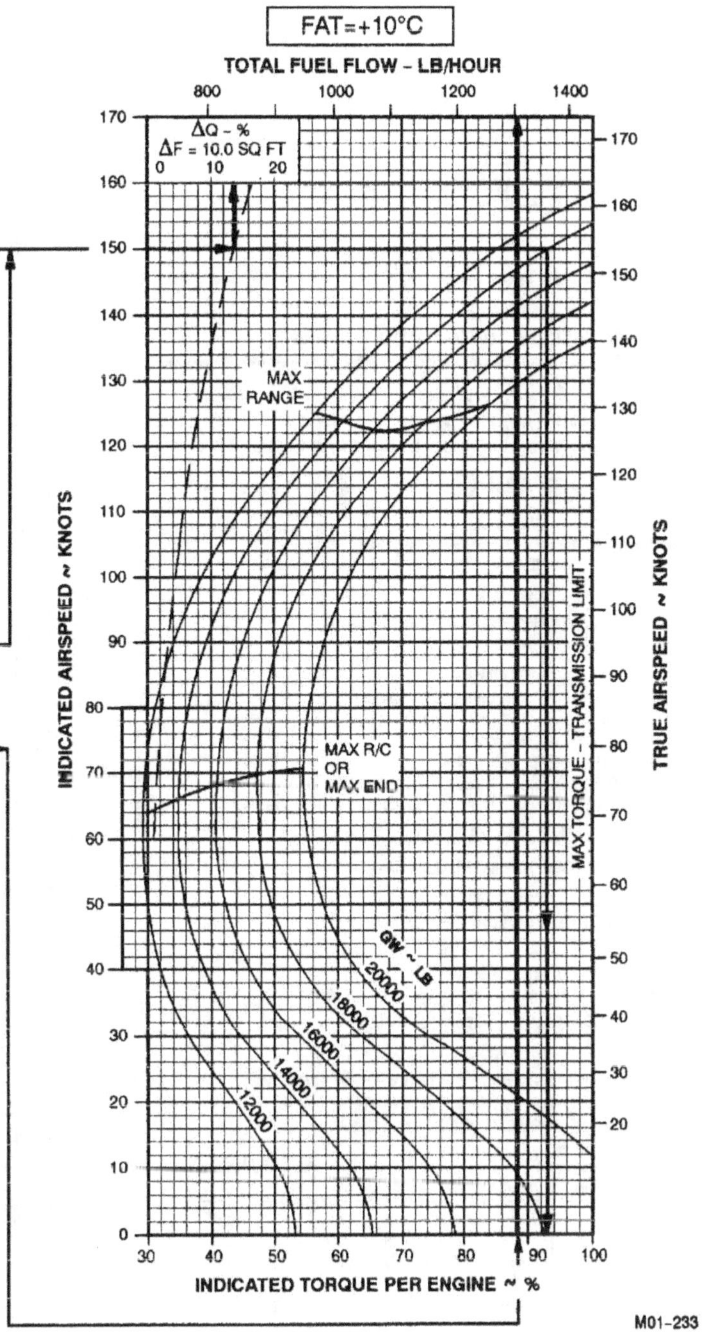

Figure 7A-7. Cruise Chart, Example 701C

CRUISE
PRESSURE ALTITUDE – SEA LEVEL
100% RPM, 8 HELLFIRE CONFIGURATION, JP-4 FUEL

CRUISE
AH-64A
T700-GE-701C

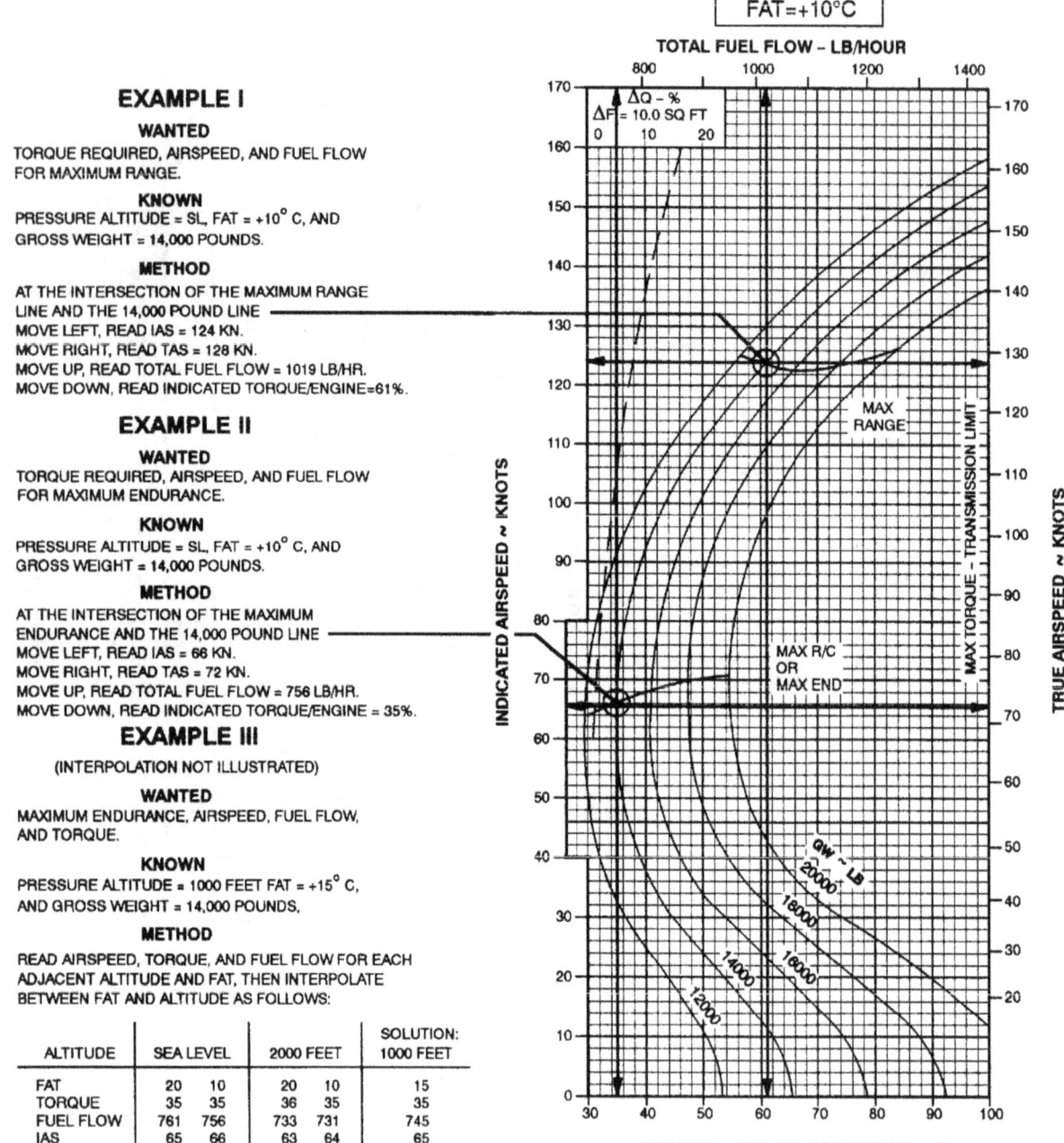

EXAMPLE I
WANTED
TORQUE REQUIRED, AIRSPEED, AND FUEL FLOW FOR MAXIMUM RANGE.
KNOWN
PRESSURE ALTITUDE = SL, FAT = +10° C, AND GROSS WEIGHT = 14,000 POUNDS.
METHOD
AT THE INTERSECTION OF THE MAXIMUM RANGE LINE AND THE 14,000 POUND LINE
MOVE LEFT, READ IAS = 124 KN.
MOVE RIGHT, READ TAS = 128 KN.
MOVE UP, READ TOTAL FUEL FLOW = 1019 LB/HR.
MOVE DOWN, READ INDICATED TORQUE/ENGINE=61%.

EXAMPLE II
WANTED
TORQUE REQUIRED, AIRSPEED, AND FUEL FLOW FOR MAXIMUM ENDURANCE.
KNOWN
PRESSURE ALTITUDE = SL, FAT = +10° C, AND GROSS WEIGHT = 14,000 POUNDS.
METHOD
AT THE INTERSECTION OF THE MAXIMUM ENDURANCE AND THE 14,000 POUND LINE
MOVE LEFT, READ IAS = 66 KN.
MOVE RIGHT, READ TAS = 72 KN.
MOVE UP, READ TOTAL FUEL FLOW = 756 LB/HR.
MOVE DOWN, READ INDICATED TORQUE/ENGINE = 35%.

EXAMPLE III
(INTERPOLATION NOT ILLUSTRATED)
WANTED
MAXIMUM ENDURANCE, AIRSPEED, FUEL FLOW, AND TORQUE.
KNOWN
PRESSURE ALTITUDE = 1000 FEET FAT = +15° C, AND GROSS WEIGHT = 14,000 POUNDS,
METHOD
READ AIRSPEED, TORQUE, AND FUEL FLOW FOR EACH ADJACENT ALTITUDE AND FAT, THEN INTERPOLATE BETWEEN FAT AND ALTITUDE AS FOLLOWS:

ALTITUDE	SEA LEVEL		2000 FEET		SOLUTION: 1000 FEET
FAT	20	10	20	10	15
TORQUE	35	35	36	35	35
FUEL FLOW	761	756	733	731	745
IAS	65	66	63	64	65
TAS	72	72	73	73	73

Figure 7A-8. Cruise Chart, Sea Level, +10 °C Example

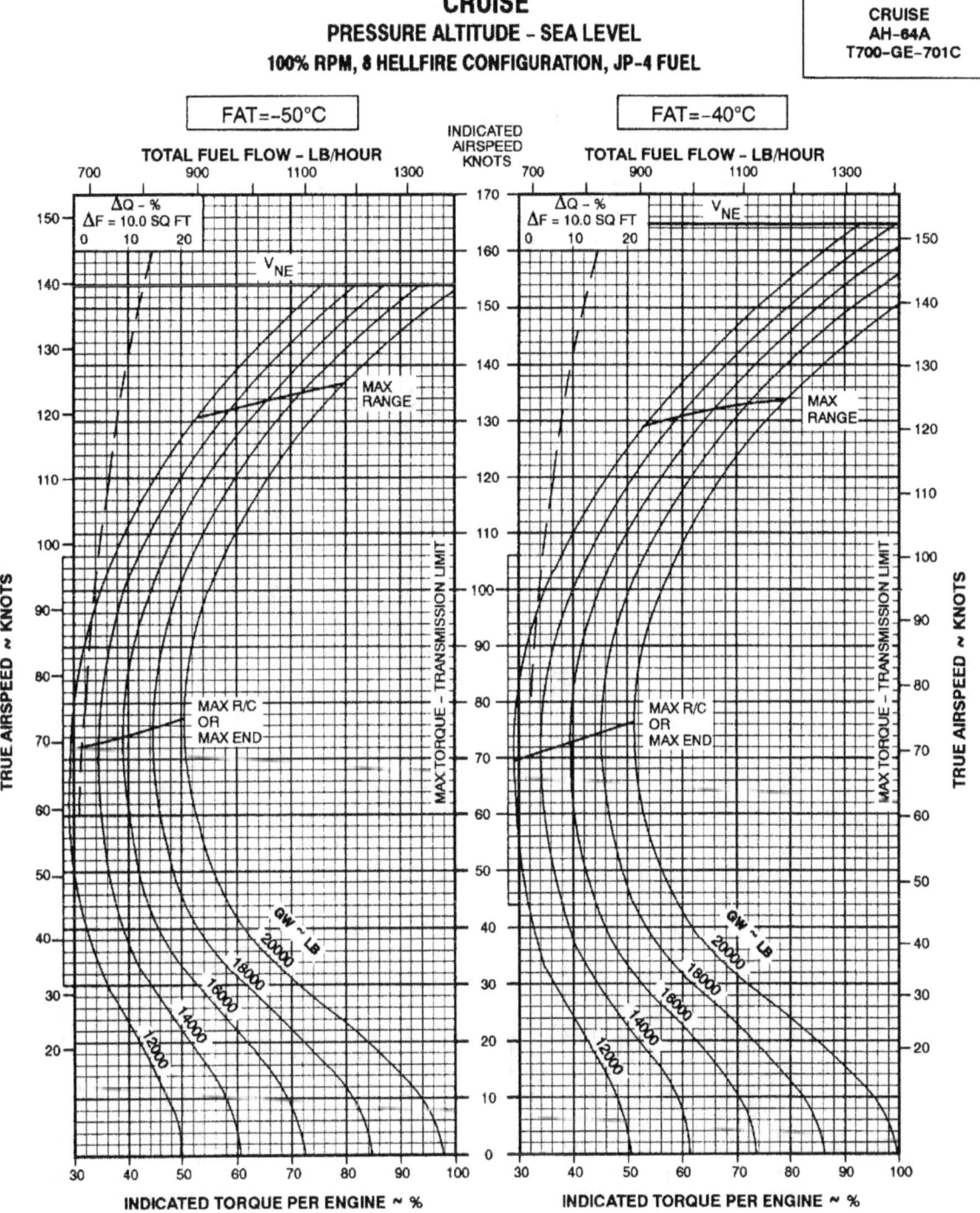

Figure 7A-9. Cruise Chart, Sea Level, -50 °C and -40 °C (Sheet 1 of 6) 701C

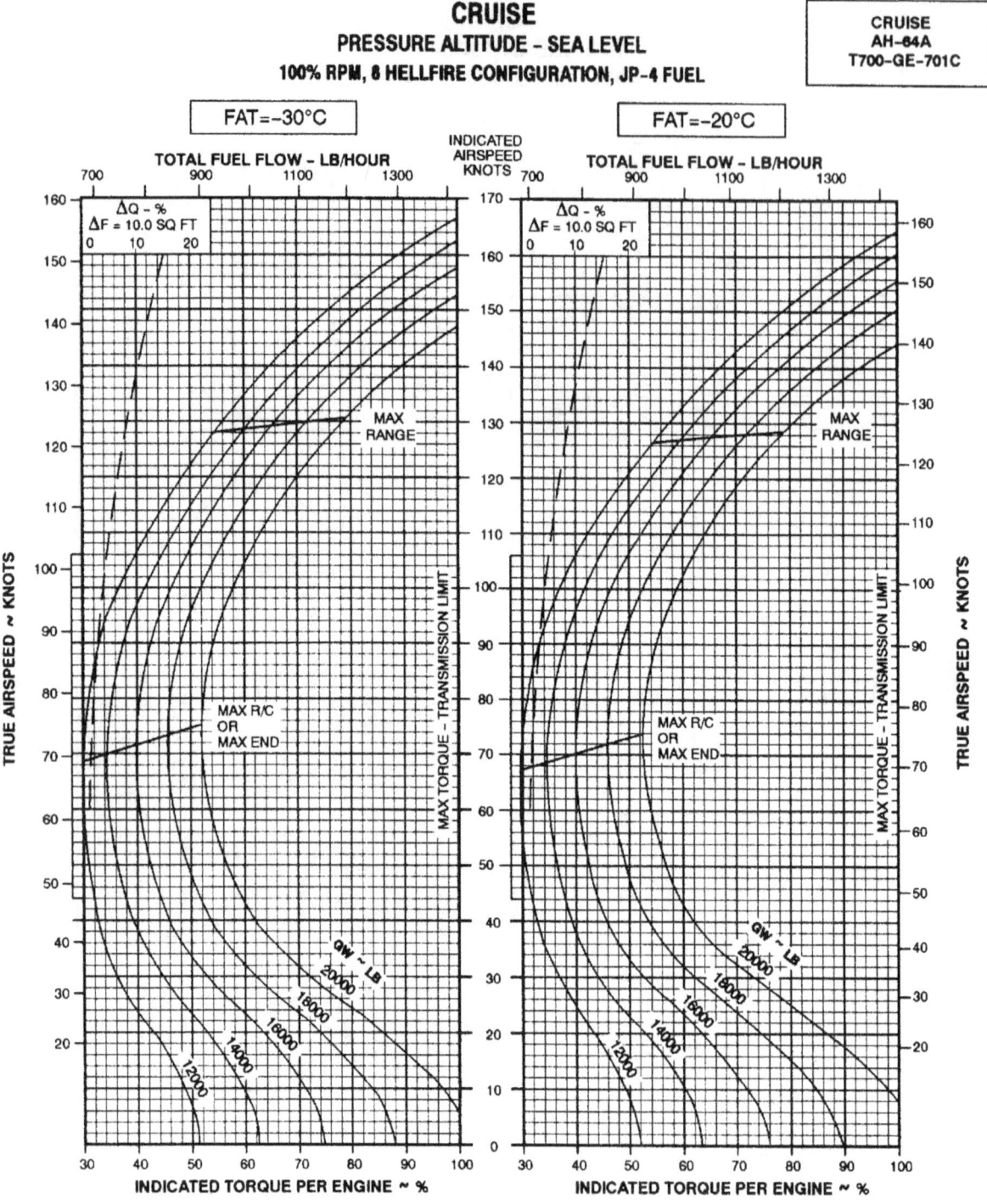

Figure 7A-9. Cruise Chart, Sea Level, -30 °C and -20 °C (Sheet 2 of 6)

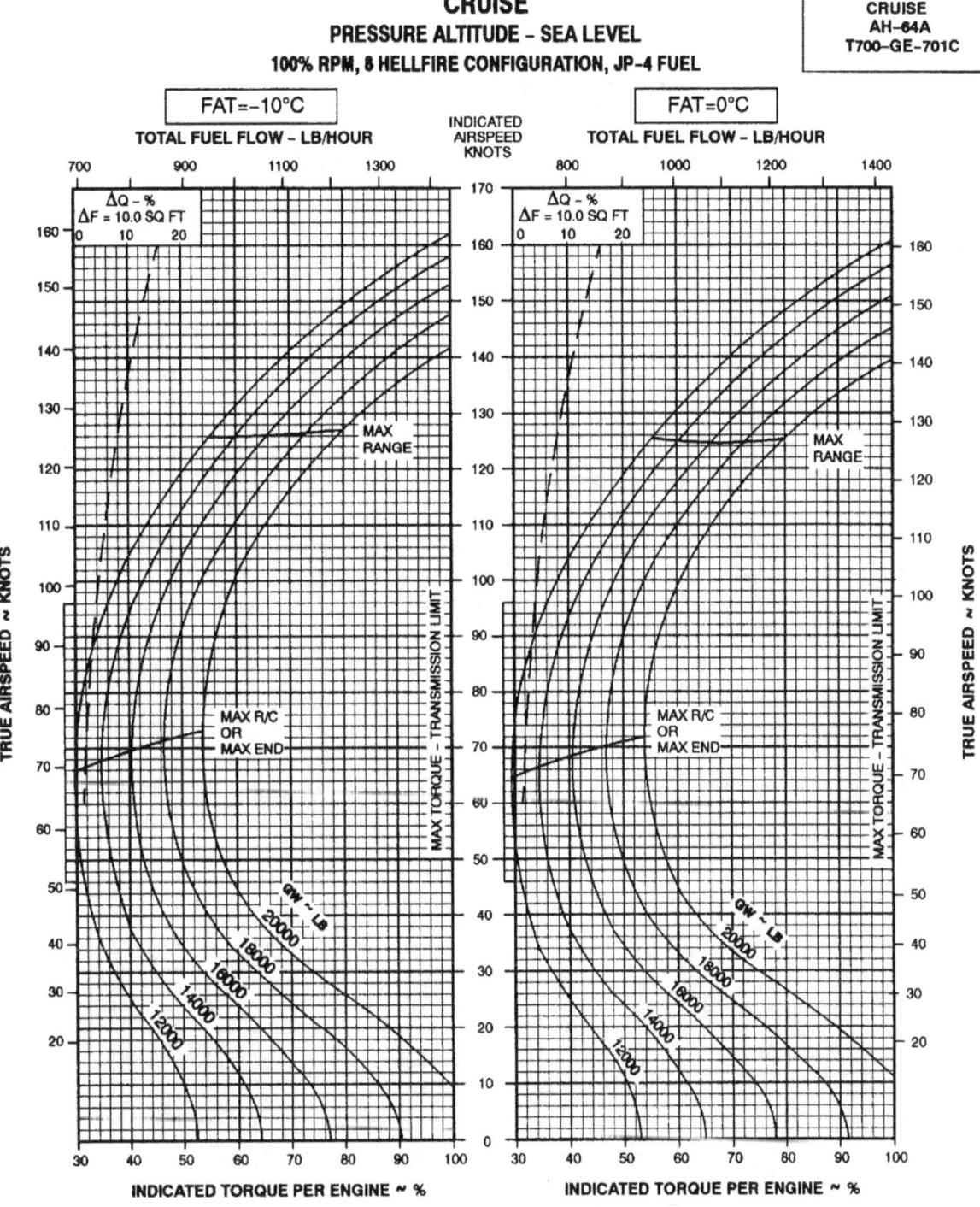

Figure 7A-9. Cruise Chart, Sea Level, -10 °C and 0 °C (Sheet 3 of 6) 701C

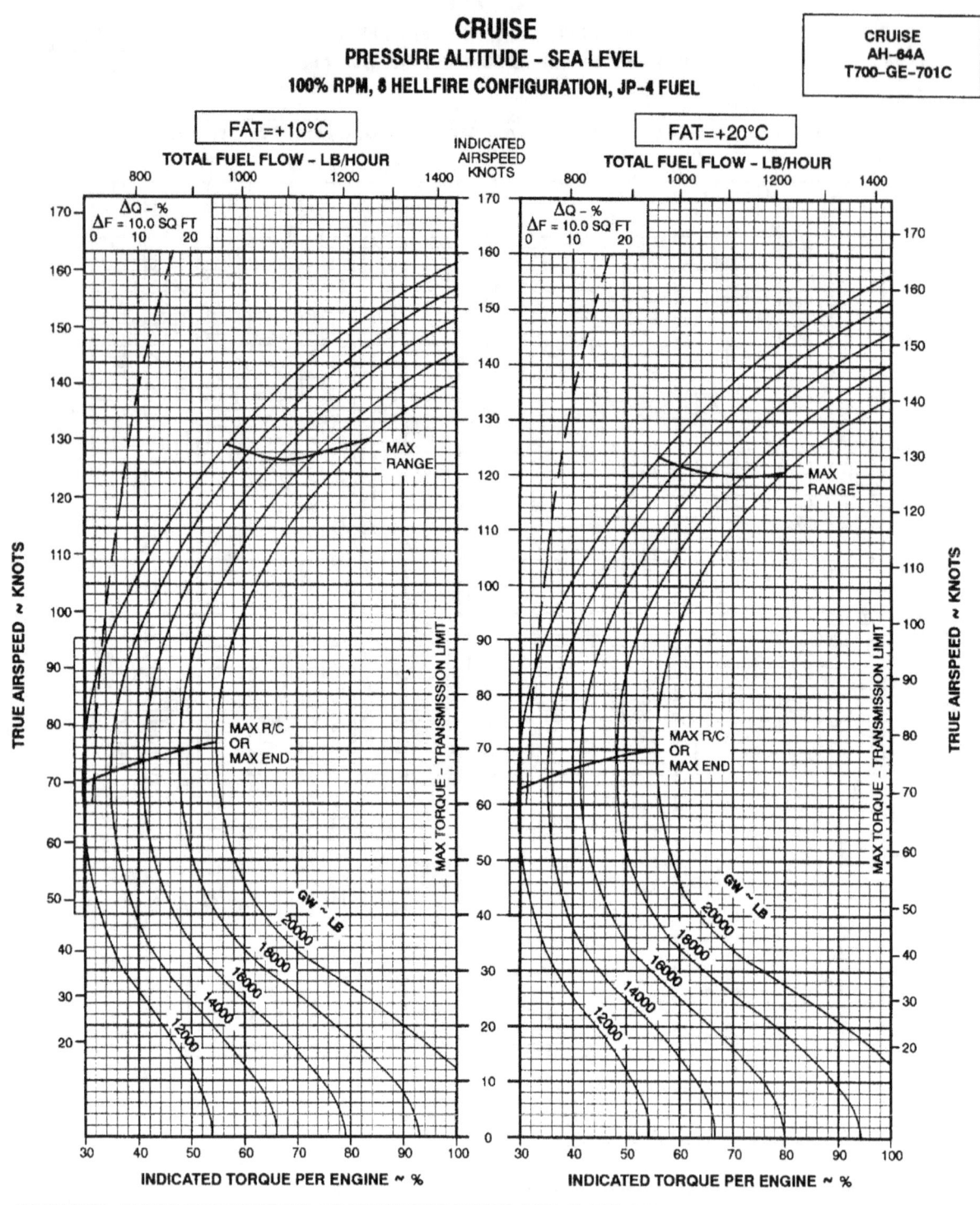

Figure 7A-9. Cruise Chart, Sea Level, +10 °C and +20 °C (Sheet 4 of 6) 701C

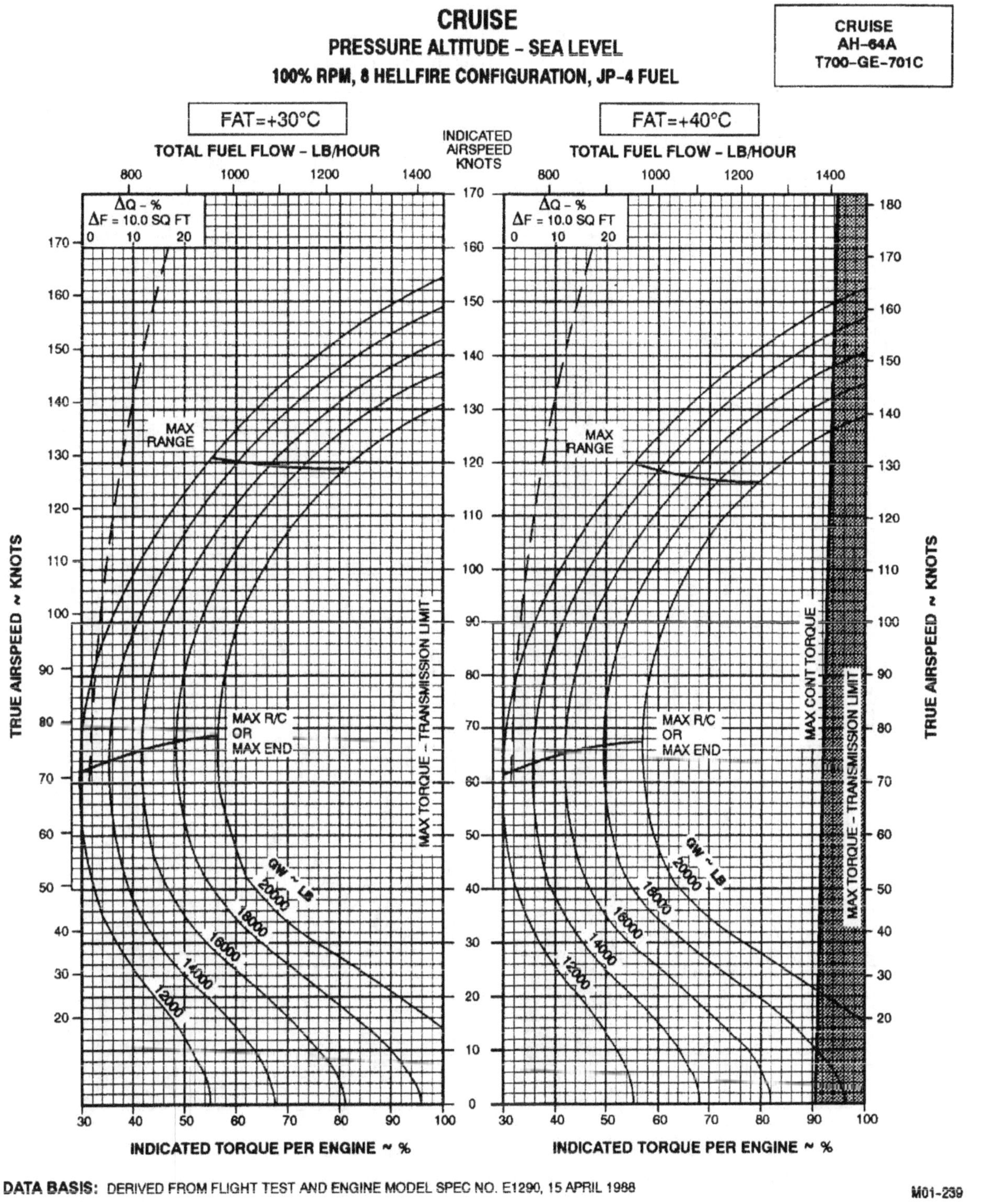

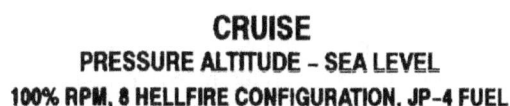

Figure 7A-9. Cruise Chart, Sea Level, +30 °C and +40 °C (Sheet 5 of 6) 701C

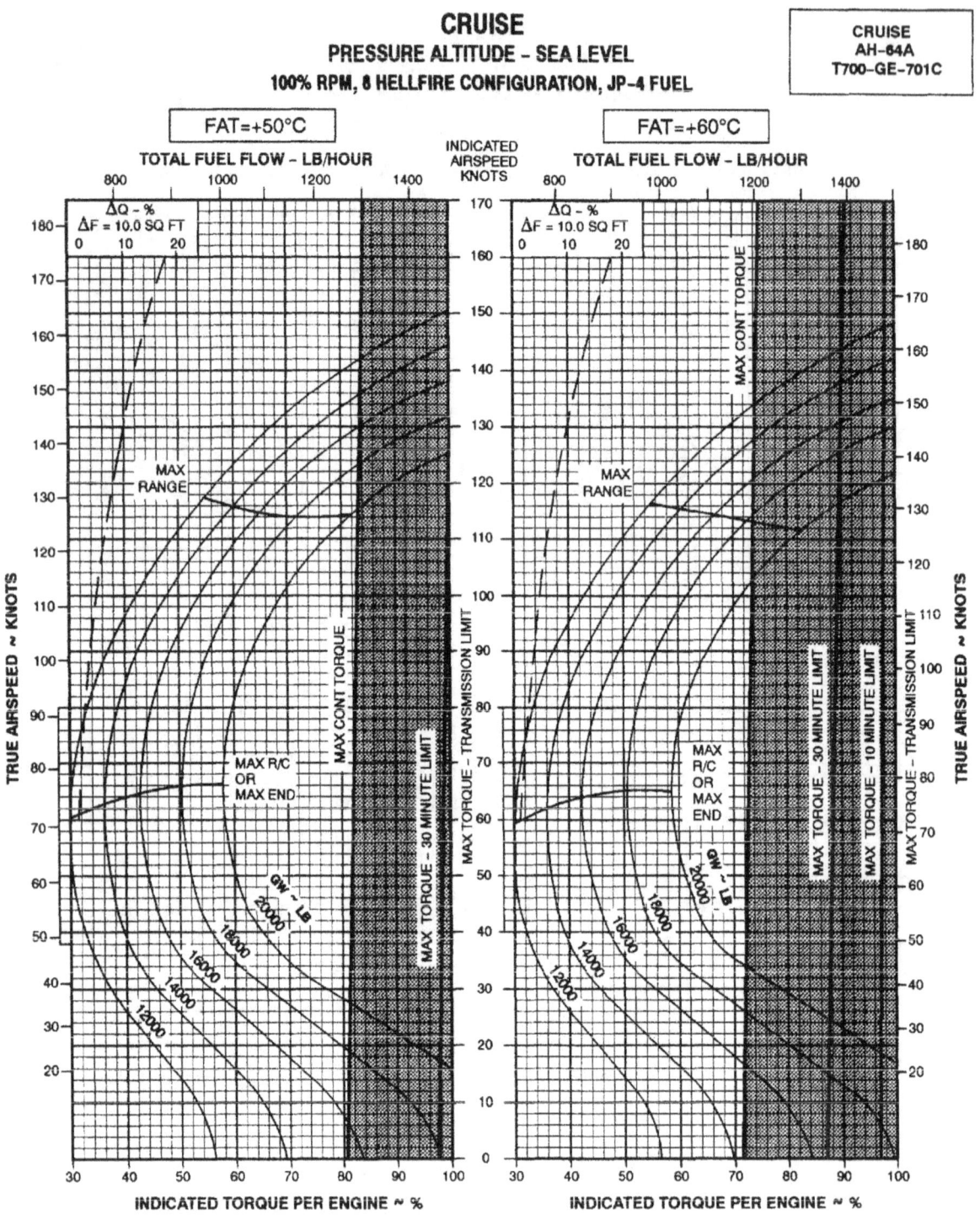

Figure 7A-9. Cruise Chart, Sea Level, +50 °C and +60 °C (Sheet 6 of 6) 701C

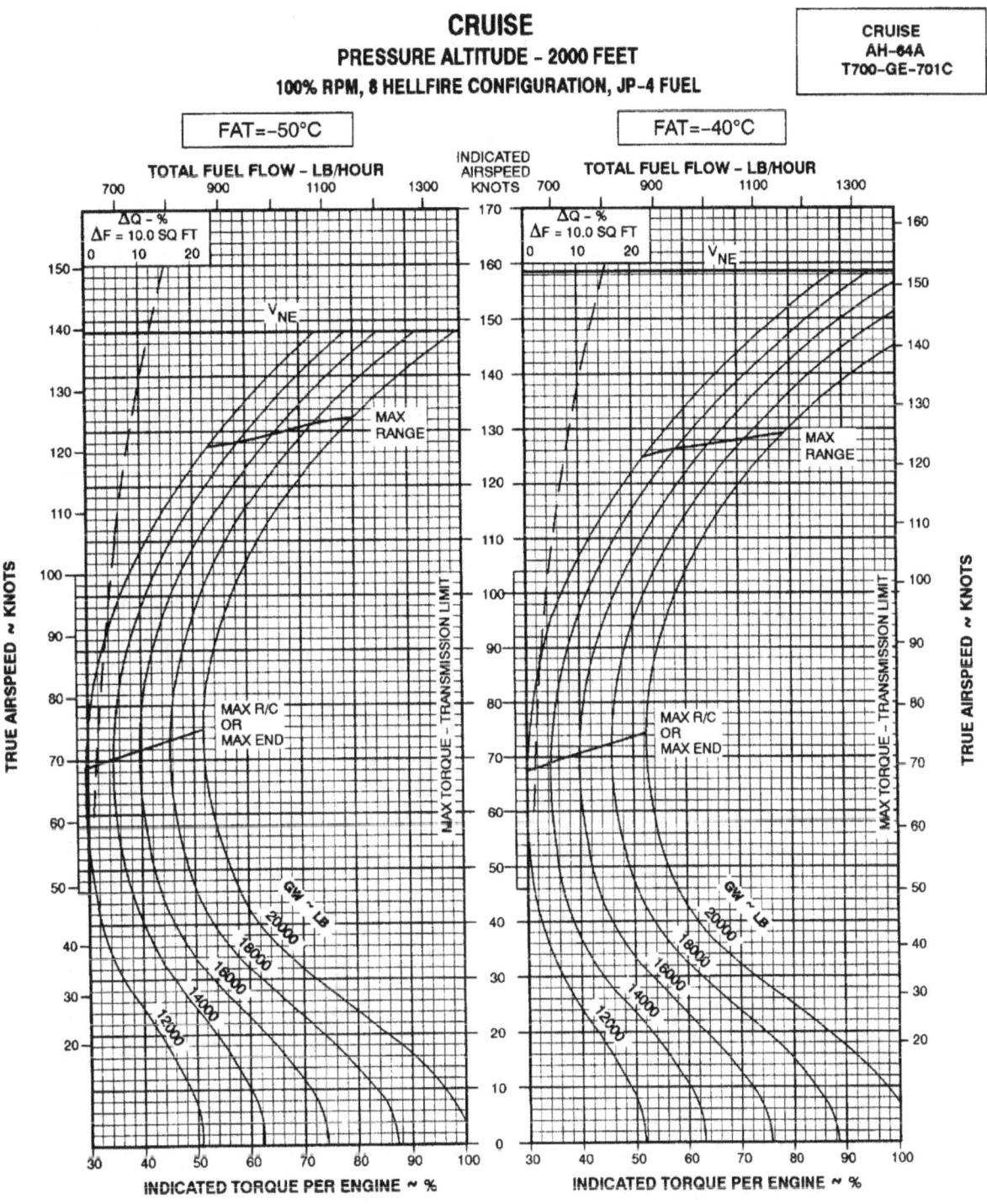

Figure 7A-10. Cruise Chart, 2,000 Feet, -50 °C and -40 °C (Sheet 1 of 6)

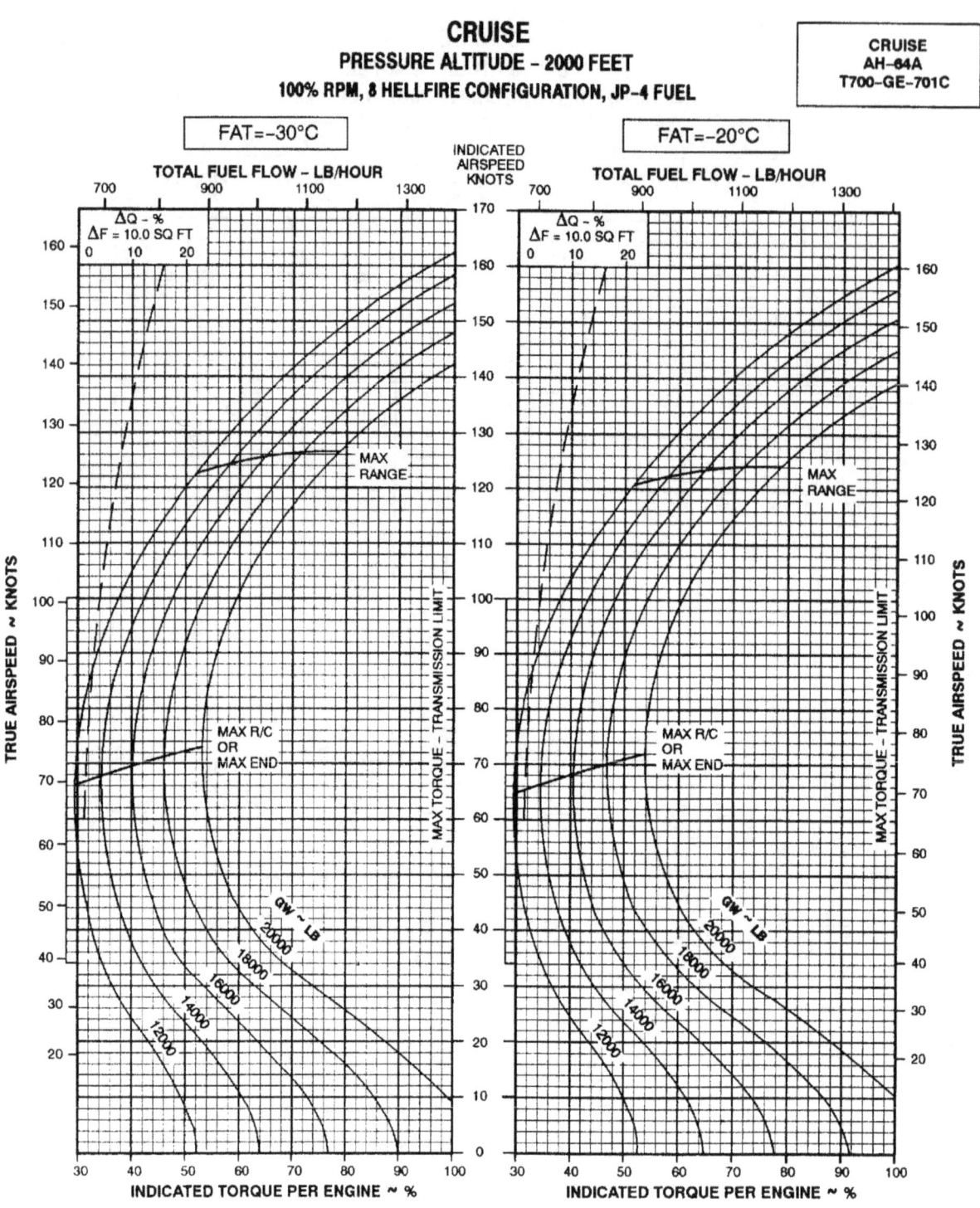

Figure 7A-10. Cruise Chart, 2,000 Feet, -30 °C and -20 °C (Sheet 2 of 6)

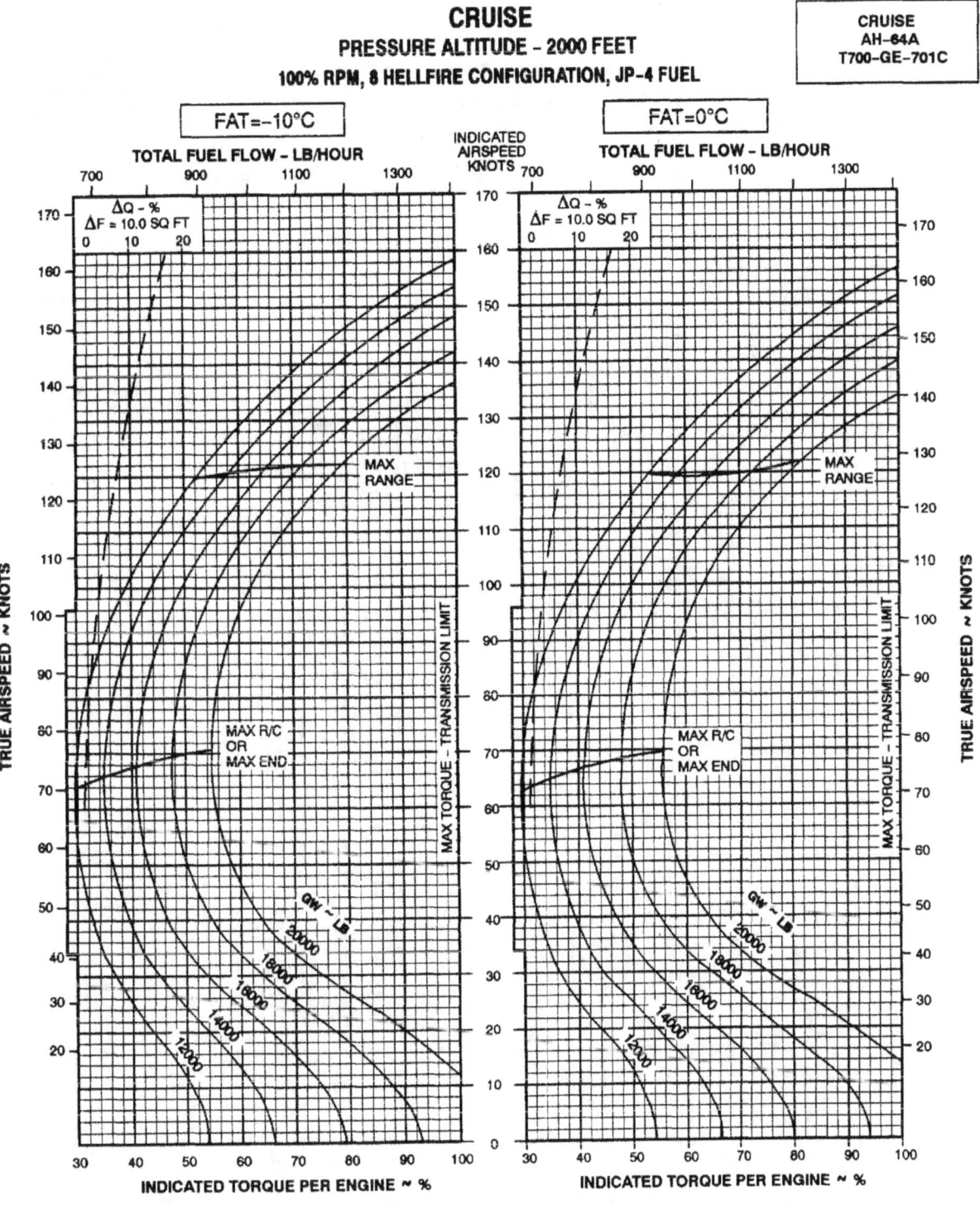

Figure 7A-10. Cruise Chart, 2,000 Feet, -10 °C and 0 °C (Sheet 3 of 6)

TM 1-1520-238-10

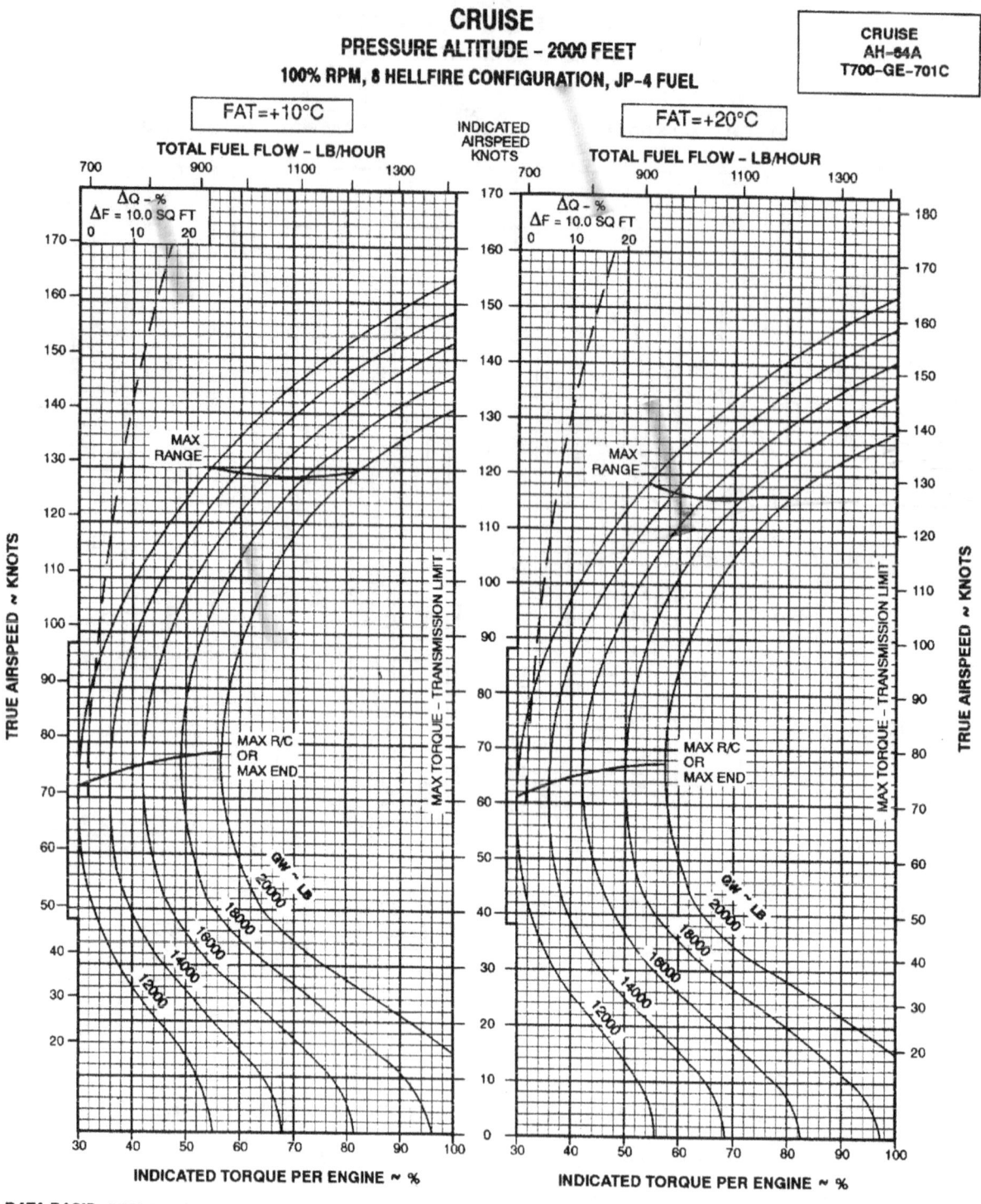

Figure 7A-10. Cruise Chart, 2,000 Feet, +10 °C and +20 °C (Sheet 4 of 6) 701C

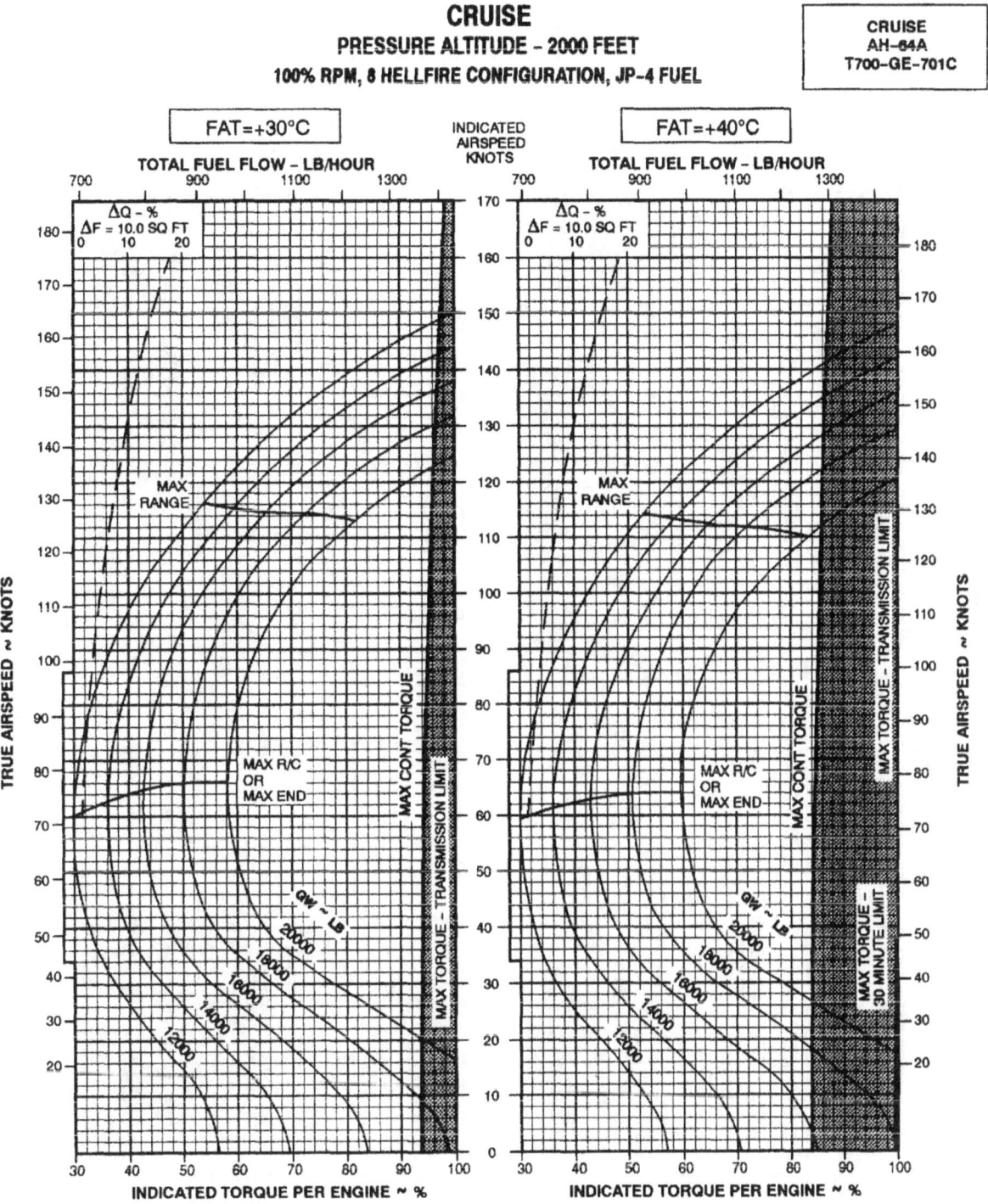

Figure 7A-10. Cruise Chart, 2,000 Feet, +30 °C and +40 °C (Sheet 5 of 6)

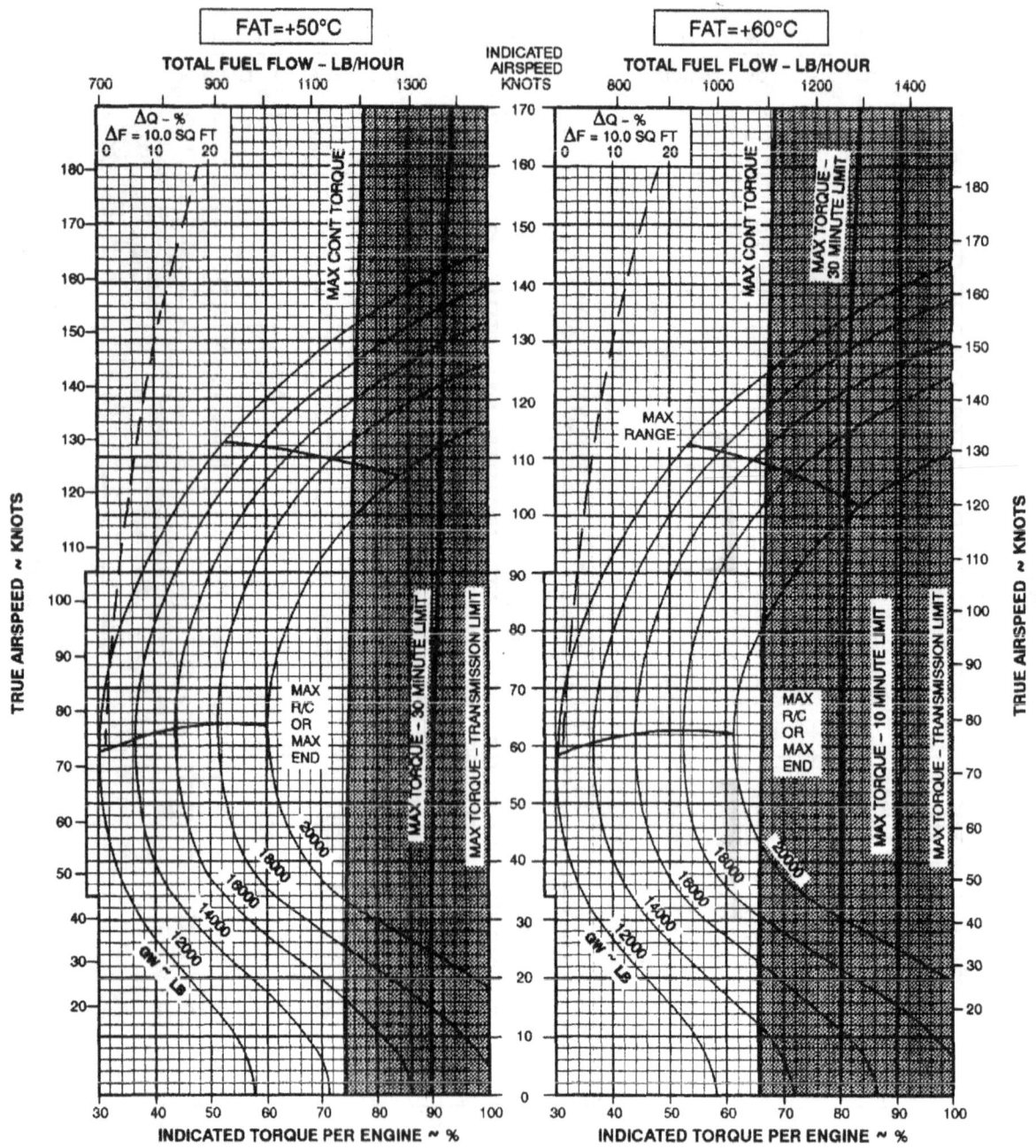

Figure 7A-10. Cruise Chart, 2,000 Feet, +50 °C and +60 °C (Sheet 6 of 6) 701C

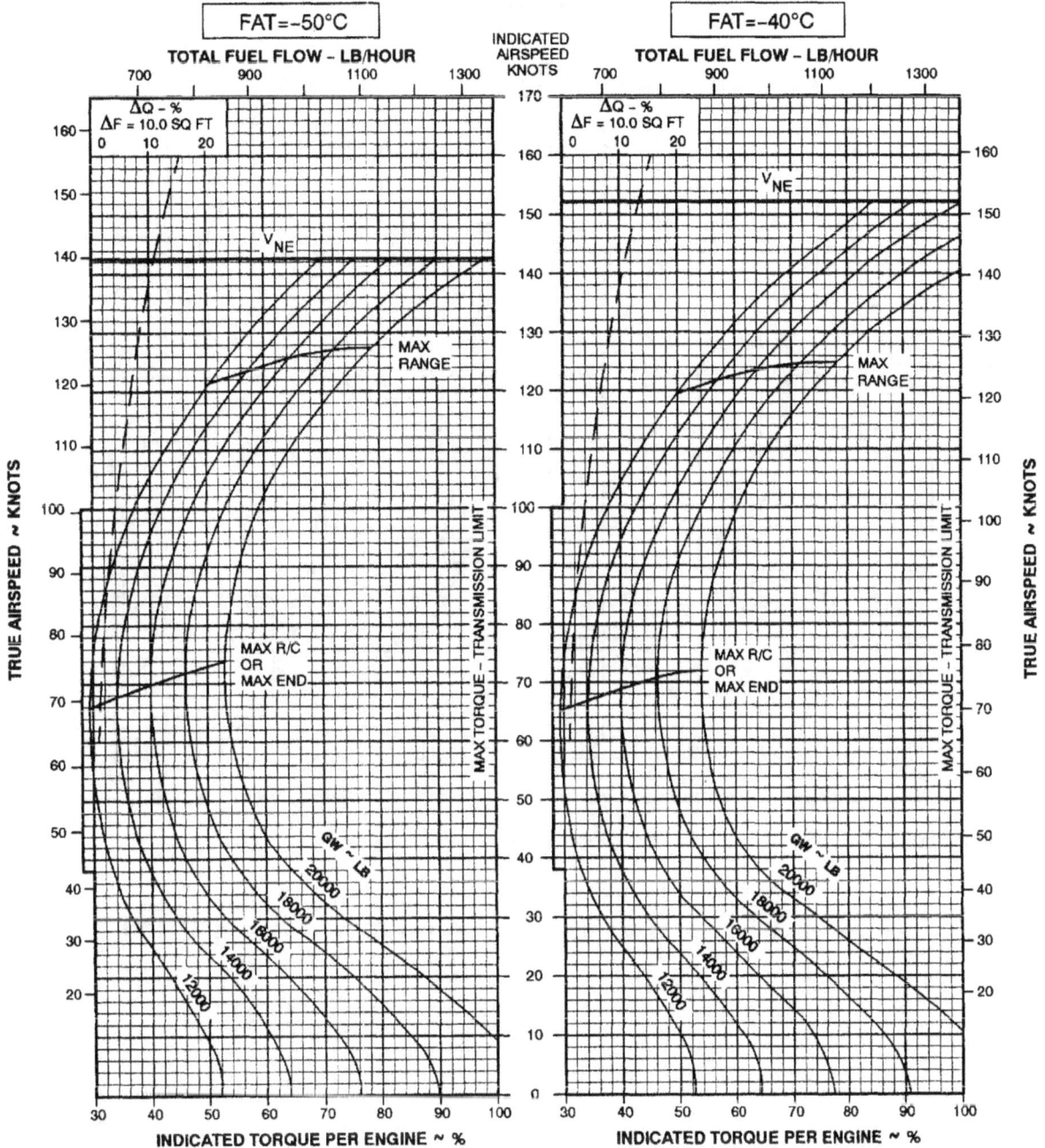

Figure 7A-11. Cruise Chart, 4,000 Feet, -50 °C and -40 °C (Sheet 1 of 6)

TM 1-1520-238-10

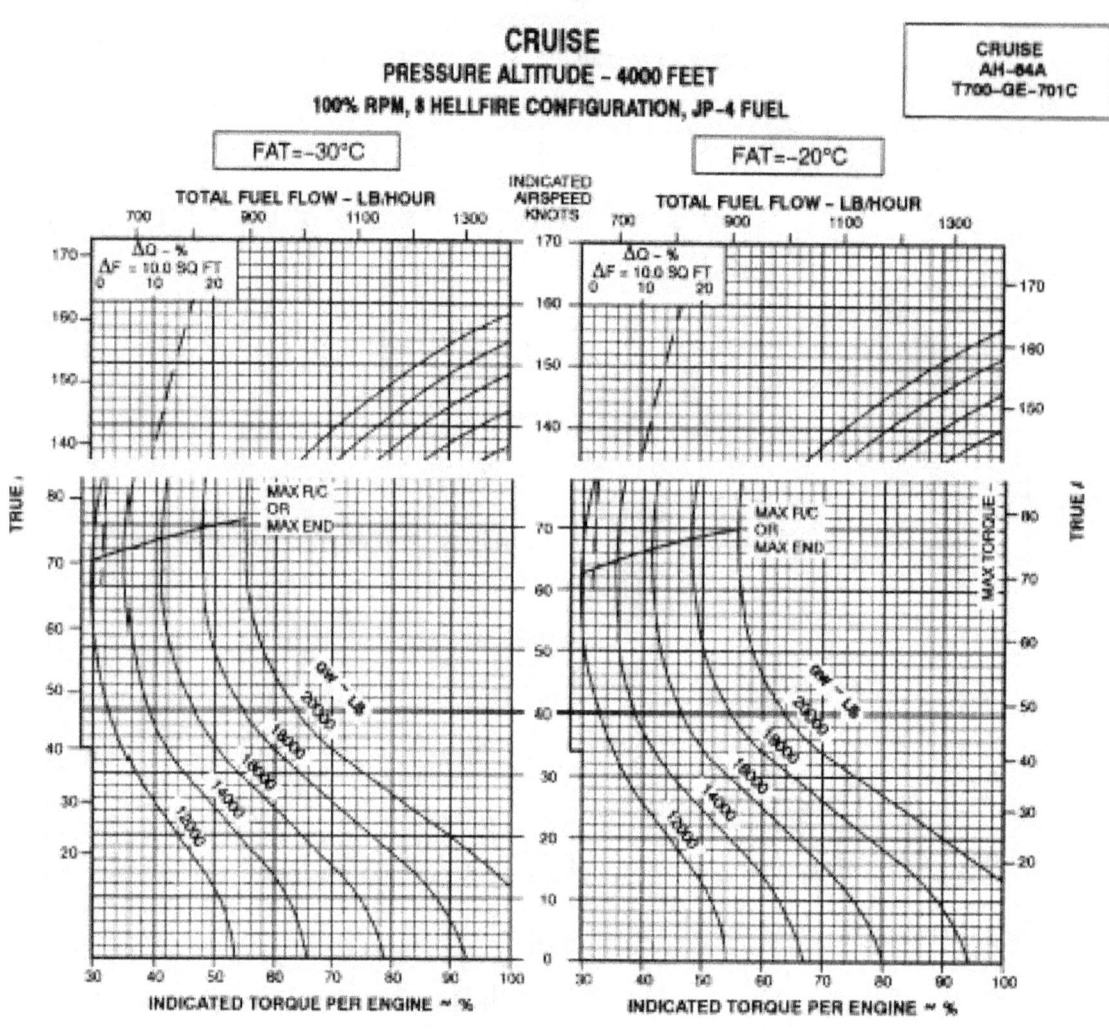

Figure 7A-11. Cruise Chart, 4,000 Feet, -30 °C and -20 °C (Sheet 2 of 6)

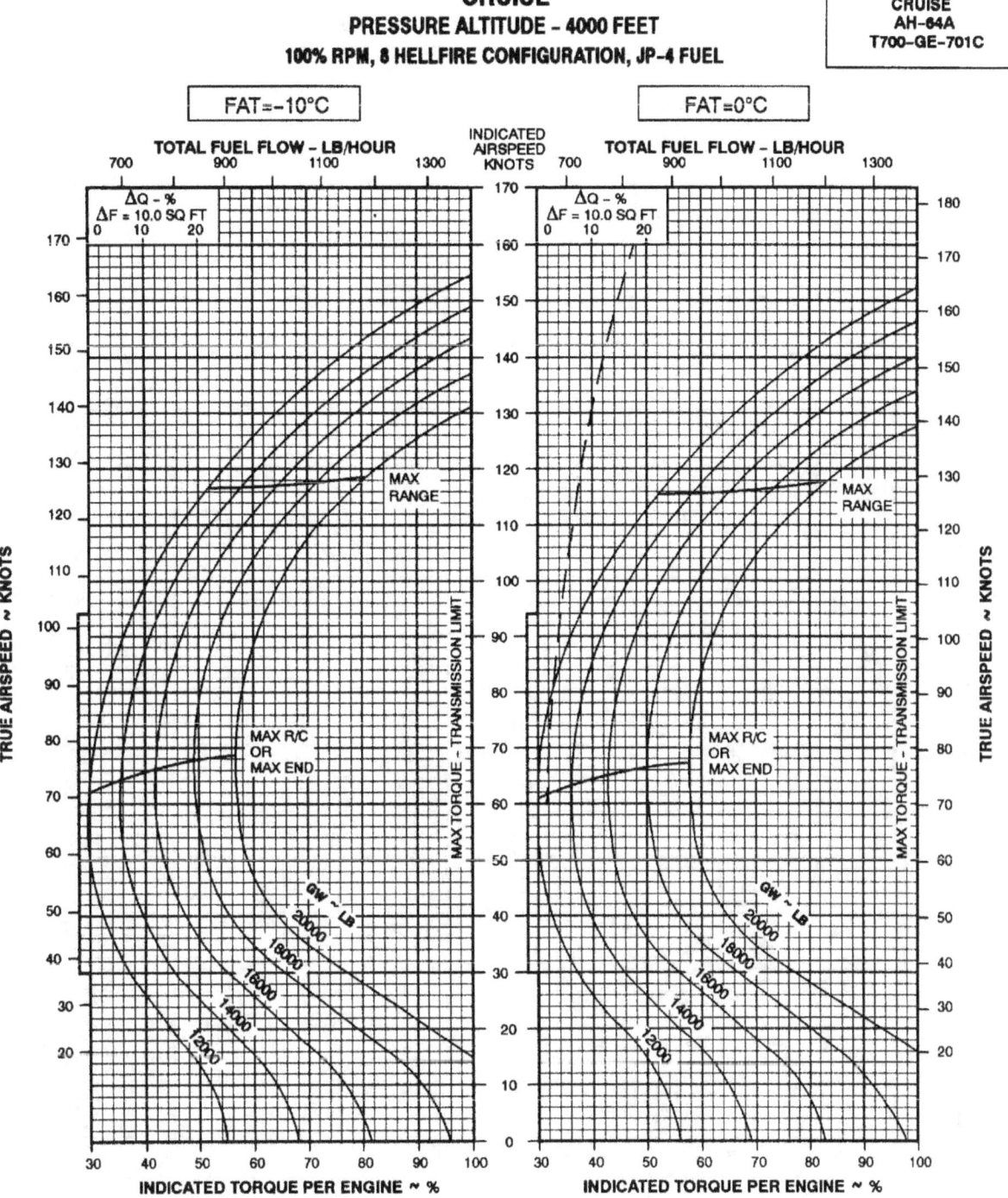

Figure 7A-11. Cruise Chart, 4,000 Feet, -10 °C and 0 °C (Sheet 3 of 6) 701C

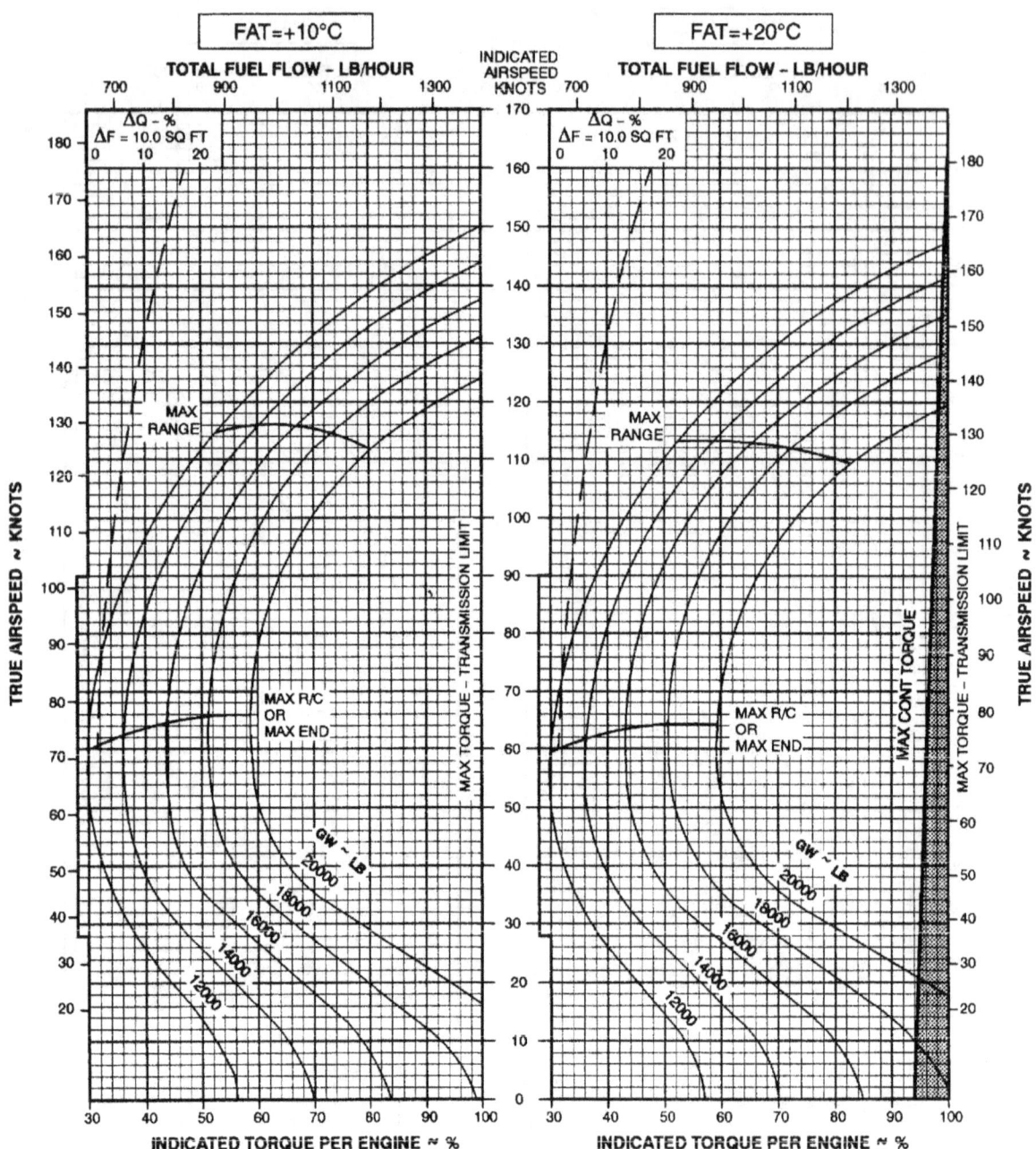

Figure 7A-11. Cruise Chart, 4,000 Feet, +10 °C and +20 °C (Sheet 4 of 6) 701C

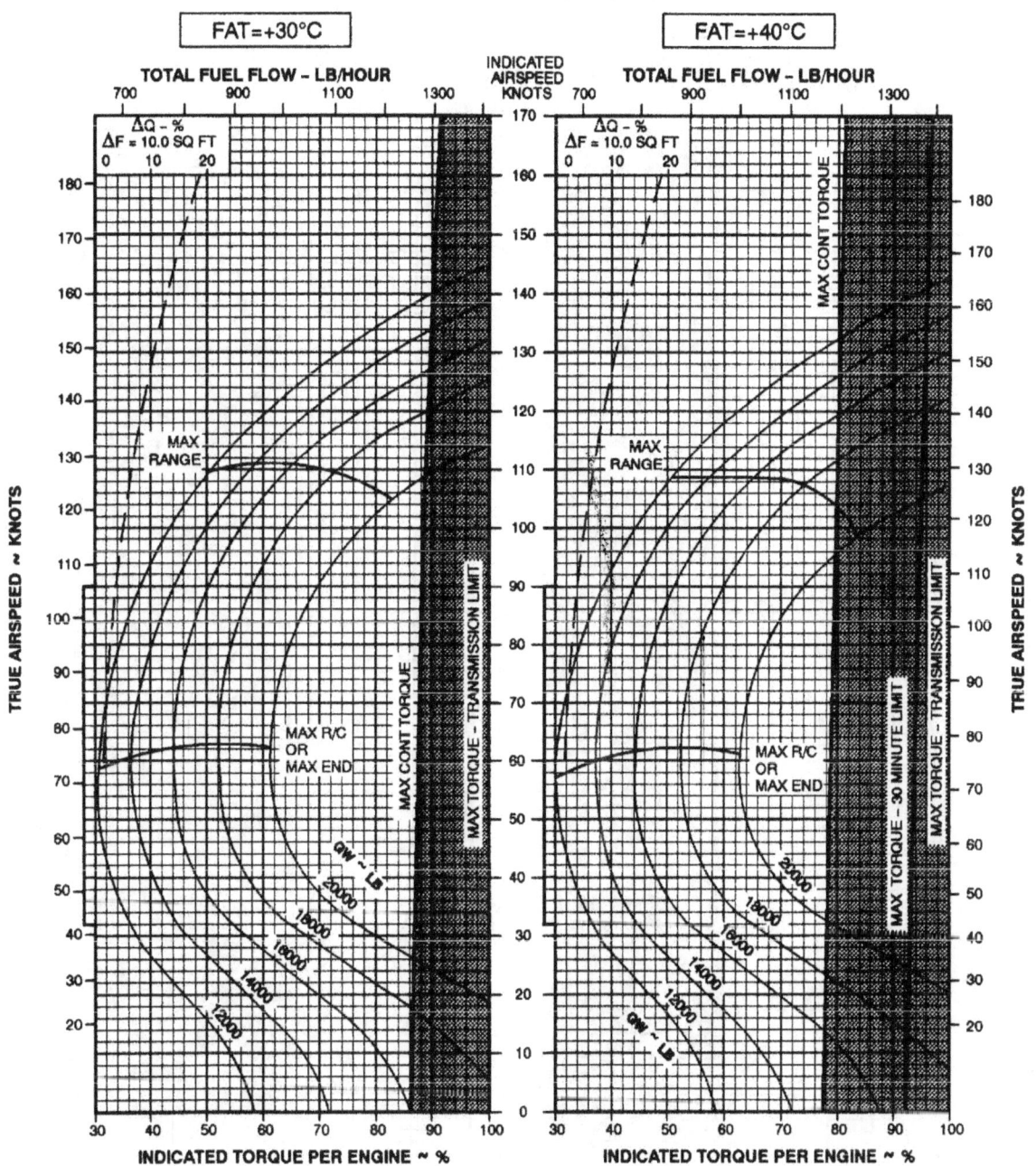

Figure 7A-11. Cruise Chart, 4,000 Feet, +30 °C and +40 °C (Sheet 5 of 6) 701C

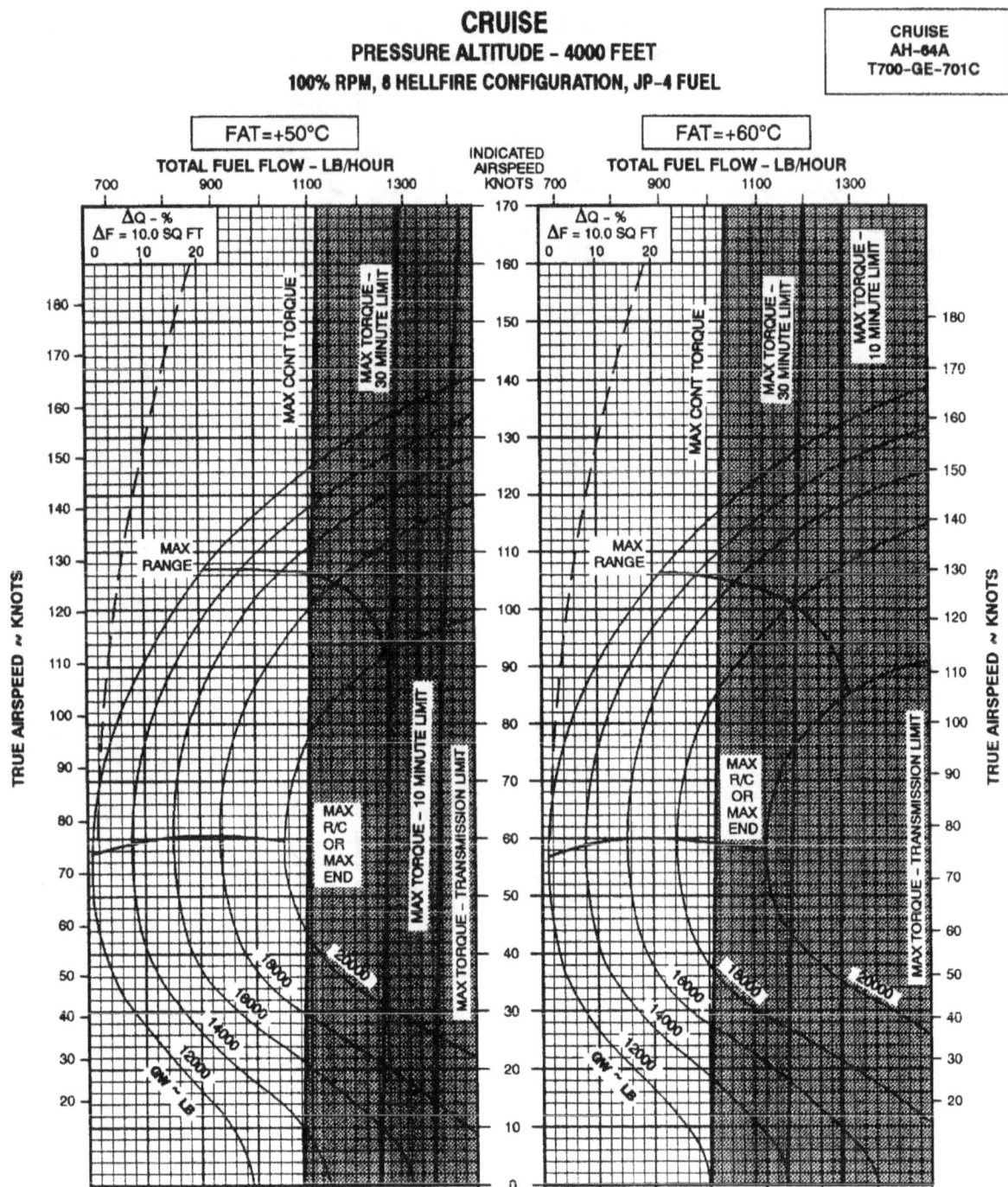

Figure 7A-11. Cruise Chart, 4,000 Feet, +50 °C and +60 °C (Sheet 6 of 6)

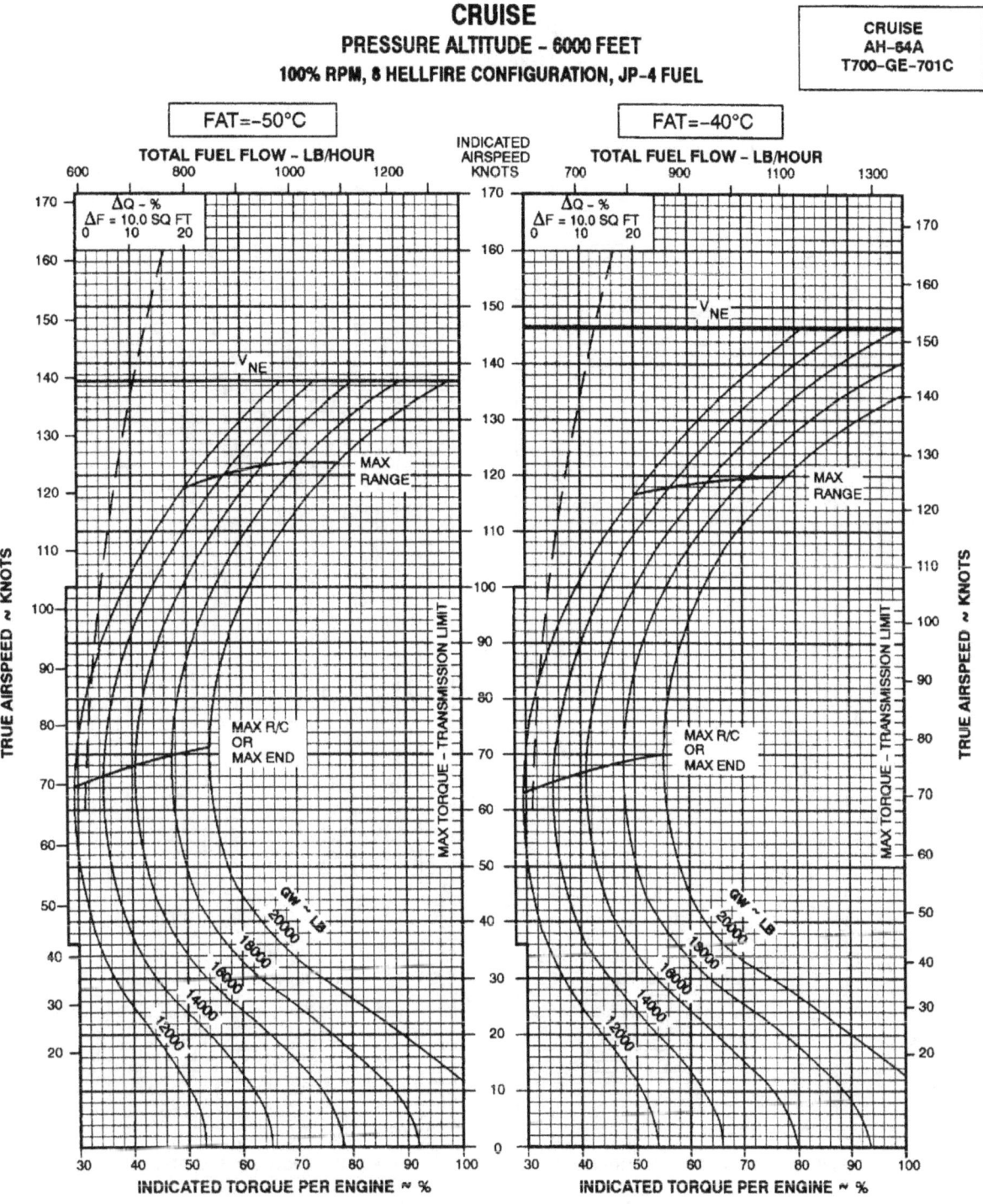

Figure 7A-12. Cruise Chart, 6,000 Feet, -50 °C and -40 °C (Sheet 1 of 5) 701C

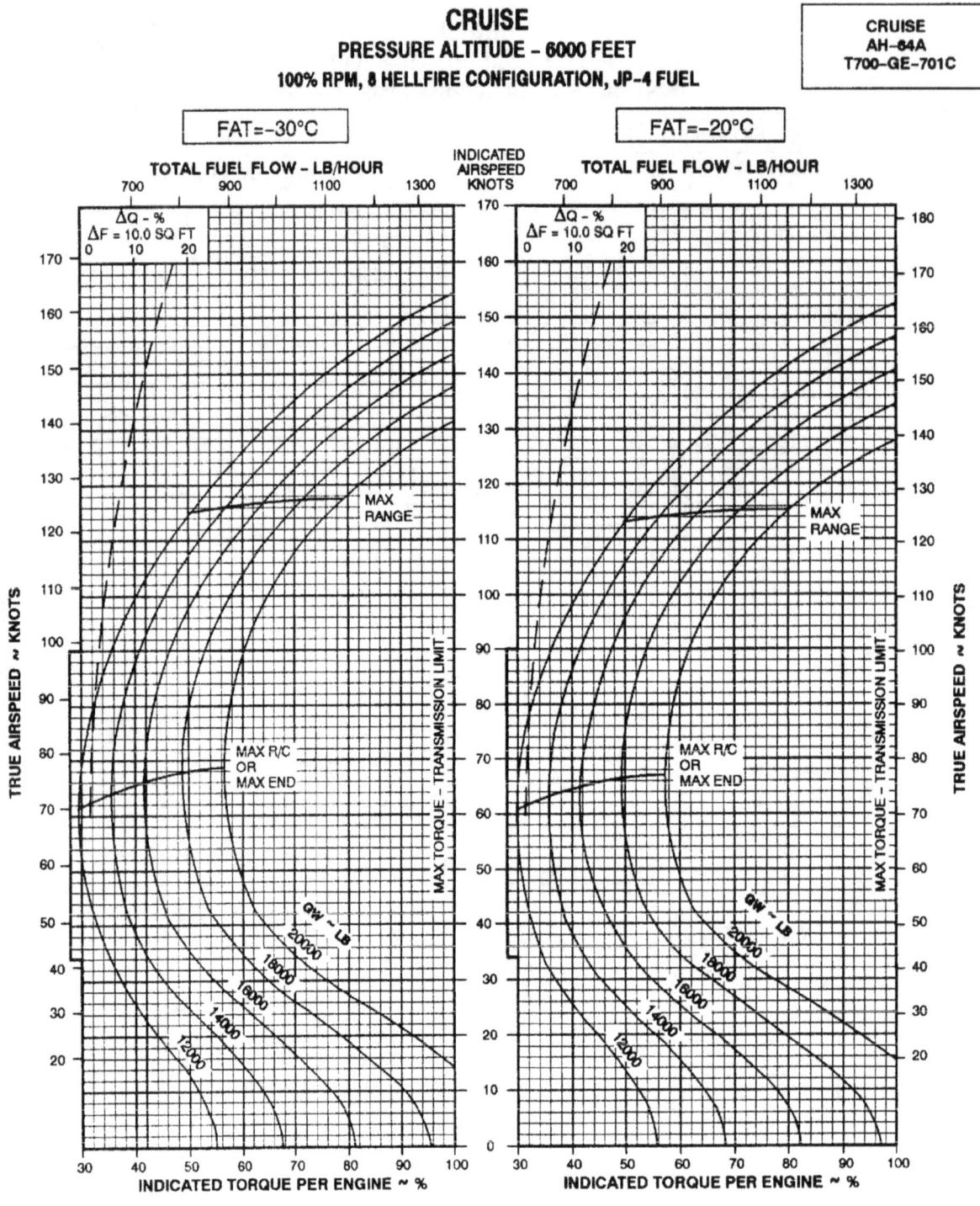

Figure 7A-12. Cruise Chart, 6,000 Feet, -30 °C and -20 °C (Sheet 2 of 5)

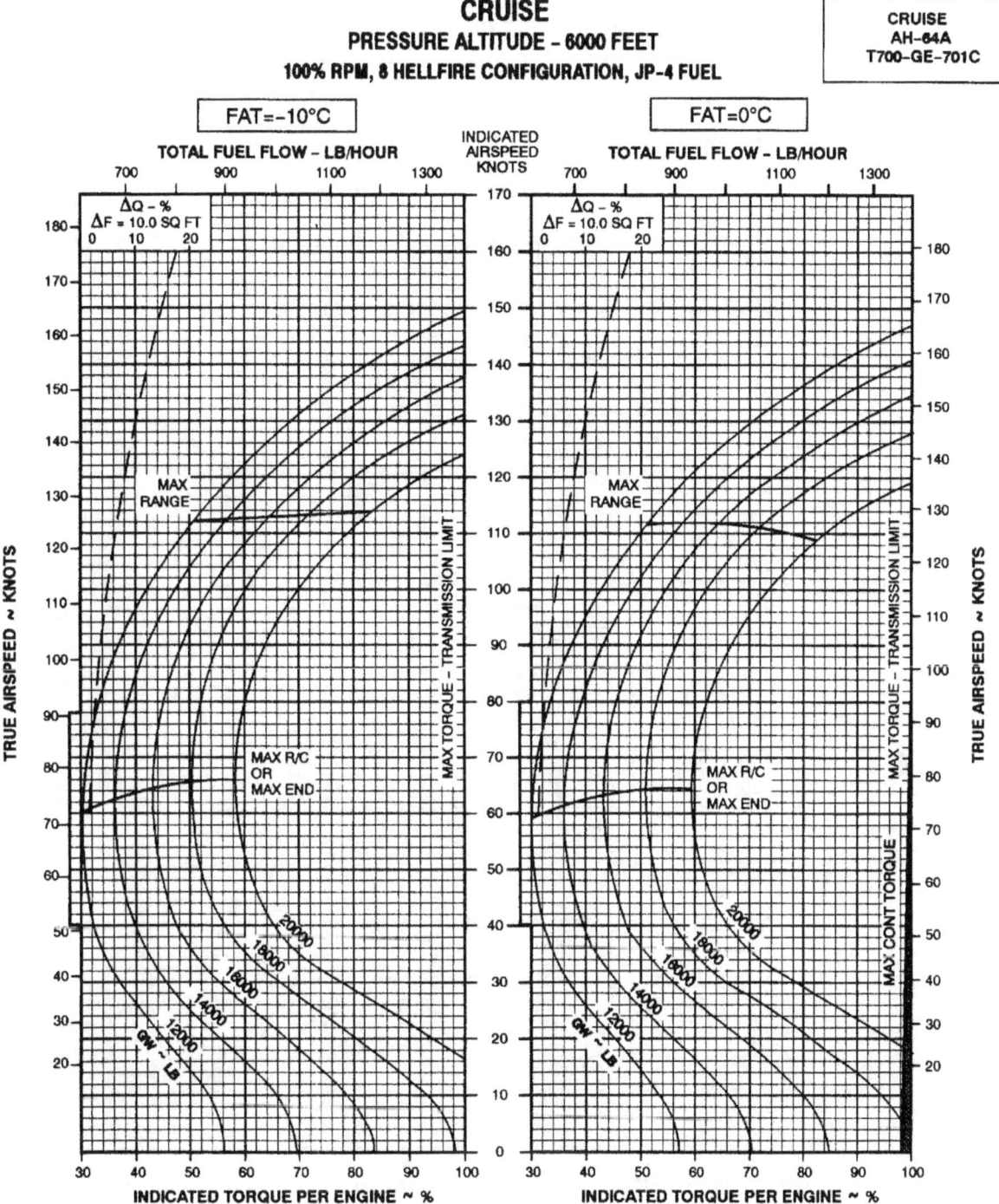

Figure 7A-12. Cruise Chart, 6,000 Feet, -10 °C and 0 °C (Sheet 3 of 5) 701C

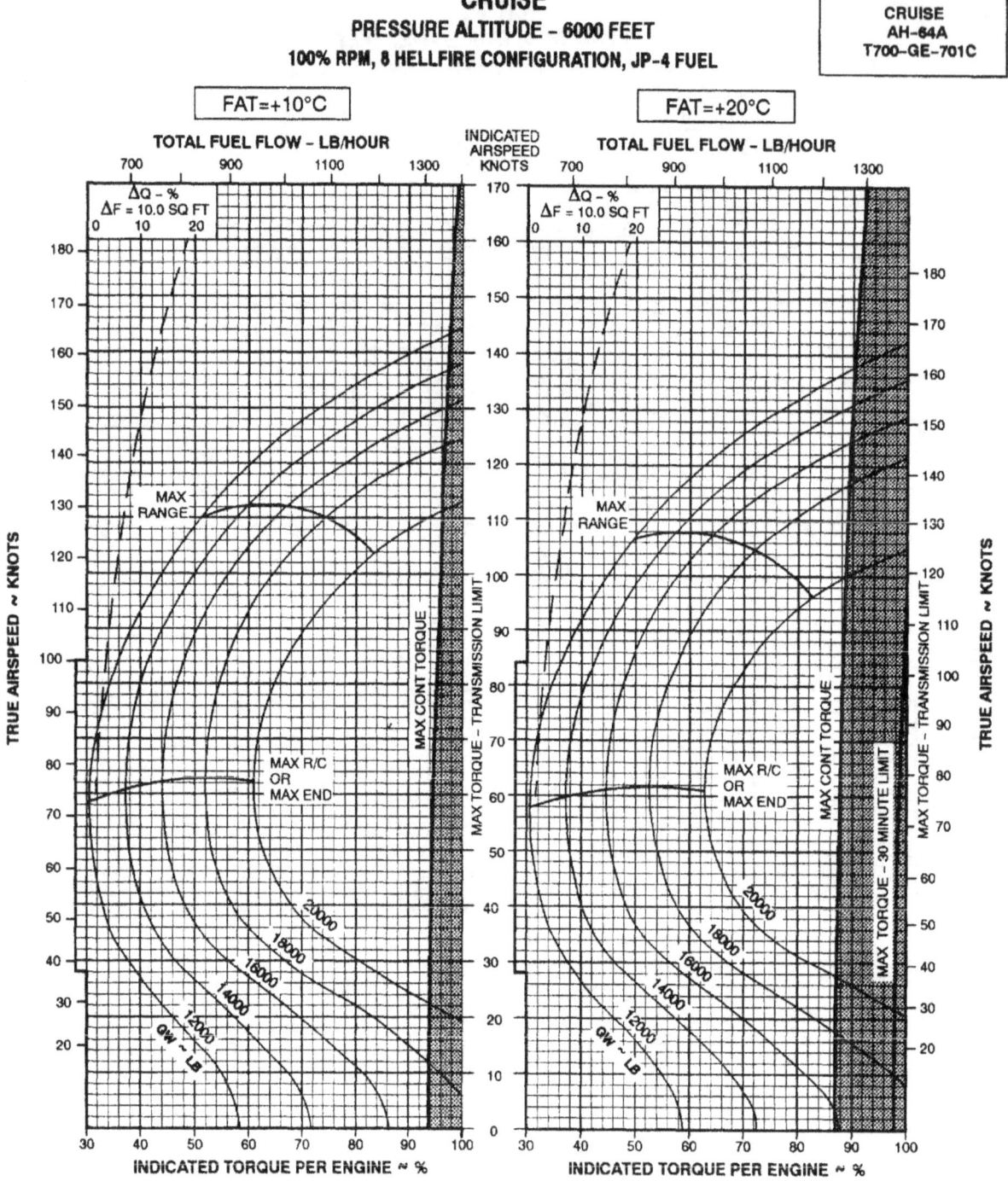

Figure 7A-12. Cruise Chart, 6,000 Feet, +10 °C and +20 °C (Sheet 4 of 5) 701C

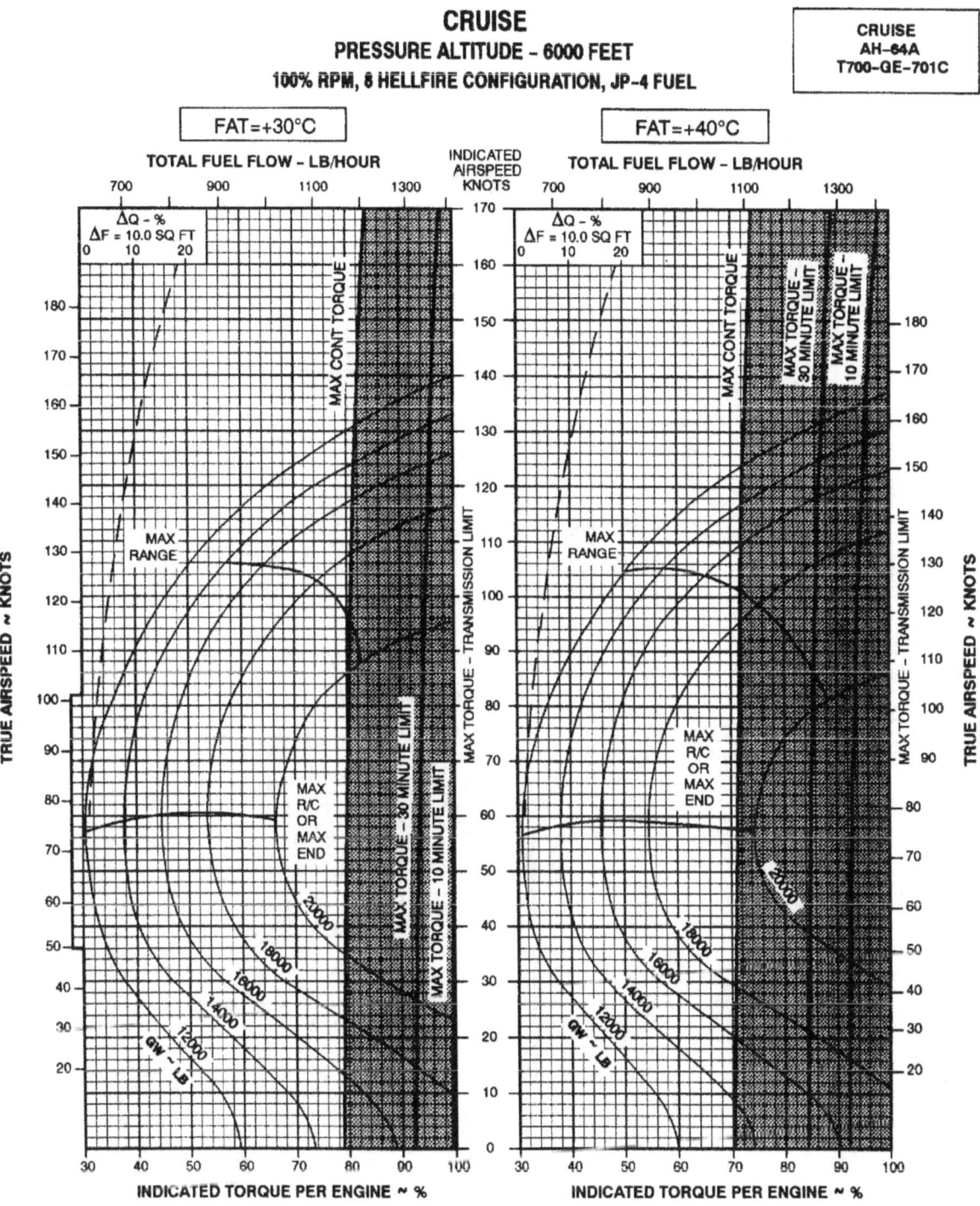

Figure 7A-12. Cruise Chart, 6,000 Feet, +30 °C and +40 °C (Sheet 5 of 5)

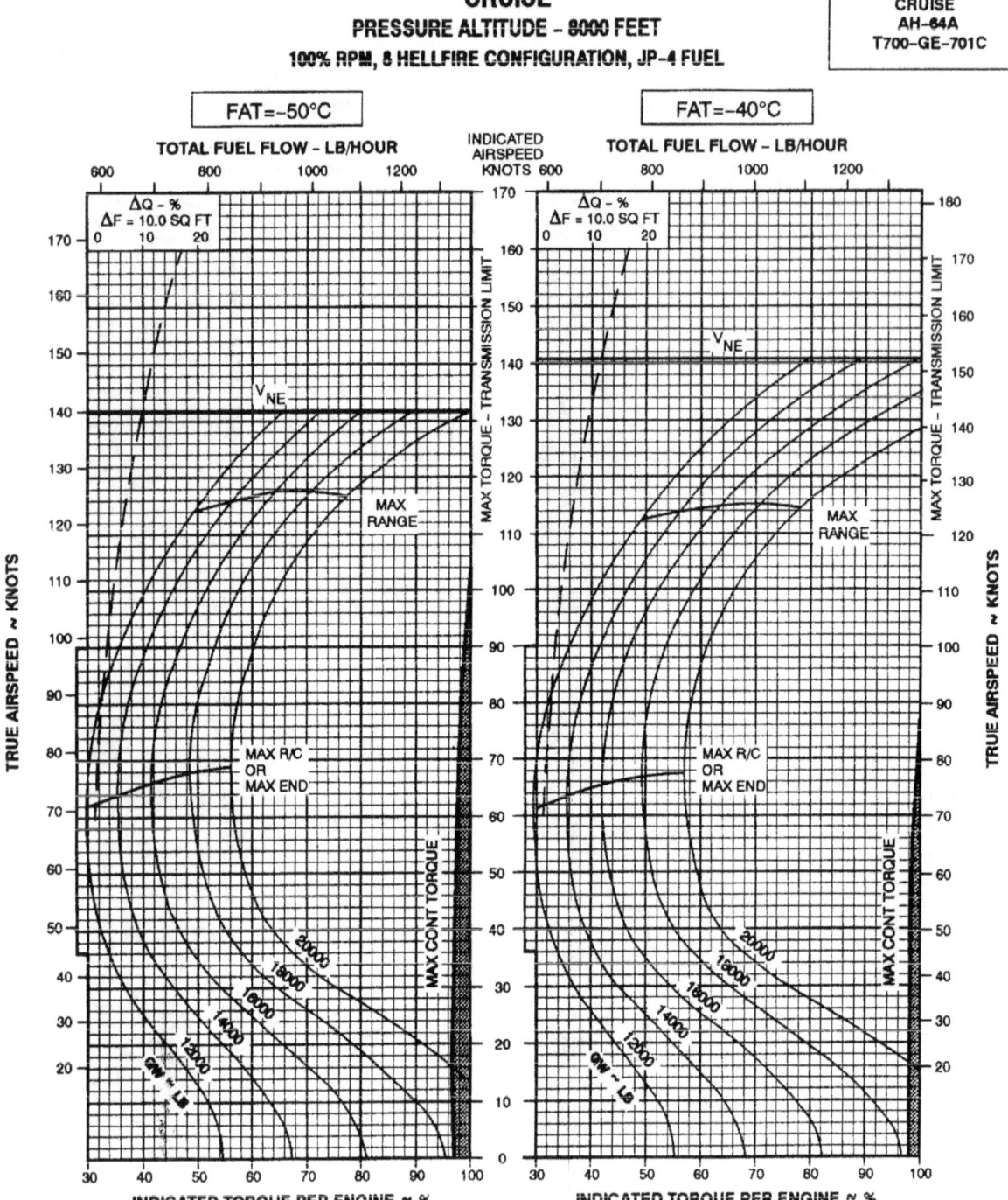

Figure 7A-13. Cruise Chart, 8,000 Feet, -50 °C and -40 °C (Sheet 1 of 5)

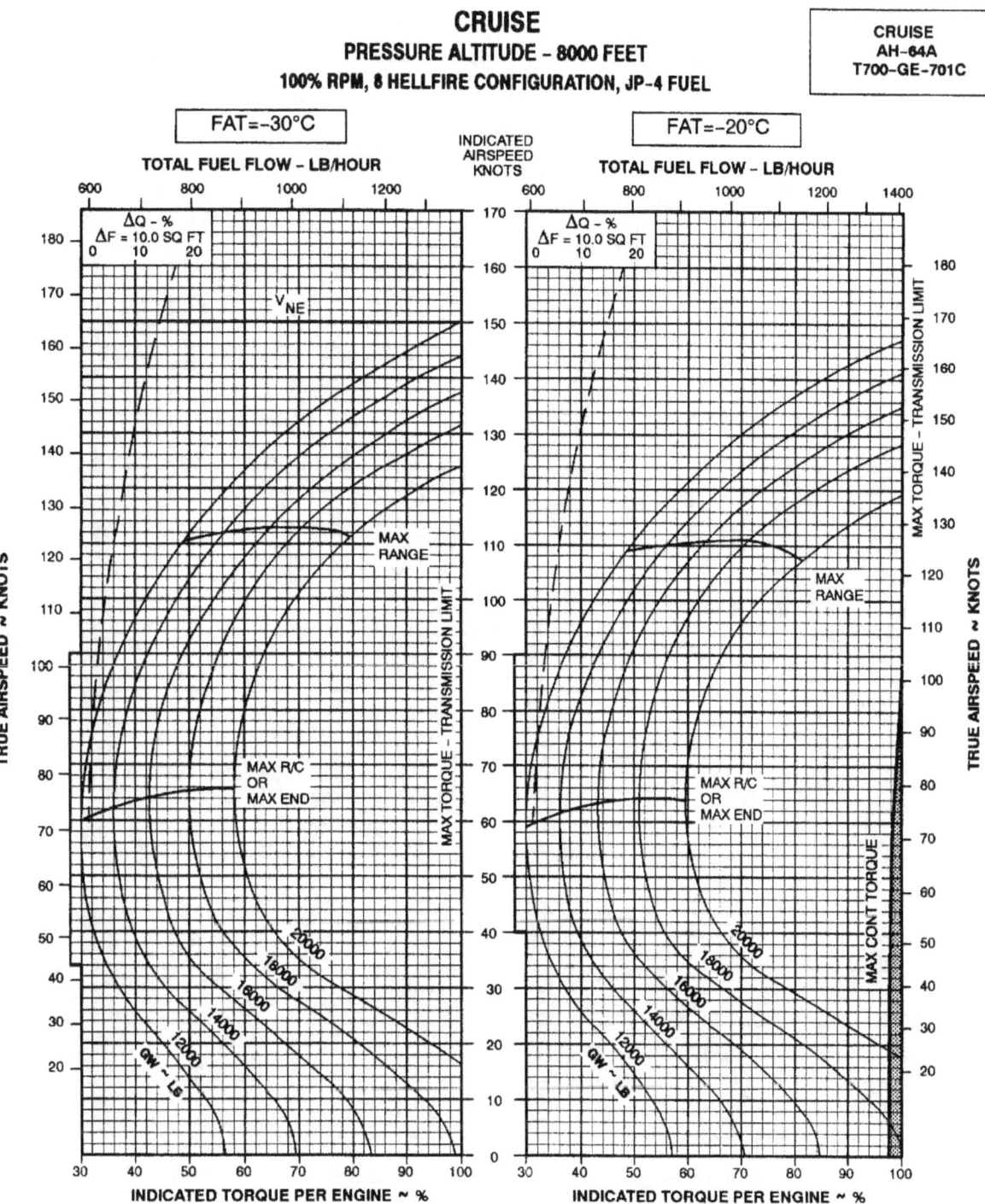

Figure 7A-13. Cruise Chart, 8,000 Feet, -30 °C and -20 °C (Sheet 2 of 5) 701C

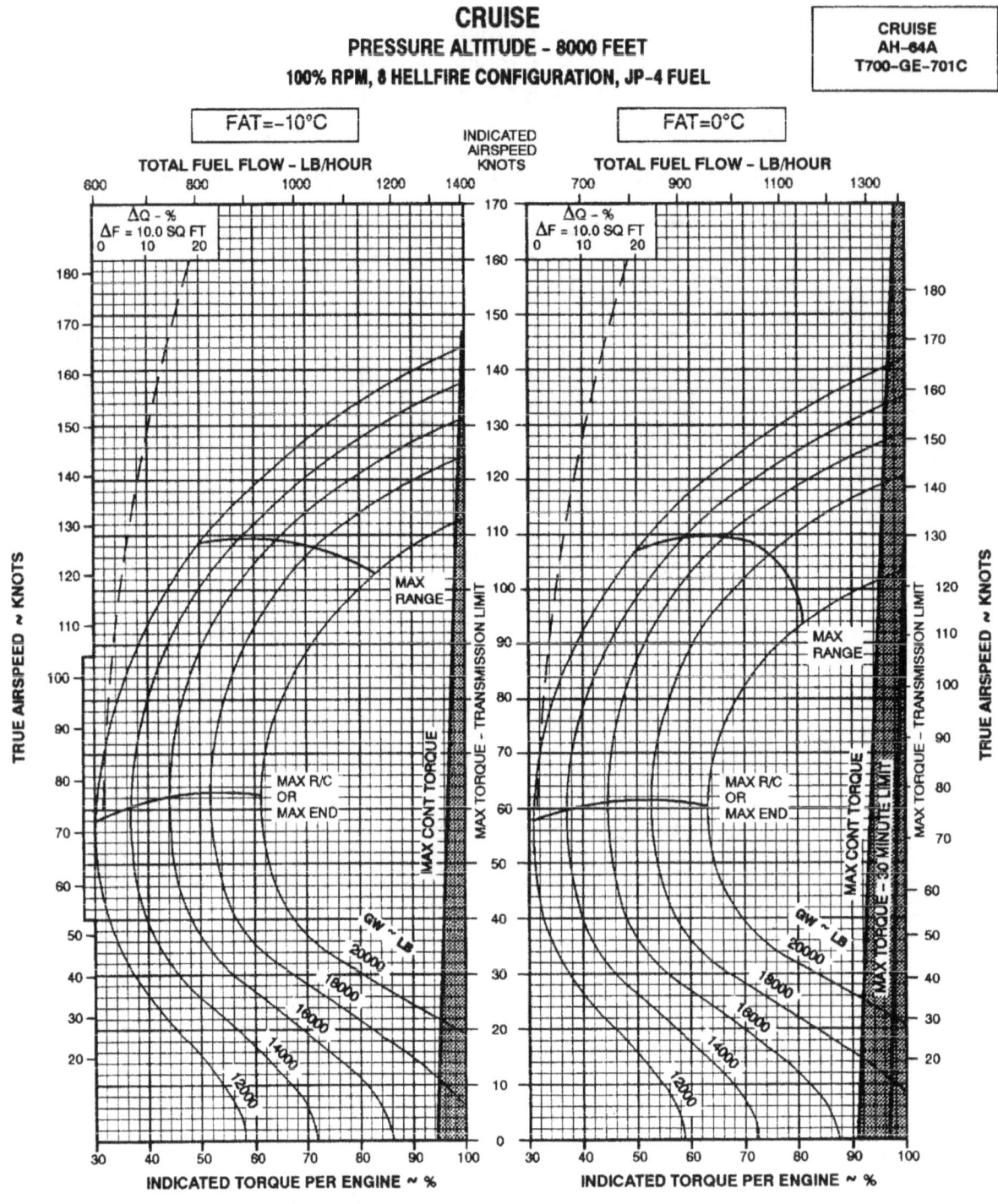

Figure 7A-13. Cruise Chart, 8,000 Feet, -10 °C and 0 °C (Sheet 3 of 5) 701C

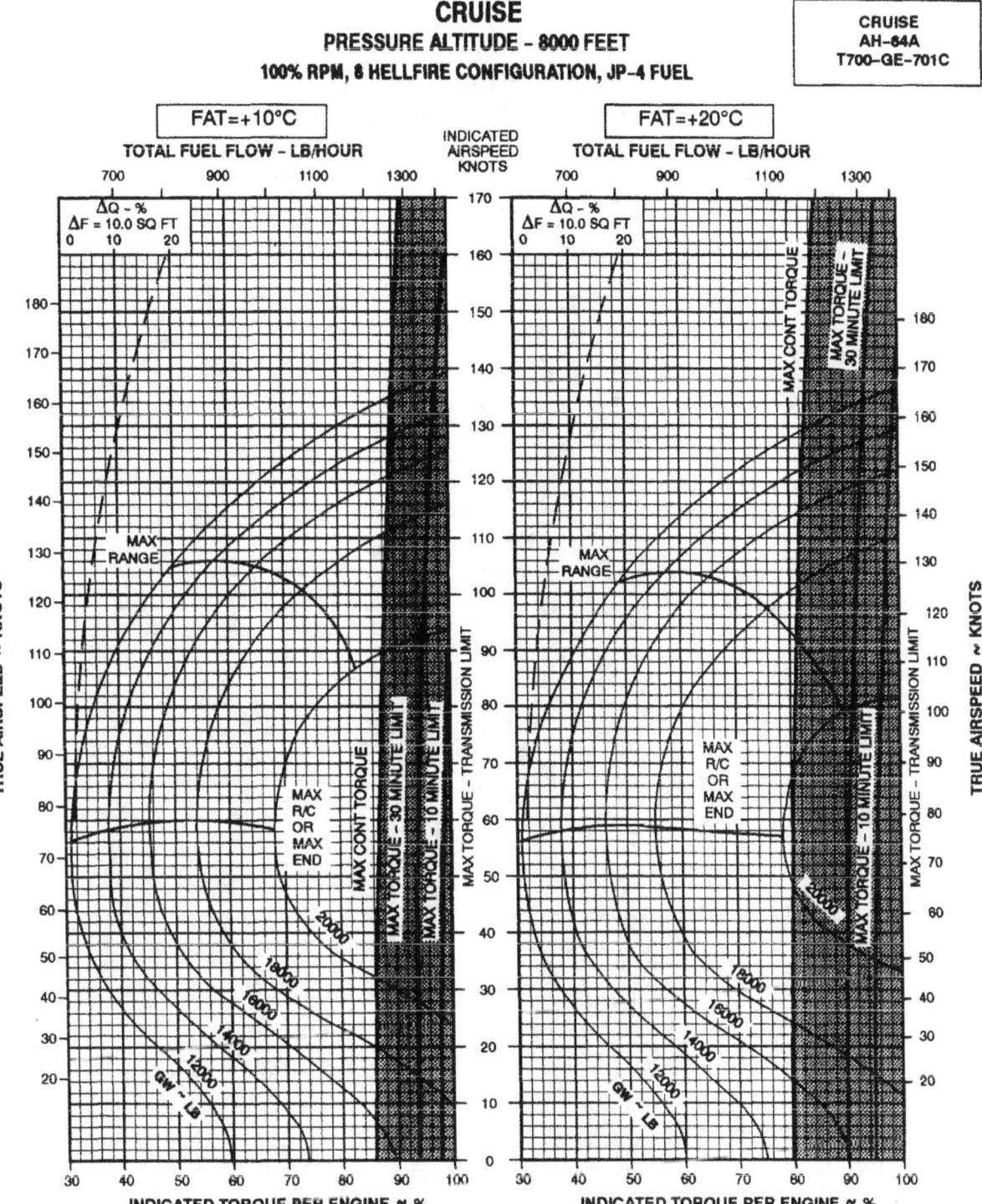

Figure 7A-13. Cruise Chart, 8,000 Feet, +10 °C and +20 °C (Sheet 4 of 5)

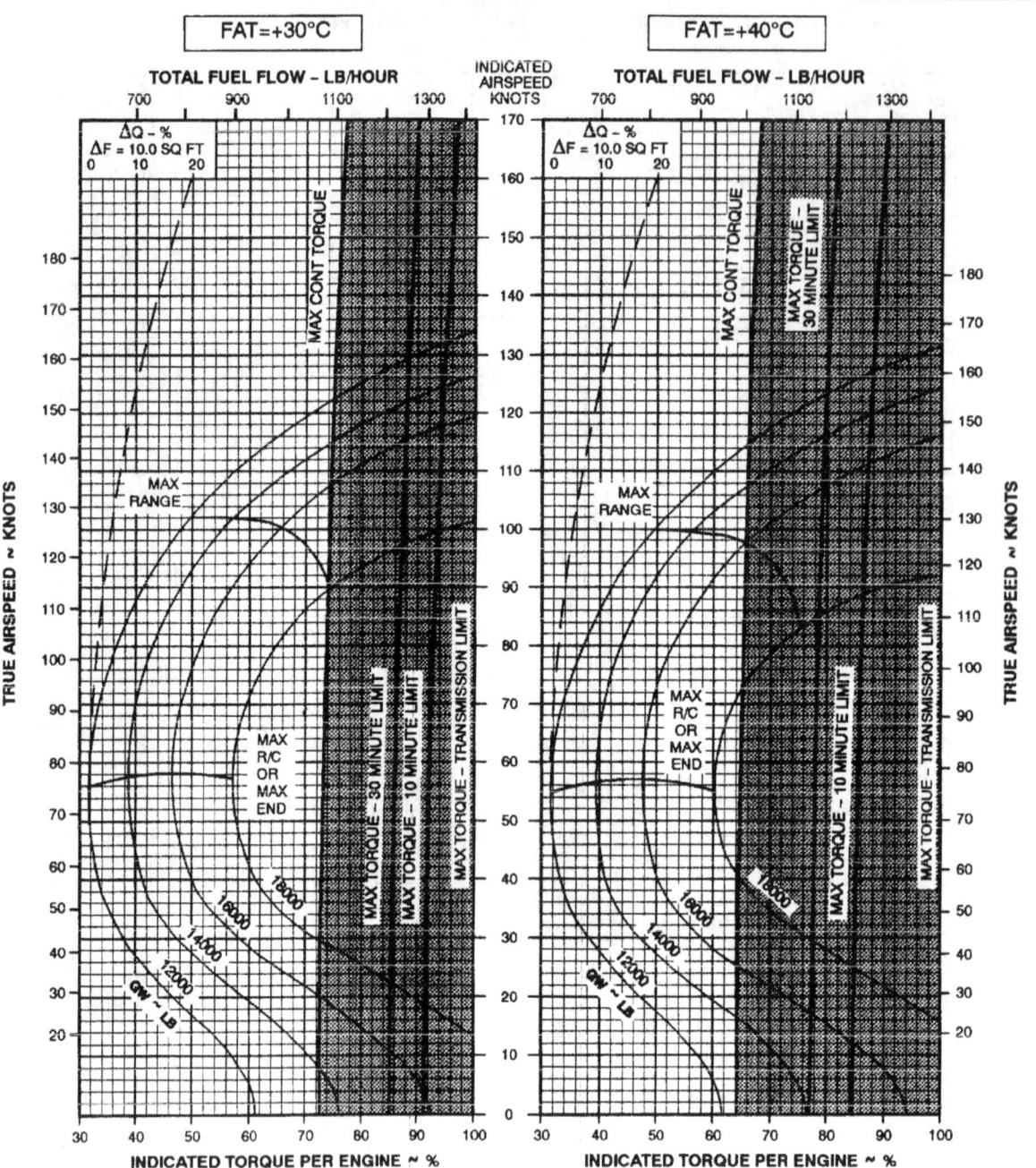

Figure 7A-13. Cruise Chart, 8,000 Feet, +30 °C and +40 °C (Sheet 5 of 5)

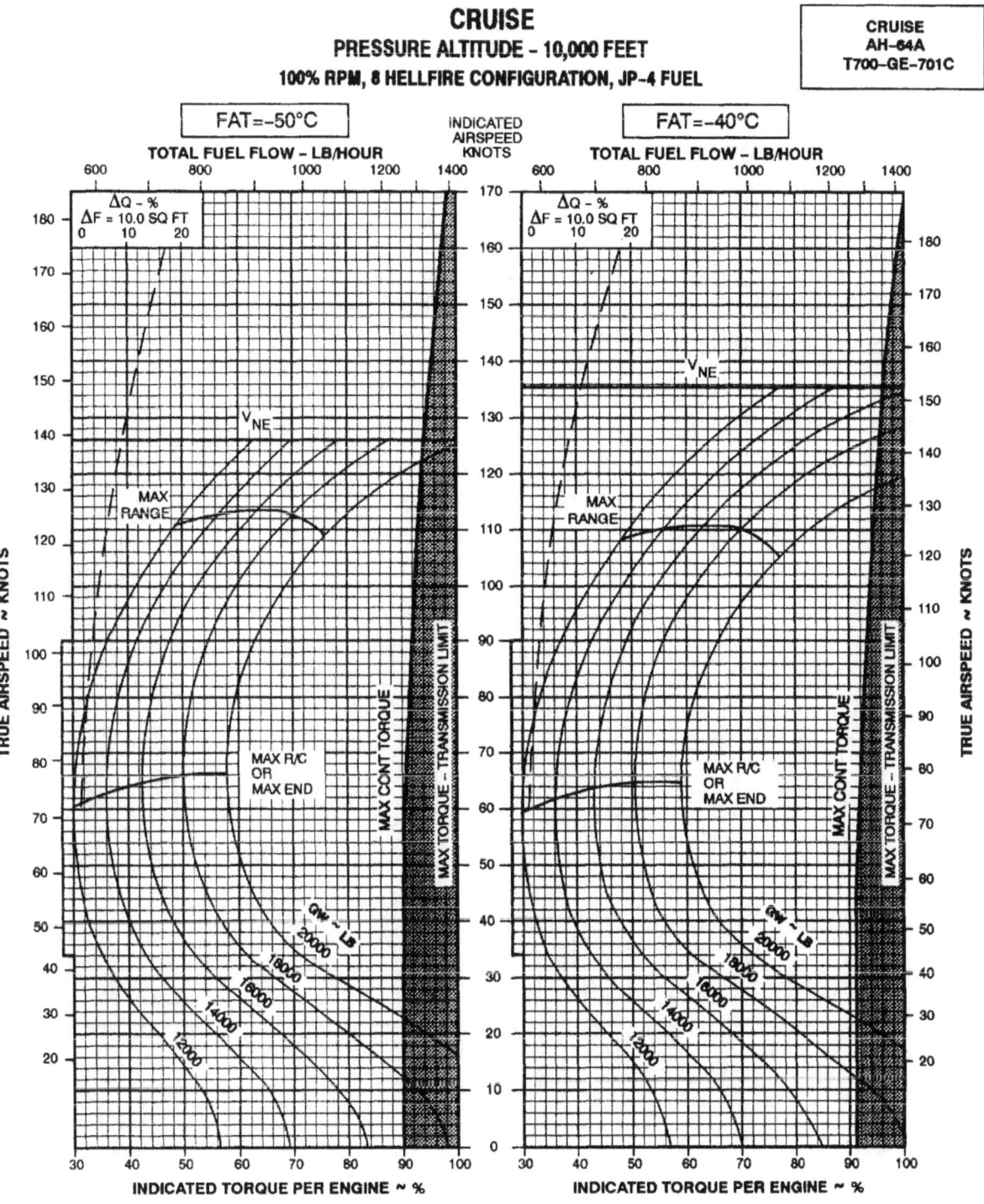

Figure 7A-14. Cruise Chart, 10,000 Feet, -50 °C and -40 °C (Sheet 1 of 5)

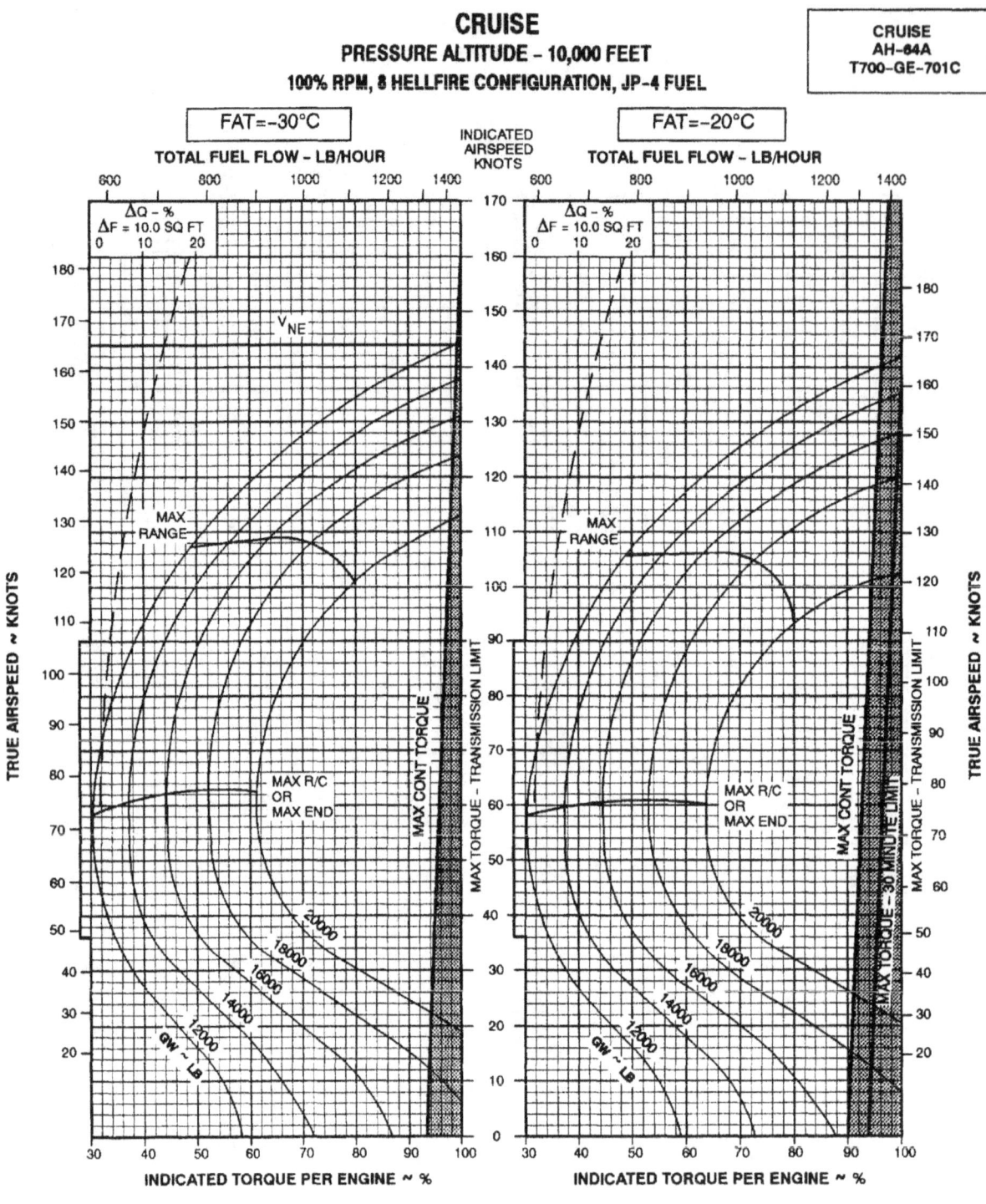

Figure 7A-14. Cruise Chart, 10,000 Feet, -30 °C and -20 °C (Sheet 2 of 5)

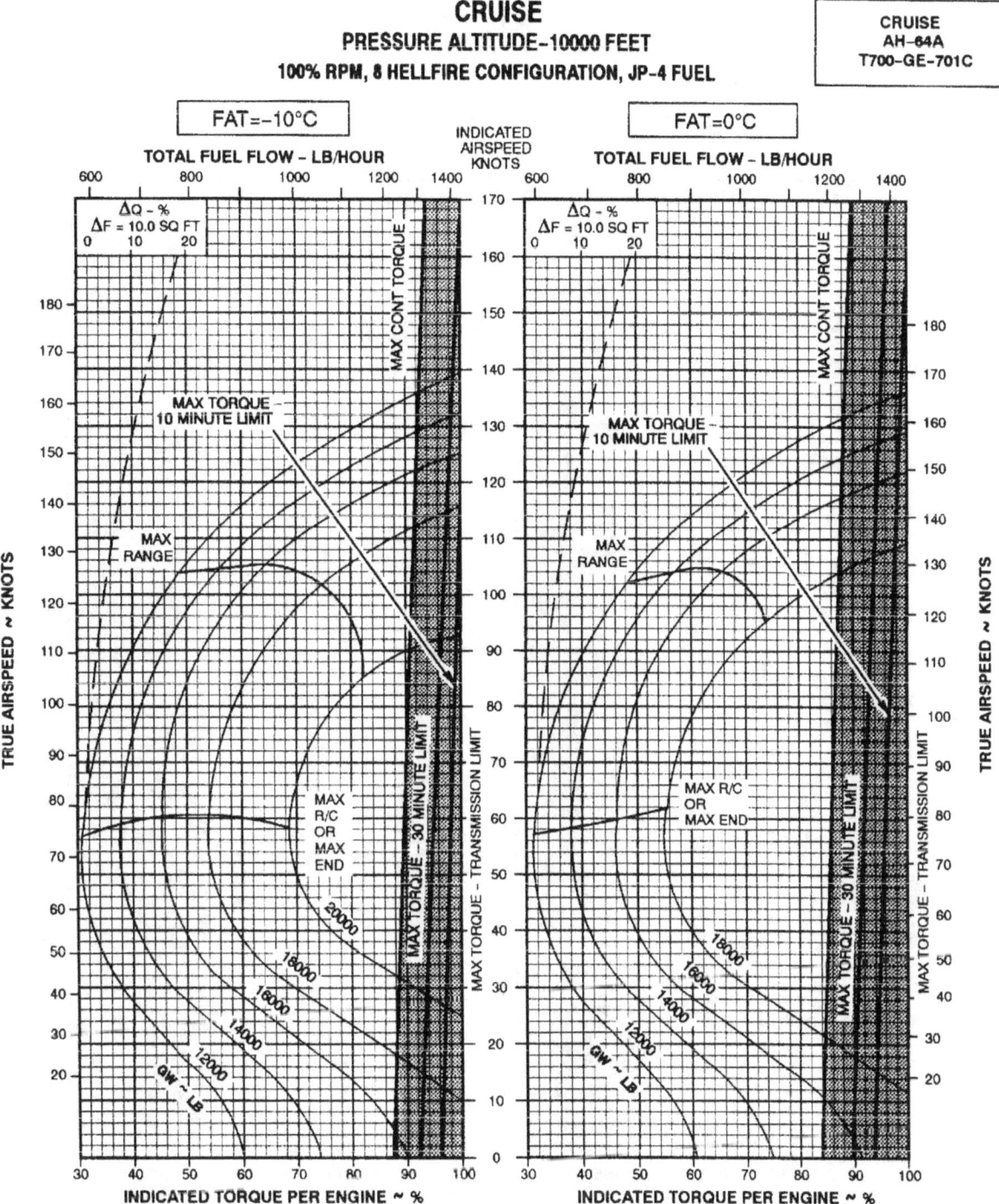

Figure 7A-14. Cruise Chart, 10,000 Feet, -10 °C and 0 °C (Sheet 3 of 5) 701C

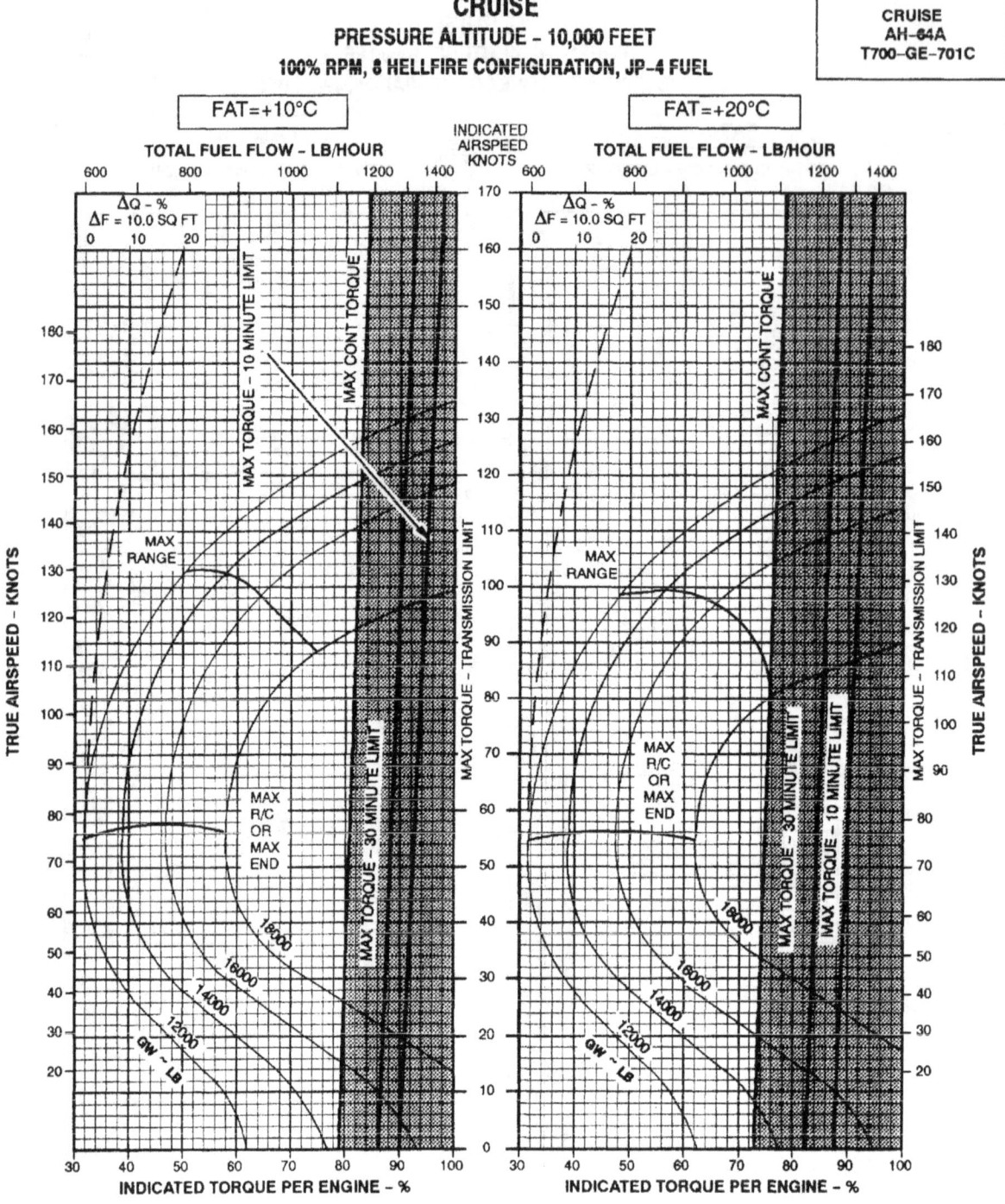

Figure 7A-14. Cruise Chart, 10,000 Feet, +10 °C and +20 °C (Sheet 4 of 5) 701C

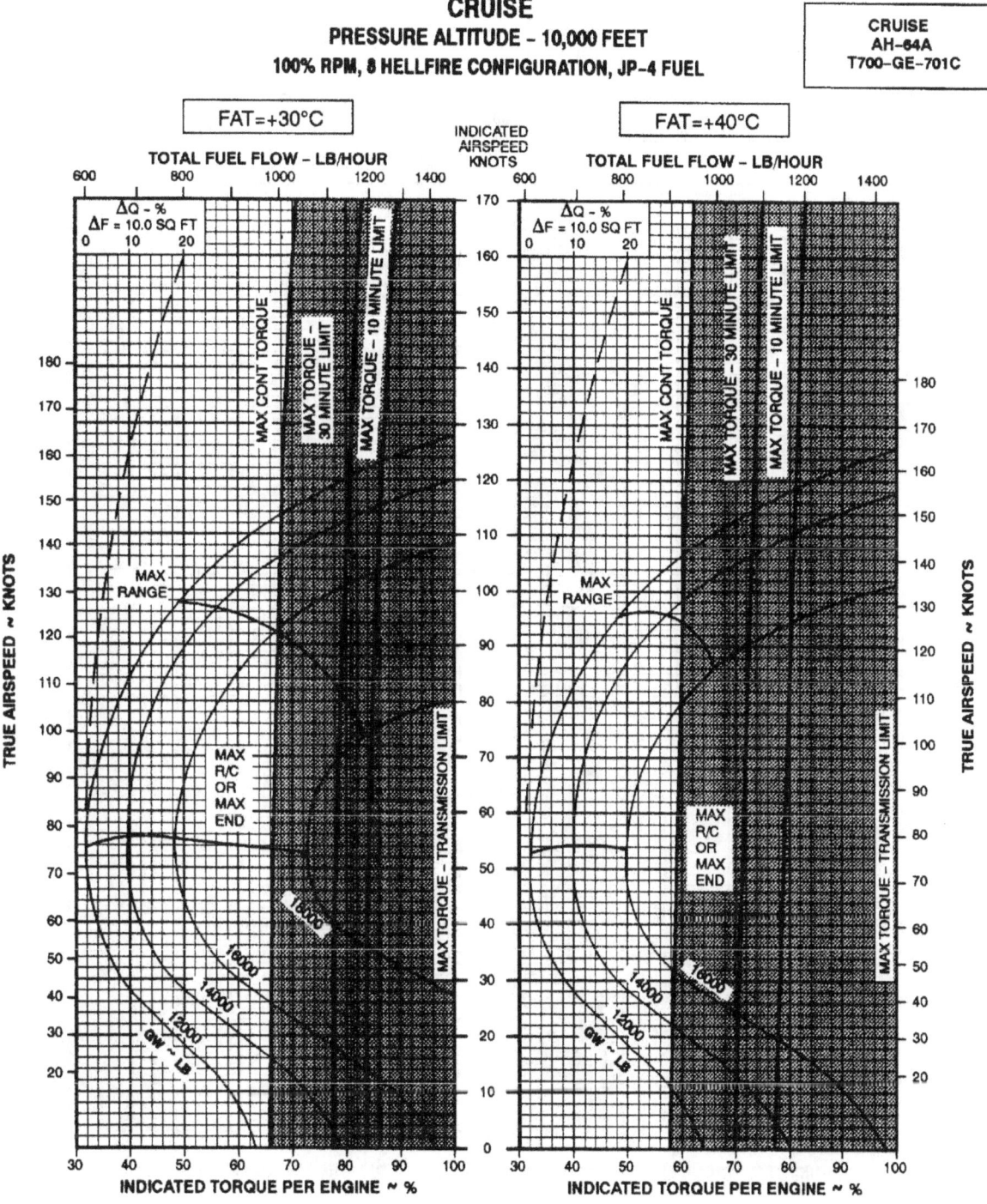

Figure 7A-14. Cruise Chart, 10,000 Feet, +30 °C and +40 °C (Sheet 5 of 5)

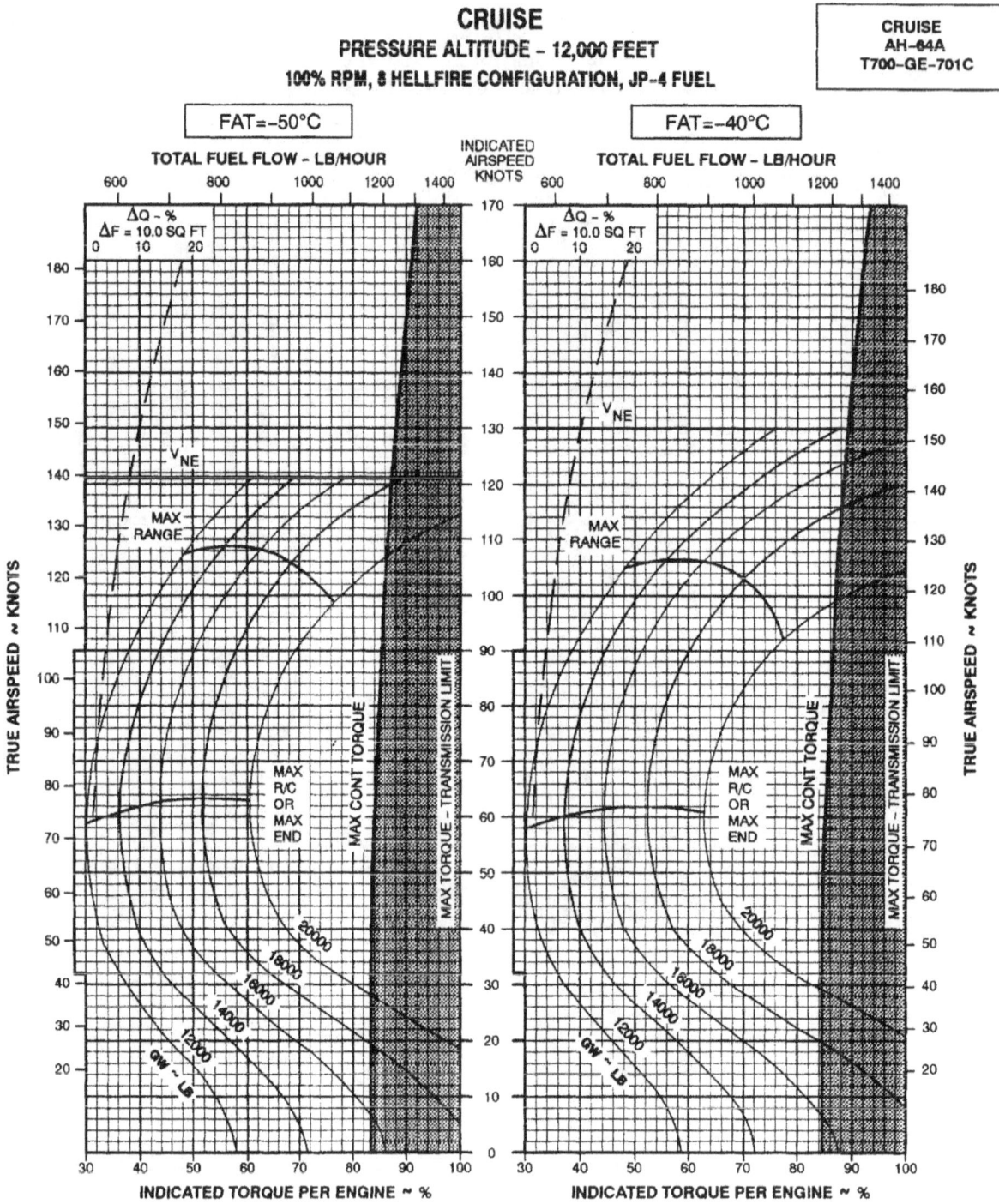

Figure 7A-15. Cruise Chart, 12,000 Feet, -50 °C and -40 °C (Sheet 1 of 5)

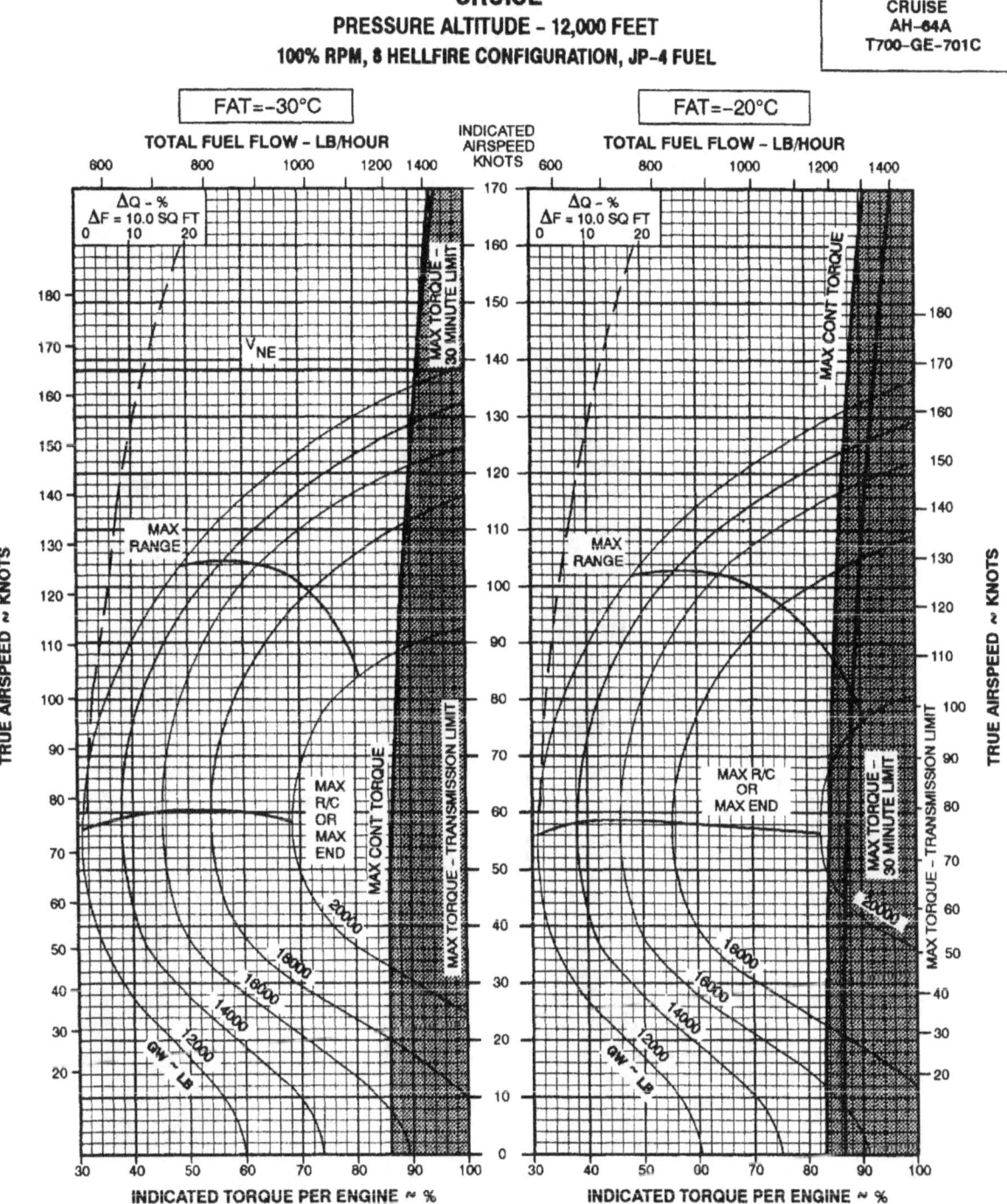

Figure 7A-15. Cruise Chart, 12,000 Feet, -30 °C and -20 °C (Sheet 2 of 5)

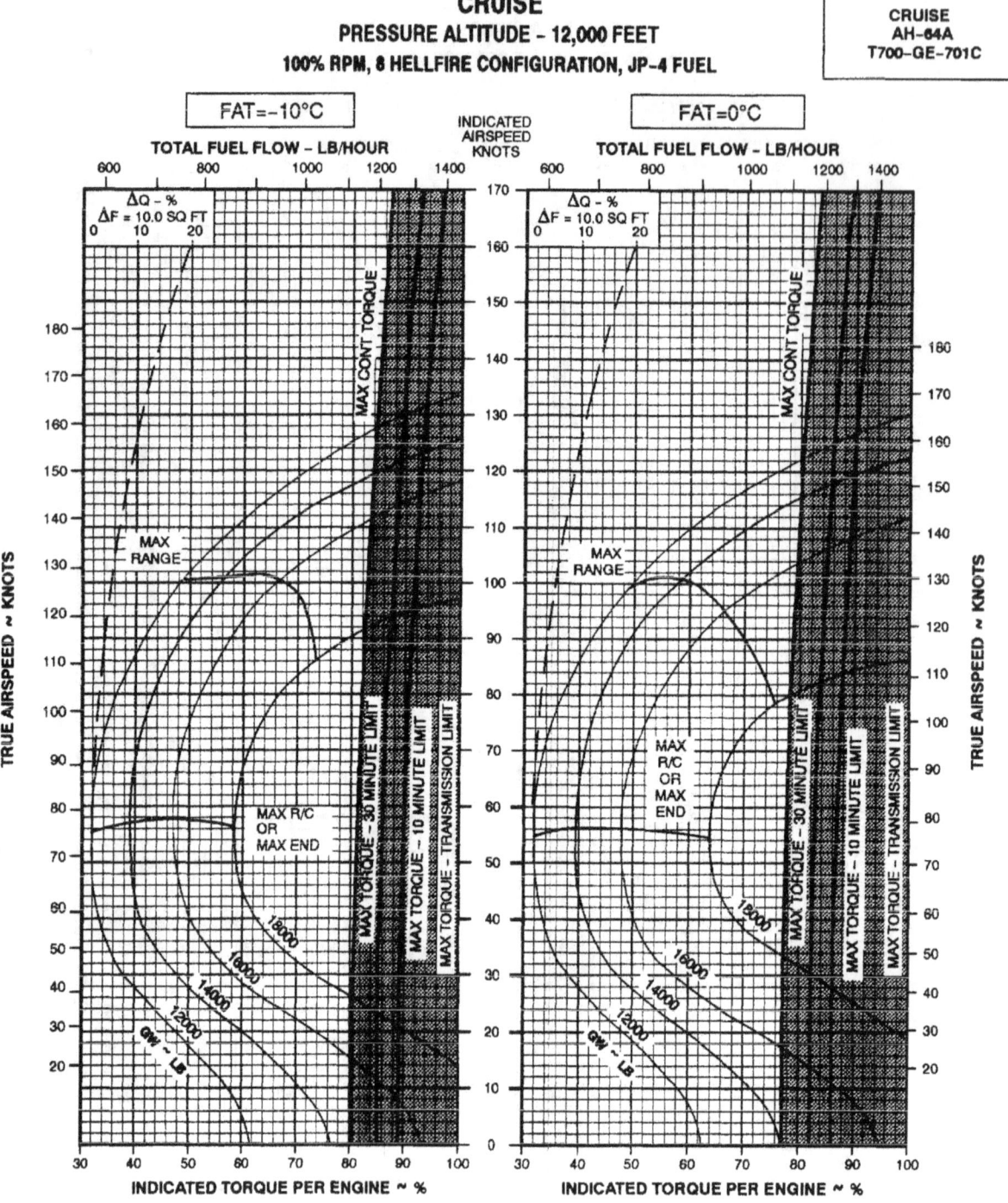

Figure 7A-15. Cruise Chart, 12,000 Feet, -10 °C and 0 °C (Sheet 3 of 5)

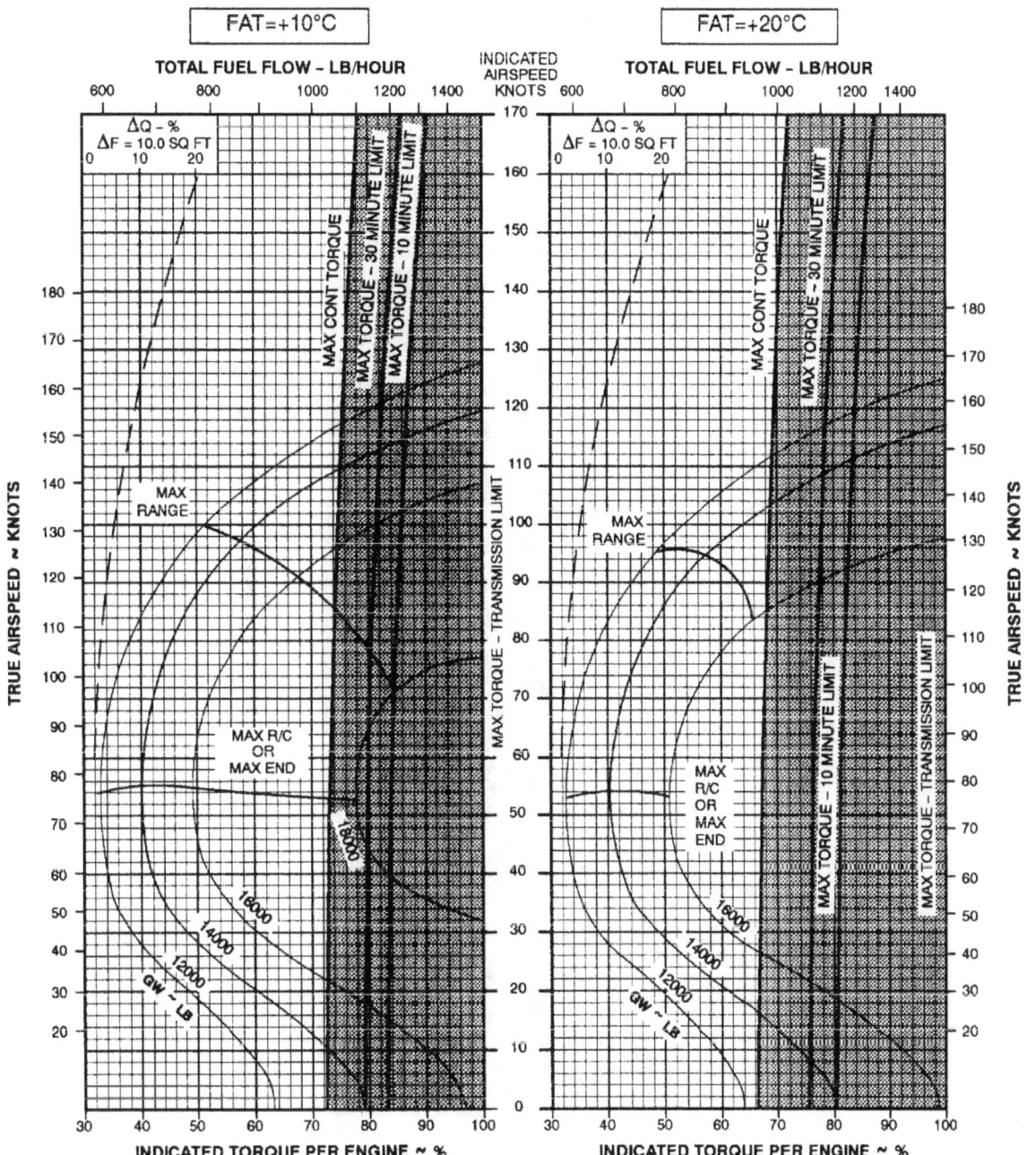

Figure 7A-15. Cruise Chart, 12,000 Feet, +10 °C and +20 °C (Sheet 4 of 5) 701C

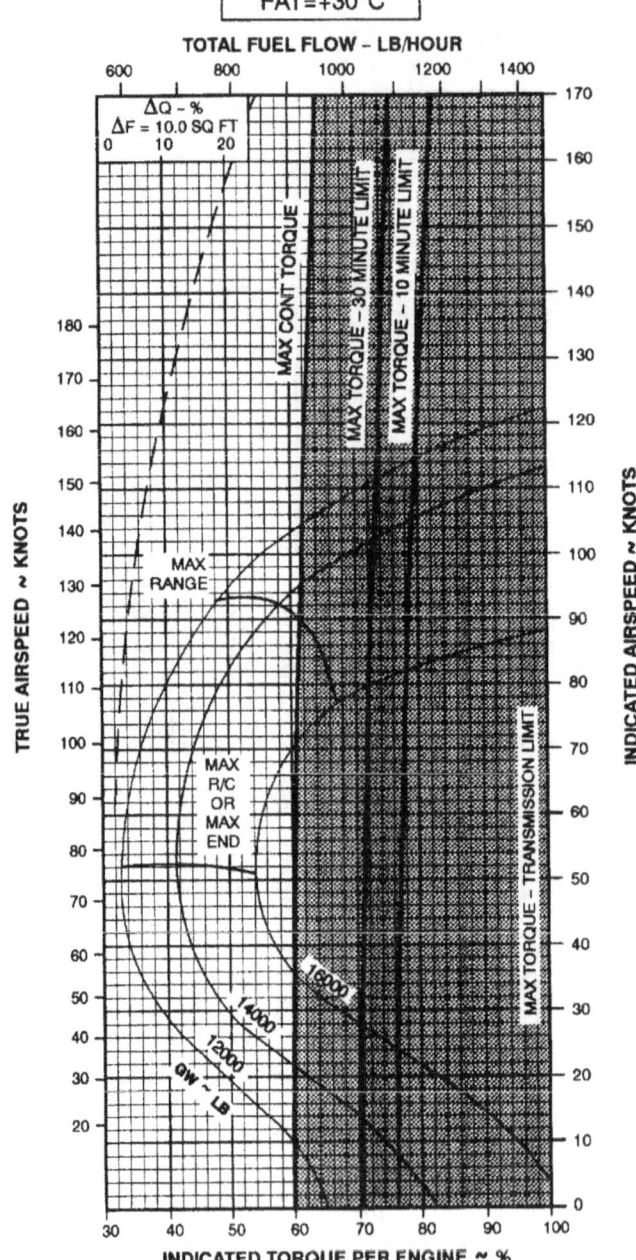

Figure 7A-15. Cruise Chart, 12,000 Feet, +30 °C (Sheet 5 of 5)

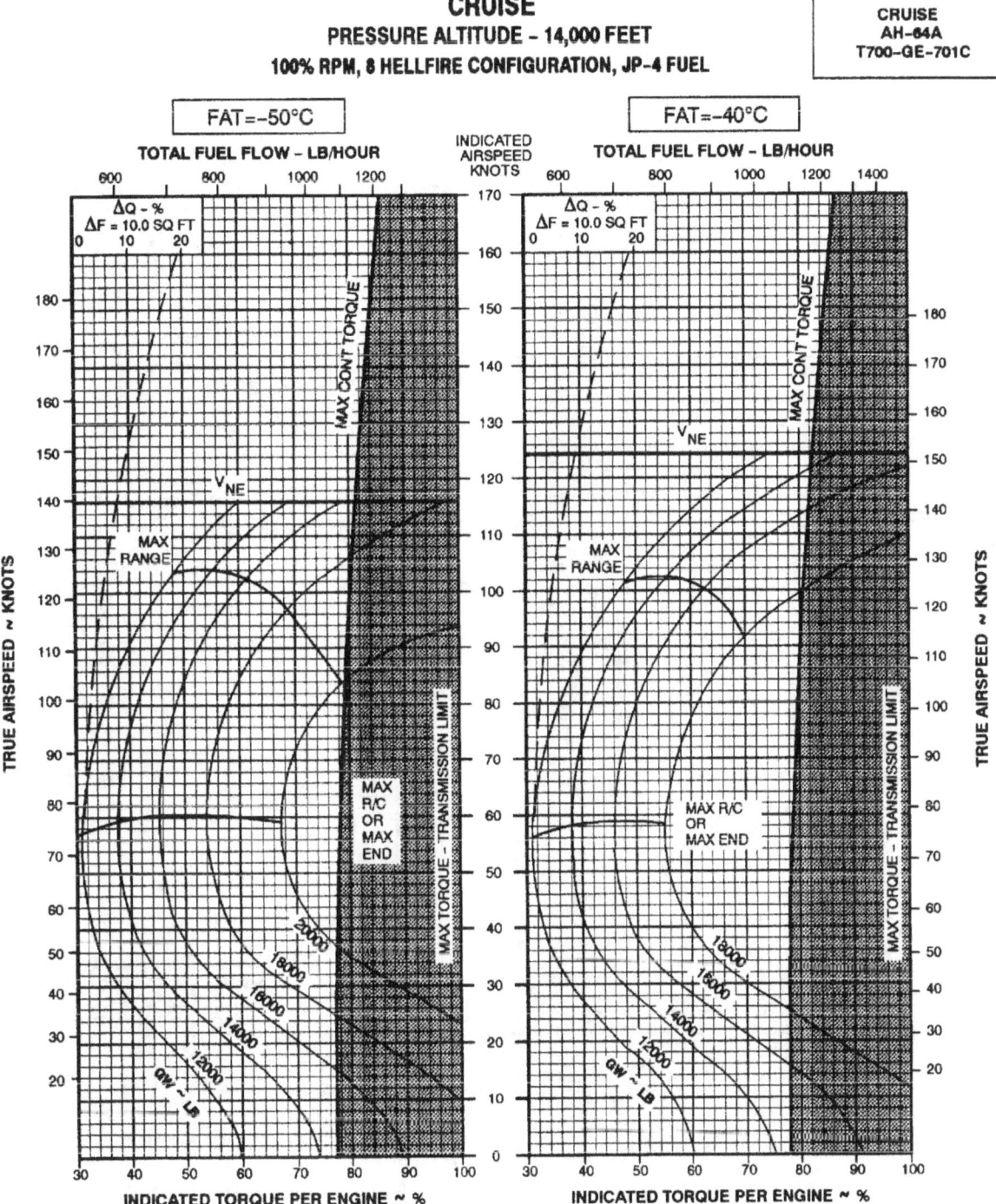

Figure 7A-16. Cruise Chart, 14,000 Feet, -50 °C and -40 °C (Sheet 1 of 5) 701C

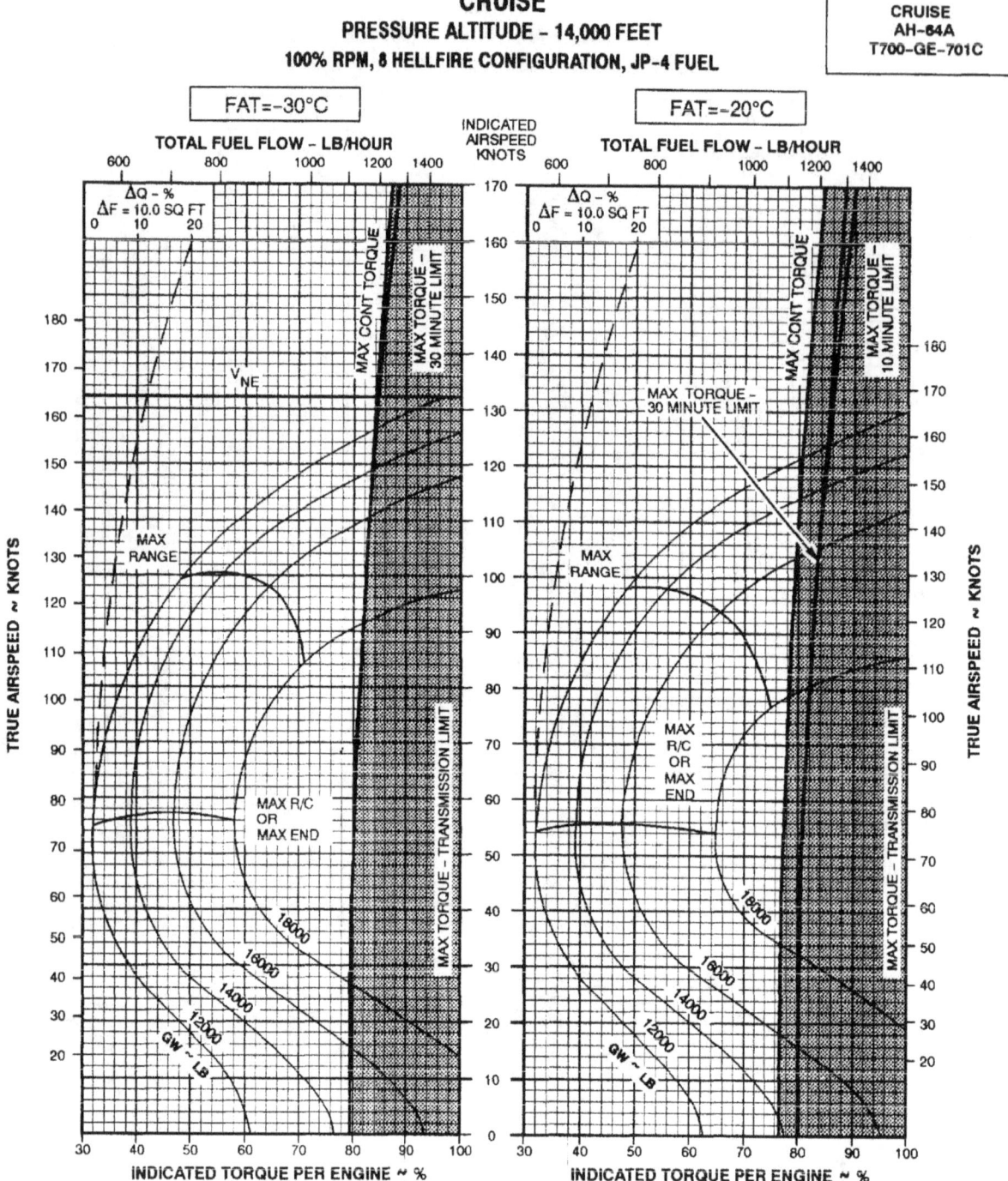

Figure 7A-16. Cruise Chart, 14,000 Feet, -30 °C and -20 °C (Sheet 2 of 5)

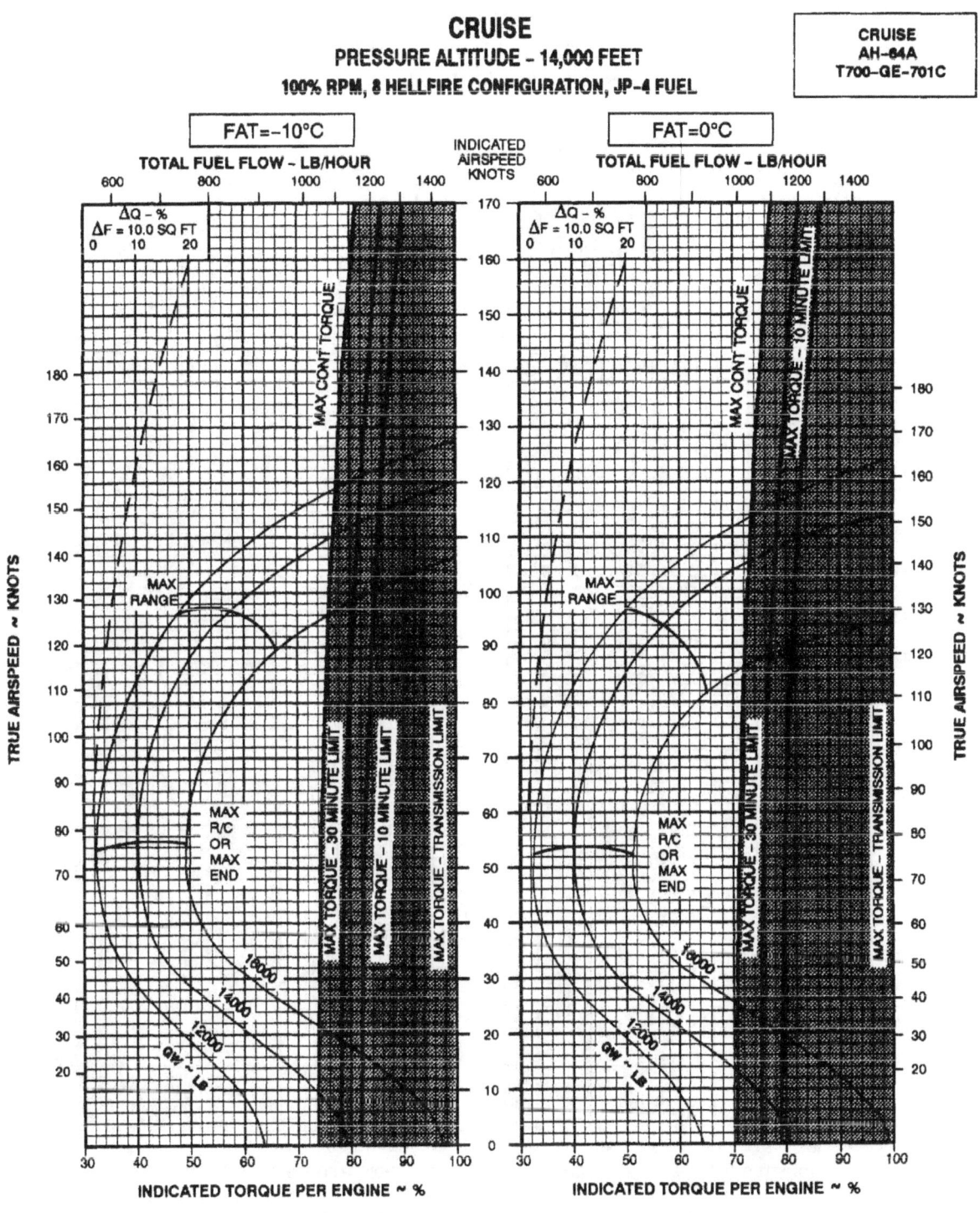

Figure 7A-16. Cruise Chart, 14,000 Feet, -10 °C and 0 °C (Sheet 3 of 5)

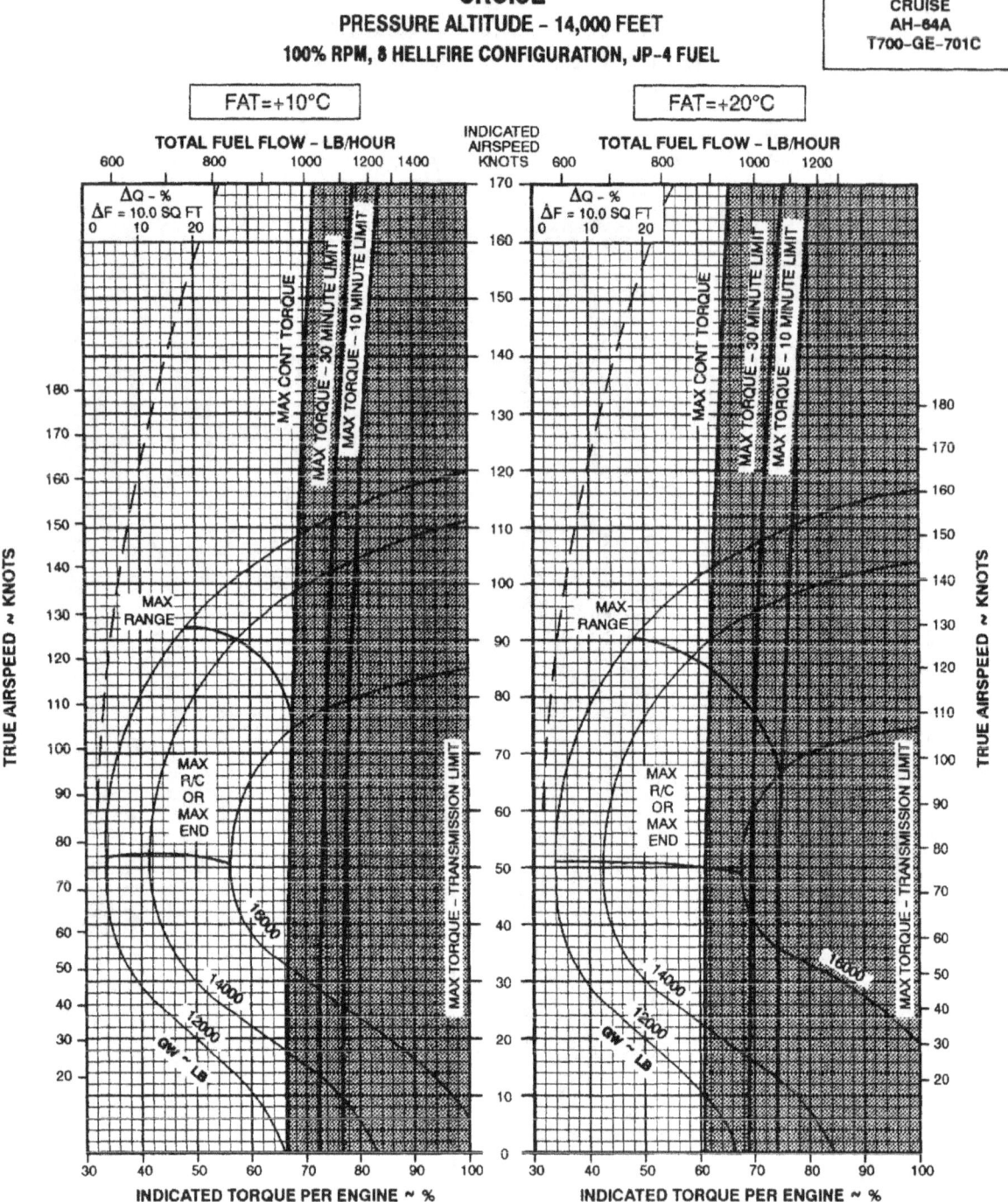

Figure 7A-16. Cruise Chart, 14,000 Feet, +10 °C and +20 °C (Sheet 4 of 5) 701C

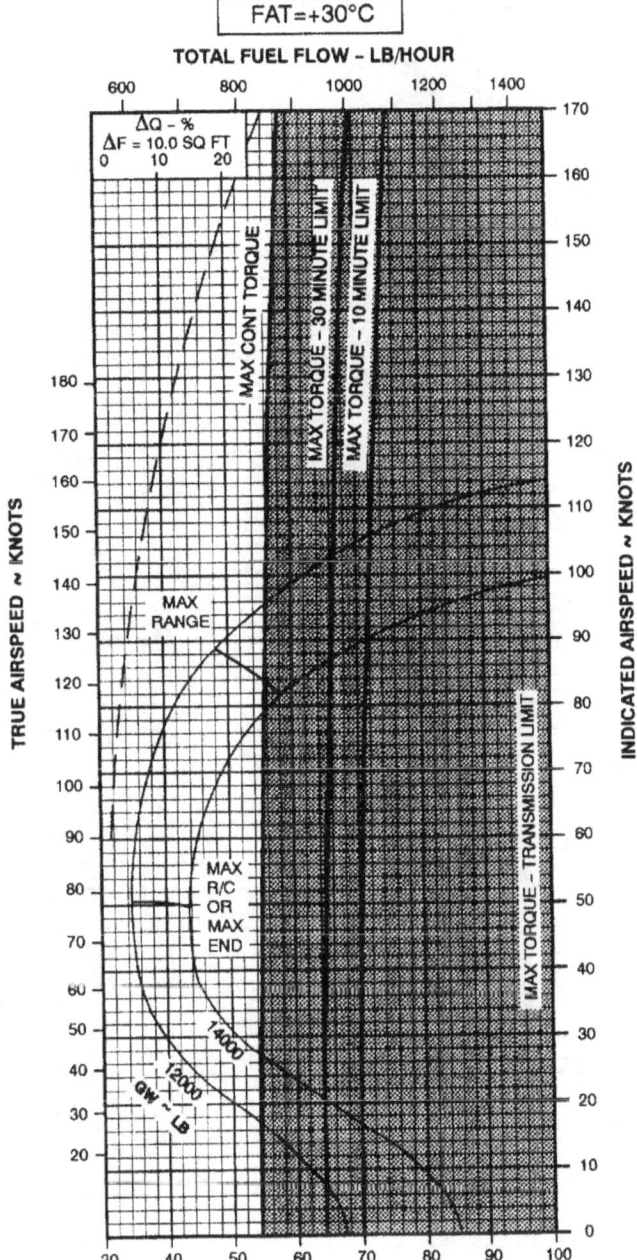

Figure 7A-16. Cruise Chart, 14,000 Feet, +30 °C (Sheet 5 of 5) 701C

CRUISE

PRESSURE ALTITUDE - 16,000 FEET

100% RPM, 8 HELLFIRE CONFIGURATION, JP-4 FUEL

CRUISE
AH-64A
T700-GE-701C

DATA BASIS: DERIVED FROM FLIGHT TEST AND ENGINE MODEL SPEC NO. E1290, 15 APRIL 1988

M01-278

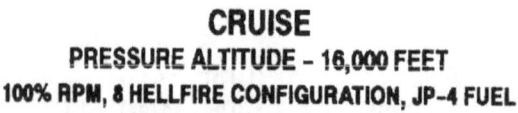

Figure 7A-17. Cruise Chart, 16,000 Feet, -50 °C and -40 °C (Sheet 1 of 4) 701C

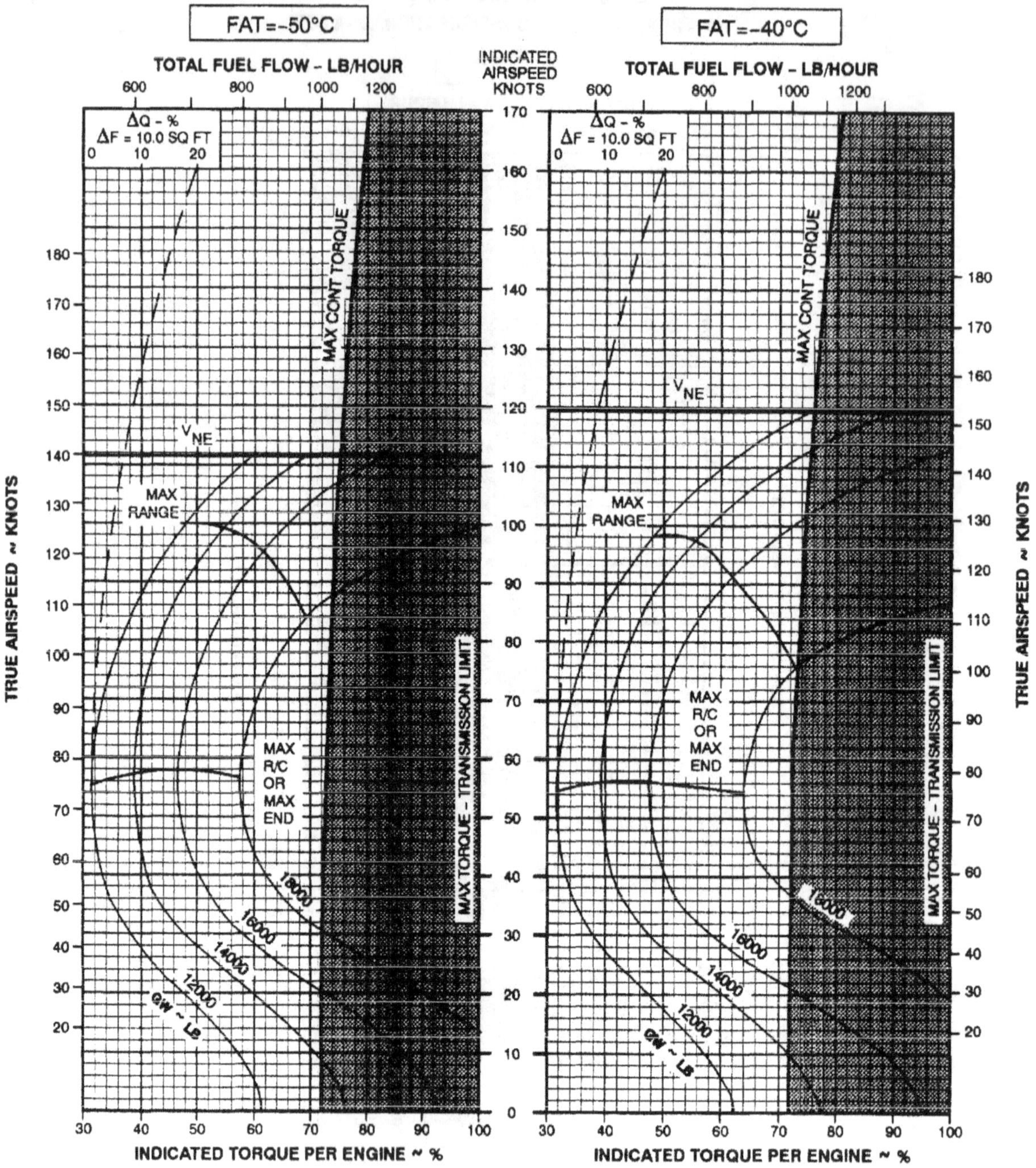

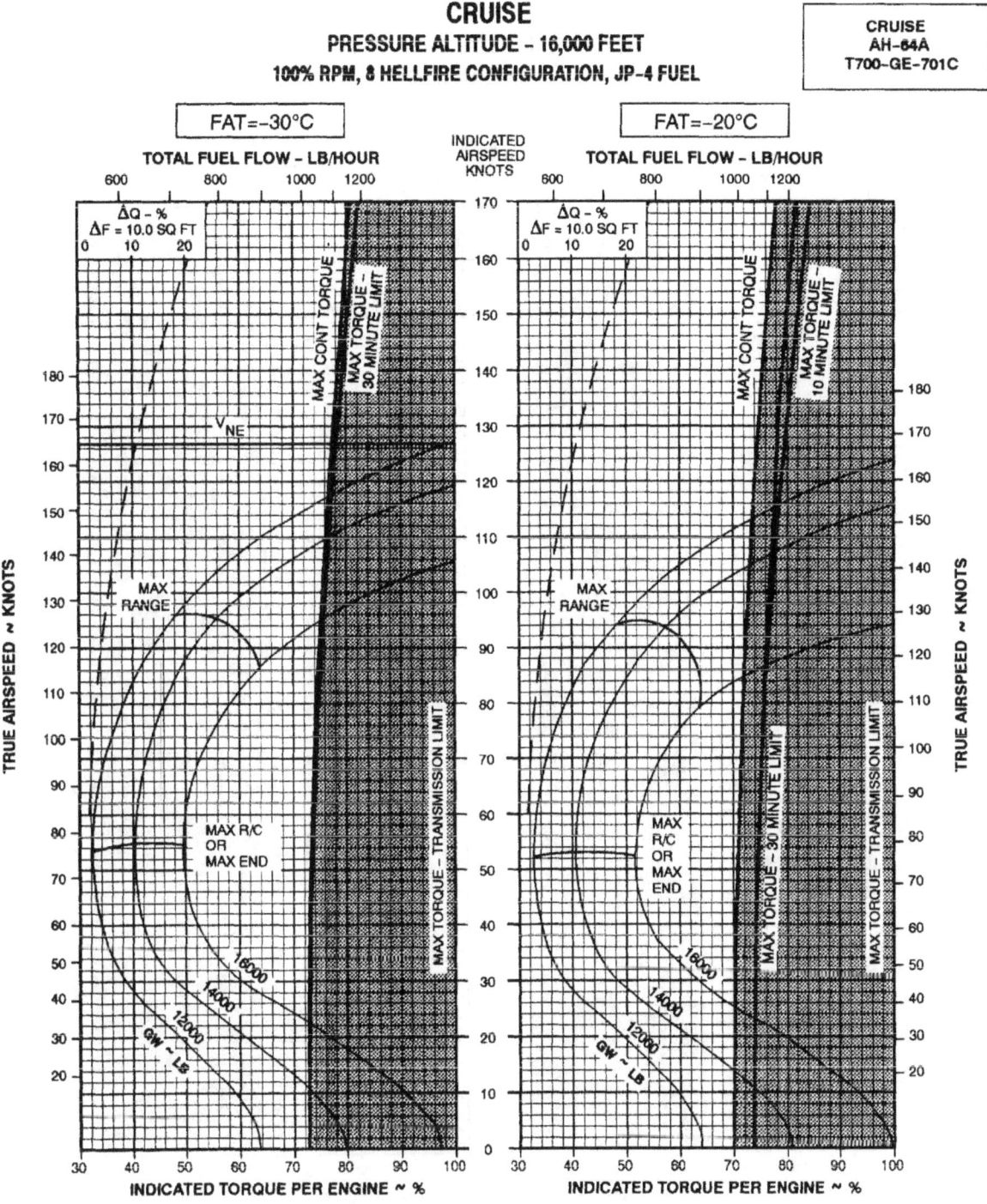

Figure 7A-17. Cruise Chart, 16,000 Feet, -30 °C and -20 °C (Sheet 2 of 4)

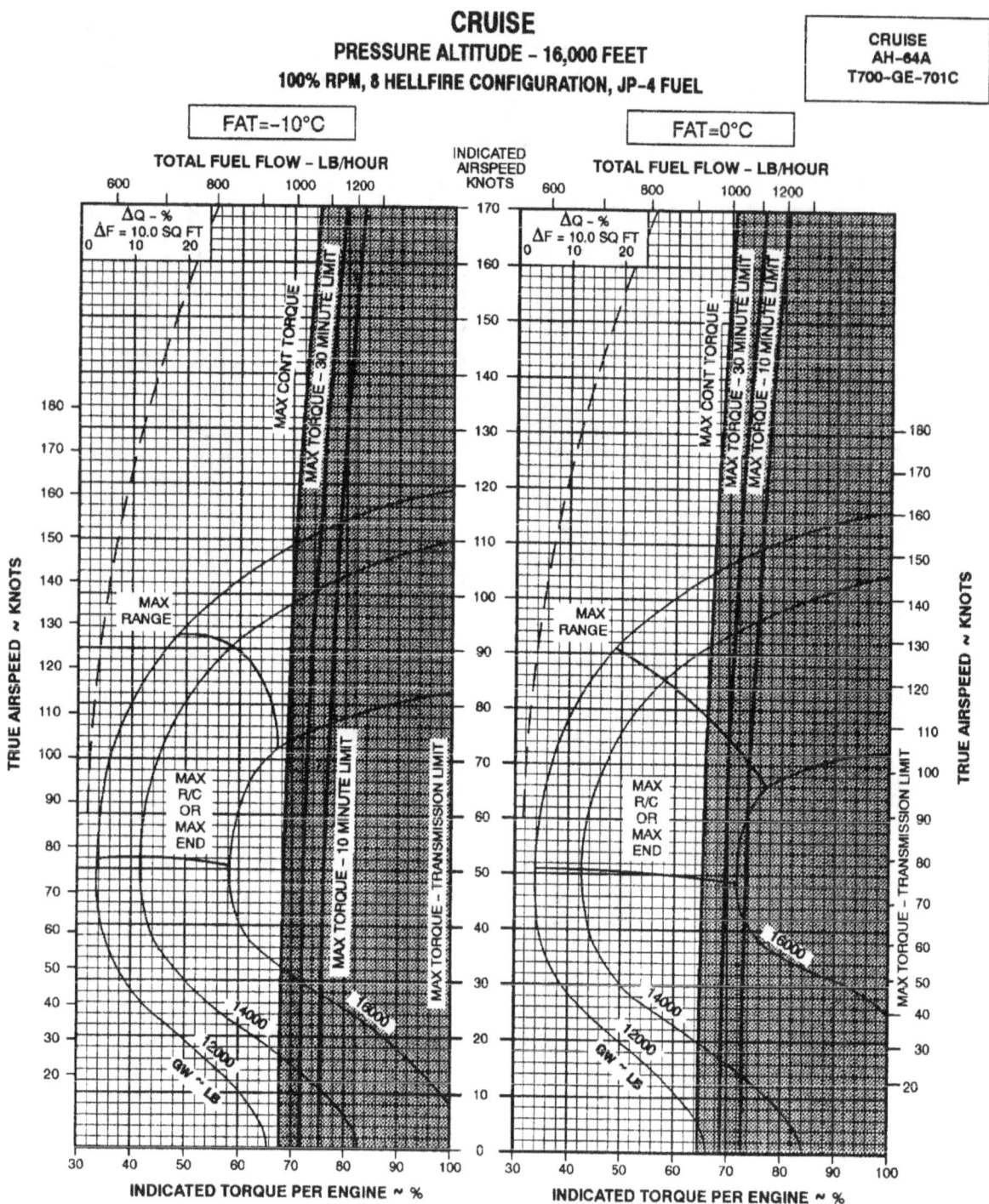

Figure 7A-17. Cruise Chart, 16,000 Feet, -10 °C and 0 °C (Sheet 3 of 4)

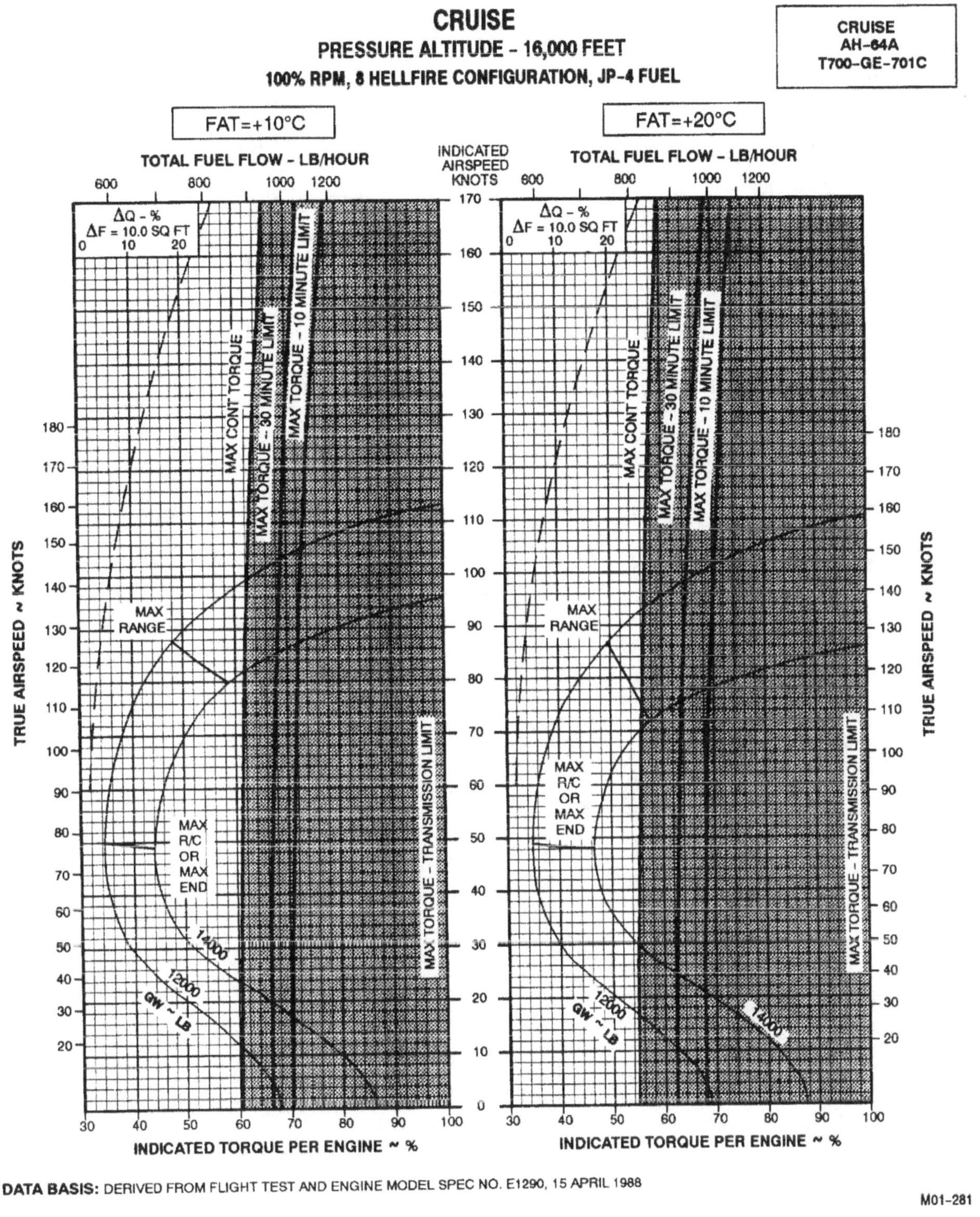

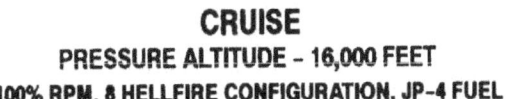

Figure 7A-17. Cruise Chart, 16,000 Feet, +10 °C and +20 °C (Sheet 4 of 4) 701C

TM 1-1520-238-10

Section VI. DRAG

7A.22. DESCRIPTION.

The drag chart (fig 7A-18) shows the change in frontal area for each wing-stores combination that can be installed on the helicopter. The baseline configuration includes the Wire Strike Protection System (WSPS).

7A.23. USE OF CHART.

To determine the change in frontal area (AF) and the associated multiplying factor, it is necessary to know what combination of stores is installed. Enter the chart at the top, move down to the illustration that matches the desired combination, and then move right and read ∧F and the multiplying factor. Use the multiplying factor and data in Section VI, Cruise to determine the resulting change in torque.

7A.24. CONDITIONS.

The drag chart is based on the mission configuration having zero change in frontal area. If the WSPS is installed, increase the frontal area for all configurations in the drag table by 1.9 sq ft and increase the multiplying factor by 0.19.

7A-66 Change 3

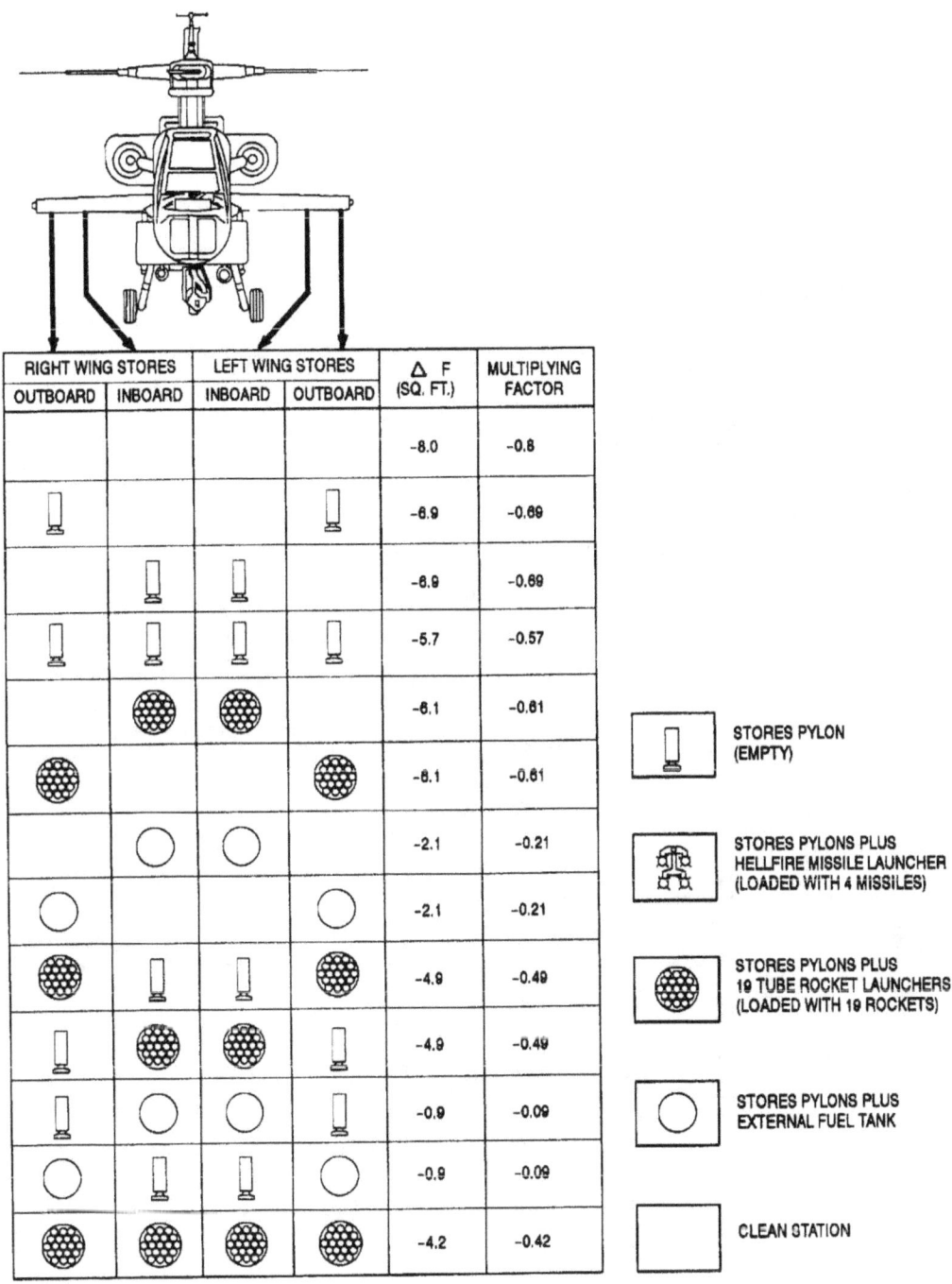

Figure 7A-18. Drag Chart and Authorized Armament Configurations (Sheet 1 of 2)

TM 1-1520-238-10

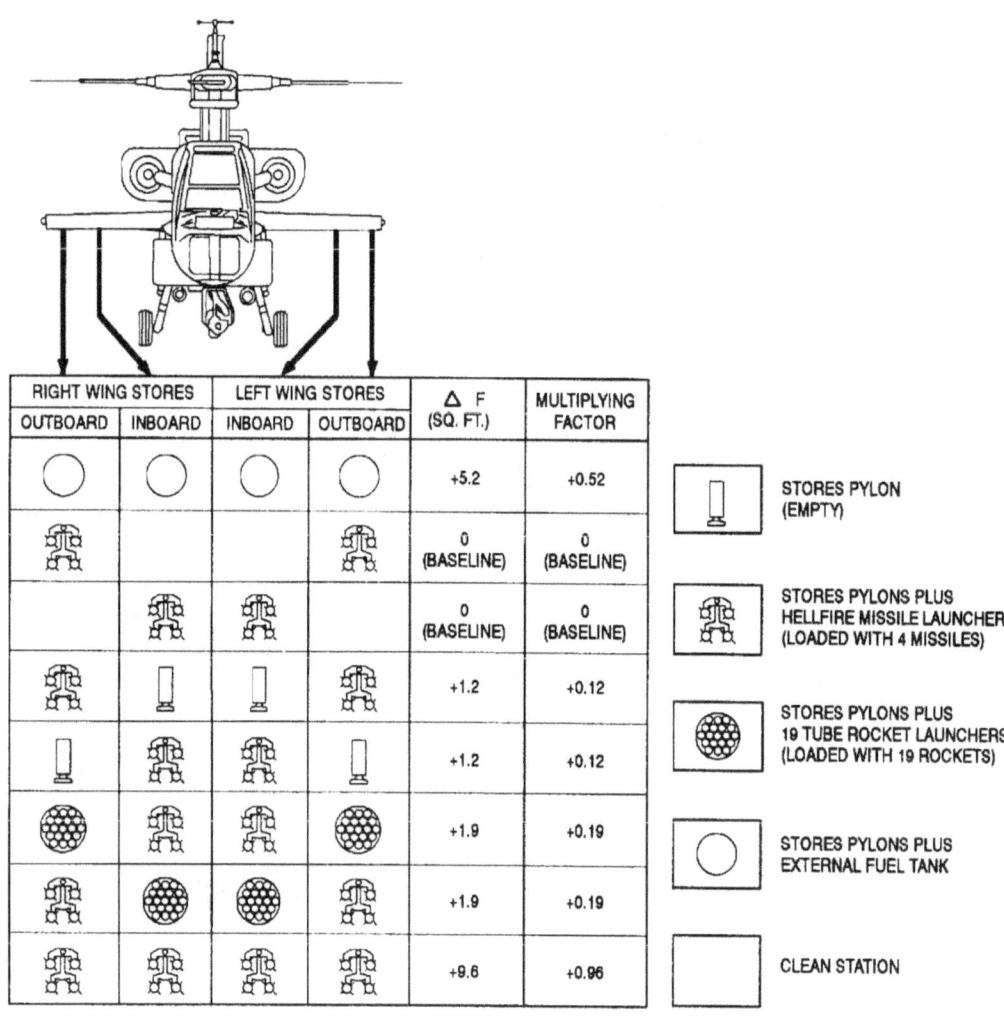

Figure 7A-18. Drag Chart and Authorized Armament Configurations (Sheet 2 of 2)

Section VII. CLIMB-DESCENT

7A.25. DESCRIPTION.

The climb descent chart (fig 7A-19) shows the change in torque (above and below the torque required for level flight under the same configuration, gross weight, and atmospheric conditions) to obtain a desired rate of climb or descent.

7A.26. USE OF CHART.

The primary use of the climb-descent chart is illustrated by the example. To determine the change in torque, it is necessary to know the gross weight and the desired rate of climb or descent. Enter the chart at the desired rate of climb or descent, move right to the known gross weight, and then move down and read the torque change. This torque change must be added to (for climb) or subtracted from (for descent) the torque required for level flight (obtained from the appropriate cruise chart) to obtain a total climb or descent torque.

By entering the chart with a known torque change, and moving up to the known gross weight and then left, the corresponding rate of climb or descent can also be obtained.

7A.27. CONDITIONS.

The climb-descent chart is based on 100% rotor rpm.

CLIMB - DESCENT

CLIMB-DESCENT
AH-64A
T700-GE-701C

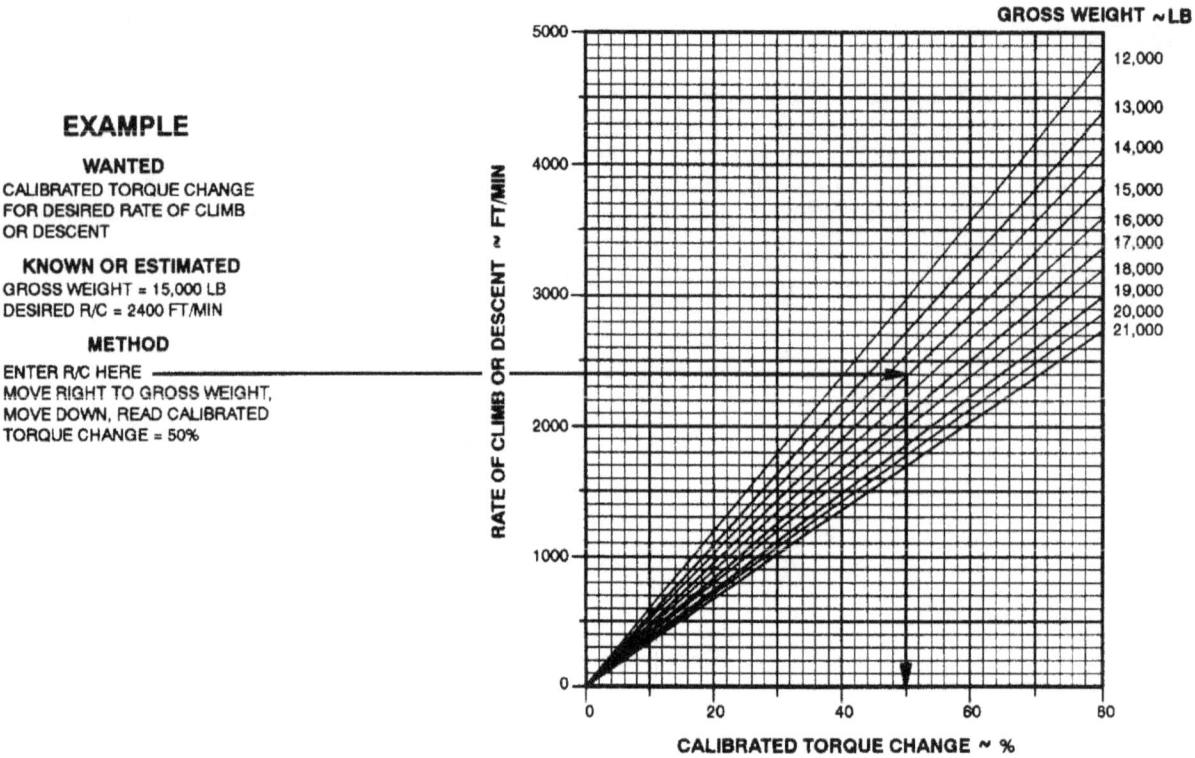

EXAMPLE

WANTED
CALIBRATED TORQUE CHANGE
FOR DESIRED RATE OF CLIMB
OR DESCENT

KNOWN OR ESTIMATED
GROSS WEIGHT = 15,000 LB
DESIRED R/C = 2400 FT/MIN

METHOD
ENTER R/C HERE
MOVE RIGHT TO GROSS WEIGHT,
MOVE DOWN, READ CALIBRATED
TORQUE CHANGE = 50%

REMARK
TORQUE CHANGE IS THE DIFFERENCE BETWEEN
TORQUE USED DURING CLIMB OR DESCENT AND
THE TORQUE REQUIRED FOR LEVEL FLIGHT AT
THE SAME CONDITIONS (ALTITUDE, TEMPERATURE,
AIRSPEED, CONFIGURATION, ETC)

DATA BASIS: DERIVED FROM FLIGHT TEST

Figure 7A-19. Climb-Descent Chart

CHAPTER 8
NORMAL PROCEDURES

Section I. CREW DUTIES

8.1 CREW DUTIES.

8.1.1 PILOT. The pilot in command is responsible for all aspects of mission planning, preflight, and operation of the helicopter. He will assign duties and functions to all other crewmembers as required. Prior to or during preflight, the pilot will brief the CPG on items pertinent to the mission; e.g., performance data, monitoring of instruments, communications, and emergency procedures.

8.1.2 CPG. The CPG must be familiar with the pilots duties. The CPG will assist the pilot as directed.

8.2 CREW BRIEFING.

A crew briefing shall be conducted to ensure a thorough understanding of individual and team responsibilities. The briefing should include, but not be limited to, CPG and ground crew responsibilities, and the coordination necessary to complete the mission in the most efficient manner. A review of visual signals is desirable when ground guides do not have direct voice communications link with the crew.

8.2.1 Crew Introduction.

8.2.2 Equipment.

  a. Personal, to include I.D. tags.

  b. Professional.

  c. Survival.

8.2.3 Flight Data.

  a. Route.

  b. Altitude.

  c. Time enroute.

  d. Weather.

8.2.4 Normal Procedures.

  a. Entry and exit from aircraft.

  b. Seating.

  c. Seat Belts.

  d. Movement in aircraft.

  e. Internal communications.

  f. Security of equipment.

  g. Smoking.

  h. Refueling.

  i. Weapons.

  j. Protective masks.

8.2.5 Emergency Procedures.

  a. Emergency exits.

  b. Emergency equipment.

  c. Emergency landing/ditching procedures.

8.3 MISSION PLANNING.

Mission planning begins when the mission is assigned, and extends to the preflight check of the helicopter. It includes, but is not limited to, checks of operating limits and restrictions; weight, balance, and loading; performance; publications; flight plan; and crew briefings. The pilot in command shall ensure compliance with the contents of this manual that are applicable to the mission.

8.4 AVIATION LIFE SUPPORT EQUIPMENT (ALSE).

All ALSE required for the mission; e.g. helmets, gloves, survival vests, survival kits, etc , shall be checked.

Section II. OPERATING PROCEDURES AND MANEUVERS

8.5 OPERATING PROCEDURES AND MANEUVERS.

This section deals with normal procedures. It includes all steps necessary for safe, efficient operation of the helicopter from the time a preflight check begins until the flight is completed and the helicopter is parked and secured. Unique feel, helicopter characteristics, and reaction of the helicopter during various phases of operation and the techniques and procedures used for taxiing, taking off, climbing, etc., are described, including precautions to be observed. Your flying experience is recognized; therefore, basic flight principles are avoided. Only the duties of the minimum crew necessary for the actual operation of the helicopter are included. Mission equipment checks are contained in Chapter 4, MISSION EQUIPMENT. Procedures specifically related to instrument flight that are different from normal procedures are covered in this section, following normal procedures. Descriptions of functions, operations, and effects of controls are covered in Section IV, FLIGHT CHARACTERISTICS, and are repeated in this section only when required for emphasis. Checks that must be performed under adverse environmental conditions, such as desert and cold-weather operations, supplement normal procedures checks in this section and are covered in Section V, ADVERSE ENVIRONMENTAL CONDITIONS.

8.6 SYMBOL DEFINITIONS.

Items which apply only to night or only to instrument flying shall have an N or an I, respectively, immediately preceding the check to which it is pertinent. The symbol O shall be used to indicate if installed. Those duties which are the responsibility of the CPG, will be indicated by a circle around the step number: i.e. The symbol star ☆ indicates an operational check is required. Operational checks are contained in the performance section of the condensed checklist. The symbol asterisk * indicates that performance of step is mandatory for all thru-flights. The asterisk applies only to checks performed prior to takeoff. Placarded items such as switch and control labels appear in boldface upper case.

8.7 CHECKLIST.

Normal procedures are given primarily in checklist form, and amplified as necessary in accompanying paragraph form, when a detailed description of a procedure or maneuver is required. A condensed version of the amplified checklist, omitting all explanatory text, is contained in the operators checklist. To provide for easier cross-referencing, the procedural steps in the checklist are numbered to coincide with the corresponding numbered steps in this manual.

8.8 PREFLIGHT CHECK.

The pilots walk-around and interior checks are outlined in the following procedures. The preflight check is not intended to be a detailed mechanical inspection. The steps that are essential for safe helicopter operation are included. The preflight may be made as comprehensive as conditions warrant at the discretion of the pilot.

8.9 BEFORE EXTERIOR CHECK.

WARNING

Do not preflight until armament systems are safe.

* 1. Helicopter covers, locking devices, tiedowns and grounding cables – removed and stowed. Pylon safety pins - installed.

2. Cockpit safety - Check as follows:

   a. **BATT** switch - **OFF** (CPG **BAT OVRD - NRML**).

   b. Interior **CANOPY JETTISON** pins - installed.

   c. **ENG FIRE PULL** fire handles - In.

   d. **APU FIRE PULL** handle - In.

   e. **APU START** switch - **OFF**.

3. Armament subsystems - Safe as follows:

   a. PILOT

      (1) **MASTER ARM/SAFE** switch - **OFF**.

      (2) **STORES JETT** switches - **OFF**.

b. CPG:

   (1) **CPG ARM/SAFE** switch - **OFF**.

   (2) **PLT/GND ORIDE** switch - **OFF**.

4. Cockpit - General.

   a. Ignition key - In and **ON**.

   b. First aid kits - Installed.

   c. Seat cushions - Conditions.

   d. Restraint system - Condition.

   e. Canopy - Check.

   f. Loose equipment - Secured.

5. Publications - As required by DA PAM 738-751; locally required forms and publications, and availability of Operator's Manual.

6. Fuel sample - Check for first flight of day.

8.10 EXTERIOR CHECK.

See figure 8-1

8.10.1 Right Side - Under Side Fuselage (Area 1).

1. 30mm gun turret - Check.

   a. Gun mounting - Check.

   b. Feed chute - Check.

2. Searchlight - Check.

8.10.2 Right Side - Lower Center Fuselage (Area 2).

1. Radar warning antenna - Check.

2. Forward avionics bay - Check.

3. Static port - Unobstructed.

4. Right main landing gear - Check.

5. Portable fire extinguisher - Check.

*6. Refueling panel - Secure door.

7. Forward gravity fuel cap - Secure.

8. Single point fuel access - Secure.

8.10.3 Right Side - Mast (Area 3).

*1. Main transmission - Check oil level.

*2. Nose gearbox - Check oil level, oil cap secured, cowling secure.

3. Engine inlet - Unobstructed.

*4. Engine oil level - Check and secure door.

5. Upper flight controls and swashplate - Check.

*6. Main rotor head and blades - Check.

8.10.4 Right Side - Wing (Area 4).

1. Wing - Check.

O 2. Pylons - Check.

NOTE

When icing conditions exist, or are predicted and HELLFIRE operations are expected, launcher arm/safe switch located on each HELLFIRE launcher must be manually placed in the **ARM** position prior to liftoff. It is possible for this switch to be rendered inoperative by icing.

O 3. HELLFIRE - Check as follows:

   a. Launcher **ARM/SAFE** switch - As required.

   b. Launcher mounting - Check that aft and forward attach lugs are secure to rack. Rack swaybrace bolts firmly against launcher swaybrace pads.

   c. Electrical connector - Check that HELLFIRE harness cannon plug is connected to launcher. Jettison quick-disconnect lanyard is attached to connector plug and rack.

TM 1-1520-238-10

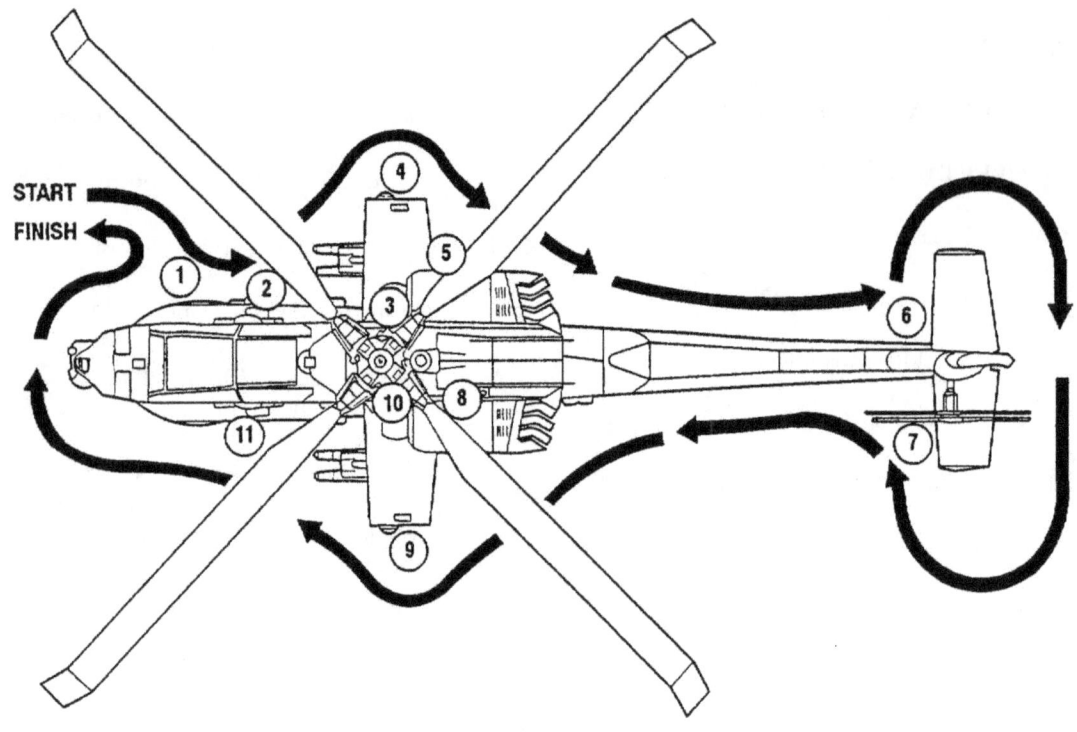

EXTERIOR CHECK
AREA 1   RIGHT SIDE – UNDERSIDE FUSELAGE
AREA 2   RIGHT SIDE – LOWER CENTER FUSELAGE
AREA 3   RIGHT SIDE – MAST AREA
AREA 4   RIGHT SIDE – LOWER CENTER FUSELAGE AND WING
AREA 5   RIGHT SIDE – REAR CENTER FUSELAGE
AREA 6   RIGHT SIDE – AFT FUSELAGE/EMPENNAGE
AREA 7   LEFT SIDE – AFT FUSELAGE/EMPENNAGE
AREA 8   LEFT SIDE – REAR CENTER FUSELAGE
AREA 9   LEFT SIDE – LOWER CENTER FUSELAGE AND WING
AREA 10   LEFT SIDE – MAST
AREA 11   LEFT SIDE – CREW STATION AND LOWER CENTER FUSELAGE

MO1O5A

Figure 8-1. Exterior Check Diagram

d. Missile installation Check that each missile is seated and hold-back latch is locked.

O 4. Rocket launcher Check as follows:

   a. Electrical connector Check that rocket harness connector plug is connected to launcher and jettison quick disconnect lanyard is attached to connector plug and rack.

   b. Launcher Check launcher exterior and tube interiors for damage and corrosion.

   c. Rocket installation Check that rocket aft end is secure in launcher tube aft detent. Note number and zones of rocket loading.

   d. Igniter arms Check for damage and corrosion. Check arms are in contact with rockets.

O 5. External fuel tanks Check.

6. Pitot tube Check tube unobstructed.

7. Wing lighting Check condition of anticollision, navigation and formation lights.

8. Ammunition bay access Secure.

8.10.5 Right Side Rear Center Fuselage (Area 5).

NOTE
Ensure that nacelle access doors are properly latched and secured.

1. Nacelle fire louvers Check open.

*2. APU oil level Check and secure door.

3. Aft gravity fuel cap Check security.

4. Aft avionics bay Secure door.

5. APU exhaust Check.

6. IR suppressor/engine exhaust Check.

*7. Utility hydraulic accumulator Check hydraulic pressure (2600 psi minimum).

8. Survival kit bay Check and secure door.

9. External power receptacle Access door closed if external power source is not used.

10. Belly antennas Check.

8.10.6 Right Side Aft Fuselage/Empennage (Area 6).

1. Aft tailboom and empennage (right side) Check.

2. Stabilator Check.

3. Tail landing gear Check.

8.10.7 Left Side Aft Fuselage/Empennage (Area 7).

1. Empennage Check.

O 2. FM-AM whip antenna Check.

O 3. GPS antenna - Check.

4. Tail rotor, controls, hub, and blades Check.

5. Stabilator Check.

8.10.8 Left Side Rear Center Fuselage (Area 8).

NOTE
Ensure that nacelle access doors are properly latched and secured.

1. Aft tailboom Check.

2. Transmission deck catwalk area for FOD, fire bottles for charge and APU enclosure for security.

3. Transmission deck catwalk doors Check security.

4. Survival kit bay Check and secure door.

5. IR suppressor/engine exhaust Check.

6. Aft storage bay Check and secure door.

7. Nacelle fire louvers Check open.

8. Fire extinguisher disc Check that yellow disc is visible.

9. Ammunition bay access Secure.

TM 1-1520-238-10

8.10.9 Left Side Wing (Area 9).

1. Wing Check.
2. Wing lighting Check anti-collision, navigation, and formation lights.
3. Pitot tube Check.
- O 4. Pylons Check.
- O 5. HELLFIRE Check, same as right side.
- O 6. Rocket launcher Same as right side.
- O 7. External fuel tanks Check.

8.10.10 Left Side Mast (Area 10).

- *1. Main transmission Check oil level.
- *2. Nose gearbox Check oil level; secure oil cap and cowling.
- *3. Primary hydraulic manifold Check oil level.
- 4. Engine inlet Unobstructed.
- *5. Engine oil level Check and secure door.
- 6. Upper flight controls and swashplate Check.
- 7. Main rotor head and blades Check.
- 8. Air data sensor Check.

8.10.11 Left Side Lower Center Fuselage and Nose (Area 11).

1. Canopy Check.
2. OAT gauge extension Check security.
3. Static port Unobstructed.
4. Utility hydraulic accumulator Check.
5. Left main landing gear Check.
6. Static ground cable Check.
7. Forward avionics bay Check.
8. Radar warning antenna Check.

NOTE
When icing conditions exist, ensure that TADS/PNVS gear teeth are free of ice.

9. TADS/PNVS turrets Check.
10. Crew briefing Complete as required.

8.11 INTERIOR CHECK PILOT.

- *1. Canopy door As desired.
- *2. Loose equipment Secured.
- 3. Seat Adjust to design eye position.
- *4. Restraint harness Fasten and adjust.

NOTE
Routing the IHADSS HDU cable under the right arm may cause entanglement during emergency egress.

5. Inertial reel lock Check.
6. Pedals Check and adjust.
- *7. **PARK BRAKE** Set.
- 8. **EDGE LT PNL** switch As desired.

NOTE
The left and right nose gearbox heater circuit breakers shall be opened unless the system is required.

- *9. Overhead circuit breakers As desired.
- 10. Collective switches As desired.

CAUTION
Physically confirm that engine chop collar is seated in its latched/centered position and safetied.

11. Auxiliary vent handle Closed.
12. Utility light As desired.
13. OAT gauge Check.
14. ANTI-ICE panel switches OFF.
- *15. EXT LT and INTR LT panel switches and controls As desired.
- *16. FUEL panel switches Set as follows: a. EXT TK switch OFF.

b. **TRANS** switch - **OFF**.

c. **CROSSFEED** switch - **NORM**.

d. **ENG 1** switch - **ON**.

e. **ENG 2** switch - **ON**.

17. **PWR** levers - **OFF**.

18. **ENG START** switch - **OFF**.

*19. **MASTER IGN** switch - Verify **ON**.

20. **RTR BK** - **OFF**.

21. **ELEC PWR** panel switches - Check switch **OFF**.

22. **STORES JETT** select switches - Guard covers down.

*23. **ROCKETS** control panel - Set.

24. **ECS** control panel - Set as follows

 a. **ENCU** - **ON**.

 b. **FAN** - **NORM**.

 c. **TEMP** control - As desired.

25. **TAIL WHEEL** switch - **LOCK**.

*26. **CANOPY JETTISON** pin - Remove and stow.

27. **FIRE BTL** select switch - Centered.

28. **FIRE CONTROL** panel - Set as follows:

 a. **SIGHT SEL** switch - **STBY**.

 b. **ACQ SEL** switch - **OFF**.

 c. **VID SEL** switch - **PLT**.

 d. **ACM** switch - **OFF**.

 e. **PNVS** switch - **OFF**.

 f. **IHADSS BRSIT** switch - **OFF**.

 g. **RKT** select switch - **OFF**.

 h. **GUN** select switch - **OFF**.

 i. **MSL** select switch - **OFF**.

29. Magnetic compass - Check.

30. BRU - Check.

31. Engine instrument test panel switch - As desired.

32. Flight instruments - Check or set as follows:

 a. Airspeed indicator.

 b. Standby attitude indicator - Cage.

 c. **VDU** - **OFF**.

 d. Radar altimeter - **OFF**.

 e. Altimeter - Check.

 f. Vertical speed indicator.

 g. HSI.

 h. Stabilator position indicator and placard.

33. Clock - Set.

34. Accelerometer - Reset.

35. **HARS** - **OFF**.

36. **EMERG HYD** switch - **OFF**:

37. **CSC** panel switches - As desired.

38. Right console avionics - OFF. Set frequencies as desired.

8.12 INTERIOR CHECK - CPG.

*1. Canopy door - As desired.

*2. Loose equipment - Secured.

3. Seat - Adjusted to design eye position.

*4. Restraint harness - Fastened and adjust.

NOTE

Routing the IHADSS HDU cable under the right arm may cause entanglement during emergency egress.

5. Inertial reel lock - Check.

6. Pedals - Check and adjust.

7. Collective switches - As desired.

CAUTION

Physically confirm that engine chop collar is seated and safetied in its latched/centered position.

*8. Circuit breakers - As desired.

9. Utility light - As desired.

*10. **INTR LT** panel - As desired.

*11. **FUEL** panel switches - Set as follows:

   a. **ORIDE** - **PLT**.

   b. **TRANS** - **OFF**.

   c. **BOOST** - **OFF**.

   d. **TK SEL** - **NORM**.

12. **PWR** levers - **OFF**.

13. **EMER HYD PWR** switch - **OFF**, guard down.

14. **BAT OVRD** switch - **NRML**, guard down.

15. **ANTI-ICE** panel switches - Set as follows:

   a. **TADS/PNVS GND** - **OFF**.

   b. **W WIPER** - **PLT**.

16. **AUX** panel switches - Set as follows:

   a. **STBY FAN** - **OFF**.

   b. **ADSS** - **OFF**.

17. **RECORDER** panel switches - As desired.

18. **MSL** panel switches - Set as follows:

   a. **TYPE** - **LASER**.

   b. **MODE** - **STBY**.

   c. **LOAL** - **OFF**.

19. **DATA ENTRY** keyboard - **OFF**.

*20. **CANOPY JETTISON** pin - Remove and stow.

21. **ENG FIRE PULL** handles - In.

22. **FIRE BTL** select switch - Centered.

23. **FIRE CONTROL** panel switches - Set as follows:

   a. **RKT** - **OFF**.

   b. **GUN** - **OFF**.

   c. **MSL** - **OFF**.

   d. **LSR** - **OFF**.

   e. **SIGHT SEL** - **STBY**.

   f. **ACQ SEL** - **FXD**.

   g. **MUX** - **PRI**.

   h. **FCC/MUX** - **ON**.

   i. **BRSIT IHADSS** and **TADS** - **OFF**.

   j. **LSR MSL CCM** - **OFF**.

   k. **PLT/GND ORIDE** - **OFF**.

   l. **LRF/D CCM** - **OFF**.

   m. **FC SYM GEN** - **OFF**.

   n. **IHADSS** - **OFF**.

   o. **TADS** - **OFF**.

24. Engine instrument test panel switch - As desired.

25. Engine instruments - Check.

26. Flight instruments - Check or set as follows:

   a. Airspeed indicator.

   b. Attitude indicator.

c. RMI.

d. Altimeter - Check.

e. Vertical velocity indicator.

f. Stabilator position indicator and placard.

27. Clock - Set.

28. **LT-OFF**.

29. **CSC** panel switches - As desired.

30. Right console avionics - OFF. Set to desired frequencies.

31. **DPLR/NAV MODE** select switch - **OFF**

8.13 BEFORE STARTING APU - PILOT.

*1. **BATT/EXT PWR** switch - **BATT** position. (**EXT PWR** if external power is to be used. Depress **RESET** button after **EXT PWR** is applied.)

2. ICS system - Check.

*3. **MASTER CAUTION** panel - Check the following are illuminated:

a. **LOW RPM ROTOR**.

b. **ENG 1 OUT**.

c. **ENG 2 OUT**

4. **MASTER CAUTION** switch - **PRESS TO TEST** - Check that all caution/warning and advisory lights illuminate.

5. Caution/warning panel - Check that the following segments illuminate:

a. **PRI HYD PSI**.

b. **UTIL HYD PSI**.

c. **OIL PSI ACC PUMP**.

d. **OIL PSI NOSE GRBX 1**.

e. **OIL PSI MAIN XMSN 1**.

f. **OIL PSI MAIN XMSN 2**.

g. **OIL PSI NOSE GRBX 2**.

h. **OIL PSI ENG 1**.

i. **OIL PSI ENG 2**.

j. **GEN 1/(RECT 1 - BATT)**.

k. **GEN 2/(RECT 2 - BATT)**

l. **FUEL PSI ENG 1**.

m. **SHAFT DRIVEN COMP.**

n. **MAN STAB** (BATT).

o. **FUEL PSI ENG 2**.

p. **ENG 1 ANTI ICE**.

q. **ENG 2 ANTI ICE**.

r. **CANOPY** (if open).

s. **EXT PWR** (EXT PWR).

t. **ADS** (EXT PVVR).

u. **CHARGER** (EXT PWR).

v. **PNVS** (EXT PWR).

w. **TADS** (EXT PWR).

6. Fire detectors - Test as follows:

a. **FIRE TEST DET** switch - lb position **1**. Check that all fire handles and APU fire warning lights in both crew stations illuminate.

b. **FIRE TEST DET** switch - lb position **2**. Check same as in step a. above.

7. Engine instrument test panel switch - **TST**. Check engine/rotor instrument vertical scale readings and segments light and digital display **(888)** illuminates.

8. Utility hydraulic pressure gauge - Check 2600 psi minimum.

8.14 BEFORE STARTING APU – CPG.

*1. **MASTER CAUTION** panel – Check that the following are illuminated.
   a. **LOW RPM ROTOR**.
   b. **ENG 1 OUT**.
   c. **ENG 2 OUT**.

2. **MASTER CAUTION** panel: **PRESS TO TEST** – Check that all caution/warning lights illuminate.

3. Caution/warning panel – Check that the following segments are illuminated.
   a. **PRI HYD**.
   b. **UTIL HYD**.
   c. **MAN STAB** (BATT).
   d. **MAIN XMSN 1**.
   e. **MAIN XMSN 2**.
   f. **ENG 1**.
   g. **ENG 2**.
   h. **ELEC SYS FAIL** (BATT).
   i. **ENG ANTI ICE**.
   j. **ADS** (EXT PWR).
   k. **TADS** (EXT PWR).

4. Engine instrument test panel switch – **TST**. Check engine/rotor instrument segment lights and digital display (**888**) illuminates.

CAUTION

If an engine is shut down from a high power setting (above 90% N_G) without being cooled for two minutes at IDLE, and it is necessary to restart the engine, the restart should be accomplished within five minutes after shutdown. If the restart cannot be accomplished within five minutes, the engine shall be allowed to cool for four hours before attempting an engine restart.

8.15 STARTING ENGINES – EXTERNAL PRESSURIZED AIR SOURCE – PILOT.

NOTE

Control sweeps and BUCS self-test are not mandatory prior to external air start. If checks are desired, connect a 3000 psi external hydraulic source to the primary hydraulic system to accomplish the checks.

1. External air source – Connected to helicopter.
2. Pressurized air – Verify available at helicopter.
3. **B** Control locks – Remove.
L4. Engines – Start same as normal (refer to paragraph 8.20).
5. **GEN 1** and **GEN 2** switches – **GEN 1** and **2**.
6. **EXT PWR/BATT** switch – To **BATT** if external power used for start.
7. External power – Disconnect.
8. External air source – Disconnect.
9. Continue with AFTER STARTING APU – Pilot and CPG (refer to paragraphs 8.17 and 8.18).

*8.16. STARTING APU – PILOT.

NOTE

During cold weather starts at temperatures below 0 °F (–18 °C), the **95%** switch is used to manually prevent PTO clutch engagement until the APU has reached 95% N_G. Prior to normal clutch engagement, place and hold the spring loaded **95%** switch to the **95%** position. When the **APU ON** advisory light illuminates, release the switch to allow clutch engagement. This procedure prevents shut down of the APU at low speeds under extreme cold conditions.

1. Fire guard – Posted, if available.
2. APU – Start as follows:

If ENG START LIGHT remains illuminated after reaching 66 – 68% N_G, set ENG START switch for affected engine to IGN ORIDE, then OFF.

   a. **APU** switch – Set to **RUN**, pause, set to **START**, then release.
   b. **APU FAIL** caution light – Within 5 seconds of start, check extinguished.
   c. **APU ON** caution light – Check that light illuminates.

3. **GEN 1** and **GEN 2** switches – **GEN 1** and **2**.

NOTE

701C ENGINES ONLY

Generators ON prior to engine start may result in an erroneous TGT indication or a mismatch between pilot and CPG TGT gauges. The indicated TGT will be accurate when the actual TGT, minus the 71 °C DECU bias equals a positive number.

4. **B** Control locks – Remove.
5. **EXT PWR/BATT** switch – Set to **BATT** if external power was used for start.
6. External power – Disconnect. **EXT PWR** caution light should be off.

*8.17. **AFTER STARTING APU – PILOT.**

| CAUTION |

Do not turn PNVS power on immediately after power was turned off. This will damage the PEU. If PNVS switch was just set to OFF, Wait a minimum of 10 seconds before doing step 1 below.

NOTE

If both TADS/PNVS are to be used simultanously, TADS must be turned on prior to PNVS power up to ensure proper operation of both systems.

1. **PNVS** switch – As required. (Verify TADS on)
2. Standby attitude indicator – Uncage.
3. VDU switch – As desired.
4. Radar altimeter – On.
5. Avionics – As desired.
6. Canopy door – Secure.

| CAUTION |

- **B** Ensure both pilot and CPG control locks are removed and that both pilot and CPG collective friction is off.
- **B** Do not execute control sweep or BUCS test unless rotor is completely stopped.

NOTE

Control motions during the BUCS self-test should be rapid and positive. Any jerkiness or oscillations during movement indicates possible fault in the subsystem.

7. FCC – Verify CPG entered present position.

8. Control sweep – **FORCE TRIM** – **OFF**. Check cyclic, collective and pedals for freedom of movement. Neutralize cyclic and pedals; place collective full down. Position indicators on base of cyclic are not used for centering. **FORCE TRIM** – **ON** and operational.
9. Stabilator – Check for full travel and position indicator function – Reset.
10. **HARS** control switch – for **-49A** and previous FCC software select **NORM**. For emerergency operations select **FAST**. For **-51** FCC software select **NORM** for stationary starts, or **FAST** for airborne or moving starts.
L11. **B** BUCS – Test as follows:
   a. **RTR BK** switch – **BRAKE**.
   b. Controls – Friction off, centered, and cleared.
   c. **BUCS TST** switch – **PLT** and hold. Warn CPG that flight controls will move.
   d. Observe **BUCS ON** caution light illuminates. Hold switch in **PLT** position until **BUCS ON** caution light is not illuminated (approximately 20 seconds). If **BUCS FAIL** warning light illuminates during the test, do not fly helicopter.
   e. If after 15 seconds, the **BUCS FAIL** warning light does not illuminate, proceed to step f.
   f. **BUCS TST** switch – **CPG** and hold. Warn CPG that flight controls will move.
   g. Observe **BUCS ON** caution light illuminates. Hold switch in **CPG** position until **BUCS ON** caution light is not illuminated (approximately 20 seconds). If **BUCS FAIL** warning light illuminates during the test, do not fly helicopter.
   h. BUCS select trigger (CPG) – Press. Verify illumination of **BUCS FAIL** warning light in both crew stations.
   i. Collective – Full down.
   j. **RTR BK** switch – As desired.
L12. IHADSS boresight – As required.

NOTE

The PNVS needs 1 minute for gyro runup before PNVS turret assembly can be commanded out of stow.

L13. FLIR operational check – As required.
14. Radar altimeter – Test.

*8.18. AFTER STARTING APU – CPG.

1. **B** Control locks – Remove.

2. Canopy door – Secure.

3. Avionics – As desired.

4. **ADSS** switch – On.

5. **FC SYM GEN** switch – As required.

6. **IHADSS** switch – As required (Announce to pilot).

NOTE

Do not turn the SYSTEM TADS/FLIR-OFF/OFF switch to TADS immediately after being set to OFF. Damage to the TADS power supply could result.

7. **TADS** switch – As required (Announce to pilot).

NOTE

If an erroneous magnetic variation is entered, the compass card in the HSI will give erroneous readings.

L8. Fire control system – Enter data/interrogate as desired for:

    a. PPOS. (Announce to pilot when present position has been entered.)

    b. Waypoint/target data.

    c. Laser codes.

NOTE

Steps 9 thru 15 may be performed in any order.

L9. IHADSS boresight – As required.

10. Doppler – Program as desired.

L11. TADS operational checks – As required.

L12. TADS internal boresight – As required.

L13. TADS outfront boresight – As required.

L14. FLIR operational check – As required.

15. Weapons Systems – As desired.

*8.19. BEFORE STARTING ENGINES – PILOT.

1. **SHAFT DRIVEN COMP** caution light – Extinguished. Verify ECS airflow from crew station vents.

2. **ANTI COL** switch – As desired.

*8.20. STARTING ENGINES – PILOT.

1. Area – Clear.

CAUTION

- During a start with RTR BK switch set to LOCK, if rotor blades begin to rotate, set RTR BK switch to OFF.

- The T700-GE-701C engine exhibits inconsistent starting capability above 6000 feet density altitude. Starts above this density altitude may be unsuccessful and require "over temperature" abort by the pilot.

2. **RTR BK** switch – **OFF** or **LOCK**.

NOTE

Use the procedures in para 3.a. thru 3.e. for COLD and WARM -701 **701** engine starts and for COLD -701C **701C** engine starts (more than 4 hours since last shutdown) and all INFLIGHT **701** and **701C** engine starts. Use the procedures in para 3.f.(1) thru 3.f.(4) for WARM **701C** engine starts on the GROUND.

L3. First engine – Start as follows:

    a. **START** switch – **START**.

    b. **PWR** lever – **IDLE** after N_G speed increases and TGT is below 150 °C **701**.

    c. **ENG OIL** pressure gauge – Monitor.

    d. **TGT** gauge – Monitor.

    e. N_G gauge – Monitor.

> **CAUTION**
>
> If ENG START LIGHT remains illuminated after reaching 66 - 68% N_G, set ENG START switch for affected engine to IGN ORIDE, then OFF.

NOTE

- The **MASTER CAUTION** and **FUEL PSI ENG 1** or **2** caution light on pilot caution/warning panel may illuminate during start. **FUEL PSI ENG 1** or **2** caution light should extinguish during start of the respective engine.

- **701** For helicopters with -701 engines, a full scale torque spike may occur during engine start.

- **701C** For helicopters with -701C engines, the torque spike will not occur.

    f. **701C** WARM ENGINE START. Less than 4 hours since last shutdown. Do not use this procedure for inflight restarts. Start procedures are as follows:

    (1) **START** switch - **IGN OVRD** until N_G reaches 18 - 20%.

    (1) **START** switch - **OFF**. Allow N_G to spool down below 5%.

    (1) **START** switch - **START**.

    (1) **PWR** lever - **IDLE** after N_G speed increases and TGT is below 80 °C.

    g. Caution/warning lights - Check.

NOTE

The second engine may be started with the **RTR BK** switch set to **LOCK**, as desired or required. Do not advance either **PWR** lever to **FLY** until **RTR BK** switch is **OFF**.

*4. Second engine - Start same as step 3, above.

5. **RTR BK** switch - **OFF**

> **CAUTION**
>
> Prior to advancing PWR levers, ensure both engines are stabilized (N_p, N_G, torque, and oil pressure). While advancing PWR levers to FLY, ensure both engines indicate a torque rise to confirm that the sprag clutches are engaged. If an engine indicates near 0 torque, retard PWR lever of affected engine to OFF and shut down the helicopter.

6. **PWR** levers - **FLY**. Advance both **PWR** levers smoothly to **FLY** and ensure both torques increase simultaneously.

7. N_p and N_r - 100%.

8. Caution/warning lights - Check.

9. **ANTI-ICE** panel/**ENG INLET** switch - As required.

10. **APU** control switch - **OFF**.

*8.21. **BEFORE TAXI CHECK.**

1. Armament and pylon safety pins - Removed.

2. Chocks and external ICS cords - Removed and disconnected.

3. **HARS** control switch - Check aligned then advance to **OPR**.

4. DASE - As desired.

NOTE

- HIT/ANTI-ICE checks while operating in adverse conditions (e.g., dust, desert, coastal beach area, dry river beds) may be deferred (maximum 5 flight hours) at the discretion of the pilot in command until a suitable location is reached.

- In sandy or dusty conditions, it is advisable to perform the HIT check while airborne.

5. HIT check - As required. Perform for first flight of day. Refer to HIT log in aircraft logbook. This step may be performed at any time prior to takeoff.

6. **ASE** panel switches - As desired.

7. **EXT LT** switches - As desired.

8. **PARK BRAKE** - Release.

9. **TAIL WHEEL** switch - **UNLOCK**. Note that green advisory light illuminates.

NOTE

If the advisory light fails to illuminate, taxi forward a short distance while making light pedal inputs.

*8.22. TAXI CHECK.

1. Wheel brakes - Check.

2. Engine/rotor instruments - Check.

3. Flight instruments - Check.

8.23 TAXI.

CAUTION

- Excessive cyclic displacement with low power settings may result in droop stop pounding.

- Excessive forward cyclic displacement with low power settings may result in high strap pack loads.

Initiate taxi by increasing collective to 20 to 24% torque and applying slight forward cyclic pressure as required to begin aircraft movement. To maintain forward movement or taxi speed, do not reduce torque below 20%. Control aircraft heading with pedals and speed with a combination of collective, cyclic and brakes. Deceleration may be controlled by using collective, aft cyclic, and if required, brakes. Forward cyclic should be limited to the position used to initiate taxi. Use slight lateral cyclic into turns to maintain a level fuselage attitude. Taxiing on soft, rough or sloping terrain may require the use of more collective than on smooth, level surfaces.

*8.24. BEFORE TAKEOFF CHECK.

1. **HARS** switch - Verify **OPR**.

2. Weapons systems - Safe.

   a. Pilot **MASTER ARM** switch - **OFF** or **SAFE**.

   b. **CPG ARM/SAFE** switch - **OFF** or **SAFE**.

   c. PLT and CPG weapons select switches - As desired.

   d. Ensure weapons not actioned.

3. **TAIL WHEEL** switch - **LOCK**.

4. **PARK BRAKE** - As desired

5. Systems check as follows:

   a. **FUEL** panel switches.

   b. Fuel quantity.

   c. Engine instruments.

   d. Caution/warning panel.

6. **PWR** levers - **FLY**.

7. If the DTC overwrites the active fly-to or target, it is necessary to de-select and re-select the active fly-to or target.

8. Power check - Perform. The power check is done by comparing indicated torque required to hover with the predicted values from performance charts in Chapter 7 **701** or Chapter 7 A **701C**.

8.25 BEFORE LANDING CHECK.

NOTE

Prior to landing, external stores may be placed in the ground stow position by placing the rocket select switch on either fire control panel in the **GND STOW** position.

1. Weapons systems - Safe.

   a. Pilot **MASTER ARM/SAFE** switch - **OFF** or **SAFE**.

   b. **CPG ARM/SAFE** switch - **OFF** or **SAFE**.

   c. PLT and CPG weapon select switches - As desired.

   d. Ensure weapons are not actioned.

2. **TAIL WHEEL** switch - **LOCK**.

3. **PARK BRAKE** - As required.

8.26 AFTER LANDING CHECK.

1. **TAIL WHEEL** switch – As required.
2. **EXT LT** controls – As required.
3. Avionics – As required.
4. **ANTI-ICE** panel **TADS/PNVS** switch – **OFF**.
5. **ASE** panel switches – As Required.

8.27 ENGINE SHUTDOWN – PILOT.

NOTE

- To ensure availability of PAS during engine shutdown, monitor ECS airflow before, during and after APU start up.
- Normal ECS airflow through crew station vents approximately one minute after APU start indicates PAS is available during engine shutdown.

1. **TAIL WHEEL** switch – **LOCK**.
2. **PARK BRAKE** – Set.
3. **APU** control switch – **START**, then release.
L 4. Weapons systems switches – Secure as follows:
   a. **SIGHT SEL** switch – **STBY**.
   b. **PNVS** – Off (announce to CPG).
   c. **ACQ SEL** – **OFF**.
   d. **VID SEL** – **PLT**.
   e. **ACM** – **OFF**.
   f. Weapons select switches – **OFF**.
   g. **MASTER ARM/SAFE** switch – **OFF**.
5. DASE release switch – Press.
6. Standby attitude indicator – Cage.
7. VDU – **OFF**.
8. Radar altimeter – **OFF**.
9. **HARS** control switch – **OFF**.
10. **APU ON**, caution/warning light – On.
11. **SHAFT DRIVEN COMP** caution/warning light – Extinguished.
12. **PWR** levers – **IDLE**.

CAUTION

If an engine is shut down from a high power setting (above 90% N_G) without being cooled for two minutes at IDLE, and it is necessary to restart the engine, the restart should be accomplished within five minutes after shutdown. If the restart cannot be accomplished within five minutes, the engine shall be allowed to cool for four hours before attempting an engine restart.

13. **PWR** levers – **OFF** after engines have cooled for two minutes.
14. **FUEL** panel switches – Set as follows:
    a. **EXT TK** switch – **OFF**.
    b. **TRANS** switch – **OFF**.
    c. **CROSSFEED** switch – **NORM**.
15. TGT – Monitor.
16. **RTR BK** switch – **BRAKE**, below 50% N_r.
17. Avionics – OFF.
18. Set stabilator to **0°** (zero).
19. Confirm with CPG that shutdown is complete.
20. **RTR BK** switch – **OFF** (when rotor stops).
21. **SRCH LT** switch – **STOW**.
22. **B** Control locks – Install.
23. Torque gages – Note DECU fault codes **701C**.
24. **GEN 2** and **GEN 1** switches – **OFF**.
25. **APU** control switch – **OFF**.
26. **BATT/EXT PWR** switch – **OFF**.
27. Ignition key – **OFF** and remove.
28. **CANOPY JETTISON** pin – Install.
29. Light switches – Off.

8.28 ENGINE SHUTDOWN – CPG.

1. **SIGHT SEL** switch – **STBY**.

   NOTE

   - For TADS OI aircraft, monitor HOD CRT as **TADS/FLIR** switch is placed in **OFF**. After approximately 15 seconds, HOD should flash and loss of SYM BRT control should be experienced. When symbol brightness is no longer adjustable, the system is in independent HOD mode and power-down can be continued.

   - Prior to turning TADS off, ensure PNVS is turned off to allow proper data transfer into the TADS non-volatile memory.

2. **TADS** switch – **OFF** (verify PNVS off).

   NOTE

   Prior to turning IHADSS off, confirm TADS power down sequence is complete by waiting 15 seconds after TADS is turned off and verifying ORT SYM BRT switch is inoperative.

3. ORT **SYM BRT** switch – Inoperative.
4. **IHADSS** switch – **OFF**.
5. **FC SYM GEN** switch – **OFF**.
6. Weapons select switches – **OFF**.
7. **CPG ARM/SAFE** switch – **OFF**.
8. **PLT/GND ORIDE** switch – **OFF**.
9. **ADSS** switch – **OFF**.
10. **MSL MODE** switch – **STBY**.
11. **RECORDER MODE** switch – **OFF**.
12. Avionics – Off.
13. **B** Control locks – Install.
14. **CANOPY JETTISON** pin – Install.
15. Light switches – Off.

8.29 BEFORE LEAVING THE HELICOPTER.

1. Armament and pylon safety pins – Installed.
2. Conduct walkaround.
3. Complete forms. An entry in DA form 2408-13 is required if helicopter was:

   a. Flown in loose grass environment.

   b. Operated within 10 NM of salt water.

   c. Operated within 200 NM of volcanic activity.

   d. Exposed to radioactivity.

4. Secure helicopter – As required.

Section III. INSTRUMENT FLIGHT

8.30 INSTRUMENT FLIGHT.

This helicopter is qualified for operation in instrument meteorological conditions.

8.31 INSTRUMENT FLIGHT PROCEDURES.

Refer to AR 95-1 and FM 1-240.

Section IV. FLIGHT CHARACTERISTICS

8.32 FLIGHT CHARACTERISTICS - GENERAL.

The safe maximum operating airspeed range is described in Chapter 5, Section V.

8.33 STABILATOR OPERATION.

The stabilator is normally operated in the automatic mode. However, the two additional modes available to the pilot can improve helicopter flight characteristics during certain maneuvers. These are:

8.33.1 NOE/APPR Mode. If the pilot desires to improve his over-the-nose visibility for landings or during NOE flight, the NOE/APPR mode may be engaged at any time. An additional benefit is improved forward speed control in NOE flight.

8.33.2 Manual Mode. The manual mode of operation is particularly useful for positioning the stabilator to help minimize airframe vibrations when hovering in crosswinds or tailwinds.

8.34 SLOPE/ROUGH TERRAIN LANDING.

CAUTION

Care shall be exercised when operating the helicopter on rough terrain. Damage to the underside antennas may result.

For slope landings and all ground operations, avoid using combinations of excessive cyclic and low collective settings. Where minimum collective is used, maintain cyclic near neutral position and avoid abrupt cyclic inputs. If cyclic pitch is required, increase collective slightly to avoid hitting the droop stops and possible rotor-blade-to-fuselage contact.

Section V. ADVERSE ENVIRONMENTAL CONDITIONS

8.35 GENERAL

This section informs the crewmembers of the special precautions and procedures to be followed during the various weather and climatic conditions that may be encountered. This material will be additional to that already covered in other chapters regarding the operation of various helicopter systems.

8.36 COLD WEATHER OPERATION.

Helicopter operation in cold weather or an arctic environment presents no unusual problems if the flight crew is aware of the various changes that occur in low temperature conditions.

8.37 PREPARATION FOR FLIGHT.

CAUTION

Ice removal shall never be done by scraping or chipping. Remove ice by applying heat or de-ice liquid (TM 1-1500-204-23).

In addition to doing a normal preflight in Section 11, the rotor head, main rotor blades, tail rotor, and flight controls should be free of all ice and snow. Failure to remove snow and ice accumulations while on the ground can result in serious aerodynamics and structural effects in flight. Check that all fuel tank vents, static ports, pitot tubes, engine inlet, APU inlet, ENCU inlet, and heat exchangers are free of snow and ice; and that tires, landing gear struts, and the hydraulic accumulator are properly serviced.

  a. If ice or snow is found in engine inlets and exhaust, remove as much as possible by hand and thaw engine out with hot air before attempting to start. Actuate **PWR** levers for freedom of movement before starting main engine.

  b. Attempt to turn rotor system by rotating APU drive shaft by hand in the direction of rotation. If rotor cannot be turned, apply heat to main transmission area.

  c. As long as fuel will flow freely from the drains in the tanks, it can be assumed that the system is free of ice. Any indication that flow is restricted is cause for application of heat.

CAUTION

Fuel draining from the affected component after several minutes of heat application does not necessarily indicate that all ice has been melted. Ice may still remain in the unit, and it could be a serious hazard to flight operations. Heat should be applied for a short time after fuel begins to flow from the drain, and the drainage should be checked frequently until it is evident that all water has been removed.

  d. If water collected in sumps has frozen (indicated by a lack of flow from drain), apply heat liberally and open drain frequently. Catch all drainage in a clear container and inspect for water globules in the fuel. Continue sampling until fuel is free of water globules.

CAUTION

Due to an elapsed time requirement, it is recommended that the tail rotor teetering bearings warmup procedure be accomplished as the last item of the exterior check during the flight crew preflight inspection. At -31 °C (-24 °F), the helicopter must be started and the tail rotor must be turning within 5 minutes of teeter bearing warmup. Below -32 °C (-25 °F), elapsed time is reduced to 2 minutes. At a temperature of -32 °C (-25 °F) or below, the tail rotor must be cycled by an applied force no greater than 75 lbs.

The tail rotor teetering bearings are made of elastomeric material which prior to certain cold weather flights require special warmup procedures as follows:

  e. One crewmember should apply a teetering motion by manually pushing back-and-forth at the tip of the tail rotor blade until the blade has reached its teetering stops. When the blade can be pushed to the top, the bearing has been sufficiently warmed up.

8-19

f. Based on unapplied force of 75 lbs, the blade must be cycled one time at -32 °C (-25 °F), five times at -42 °C (-44 °F), and ten times at -54 °C (-65 °F).

8.38 ENGINE STARTING.

When starting in cold weather below -40 °C (-40 °F), if light-off does not occur within 10 seconds after placing the **PWR** lever to **IDLE**, quickly move engine **PWR** lever for the affected engine to **OFF** and then to **IDLE** three times. Then, leave the **PWR** lever in **IDLE**. If light-off still does not occur within 40 seconds, abort start and do the following:

1. Engine **PWR** lever (affected engine) - Hold at **LOCKOUT**.

2. Fuel **CROSSFEED** switch - **AFT TK**.

3. Fuel **BOOST** switch **ON**. Check that ground crew verifies fuel flow from upper drain.

4. Engine **PWR** lever (affected engine) - **OFF**.

5. Attempt another start.

6. After engine start - **CROSSFEED** switch - **NORM**.

8.39 WARMUP AND GROUND TESTS.

It is normal for engine oil pressure to be high during initial starts when oil is cold. Run engine at idle until oil pressure is within normal operating limits. Oil pressure should return to the normal range within five minutes. However, time required for warmup will depend on temperature of the engine and lubrication system before start.

During starts in extremely cold weather (near -54 °C, -65 °F), the following engine oil pressure characteristics are typical:

a. Oil pressure may remain at zero from 20 to 30 seconds after initiating the start. Abort start if oil pressure does not register within one minute of initiating start.

b. Once oil pressure begins to indicate on the gauge, it will increase rapidly and go over the 100 psig limit. The pressure will decrease as oil temperature rises and return to within the green band on the gauge. This condition is considered normal. The time for oil pressure to decrease to 100 psig or below will depend on the severity of the ambient temperature, but it should be inside the green band within five minutes of starting the engine.

c. Oil pressure may increase above the maximum pressure limit of 100 psig if the engine is accelerated above idle while oil temperature is below normal operating range. The pressure will decrease to within the normal operating range as oil temperature increases.

d. The **OIL BYP ENG 1** or **2** caution light normally goes on when starting an engine with oil below normal operating temperatures because of the relatively high oil viscosity and the amount of contamination in the oil filter. When oil temperature reaches about 38 °C (100 °F) during engine warmup, the light should go off.

e. At temperatures between -17 °C and -43 °C (1 °F and -45 °F), warmup engines during engine run-up for three minutes.

To eliminate any possibility of main rotor droop stop wear should the main rotor blades move through a pitch change angle while resting on the droop stop, the flight crew should observe the following:

At a temperature of -42 °C (-44 °F) or below, and with a rotor speed of 100% Nr, maintain neutral cyclic position for one minute. Then move the cyclic forward one inch and hold for one minute. Move the cyclic forward one additional inch and hold for one minute. The total procedure requires three minutes after reaching normal rotor RPM and can be accomplished simultaneously with the engine warmup procedures.

8.40 DESERT AND HOT WEATHER OPERATIONS.

a. In sandy or dusty conditions, it is advisable to perform the HIT check while airborne.

b. Refer to FM 1-202, Environmental Flight.

8.41 TURBULENCE AND THUNDERSTORM OPERATION.

8.41.1 Turbulence Operation:

a. For moderate turbulence, airspeed should be less than 150 knots.

b. For light turbulence, reduce airspeed, if desired, to minimize vibration.

8.41.2 Thunderstorm Operation:

a. Lightning strikes may result in the loss of the digital automatic stabilization equipment (DASE), stabilator control, engine electronic control units, and helicopter electrical power. The high voltages passing through the helicopter structure are expected to couple into the helicopter wiring, producing secondary effects which cause degradation to the mission equipment.

b. If a lightning strike occurs where all helicopter electrical power and engine electrical control units are lost, both engines will go immediately to maximum power output, as if in **LOCKOUT**. The flight crew shall have to react immediately to retard the **PWR** levers to **IDLE** and enter autorotation. The pilot could then advance the **PWR** levers, restoring power, relying solely on rotor and engine sounds and general helicopter handling because of the high probability that all engine instruments would be inoperative.

8.42 ICE AND RAIN.

CAUTION

Prolonged operation of the anti-ice/de-ice systems while on the ground may result in damage.

8.42.1 Preflight/Runup. Prior to flight in icing conditions (visible moisture and below freezing temperatures), special care should be taken to ensure that all necessary anti-ice/de-ice systems are operational. The blade de-ice system may be checked by holding the **BLADE** de-ice switch, on the pilots **ANTI-ICE** panel (fig 2-17), to the **TEST** position. The **BLADE ON** advisory light, on the same panel, will illuminate for approximately 3 to 4 seconds. Additionally, the on-command FD/LS may be used to ensure proper system operation. The engine inlet and nose gearbox anti-ice system may be checked by moving the **ENG INLET** switch, on the pilots **ANTI-ICE** panel, to the **ON** position. The **ENG 1** and **ENG 2 ANTI-ICE** lights on the pilots caution/warning panel will illuminate and remain on until the fairing heaters reach operating temperature. After approximately 30 to 40 seconds, the **ENG 1** and **ENG 2 ANTI-ICE** lights on the pilots caution/warning panel will extinguish and the **ENG 1** and **ENG 2** advisory lights on the pilots **ANTI-ICE** panel, will illuminate.

8.42.2 Inflight. Anti-ice/de-ice systems should be activated prior to flight in potential icing conditions (visible moisture and below freezing conditions).

a. The green **BLADE**, **ENG 1**, and **ENG 2** advisory lights on the pilots **ANTI-ICE** panel should be illuminated. When actual icing conditions are encountered, the **ENG ICE** light on the caution/warning panel will illuminate and remain on during the icing encounter. The main and tail rotor blades will accumulate ice between heating cycles. This will result in approximately 6 to 10% increase in indicated torque (without collective movement). When the main and tail rotor blades heat and shed ice, there will be a slight momentary increase in airframe vibrations and the indicated torque will drop to approximately the original reading.

b. Particular attention should be devoted to unusual torque rises or persistent airframe vibrations as these may be the first indications of a blade de-ice system malfunction. Icing rates tend to vary, even over short distances, and may result in continuous changes in the indicated icing rate on the ice rate meter. If continuous large erratic rate needle movements of more than 0.3 gm/m3 and/or large indicated torque rises are observed, recommend moving the de-ice mode switch (fig 2-17) to the manual MOD position, and depart the icing conditions.

c. Prolonged flight in icing conditions will reduce aircraft maximum range. For fuel consumption, refer to Chapter 7 **701** or Chapter 7A **701C**.

d. After departing the icing conditions, recommend minimum use of the windshield wipers as shed ice may damage main rotors, tail rotors, and engines.

8.43 GROUND OPERATIONS DURING HIGH WINDS.

WARNING

The maximum wind velocity for rotor start or stops is 45 knots from any direction. Ground operation of the helicopter in winds greater than 45 knots may cause the main rotor blades to contact the fuselage or the helicopter to roll over.

a. If surface winds above 45 knots are anticipated, ground operations should cease and the helicopter should be hangered or moored in accordance with TM 1-1520-238-23. If the helicopter cannot be hangered or moored, and sufficient time exists to shut the helicopter down prior to winds exceeding 45 knots, do so. Ensure rotor blades are positioned to the 45° point displacement and place **RTR BK** switch - **LOCK** prior to shutting down the APU.

b. If surface winds above 45 knots are inadvertently encountered during ground operations, head aircraft into the wind and maintain 100% Np/Nr, **TAIL WHEEL** switch - **LOCK, PARK BRAKE** - set, and operate the stabilator in automatic. Adjust the cyclic and collective as necessary to prevent droop 'stop pounding and keep the helicopter upright.

CHAPTER 9
EMERGENCY PROCEDURES

Section I. AIRCRAFT SYSTEMS

9.1 AIRCRAFT SYSTEMS.

This section describes the aircraft system emergencies that may reasonably be expected to occur and presents the procedures to be followed. Table 9-1 shows the messages displayed on the **MASTER CAUTION** panels, the caution/warning panels, and the corrective action required. Emergency procedures are given in checklist form when applicable. A condensed version of these procedures is contained in the condensed checklist, TM 1-1520-238-CL.

NOTE

- The urgency of certain emergencies requires immediate and instinctive action by the pilot. The most important single consideration is helicopter control. All procedures are subordinate to this requirement.

- The **MASTER CAUTION** segment should be reset after each malfunction to allow systems to respond to possible subsequent malfunctions. If time permits during a critical emergency, transmit a MAYDAY call.

9.2 DEFINITION OF EMERGENCY TERMS.

Those steps that must be performed immediately in an emergency situation are underlined. These steps must be performed without reference to the checklist. When the situation permits, non-underlined steps will be accomplished with use of the checklist.

　a. The term LAND AS SOON AS POSSIBLE is defined as landing at the nearest suitable landing area (e.g., open field) without delay. (The primary consideration is to ensure the survival of occupants.)

　b. The term "LAND AS SOON AS PRACTICABLE" is defined as landing at a suitable landing area. (The primary consideration is the urgency of the emergency.)

　c. The term AUTOROTATE is defined as adjusting the flight controls as necessary to establish an autorotational descent.

CAUTION

- When shutting down an engine that has malfunctioned in flight, it is important to identify the malfunctioning engine to avoid shutting the wrong engine down.

- Monitor TGT after shutdown. If TGT rises above 540 °C, or there is evidence of combustion as indicated by a rapid rise in TGT, place the engine START switch in IGN OVRD position and motor engine until TGT decreases below 540 °C.

　d. The term "EMER ENG SHUTDOWN" is defined as engine shutdown without delay. Engine shutdown in flight is usually not an immediate-action item unless a fire exists. Before attempting an engine shutdown, identify the affected engine by checking engine-out warning lights, torque meters, TGT indicators, N_G, N_p, and engine oil pressure indicators.

　e. Other terms may be defined, as necessary, to simplify the procedural memory steps within the existing emergency procedures. The term can then be used as an emergency procedure step instead of the steps used to define it. EXAMPLE: The term EMER ENG SHUTDOWN is defined as engine stoppage without delay and is accomplished as follows:

　　1. PWR lever (affected engine) – OFF.

　　2. FUEL switch (affected engine) – OFF

9.3 AFTER-EMERGENCY ACTION.

After a malfunction of equipment has occurred, appropriate emergency actions have been taken and the helicopter is on the ground, an entry shall be made in the Remarks Section of DA Form 2408-13-1 describing the malfunction. Ground and flight operations shall be discontinued until corrective action has been taken.

9.4 EMERGENCY EXITS AND EQUIPMENT.

Emergency exits and equipment are shown in figure 9-1.

WARNING

- Activation of the canopy removal system when combustible fuel/vapors are present in the cockpit can result in an explosion/fire. An explosion/fire can also occur if the aircraft has rolled on its side and fuel/vapors have gathered on the ground adjacent to the canopy side panels. The crewmembers survival knife may be used to fracture the canopy side panels as an alternate means of egress.

- Rotate the CANOPY JETTISON handle to the ARM (90°) position and release. Push the jettison handle to actuate the canopy jettison. Continuing to twist the jettison handle while trying to push may cause the actuator piston to jam and thereby prevent operation of the canopy severance system. If canopy jettison does not occur on the first attempt, ensure the jettison handle is in the 90° position, and push again. A push force of 140 – 150 lbs may be required to overcome the jam and initiate canopy jettison.

- In the event that canopy jettison does not occur when the canopy removal system is actuated, the personal survival knife should be used to fracture the canopy panel and permit egress.

9.4.1 Emergency Egress.

WARNING

- If emergency egress is required before the rotor blades have stopped, ensure cyclic remains centered. BATT and FORCE TRIM switches shall be left on to prevent rotors from striking the aircraft or personnel.

- In all cases of canopy jettison, remain clear of canopy side panels to avoid high velocity canopy fragments.

Emergency egress is accomplished by exiting through the emergency exits. If possible, use the manual canopy opening handles to exit the aircraft. To permit emergency egress by the pilot and CPG from the helicopter, the transparent portion of the four canopy side panels can be jettisoned by activating a detonation cord. Turning 90° and pushing any of three **CANOPY JETTISON** handles will initiate canopy side panel jettison. If emergency egress becomes necessary, proceed as follows:

1. Helmet visors – Down.

2. **CANOPY JETTISON** handle – Either crew member turn 90° release, then push.

9.4.2 Emergency Entrance. If it becomes necessary to jettison the canopy to gain entrance in case of an emergency.

1. Canopy emergency release door – Open.

2. Area around helicopter – Clear of personnel at least 50 feet from all canopy side panels.

3. **CANOPY JETTISON** handle – Turn 90° release, then push to jettison canopy.

TM 1-1520-238-10

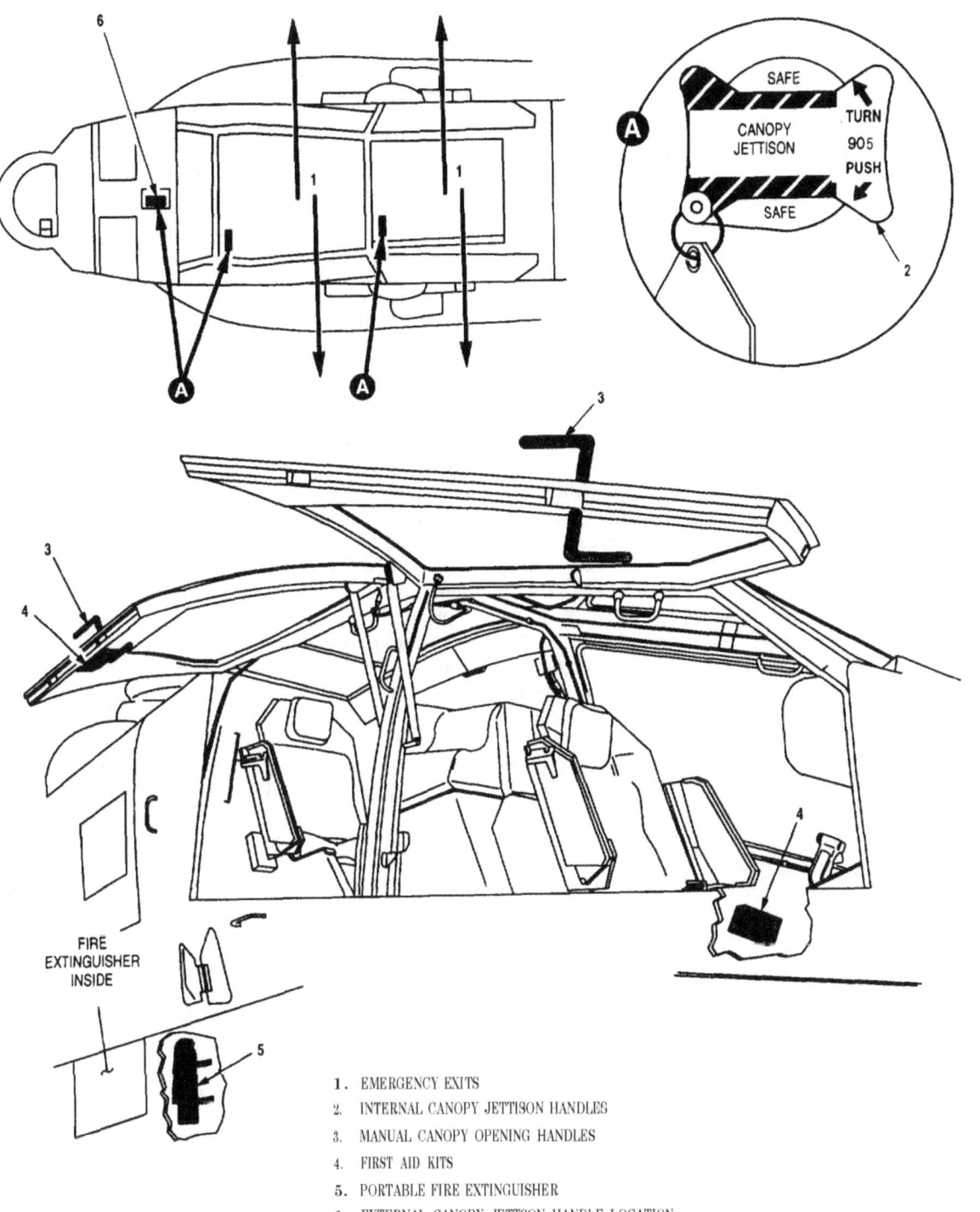

1. EMERGENCY EXITS
2. INTERNAL CANOPY JETTISON HANDLES
3. MANUAL CANOPY OPENING HANDLES
4. FIRST AID KITS
5. PORTABLE FIRE EXTINGUISHER
6. EXTERNAL CANOPY JETTISON HANDLE LOCATION

M01-096

Figure 9-1. Emergency Exits and Equipment

9-3

9.5 ENGINE FAILURE.

The various conditions under which an engine may fail prevent a standard procedure for all cases. A thorough knowledge of both emergency procedures and flight characteristics will enable the pilot to respond correctly and automatically in an emergency. The engine instruments often provide ample warning of an impeding failure by deviating from normal behavior. Engine failure is normally indicated by a rapid drop in N_G, N_p, torque, TGT, oil pressure and the engine and symbolic torque value will flash for the affected engine. The **ENGINE OUT 1 or 2** warning lights will illuminate and an audio signal will be heard through both headsets. Engines may fail only partially, and the degree of failure (amount of power loss) is another factor affecting crewmember response.

When an engine fails completely, the engine **PWR** lever and **FUEL** panel switch of the failed engine should be turned **OFF**. The reduction required in collective after engine failure will vary with altitude and airspeed at the time of failure. For example, the collective should not be reduced when an engine fails while the helicopter is hovering below 15 feet. During cruise flight, when altitude and airspeed permit a significant reduction in collective pitch, N_r can be restored to 100% before landing. During single-engine flight or during autorotation airspeed should be kept at the optimum. Optimum autorotation airspeeds are shown in figure 9-3. In autorotation, as airspeed increases above 70 - 80 KIAS, the rate of descent and glide distance increase significantly. As airspeed decreases below 64 KIAS, the rate of descent will increase and glide distance will decrease. Autorotation during an out-of-trim condition will increase the rate of decent and decrease the glide distance. Engine failure accompanied by an explosion or loud noise would indicate engine damage, and there is a possibility that an attempt of restart the engine would result in a fire.

9.51 Engine Failure Flight Characteristics. The flight characteristics and the required crewmember control response after a dual engine failure are similar to those during a normal power-on descent. Full control of the helicopter can be maintained during autorotational descent. When one engine has failed, the helicopter can often maintain altitude and airspeed until a suitable landing site can be selected. Whether or not this is possible becomes a function of such combined variables as aircraft weight, density altitude, and altitude and airspeed at the time of the engine failure.

Crewmember response time and control technique may be additional factors.

9.5.2 Single Engine Failure.

> **CAUTION**
>
> Prior to movement of either PWR lever, it is imperative that the malfunctioning engine and the corresponding PWR lever be identified.

Proper response to an engine failure depends on various factors: density altitude, airspeed, aircraft weight, single engine performance, and environmental conditions. The SAFE region in the height velocity diagram (fig 9-2) defines the airspeed and wheel-height combinations at various gross weight and density altitude combinations that will permit a safe landing in event of an engine failure. Crewmember recognition and subsequent action are essential and should be based on the following general guidelines: At low altitude and low airspeed, it may be necessary to lower the collective only enough to maintain N_r normal range. At higher density altitude, however, the collective may be lowered significantly to increase N_r to 100%. When hovering in ground effect, the collective should be used only as required to cushion the landing, and the primary consideration is in maintaining a level attitude. In forward flight at low altitude (as in takeoff), when a single-engine capability to maintain altitude does not exist, a decelerating attitude will initially be required to prepare for landing. Conversely, if airspeed is low and altitude sufficient, the helicopter should be placed in an accelerating attitude to gain sufficient airspeed for single-engine fly-away to a selected landing site. When the power available during single-engine operation is marginal or less; consideration should be given to jettisoning the external wing stores.

9.5.3 Single Engine Failure Low Altitude/Low Airspeed and Cruise.

Continued flight is possible.

1. LAND AS SOON AS PRACTICABLE.

Continued flight is <u>not</u> possible.

2. **STORES JETT** switches - Activate if required.

3. <u>LAND AS SOON AS POSSIBLE.</u>

TM 1-1520-238-10

HEIGHT – VELOCITY
SINGLE-ENGINE FAILURE
STD TEMP, ZERO WIND
GROSS WEIGHT = 14,550 LB OR LESS

NOTE: THERE IS NO AVOID AREA AT SEA LEVEL

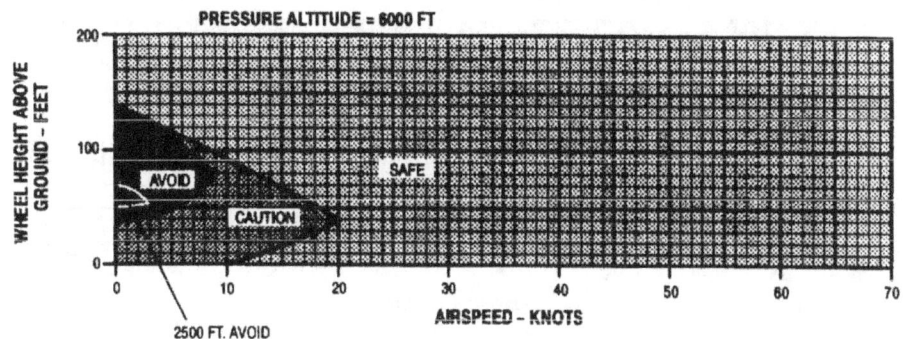

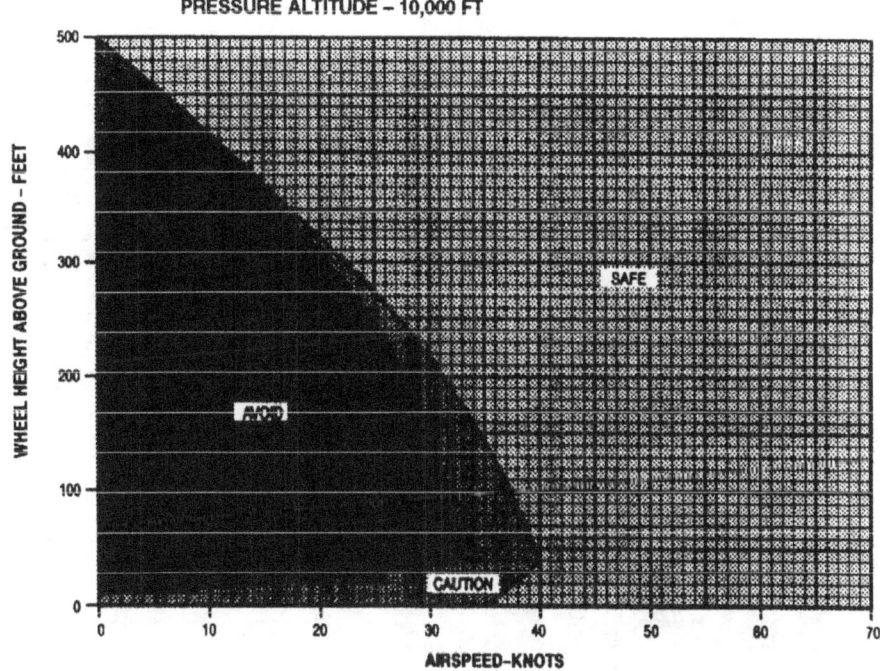

Figure 9-2. Height Velocity Plots (Sheet 1 of 2)

TM 1-1520-238-10

HEIGHT – VELOCITY
SINGLE-ENGINE FAILURE
STD TEMP, ZERO WIND
GROSS WEIGHT = 17,650 LB

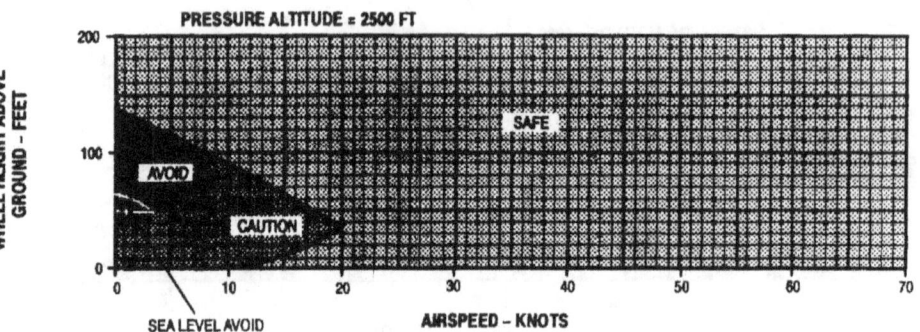

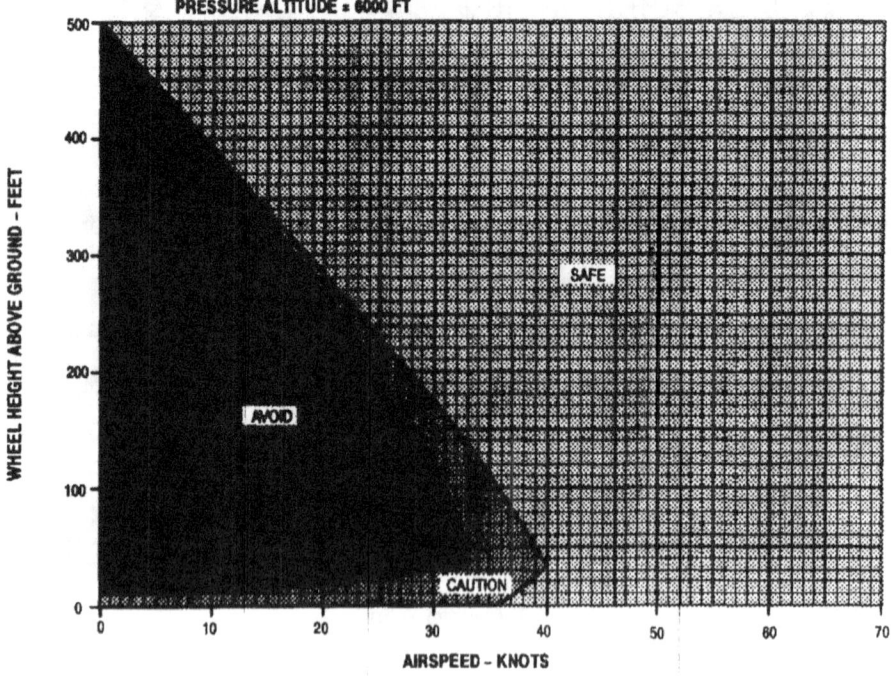

Figure 9-2. Height Velocity Plots (Sheet 2 of 2)

9.5.4 Dual Engine Failure.

WARNING

In the event of an inadvertent activation of the engine chop collar, initial indications from Np and Nr could be interpreted as a dual engine failure. Indications of engine chop collar activation are the ENGINE CHOP light illuminates and engine idle indications on the NG, TGT and Np There is NO illumination of the ENGINE OUT 1 or 2 lights on the MASTER CAUTION panel.

If both engines fail, immediate action is required to make a safe autorotative descent. The altitude and airspeed at which a two engine failure occurs will dictate the action to be taken. After the failure, main rotor rpm will decay rapidly and the aircraft will yaw to the left. Unless a two-engine failure occurs near the ground, it is mandatory that autorotation be established immediately. At gross weights above 15,000 lbs, immediate consideration should be given to jettisoning the external wing stores. During cruise at airspeeds to Vne, reduce collective immediately to regain Nr and then adjust as required to maintain rpm. The cyclic should be adjusted as necessary to attain and maintain airspeed in the optimum range (fig 9-3) An airspeed between 64 and 98 KIAS should be maintained except for high altitude high gross weight conditions where the maximum allowable autorotation airspeed should be maintained if it is less than 64 KIAS. A landing area must be selected immediately after both engines fail, and control inputs must be made to fly to the intended site. Throughout the descent, adjust collective as necessary to maintain Nr within normal range. At high gross weights, the rotor may tend to overspeed and the collective must be used to maintain the desired rotor rpm (fig 5-1). Nr should be maintained at or slightly above 100% to allow ample rpm before touchdown, and heading maintained by pedals.

Main rotor rpm will increase momentarily when the cyclic is moved aft with no change in collective pitch setting. An autorotative rpm of approximately 100% provides for a good rate of descent. Nr above 100% may result in a higher than desired rate of descent. At 100 to 125 feet AGL, use aft cyclic to decelerate. This reduces airspeed and rate of descent and causes an increase in Nr. The degree of increase depends upon the amount and rate of deceleration. An increase in Nr can be desirable in that more inertial energy in the rotor system will be available to cushion the landing. Ground contact should be made with some forward speed. If a rough landing is selected, a more pronounced deceleration is necessary and touchdown speed should approach zero. It is possible that during the autorotative approach, the situation may require additional deceleration. In that case, it is necessary to assume a landing attitude at a higher altitude than normal. Should both engines fail at low airspeed, initial collective reduction may vary widely. The objective is to reduce collective as necessary to maintain Nr within normal range. In some instances at low altitude or low airspeed, settling may be so rapid that little can be done to avoid a hard-impact landing.

In that case, it is critical to maintain a level landing attitude. Cushion the landing with remaining collective as helicopter settles to the ground.

9.5.5 Dual Engine Failure Low Altitude/Low Airspeed and Cruise.

    1. AUTOROTATE. ∎

    2. Chop Collar - Reset.

    3 **STORES JETT** switches - Activate as desired (time permitting)

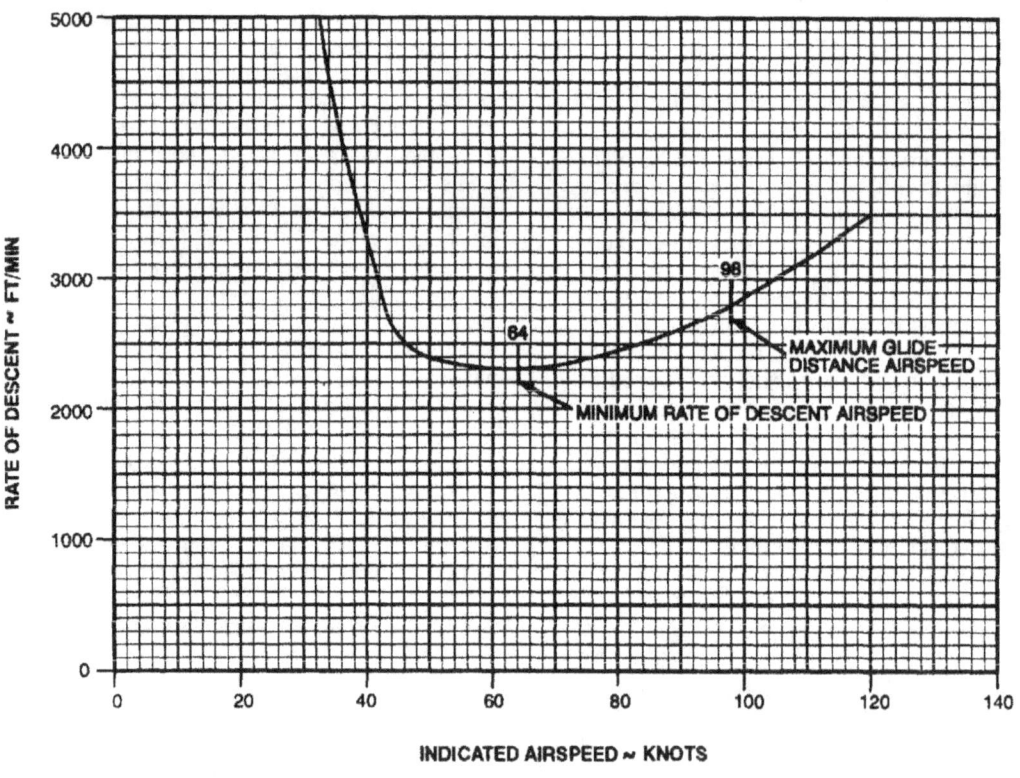

Figure 9-3. Autorotative Glide Chart

9.6 ENGINE COMPRESSOR STALL.

An engine compressor stall is normally recognized by a noticeable bang or popping noise and possible aircraft yaw. These responses are normally accompanied by a rapid increase in TGT and fluctuations in N_G, torque, and N_p readings for the affected engine. In the event of a compressor stall:

1. Collective - Reduce.

If condition persists:

2. **PWR** lever (affected engine) - Retard. (TGT should decrease.)

3. **PWR** lever (affected engine) - **FLY**.

If stall condition reoccurs:

4. **PWR** lever (affected engine) - **IDLE**.

5. LAND AS SOON AS PRACTICABLE.

9.7 ENGINE RESTART DURING FLIGHT.

After an engine failure in flight, an engine restart may be attempted. A failed engine should not be restarted unless it can be determined that it is reasonably safe to do so.

9.8 ENGINE EMERGENCY START (DECU INSTALLED ONLY 701C)

CAUTION

- Engine overspeed and overtemperature protection are not provided when the ENG 1 or ENG 2 (as appropriate) circuit breaker is out.

- The T700-GE-701C engine exhibits inconsistent starting capability above 6000 feet density altitude. Starts above this density altitude may be unsuccessful and require "over temperature" abort by the pilot.

The following emergency start may be attempted if the hot start preventer will not permit a normal engine start and starting is necessary because of a tactical emergency situation.

1. **ENG 1** or **ENG 2** (engine to be started) circuit breaker - OUT.

2. Start engine using normal starting procedure.

3. **ENG 1** or **ENG 2** circuit breaker - IN, prior to advancing **PWR** lever - **FLY**.

9.9 ROTORS, TRANSMISSIONS, AND DRIVE SYSTEM FAILURES AND MALFUNCTIONS.

9.9.1 Tail Rotor Malfunctions.

WARNING

Pilot situational awareness and correct analysis of the helicopters condition and operational environment are critical in the successful accomplishment of these procedures. The low inertia rotor system, coupled with high rates of descent during vertical autorotations, may not provide the pilot with adequate reaction time and cushioning pitch. Activation of the CHOP collar or retarding the PWR levers prior to reduction of the collective will result in rapid decay of rotor RPM. Successful completion of an out-of-ground-effect hovering autorotation is doubtful.

These procedures represent a best estimate of helicopter reactions and crewmember procedures. The most critical consideration in responding to any tail rotor malfunction is that the crewmember correctly interprets the nature and extent of the problem.

NOTE

Tail wheel should be locked during all landings.

9.9.2 Loss of Tail Rotor Thrust.

CAUTION

If engine chop is used to minimize main rotor torque, increasing collective pitch without first retarding PWR levers to IDLE may cause an uncommanded yaw.

Loss of tail rotor thrust occurs when there is a break in the drive system; for example, a severed drive shaft. The nose of the helicopter will turn to the right. If the helicopter is in forward flight, there will be a right roll of the fuselage along the longitudinal axis, and the nose of the helicopter may pitch downward. Normal pilot input to compensate for the right roll will produce a left side-slip. This downward pitch will be more pronounced if a tail rotor component has separated from the helicopter. In some cases, depending on the severity of the right rotation, powered flight to a suitable landing area can be accomplished by maintaining or increasing airspeed. The degree of side-slip and the amount of roll may be varied by changing airspeed and by varying collective pitch. Neither, however, can be completely eliminated.

9.9.3 Loss of Tail Rotor Thrust in Cruise Flight.

> **WARNING**
>
> **If the airspeed is allowed to approach effective translational lift, the sideslip angle may become quite severe and helicopter control may be lost.**

a. Continued Flight Possible. At cruise airspeeds, it may be possible that level flight at some stabilized yaw angle can be maintained. The degree of sideslip will depend on the airspeed and power required to maintain flight. Some left cyclic should be used to stop the slow right turn induced by loss of thrust. Care should be taken to avoid slowing the helicopter. The airspeed indicator may not provide useful information once the sideslip is established, but true airspeed, yaw angle, engine torque, and rate of climb or descent should provide the cues necessary to maintain flight. If yaw angle becomes excessive, reduce power and lower the nose to regain adequate airspeed. A minimum of 80 knots during a shallow approach to a roll-on landing should be maintained until approximately 10 to 20 feet above the touchdown point. Begin a gradual deceleration to arrive at approximately 5 to 10 feet above touchdown as the yaw angle begins to increase (to the right). At this point, retard the **PWR** levers as necessary to align the helicopter fuselage with the landing direction. Care should be taken to use minimum collective pitch to cushion the landing during touchdown.

After touchdown, the wheel brakes should be used to maintain heading and the collective should be lowered to minimize torque.

1. Airspeed - 80 KIAS minimum (until 10 to 20 feet above toucdown

2. PWR levers - Reduce as necessary (5 to 10 feet above touchdown).

b. Continued Flight Not Possible. If powered flight at an airspeed sufficient to maintain helicopter control is not possible, enter autorotation, and shutdown both engines using the **CHOP** collar (altitude and airspeed permitting). In autorotation, the sideslip and roll angles may be significantly reduced by maintaining a sufficiently high airspeed to allow the fuselage to streamline. A roll-on landing during touchdown will minimize the required pitch application and should be used if terrain permits.

Before touchdown, time permitting, the engine **PWR** levers should be retarded to **OFF**.

1. AUTOROTATE.

2. CHOP collar - CHOP.

3. PWR levers - OFF (time permitting).

9.9.4 Loss of Tail Rotor Thrust At Low Airspeed/Hover.

> **CAUTION**
>
> **Continuous right rotation during descent and touchdown can be expected.**

Loss of tail rotor thrust at low speed may result in extreme yaw angles and uncontrolled rotation to the right. Immediate collective pitch reduction should be initiated to reduce the yaw and begin a controlled rate of descent. If the helicopter is high enough above the ground, an attempt should be made to increase airspeed to streamline the helicopter. This may permit continued flight with a stabilized and manageable yaw angle. If this increase in airspeed does reduce yaw angle, proceed as outlined in LOSS of TAIL ROTOR THRUST in CRUISE FLIGHT (Continued Flight Possible) paragraph 9.9.3. If the aircraft cannot be accelerated into forward flight, initiate a power-on descent. Collective should be adjusted so that an acceptable compromise between rate of turn and rate of descent is maintained. At approximately 5 to 10 feet above touchdown, perform a hovering autorotation by CHOP collar - **CHOP** or PWR levers - **OFF**.

1. Collective - Reduce.

2. PWR levers or chop collar - **OFF** or **CHOP** (5 to 10 feet above touchdown).

9-10   Change 3

9.9.5 Tail Rotor Fixed Pitch Malfunction.

A fixed pitch failure may be evident by slow, intermittent, or no response to pedal input or no pedal movement. A left or right yaw may be apparent.

a. In Ground Effect. If a failure occurs during in-ground-effect hover, reaction may vary from adjusting collective and **PWR** levers during a left rotation to activating the **CHOP** collar to stop a right rotation. In any case, the primary concern should be to land the aircraft with as little yaw rate as possible.

1. If the aircraft has an uncontrolled turn to the left, a reduction in the **PWR** levers coordinated with an increase in collective may slow or stop the rotation so that a controlled power on descent to landing can be accomplished.

2. If the aircraft is not turning, a slight reduction in collective pitch will begin a descent. During the descent, a slight rotation to the left may be present; increasing collective just prior to touchdown should stop the rotation.

3. If the aircraft has an uncontrolled turn to the right, reduce collective to begin descent. At approximately 5 to 10 feet AGL perform a hovering autorotation by **CHOP** Collar – **CHOP** or **PWR** Levers – **OFF**.

b. Out-Of-Ground Effect.

1. If little or no right rotation or if left rotation is experienced and control can be maintained, the aircraft should be accelerated into forward flight and perform approach and landing appropriate to power setting and condition of flight at time of failure.

2. If the aircraft cannot be accelerated into forward flight, initiate a power-on descent. Collective should be adjusted so that an acceptable compromise between rate of turn and rate of descent is maintained. At approximately 5 to 10 feet above touchdown, perform a hovering autorotation by **CHOP** Collar – **CHOP** or **PWR** Levers – **OFF**.

9.9.6 Main Transmission Input Drive Clutch Failure.

An input drive clutch malfunction is most likely to occur during engine start or when an engine power lever is advanced. Indications may include: erratic torque indication on the affected engine, or a complete loss of torque indication on the affected engine, and/or N_p of the affected engine exceeding N_r. If the failure is a sudden disengagement, the torque of the opposite engine will double as it attempts to carry the load. A sudden high torque input drive clutch engagement may cause severe engine and/or drive train damage. A sudden engagement is indicated by a loud noise and/or a sudden increase in engine torque. Should an input drive clutch fail to engage, perform the following:

a. In Flight.

1. <u>**PWR** lever (affected engine) – **IDLE**.</u>

If N_P (affected engine) does not drop below N_R:

2. <u>LAND AS SOON AS POSSIBLE.</u>

If N_P (affected engine) goes below N_R:

3. <u>EMER ENG SHUTDOWN (affected engine only).</u>

4. LAND AS SOON AS PRACTICABLE.

b. On Ground.

> **CAUTION**
>
> Do not shutdown both engines simultaneously.

1. <u>EMER ENG SHUTDOWN (affected engine only)</u>

2. <u>Check N_C is less than 10% (affected engine)</u>

3. <u>Perform normal engine shutdown.</u>

9.9.7 High RPM Rotor (Warning Light On) N_P Failed High.

1. <u>Collective – Adjust</u> to Maintain N_r within limits.

If condition persists:

2. <u>**PWR** lever (affected engine) – Retard</u> to equalize torque on both engines.

3. LAND AS SOON AS PRACTICABLE.

9.9.8 Low RPM Rotor (Warning Light On) Np Failed Low.

1. <u>Collective - Adjust</u> to maintain Nr within limits.

   If condition persists:

2. **PWR** lever (affected engine) **LOCKOUT** then retard to equalize torque output of both engines.

NOTE

Advancing the **PWR** lever of the engine with low torque and TGT to LOCKOUT locks out the signal from the ECU **701** or DECU **701C**. The engine must be controlled manually to ensure that it does not exceed operating limits.

If manual control is not possible:

3. PWR lever (affected engine) IDLE.

4. LAND AS SOON AS PRACTICABLE.

9.10 FIRES.

The safety of the helicopter occupants is the primary consideration when a fire occurs. On the ground, it is essential that the engine(s), and APU be shut down, the crew evacuated, and fire fighting begin immediately. If time permits, a MAYDAY radio call should be made before electrical power is OFF to expedite assistance from fire fighting equipment and personnel. If airborne, the most important single action that can be taken by the crew is to land the helicopter. Consideration should be given to jettisoning external stores prior to landing.

NOTE

- The **PRI** bottle should be selected first; in the event of a malfunction or failure to extinguish the fire, select **RES**.

- The fire bottle discharge switch directly associated with the fire handle pulled must be used to dispense extinguishing agent.

- If the APU is running, accomplish an APU shutdown prior to evacuating the aircraft.

9.10.1 Engine/Fuselage Fire on Ground.
If ENG FIRE PULL handle illuminates or if fire is observed:

1. <u>**PWR** levers OFF.</u>

2. <u>Illuminated **ENG FIRE PULL** handle Pull</u> if applicable.

3. <u>**FIRE BTL** switch Activate if applicable.</u>

9.10.2 APU FIRE PULL Handle Illumination in Flight.
An **APU FIRE PULL** handle illumination in flight may be an indication of a fire in the transmission area. If the fire is in the transmission area, pulling the **APU FIRE PULL** handle and discharging the fire bottles may have little or no effect on the fire.

1. <u>**APU FIRE PULL** handle - Pull</u>

2. <u>**ECS** Off</u>

3. <u>LAND AS SOON AS POSSIBLE.</u>

9.10.3 APU Compartment Fire.
If fire is observed in APU compartment or if **APU FIRE PULL** handle on pilot right console illuminates:

1. <u>**APU FIRE PULL** handle - Pull</u>

2. <u>**FIRE BTL** switch Activate.</u>

3. <u>**ECS** Off</u>

9.10.4 Engine Fire in Flight.

1. PWR lever (affected engine) OFF.

2. Illuminated ENG FIRE PULL handle Pull.

3. FIRE BTL switch Activate.

4. LAND AS SOON AS POSSIBLE.

9.10.5 Electrical Fire in Flight.
Prior to shutting off all electrical power, the pilot must consider the equipment that is essential to a particular flight environment which will be affected; e.g., flight instruments, flight controls, etc. With electrical power off, engine anti-ice is automatically on. If an immediate landing cannot be made, the defective circuit may be isolated by selectively turning off electrical equipment and/or pulling circuit breakers.

1. <u>**GEN 1** and **2** switches **OFF**.</u>

2. <u>LAND AS SOON AS POSSIBLE.</u>

2. LAND AS SOON AS POSSIBLE.

9.10.6 Smoke and Fume Elimination.

1. Airspeed - Slow to 20 KIAS maximum.

2. Canopy door (affected crew position) - Open to intermediate position.

3. LAND AS SOON AS POSSIBLE.

9.10.7 Aborting Engine Start.

WARNING

701C Aborted engine starts may cause fuel to collect in the engine nacelle. Subsequent engine starts may be attempted only after the nacelle door/work platform is opened and the nacelle inspected for fuel. If during the initial start an abnormal TGT rise was evident, or fuel is evident in the nacelle, the ignition system shall be checked IAW standard maintenance procedures.

CAUTION

Abort Start for any of the following reasons:

- If it becomes apparent that TGT will exceed **852 °C** before N_G idle speed (63% or more) is attained.
- If TGT does not increase within 45 seconds after moving PWR lever to IDLE.
- If no N_p within 45 seconds after moving PWR lever to IDLE unless rotor is locked.
- If **positive** oil pressure indication does not occur within 45 seconds *after moving* PWR lever to IDLE.
- ENG START light extinguishes **prior** to 52% N_G.

Abort start procedures are as follows:

1. PWR lever - OFF

2. ENG START switch - IGN OVRD for 30 seconds or until TGT is below 540 °C.

9.11 ELECTRICAL SYSTEM FAILURES.

CAUTION

In the event any circuit breaker opens for unknown reasons, do not attempt to reset the breaker more than one time. Repeated tripping of a circuit breaker is an indication of a possible problem with equipment or electrical wiring. Multiple attempts to reset the circuit breaker may result in equipment damage and/or an electrical fire.

NOTE

- If generator fails in flight do not place generator switch of failed generator in test position.
- During an electrical system malfunction and operating on **EMERG** BATT power, the HSI/RMI will not provide adequate indications to the station.

9.11.1 GEN 1 and GEN 2 Caution Light On.

NOTE

In the event both **generators** are lost, turn off all unnecessary equipment on the emergency bus to conserve battery power. Battery power, assuming a 90% charge, is normally sufficient for 12 minutes emergency bus operation. When the battery heater is inoperative/disconnected and operating at temperatures below -30 °C (-22 °F), power may only be available for one minute of emergency operation.

1. GEN switches 1 and 2 - OFF/RESET - GEN.

   If power is not restored:

2. GEN switches - OFF

3. LAND AS SOON AS PRACTICABLE.

9.12 HYDRAULIC SYSTEM FAILURES.

9.12.1 PRI HYD PSI and UTIL HYD PSI Light On.

1. **EMER** HYD switch - On.

2. **LAND** WITHOUT DELAY.

Change 7   9-13

WARNING

Immediate emergency action must follow failure of both hydraulic systems. Any hesitation could result in loss of helicopter control. With emergency hydraulic power in use, flight control inputs must be kept to **a minimum as hydraulic pressure may only be available for 30 to 41 seconds. It is imperative that a landing be executed without delay.**

9.12.2 PRI HYD PSI and OIL LOW UTIL HYD Light On. In the event of a **PRI HYD** PSI failure and OIL LOW UTIL **HYD** condition, hydraulic power to the tail rotor servo may be lost. This may require a landing in accordance with **TAIL ROTOR FIXED PITCH MALFUNCTION** paragraph **9.9.5.**

1. <u>LAND AS SOON AS POSSIBLE.</u>

9.13 LANDING AND DITCHING.

9.13.1 Emergency Landing In Wooded Areas (Power Off).
1. <u>AUTOROTATE.</u>
2. <u>Collective</u> - **Adjust** <u>to maximum before main</u> **rotor** <u>contacts</u> tree <u>branches.</u>

9.13.2 Ditching (Power On). The decision to ditch the helicopter shall be made by the pilot when an emergency makes further flight unsafe.

1. Approach to hover.
2. **Canopies - <u>Jettison prior to entering water.</u>**
3. Pilot shoulder harness - Lock.
4. CPG - Exit helicopter.
5. Hover downwind a safe distance.
6. PWR levers - **OFF**.
7. Perform hovering autorotation - Apply full collective to decay RPM as helicopter settles.
8. Cyclic - Position in direction of roll.
9. Exit when main rotor has stopped.

9.13.3 Ditching (Power Off**).** If autorotational landing over water becomes necessary

1. <u>AUTOROTATE</u> - Apply full collective to decay rotor RPM as helicopter settles.
2. <u>Canopies - Jettison prior to entering water.</u>
3. <u>Cyclic</u> - <u>Position in direction of roll</u>

4. <u>Exit when main rotor has stopped.</u>

9.14 FLIGHT CONTROL FAILURES AND MALFUNCTIONS.

a. Failure of components within the flight control system may be indicated through varying degrees of feedback, binding, resistance, sloppiness or abnormal control response. These conditions should not be mistaken for the malfunction of the DASE.

b. Imminent failure of main rotor components may be indicated by a sudden increase in main rotor vibration and/or unusual noise. Severe changes in **lift** characteristics **and/or** balance condition can occur due to blade strikes, skin separation, **shift** or loss of balance weights or other material. Malfunctions may result in severe main-rotor flapping. If the main rotor system malfunctions, proceed as follows:

WARNING

Danger exists that the main rotor system could **collapse** or separate from the aircraft after landing. A decision must be made whether occupant egress occurs before or after the rotor has stopped.

1. <u>LAND AS SOON AS POSSIBLE.</u>
2. <u>EMER ENG(S) SHUTDOWN</u> after <u>landing.</u>

c. During ground operations any abnormal control inputs required to maintain desired fuselage attitude may be indicative of a problem. If **this** condition occurs, complete a normal engine shutdown.

9.14.1 Stabilator Automatic Mode Failure.
1. Stabilator **RESET** button - Press. If automatic mode is not restored:
2. Use manual stabilator.

9.14.2 Stabilator Auto/Manual Mode Failure.
1. Airspeed - Use placard limits. If both crew-station indicators are inoperative (90 knots maximum):
2. LAND AS SOON AS PRACTICABLE.

9.14.3 DASE Malfunction. DASE malfunctions may manifest themselves as uncommanded control inputs, which may cause unusual rotor disc movement, or aircraft attitude/heading changes.

1. <u>ASE release</u> switch - <u>Press.</u>
2. SAS - **Re-engage** unaffected axes.

9.14.3 DASE Malfunction. DASE malfunctions may manifest themselves as uncommanded control inputs, which may cause unusual rotor disc movement, or aircraft attitude/heading changes.

1. ASE release switch - Press.
2. SAS - Re-engage unaffected axes.

9.14.4 B BUCS Failure.

WARNING

Illumination of the BUCS FAIL warning light in flight shall be treated as a flight control system emergency. Exercise extreme care in making large or rapid control inputs.

1. LAND AS SOON AS POSSIBLE.

9.15 CAUTION/WARNING LIGHT EMERGENCY PROCEDURES.

For caution/warning light emergency procedures, reference table 9-1.

NOTE
- During maneuvering flight, a momentary reduction of oil pressure may occur in the main transmission and engine nose gearboxes causing a low oil pressure caution/warning light. As long as the caution/warning light goes out within 10 seconds, no action is required.
- During approaches, maneuvering flight and other areas of 4/rev vibration, the **VIB GRBX** caution light may illuminate. As long as the caution light extinguishes within 5-10 seconds after exiting the 4/rev environment, no action is required.
- FD/LS 19 Pressure indications of 50 or 99 lbs are normally wiring problems. If these indications are observed, make appropriate entry on DA form 2408-13 series forms.

Table 9-1. Caution/Warning Light Corrective Actions

WORD SEGMENT	CORRECTIVE ACTION
NOTE	
• Illumination of a light that is *information/system status* shows condition of system components. Mission accomplishments may be degraded. Mission requirements will dictate further actions.	
• For conditions which cause the lights to illuminate, see Tables 2-3, 2-4, and 2-5.	
MASTER CAUTION PANELS:	
MASTER CAUTION	Check Master/Caution Warning and Caution Panels for other lights. If no other light is flashing LAND AS SOON AS POSSIBLE.
LOW RPM ROTOR	See LOW RPM ROTOR (Warning Light On) Np Failed Low (paragraph 9.98).
FIRE APU	See APU Compartment Fire (paragraph 9.10.3).
ENGINE 1 OUT	See Single-Engine Failure Low Altitude/Low Airspeed and Cruise (paragraph 95.3).
ENGINE CHOP	Chop collar - Reset
ENGINE 2 OUT	See Single-Engine Failure Low Altitude/Low Airspeed and Cruise (paragraph 9.5.3).

Change 4   9-15

Table 9-1. Caution/Warning Light Corrective Actions - continued

WORD SEGMENT	CORRECTIVE ACTION
NOTE	
ENG 1 and 2 out. See Dual-Engine Failure Low Altitude/Low Airspeed and Cruise.	
HIGH RPM ROTOR	See HIGH RPM ROTOR (Rotor Light On) Np Failed High (paragraph 9.9.7).
B BUCS FAIL	LAND AS SOON AS POSSIBLE.
CAUTION/WARNING PANELS:	
FUEL LOW FWD	LAND AS SOON AS PRACTICABLE.
EXT EMP	Information/system status.
FUEL XFR	Information/system status.
FUEL XFR (Green) Modified C/W panel	Information/system status.
FUEL XFR (Amber) Modified C/W panel	Evaluate remaining fuel per cell.
X FEED (Green) Modified C/W panel	Information/system status.
X FEED (Amber) Modified C/W panel	Achieve safe single engine airspeed and restore previous switch position. If caution segment extinguishes, continue mission. If caution segment remains illuminated be prepared for a single engine flame-out. LAND AS SOON AS PRACTICABLE.
PRI HYD PSI	LAND AS SOON AS POSSIBLE.
UTIL HYD PSI	LAND AS SOON AS POSSIBLE.
MAN STAB	With audio - Refer to emergency procedures. Light only - Information/system status.
BUCS ON	Information/system status.
B BUCS ON	LAND AS SOON AS PRACTICABLE.
ADS	Information/system status.
FUEL LOW AFT	LAND AS SOON AS PRACTICABLE.
BOOST PUMP ON	Information/system status.
OIL LOW PRI HYD	LAND AS SOON AS PRACTICABLE.
OIL LOW UTIL HYD	LAND AS SOON AS PRACTICABLE.
PRI HYD PSI and UTIL HYD PSI	Refer to emergency procedures.

Table 9-1. Caution/Warning Light Corrective Actions - continued

WORD SEGMENT	CORRECTIVE ACTION
NOTE	
OIL LOW PRI HYD and UTIL HYD PSI - <u>LAND AS SOON AS POSSIBLE</u>.	
OIL LOW UTIL HYD and PRI HYD PSI - <u>LAND AS SOON AS POSSIBLE</u>.	
OIL PSI ACC PUMP	IN FLIGHT - <u>LAND AS SOON AS POSSIBLE</u>. APU ONLY OPERATION - Shutdown APU immediately.
ASE	LAND AS SOON AS PRACTICABLE.
REFUEL VALVE OPEN	Information/system status.
CHIPS NOSE GRBX 1	PWR lever - IDLE when conditions permit.
	LAND AS SOON AS PRACTICABLE.
OIL BYP PRI HYD	LAND AS SOON AS PRACTICABLE.
OIL BYP UTIL HYD	LAND AS SOON AS PRACTICABLE.
CHIPS NOSE GRBX 2	PWR lever - IDLE when conditions permit.
	LAND AS SOON AS PRACTICABLE.
CHIPS ENG 1	PWR lever - IDLE when conditions permit.
	LAND AS SOON AS PRACTICABLE.
OIL PSI NOSE GRBX 1	PWR lever - IDLE when conditions permit.
	LAND AS SOON AS PRACTICABLE.
OIL PSI MAIN XMSN 1	LAND AS SOON AS PRACTICABLE.
OIL PSI MAIN XMSN 2	LAND AS SOON AS PRACTICABLE.
NOTE	
OIL PSI MAIN XMSN 1 and 2 - <u>LAND AS SOON AS POSSIBLE</u>.	
OIL PSI NOSE GRBX 2	PWR lever - IDLE when conditions permit. LAND AS SOON AS PRACTICABLE.
CHIPS ENG 2	PWR lever - IDLE when conditions permit. LAND AS SOON AS PRACTICABLE.
OIL PSI ENG 1	<u>EMER ENG SHUTDOWN</u> when conditions permit. LAND AS SOON AS PRACTICABLE. PWR lever - IDLE when conditions permit. OIL HOT NOSE GRBX 1
OIL HOT MAIN XMSN 1	LAND AS SOON AS PRACTICABLE.
OIL HOT MAIN XMSN 2	LAND AS SOON AS PRACTICABLE.
NOTE	
OIL HOT MAIN XMSN 1 and 2 - <u>LAND AS SOON AS POSSIBLE</u>.	
OIL HOT NOSE GRBX 2	PWR lever - IDLE when conditions permit. LAND AS SOON AS PRACTICABLE.

TM 1-1520-238-10

Table 9-1. Caution/Warning Light Corrective Actions - continued

WORD SEGMENT	CORRECTIVE ACTION
OIL PSI ENG 2	<u>EMER ENG SHUTDOWN</u> when conditions permit. LAND AS SOON AS PRACTICABLE.
OIL BYP ENG 1	PWR lever - IDLE when conditions permit. LAND AS SOON AS PRACTICABLE.
GEN 1 and GEN 2	GEN switches 1/2 - OFF/RESET - GEN. If power not restored, GEN switches - OFF/RESET. LAND AS SOON AS PRACTICABLE.
RECT 1 and RECT 2	LAND AS SOON AS PRACTICABLE.
GEN 1	GEN switch - OFF/RESET - GEN. If power is not restored - GEN switch - OFF/RESET.
RECT 1	Information/system status.
GEN 2	GEN switch - OFF/RESET - GEN. If power is not restored - GEN switch - OFF/RESET.
RECT2	Information/system status.
OIL BYP ENG 2	PWR lever - IDLE when conditions permit. LAND AS SOON AS PRACTICABLE.
FUEL BYP ENG 1	PWR lever - IDLE when conditions permit. LAND AS SOON AS PRACTICABLE.
HOT RECT 1	POWER XFMR RECT 1 circuit breaker - Out.
CHIPS MAIN XMSN	<u>LAND AS SOON AS POSSIBLE</u>.
TEMP INT	LAND AS SOON AS PRACTICABLE.
TEMP TR	LAND AS SOON AS PRACTICABLE.
HOT RECT 2	POWER XFMR RECT 2 circuit breaker - Out
FUEL BYP ENG 2	PWR lever - IDLE when conditions permit. LAND AS SOON AS PRACTICABLE.
FUEL PSI ENG 1	Achieve safe single-engine airspeed. LAND AS SOON AS PRACTICABLE.
PRI MUX	Information/system status.
RDR JAM	Information/system status.

TM 1-1520-238-10

Table 9-1. Caution/Warning Light Corrective Actions - continued

WORD SEGMENT	CORRECTIVE ACTION
SHAFT DRIVEN COMP	LAND AS SOON AS POSSIBTE.
VIB GRBX	LAND AS SOON AS POSSIBLE.
HOT BAT LAND AS SOON AS PRACTICABLE.	BATT switch - OFF.
CHARGER	Information/system status.
FUEL PSI ENG 2 LAND AS SOON AS PRACTICABLE.	Achieve safe single-engine airspeed.
FUEL PSI ENG 1 AND ENG 2	Fuel CROSSFEED switch - AFT TK - Fuel BOOST switch - ON LAND AS SOON AS POSSIBLE.
GUN	Information/system status.
ROCKET	Information/system status.
IR JAM	Information/system status.
PNVS	See PNVS failure.
BLADE ANTI ICE FAIL ENG ICE	Information/system status. ANTI-ICE panel switches - As desired.
RTR BK	RTR BK switch - OFF. LAND AS SOON AS POSSIBLE.
CANOPY	Information/system status.
EXT PWR	Information/system status
MISSILE	Information/system status.
IFF	Information/system status.
ECS	Information/system status.
TADS	Information/system status.
CANOPY ANTI ICE FAIL	Information/system status.
ENG 1 ANTI ICE	Information/system status.
ENG 2 ANTI ICE	Information/system status.

TM 1-1520-238-10

Table 9-1. Caution/Warning Light Corrective Actions - continued

WORD SEGMENT	CORRECTIVE ACTION
APU ON	Information/system status.
APU FAIL	Information/system status.
- (SPARE)	LAND AS SOON AS POSSIBLE.
CPG CAUTION WARNING PANEL ONLY:	
PRI HYD	Coordinate actions with pilot.
UTIL HYD	Coordinate actions with pilot.
MAIN XMSN 1	Coordinate actions with pilot.
MAIN XMSN 2	Coordinate actions with pilot.
ENG 1	Coordinate actions with pilot.
ENG 2	Coordinate actions with pilot.
ELEC SYS FAIL	Coordinate actions with pilot.
ENG ANTI ICE	Coordinate actions with pilot.
VOICE CIPHER	Information/system status.
FUEL XFR (Green) Modified C/W panel	Information/system status.
FUEL XFR (Amber) Modified C/W panel	Evaluate remaining fuel per cell.
X FEED (Green) Modified C/W panel	Information/system status.
X FEED (Amber)	Achieve safe single engine airspeed and restore previous switch Modified CIW panel position. If caution segment extinguishes, continue mission. If caution segment remains illuminated be prepared for a single engine flame out. LAND AS SOON AS PRACTICABLE.

9-20 Change 3

Section II. MISSION EQUIPMENT

9.16 MISSION EQUIPMENT FAILURES AND MALFUNCTIONS.

Emergency operation of mission equipment is contained in this section, insofar as its use affects safety of flight.

9.16.1 Wing Stores Jettison.

> **CAUTION**
>
> Do not jettison Hellfire missile if hangfire is in progress (i.e. missile is still ignited).

a. **Armament Wing Stores.**
  1. Airspeed – 120 KIAS maximum.
  2. Selected **STORES JETT** switches – Activate.

     OR

  3. **ST JETT** – Press.

b. **External Fuel Wing Stores.**
  1. Airspeed – 100 KIAS maximum.
  2. Selected **STORES JETT** switches – Activate.

     OR

  3. **ST JETT** – Press.

9.16.2 ENCU Malfunction.

1. **ENCU** switch – **OFF**.
2. **ECS** panel – **STBY FAN** if desired.
3. Emergency crew station ventilator door – Open if desired.

9.16.3 PNVS/IHADSS Failure.

> **WARNING**
>
> If night or simulated night NOE, reaction to the following malfunctions must be immediate. Exit the NOE environment immediately.

> **NOTE**
>
> During rolling maneuvers, a phenomena known as AC coupling may degrade the PNVS imagery. Generally, this image degradation will worsen as the bank angle is increased. To reduce the adverse affects of AC coupling, the pilot should reduce the amount of sky visible within the PNVS field of view by viewing the terrain below the horizon.

a. **PNVS Failure.**

Pilot **NVS select switch** – **TADS**. Switch over to TADS WFOV. FLIR image should occur in about 3 seconds. TADS slew rates are noticeably slower in azimuth than PNVS. Some gain and level adjustment is usually necessary for optimum image.

> **NOTE**
>
> In the event of a GEN-1 failure, the automatic power transfer to GEN-2 will result in gray-scale selection in the HDU on some aircraft. Should this condition occur, resetting the **VID SEL** switch to **TADS** will restore normal video.

b. **IHADSS/HDU Failure.**

> **WARNING**
>
> In the event of IHADSS failure with gun selected, the gun will be commanded to the fixed forward position. Once in this position, the gun can still be fired without having to re-action the gun.

Pilot

  1. Establish visual flight.

VDU control switch – **PLT**.

ACQ SEL – **NVS FXD**.

   OR

CPG

CPG **SIGHT SEL** switch – **NVS**.

PLT/GND ORIDE switch – **ORIDE**.

NVS select switch – **PNVS** or **TADS**.

CPG assume helicopter control.

9.16.4 Dual IHADSS/HDU Failure.

Pilot

Establish visual flight.

VDU control switch – **PLT**.

  3. **ACQ SEL** – **NVS FXD**.

     OR

CPG

  1. Establish visual flight.

  2. CPG assume helicopter control.

9.16.5 Engine Alternator Malfunction.

> **CAUTION**
>
> - **[701]** Complete failure of the alternator or of the winding provided N_G speed indicated signal will activate the ENG OUT warning light and audio. The pilot shall check the N_r gauge indication and be prepared to carry out the actions for a high side failure. Thereafter, reference TGT and OIL PSI ENG, along with FUEL PSI ENG and OIL PSI NOSE GRBX segment lights of the affected engine.
> - **[701C]** Following a complete failure of an alternator, operation of the corresponding engine and all indications from engine instruments will be normal, except that N_G indications will be lost and the corresponding ENG OUT warning light and audio will be activated.

The engine alternator has three windings, providing power for engine ignition, aircraft N_G speed indication, and electrical control system operation. A failure of the ignition winding would result in loss of electrical power to the ignition circuitry which would be detected by inability to start the engine. A failure of the winding providing N_G speed indication signal would not affect actual engine operation; however, the pilot would have no N_G indication. A failure of the winding providing electrical power to the engine ECU **[701]** is the most severe and requires immediate action by the pilot. The immediate indication to the pilot may be the affected engine accelerating to maximum power. There will also be a loss of N_p and torque indication. If this failure is due to a complete loss of the alternator, then no N_G indication will be present either. The engine TGT will still be indicating because it is not acted upon by the ECU **[701]**; however, TGT limiting will no longer be available. N_p overspeed protection will be present because aircrarft power is supplied for that function. If an alternator failure is suspected and the affected engine accelerates to maximum power, proceed as in **HIGH RPM ROTOR** (warning light ON) N_P FAILED HIGH.

> **WARNING**
>
> If night or simulated night NOE, reaction to the following malfunctions must be immediate. Exit the NOE environment immediately.

9.16.6 Symbol Generator Failure.
Failure of the symbol generator will result in the loss of video to the VDU and video recorder. Both the PLT and CPG HDUs will automatically revert to PNVS FLIR-2 (PNVS video with no symbology). The ORT, HOD or HDD, as selected, will automatically revert to TADS FLIR-2 (TADS FLIR video). If the CPG **VID SEL** switch is positioned to **TADS**, the LOS reticle and the IAT gates (if operational) symbology from the TEU will be displayed. If the CPG **VID SEL** switch is set to **PNVS**, these symbols will not be displayed. The AND display will continue to operate normally. The pilot may continue flight in a degraded mode without flight or weapon symbology. The CPG may continue weapons engagement in a degraded mode without range information.

9.16.7 TADS Electronic Unit (TEU) Failure.
Complete failure of the TEU will cause the CPG's displays to lose imagery and the AND to blank. The pilot may be limited to ±75 degrees in azimuth on the PNVS and the message **PNVS DIRECT** will be displayed in the HAD sight status section.

9.16.8 FLIR Cooler Failure.
Failure of the PNVS or TADS FLIR cooler normally occurs during FLIR cooldown prior to flight. However, the cooler may fail in flight. Cooler failures are indicated by a loss of resolution on the FLIR. Resolution loss occurs gradually initially, and becomes more rapid as detector temperatures increase until resolution is lost completely. In the event of a PNVS cooler failure, proceed as indicated in PNVS failure emergency.

9.16.9 FCC Failure.
Failure of the FCC (primary MUX BUS Controller) will be indicated by illumination of the **MASTER CAUTION** and **PRI MUX** caution segment lights in the pilot and CPG stations. MUX BUS control will be automatically assumed by the BBC. In addition to the degradations listed in Chapter 4, the following effects will be present: DASE channels may disengage causing temporary instability; range inputs thru the DEK range page will be accepted but will have no effect on ordnance delivery; laser designator codes may not transfer to BBC registers and boresight correction data will not be available in the BBC. In the event of a **PRI MUX** failure, DASE channels should be re-engaged. The **MUX** switch should be placed in **SEC**. If weapons system use is necessary, correct laser codes should be confirmed and outfront boresight alignment should be checked prior to weapons systems engagements.

APPENDIX A
REFERENCES

AR 70-50	Designating and Naming Military Aircraft, Rockets, and Guided Missiles
AR 95-1	Flight Regulations
AR 95-13	Safety Procedures for Operation and Movement of Army Aircraft on the Ground
AR 385-40	Accident Reporting and Records
AR 385-63	Policies and Procedures for Firing Ammunition for Training, Target Practice and Combat
FLIP	Flight Information Publication
FM 1-202	Environmental Flight
FM 1-203	Fundamentals of Flight
FM 1-230	Meteorology for Army Aviators
FM 1-240	Instrument Flying and Navigation for Army Aviators
TB MED 501	Noise and Conservation of Hearing
TB MED 524	Occupational and Environmental Health: Control of Hazards from Laser Radiation
TM 1-1520-238-CL	Operator's Checklist for Army AH-64A Helicopter
TM 3-4240-312-12&P	Operator's and Unit Maintenance Manual Including Repair Parts and Special Tools List for Mask, Chemical - Biological: Aircraft, M43
TM 750 244- 1-5	Procedures for the Destruction of Aircraft and Associated Equipment to Prevent Enemy Use
TM 9-1095-206-13&P	Operator's Aviation Unit Maintenance and Aviation Intermediate Maintenance Manual Including Repair Parts and Special Tools Lists for Dispenser, General Purpose, Aircraft: M130
TM 9-4935-476-13	Operator's, Aviation Unit Maintenance and Aviation Intermediate Maintenance Manual for Captive Boresight Harmonization Kit.
TM 11-5810-262-OP	Operating Procedures for Communications Security Equipment TSEC/KY-58 in Aircraft Operations
TM 11-5841-281-12	Operator's and Organizational Maintenance Manual: Doppler Navigation Set
TM 11-5841-283-12	Aviation Unit Maintenance Manual for Radar Signal Detecting Set

TM 11-5841-294-30-2	Aviation Intermediate Maintenance Manual for Radar Signal Detecting Set. AN/APR-39(V).
TM 11-5865-200-12	Operator's and Aviation Unit Maintenance Manual Aviation Unit Maintenance (AVUM) Countermeasures Sets
TM 11-5895-1199-12	Operator's and Organizational Maintenance for Mark-XII IFF System
TM 55-1500-342-23	Army Aviation Maintenance Engineering Manual: Weight and Balance
TM 55-2840-248-23	Aviation Unit and Intermediate Maintenance Instructions. Engine, Aircraft Turboshaft, Models T700-GE-700, T700-GE-701, and T700-GE-701C.
TM 55-6600-200-20	Marking of Instruments and Interpretation of Markings
TM 55-9150-200-25	Engine and Transmission Oils, Fuels, and Additives for Army Aircraft

APPENDIX B
ABBREVIATIONS AND TERMS

SYMBOLS

*	Used as a decimal
°C	Degrees Celsius
°F	Degrees Fahrenheit
AF	Change in Frontal Area
AQ	Torque change for AF of 5 sq. ft.

A

A	Ampere
AH	Attack Helicopter
AC, ac	Alternating Current
A/C	Aircraft
ACC	Accumulator
ACM	Auto Control Module
ACQ	Acquire
ACT, ACTN	Action
ADF	Automatic Direction Finder
ADMIN	Administration
ADS	Air Data Sensor
AGL	Above Ground Level
ALT	Altimeter
AM	Amplitude Modulation
AMMO	Ammunition
AMP	Ampere
AND	Alphanumeric Display

TM 1-1520-238-10

Appendix B - Abbreviations and Terms (cont)

ANT	Antenna
ANTI COL	Anticollision
APU	Auxiliary Power Unit
ARCS	Aerial Rocket Control System
ATTD	Attitude
AUTO	Automatic
AUX	Auxiliary
AVAIL	Available
AWS	Area Weapon System
AZ	Azimuth

B

BATT (or BAT)	Battery
BITE	Built-in Test Equipment
BRK	Brake
BL	Blade (rotor)
BNK	Bunker
BOT	Beginning of Tape
BRG	Bearing
BRK	Brake
BRSIT	Boresight
BRT	Bright (switch position)
BRU	Boresight Reticle Unit
BST	Boost
BST	Boresight
BTL	Bottle (fire extinguisher)
BUCS	Backup Control System
BYP	Bypass

B-2 Change 3

Appendix B - Abbreviations and Terms (cont)

C

C	Celsius
CAN	Canopy
CAS	Command Augmentation System
CAUT	Caution
CBHK	Captive Boresight Harmonization Kit
CDU	Computer Display Unit
CG, cg	Center of Gravity
CHK	Check
CHRGR	Charger (battery)
CK	Check
CKT	Circuit
CL	Clear
CM, cm	Centimeter
CMD	Command
CNTST	Contrast
CNV	Crypto-Net Variables
COL	Collision
COLL, coll	Collective
COMM	Communication
COMM CONT	Communication Control
COMP	Compressor, Compass
CONT	Continuous, Control
COORD	Coordinate
CPG	Copilot/Gunner
CRS	Course
CRT	Cathode Ray Tube

Appendix B - Abbreviations and Terms (cont)

CSC	Communication System Control
CSL	Console
CW	Continuous Wave

D

DASE	Digital Automatic Stabilization Equipment
DC, dc	Direct Current
DECR	Decrease
DECU	Digital Electronic Control Unit
DEK	Data Entry Keyboard
DEST	Destination
DET	Detector (fire detector)
D/F	Direction Finding
DG	Directional Gyro
DIA, dia	Diameter
DIR	Directional
DISP, DSP	Display
DIST	Distance
DNS	Doppler Navigation Set
DTU	Data Transfer Unit
DTV	Day Television
DVO	Direct View Optics

E

ECS	Environmental Control System
ECU	Electrical Control Unit (for engine)
EDGE LT PNL	Edge Light Panel (switch)
EEPROM	Electrically-Eraseable Programmable Read-Only Memory
EID	Emitter Identification Data
EGI	Embedded Global Positioning System (GPS) Inertial
EL	Elevation

Appendix B - Abbreviations and Terms (cont)

ELEC, ELECT	Electrical, Electric
ELEV	Elevation
ELEX	Electronics
EMER	Emergency
EMERG	Emergency
EN, ENT	Enter, Entry
ENCU	Environmental Control Unit
END	Endurance
ENG	Engine
ENG CUT	Engine cut
ENG INST	Engine Instrument (lighting)
ENS	Environment Control System
ESS	Essential
EXT	Extend, External, Extinguisher
EXT LTS	External Lights
EXT-RET	Extend, Retract
EXT TK	External Tank, Auxiliary Fuel Tank

F

FAB	Fixed Action Button
FAB	Forward Avionics Bay
FAT	Free Air Temperature
FCC (AC)	Fire Control Computer
FCS	Fire Control System
FD/LS	Fault Detection and Location System
FFAR	Folding-fin Aerial Rocket
FIRE BTL	Fire Bottle
FLIR	Forward Looking Infrared
FLPN	Flight Planning

Appendix B - Abbreviations and Terms (cont)

FLT INST	Flight Instrument (lighting)
FLTR	Filter
FM	Frequency Modulation
FOC	Focus
FOD	Foreign Object Damage
FORM LT	Formation Light
FOV	Field of View
FRL	Fuselage Reference Line
FT, ft	Foot, Feet
FT/MIN	Feet per Minute
FTR	Filter
FWD	Forward
FXD	Fixed

G

G, g	Gravity
GEN	Generator
GND	Ground
GPM, gpm	Gallons per Minute
GRB, GRBX	Gearbox
GRWT	Gross Weight
GS	Ground Speed, Gray Scale
GSE	Ground Support Equipment
GUN MTR	Gun Motor
GW	Gross Weight

Appendix B - Abbreviations and Terms (cont)

H

H	Home
HAD	High Action Display
HARS	Heading and Attitude Reference System (set)
HAS	Hover Augmentation System
HBCM	Hover Bias Calibration Mode
HDG	Heading, Heading Set
HE	High Explosive
HF	High Frequency, Hellfire
HIGE	Hover In Ground Effect
HIT	Health Indicator Test
HMD	Helmet-Mounted Display
HMMS	Hellfire Module Missile System
HMS	Helmet-Mounted Sight
HMU	Hydromechanical Unit
HOGE	Hover Out of Ground Effect
HOV	Hover
HSI	Horizontal Situation Indicator
HSP	Hot Start Preventor
HT	Heat, Heater, Height
HTR	Heater
HYD	Hydraulic
Hz	Hertz

I

IAS	Indicated Airspeed
ICS	Intercommunication System, Internal Communication System
IDENT	Identification

Appendix B - Abbreviations and Terms (cont)

IFF	Identification Friend or Foe (transponder)
IFR	Instrument Flight Rules
IGE	In Ground Effect
IGN	Ignition
IHADSS	Integrated Helmet and Display Sight System
IMC	Instrument Meteorological Conditions
INBD	Inboard
INCR	Increase
IND	Indicator
INST, INSTR	Instrument
INTMD	Intermediate
INTR LTS	Interior Lights
IP, I/P	Identification of Position
IPAS	Integrated Pressurized Air System
IR	Infrared
IRIS	Infrared Imaging Seeker
IR JMR	Infrared Jammer
IRP	Intermediate Rated Power

J

JETT	Jettison
JMR	Jammer

K

KG, kg	Kilogram
KHz	Kilohertz
KIAS	Knots, Indicated Airspeed
KTAS	Knots, True Airspeed

Appendix B - Abbreviations and Terms (cont)

KVA, kva	Kilovolt Ampere
KYBD	Keyboard

L

L	Left
LAT	Latitude
LB	Pound
LB/HR	Pounds per Hour
LCHR	Launcher
LCF	Low Cycle Fatigue
LDG	Landing
LDGLT	Landing Light
LDNS	Lightweight Doppler Navigation System
LDS	Load Demand Spindle
LF	Low Frequency
LH	Lefthand
LOAL	Lock On After Launch
LOBL	Lock On Before Launch
LONG	Longitude
LOS	Line Of Sight
LRF/D	Laser Range Finder Designator
LRU	Line Replaceable Unit
LSR	Laser
LT	Light, Laser Tracker
LTS	Lights
LVDT	Linear Variable Differential Transducer (position sensor)
LVR	Lever

Appendix B - Abbreviations and Terms (cont)

M

M, m	Meter
MAL	Malfunction
MAN	Manual
MAX	Maximum
MCP	Maximum Continuous Power
MEM	Memory
MHz	Megahertz
mHz	Millihertz
MIC	Microphone
MIKE	Microphone
MIN	Minimum, Minutes
MLG	Main Landing Gear
MM, mm	Millimeter
MON	Monitor
MRP	Maximum Rated Power
MRTU	Multiplex Remote Terminal Unit
M/R	Main Rotor
MSL	Missile, Mean Sea Level
MTR	Motor
MUX	Multiplexer, Multiplex
M230E1	30mm Gun

N

NAV	Navigation
NAV LT	Navigation Light
NDB	Nondirectional Beacon

Appendix B - Abbreviations and Terms (cont)

NFOV	Narrow Field of View
NG, NG	Engine Gas Generator Speed
NM	Nautical Mile
NOE	Nap Of the Earth
NP, Np	Engine Power Turbine Speed
NR, Nr	Main Rotor Speed
NRML, NORM	Normal
NVS	Night Vision System

O

OAT	Outside Air Temperature
ODV	Overspeed and Drain Valve
OEI	One Engine Inoperative
OFP	Operational Flight Program
OFS	Offset
OGE	Out of Ground Effect
OI	Optical Improvement
OPR	Operator
ORT	Optical Relay Tube
OTBD	Outboard
Overtemp	Overtemperature
ORIDE	Override
OVSP	Overspeed

P

para	Paragraph
PAS	Pressurized Air System, Power Available Spindle
PCT	Percent
PEN	Penetration

Appendix B - Abbreviations and Terms (cont)

PGM	Program
PLT	Pilot
PNL	Panel
PNVS	Pilots Night Vision System
PP	Present Position
ppm	Pounds per Minute
PRES, PRESS	Pressure
PRI	Primary
PSI, psi	Pounds per Square Inch
PSID	Pounds per Square Inch Delta
PSIG, psig	Pounds per Square Inch Gage
PTO	Power Takeoff
PWR	Power
PYL	Pylon

Q

Q	Torque
QTY	Quantity

R

R	Right, Red
RAD	Radio, Radar
RAD ALT	Radar Altimeter
RAI	Remote Attitude Indicator
R/C	Rate of Climb
RCDR	Recorder
RCVR	Receiver
RDR	Radar
READ	Read Page
RNDS	Rounds

Appendix B - Abbreviations and Terms (cont)

REC	Recorder
RECT	Rectifier, Transformer-Rectifier
REF	Reference
REL	Release
REM	Remote
RES	Reserve
RET, RETR	Retract
RETRAN	Retransmit
RF	Radio Frequency
RH	Righthand
RHE	Remote Hellfire Electronics
RIPL	Ripple
RKT	Rocket
RMI	Radio Magnetic Indicator
RNG	Range
RPM, rpm	Revolutions per Minute
RTR	Rotor
RTR BK	Rotor Brake
RTR BL	Rotor Blade
RTSS	Radio Transmitter Select System

S

SAS	Stability Augmentation Subsystem
SCAS	Stability and Command Augmentation System
SCL	Scale
SDC	Shaft Driven Compressor
SDD	Selectable Digital Display

Appendix B - Abbreviations and Terms (cont)

SEC	Secondary
SEL	Select
SEU	Sight Electronics Unit
SG	Symbol Generator
SHP	Shaft Horsepower
SKR	Seeker
SL	Sea Level
SN	Serial Number
SPAD(S)	Shear-pin-actuated Decoupler (system)
SPH	Spheroid
SQ	Super Quick
SQ FT, sq ft	Square Foot, Square Feet
SCHLT	Searchlight
SRCH PWR	Searchlight Power
ST	Stores
STA, STN	Station
STAB	Stabilizer, Stabilator
STBY	Standby
STBY ATT	Standby Attitude
STD	Standard
STR	Storage, Store
STRS	Stores
SYM GEN	Symbol Generator
SYS	System

Appendix B - Abbreviations and Terms (cant)

T

T1, T2	Test patterns 1 and 2
T4.5	Five-probe harness measuring gas temperature at power turbine inlet
TADS	Target Acquisition Designation System
TAS	True Airspeed
TBD	To Be Determined
TBS	To Be Supplied
TEMP	Temperature
TGT	Turbine Gas Temperature, Target
TGT	Target
TK	Tank, Track
TK SEL	Tank Select (switch)
TKE	Tracking Angle Error
TLG	Tail Landing Gear
TNK	Tank
T/R	Transmitter-receiver, Transmit-receive, Transformer-rectifier
TRANS	Transfer
TST	Test

U

UHF	Ultra-high Frequency
UIL	Utility
UTM	Universal Transverse Mercator

Appendix B - Abbreviations and Terms (cent)

V

V, v	Volt
VAB	Variable Action Button
VAC, vac	Volts Alternating Current
VAR	Variation, Variable
VDC, vdc	Volts Direct Current
VDU	Video Display Unit
VHF	Very High Frequency
VID	Video
VID RCDR	Video Recorder
Vne, VNE	Velocity, Never Exceed (airspeed limit)
VOL	Volume
VROC	Vertical Rate of Climb
VSI	Vertical Speed Indicator
VSWR	Voltage Standing Wave Ratio

W

W	Watt
WAS	Weapon Action Select
WE	Weight Empty
WHL	Wheel
WHT	White (Switch position)
WPN	Weapon
WPT	Waypoint
WRP	Wiper (windshield)
WSHLD	Windshield
WSHLD WPR	Windshield Wiper
WSPS	Wire Striker Protection System

Appendix B - Abbreviations and Terms (cont)

X

XFEED	Crossfeed
XFER	Transfer
XMSN	Transmission
XTK	Crosstrack

Y

Y	Yellow

Z

ALPHABETICAL INDEX

Subject	Page No.
A	
Abbreviations and Terms, Appendix B, Description	1-1
AC Power Supply System	2-66
Accessory Section Module, Engine	2-22
ADF Operations	3-34
Adjustment, Seat Height	2-16
ADMIN Display, CDU (EGI)	3-64.6
Adverse Environmental Conditions, General	8-19
Aerial Rocket Control Panel (ARCS), Pilot	4-24
Aerial Rocket Control Panel Pilot (figure)	4-24
Aerial Rocket Control Panel/Indicator Functions, Pilot (table)	4-24
Aerial Rocket Control System (ARCS) 2.75	4-12
Aerial Rocket Control System, In Flight Procedures	4-66.3
Aerial Rocket Delivery System (figure)	4-12
After Landing Check	8-15
After Starting APU, CPG	8-12
After Starting APU, Pilot	8-11
Air Data Sensor Subsystem (ADSS)	4-17
Air Induction, Engine	2-22
Aircraft and Systems Description and Operation, General	2-1
Aircraft General Arrangement	2-1
Aircraft General Arrangement (figure)	2-3
Aircraft Systems Emergency Procedures, General	9-1
Airspeed Operating Limits	5-11
Airspeed Operating Limits Chart	5-11
Airspeed Operating Limits Chart (figure)	5-12
Alpha/Numeric Display (AND)	4-54
Alpha/Numeric Display (AND) (figure)	4-54

Subject	Page No.
Alpha/Numeric Display Message Location and Description (table)	4-54
Alteration of CBHK Values	5-17
Alternator, Engine	2-28
Alternator, Engine -701	2-28
Alternator, Engine -701C	2-28
Altimeter, AN/APN-209(V)	3-71
Altimeter, AN/APN-209(V) (figure)	3-71
Ammunition (30mm)	6-11
Ammunition Loading for 30mm Rounds (table)	6-14
AN/ALQ-136 Radar Countermeasures Set	4-69
AN/ALQ-144A,-144(V)3 Infrared Countermeasures Set	4-69
AN/APN-209(V) Altimeter	3-71
AN/APN-209(V) Altimeter Control and Indicator Functions (table)	3-72
AN/APN-209(V) Antenna	3-71
AN/APN-209(V) Controls, Indicators, and Functions	3-71
AN/APN-209(V) Operation	3-72
AN/APN-209(V) Stopping Procedure	3-72
AN/APR-39(V)1 (CRT) Control/Indicator Functions (table)	4-72
AN/APR-39(V)1 Control/Indicator Functions (table)	4-72
AN/APR-39(V)1 Operating Procedures	4-72
AN/APR-39(V)1 Operation Modes	4-73
AN/APR-39(V)1 Radar Warning System	4-71
AN/APR-39(V)1 Radar Warning Display Controls	4-72
AN/APR-3901 Self-Test Operation	4-74
AN/APR-39A(V)1 (CRT) Control/Indicator Functions (table)	4-76
AN/APR-39A(V)1 Control/Indicator Functions (table)	4-75
AN/APR-39A(V)1 Operating Procedures	4-75
AN/APR-39A(V)1 Operation Modes	4-76

Subject	Page No.
AN/APR-39A(V)1 Radar Warning Display Controls	4-76
AN/APR-39A(V)1 Radar Warning System	4-75
AN/APR-39A(V)1 Self-Test Operation	4-76
AN/ARC-186(V) Antennas	3-10
AN/ARC-186(V) Controls and Functions	3-10
AN/ARC-186(V) Controls and Functions (table)	3-11
AN/ARC-186(V) Control Panel (figure)	3-10
AN/ARC-186(V) Modes of Operation	3-12
AN/ARC-186(V) Operating Procedures	3-12
AN/ARC-186(V) Radio Set	3-9
AN/ARC-201 Control Functions (table)	3-25
AN/ARC-201 Control Panel (figure)	3-24
AN/ARC-201 Operating Procedures	3-27
AN/ARC-201 Radio Set	3-24
AN/ARN-89 Control Functions (table)	3-30
AN/ARN-89 Control Panel (figure)	3-30
AN/ARN-89 Direction Finder Set	3-30
AN/ARN-89 Operating Procedures	3-31
AN/ARN-149(V)3 ADF Control Panel (figure)	3-32
AN/ARN-149(V)3 ADF Control Panel Control Functions (table)	3-33
AN/ARN-149(V)3 Automatic Direction Finder Set	3-32
AN/ASN-128 Antenna	3-37
AN/ASN-128 Controls and Displays	3-37
AN/ASN-128 Control and Indicator Functions (table)	3-38
AN/ASN-128 DNS CDU CP1252 (figure)	3-38
AN/ASN-128 Doppler Navigation Set	3-37
AN/ASN-128 Methods of Operation	3-41
AN/ASN-128 Modes of Operation	3-37
AN/ASN-128 Operating Procedures	3-41
AN/ASN-128 Window Displays (table)	3-42
AN/ASN-137 ADMIN Page - PWR OFF (figure)	3-50
AN/ASN-137 ADMIN Page - PWR ON (figure)	3-50
AN/ASN-137 Antenna	3-37
AN/ASN-137 CDU Displays	3-49

Subject	Page No.
AN/ASN-137 Control and Display Functions	3-46
AN/ASN-137 Control and Display Functions (table)	3-47
AN/ASN-137 Data Entry Procedures	3-52
AN/ASN-137 DNS CDU IP1552 (figure)	3-47
AN/ASN-137 Doppler Navigation Set	3-46
AN/ASN-137 FDLS Page (figure)	3-51
AN/ASN-137 FDLS Page - Continuous Test Results (figure)	3-54
AN/ASN-137 FDLS Status Page - On Command Test Results (figure)	3-53
AN/ASN-137 FDLS Status Page - On Command (figure)	3-62
AN/ASN-137 FDLS Status Page - On Command Test Results NO-GO (figure)	3-61
AN/ASN-137 FDLS Test in Progress (figure)	3-53
AN/ASN-137 FPLN Dictionary Page (figure)	3-51
AN/ASN-137 HBCM Page (figure)	3-51
AN/ASN-137 Modes of Operation	3-48
AN/ASN-137 NAV Top Level Page - PWR ON (figure)	3-49
AN/ASN-137 Power Up Display (figure)	3-47
AN/ASN-137 Power Up Procedures	3-52
AN/ASN-137 Stopping Procedures	3-62
AN/ASN-137 CDU Displays (EGI)	3-64.5
AN/ASN-137 Computer Display Unit (CDU) IP-1552G (EGI)	3-64.1
AN/ASN-137 Control and Display Functions (table) (EGI)	3-64.3
AN/ASN-137 Modes of Operation (EGI)	3-64.2
AN/ASN-137 CDU Displays (EGI)	3-64.5
AN/ASN 137 NAV Top Level Page (EGI)	3-64.5
AN/ASN 137 ADMIN Page (EGI)	3-64.6
AN/ASN 137 ADMIN Page Variable Action Buttons (VAB) (EGI) (table)	3-64.7
AN/ASN 137 Alphanumeric Display (AND) Page (EGI)	3-64.24
AN/ASN-137 Alphanumeric Display (AND) Page (EGI) (figure)	3-64.25

Index 2 Change 3

Subject	Page No.	Subject	Page No.
AN/ASN-137 Area Weapons Subsystem (AWS HARMONIZATION) Page (EGI)	3-64.24	AN/ASN-137 GPS STATUS Page (EGI) (figure)	3-64.20
AN/ASN-137 Area Weapons Subsystem (AWS HARMONIZATION) Page (EGI) (figure)	3-64.24	AN/ASN-137 HBCM Page (EGI)	3-64.8
		AN/ASN-137 HBCM Page (EGI) (figure)	3-64.8
AN/ASN-137 Area Weapons Subsystem (AWS HARMONIZATION) VAB operations (EGI) (table)	3-64.24	AN/ASN-137 HBCM Page Variable Action Buttons (VAB) (EGI) (table)	3-64.8
AN/ASN-137 Boresight EGI (BST EGI) Page (EGI)	3-64.221	AN/ASN-137 Missile Laser Codes (CODES) Page (EGI)	3-64.15
AN/ASN-137 Boresight EGI (BST EGI) Page (EGI) (figure)	3-64.22	AN/ASN-137 Missile Laser Codes (CODES) Page (EGI) (figure)	3-64.15
AN/ASN-137 Boresight EGI (BST EGI) Page VAB operations (EGI) (table)	3-64.22	AN/ASN-137 NAV SENSOR CONTROL Page (EGI)	3-64.16
AN/ASN-137 CO-Pilot High Action Display (HAD) (EGI)	3-64.26	AN/ASN-137 NAV SENSOR CONTROL Page (EGI) (figure)	3-64.17
AN/ASN-137 DATA Menu Page (EGI)	3-64.15	AN/ASN-137 NAV SENSOR CONTROL Page (EGI) VAB operations (table)	3-64.17
AN/ASN-137 DATA Top Leve Menu Page (EGI) (Figure)	3-64.15	AN/ASN-137 NAV STATUS Page (EGI) (figure)	3-64.16
AN/ASN-137 Data Transfer Unit (DTU) Page (EGI)	3-64.18	AN/ASN-137 NAV Top Level Page (EGI) (figure)	3-64.5
AN/ASN-137 DTU Page (EGI) (figure)	3-64.18	AN/ASN-137 NAV Top Level Page Variable Action Buttons (VAB) (EGI) (table)	3-64.4
AN/ASN-137 DTU Page (EGI) VAB operations (table)	3-64.19		
AN/ASN-137 FDLS Page (EGI)	3-64.9	AN/ASN-137 Navigation Status (NAV STATUS) (EGI)	3-64.16
AN/ASN-137 FDLS Page (EGI) (figure)	3-64.10	AN/ASN-137 Offset Update (OFS UPDATE) Page (EGI)	3-64.9
AN/ASN-137 FLPN CDU Displays (EGI)	3-64.10		
AN/ASN-137 FLPN Waypoint Coordinate (WPT COORD) Page (EGI)	3-64.11	AN/ASN-137 OFS UPDATE Page (EGI) (figure)	3-64.9
AN/ASN-137 FLPN Waypoint Coordinate (WPT COORD) Selected Waypoint Coordinates (EGI) (figure)	3-64.11	AN/ASN-137 OFS UPDATE Page (EGI) Variable Action Buttons (VAB) (table)	3-64.9
		AN/ASN-137 Pilot High Action Display (HAD) (EGI)	3-64.25
AN/ASN-137 FLPN Waypoint Coordinate (WPT COORD) VAB operations with Scratchpad Empty (EGI) (table)	3-64.12	AN/ASN-137 Program Menu (PGM MENU) Page	3-64.21
		AN/ASN-137 Program Menu (PGM MENU) Page (figure)	3-64.21
AN/ASN-137 FLPN Waypoint Coordinate (WPT COORD) VAB operations (EGI) (table)	3-64.12	AN/ASN-137 Program Menu (PGM MENU) Page VAB operations (EGI) (table)	3-64.21
AN/ASN-137 FLPN Waypoint List (WPT LIST) Page (EGI)	3-64.10		
AN/ASN-137 FLPN Waypoint List (WPT LIST) Second Level Page (EGI) (figure)	3-64.10	AN/ASN-137 READ Page (EGI)	3-64.23
		AN/ASN-137 READ Page (EGI) (figure)	3-64.23
AN/ASN-137 FLPN Waypoint List (WPT LIST) Top Level Page (EGI) (figure)	3-64.10	AN/ASN-137 READ Page (EGI) VAB operations (table)	3-64.23
AN/ASN-137 GPS STATUS Page (EGI)	3-64.20		

Change 3   Index 2.1

TM 1-1520-138-10

Subject	Page No.
AN/ASN-137 Weapon Control (WPN CONTROL) Page (EGI)	3-64.13
AN/ASN-137 Weapon Control (WPN CONTROL) Page (EGI) (figure)	3-64.13
AN/ASN-137 Weapon Control (WPN CONTROL) Page CPG Range Source (EGI) (table)	3-64.14
AN/ASN-137 Weapon Control (WPN CONTROL) Page CPG Video Target Report Data (EGI) (figure)	3-64.15
AN/ASN-137 Weapon Control (WPN CONTROL) Page VAB operations (EGI) (table)	3-64.14
AN/ASN-137 Zeroize Page (EGI)	3-64.19

Subject	Page No.
AN/ASN-137 Zeroize Page (EGI) (figure)	3-64.19
AN/ASN-137 Zeroize Page (EGI) VAB operations (table)	3-64.20
AN/AVR-2A(V)1 Laser Detecting Set	4-78
Annunciators, Pilot and CPG	2-9
Antenna, RT-1167C/ARC-164(V) and HAVE QUICK Radios	3-14
Antenna, RT-1296/APX-100(V)1(IFF)	3-66
Antenna, RT-1557/APX-l00(V)1(IFF)	3-66
Antennas, AN/ARC-186(V)	3-10
Antennas, AN/ARC-201	3-24
Antennas, AN/ARN-89	3-30

Index 2.2   Change 5

Subject	Page No.
Anti-Ice Control Panels, Pilot and CPG (figure)	2-60
Anti-Ice Operation, Engine and Engine Inlet	2-24
Anti-Ice System, Engine and Engine Inlet	2-24
Anti-Ice, Engine Inlet and Nose Gearbox	2-61
Anti-Ice/De-Ice, Pitot Tube and Air Data Sensor	2-60
Anti-Ice/De-Ice, PNVS and TADS	2-61
Anti-Ice/De-Ice, Windshield	2-60
Anti-Icing and De-Icing	2-60
Anticollision Lights	2-74
Appendix A, References, Description	1-1
Appendix B, Abbreviations and Terms, Description	1-1
Approved Fuels (table)	2-93
Approved Oils (table)	2-95
APU (Auxiliary Power Unit), General	2-72
APU and Engine Fire Extinguishing System	2-19
APU Compartment Fire	9-12
APU Control Panel	2-72
APU Control Panel (figure)	2-73
APU Controller	2-72
APU Electrical System	2-72
APU FIRE PULL Handle Illumination in Flight	9-12
APU Fuel Supply	2-40
APU Fuel System	2-72
APU Lubrication System	2-72
APU Operational Limits	5-9
Area Weapon System, 30mm	4-10
Area Weapon System, 30mm (figure)	4-11
Arm, (Weight and Balance Definition)	6-3
Armament Configurations, Authorized	4-10
Armament Control Panels, CPG	4-29
Armament Control Panels, Pilot	4-24
Armament Inflight Procedures	4-66
Armament Postflight Procedures	4-68
Armament Preflight Procedures	4-61
Armor Protection (figure)	2-17
Army Aviation Safety Program	1-1
ASE Control Panel (figure)	2-46

Subject	Page No.
Audio Warning Signals (table)	2-85
Audio Warning System, Headset	2-85
Authorized Armament Configurations	4-10
Automatic Direction Finder Set, AN/ARN-149(V)3	3-32
Automatic Direction Finder Set, AN/ARN-149(V)3 Antenna	3-32
Automatic Direction Finder Set, AN/ARN-149(V)3 Controls and Functions	3-32
Automatic Direction Finder Set, AN/ARN-149(V)3 Modes of Operation	3-33
Automatic Direction Finder Set, AN/ARN-149(V)3 Operating Procedures	3-34
Autorotation Glide Chart (figure)	9-8
Auxiliary Fuel Tank	2-40
Auxiliary Fuel Tanks	6-6
Auxiliary Fuel Tanks, Fuel Loading (table)	6-9
Average Arm, (Weight and Balance Definition)	6-3
Aviation Life Support Equipment (ALSE)	8-1
Avionics Description	3-1
Avionics Equipment Configurations	3-1
Avionics Power Supply	3-1
AWS Dynamic Harmonization	4-10
AWS Harmonization Procedures	4-66
AWS Harmonization Wide FOV Corrector Guide (figure)	4-66.1
AWS Harmonization Narrow FOV Corrector Guide (figure)	4-66.2

B

Subject	Page No.
Backup Bus Controller (BBC)	4-14
Backup Control System (BUCS)	2-47
Backup Control System Failure	9-14
Balance Definitions	6-3
Basic Moment, (Weight and Balance Definition)	6-3
Basic Weight, (Definition)	6-3
Battery	2-66
Bay, Survival Kit	6-1
Before Exterior Check	8-2

Subject	Page No.
Before Landing Check	8-14
Before Leaving the Helicopter	8-16
Before Starting APU, CPG	8-10
Before Starting APU, Pilot	8-9
Before Starting Engines, Pilot	8-12
Before Takeoff Check	8-14
Before Taxi Check	8-13
Bleed Air, Engine No. 1	2-55
Boresight, TADS	4-63
Brake, Rotor	2-57
Brakes, Landing Gear	2-8
Briefing, Crew	8-1

C

Subject	Page No.
C-10414(V)3/ARC or C-11746(V)4/ARC Controls and Functions	3-7
C-10414(V)3/ARC or C-11746(V)4/ARC, Intercommunication System (ICS)	3-7
C-10414(V)3/ARC or C-11746(V)4/ARC Modes of Operation	3-9
C-10414(V)3/ARC and Remote Transmitter Selector, Control Panel (figure)	3-7
C-11746(V)4/ARC and Remote Transmitter Indicator Panel, Control Panel (figure)	3-7
Canopy and Windshield Cleaning	2-96
Canopy Jettison Handle (figure)	2-2
Canopy Jettison System	2-2
Canopy Panels and Windshield	2-2
Canopy Panels, General	2-2
Cargo Loading	6-16
Cargo, Extra	6-16
Caution/Warning Annunciators, Engine	2-33
Caution/Warning Light Corrective Actions (table)	9-15
Caution/Warning Light Emergency Procedures	9-15
Caution/Warning Light Segments, CPG (table)	2-83
Caution/Warning Light Segments, Pilot (table)	2-80.1
Caution/Warning Lights, Fuel System	2-40

Subject	Page No.
Caution/Warning Panels, Pilot and CPG	2-78
Caution/Warning Panels, Pilot and CPG (figure)	2-79
Caution/Warning Panels, Pilot and CPG (Modified) (figure)	2-80
CBHK Data Validation	4-60
CBR Filters/Blowers (table)	6-15
CDU CP1252, Used with AN/ASN-128 DNS (figure)	3-35
CDU Displays, AN/ASN-137	3-47
CDU Displays, AN/ASN-137 (EGI)	3-64.5
CDU IP1552, Used with AN/ASN-137 DNS (figure)	3-45
Center of Gravity (Computation)	6-17
Center of Gravity (Definition)	6-3
Center of Gravity Limits	5-10
Center of Gravity Limits (figure)	6-18
Center of Gravity Limits Chart	6-17
Center of Gravity Limits, (Definition)	6-3
Center of Gravity Management	6-4
Chaff Cartridges	6-11
Chaff Dispenser and Cartridges (table)	6-15
Chaff Dispenser and Countermeasures Control Panels (figure)	4-70
Chart C-Basic Weight and Balance Record	6-3
Chart, Airspeed Operating Limits (figure)	5-12
Chart, Autorotation Glide (figure)	9-8
Chart, Climb-Descent -701 Engine (figure)	7-73
Chart, Climb-Descent -701C Engine (figure)	7A-70
Chart, Cruise Example -701 Engine (figure)	7-15
Chart, Cruise Example -701C Engine (figure)	7A-17
Chart, Cruise Example, Sea Level -701 Engine (figure)	7-16
Chart, Cruise Example, Sea Level -701C Engine (figure)	7A-18
Chart, Drag and Authorized Armament Configurations -701 Engine (figure)	7-70

Subject	Page No.
Chart, Drag and Authorized Armament Configurations -701C Engine (figure)	7A-67
Chart, Flight Envelope (figure)	5-15
Chart, Hover Ceiling, 10 Min Limit -701C Engine (figure)	7A-12
Chart, Hover Ceiling, 30 Min Limit -701 Engine (figure)	7-10
Chart, Hover Ceiling, 30 Min Limit -701C Engine (figure)	7A-11
Chart, Hover, -701 Engine (figure)	7-12
Chart, Hover, -701C Engine (figure)	7A-14
Chart, Maximum Torque Available 10 Min Limit -701C Engine (figure)	7A-6
Chart, Maximum Torque Available 2.5 Min Limit -701 Engine (figure)	7-6
Chart, Maximum Torque Available 2.5 Min Limit -701C Engine (figure)	7A-7
Chart, Maximum Torque Available 30 Min Limit -701 Engine (figure)	7-5
Chart, Maximum Torque Available 30 Min Limit -701C Engine (figure)	7A-5
Chart, Temperature Conversion -701 Engine (figure)	7-3
Chart, Temperature Conversion -701C Engine (figure)	7A-3
Chart, Torque Conversion -701 Engine (figure)	7-8
Chart, Torque Conversion -701C Engine (figure)	7A-9
Chart, Torque Factor -701 Engine (figure)	7-7
Chart, Torque Factor -701C Engine (figure)	7A-8
Charts, Cruise, 10,000 Feet -701 Engine (figure)	7-48
Charts, Cruise, 10,000 Feet -701C Engine (figure)	7A-47
Charts, Cruise, 12,000 Feet -701 Engine (figure)	7-54
Charts, Cruise, 12,000 Feet -701C Engine (figure)	7A-52
Charts, Cruise, 14,000 Feet -701 Engine (figure)	7-59
Charts, Cruise, 14,000 Feet -701C Engine (figure)	7A-57

Subject	Page No.
Charts, Cruise, 16,000 Feet -701 Engine (figure)	7-64
Charts, Cruise, 16,000 Feet -701C Engine (figure)	7A-62
Charts, Cruise, 2,000 Feet -701 Engine (figure)	7-24
Charts, Cruise, 2,000 Feet -701C Engine (figure)	7A-25
Charts, Cruise, 4,000 Feet -701 Engine (figure)	7-31
Charts, Cruise, 4,000 Feet -701C Engine (figure)	7A-31
Charts, Cruise, 6,000 Feet -701 Engine (figure)	7-38
Charts, Cruise, 6,000 Feet -701C Engine (figure)	7A-37
Charts, Cruise, 8,000 Feet -701 Engine (figure)	7-43
Charts, Cruise, 8,000 Feet -701C Engine (figure)	7A-42
Charts, Cruise, Sea Level -701 Engine (figure)	7-17
Charts, Cruise, Sea Level -701C Engine (figure)	7A-19
Check, After Landing	8-15
Check, Before Exterior	8-2
Check, Before Landing	8-14
Check, Before Takeoff	8-14
Check, Before Taxi	8-13
Check, Interior, CPG	8-7
Check, Interior, Pilot	8-6
Check, Preflight	8-2
Check, Taxi	8-14
Checklist, General	8-2
Chemical, Biological, and Radiological (CBR) Filter/Blower	2-21
Chemical, Biological, and Radiological (CBR) Filters/Blowers	6-11
Chemical, Biological, Radiological (CBR) Filter/Blower Mounting Bracket	2-16
Chip Detector, Engine	2-28
Circuit Breaker Panels, CPG (figure)	2-71

Subject	Page No.
Circuit Breaker Panels, Pilot (figure)	2-70
Climb-Descent Chart -701 Engine (figure)	7-73
Climb-Descent Chart -701C Engine (figure)	7A-70
Climb-Descent Performance, Use of Chart -701 Engine	7-72
Climb-Descent Performance, Use of Chart -701C Engine	7A-69
Climb-Descent, Performance Conditions -701 Engine	7-72
Climb-Descent, Performance Conditions -701C Engine	7A-69
Climb-Descent, Performance Description -701 Engine	7-72
Climb-Descent, Performance Description -701C Engine	7A-69
Cold Section Module, Engine	2-22
Cold Weather Operations	8-19
Collective and Cyclic Mission Equipment Switches	4-22
Collective Stick and Cyclic Stick Grip Controls (figure)	2-44
Collective Stick Switch Functions (table)	4-23
Collective Sticks	2-42
Collective Switches	4-23
Common Flight Instruments	2-76
Communication/Navigation Equipment (table)	3-2
Compartments and Stations, Helicopter	6-1
Compartments, Crew	2-16
Compartments, Equipment Stowage	2-2
Compressor Stall, Engine	9-9
Conference Capability, AN/ARC-164(V) and HAVE QUICK Radios	3-17
Configuration, Wing Stores	5-17
Consoles, Pilot and CPG	2-9
Control and Display Functions, AN/ASN-137	3-45
Control Consoles, CPG (figure)	2-15
Control Consoles, Pilot (figure)	2-14
Control Panel, AN/ARC-186(V) (figure)	3-10
Control Panel, AN/ARC-201 (figure)	3-24
Control Panel, AN/ARN-89 (figure)	3-30
Control Panel, APU (figure)	2-73

Subject	Page No.
Control Panel, ASE (figure)	2-46
Control Panel, C-10414(V)3/ARC and Remote Transmitter Selector (figure)	3-7
Control Panel, C-11746(V)4/ARC and Remote Transmitter Indicator Panel (figure)	3-7
Control Panel, Countermeasures	4-70
Control Panel, RT-1296/APX-100(V)(IFF) (figure)	3-67
Control Panel, RT-1557/APX-100(V)1(IFF) (figure)	3-67
Control Panels, Chaff Dispenser and Countermeasures (figure)	4-70
Control System, Backup (BUCS)	2-47
Control System, Flight	2-42
Control System, Primary Flight (figure)	2-43
Controls and Displays, AN/ASN-128	3-37
Controls and Displays, AN/ASN-137	3-46
Controls and Displays, AN/ASN-137 (EGI)	3-64.1
Controls and Function, AN/ARN89	3-30
Controls and Functions, AN/ARC-186(V)	3-10
Controls and Functions, AN/ARC-201	3-24
Controls and Functions, AN/ARN-149(V)3	3-32
Controls and Functions, C-10414(V)3/ARC or C-11746(V)4/ARC	3-7
Controls and Functions, HARS	3-35
Controls and Functions, RT-1167C/ARC-164(V) and HAVE QUICK Radios	3-14
Controls and Functions, RT-1296/APX-100(V)1(IFF)	3-66
Controls and Functions, RT-1557/APX-100(V)1(IFF)	3-66
Controls and Functions, TSEC/KY-58	3-20
Controls, Indicators, and Functions Horizontal Situation Indicator	3-62
Controls, Indicators, and Functions, AN/APN-2090	3-71
Cooling, Engine	2-22
Countermeasure Control Panels	4-70
Covers, Protective	2-96
CPG - After Starting APU	8-12

Subject	No. Page	Subject	Page No.
CPG - Armament Control Panels	4-29	Cruise Chart, Example -701 Engine (figure)	7-15
CPG - Before Starting APU	8-10	Cruise Chart, Example -701C Engine (figure)	7A-17
CPG - Caution/Warning Light Segments (table)	2-83	Cruise Chart, Example, Sea Level -701 Engine (figure)	7-16
CPG - Circuit Breaker Panels (figure)	2-71	Cruise Chart, Example, Sea Level -701C Engine (figure)	7A-18
CPG - Control Consoles (figure)	2-15	Cruise Charts, 10,000 Feet -701 Engine (figure)	7-48
CPG - Crew Duties	8-1	Cruise Charts, 10,000 Feet -701C Engine (figure)	7A-47
CPG - Crew Station	6-1	Cruise Charts, 12,000 Feet -701 Engine (figure)	7-54
CPG - Engine Instruments	2-33	Cruise Charts, 12,000 Feet -701C Engine (figure)	7A-52
CPG - Engine Shutdown	8-16	Cruise Charts, 14,000 Feet -701 Engine (figure)	7-59
CPG - Fire Control Panel	4-30	Cruise Charts, 14,000 Feet -701C Engine (figure)	7A-57
CPG - Fire Control Panel (figure)	4-31	Cruise Charts, 16,000 Feet -701 Engine (figure)	7-64
CPG - Fire Control Panel Functions (table)	4-32	Cruise Charts, 16,000 Feet -701C Engine (figure)	7A-62
CPG - Flight Instruments	2-77	Cruise Charts, 2,000 Feet -701 Engine (figure)	7-24
CPG - Fuel Control Panel	2-39	Cruise Charts, 2,000 Feet -701C Engine (figure)	7A-25
CPG - Fuel Control Panel (figure)	2-39	Cruise Charts, 4,000 Feet -701 Engine (figure)	7-31
CPG - Fuel Control Panel Switch Functions (table)	2-39	Cruise Charts, 4,000 Feet -701C Engine (figure)	7A-31
CPG - Instrument Panel (figure)	2-13	Cruise Charts, 6,000 Feet -701 Engine (figure)	7-38
CPG - Interior Check	8-7	Cruise Charts, 6,000 Feet -701C Engine (figure)	7A-37
CPG - Missile Control Panel	4-29	Cruise Charts, 8,000 Feet -701 Engine (figure)	7-43
CPG - Missile Control Panel (figure)	4-29	Cruise Charts, 8,000 Feet -701C Engine (figure)	7A-42
CPG - Missile Control Panel Functions (table)	4-30	Cruise Charts, Additional Uses -701 Engine	7-14
CPG - Station Diagram (figure)	2-11	Cruise Charts, Additional Uses -701C Engine	7A-16
CPG and Pilot Master Caution Panel	2-77		
Crew Briefing	8-1		
Crew Compartments	2-16		
Crew Duties	8-1		
Crew Duties CPG	8-1		
Crew Duties Pilot	8-1		
Crew Introduction	8-1		
Crew Moments (table)	6-10		
Crew Requirements, Minimum	5-1		
Crew Station, CPG	6-1		
Crew Station, Pilot	6-1		
Crewmember Seat - Both Crew Stations (figure)	2-18		
Crewmember Seats	2-16		

TM 1-1520-238-10

Subject	Page No.
Cruise Charts, Airspeed -701 Engine	7-13
Cruise Charts, Airspeed -701C Engine	7A-15
Cruise Charts, Change in Frontal Area -701 Engine	7-13
Cruise Charts, Change in Frontal Area -701C Engine	7A-15
Cruise Charts, Fuel Flow -701 Engine	7-13
Cruise Charts, Fuel Flow -701C Engine	7A-15
Cruise Charts, Maximum Endurance and Rate of Climb -701 Engine	7-13
Cruise Charts, Maximum Endurance and Rate of Climb -701C Engine	7A-15
Cruise Charts, Maximum Range -701 Engine	7-13
Cruise Charts, Maximum Range -701C Engine	7A-15
Cruise Charts, Sea Level -701 Engine (figure)	7-17
Cruise Charts, Sea Level -701C Engine (figure)	7A-19
Cruise Charts, Torque -701 Engine	7-13
Cruise Charts, Torque -701C Engine	7A-15
Cruise Performance, Use of Charts -701 Engine	7-13
Cruise Performance, Use of Charts -701C Engine	7A-15
Cruise, Performance Conditions -701 Engine	7-14
Cruise, Performance Conditions -701C Engine	7A-16
Cruise, Performance Description -701 Engine	7-13
Cruise, Performance Description -701C Engine	7A-15
CSC Panel Control and Indicator Functions (table)	3-8
Cyclic and Collective Mission Equipment Switches	4-22
Cyclic Stick Grip and Collective Stick Controls (figure)	2-44
Cyclic Stick Switch Functions (table)	4-22
Cyclic Sticks	2-42
Cyclic Switches	4-22

D

Subject	Page No.
Danger Areas	2-1
Danger Areas (figure)	2-7
Danger Areas, Air Flow	2-1
Danger Areas, Canopy Jettison	2-1
Danger Areas, Exhaust Gases	2-1
Danger Areas, Illustrated	2-1
Danger Areas, Laser	2-2
DASE Malfunction	9-14
Data Entry Keyboard (DEK)	4-8
Data Entry Keyboard (figure)	4-8
Data Entry Keyboard Control Functions (table)	4-8
Data Entry Keyboard Operation	4-8
Data Entry Procedures, AN/ASN-137	3-60
Data Transfer Unit (DTU)	3-64
DC Power Supply System	2-66
De-Icing and Anti-Icing	2-60
De-Icing System, Rotor Blade	2-61
Definition of Emergency Terms	9-1
Definitions, Balance	6-3
Definitions, Symbol	8-2
Definitiona, Weight	6-3
Defogging System	2-60
Description, Avionics	3-1
Description, Operators Manual	1-1
Desert and Hot Weather Operations	8-20
Destruction of Army Material to Prevent Enemy Use	1-1
Detecting Set, Laser AN/AVR-2A(V)1	4-78
Digital Automatic Stabilization Equipment (DASE)	2-45
Digital Electronic Control Unit (DECU)	2-26
Dimensions, Principal	2-1
Dimensions, Principal, (figure)	2-6
Dimming, Master Caution Panel and Caution/Warning Panel Segment Lights	2-75
Direction Finder Set, AN/ARNSS	3-30
Direction Finder Set, AN/ARN-89 Antennas	3-30

Index 8 Change 5

Subject	Page No.
Direction Finder Set, AN/ARN-89 Controls and Function	3-30
Directional Control Pedals	2-42
Discrepancies, Performance -701 Engine	7-2
Discrepancies, Performance -701C Engine	7A-2
Dispenser Control Panel Control/Indicator Function (table)	4-70.1
Dispenser Kit M-130	4-70
Ditching (Power Off)	9-14
Ditching (Power On)	9-14
Ditching and Landing	9-13
Doppler Navigation Set, AN/ASN-128	3-37
Doppler Navigation Set, AN/ASN-137	3-46
Drag Chart and Authorized Armament Configurations -701 Engine (figure)	7-70
Drag Chart and Authorized Armament Configurations -701C Engine (figure)	7A-67
Drag Performance, Use of Chart -701 Engine	7-69
Drag Performance, Use of Chart -701C Engine	7A-66
Drag, Performance Conditions -701 Engine	7-69
Drag, Performance Conditions -701C Engine	7A-66
Drag, Performance Description -701 Engine	7-69
Drag, Performance Description -701C Engine	7A-66
Drive Shafts, Tail Rotor	2-58
Dual Engine Failure	9-7
Dual Engine Failure Low Altitude/Low Airspeed and Cruise	9-7
Dual IHADSS/HDU Failure	9-21
During Flight, Engine Restart	9-9
Duties, Crew	8-1

E

Subject	Page No.
ECS Control Panel (figure)	2-64
ECS Emergency Operation	2-65
ECS Normal Operation	2-64
Effectivity and Series Codes	1-2
Egress, Emergency	9-2

Subject	Page No.
Electrical Control Unit (ECU)	2-26
Electrical Fire in Flight	9-12
Electrical Power Distribution System (figure)	2-68
Electrical Power Sources, Fuel System	2-41
Electrical Power System	2-66
Electrical System Failures, General	9-13
Electrical System, APU	2-72
Electronic Module Switch Functions (table)	4-71
Embedded GPS Inertial (EGI)	3-63
Emergency Action, After	9-2
Emergency Egress	9-2
Emergency Engine Start, DECU Installed Only	9-9
Emergency Entrance	9-2
Emergency Equipment, General	2-19
Emergency Exits and Equipment	9-2
Emergency Exits and Equipment (figure)	9-3
Emergency Floodlight System	2-75
Emergency Landing in Wooded Areas (Power Off)	9-13
Emergency Oil System, Engine	2-28
Emergency Operation, (Avionics)	3-1
Emergency Operation, ECS	2-65
Emergency Procedures	2-21
Emergency Procedures, (Crew Briefing)	8-1
Emergency Procedures, Aircraft Systems, General	9-1
Emergency Procedures, Caution/Warning Light	9-15
Emergency Terms, Definition of	9-1
ENCU Malfunction	9-21
Engine Accessory Section Module	2-22
Engine Air Induction	2-22
Engine Alternator	2-28
Engine Alternator -701	2-28
Engine Alternator -701C	2-28
Engine Alternator Malfunction	9-22
Engine and APU Fire Detection	2-19
Engine and APU Fire Detector Testing	2-19

Subject	Page No.
Engine and APU Fire Extinguishing System..	2-19
Engine and Engine Inlet Anti-Ice Operation	2-24
Engine and Engine Inlet Anti-Ice System	2-24
Engine Anti-Ice Panel (figure)	2-25
Engine Bleed Air, No. 1	2-55
Engine Caution/Warning Annunciators	2-33
Engine Chip Detector	2-28
Engine Chop Control, Pilot and CPG	2-30
Engine Cold Section Module	2-22
Engine Compressor Stall	9-9
Engine Control System	2-30
Engine Cooling	2-22
Engine Emergency Oil System	2-28
Engine Emergency Start, DECU Installed Only	9-9
Engine Failure Flight Characteristics	9-4
Engine Failure, General	9-4
Engine Fire in Flight	9-12
Engine Fuel Boost Pump	2-25
Engine Fuel Control System	2-25
Engine Fuel Filter	2-25
Engine Fuel Pressure Warning System	2-28
Engine History Counter -701C	2-33
Engine History Recorder -701	2-33
Engine Hot Section Module	2-22
Engine Ignition System	2-28
Engine Inlet and Nose Gearbox Anti-Ice	2-61
Engine Instrument Test Panel (figure)..	2-32
Engine Instruments	2-31
Engine Load Demand System	2-31
Engine Nose Gearboxes	2-56
Engine Oil Cooler and Filter	2-28
Engine Oil System	2-28
Engine Oil Tank	2-28
Engine Overspeed Check Limitations	5-9
Engine Power Lever Quadrants	2-30
Engine Power Limitations	5-9
Engine Power Turbine Section Module	2-22

Subject	Page No.
Engine Restart During Flight	9-9
Engine Shutdown, CPG	8-16
Engine Shutdown, Pilot	8-15
Engine Start Limits	5-9
Engine Start Using APU	2-29
Engine Start Using Engine Bleed Air Source	2-29
Engine Start Using External Source	2-29
Engine Starter Limitations	5-9
Engine Starting (cold weather)	8-20
Engine Starting System	2-29
Engine T700-GE-701/T700-GE-701C (figure)	2-23
Engine Temperature Limitations	5-9
Engine/Fuselage Fire on Ground	9-12
Engines	2-22
Engines, General	2-1
Entrance, Emergency	9-2
Environmental Control System (ECS)	2-64
Environmental Restrictions	5-16
Equipment, (Crew Briefing)	8-1
Equipment, Aviation Life Support (ALSE)	8-1
Equipment, Mission, (weight/balance)	6-11
Equipment, Stowage Compartments	2-2
Exceeded Operational Limits	5-1
Exits and Equipment, Emergency	9-2
Exits and Equipment, Emergency (figure)	9-3
Explanation of Change Symbols	1-2
Exterior Check	8-3
Exterior Check - Left Side - Aft Fuselage/Empennage (Area 7)	8-5
Exterior Check - Left Side - Lower Center Fuselage and Nose (Area 11)	8-6
Exterior Check - Left Side - Mast (Area 10)	8-6
Exterior Check - Left Side - Rear Center Fuselage (Area 8)	8-5
Exterior Check - Left Side - Wing (Area 9)	8-6
Exterior Check - Right Side - Aft Fuselage/Empennage (Area 6)	8-5

Subject	Page No.
Exterior Check - Right Side - Lower Center Fuselage (Area 2)	8-3
Exterior Check - Right Side - Mast (Area 3)	8-3
Exterior Check - Right Side - Rear Center Fuselage (Area 5)	8-5
Exterior Check - Right Side - Wing (Area 4)	8-3
Exterior Check - Right Side - Under Side Fuselage (Area 1)	8-3
Exterior Check Diagram (figure)	8-4
Exterior Lighting System	2-74
External Air Source Receptacles	2-55
External Power Receptacle	2-67
External Stores and Stations (figure)	6-12
External Stores Subsystem (ESS)	4-35
Extinguishing and Fire Detection Controls (figure)	2-20
Extra Cargo	6-16

F

Subject	Page No.
Failure, BUCS	9-14
Failure, Dual Engine	9-7
Failure, Dual Engine, Low Altitude/Low Airspeed and Cruise	9-7
Failure, Dual IHADSS/HDU	9-21
Failure, Engine, Flight Characteristics	9-4
Failure, Engine, General	9-4
Failure, FCC	9-22
Failure, FLIR Cooler	9-22
Failure, IHADSS	4-43
Failure, IHADSS/HDU	9-21
Failure, Main Transmission Input Drive Clutch	9-11
Failure, PNVS	9-21
Failure, PNVS FLIR	4-43
Failure, Single Engine	9-4
Failure, Single Engine, Low Altitude/Low Airspeed and Cruise	9-4
Failure, Stabilator Auto/Manual Mode	9-14
Failure, Stabilator Automatic Mode	9-14
Failure, Symbol Generator	4-43

Subject	Page No.
Failure, Symbol Generator	9-22
Failure, TADS Electronic Unit	4-43
Failure, TADS Electronic Unit (TEU)	9-22
Failure, Turret	4-43
Failures and Malfunctions, Flight Control, General	9-14
Failures and Malfunctions, Mission Equipment	9-21
Failures and Malfunctions, -Rotors, Transmissions, and Drive System, General	9-9
Failures, Electrical System, General	9-13
Failures, Hydraulic System, General	9-13
Fault Codes - Signal Validation (figure)	2-27
Fault Detection/Location System (FD/LS)	4-16
Fault Detection/Location System On-Command Tests	4-16
FDLS Display, CDU (EGI)	3-64.9
Fire Control Computer (FCC)	4-14
Fire Control Computer Failure	9-22
Fire Control Panel Functions, CPG (table)	4-32
Fire Control Panel, CPG	4-30
Fire Control Panel, CPG (figure)	4-31
Fire Control Panel, Pilot	4-27
Fire Control Panel, Pilot (figure)	4-27
Fire Control Panel/Indicator Functions Pilot (table)	4-27
Fire Control System Messages	4-46
Fire Detection and Extinguishing Controls (figure)	2-20
Fire Detection, Engine and Auxiliary Power Unit (APU)	2-19
Fire Detector Testing, Engine and APU	2-19
Fire Extinguisher, Portable	2-19
Fire in Flight, Electrical	9-12
Fire in Flight, Engine	9-12
Fire on Ground, Engine/Fuselage	9-12
Fire, APU Compartment	9-12
Fires, General	9-12
First Aid Kits	2-21
Flight Characteristics, Engine Failure	9-4

Subject	Page No.
Flight Characteristics, General	8-18
Flight Control Failures and Malfunctions, General	9-14
Flight Control Servos	2-45
Flight Control System	2-42
Flight Controls	2-9
Flight Data, (Crew Briefing)	8-1
Flight Envelope Chart (figure)	5-15
Flight Instrument Procedures	8-17
Flight Instruments	2-76
Flight Preparation for (cold weather)	8-19
Flight Symbology	4-2
Flight Symbology Definitions (table)	4-3
Flight Symbology Modes (figure)	4-2
FLIR Cooler Failure	9-22
FLIR Operational Check	4-61
FLPN Display, CDU (EGI)	3-64.10
Formation Lights	2-74
Forms and Records	1-1
Fuel and Lubricant Specifications and Capacities (table)	2-92
Fuel Boost Pump, Engine	2-25
Fuel Control Panel, CPG	2-39
Fuel Control Panel, CPG (figure)	2-39
Fuel Control Panel, Pilot	2-34
Fuel Control Panel, Pilot (figure)	2-34
Fuel Control Panel, Pilot (Modified) (figure)	2-34
Fuel Control System, Engine	2-25
Fuel Filter, Engine	2-25
Fuel Loading	6-4
Fuel Loading, Aft Tank (table)	6-8
Fuel Loading, Forward Tank (table)	6-7
Fuel Management	6-4
Fuel Moments	6-6
Fuel Pressure Warning System, Engine	2-28
Fuel Quantity Indicators	2-40
Fuel System	2-34
Fuel System (figure)	2-37
Fuel System Caution/Warning Lights	2-40
Fuel System Electrical Power Sources	2-41

Subject	Page No.
Fuel System Servicing	2-86
Fuel System, APU	2-72
Fuel Tank, Auxiliary	2-40
Fuel Tanks, Auxiliary	6-6
Fuel Tanks, Auxiliary, Fuel Loading (table)	6-9
Fuel Weight and Moment	6-6
Fuels, Approved (table)	2-93
Fuselage, General	2-1

G

Subject	Page No.
Gearbox, Intermediate	2-58
Gearboxes, Engine Nose	2-56
GEN 1 and GEN 2 Caution Light On	9-13
General Arrangement, Aircraft	2-1
General Arrangement, Aircraft (figure)	2-3
General Conditions, Performance -701 Engine	7-2
General Conditions, Performance -701C Engine	7A-2
Generator, Symbol	4-16
Gravity Filling, Fuel	2-40
Gross Weight, (Definition)	6-3
Ground Air Source	2-96
Ground Air Source (figure)	2-96
Ground Clearance and Turning Radius	2-1
Ground Clearance and Turning Radius (figure)	2-6
Ground Operations During High Winds	8-21
Ground Test and Warmup (cold weather)	8-20
Guard Operation, RT-1167C/ARC-164(V) and HAVE QUICK Radios	3-17
Gun System, 30mm	4-66

H

Subject	Page No.
HARS (Heading and Attitude Reference System)	3-34
HARS Alignment Methods (table)	3-36
HARS Control and Functions (table)	3-36
HARS Control Panel (figure)	3-35
HARS Controls and Functions	3-35
HARS System Description	3-34

Subject	Page No.
HAVE QUICK Emergency Startup of TOD Clock RT-1167C/ARC-164(V)	3-18
HAVE QUICK Mode Operating Procedures RT-1167C/ARC-164(V)	3-18
HBCM Display, CDU (EGI)	3-64.8
Height Velocity Plots (figure)	9-5
Helicopter CG Movement When Loaded Items Are Expended (table)	6-5
Helicopter Class	6-1
Helicopter Compartments and Stations	6-1
Hellfire Launchers	6-11
Hellfire Missiles	6-11
Hellfire Missiles Loading (table)	6-13
High Action Display (figure)	4-47
High Action Display (HAD)	4-47
High Action Display Opposite Crew Station Weapon Control (table)	4-54
High Action Display Range and Range Source (table)	4-50
High Action Display Sight Status (table)	4-48
High Action Display Weapons Status (table)	4-51
High RPM Rotor (Warning Light On) Nr Failed High	9-11
History Counter, Engine -701C	2-33
History Recorder, Engine -701	2-33
Horizontal Situation Indicator	3-62
Horizontal Situation Indicator (figure)	3-62
Horizontal Situation Indicator Controls, Indicators, and Functions	3-62
Hot Section Module, Engine	2-22
Hover Ceiling Chart, 10 Min Limit -701C Engine (figure)	7A-12
Hover Ceiling Chart, 30 Min Limit -701 Engine (figure)	7-10
Hover Ceiling Chart, 30 Min Limit -701C Engine (figure)	7A-11
Hover Ceiling Performance, Use of Chart -701 Engine	7-9
Hover Ceiling Performance, Use of Chart -701C Engine	7A-10
Hover Ceiling, Performance Conditions -701 Engine	7-9
Hover Ceiling, Performance Conditions -701C Engine	7A-10

Subject	Page No.
Hover Ceiling, Performance Description -701 Engine	7-9
Hover Ceiling, Performance Description -701C Engine	7A-10
Hover Chart -701 Engine (figure)	7-12
Hover Chart -701C Engine (figure)	7A-14
Hover Performance, Use of Chart -701 Engine	7-11
Hover Performance, Use of Chart -701C Engine	7A-13
Hover, Performance Conditions -701 Engine	7-11
Hover, Performance Conditions -701C Engine	7A-13
Hover, Performance Description -701 Engine	7-11
Hover, Performance Description -701C Engine	7A-13
HSI Controls and Indicators (table)	3-62
Hydraulic Hand Pump	2-54
Hydraulic System Failures, General	9-13
Hydraulic System, Primary	2-50
Hydraulic System, Primary (figure)	2-51
Hydraulic System, Utility	2-52
Hydraulic System, Utility (figure)	2-53
Hydraulic Systems	2-50
Hydromechanical Unit (HMU)	2-25

I

Subject	Page No.
Ice and Rain (operation)	8-21
Icing Severity Meter	2-62
Icing Severity Meter and Press to Test Switch (figure)	2-62
Ignition System, Engine	2-28
IHADSS Boresight - CPG	4-61
IHADSS Boresight - Pilot	4-61
IHADSS Failure	4-43
IHADSS Subsystem	4-20
IHADSS Subsystem (figure)	4-21
IHADSS Subsystem Operation	4-20
IHADSS/HDU Failure	9-21
Index, Description	1-1

Subject	Page No.
Indicator, Stabilator Position (figure)	2-49
Indicators, Fuel Quantity	2-40
Indicators, Pilot and CPG	2-9
Inertia Reel Lock Lever, Shoulder Harness	2-16
Inflight Ice and Rain Operation	8-21
Infrared (IR) Suppression System	2-22
Infrared Countermeasures Set, AN/ALQ-144A,-144(V)3	4-69
Infrared Countermeasures Set Control Panel, AN/ALQ-144A,-144(V)3	4-69
Inspection and Maintenance Light	2-75
Instrument Flight	8-17
Instrument Flight Procedures	8-17
Instrument Marking Color Code	5-2
Instrument Markings (figure)	5-3
Instrument Operating Ranges and Markings	5-2
Instrument Panel, CPG (figure)	2-13
Instrument Panel, Pilot (figure)	2-12
Instrument Panels, Pilot and CPG	2-9
Instrument Test Panel, Engine (figure)	2-32
Instruments, Engine	2-31
Instruments, Engine, CPG	2-33
Instruments, Engine, Pilot	2-32
Instruments, Flight, Common	2-76
Instruments, Flight, CPG	2-77
Instruments, Flight, Pilot	2-76
Integrated Navigation System	3-63
Intercommunication System (ICS), C-1041(V)3/ARC or C-11746(V)4/ARC	3-7
Interior Check, CPG	8-7
Interior Check, Pilot	8-6
Interior Lighting System	2-75
Intermediate Gearbox	2-58
Introduction to Operators Manual, General	1-1

J

Subject	Page No.
Jettison Handle, Canopy (figure)	2-2
Jettison System, Canopy	2-2
Jettison, Wing Stores	9-21

K

Subject	Page No.
Kit, Dispenser, M-130	4-70
Kits, First Aid	2-21

L

Subject	Page No.
Landing and Ditching	9-13
Landing Gear	2-8
Landing Gear, Brakes	2-8
Landing Gear, Main	2-8
Landing Gear, Tail	2-8
Landing Limits	5-14
Landing, Slope/Rough Terrain	8-18
LAT/LONG Coordinate Data (table)	3-59
Launcher, Hellfire	6-11
Launchers, Rocket (2.75)	6-11
Left Aft Storage Bay	6-1
Light, Inspection and Maintenance	2-75
Light, Utility	2-75
Lighting Control Panels, Pilot and CPG (figure)	2-74
Lighting Equipment	2-74
Lighting System, Exterior	2-74
Lighting System, Interior	2-75
Lights, Anticollision	2-74
Lights, Dimming, Master Caution Panel and Caution/Warning Segments	2-75
Lights, Formation	2-74
Lights, Navigation	2-74
Limit Chart, Center of Gravity	6-17
Limitations, Engine Overspeed Check	5-9
Limitations, Engine Power	5-9
Limitations, Engine Starter	5-9
Limitations, Engine Temperature	5-9
Limitations, Rotor	5-2
Limitations, Weight	5-10
Limits, Airspeed Operating	5-11
Limits, Center of Gravity	5-10
Limits, Engine Start	5-9
Limits, Landing	5-14

Subject	Page No.
Limits, Maneuvering	5-14
Limits, Operational -701 Engine	7-1
Limits, Operational -701C Engine	7A-1
Limits, Pneumatic Source Inlet	5-9
Loading Cargo	6-16
Loading Data	6-3
Loss of Tail Rotor Thrust	9-9
Loss of Tail Rotor Thrust at Low Airspeed/Hover	9-10
Loss of Tail Rotor Thrust in Cruise Flight	9-10
Low RPM Rotor (Warning Light On) Nr Failed Low	9-12
Lubrication System, APU	2-72

M

Subject	Page No.
M-130 Chaff Dispenser Assembly	4-70.1
M-130 Chaff Dispenser Kit Electronic Module	4-70.1
M-130 Chaff Dispenser Kit Electronic Module Controls	4-70.1
M-130 Chaff Dispenser Kit Operation	4-71
M-130 Chaff Dispenser Kit Payload Module Assembly	4-70.1
M-130 Chaff Dispenser Kit Remote Safety Switch	4-70.1
M-130 Chaff Dispenser Kit Safety Procedures	4-71
Main Landing Gear	2-8
Main Rotor	2-59
Main Transmission	2-56
Main Transmission Input Drive Clutch Failure	9-11
Malfunction, DASE	9-14
Malfunction, ENCU	9-21
Malfunction, Engine Alternator	9-22
Malfunctions, Tail Rotor	9-9
Malfunction, Tail Rotor Fixed Pitch	9-11
Maneuvering Limits	5-14
Master Caution Panel Indications (table)	2-78
Master Caution Panel, Pilot and CPG	2-77
Master Caution Panel, Pilot and CPG (figure)	2-77

Subject	Page No.
Maximum Torque Available Chart, 30 Min Limit -701 Engine (figure)	7-5
Maximum Torque Available Chart, 30 Min Limit -701C Engine (figure)	7A-5
Maximum Torque Available Performance, Use of Chart -701 Engine	7-4
Maximum Torque Available Performance, Use of Chart -701C Engine	7A-4
Maximum Torque Available, 10 Min Limit -701C Engine (figure)	7A-6
Maximum Torque Available, 2.5 Min Limit -701 Engine (figure)	7-6
Maximum Torque Available, 2.5 Min Limit -701C Engine (figure)	7A-7
Maximum Torque Available, Performance Conditions -701 Engine	7-4
Maximum Torque Available, Performance Conditions -701C Engine	7A-4
Maximum Torque Available, Performance Description -701 Engine	7-4
Maximum Torque Available, Performance Description -701C Engine	7A-4
Method, Torque Factor -701 Engine	7-4
Method, Torque Factor -701C Engine	7A-4
Methods of Operation, AN/ASN-128	3-39
Minimum Crew Requirements	5-1
Missile Control Panel Control/Indicator Functions Pilot (table)	4-26
Missile Control Panel Functions, CPG (table)	4-30
Missile Control Panel, CPG	4-29
Missile Control Panel, CPG (figure)	4-29
Missile Control Panel, Pilot	4-26
Missile Control Panel, Pilot (figure)	4-26
Mission Equipment (weight/balance)	6-11
Mission Equipment Failures and Malfunctions	9-21
Mission Kits, Special	2-1
Mission Planning	8-1
Modes of Operation, AN/ARC-186(V)	3-12
Modes of Operation, AN/ARC-201	3-27
Modes of Operation, AN/ARN-149(V)3	3-33

Subject	Page No.
Modes of Operation, AN/ASN-128	3-38
Modes of Operation, AN/ASN-137	3-47
Modes of Operation, C-10414(V)3/ARC or C-11746(V)4/ARC	3-9
Modes of Operation, RT-1167C/ARC-164(V) and HAVE QUICK Radios	3-15
Moment, (Weight and Balance Definition)	6-3
Moment, Fuel Weight	6-6
Moment, Oil Weight	6-6
Moment, Personnel	6-10
Moments, Crew (table)	6-10
Mooring	2-96
Mounting Bracket, Chemical, Biological, Radiological (CBR) Filter/Blower	2-16
Multiplex Bus (1553) (figure)	4-15
Multiplex Bus Subsystem (MUX)	4-14

N

Subject	Page No.
Navigation Lights	2-74
Navigation/Communication Equipment	3-2
Nitrogen Inerting Unit (NIU)	2-41
Normal Mode Operating Procedures, RT-1167/ARC-164(V)	3-17
Normal Operation, RT-1296/APX-100(V)1(IFF)	3-69
Normal Operation, RT-1557/APX-100(V)1(IFF)	3-69
Normal Procedures, (Crew Briefing)	8-1

O

Subject	Page No.
OI Configuration	4-40
Oil Cooler and Filter, Engine	2-28
Oil System Servicing	2-87
Oil System, Engine	2-28
Oil Tank, Engine	2-28
Oil Weight and Moment	6-6
Oils, Approved (Table)	2-95
Operating Limits and Restrictions, General	5-1
Operating Procedures and Maneuvers	8-2

Subject	Page No.
Operating Procedures, AN/APR-39(V)1	4-72
Operating Procedures, AN/APR-39A(V)1	4-75
Operating Procedures, AN/ARC-186(V)	3-12
Operating Procedures, AN/ARC-201	3-27
Operating Procedures, AN/ARN-89	3-31
Operating Procedures, AN/ARN-149 ADF	3-34
Operating Procedures, AN/ASN-128	3-42
Operating Procedures, AN/ASN-137	3-53
Operating Procedures, TSEC/KY-58	3-23
Operating Ranges and Markings, Instrument	5-2
Operating Weight, (Definition)	6-3
Operation, AN/APN-209(V)	3-72
Operation, Cold Weather	8-19
Operation, HARS	3-64
Operation, RT-1296/APX-100(V)1(IFF)	3-68
Operation, RT-1557/APX-100(V)1(IFF)	3-68
Operation, Stabilator	8-18
Operation, Turbulence and Thunderstorm	8-20
Operational Limits -701 Engine	7-1
Operational Limits -701C Engine	7A-1
Operational Limits, APU	5-9
Operational Limits, Exceeded	5-1
Operations, Desert and Hot Weather	8-20
Operations, Ground, During High Winds	8-21
Optical Relay Tube (ORT) Controls and Displays	4-36
ORT and Hand Controls (figure)	4-36
ORT and Hand Controls/Display Functions (table)	4-37
Overspeed and Drain Valve (ODV)	2-25

P

Subject	Page No.
Parking	2-96
Performance Charts, Chart Explanation -701 Engine	7-2
Performance Charts, Chart Explanation -701C Engine	7A-2
Performance Charts, Data Basis -701 Engine	7-2

Subject	Page No.
Performance Charts, Data Basis -701C Engine	7A-2
Performance Charts, Reading the Charts -701 Engine	7-2
Performance Charts, Reading the Charts -701C Engine	7A-2
Performance Charts, Use of -701 Engine	7-2
Performance Charts, Use of -701C Engine	7A-2
Performance Conditions, Aircraft Variation -701 Engine	7-2
Performance Conditions, Aircraft Variation -701C Engine	7A-2
Performance Conditions, Climb-Descent -701 Engine	7-72
Performance Conditions, Climb-Descent -701C Engine	7A-69
Performance Conditions, Configurations -701 Engine	7-2
Performance Conditions, Configurations -701C Engine	7A-2
Performance Conditions, Cruise -701 Engine	7-14
Performance Conditions, Cruise -701C Engine	7A-16
Performance Conditions, Drag -701 Engine	7-69
Performance Conditions, Drag -701C Engine	7A-66
Performance Conditions, Hover -701 Engine	7-11
Performance Conditions, Hover -701C Engine	7A-13
Performance Conditions, Hover Ceiling -701 Engine	7-9
Performance Conditions, Hover Ceiling -701C Engine	7A-10
Performance Conditions, Instrument Variation -701 Engine	7-2
Performance Conditions, Instrument Variation -701C Engine	7A-2
Performance Conditions, Maximum Torque Available -701 Engine	7-4
Performance Conditions, Maximum Torque Available -701C Engine	7A-4

Subject	Page No.
Performance Conditions, Pilot Technique -701 Engine	7-2
Performance Conditions, Pilot Technique -701C Engine	7A-2
Performance Conditions, Rigging -701 Engine	7-2
Performance Conditions, Rigging -701C Engine	7A-2
Performance Data -701 Engine	7-1
Performance Data -701C Engine	7A-1
Performance Description, Climb-Descent -701 Engine	7-72
Performance Description, Climb-Descent -701C Engine	7A-69
Performance Description, Cruise -701 Engine	7-13
Performance Description, Cruise -701C Engine	7A-15
Performance Description, Drag -701 Engine	7-69
Performance Description, Drag -701C Engine	7A-66
Performance Description, Hover -701 Engine	7-11
Performance Description, Hover -701C Engine	7A-13
Performance Description, Hover Ceiling -701 Engine	7-9
Performance Description, Hover Ceiling -701C Engine	7A-10
Performance Description, Maximum Torque Available -701 Engine	7-4
Performance Description, Maximum Torque Available -701C Engine	7A-4
Performance Discrepancies -701 Engine	7-2
Performance Discrepancies -701C Engine	7A-2
Performance General Conditions -701 Engine	7-2
Performance General Conditions -701C Engine	7A-2
Performance Specific Conditions -701 Engine	7-2
Performance Specific Conditions -701C Engine	7A-2

Subject	Page No.
Personnel General	6-10
Personnel Moment	6-10
Personnel Weight	6-10
Pilot - Aerial Rocket Control Panel (ARCS)	4-24
Pilot - Aerial Rocket Control Panel (figure)	4-24
Pilot - Aerial Rocket Control Panel/Indicator Functions (table)	4-24
Pilot - After Starting APU	8-11
Pilot - Armament Control Panels	4-24
Pilot - Before Starting APU	8-9
Pilot - Before Starting Engines	8-12
Pilot - Caution/Warning Light Segments (table)	2-80.1
Pilot - Circuit Breaker Panels (figure)	2-70
Pilot - Control Consoles (figure)	2-14
Pilot - Crew Duties	8-1
Pilot - Crew Station	6-1
Pilot - Electrical Power Control Panel (figure)	2-67
Pilot - Engine Shutdown	8-15
Pilot - Fire Control Panel	4-27
Pilot - Fire Control Panel (figure)	4-27
Pilot - Fire Control Panel/Indicator Functions (table)	4-27
Pilot - Flight Instruments	2-76
Pilot - Fuel Control Panel	2-34
Pilot - FuelControl Panel (figure)	2-34
Pilot - Fuel Control Panel (Modified) (figure)	2-34
Pilot - Fuel Control Panel Switch Functions (table)	2-36
Pilot - Instrument Panel (figure)	2-12
Pilot - Interior Check	8-6
Pilot - Missile Control Panel	4-26
Pilot - Missile Control Panel (figure)	4-26
Pilot - Missile Control Panel Control/Indicator Functions (table)	4-26
Pilot - Night Vision Sensor (PNVS) AN/AAQ-11	4-42
Pilot - Starting APU	8-10
Pilot - Starting Engines	8-12

Subject	Page No.
Pilot - Starting Engines - External Pressurized Air Source	8-10
Pilot - Station Diagram (figure)	2-10
Pilot and CPG Anti-Ice Control Panels (figure)	2-60
Pilot and CPG Caution/Warning Panels	2-78
Pilot and CPG Caution/Warning Panels (figure)	2-79
Pilot and CPG Caution/Warning Panels (Modified) (figure)	2-80
Pilot and CPG Engine Chop Control	2-30
Pilot and CPG Indicators, Instrument Panels, Consoles, and Annunciators	2-9
Pilot and CPG Lighting Control Panels (figure)	2-74
Pilot and CPG Master Caution Panel	2-77
Pilot and CPG Master Caution Panel (figure)	2-77
Pilot Emergency Check Overspeed Test Panel Power Lever Quadrant and CPG Power Lever Quadrant (figure)	2-30
Pilot Engine Instruments	2-32
Pitot Static System	2-77
Pitot Tube and Air Data Sensor Anti-Ice/De-Ice	2-60
Planning, Mission	8-1
Pneumatic Source Inlet Limits	5-9
PNVS and TADS Anti-Ice/De-Ice	2-61
PNVS Electrical Power	4-43
PNVS Equipment Data (table)	4-43
PNVS Failure	9-21
PNVS FLIR Failure	4-43
Point Target (HELLFIRE) Perferred Missile Firing Order (figure)	4-13
Point Target Weapon System, In Flight Procedures	4-66.3
Point Target Weapons System	4-13
Portable Fire Extinguisher	2-19
Power Supply, Avionics	3-1
Power Train	2-56
Power Train (figure)	2-57
Power Turbine Section Module, Engine	2-22
Power Up Procedures, AN/ASN-137	3-51

Subject	Page No.
PPOS - Partial Entry Rules (table)	3-57
Preflight Check	8-2
PreflightRunup Ice and Rain Operation	8-21
Preparation for Flight (cold weather)	3-19
Pressure Refueling	2-40
Pressurized Air System (PAS)	2-54
PRI HYD PSI and UTIL HYD PSI Light On	9-13
Primary Flight Control System (figure)	2-43
Primary Hydraulic System	2-50
Primary Hydraulic System (figure)	2-51
Principal Dimensions	2-1
Principal Dimensions (figure)	2-5
Procedure, Torque Factor -701 Engine	7-4
Procedure, Torque Factor -701C Engine	7A-4
Procedures and Maneuvers, Operating	8-2
Procedures, Emergency	2-21
Protective Covers	2-96
Protective Covers, Mooring, Towing (figure)	2-97
Pump, Hand, Hydraulic	2-54
Pylon Assemblies	4-35

Q

R

Subject	Page No.
Radar Countermeasures Set, AN/ALQ-136	4-69
Radar Countermeasures Set Control Panel, AN/ALQ-136	4-2
Radar Warning Discriminator Off Mode Display (figure)	4-73
Radar Warning Discriminator On Mode Display (figure)	4-74
Radar Warning Display (figure)	4-76
Radar Warning OFD and EID Display (figure)	4-77
Radar Warning Receiver Check Display (figure)	4-77
Radar Warning Self-Test Mode Display (figure)	4-74

Subject	Page No.
Radar Warning System, AN/APR-39(V)1	4-71
Radar Warning System, AN/APR-39A(V)1	4-75
Radio Set, AN/ARC-164(V) and HAVE QUICK Radios Conference Capability	3-17
Radio Set, AN/ARC-186(V)	3-9
Radio Set, AN/ARC-201	3-24
Radio Set, AN/ARC-201 Antennas	3-24
Radio Set, AN/ARC-201 Controls and Functions	3-24
Radio Set, AN/ARC-201 Modes of Operation	3-27
Radio Set, RT-1167C/ARC-164(V) and HAVE QUICK Radios	3-13
Radio Set, RT-1167C/ARC-164(V) and HAVE QUICK Radios Antenna	3-14
Radio Set, RT-1167C/ARC-164(V) and HAVE QUICK Radios Controls and Functions	3-14
Radio Set, RT-1167C/ARC-164(V) and HAVE QUICK Radios Guard Operation	3-17
Radio Set, RT-1167C/ARC-164(V) and HAVE QUICK Radios Modes of Operation	3-15
Radio Set, RT-1167C/ARC-164(V) and HAVE QUICK Radios Secure Communication	3-17
Radio Set, RT-1167C/ARC-164(V) and HAVE QUICK Radios Warning Tones	3-17
Rain and Ice (operation)	8-21
Rain Removal	2-62
Rapid Rearming	4-65
Rapid Refueling	2-86
Receiver Transmitter Radio, RT-1167C/ARC-164(V) (figure)	3-13
Receptacles, External Air Source	2-55
Reference Datum, (Definition)	6-3
References, Appendix A, Description	1-1
Refueling Provisions	2-40
Restrictions, Environmental	5-16
Rocket Launchers (2.75)	6-11

Subject	Page No.
Rockets (2.75)	6-11
Rockets (2.75) Loading for H519 or Dummy (table)	6-13
Rockets (2.75) Loading for MPSM Warhead with MK66 Motor (table)	6-14
Rotor Blade De-Icing System	2-61
Rotor Brake	2-57
Rotor Limitations	5-2
Rotor System	2-59
Rotors, General	2-1
Rotors, Transmissions, and Drive System Failures and Malfunctions, General	9-9
RT-1167/ARC-164(V) Normal Mode Operating Procedures	3-17
RT-1167C/ARC-164(V) and HAVE QUICK Radios	3-13
RT-1167C/ARC-164(V) Controls and Functions (table)	3-14
RT-1167C/ARC-164(-V) HAVE QUICK Emergency Startup of TOD Clock	3-18
RT-1167C/ARC-164(V) HAVE QUICK Mode Operating Procedures	3-18
RT-1296/APX-100(V)1(IFF) Antenna	3-66
RT-1296/APX-l00(V)1(IFF) Controls and Functions	3-66
RT-1296/APX-100(V)1(IFF) Control and Indicator Functions (table)	3-67
RT-1296/APX-100(V)1(IFF) Control Panel (figure)	3-67
RT-1296/APX-100(V)1(IFF) Normal Operation	3-69
RT-1296/APX-100(V)1(IFF) Operation	3-68
RT-1296/APX-100(V)1(IFF) Stopping Procedure	3-70
RT-1296/APX-100(V)1(IFF) Transponder	3-66
RT-1557/APX-100(V)1(IFF) Antenna	3-66
RT-1557/APX-100(V)1(IFF) Control and Indicator Functions (table)	3-67
RT-1557/APX-100(V)1(IFF) Control Panel (figure)	3-67
RT-1557/APX-100(V)1(IFF) Controls and Functions	3-66
RT-1557/APX-100(V)1(IFF) Normal Operation	3-69

Subject	Page No.
RT-1557/APX-100(V)1(IFF) Operation	3-68
RT-1557/APX-100(V)1(IFF) Stopping Procedure	3-70
RT-1557/APX-100(V)l(IFF) Transponder	3-66

S

Subject	Page No.
Searchlight	2-74
Seat Height Adjustment	2-16
Seat, Crewmember - Both Crew Stations (figure)	2-18
Seats, Crewmember	2-16
Secure Communication, RT-1167C/ARC-164(V) and HAVE QUICK Radios	3-17
Selectable Digital Display Panel (figure)	2-33
Self-Test HARS	3-64
Series and Effectivity Codes	1-2
Servicing, Diagram (figure)	2-88
Servicing, Fuel System	2-86
Servicing, General	2-86
Servicing, Oil System	2-87
Shaft Driven Compressor (SDC)	2-55
Shall, Should, and May, Use of	1-2
Shoulder Harness Inertia Reel Lock Lever	2-16
Signal Validation - Fault Codes (figure)	2-27
Single Engine Failure	9-4
Single Engine Failure Low Altitude/Low Airspeed and Cruise	9-4
Slope/Rough Terrain Landing	8-18
Smoke and Fume Elimination	9-13
Special Mission Kits	2-1
Specific Conditions, Performance -701 Engine	7-2
Specific Conditions, Performance -701C Engine	7A-2
Spheroid String Enties (table)	3-56
Stabilator Auto/Manual Mode Failure	9-14
Stabilator Automatic Mode Failure	9-14
Stabilator Operation	8-18
Stabilator Operation Manual Mode	8-18

Subject	Page No.
Stabilator Operation NOE/APPR Mode	8-18
Stabilator Position Indicator (figure)	2-49
Stabilator System	2-49
Start, Using APU, Engine	2-29
Start, Using Engine Bleed Air Source, Engine	2-29
Start, Using External Source, Engine	2-29
Starting APU, Pilot	8-10
Starting Engines (cold weather)	8-20
Starting Engines, External Pressurized Air Source, Pilot	8-10
Starting Engines, Pilot	8-12
Starting System, Engine	2-29
Station Diagram (figure)	6-2
Stopping Procedure, AN/APN-209(V)	3-72
Stopping Procedure, RT-1296/APX-100(V)1(IFF)	3-70
Stopping Procedure, RT-1557/APX-100(V)1(IFF)	3-70
Stopping Procedures, AN/ASN-137	3-61
Storage Bay and Survival Kit Bay Equipment Weights and Moments (table)	6-16
Storage Bay, Left Aft	6-1
Survival Kit Bay	6-1
Switches, Collective	4-23
Switches, Cyclic	4-22
Symbol Definitions	8-2
Symbol Generator	4-16
Symbol Generator Failure	4-43
Symbol Generator Failure	9-22
Symbology, Weapons	4-43

T

Subject	Page No.
TADS Foresight	4-63
TADS Electronic Unit (TEU) Failure	9-22
TADS Electronic Unit Failure	4-43
TADS Equipment Data	4-41
TADS Equipment Data (table)	4-41
TADS Manual Servo Drift Null	4-63
TADS Operation Data	4-41
TADS Operational Checks - CPG	4-62
TADS Pechan Alignment	4-62
TADS/PNVS Equipment Data (figure)	4-43
Tail Landing Gear	2-8
Tail Rotor	2-59
Tail Rotor Drive Shafts	2-58
Tail Rotor Fixed Pitch Malfunction	9-11
Tail Rotor Gearbox	2-58
Tail Rotor Malfunctions	9-9
Tail Rotor Thrust at Low Airspeed/Hover, Loss of	9-10
Tail Rotor Thrust in Cruise Flight, Loss of	9-10
Tail Rotor Thrust, Loss of	9-9
Tail Wheel Lock Panel (figure)	2-8
Target Acquisition Designation Sight (TADS), AN/ASQ-170	4-40
Taxi	8-14
Taxi Check	8-14
Temperature Conversion -701 Engine	7-2
Temperature Conversion -701C Engine	7A-2
Temperature Conversion Chart -701 Engine (figure)	7-3
Temperature Conversion Chart -701C Engine (figure)	7A-3
Thunderstorm Operation	8-21
Torque Conversion Chart -701 Engine (figure)	7-8
Torque Conversion Chart -701C Engine (figure)	7A-9
Torque Factor Chart -701 Engine (figure)	7-7
Torque Factor Chart -701C Engine (figure)	7A-8
Torque Factor Method -701 Engine	7-4
Torque Factor Method -701C Engine	7A-4
Torque Factor Procedure -701 Engine	7-4
Torque Factor Procedure -701C Engine	7A-4
Torque Factor Terms -701 Engine	7-4
Torque Factor Terms -701C Engine	7A-4
Towing	2-96

Subject	Page No.
Transponder Computer, KIT-1A/TSEC and KIT-1C/TSEC	3-70
Transponder, RT-1296/APX-100(V)1(IFF)	3-66
Transponder, RT-1557/APX-100(V)1(IFF)	3-66
Trim and Force Feel	5-17
Trim Feel	2-45
TSEC/KY-28 Voice Security System	3-24
TSEC/KY-58 Control and Indicator Functions (table)	3-21
TSEC/KY-58 Controls and Functions	3-20
TSEC/KY-58 Operating Procedures	3-23
TSEC/KY-58 Voice Security System	3-20
Turbulence	5-10
Turbulence Operation	8-20
Turning Radius and Ground Clearance	2-1
Turning Radius and Ground Clearance (figure)	2-6
Turret Failure	4-43

U

Subject	Page No.
Use of Chart, Climb-Descent Performance -701 Engine	7-72
Use of Chart, Climb-Descent Performance -701C Engine	7A-69
Use of Chart, Drag Performance -701 Engine	7-69
Use of Chart, Drag Performance -701C Engine	7A-66
Use of Chart, Hover Ceiling Performance -701 Engine	7-9
Use of Chart, Hover Ceiling Performance -701 C Engine	7A-10
Use of Chart, Hover Performance -701 Engine	7-11
Use of Chart, Hover Performance -701C Engine	7A-13
Use of Chart, Maximum Torque Available Performance -701 Engine	7-4
Use of Chart, Maximum Torque Available Performance -701C Engine	7A-4
Use of Charts, Cruise Performance -701 Engine	7-13
Use of Charts, Cruise Performance -701C Engine	7A-15

Subject	Page No.
Use of Performance Charts -701 Engine	7-2
Use of Performance Charts -701C Engine	7A-2
Use of Shall, Will, Should, and May	1-2
Utility Hydraulic System	2-52
Utility Hydraulic System (figure)	2-53
Utility Light	2-75
Utility Manifold (figure)	2-54
UTM Coordinate Data - Valid Entries (table)	3-57

V

Subject	Page No.
Video Display Unit (figure)	4-5
Video Display Unit (VDU)	4-5
Video Display Unit Control/Indicator Functions (figure)	4-6
Video Recorder Control Panel/Indicator Functions (table)	4-7
Video Recorder Subsystem Control Panel (figure)	4-6
Video Recorder Subsystems (VRS)	4-6
Video Recorder Subsystems Operation	4-7
Voice Security System Equipment (figure)	3-22
Voice Security System, TSEC/KY-28	3-24
Voice Security System, TSEC/KY-58	3-20

W

Subject	Page No.
Warmup and Ground Tests (cold weather)	8-20
Warning Tones, RT-1167C/ARC-164(V) and HAVE QUICK Radios	3-17
Warnings, Cautions, and Notes	1-1
Waypoint Dictionary - Partial Entry Rules (table)	3-56
Waypoint/Targeting	4-17
Waypoint/Targeting Coordinate Data Storing	4-18
Waypoint/Targeting Cueing	4-19
Waypoint/Targeting Data Entry	4-17
Waypoint/Targeting Position Update	4-18
Waypoints (figure)	3-50
Weapon Symbology Definitions (table)	4-44
Weapon Symbology Modes (figure)	4-44

Subject	Page No.
Weapons Symbology	4-43
Weight and Balance	6-3
Weight and Balance Clearance Form F	6-4
Weight Definitions	6-3
Weight Limitations	5-10
Weight, Basic, (Definition)	6-3
Weight, Gross, (Definition)	6-3
Weight, Operating, (Definition)	6-3
Weight, Personnel	6-10
Weight/Balance and Loading, Introduction	6-1
Weights and Moments, Storage Bay and Survival Kit Bay Equipment (table)	6-16
Windshield and Canopy Cleaning	2-96
Windshield and Canopy Panels	2-2
Windshield Anti-Ice/De-Ice	2-60

Subject	Page No.
Windshield, General	2-2
Wing Stores Configuration	5-17
Wing Stores Jettison	9-21
Wings, General	2-1
Wire Strike Protection System (figure)	2-63
Wire Strike Protection System (WSPS)	2-62

X

Y

Z

Z-AHP Remote Control Unit Control and Indicator Functions (table)	3-21
Z-AHQ Power Interface Adapter Control and Indicator Functions (table)	3-20

By Order of the Secretary of the Army:

GORDON R. SULLIVAN
*General, United States Army
Chief of Staff*

Official:

Milton H. Hamilton
MILTON H. HAMILTON
*Administrative Assistant to the
Secretary of the Army*
07462

DISTRIBUTION:
To be distributed in accordance with DA Form 12-31-E, block no. 0293, requirements for TM 1-1520-238-10.

* U.S. GOVERNMENT PRINTING OFFICE: 1994-555-121/00026

TM 55-1520-238-10

THESE ARE THE INSTRUCTIONS FOR SENDING AN ELECTRONIC 2028

The following format must be used if submitting an electronic 2028. The subject line must be exactly the same and all fields must be included; however only the following fields are mandatory: 1, 3, 4, 5, 6, 7, 8, 9, 10, 13, 15, 16, 17, and 27.

From: 'Whomever' <whomever@avma27.army.mil>

To: ls-lp@redstone.army.mil

Subject: DA Form 2028

1. **From:** Joe Smith
2. Unit: Home
3. **Address:** 4300 Park
4. **City:** Hometown
5. **St:** MO
6. **Zip:** 77777
7. **Date Sent:** 19-Oct-93
8. **Pub no:** 55-2840-229-23
9. **Pub Title:** TM
10. **Publication Date:** 04-Jul-85
11. Change Number: 7
12. Submitter Rank: MSG
13. **Submitter Fname:** Joe
14. Submitter Mname: T
15. **Submitter Lname:** Smith
16. **Submitter Phone:** (123) 123-1234
17. **Problem:** 1
18. Page: 2
19. Paragraph: 3
20. Line: 4
21. NSN: 5
22. Reference: 6
23. Figure: 7
24. Table: 8
25. Item: 9
26. Total: 123
27. Text:

This is the text for the problem below line 27.

Change 5

Warships DVD Series

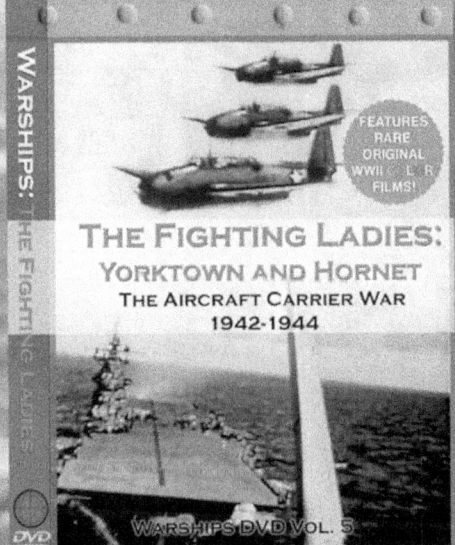

Now Available!

WARSHIPS DVD SERIES

WARSHIPS: CARRIER MISHAPS

AIRCRAFT CARRIER MISHAPS
SAFETY AND TRAINING FILMS

-PERISCOPEFILM.COM-

NOW AVAILABLE ON DVD!

Aircraft At War DVD Series

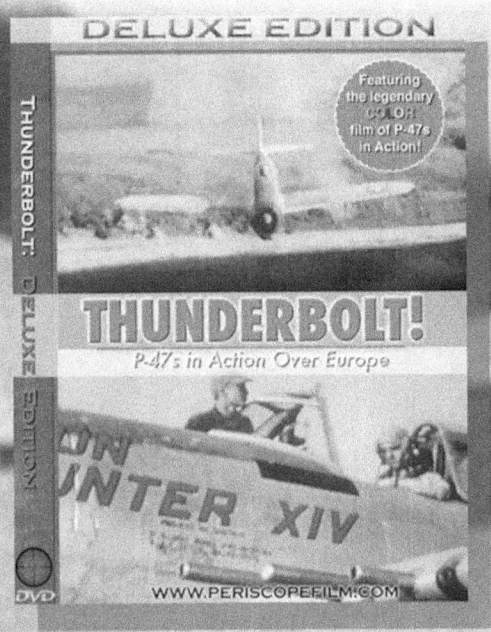
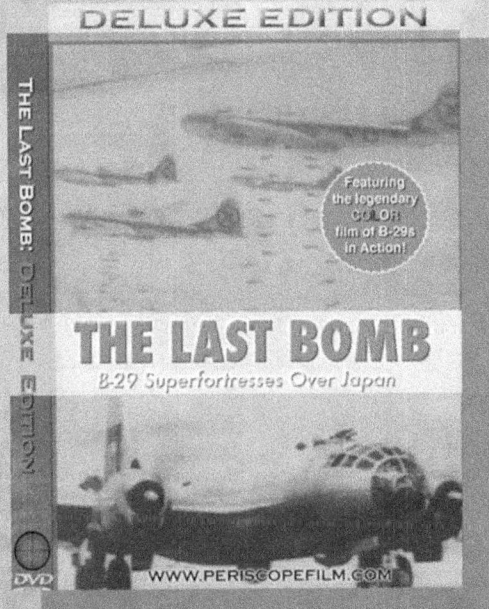
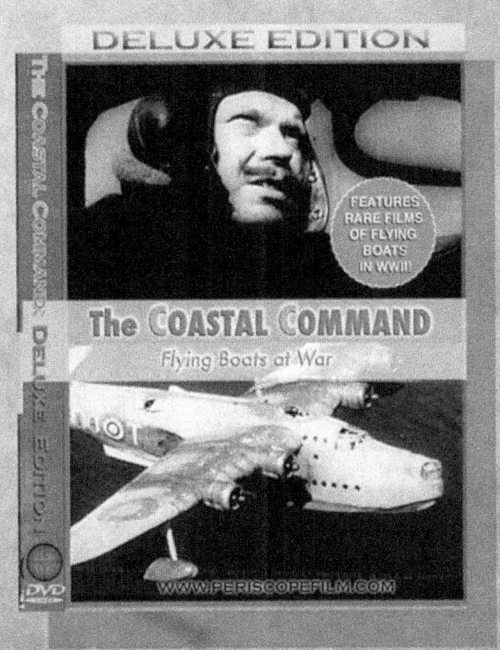

Now Available!

Epic Battles of WWII

Now Available on DVD!

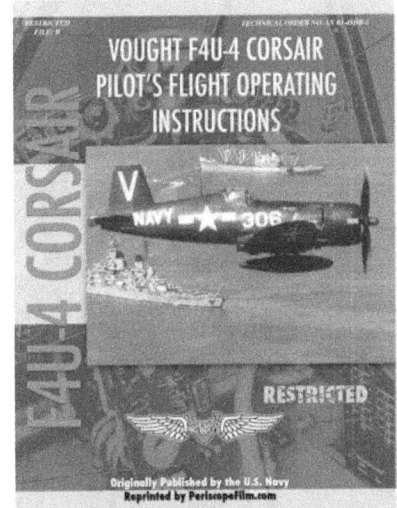
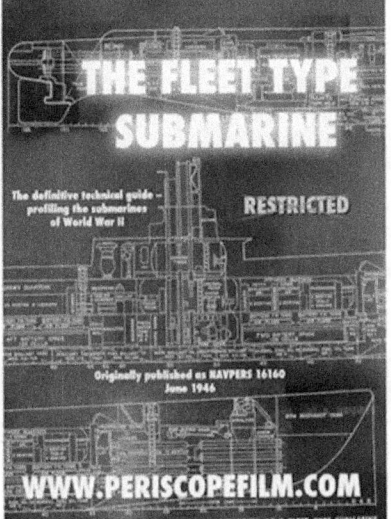

ALSO NOW AVAILABLE
FROM PERISCOPEFILM.COM

©2011 Periscope Film LLC
ISBN #978-1-935700-67-8
www.PeriscopeFilm.com

www.ingramcontent.com/pod-product-compliance
Lightning Source LLC
Chambersburg PA
CBHW080527300426
44111CB00017B/2639